AF577036

SUPEROXIDE
AND
SUPEROXIDE DISMUTASES

SUPEROXIDE AND SUPEROXIDE DISMUTASES

Edited by

A. M. MICHELSON
Institut de Biologie Physico-Chimique
Fondation Edmond de Rothschild
Paris, France

J. M. McCORD
Duke University Medical Center
Durham, North Carolina
USA

I. FRIDOVICH
Duke University Medical Center
Durham, North Carolina
USA

1977

ACADEMIC PRESS
LONDON · NEW YORK · SAN FRANCISCO
A Subsidiary of Harcourt Brace Jovanovich, Publishers

ACADEMIC PRESS INC. (LONDON) LTD.
24/28 Oval Road,
London NW1

United States Edition published by
ACADEMIC PRESS INC.
111 Fifth Avenue
New York, New York 10003

Library of Congress Catalog Card Number: 77–71799
ISBN: 0–12–494050–1

CONTRIBUTORS

Allen, J.F., Botany School, South Parks Road, Oxford OX1 3RA, England.

Anastasi, A., Department of Physiology and Biochemistry, Royal University of Malta, Msida, Malta.

Azzi, A., Instituto di Patologia Generale, Universita di Padova, Padova, Italy.

Babior, B.M., Hematology Service, New England Medical Center Hospital and Department of Medicine, Tufts University School of Medicine, Boston, Massachusetts 02111, U.S.A.

Bannister, J.V., Department of Physiology and Biochemistry, Royal University of Malta, Msida, Malta.

Bannister, W.H., Department of Physiology and Biochemistry, Royal University of Malta, Msida, Malta.

Bell, G.R., Department of Biochemistry, The University of Georgia, Athens, GA 30601, U.S.A.

Bielski, B.H.J., Department of Chemistry, Brookhaven National Laboratory, Upton, New York 11973, U.S.A.

Bohnenkamp, W., Anorganische Biochemie, Physiologisch-Chemisches Institut der Universität Tübingen, D-7400 Tübingen, W. Germany.

Bonneau, J.C., Centre Régional de Transfusion Sanguine et de Génétique Humaine, 609 Chemin de la Bretèque, 76230 Bois-Guillaume, France.

Boyle, J.A., Department of Medicine and Biochemistry, Duke University Medical Center, Durham, North Carolina 27710, U.S.A.

Bray, R.C., School of Molecular Sciences, University of Sussex, Brighton BN1 9QJ, England.

Brunori, M., Institute of Chemistry, Faculty of Medicine, University of Rome, Rome, Italy.

Calabrese, L., C.N.R. Center for Molecular Biology, Rome, Italy.

Carle, I., Physics Division, Institute of Cancer Research, Sutton, Surrey, England.

Chan, P.C., Department of Biochemistry, State University of New York, Downstate Medical Center, Brooklyn, New York 11203, U.S.A.

Charbonneau, H., Bioluminescence Laboratory, Department of Biochemistry, University of Georgia, Athens, GA 30602, U.S.A.

Cockle, S.A., Department of Biochemistry, McMaster University, Health Sciences Centre, Hamilton, Ontario, Canada.

Cohen, G., Department of Neurology, Mount Sinai School of Medicine, New York, New York 10029, U.S.A.

Cormier, M.J., Bioluminescence Laboratory, Department of Biochemistry, University of Georgia, Athens, GA 30602, U.S.A.

Crapo, J.D., Department of Medicine, Duke University Medical Center, Durham, North Carolina 27710, U.S.A.

CONTRIBUTORS

Crichton, R.R., Unité de Biochimie, Université Catholique de Louvain, 1348 Louvain-la-Neuve, Belgique.

Day, E.D., Department of Medicine and Biochemistry, Duke University Medical Center, Durham, North Carolina 27710, U.S.A.

Diliberto, E.J. Jr., Laboratory of Neurochemistry, NIMH, Bethesda, Maryland 20014, U.S.A.

Durosay, P., Institut de Biologie Physico-Chimique, Service de Biochimie-Physique, 13, rue P. et M. Curie, 75005 Paris, France.

Falconi, G., Laboratory of Molecular Biology, University of Camerino, Camerino, Italy.

Fee, J.A., Biophysics Research Division and Department of Biological Chemistry, University of Michigan, Ann Arbor, MI 48109, U.S.A.

Fielden, E.M., Physics Division, Institute of Cancer Research, Sutton, Surrey, England.

Fioretti, E., Laboratory of Molecular Biology, University of Camerino, Camerino, Italy.

Flohe, L., The Research Laboratories of Chemie Grünenthal GmbH, 519 Stolberg, W. Germany.

Fridovich, I., Department of Biochemistry, Duke Medical Center, Durham, North Carolina 27710, U.S.A.

Hall, D.O., University of London, King's College, 68 Half Moon Lane, London SE24 9JF, England.

Halliwell, B., Department of Biochemistry, King's College, Strand, London WC2R 2LS, England.

Harris, J.I., MRC Laboratory of Molecular Biology, Hills Road, Cambridge, CB2 2QH, England.

Hatchikian, C.E., Laboratoire de Chimie Bactérienne, CNRS, 13274 Marseille, Cedex 2, France.

Hayaishi, O., Department of Medical Chemistry, Kyoto University Faculty of Medicine, Kyoto 606, Japan.

Heikkila, R.E., Department of Neurology, Mount Sinai School of Medicine, New York, New York 10029, U.S.A.

Henry, J.P., Institut de Biologie Physico-Chimique, Service de Biochimie-Physique, 13, rue P. et M. Curie, 75005 Paris, France.

Henry, L., University of London, King's College, 68 Half Moon Lane, London SE24 9JF, England.

Hirata, F., Department of Medical Chemistry, Kyoto University Faculty of Medicine, Kyoto 606, Japan.

Huber, W., Diagnostic Data Inc., 518 Logue Av, Mountain View, California 94043, U.S.A.

Johnston, R.B., Jr., Departments of Pediatrics and Microbiology and the Compréhensive Cancer Center, The University of Alabama Medical Center, Birmingham, Alabama 35294, U.S.A.

Kaufman, S., Laboratory of Neurochemistry, NIMH, Bethesda, Maryland 20014, U.S.A.

Lavelle, F., Institut de Biologie Physico-Chimique, Service de Biochimie-Physique, 13, rue P. et M. Curie, 75005 Paris, France.

Le Gall, J., Laboratorie de Chimie Bactérienne, CNRS, 13274 Marseille, Cedex 2, France.

Lehmeyer, J.E., Departments of Pediatrics and Microbiology and the Compréhensive Cancer Center, The University of Alabama Medical Center, Birmingham, Alabama 35294, U.S.A.

CONTRIBUTORS

Loschen, G., Research Laboratories of Chemie Grünenthal GmbH, 519 Stolberg, W. Germany.

Lumsden, J., University of London, King's College, 68 Half Moon Lane, London SE24 9JF, England.

*McCord, J.**, Box 3206, Duke University Medical Center, Durham, North Carolina 27710, U.S.A.

Matkovics, B., Biological Isotop Laboratory, "A.J." University, Szeged, Hungary.

Mautner, G.N., School of Molecular Sciences, University of Sussex, Brighton BN1 9QJ, England.

Menander-Huber, K.B., Diagnostic Data Inc., 518 Logue Av., Mountain View, California 94043, U.S.A.

Michelson, A.M., Institut de Biologie Physico-Chimique, Service de Biochimie-Physique, 13, rue P. et M. Curie, 75005 Paris, France.

Puget, K., Institut de Biologie Physico-Chimique, Service de Biochimie-Physique, 13, rue P. et M. Curie, 75005 Paris, France.

Richardson, D.C., Department of Biochemistry, Duke University Medical Center, Durham, North Carolina 27710, U.S.A.

Richter, C., Instituto di Patologia Generale, Universita di Padova, Padova, Italy.

Rigo, A., Institute of Physical Chemistry, University of Venice, Venice, Italy.

Rizzolo, L.J., Department of Medicine and Biochemistry, Duke University Medical Center, Durham, North Carolina 27710, U.S.A.

Roos, D., Central Laboratory of the Netherlands Red Cross Blood Transfusion Service, Amsterdam, The Netherlands.

Rotilio, G., C.N.R. Center for Molecular Biology, Rome, Italy.

Saifer, M.G.P., Diagnostic Data Inc., 158 Logue Av., Mountain View, California 94043, U.S.A.

Salin, M.L., Department of Medicine and Biochemistry, Duke University Medical Center, Durham, North Carolina 27710, U.S.A.

Sinet, P.M., Hôpital Necker, Service de Biochimie Génétique, 149 rue de Sèvres, 75015 Paris, France.

*Steinman, H.M.***, Department of Biochemistry, Duke University Medical Center, Durham, North Carolina 27710, U.S.A.

Valentine, J.S., Department of Chemistry, Douglass College, Rutgers, The State University, New Brunswick, N.J. 08903, U.S.A.

Van Schaik, M.L.J., Pediatric Clinic, Binnen·Gasthuis, University of Amsterdam, Amsterdam, The Netherlands.

Vanopdenbosch, B., Université Catholique de Louvain, 1348 Louvain-la-Neuve, Belgique.

Viglino, P., Institute of Physical Chemistry, University of Venice, Venice, Italy.

Ward, W.W., Bioluminescence Laboratory, Department of Biochemistry, University of Georgia, Athens, GA 30602, U.S.A.

Weening, R.S., Pediatric Clinic, Binnen Gasthuis, University of Amsterdam, Amsterdam, The Netherlands.

Wendal, A., Anorganische Biochemie, Physiologisch-Chemisches Institut der Universität Tübingen, D-74 Tübingen, Hoppe-Seylerstrasse 1, W. Germany.

Weser, U., Anorganische Biochemie, Physiologisch-Chemisches Institut der Universitat Tubingen, D-74 Tubingen, Hoppe-Seylerstrasse 1, W. Germany.

Wever, R., Laboratory of Biochemistry, B.C.P., Jansen Instituut, University of Amsterdam, Amsterdam, The Netherlands.

CONTRIBUTORS

Current Addresses

**McCord, J.*, Associate Professor of Biochemistry, College of Medicine, University of South Alabama, Mabile, Alabama 36688, U.S.A.

***Steinman, H.M.*, Department of Biochemistry, Albert Einstein College of Medicine, 1300 Morris Park Avenue, Bronx, New York 10461, U.S.A.

PARTICIPANTS

EMBO WORKSHOP ON SUPEROXIDE AND SUPEROXIDE DISMUTASE

20 - 26 June 1976, Banyuls, France

Allen, J.F., Botany School, South Parks Road, Oxford OX1 3RA, U.K.

Babior, B.M., New England Medical Center Hospital, 171, Harrison Av., Boston, Mass. 02111, U.S.A.

Bannister, J.V., The Royal University of Malta, Department of Physiology, Malta.

Bannister, W.H., The Royal University of Malta, Department of Physiology, Malta.

Bray, R.C., University of Sussex, The School of Molecular Sciences, Falmer Brighton BN1 9QJ, U.K.

Bruneau, P., ICI Pharma, Zone Industrielle Sud-Est Reims, Cedex 51064, France.

Cohen, G., Mount Sinai School of Medicine of the City University of New York, Department of Neurology, Fifth Av. and 100th Street, New York, N.Y. 10029, U.S.A.

Cohen, H.J., Division of Hematology, Oncology, The Children's Hospital Medical Center, 300, Longwood Av., Boston, Mass. 02115, U.S.A.

Cormier, M.J., Department of Biochemistry, University of Georgia, Athens, Georgia 30602, U.S.A.

Crichton, R.R., Université Catholique de Louvain, Institut Lavoisier, Place Louis Pasteur, 1348 Louvain-la-Neuve, Belgium.

De Ryker, J., Institut voor Molekulaire Biologie, Vrije Universiteit Brussel, Paardenstraat 65, 1640 St. Genesius-Rode, Belgium.

Fee, J.A., The University of Michigan, Institute of Science and Technology, Biophysic Research Division, 2200 North Campus Boulevard, Ann Arbor, Michigan 48105, U.S.A.

Ferradini, C., Faculté de Médecine, Service de Chimie-Physique, 45, rue des St. Pères, 75006 Paris, France.

Fielden, M., Institut of Cancer Research, Physics Dept., Clifton Av., Belmont, Sutton, Surrey, U.K.

Fitzsimons, D.W., The Ciba Foundation, 41, Portland Place, London W1N 4BN, U.K.

Flohe, L., Chemie Grünenthal GMBH, 519 Stolberg/RHL Zweifaller Strasse, Germany.

Halliwell, B., Dept. of Biochemistry, King's College, Strand, London WC2R 2LS, U.K.

Harris, J.I., MRC Laboratory of Molecular Biology, Hills Road, Cambridge CB2 2QH, U.K.

Hatchikian, C., Laboratoire de Chimie Bactérienne, 31, Chemin Joseph Aiguier, 13274 Marseille, Cedex 2, France.

PARTICIPANTS

Heikkila, R., Mount Sinai School of Medicine of the City University of New York, Dept. of Neurology, Fifth Av. and 100th Street, New York, N.Y. 10029, U.S.A.

Henry, J.P., Institut de Biologie Physico-Chimique, Service de Biochimie-Physique, 13, rue P. et M. Curie, 75005 Paris, France.

Huber, W., Diagnostic Data Inc., 518 Logue Av., Mountain View, California 94043, U.S.A.

Jacquet, B., L'Oreal, 1, Av. de St. Germain, 93601 Aulnay-sous-Bois, France.

Johnston, R.B., The Rockfeller University, Lab. of Cellular Physiology and Immunology, 1230 York Av., New York, N.Y. 10021, U.S.A.

Kauffman, S., Lab. of Neurochemistry, National Institute of Mental Health, 9000 Rockville Pike, Bethesda, Maryland 20014, U.S.A.

Labeyrie, F., Institut de Génétique Moléculaire, CNRS 91, Gif/Yvette, France.

Lavelle, F., Institut de Biologie Physico-Chimique, Service de Biochimie-Physique, 13, rue P. et M. Curie, 75005 Paris, France.

Legall, M.J., Laboratoire de Chimie Bactérienne, 31, Chemin Joseph Aiguier, 13274 Marseille, Cedex 2, France.

Lumsden, J., University of Glasgow, Dept. of Biochemistry, Glasgow G12 8QQ, U.K.

McAdam, M.E., Institut of Cancer Research, Royal Cancer Hospital, Dept. of Physics, Clifton Av., Sutton, Surrey SM2 5PX, U.K.

McCord, J., Box 3206, Duke University Medical Center, Durham, N.C. 27710, U.S.A.

Menander-Huber, K.B., Diagnostic Data Inc., 518, Logue Av., Mountain View, California 94043, U.S.A.

Michelson, A.M., Institut de Biologie Physico-Chimique, Service de Biochimie-Physique, 13, rue P. et M. Curie, 75005 Paris, France.

Moore, G.G.I., Riker Laboratories, Inc., 3 M Center, Building 218, 1, St. Paul, Minnesota 55101, U.S.A.

Puget, K., Institut de Biologie Physico-Chimique, Service de Biochimie-Physique, 13, rue P. et M. Curie, 75005 Paris, France.

Puig-Muset, P., José Bertrand 7, Barcelona 6, Spain.

Richardson, D.C., Duke University Medical Center, Dept. of Biochemistry, 213, Medical Sciences I, Durham, N.C. 27710, U.S.A.

Rigo, A., Institute of Physical Chemistry, University of Venice, Venice, Italy.

Roos, D., Central Laboratory of the Netherlands, Red Cross Blood Transfusion Service, Plesmanla 125, Amsterdam W., Holland.

Rotilio, G., Instituto di Chimica Biologica, Università di Roma, Città Universitarià, 00185 Roma, Italy.

Salin, M.L., Duke University Medical Center, Dept. of Biochemistry, Durham, N.C. 27710, U.S.A.

Signour, M., Institut de Biologie Physico-Chimique, Service de Biochimie-Physique, 13, rue P. et M. Curie, 75005 Paris, France.

Sinet, P.M., Hôpital Necker, Service de Biochimie Génétique, 149 rue de Sèvres, 75015 Paris, France.

Vigny, A., Institut de Biologie Physico-Chimique, Service de Biochimie-Physique, 13, rue P. et M. Curie, 75005 Paris, France.

Weser, U., Anorganische Biochemie, Physiologish-Chemisches Institut der Universität Tübingen, 7400 Tübingen 1, Hoppe Seyler Strasse 1, Germany.

PREFACE

This book is a result of the first EMBO Workshop on "Superoxide and Superoxide Dismutase" held at Banyuls, France, June 21st - 26th 1976. Apart from financial support from EMBO (European Molecular Biology Organisation) we are extremely grateful to Prof. P. DRACH (Director of the Laboratoire Arago, Banyuls), for the facilities and support graciously accorded and without which this meeting could not have taken place.

Biological aspects of superoxide radical anions were initiated by a single publication in 1969 by McCord and Fridovich, identifying the enzymic activity of erythrocuprein. Since then, the intellectual and creative explosion which has followed (as evidenced by the contents of this book, by no means comprehensive) with respect to the metabolism and molecular biology of oxygen, can only be compared with that occurring as a result of a well known publication in 1953 by Watson and Crick on the double helical structure of DNA. From one who has had the good scientific fortune to be at least a close spectator of both events the comparison is not unjustified. Some of the same kind of excitement generated twenty odd years ago in a completely different area could be felt at the Banyuls meeting. For the past few years, the modern history of DNA has been enshrined via migrant tape recorder equipped scientific historians or journalists. We are no doubt fortunate that the history of superoxide dismutase began only seven years ago and can be represented with perhaps a greater verity than would otherwise be the case.

A.M. MICHELSON

CONTENTS

CONTENTS

CONTENTS

SUPEROXIDE DISMUTASES: A HISTORY

J.M. McCORD and I. FRIDOVICH

Departments of Medicine and Biochemistry
Duke University Medical Center
Durham, North Carolina 27710, U.S.A.

One minor product of the biological reduction of molecular oxygen is a free radical of such great self-reactivity that it has but a fleeting lifetime in neutral aqueous solutions. Incredibly, the production of this free radical has necessitated the evolution of a class of enzymes whose function is to facilitate that very reaction which occurs rapidly even in the absence of catalysis. These enzymes are ubiquitous among respiring organisms and appear to provide an indispensable defense against the toxicity of oxygen. The free radical alluded to is the superoxide anion, O_2^-, formed by the univalent reduction of oxygen. The self reaction is the dismutation, $O_2^- + O_2^- + 2H^+ \rightarrow H_2O_2 + O_2$, and the enzymes which catalyze this reaction are therefore called superoxide dismutases.

Obviously, enzymes which act upon an unstable substrate could not have been discovered by design. All the more so since this substrate radical was not really known to be a normal product of biological oxidations prior to the discovery of the superoxide dismutase activity. In fact, one of the superoxide dismutases was isolated (Mann and Keilin, 1938), crystallized (Mohamed and Greenberg, 1953) and extensively studied (Markowitz *et al.*, 1959; Carrico and Deutsch, 1969 and references therein) as a copper-containing protein of unknown function, during the thirty years preceding the discovery of its superoxide dismutase activity. The story of superoxide dismutase must be and is a tale of fortunate happenstance, persistent research and liberal and far-sighted granting agencies. In this chapter we shall try to reconstruct the chain of events which led to our present comprehension.

In 1941 Handler *et al.* (1941) described a sarcosine oxidase activity in rat liver homogenates. They found that sarcosine (N-methylglycine) was oxidized to formaldehyde and glycine. In 1951 Murray Heimberg, working as a graduate student under the direction of Philip Handler, was studying sarcosine oxidase. One of his approaches involved the use of sulfonic acid analogues of sarcosine as possible competitive inhibitors. Surprisingly α-amino and α-hydroxy sulfonic acids did not inhibit oxygen consumption in his assay system, but rather augmented it. These sulfonic acids were, in fact, themselves oxidized by the liver preparations and one of the products was sulfate. In 1952 Fridovich joined the Department of Biochemistry at Duke University as a graduate student and chose to work under the tutelage of P. Handler. In practice this meant that he joined M. Heimberg in his studies of sarcosine oxidase. The two of them kept the Warburg microrespirometer shaking almost continuously, both day and night. It soon became apparent that α-amino and α-hydroxy sulfonic acids were substrates because they dissociated and because the sulfite produced by that dissociation was rapidly oxidized to sulfate in the presence of liver extracts. The existence of an enzyme catalyzing the oxidation of inorganic sulfite was thus exposed and a four fold purification of this activity was achieved (Heimberg *et al.*, 1953). Heimberg wrote his thesis and turned his attention to other endeavors in other places, while Fridovich continued to study sulfite oxidase. This activity was found to be localized in liver mitochondria and a modest purification was achieved (Fridovich and Handler, 1954).

The aerobic oxidation of sulfite is complicated by the tendency of sulfite to undergo a free radical chain oxidation. This leads to a great sensitivity to chain initiators, which augment the oxidation, and to chain breakers, which inhibit the oxidation (Bäckström, 1927). This information would have been invaluable to Fridovich and Handler in explaining the strange behavior of the sulfite oxidase system. Unfortunately, they were unaware of the earlier work on the sulfite-oxygen chain reaction and had laboriously to discover the phenomenon all over again. Thus the sulfite oxidase could be assayed in terms of the reduction of methylene blue in evacuated Thunberg tubes. Using this assay it was seen that dialysis of the enzyme caused a loss of activity which could be restored by a heat-stable, dialyzable component of the enzyme preparation (Fridovich and Handler, 1955). This substance which behaved like a classical enzyme cofactor was unimaginatively named cofactor x and was isolated from bovine liver. It proved to be hypoxanthine. The general excitement over this apparent discovery of a new cofactor led one of Handler's colleagues to order for him a set of silver cufflinks upon which the structure of hypoxanthine was engraved.

The facilitation of the enzymatic oxidation of sulfite by hypoxanthine led naturally to a suspicion of the involvement of xanthine oxidase. In the interim the free radical chain oxidation of sulfite was exposed by an effect of cytochrome *c*. Thus sulfite slowly reduced ferricytochrome *c*. Ferricytochrome *c*, but not ferrocytochrome *c*, caused the aerobic oxidation of sulfite. The relative rates of these reactions were such that thousands of sulfite ions were oxidized per cytochrome *c* reduced. It was apparent that the oxidation of one sulfite by cytochrome *c* generated a species, probably a radical (SO_3^-), which then caused the aerobic oxidation of thousands of sulfites. Xanthine oxidase, when oxidizing its substrates xanthine or hypoxanthine, was then found to initiate this sulfite-oxygen free radical reaction (Fridovich and Handler, 1958). Hypoxanthine, originally thought of as a cofactor, was thus seen to act as a cosubstrate. The great length of the sulfite-oxygen free radical chain reaction provided amplification of the initiating event and sulfite oxidation was exploited as an ultrasensitive assay for xanthine oxidase.

Numerous other oxidases, including glucose oxidase, D-amino acid oxidase and uricase were found unable to initiate sulfite oxidation. One had to suppose that the xanthine oxidase reaction produced some special product which was reactive enough to start the sulfite chain reaction. Since uric acid and H_2O_2 were not able to do so, an intermediate of oxygen reduction was suspected. Ferrous salts were seen to initiate sulfite oxidation, during their autoxidation to the ferric state, but ferric salts were much less effective. It appeared likely that O_2^-, made during the oxidation of ferrous salts, $Fe^{++} + O_2 \rightarrow Fe^{+++} + O_2^-$, was the initiator of sulfite oxidation. The ability of O_2^- to initiate directly sulfite oxidation was supported by the observation that the cathodic reduction of oxygen initiated sulfite oxidation (Fridovich and Handler, 1961). The initiation of sulfite oxidation was seen as a sensitive test for the production of free radicals and other systems were explored. Rat liver microsomes oxidizing NADPH were seen to initiate sulfite oxidation. The production of reactive radicals in photochemical systems were also explored (Fridovich and Handler, 1960) and an ultra-sensitive manometric actinometer was developed (Fridovich and Handler, 1959). It appeared likely at this time that xanthine oxidase initiated sulfite oxidation by virtue of the production of O_2^-. If the production of superoxide radicals by xanthine oxidase raised concerns over the stability of the enzyme in the presence of such free radicals, one could always propose that the O_2^- remained bound to the enzyme.

Our infatuation with the free radical chain oxidation of sulfite did not lead to total neglect of the enzymatic oxidation of sulfite. That was pursued as an independent effort. Sulfite is an intermediate in the oxidative catabolism of the sulfur-containing amino acids and there is a biological need for a sulfite oxidase which acts, by a non-radical mechanism, at low concentrations of sulfite. This true sulfite oxidase was studied (MacLeod *et al.*, 1961) and was eventually characterized as a molybdohemoprotein (Cohen and Fridovich, 1971; Cohen *et al.*, 1971). This is a fascinating enzyme in its own right and it continues to be studied in the laboratory of K.V. Rajagopalan.

The literature concerning the reduction of cytochrome *c* by xanthine oxidase was then confusing and our attempts to lessen this confusion also contributed to the discovery of superoxide dismutase. Cytochrome *c* was reported to be both a good electron acceptor and a poor electron acceptor. The reduction of cytochrome *c* was reported to be faster in the absence of oxygen (Morell, 1952) and dependent upon the presence of oxygen (Horecker and Heppel, 1949). In our hands xanthine oxidase did reduce some samples of cytochrome *c* very rapidly, while other samples were hardly reduced at all. Furthermore those preparations of cytochrome *c* which were not reducible by xanthine oxidase were found to contain a potent heat-labile and non-dialyzable inhibitor of cytochrome *c* reduction. This was a very specific sort of inhibitor, since it prevented the reduction of cytochrome *c* by xanthine oxidase acting upon xanthine, while having no effect upon the rate of reduction of oxygen. Since the globin of myoglobin was a frequent contaminant of the horse heart cytochrome *c* being used, myoglobin and the corresponding globin were prepared and were found to be powerful competitive inhibitors of cytochrome *c* reduction by xanthine oxidase (Fridovich, 1962). Figure 1 illustrates that this inhibition of cytochrome *c* reduction by horse heart globin was classically competitive with respect to cytochrome *c*. The specificity inherent in this inhibition was demonstrated by the failure of many other proteins to inhibit the reduction of cytochrome *c* by xanthine oxidase and by the failure of horse heart globin to inhibit the reduction of cytochrome *c* by the NADH-cytochrome *c* reductase or by the sulfite-cytochrome *c* reductase.

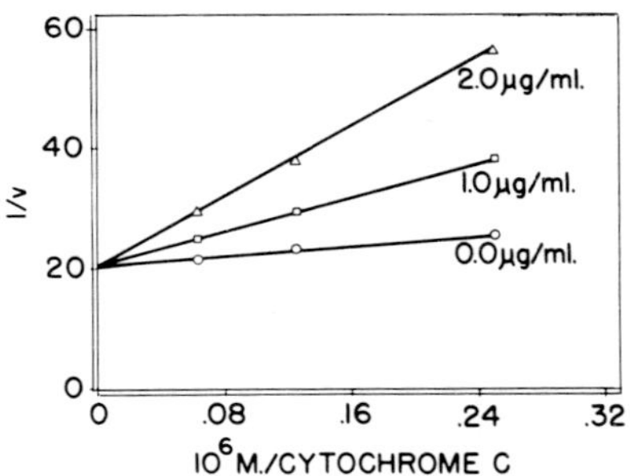

Fig.1. *Inhibition by horse heart globin of the reduction of cytochrome* c *by xanthine oxidase. Each cuvette contained 0.75 μmoles of xanthine, 600 Worthington units of catalase, 8 μg of xanthine oxidase, and the indicated amounts of cytochrome* c *and horse heart globin in a final volume of 10.0 ml buffered at pH 7.8 by 0.05 M potassium phosphate. (Fridovich, 1962).*

Oxygen was indeed essential for the reduction of cytochrome *c* by xanthine oxidase and H_2O_2 was excluded as the responsible reductant which was made from oxygen (Fridovich and Handler, 1962). Thus the concentration of oxygen needed to saturate the rate of cytochrome *c* reduction was much greater than that needed to saturate the oxidation of xanthine to urate. Furthermore, enzymatic scavengers of H_2O_2 did not prevent the reduction of cytochrome *c* and alternate enzymatic sources of H_2O_2 did not support the reduction of cytochrome *c*. In addition a catechol disulfonate (tiron) was found to inhibit the reduction of cytochrome *c*, without affecting the enzymatic turnover of xanthine oxidase. It appeared most probable at this time that O_2^-, generated on the enzyme and bound to it, was responsible for the reduction of cytochrome *c* by xanthine oxidase and that myoglobin or its globin had inhibited by binding to this O_2^- site thereby preventing access by cytochrome *c*. The other oxygen-dependent activities of xanthine oxidase, such as the initiation of sulfite oxidation and the induction of luminescence from luminol and lucigenin (Greenlee *et al.*, 1962) were also related to the bound O_2^- and were also inhibited by horse heart myoglobin or globin and by tiron.

Given the marvelous clarity of hindsight, all of these observations make perfect sense. Thus xanthine oxidase, acting upon its substrates in the presence of oxygen, produces O_2^-. This O_2^- can reduce cytochrome *c*, initiate the free radical chain oxidation of sulfite or cause the luminescence of luminol or lucigenin. The preparations of myoglobin, then in hand, contained a small amount (less than 1%) of superoxide dismutase which, by catalytically scavenging O_2^-, prevented all of those reactions dependent upon this radical. Tiron, like other catechols, is oxidized by O_2^- and thus also scavenges it and so acted like the myoglobin did in specifically inhibiting the O_2^--dependent reactions, without inhibiting the enzymatic turnover of xanthine oxidase. The oxygen requirement for cytochrome *c* reduction has an obvious explanation since in the absence of oxygen there could be no production of O_2^-. The striking difference in oxygen saturation curves for cytochrome *c* reduction and for xanthine oxidation has a more subtle explanation. Thus xanthine oxidase can carry out both univalent and divalent reductions of oxygen and the ratio of the univalent to the divalent process increases with the concentration of oxygen. Because of this, the rate of production of O_2^- and therefore the rate of reduction of cytochrome *c* would continue to increase with increasing oxygen concentration, even though the rate of oxidation of xanthine to urate had already been saturated. Thus the partition of substrate electrons between univalent and divalent reductions of oxygen depends upon the concentration of oxygen (Fridovich, 1970).

In 1962 our appreciation of these matters was in a much less satisfactory state. In that year Fridovich moved his wife and two infant daughters from Durham to Boston and spent a sabbatical year at Harvard in the Department of Chemistry. Professor F.W. Westheimer was so helpful and the atmosphere in his department was so congenial, that the year proved both pleasant and productive, despite the chaotic Boston traffic and the often inclement weather. Acetoacetic decarboxylase and its mechanism of action was the focus of attention during that year and we did learn a few things about it, which were duly reported (Fridovich and Westheimer, 1962; and Fridovich, 1963). The year at Harvard broke the thread of the studies of oxygen reduction by xanthine oxidase and even after the return to Duke University in the fall of 1963, this problem lay fallow for several years. Other problems concerning other enzymes claimed center stage and we seldom thought of xanthine oxidase, sulfite oxidation or superoxide radicals. However chance events were ultimately to bring us back onto the track of superoxide dismutase.

The triggering factor was a paper by Pocker and Meany (1965) which dealt with the specificity of carbonic anhydrase. These authors demonstrated that this enzyme was not restricted to catalyzing the hydration of carbon dioxide but could act upon a variety of aldehydes as well. At that time, textbooks of biochemistry depicted the enzymatic oxidation of aldehydes as the dehydrogenation of the corresponding diol hydrates. Was this actually the case or was the true substrate for aldehyde oxidases the anhydrous aldehyde? The relatively slow rate of the spontaneous hydration of aldehydes and the catalysis of this hydration by carbonic anhydrase was seen as providing the opportunity for a simple yet decisive experiment. Thus we could rapidly add the anhydrous aldehyde to a buffered solution of the aldehyde oxidase and immediately measure the rate of its oxidation. The concentration of the anhydrous aldehyde would be highest at the outset and would fall as some of it was converted to the hydrate. If the amount of aldehyde added was less than that needed to saturate the enzyme, then the enzymatic rate would reflect the concentration of true substrate. If the aldehyde was the substrate the rate would be fastest at the outset and would then fall as the hydration equilibrium was established. Conversely, if the hydrate was the substrate then the rate would increase with time.

There were some technical problems but they were easily circumvented. Distillation of acetaldehyde over drying agents followed by storage in vacuo at -20° provided the needed anhydrous aldehyde substrate. The need for recording reaction rates within a short time after mixing was met by fitting our recording spectrophotometer with a motor which would spew out six inches of chart paper every minute and the reaction was slowed to a manageable degree by working at 15°. Because xanthine oxidase has a substrate specificity which encompasses a variety of aldehydes and because we were thoroughly familiar with this enzyme, it was the aldehyde oxidase of choice. We wished to follow the oxidation of aldehydes spectrophotometrically so we used cytochrome *c* as the terminal electron acceptor. The experiment was done as originally conceived and the results left no doubt that the aldehyde, rather than the hydrate, was the true substrate (Fridovich, 1966). Carbonic anhydrase was then added to accelerate the hydration reaction. We expected that it would have no effect on the initial rate but would accelerate the rate decrease which accompanied the establishment of the aldehyde hydration reaction. The actual result was striking and surprising. Carbonic anhydrase acted as a competitive inhibitor of the reduction of cytochrome *c*, whether the substrate was acetaldehyde or xanthine, and its inhibitory K_i was an astonishing 2×10^{-9} M. This was reminiscent of the behavior previously seen with myoglobin and ignited our interest once again in the mechanism of the oxygen-dependent reduction of cytochrome *c*, by xanthine oxidase.

This inhibition by carbonic anhydrase was carefully explored. The carbonic anhydrase was strictly competitive with respect to ferricytochrome *c*. Sulfonamide inhibitors of carbonic anhydrase did not interfere with its ability to inhibit the reduction of cytochrome *c*. Indeed removal of the Zn(II) from carbonic anhydrase was without effect. The competitive K_i decreased as the pH increased but was unaffected by temperature and a variety of other proteins, chosen at random, did not inhibit cytochrome *c* reduction (Fridovich, 1967). We now know that superoxide dismutase, present as a minor impurity in the carbonic anhydrase, was responsible for its inhibitory action. At the time, however, it seemed most likely that a high affinity, high specificity association between carbonic anhydrase and xanthine oxidase was the basis of this inhibition. Given this point of view, it was natural to try to demonstrate a physical association between carbonic anhydrase and xanthine oxidase. This project was assigned to J.M. McCord when he entered the laboratory as a graduate student. His diligent attempts to demonstrate binding between these two proteins yielded uniformly negative results. After each experiment Fridovich suggested refinements in technique which might yield positive results and the manipulations grew ever more difficult and tedious.

It was suggested, for example, that carbonic anhydrase might bind only to the reduced form of the enzyme. Thereafter, binding experiments were repeated under strictly anaerobic conditions to prevent the air oxidation of substrate-reduced xanthine oxidase. Still, there was no evidence for binding. Finally, McCord became convinced that the accumulation of negative data was, in fact, saying something other than that the experiments were not properly designed. He wondered whether cytochrome *c* interacted with a binding site on the xanthine oxidase surface. This had been assumed, due to the apparently classical Michaelis-Menten kinetic behavior, showing a rectangular hyperbolic saturation plot when cytochrome *c* concentration was varied, and a typical Lineweaver-Burk plot. But could this behavior also be observed in a completely atypical situation - one not involving binding at all?

On April 2, 1968, McCord observed that when methylene blue was added to an aerobic xanthine oxidase-xanthine-cytochrome *c* reaction, the dye-mediated cytochrome *c* reduction yielded different kinetic constants, but still showed classical hyperbolic saturation despite the fact that the enzyme-reduced dye was thought to transfer electrons to cytochrome *c* in free solution. At high cytochrome *c* concentrations the reduced dye efficiently passed electrons between xanthine oxidase and the cytochrome. At low cytochrome concentrations, however, the reduced dye, unable to find cytochromes to accept the electrons, could simply autoxidize. That evening it occurred to him that oxygen could carry electrons in a manner completely analogous to methylene blue, through free solution. Saturation kinetics would still obtain, because at low cytochrome concentrations superoxide could largely "autoxidize" via its dismutation reaction: $O_2^- + O_2^- + 2H^+ \rightarrow H_2O_2 + O_2$. At high cytochrome *c* concentrations the complete scavenging of O_2^- by the cytochrome would be approached. At half-saturation (the apparent K_m), half the superoxide produced would disappear via the second-order dismutation, and half via first-order reaction with cytochrome *c*. If this were true, some kinetic predictions could be made. The apparent K_m for cytochrome *c* would reflect the steady-state concentration of superoxide, which in turn would reflect the rate of production of the radical. The analogy that came to mind was that of a bucket with a hole in the bottom. If a stream of water is directed into the bucket, the level of water will rise, increasing the rate of outflow at the bottom until influx and eflux are equal. If the rate of influx is changed, a new steady-state level will result. By analogy, the K_m for cytochrome *c* in the xanthine oxidase system should be a function of the concentration of xanthine oxidase, since this is the factor that determines the rate at which superoxide is produced and, hence, the steady-state concentration. The next day he asked Fridovich if it was indeed true that, in the case of binding of a substrate to an enzyme, K_m for that substrate was independent of the concentration of the enzyme. The response was affirmative, within the limitation that the concentration of the substrate always exceeds that of the enzyme. It seemed satisfactory to McCord and nothing more was said on this subject for 24 hours, during which time he produced figures 2 and 3. The mathematical relationship shown in Fig.3, $v = K\sqrt{\text{xanthine oxidase}}$, is predicted from a straightforward derivation based on the assumptions that superoxide production is first order in xanthine oxidase while its disappearance via dismutation in free solution is second order in the radical (McCord and Fridovich, 1968). These two figures provided McCord and Fridovich an understanding of the mechanism of cytochrome *c* reduction by xanthine oxidase, but more importantly they forced a breakthrough in the understanding of how the various protein preparations inhibited the phenomenon. No longer could a binding mechanism be invoked. The inhibitory proteins *had to intercept superoxide in free solution,* preventing it from reducing cytochrome *c*. There were two conceivable ways of accomplishing this feat: 1) a stoichiometric consumption of the radical, or 2) a catalytic elimination of the radical. The first possibility was immediately discarded, since carbonic anhydrase appeared to efficiently inhibit at concentrations as low as 10^{-8} M. On a mole-for-mole basis, this concentration of inhibitor would be consumed in less than a second in the assay used, whereas the observed inhibition continued indefinitely. The only possibility was a catalytic decomposition of the radical - a *superoxide dismutase.* Because the phenomenon had been thought of for so long in terms of a protein-protein interaction, the first order of business was to confirm the activity in a system free of xanthine oxidase and cytochrome *c*. McCord quickly assembled such a system based on some studies by Fridovich and Handler (1961) several years earlier on the mechanism of initiation of sulfite free radical chain oxidation by superoxide radicals generated electrochemically at a platinum electrode. As predicted, the system was quite susceptible to inhibition by this new "superoxide dismutase". When the manuscript was submitted for publication we were still ascribing this bizarre superoxide dismutase activity to the proteins carbonic anhydrase and myoglobin. A note was added in proof, however, stating that carbonic anhydrase could be cleanly separated from the superoxide dismutase activity chromatographically. With these findings in print (McCord and Fridovich, 1968), the superoxide dismutase field was launched.

The way was now clear. Starting with bovine erythrocytes, McCord isolated the superoxide dismutase on the basis of its enzymic activity. Once it was pure and in hand, the identity between this blue-green, copper-containing bovine superoxide dismutase and the previously described bovine

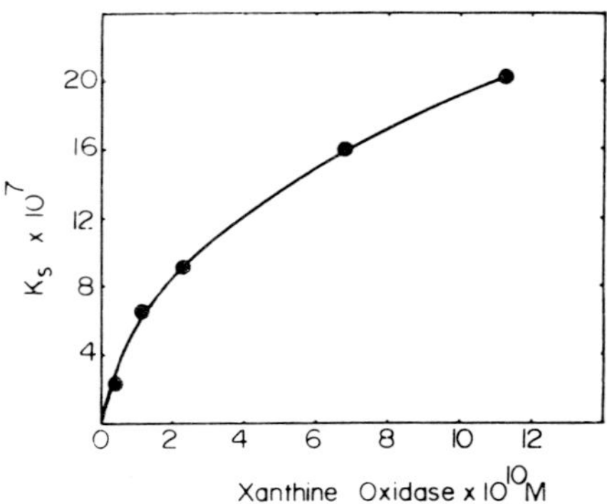

Fig.2. *Effect of enzyme concentration on K_m for cytochrome* c. *Initial rates of cytochrome* c *reduction were measured as a function of the concentration of ferricytochrome* c *at the indicated concentrations of xanthine oxidase. Plots of the data on reciprocal coordinates defined straight lines, from which the values of K_m were derived. All reactions were performed at 25° in 0.05 M potassium phosphate, pH 7.8, containing 0.1 mM EDTA. Cuvettes with a 10.0 cm light path were used. (McCord and Fridovich, 1968).*

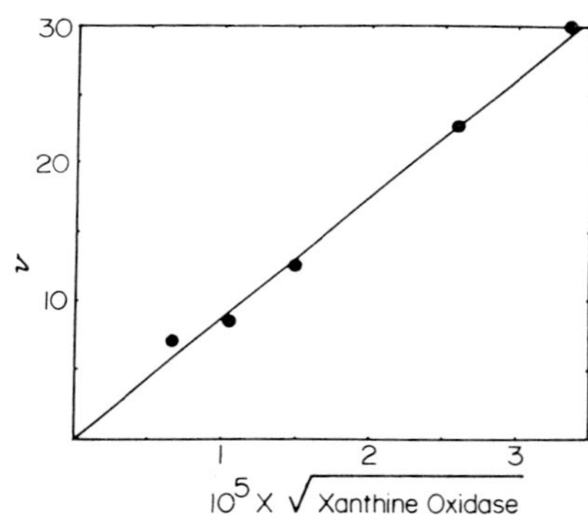

Fig.3. *Relationship between the rate of reduction of cytochrome* c *at 5×10^{-8} M and the square root of the concentration of xanthine oxidase. Reactions were carried out as in Fig.2. Initial rates which would obtain at 5×10^{-8} M cytochrome* c *were calculated by extrapolation from data at higher concentrations of cytochrome* c. *These calculated initial rates are here seen to be a linear function of the square root of concentration of xanthine oxidase. (McCord and Fridovich, 1968).*

hemocuprein (Mann and Keilin, 1938) and human erythrocuprein (Markowitz *et al.*, 1959) was apparent. These results were presented orally at the April, 1969, meeting of the American Society of Biological Chemists, and were published in the Journal of Biological Chemistry in November of that year (McCord and Fridovich, 1969). Almost immediately, scores of research publications began appearing on the subjects of superoxide and superoxide dismutases from a large number of investigators - some wildly enthusiastic, and a few unabashedly antagonistic. The chapters which have been contributed to this volume illustrate the broad spectrum disciplines which have been involved, from the physical-chemical to the human-clinical. Some studies have produced "hard" data and definitive, unequivocal results. Other studies have produced mere suggestions to stir the imagination as to the biological roles of the superoxide radical and of superoxide dismutases.

A recent survey of the literature indicated that there are nearly 500 reports dealing with the biology of superoxide and superoxide dismutases. Fortunately, there are several reviews which summarize this mass of information (Fridovich, 1972; McCord *et al.*, 1973; Halliwell, 1974; Bors *et al.*, 1974; Fridovich, 1974; Henry, 1975; Fridovich, 1975, 1975a; Michalski, 1975).

REFERENCES

1. Bäckström, H.L.J. (1927). *J. Am. Chem. Soc. 49*, 1460-1472.
2. Bors, W., Saran, M., Lengfelder, E., Stöttl, R. and Michel, C. (1974). *Curr. Top. Rad. Res. Quart. 9*, 247-309.
3. Carrico, R.J. and Deutsch, H.F. (1969). *J. Biol. Chem. 244*, 6087-6093.
4. Cohen, H.J. and Fridovich, I. (1971). *J. Biol. Chem. 246*, 359-366, 367-373.
5. Cohen, H.J., Fridovich, I. and Rajagopalan, K.V. (1971). *J. Biol. Chem. 246*, 374-382.
6. Fridovich, I. (1962). *J. Biol. Chem. 237*, 584-586.
7. Fridovich, I. (1963). *J. Biol. Chem. 238*, 592-598.
8. Fridovich, I. (1966). *J. Biol. Chem. 241*, 3126-3128.
9. Fridovich, I. (1967). *J. Biol. Chem. 242*, 1445-1449.
10. Fridovich, I. (1970). *J. Biol. Chem. 245*, 4053-4057.
11. Fridovich, I. (1972). *Accounts Chem. Res. 5*, 321-326.
12. Fridovich, I. (1974). *Advan. Enzymol. 41*, 35-97.
13. Fridovich, I. (1975). *Ann. Rev. Biochem. 44*, 147-159.
14. Fridovich, I. (1975a). *Ann. Rep. Med. Chem. 10*, 257-264.
15. Fridovich, I. and Handler, P. (1954). *Federation Proc. 13*, 212-213.
16. Fridovich, I. and Handler, P. (1955). *Federation Proc. 14*, 214.
17. Fridovich, I. and Handler, P. (1958). *J. Biol. Chem. 233*, 1578-1580, 1581-1585.
18. Fridovich, I. and Handler, P. (1959). *Biochim. Biophys. Acta 35*, 546-547.
19. Fridovich, I. and Handler, P. (1960). *J. Biol. Chem. 235*, 1835-1838.
20. Fridovich, I. and Handler, P. (1961). *J. Biol. Chem. 236*, 1836-1840.
21. Fridovich, I. and Handler, P. (1962). *J. Biol. Chem. 237*, 916-921.
22. Fridovich, I. and Westheimer, F.H. (1962). *J. Am. Chem. Soc. 84*, 3208-3209.
23. Greenlee, L.L., Fridovich, I. and Handler, P. (1962). *Biochemistry 1*, 779-783.
24. Halliwell, B. (1974). *New Phytol. 73*, 1075-1086.
25. Handler, P., Bernheim, M.L.C. and Klein, J.R. (1941). *J. Biol. Chem. 138*, 211-218.
26. Heimberg, M., Fridovich, I. and Handler, P. (1953). *J. Biol. Chem. 204*, 913-926.
27. Henry, J.P. (1975). *La Recherche 6*, 370-372.
28. Horecker, B.L. and Heppel, L.A. (1949). *J. Biol. Chem. 178*, 683-690.
29. MacLeod, R.M. Farkas, W., Fridovich, I. and Handler, P. (1961). *J. Biol. Chem. 236*, 1841-1846.
30. Mann, T. and Keilin, D. (1938). *Proc. Roy. Soc. (London) B126*, 303-315.
31. Markowitz, H., Cartwright, G.E. and Wintrobe, M.M. (1959). *J. Biol. Chem. 234*, 40-45.

32. McCord, J.M., Beauchamp, C.O., Goscin, S., Misra, H.P. and Fridovich, I. (1973). In "Proc. 2nd Int. Symp. Oxidases and Related Redox Systems", Memphis, Tenn. 1971 (King, T.E., Mason, H.S. and Morrison, M., eds.) Univ. Park Press, Baltimore, pp. 51-76.
33. McCord, J.M. and Fridovich, I. (1968). *J. Biol. Chem. 243*, 5753-5760.
34. McCord, J.M. and Fridovich, I. (1969). *J. Biol. Chem. 244*, 6049-6055.
35. Michalski, W. (1975). *Postepy. Biochem. 21*, 295-317.
36. Mohamed, M.S. and Greenberg, D.M. (1953). *J. Gen. Physiol. 37*, 433-439.
37. Morell, D.B. (1952). *Biochem. J. 51*, 666-669.
38. Pocker, Y. and Meany, J.E. (1965). *J. Am. Chem. Soc. 87*, 1809-1811.

SUPEROXIDE DISMUTASE ASSAYS: A REVIEW OF METHODOLOGY

J.M. McCORD, J.D. CRAPO and I. FRIDOVICH

Departments of Medicine and Biochemistry
Duke University Medical Center
Durham, North Carolina 27710, U.S.A.

Superoxide is the conjugate base of the weak acid $HO_2^{\cdot}$, which has a pK_a of 4.8 (Rabani and Nielsen, 1969; Behar *et al.*, 1970). Its well studied dismutation reaction occurs as follows:

$$HO_2^{\cdot} + O_2^{\bar{\cdot}} \rightarrow HO_2^{-} + O_2 \,, \quad k = 8.5 \times 10^{7}\ M^{-1}sec^{-1} \,.$$

In contrast, the reaction

$$H^{+} + O_2^{\bar{\cdot}} + O_2^{\bar{\cdot}} \rightarrow HO_2^{-} + O_2$$

proceeds very slowly if at all in the absence of catalysis, probably because of electrostatic repulsion between the radical anions. The reaction between two molecules of $HO_2^{\cdot}$ is also slower than the reaction between $HO_2^{\cdot}$ and $O_2^{\bar{\cdot}}$. As a consequence, the dismutation of these radicals is most rapid at pH 4.8 and falls an order of magnitude for each unit the pH is raised due to the progressive decline in the concentration of $HO_2^{\cdot}$. At physiologic pH the rate constant is about $1 \times 10^{5}\ M^{-1}sec^{-1}$. While this appears to be a fast reaction, we must recognize that *fast* is a relative term. The rate constant for the superoxide dismutase (SOD) catalyzed reaction under the same conditions is $2 \times 10^{9}\ M^{-1}sec^{-1}$, being first order in enzyme and first order in the radical, for a catalytic efficiency factor of 2×10^{4}. Coupled with the fact that the SOD concentration in a liver cell is greater than 10^{-5} M, whereas the steady-state superoxide concentration is probably at least four or five orders of magnitude lower, the overall gain in the rate of the dismutation *in vivo* may exceed a factor of 10^{9}. Because of the relative instability of its substrate, assaying SOD activity is less than straightforward. Direct assay has been achieved by optically monitoring the SOD-catalyzed disappearance of electron pulse-generated superoxide free radical over a millisecond time-scale (Klug *et al.*, 1972; Rotilio *et al.*, 1972), but this assay is hardly suited for general or routine laboratory use since linear electron accelerators are not commonly available. Consequently, to work in a more easily handled time-scale and with ordinary equipment, assays of SOD activity have had to rely on enzyme catalyzed alterations in *steady-state* concentrations of superoxide. Thus, all the commonly used assays which have been devised consist of two components: a superoxide generator, and a superoxide detector (see Fig.1). The generator serves to produce the radical at a controlled, constant rate. In the absence of SOD the radical accumulates to a concentration such that its rate of reaction with the detector equals its rate of generation. Typically, this steady-state is achieved in a second or so. If SOD is present, it competes with the detector for superoxide, resulting in a lowered steady-state concentration of the radical. This decreased concentration of superoxide is manifested in a decreased or inhibited rate of occurrance of the detection phenomenon being monitored.

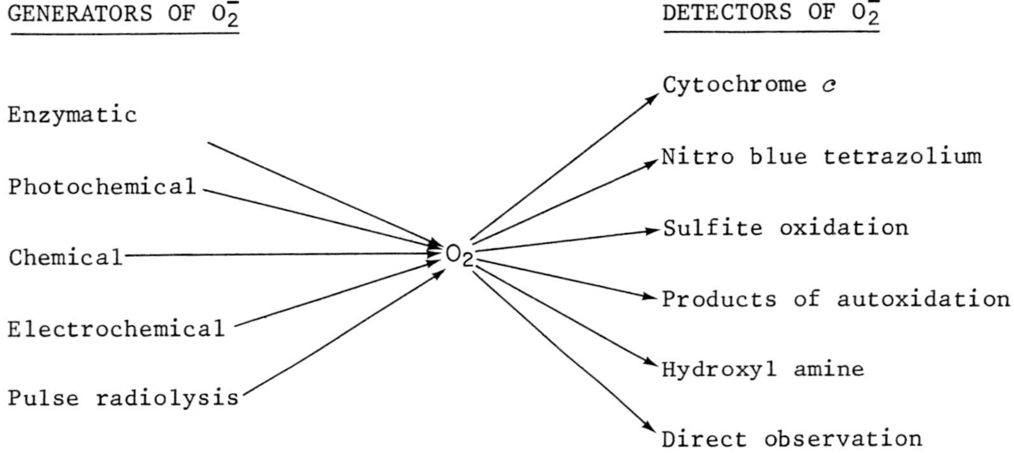

Fig.1. *Assay methods for superoxide dismutase*

One fault shared by all the assays is that saturation of the enzyme with substrate is never even approached. In the commonly used assays, the steady-state concentrations of superoxide are minuscule. Except for the assay employing pulse radiolysis where the rate constant between enzyme and substrate is measured directly, all other assays rely on a competition between SOD and the detector. Hence, the ratio of the rate constants and the concentration of the detector determine the sensitivity of the assay to SOD. Since a "unit" is defined as the quantity of SOD required to produce a given degree of inhibition under a given set of conditions, a "unit" is actually a *concentration* of SOD, not an absolute amount of the enzyme.

SOD was discovered as an inhibitor of cytochrome *c* reduction by xanthine oxidase (McCord and Fridovich, 1968). This system provided the basis for the first assay, the "standard assay", for the enzyme (McCord and Fridovich, 1969). This assay, with various modifications, has probably remained the most generally useful, versatile, and reliable, and will be discussed in some detail below. First, let us examine the relative merits and disadvantages of the various superoxide-generating systems and superoxide-detecting systems which have been coupled in various combinations to create SOD assays.

GENERATORS OF O_2^-

Enzymatic

Xanthine oxidase, as already mentioned, is a commercially available

(Sigma Chemical Co., Grade I), very well studied and well characterized source of the radical (Fridovich, 1970). While xanthine is the substrate most commonly employed, hypoxanthine and purine may also be used if long-term radical generation is desired. In addition to being more soluble, hypoxanthine provides twice as much superoxide per mole as does xanthine, and purine provides three times as much. Aliphatic aldehydes also serve as substrates for xanthine oxidase and have the advantage that neither they nor their products absorb in the near ultraviolet region of the spectrum.

To produce usable assays, the enzymatic source of superoxide has been coupled with the following detectors: cytochrome *c* (McCord and Fridovich, 1969; Salin and McCord, 1974); nitroblue tetrazolium (Beauchamp and Fridovich, 1971); sulfite oxidation (Tyler, 1975); and nitrite formation from hydroxylammonium chloride (Elstner and Heupel, 1976).

Photochemical

Massey *et al.* (1969) first reported the production of superoxide by the spontaneous re-oxidation of photo-reduced flavin, and coupled the system to cytochrome *c* reduction. Beauchamp and Fridovich (1971) coupled this photochemical generation of superoxide to the reduction of nitroblue tetrazolium (NBT) for a spectrophotometric assay, but the major contribution of this combination has been its application to polyacrylamide and agarose gel electrophoresis as a means of localizing SOD activity bands on the gels. It is ideally suited for this purpose. The reactants (riboflavin, an oxidizable substance such as tetramethylethylenediamine, and NBT) may be mixed together in subdued light and allowed to soak into the gel (see Salin and McCord, 1974). The gels may then be removed from the solution and the reaction started by exposure to light. Superoxide is photochemically generated throughout the gel, reducing NBT to the insoluble purple formazan. Achromatic zones appear where SOD scavenges the radical, preventing formazan formation. After color development, excess reactants may be soaked out of the gels, leaving the insoluble formazan conveniently and permanently locked in place.

Chemical

Several assays based on chemical reactions that produce superoxide have been proposed. Misra and Fridovich (1971) reported that the autoxidation of epinephrine to adrenochrome proceeds along at least two distinct pathways which are quite pH-dependent. At high pH the autoxidation primarily involves a free radical chain involving superoxide as a chain-mediating species. The assay is initiated by adding epinephrine to a pH 10.2 buffer and then monitoring the appearance of adrenochrome at 480 nm. SOD inhibits this reaction, and over a certain usable range the degree of inhibition is a function of SOD concentration. This assay was initially reported as convenient and sensitive (about twice the sensitivity of the standard assay) and particularly applicable to systems demanding multiple routine assays. This is true, especially when one is dealing with relatively pure preparations of SOD. However, in working with crude homogenates of tissues a great deal of assay variability is found which is frequently difficult to explain. The autoxidation of epinephrine at high pH involves a poorly understood set of free radical reactions. The system becomes even more difficult to interpret when one deals with all of the possible reactants, chain-breakers, etc., in a crude homogenate.

The major advantage to this assay over most of the others is that the superoxide involved in mediating the process appears to function as an oxidant, not a reductant. Hence, this assay may be less prone to specific interference by endogenous reductants than an assay employing a detector which must be reduced by superoxide.

Certain other substances have been found to autoxidize by mechanisms which are dependent on superoxide formation. Any of these could potentially be used as the basis of an assay for SOD. One such example is pyrogallol (Marklund and Marklund, 1974; Puget and Michelson, 1974). Pyrogallol autoxidation also proceeds by more than one mechanism, but at pH 8.2 it is essentially fully inhibitable by SOD, and hence can be used as an effective assay for the enzyme.

A purely chemical assay system described by Fried (1975) has some interesting features. The assay relies on the phenazine methosulfate catalyzed oxidation of NADH to produce superoxide, and utilizes NBT as the detector. The total quantity of superoxide produced depends on the amount of NADH present, and the rate of the oxidation depends on the concentration of phenozine methosulfate present. As described by Fried, the assay may be set up as an "end-point" assay. With no SOD present, all the superoxide produced is detected by the NBT, present in excess. With SOD present, a fraction of the superoxide produced will be catalytically dismuted, resulting in less NBT reduced when the reaction ceases. An end-point assay is particularly useful when large numbers of assays must be performed, as in screening column effluents.

Electrochemical

Superoxide may be generated at the surface of a platinum cathode in contact with aqueous solution (Fridovich and Handler, 1961). Such a system was used by McCord and Fridovich (1968) to verify SOD activity in a system not involving xanthine oxidase. Superoxide may also be generated electrochemically in aprotic solvents to produce stable, reasonably concentrated solutions of the radical. Such a solution may be slowly infused into stirred aqueous solutions as a source of superoxide. This system has been coupled with the superoxide-mediated reduction of tetranitromethane to nitroform (McCord and Fridovich, 1969). Electrochemically generated superoxide is not a very practical source of the radical for routine use.

A novel SOD assay has been described by Rigo *et al.* (1975) which is based on the polarographic reduction of molecular oxygen at a dropping mercury electrode. Normally, oxygen is reduced univalently at the electrode surface to form superoxide. The radical immediately becomes protonated and is thus able to accept a second electron, forming hydrogen peroxide. If, however, a compound such as triphenylphosphine oxide is present, it forms an aprotic molecular film over the growing mercury drop through which the nascent superoxide radical may diffuse, thereby escaping protonation and further reduction. Thus, at a given concentration of molecular oxygen in solution, triphenylphosphine oxide decreases the value of the limiting current observed to a value approaching half that observed in its absence. If SOD is present in the solution, the nascent superoxide is quickly dismuted near the surface of the electrode, regenerating one molecule of oxygen and one molecule of hydrogen peroxide for every two superoxide radicals produced. This effectively increases the concentration of oxygen in the immediate vicinity of the electrode surface, which in turn increases the value of the limiting current until it approaches that observed in the absence of triphenylphosphine oxide. Hence, over a limited range, the value of the limiting current obtained in the presence of triphenylphosphine oxide is a function of the SOD concentration in the solution. The assay is reportedly free from interferences and may be used to assay biological fluids and crude homogenates. It has the advantage of not employing a detector substance to compete with SOD and thus may be considered a *direct* assay. Potential disadvantages are that it is applicable only above pH 9 and requires equipment that may not be readily available.

DETECTORS OF $O_2^{\overline{\cdot}}$

Cytochrome c

The rate constant for the reaction between ferricytochrome *c* and superoxide at pH 7.8 is about $6 \times 10^5\ M^{-1}sec^{-1}$. The reduction of cytochrome *c* is accompanied by an increase in absorbance at 550 nm, $\Delta\varepsilon_{550} = 20{,}000\ M^{-1}cm^{-1}$ (Massey, 1959). Due to a shift in the Sorêt absorption, an even greater increase in absorbance occurs at 418 nm, $\Delta\varepsilon_{418} = 70{,}000\ M^{-1}cm^{-1}$ (Salin and McCord, 1974). Ferricytochrome *c* reduction is not, of course, a specific detector of superoxide. Tissue homogenates often contain low molecular weight compounds capable of chemically reducing cytochrome *c*, and occasionally they are abundant enough to interfere with SOD assays. Dialysis of the homogenate usually eliminates the problem. Cytochrome oxidase is another common source of frustration when assaying tissue extracts via cytochrome *c* reduction. To the uninitiated, cytochrome oxidase activity may be mistaken for SOD activity, since its re-oxidation of reduced cytochrome *c* mimics an inhibition of cytochrome *c* reduction. If cytochrome oxidase is present, however, reaction rates become distinctly non-linear, with more apparent inhibition developing as the concentration of ferrocytochrome *c* increases. Its interference may be eliminated by a concentration of cyanide low enough (10^{-5} M) not to inhibit seriously the cuprozinc SOD. A more novel and successful approach was suggested by Azzi *et al.* (1975). If cytochrome *c* is heavily acetylated it is no longer recognized as a substrate by cytochrome oxidase. It still reacts with superoxide, however, although the rate constant is approximately halved. The decreased rate constant leads to a two-fold increase in the sensitivity of assays employing acetylated cytochrome *c*.

Cytochrome *c* is commercially available (Sigma Chemical Co.) but frequently contains trace amounts (<0.01%) of cuprozinc SOD. At pH 7.8, this is usually negligible. At pH 10, however, the rate constant for the reduction of cytochrome *c* by superoxide falls an order of magnitude to about $6 \times 10^4\ M^{-1}sec^{-1}$, while the enzyme's rate constant remains at $2 \times 10^9\ M^{-1}sec^{-1}$. Under these conditions, one part SOD competes equally with 33,000 parts cytochrome *c*. If the cytochrome *c* preparation is completely free of SOD, the addition of 1 mM potassium cyanide will not affect its rate of reduction. If cyanide enhances this rate, the difference is a measure of the amount of SOD present.

Nitroblue Tetrazolium

The unique feature of this freely soluble yellow compound is that upon reduction it forms an intensely blue product (formazan) which is virtually insoluble in aqueous solution and quickly precipitates. While this property makes it ideally suited for the gel stain discussed above, its utility in spectrophotometric assays is rather limited. Fried (1975), however, reports that the inclusion of 0.33 mg gelatin per ml in the assay medium prevents the precipitation of formazan.

The rate constant for the reduction of NBT by superoxide is about 5×10^4 $M^{-1}sec^{-1}$ at pH 7.8, and drops to about 1×10^4 $M^{-1}sec^{-1}$ at pH 10.2. NBT, like cytochrome *c*, is susceptible to direct reduction by certain reductants likely to be found in crude homogenates, and can be reduced by a number of dehydrogenases via mechanisms not inhibitable by SOD.

Sulfite Oxidation

Sulfite autoxidizes by a free radical chain reaction which may be initiated by superoxide, but in which superoxide is not a chain propagating species (McCord and Fridovich, 1968). The chain length is a function of sulfite concentration and other factors, and can be manipulated such that a single initiating event results in the conversion of thousands of molecules of sulfite to sulfate with a corresponding uptake of molecular oxygen. This phenomenon allows an assay based on oxygen consumption, easily measured by microrespirometry or by an oxygen electrode. Tyler (1975) has used xanthine oxidase as a generator and sulfite oxidation, monitored polarographically, as a detector to assay SOD in rat liver fractions. The assay should be considered if excessively turbid samples rule out spectrophotometric assays.

SIMULTANEOUS ASSAY OF MANGANESE AND CUPROZINC SUPEROXIDE DISMUTASES

Since most mammalian tissues contain two types of SOD, a manganese enzyme, it is frequently desirable to assay specifically for one in the presence of the other. This can easily be accomplished with the xanthine oxidase/cytochrome *c* assay, making use of two distinguishing features of the two families of enzymes. The cuprozinc enzymes are all susceptible to inhibition by cyanide, whereas the manganese enzymes are unaffected by this ligand. (For several examples see Beauchamp and Fridovich, 1973.) The assay is unaffected by the presence of cyanide if monitored at 550 nm. (Cytochrome *c* and cyanide slowly form a complex which perturbs the 418 nm absorption, but reducibility by superoxide is unaffected.) Thus, in the presence of 1 or 2 nM cyanide, any SOD activity observed will be due to manganese (or iron) SOD. Cyanide solutions are not stable indefinitely. Solutions should be at alkaline pH and freshly prepared.

The other useful difference for distinguishing between the two types of enzymes is their variation in activity versus pH. (Again, for several specific examples see Beauchamp and Fridovich, 1973.) Typically, the ratio of activity at pH 10.2 to activity at pH 7.8 is around 10 for the cuprozinc enzymes and around 2 for the manganese enzymes. The percentage of cuprozinc SOD in a mixture of the two types of enzyme which gave a pH ratio of 5, for example, can be calculated as follows:

$$\%\ \text{cuprozinc SOD} = \frac{5 - 2}{10 - 2} \times 100 = 37.5\%$$

to use this method with accuracy, the "pH ratios" must be known for the purified enzymes from the species being assayed.

ACKNOWLEDGEMENTS

This work was supported by U.S. Public Health Service grants AM 17091 and HL 17603.

REFERENCES

1. Azzi, A., Montecucco, C. and Richter, C. (1975). *Biochem. Biophys. Res. Commun. 65*, 597-603.
2. Beauchamp, C. and Fridovich, I. (1971). *Anal. Biochem. 44*, 276-287.
3. Beauchamp, C. and Fridovich, I. (1973). *Biochim. Biophys. Acta 317*, 50-64.
4. Behar, D., Czapski, G., Dorfman, L., Rabani, J. and Schwarz, H. (1970). *J. Phys. Chem. 74*, 3209.
5. Elstner, E. and Heupel, A. (1976). *Anal. Biochem. 70*, 616-620.
6. Fridovich, I. (1970). *J. Biol. Chem. 245*, 4053-4057.
7. Fridovich, I. and Handler, P. (1961). *J. Biol. Chem. 236*, 1836-1840.
8. Fried, R. (1975). *Biochimie 57*, 657-660.
9. Klug, D., Rabani, J. and Fridovich, I. (1972). *J. Biol. Chem. 247*, 4839-4842.
10. Marklund, S. and Marklund, G. (1974). *Eur. J. Biochem. 47*, 469-474.
11. Massey, V. (1959). *Biochim. Biophys. Acta 34*, 255.
12. Massey, V., Strickland, S., Mayhew, S., Howell, L., Engel, P., Matthews, R., Schuman, M. and Sullivan, P. (1969). *Biochem. Biophys. Res. Commun. 36*, 891-897.
13. McCord, J. and Fridovich, I. (1968). *J. Biol. Chem. 243*, 5753-5760.
14. McCord, J. and Fridovich, I. (1969). *J. Biol. Chem. 244*, 6049-6055.
15. Misra, H.P. and Fridovich, I. (1972). *J. Biol. Chem. 247*, 3170-3175.
16. Puget, K. and Michelson, A.M. (1974). *Biochimie, 56*, 1255-1267.
17. Rabani, J. and Nielsen, S.O. (1969). *J. Phys. Chem. 72*, 3836.
18. Rigo, A., Viglino, P. and Rotilio, G. (1975). *Anal. Biochem. 68*, 1.
19. Rotilio, G., Bray, R.C. and Fielden, E.M. (1972). *Biochim. Biophys. Acta 268*, 605-609.
20. Salin, M.L. and McCord, J.M. (1974). *J. Clin. Invest. 54*, 1005-1009.
21. Tyler, D.D. (1975). *Biochem. J. 147*, 493-504.

CHEMICAL AND PHYSICAL PROPERTIES OF SUPEROXIDE

J.A. FEE and J.S. VALENTINE

Biophysics Research Division and Department of Biological Chemistry
University of Michigan
Ann Arbor, MI 48109, U.S.A.

and

Department of Chemistry
Douglass College, Rutgers
The State University
New Brunswick, NJ 08903, U.S.A.

I. INTRODUCTION

The chemistry and biochemistry of superoxide is currently an active area of biomedical research. This is due largely to the discovery by McCord and Fridovich (1) in 1969 that a copper containing protein, erythrocuprein, isolated from red blood cells could catalyze the dismutation of superoxide ions. This protein was renamed superoxide dismutase and has attracted widespread interest. At the same time, study of the biochemistry of superoxide has been stimulated by the suggestion (2,3) that superoxide toxicity may be an important factor in the etiology of oxygen poisoning (4). Moreover it was found that the presence of superoxide dismutases affected the course of many biological reactions involving dioxygen. As yet, few chemical or biochemical reactions of superoxide itself have been discovered which are of a type generally acknowledged to be deleterious to the living cell. Indeed, if superoxide ion is a significant cytotoxin, the specific reactions involved remain to be discovered.

The purposes of this review are to outline some of the physical properties of the superoxide ion, to examine its chemical properties in some detail, to compare them with those of dioxygen, hydrogen peroxide and hydroxyl radical, and to suggest some hitherto unexplored avenues of research into the biological properties of superoxide ion.

Our discussions of the chemical reactions of superoxide will not include reactions where the only evidence for the involvement of superoxide comes from the use of superoxide dismutase enzymes. This is not to imply that these proteins are never a reliable probe for superoxide involvement, but it serves to limit the scope of the review to physically and chemically well defined systems. This review also will not consider the purely inorganic aspects of superoxide chemistry as these are discussed rather fully in the monograph of Vol'nov (5) and the review of Vannerberg (6). No attempt will be made to review the reactivity and spectroscopic properties of superoxide in the gas phase (7) or when associated with surfaces or of the several detailed molecular orbital calculations which have been carried out on this ion and the perhydroxyl radical (9).

II. BONDING AND ELECTRONIC STRUCTURE

A. *Bonding in Dioxygen and Related Molecules*

Before beginning a discussion of the bonding and electronic structure of dioxygen and related species, it is appropriate to review the nomenclature of these species. Elemental oxygen, O_2, is properly referred to as dioxygen, although the term molecular oxygen is often used. One electron reduction gives superoxide, O_2^-, which is also referred to as superoxide anion or superoxide radical. The product of protonation of superoxide, HO_2, will here be referred to as perhydroxyl radical although the terms hydroperoxy or hydroperoxyl radical are often seen. Two electron reduction of O_2 gives peroxide, O_2^{2-}; peroxide or peroxy anion, HO_2^-; or hydrogen peroxide, H_2O_2, depending on the state of protonation. Related organic derivatives are ROOR, dialkyl peroxides, and ROOH, alkyl hydroperoxides. Other species that are mentioned in this review are HO·, hydroxyl radical, and O_2^+, dioxygenyl cation. Bond lengths and energies of dioxygen and related molecules are presented in Table I.

TABLE I

Oxygen-Oxygen Bond Distances and Strengths

	Example	O-O(Å)	D(kcal/mole)	Reference
O_2^+ dioxygenyl	O_2PtF_6	1.12	-	36
O_2 dioxygen		1.21	118	19,36
O_2^- superoxide	LiO_2	1.33		143
	HO_2	-	64	39
O_2^{2-} peroxide	Na_2O_2	1.49	-	143
	H_2O_2	1.49	51	24,37,39
	ROOR[a]	-	38	38,39

[a]R = alkyl

It can be seen that the oxygen-oxygen bond length becomes progressively greater on going from O_2^+ to H_2O_2 and that there is a decrease in the strength of the O-O bond accompanying this serial reduction. It is also notable that the bond strength of the O-O bond in peroxides is dependent on the nature of the molecular fragment bonded to each oxygen atom.

A general explanation of these facts is offered by study of the molecular orbital diagram shown in Fig.1a. The $2p\sigma_g$ and $2p\pi_u$ bonding orbitals of dioxygen are fully occupied with two additional electrons in the degenerate $2p\pi_g^*$ antibonding orbitals. For such an orbital occupancy, a bond order of 2.0 is predicted. Oxidation of O_2 to form O_2^+ removes an antibonding electron, increases the bond order to 2.5, and thus causes a

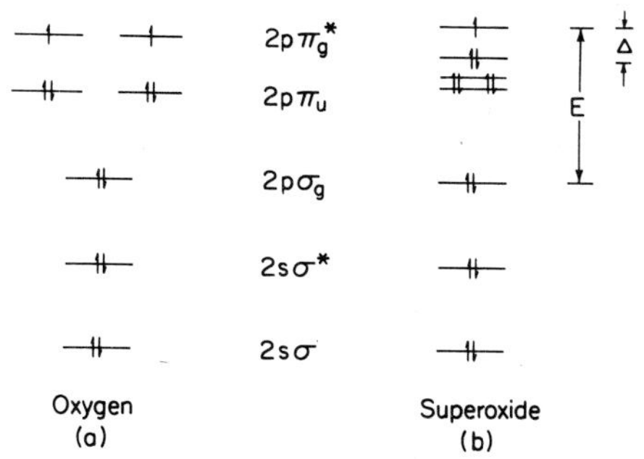

Oxygen (a)		Superoxide (b)
⊣⊢ ⊣⊢ (one electron each)	$2p\pi_g^*$	split levels, Δ
⫲ ⫲	$2p\pi_u$	⫲⫲
⫲	$2p\sigma_g$	⫲
⫲	$2s\sigma^*$	⫲
⫲	$2s\sigma$	⫲

Fig.1. *Qualitative molecular orbital scheme for dioxygen and superoxide radical.*

shortening of the O-O bond. Similarly, O-O bond orders are predicted to decrease for O_2^- (1.5) and O_2^{2-} (1.0) because electrons are added to anti-bonding orbitals. The bond lengths and energies of these species are in accord with such predictions (Table I). The O-O bond in dioxygen may be thought of as one sigma bond and one π bond, in superoxide as one sigma bond and one-half π bond, while in peroxide the one bond is entirely sigma in character. In hydrogen peroxide and organic derivatives, each oxygen atom is sigma bonded to a hydrogen atom or a carbon atom and also has two non-bonded electron pairs. Repulsive interactions between these non-bonded pairs along the O-O axis are responsible in part for the relative weakness of the O-O bond in such compounds.

B. *Spectroscopic and Magnetic Properties*

i. *Dioxygen*

The above molecular orbital description of the dioxygen molecule is useful in understanding the spectral and magnetic properties. Thus ground state dioxygen is predicted to contain two unpaired electrons and is in fact found to be a triplet. Potential energy curves for dioxygen are shown in Fig.2. Transitions have been shown to occur between the ground state and the indicated excited states. The three lowest energy states result from the fact that the two electrons can be placed in the two degenerate $2p\pi_g^*$ orbitals in three different ways as shown in order of increasing energy in Fig.3. The transitions from the $^1\Delta_g$ or $^1\Sigma_g^+$ excited states to $^3\Sigma_g^-$ ground state are forbidden; this accounts both for their extremely low transition probabilities and for the fact that the lifetimes of these excited "singlet" states are long compared to excited states which can decay via allowed transitions. The observation of "singlet oxygen chemistry" is due to the relatively long lifetimes of these excited states (8).

ii. *Superoxide*

The spectral and magnetic properties of superoxide are also best discussed with reference to a simplified molecular orbital scheme as shown in Fig.1b. The superoxide ion is formed by adding one electron to one of the degenerate $2p\pi_g^*$ orbitals. The unequal occupancy of these orbitals causes a Jahn-Teller distortion which leads to splitting of these levels (9). This qualitative molecular orbital scheme is useful in predicting some of the properties of the superoxide ion. The paramagnetism of superoxide, for example, allows study by several techniques, and the fact that an electronic transition between the $2p\sigma_g$ and $2p\pi_g{}^*$ states is possible suggests that this transition may be observable by spectroscopic techniques. These properties are discussed in the following sections.

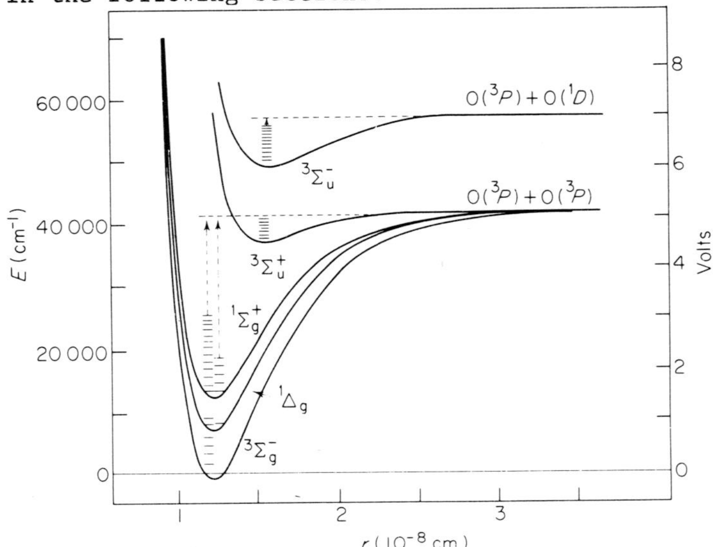

Fig.2. *Potential energy surfaces of ground and excited states of dioxygen. Taken from Reference 19.*

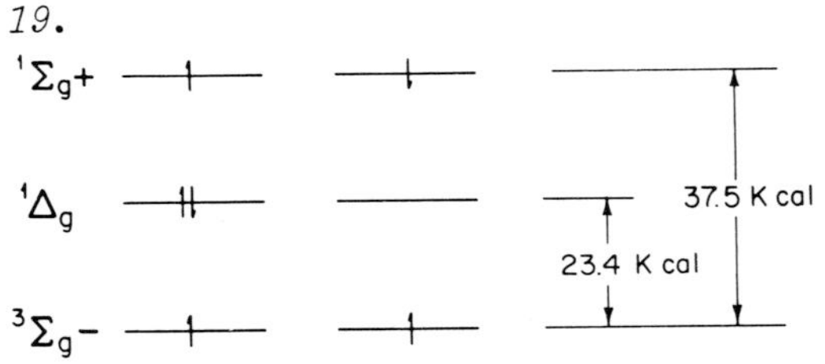

$^1\Sigma_g+$

$^1\Delta_g$

$^3\Sigma_g-$

37.5 K cal

23.4 K cal

Fig.3. π^* *orbital occupancy and energies of triplet and singlet dioxygen.*

Paramagnetism

Early work on superoxide showed that it was indeed paramagnetic containing one unpaired electron (10). Most systems containing one unpaired electron are readily observable by electron spin resonance techniques and for this reason superoxide has been widely studied. A typical e.p.r. spectrum is shown in Fig.4. There can be rather substantial variation of the g-values when the ion is bound to cations (see below). Känzig and Cohen (11) have offered theoretical expressions for the g-values of superoxide ions (equations 1-3,

$$g_{zz} = g_e + 2\,[\lambda^2/(\lambda^2 + \Delta^2)]^{\frac{1}{2}}\ell \qquad 1$$

$$g_{yy} = g_e\,\left(\frac{\Delta^2}{\lambda^2+\Delta^2}\right)^{\frac{1}{2}} - \frac{\lambda}{E}\,\left[-\left(\frac{\lambda^2}{\lambda^2+\Delta^2}\right)^{\frac{1}{2}} - \frac{\Delta}{(\lambda^2+\Delta^2)^{\frac{1}{2}}} + 1\right] \qquad 2$$

$$g_{xx} = g_e\,\left(\frac{\Delta^2}{\lambda^2+\Delta^2}\right)^{\frac{1}{2}} - \frac{\lambda}{E}\,\left[\left(\frac{\Delta^2}{\lambda^2+\Delta^2}\right)^{\frac{1}{2}} - \frac{\Delta}{(\lambda^2+\Delta^2)^{\frac{1}{2}}} - 1\right] \qquad 3$$

where Δ and E are defined as in Fig.1, λ is the spin orbit coupling constant taken to be 226 cm^{-1}, and ℓ is the orbital reduction factor which has a value less than but near unity).

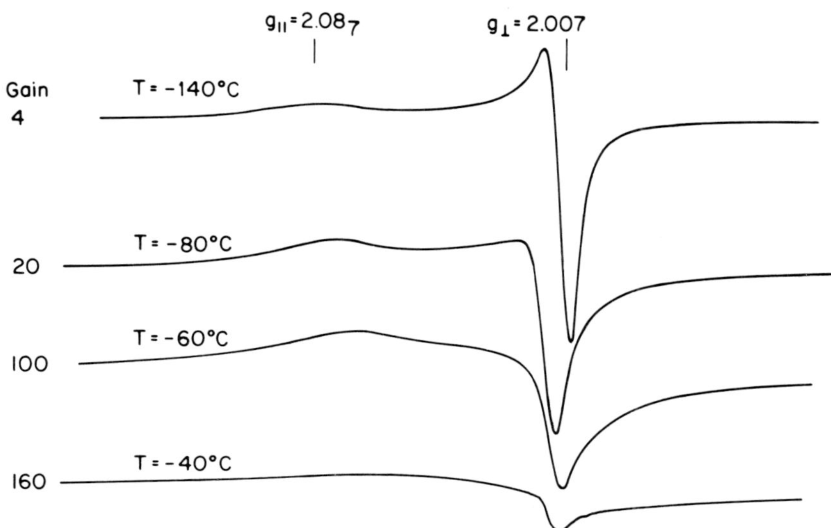

Fig.4. *E.p.r. spectrum of superoxide ion in DMSO at various temperatures.*

The value of Δ will be determined by the ionic environment of the superoxide ion, i.e. by the "crystal field" in which the ion finds itself. Lunsford (12) has reviewed the literature concerned with superoxide adsorbed to surfaces and has noted a correlation between the value of g_z ($=g_{zz}=g_{\parallel}$) and the oxidation state of the metal ion with which the superoxide is associated on the solid. Thus the value of g_z decreased from 2.436 to 2.015, corresponding to changes in Δ from 480 cm^{-1} to 17,300 cm^{-1}, when the superoxide ion is observed in bulk KCl and in solid Mo(VI)/Al_2O_3, respectively.

It has been known for some time that the value of $g_{\parallel}$ in the e.p.r. spectrum of O_2^- is variable, apparently depending on the nature of the solvent. This may now be understood as an extreme sensitivity of Δ toward the environment of the superoxide ion. Recent results of Bray and co-workers (13) show that the $g_{\parallel}$-value of O_2^- is markedly shifted toward $g_{\perp}$ in the presence of cations such as Ca^{2+}, Ba^{2+}, and even high concentrations of Na^+, indicating a complexation of O_2^- with these ions.

The e.p.r. spectrum of superoxide exhibits an unusual temperature dependence as shown in Fig.4. It is clear that at approximately -80°C the spectrum begins to broaden and at -40°C it is barely detectable, but on lowering the temperature the original signal reappears. This result suggests that superoxide will not be observable by e.p.r. in aqueous solutions at room temperature as has been discussed by Czapski (14). When superoxide is protonated to form HO_2, however, a signal having a linewidth of 27 gauss is observable at room temperature (Section IV.C.iii). When HO_2 binds to diamagnetic metal ions a narrow e.p.r. signal can be observed in the fluid state which is typical of most free radicals. This unusual behavior has not been subjected to theoretical analysis; it probably has its origin in some type of dynamic Jahn-Teller effect which is suppressed when the value of Δ in Fig.1

becomes very large, as, for example, when superoxide is bound to a metal ion.

Pshezhetskii *et al.* (15) have reviewed the literature relevant to the e.p.r. spectrum of HO_2 at low temperatures. This radical yields a spectrum similar to that of O_2^- except for a hyperfine splitting, A, of 10-12 gauss arising from a single proton. The reported values of g and A obtained under various environmental conditions are tabulated in reference 15.

Optical Properties

The molecular orbital scheme of Fig.1 indicates a transition between the $2p\sigma_g$ and $2p\pi_g^*$ orbitals (11). This transition has been observed in the near ultraviolet region, and the literature concerning this transition in aqueous and non-aqueous solutions of superoxide has been reviewed by Czapski (14) and Bielski and Gebicki (16). While there is a great deal of controversy concerning the intensity of the absorption, there is no question that superoxide does absorb light maximally in the 240 to 260 nm range. Bielski and Gebicki (16) concluded from their study of the literature that the absorption maxima of superoxide and the perhydroxyl radical were both at 240 nm and that the extinction coefficients were within error identical, 1000 ± 100 $M^{-1}cm^{-1}$. They report that the spectrum of O_2^- does differ from that of HO_2, however, in that it tends to much longer wavelengths, being significant even well beyond 300 nm. Czapski (14) on the other hand has concluded that the absorption maximum of the HO_2 radical is at 230 nm and the extinction coefficient is 1100-1200 $M^{-1}cm^{-1}$. Moreover he concludes that the spectrum of O_2^- is shifted to 240 nm, its overall spectrum is broader than that of HO_2, and the best value of the extinction coefficient at 240 nm is 1950 ± 50 $M^{-1}cm^{-1}$. We offer no clarification.

There have been conflicting reports in the literature concerning the possibility that superoxide possesses an absorption band in the visible region. Many solutions of superoxide in aprotic solvents have been reported to be colored, but it is by no means certain that this absorption arises from superoxide itself. Indeed it is very likely that superoxide ion can enter into some interaction with solvent molecules to form highly colored species (cf. References 14 and 15 for further discussion), but this has not been investigated thoroughly. The present authors have found that 2 mM solutions of superoxide prepared electrochemically in dry acetonitrile (17) and 0.2 M solutions of KO_2 in DMSO/crown ether (64) show no bands with maxima in the visible region. The latter solution is yellow due to the fact that the intense u.v. band tails into the visible, but no new transitions in this region are apparent. It is possible that solvent interactions influence the shape of the u.v. band and may cause it to have some intensity in the visible region. This has not been systematically investigated.

It is possible that the Δ transition may be observable in the visible region in ionic complexes such as, for example, solid salts of superoxide, but an explicit assignment of this transition does not appear to have been reported for O_2^- in solution (16).

Detection of Superoxide

From the above discussion it can be seen that there are two physical properties of superoxide which can be used in its detection. The ultraviolet absorption will of course be useful only in the absence of interfering chromophores, but the e.p.r. spectrum is unique and could be extracted even from a complicated set of overlapping signals. In this regard, the temperature dependence of this signal (see Fig.4) may be useful in detecting and assigning the superoxide signal. It is noteworthy that e.p.r. was used in the first physical demonstration of the formation of O_2^- in a biological system (18). Another procedure involving e.p.r. spectroscopy but useful for detecting very low concentrations of O_2^-, i.e. spin-trapping, will be discussed below (Section IV.C.vii).

iii. *Hydrogen Peroxide*

Hydrogen peroxide is diamagnetic. It absorbs light continuously from 410 nm (ε = 0.001) to 200 nm (ε = 140 $M^{-1}cm^{-1}$). Absorption of light results in decomposition into two ·OH radicals (19).

III. IONIZATIONS AND REDOX POTENTIALS OF DIOXYGEN AND RELATED SPECIES

A. *Ionizations*

The discussion of the redox potentials of the dioxygen, superoxide and hydrogen peroxide couples must be prefaced with some knowledge of the proton binding properties of the various species in water. These are summarized in Fig.5, where the values on the horizontal correspond to the observed pK_a values for that ionization. Dioxygen is not protonated even in strong acids, superoxide accepts a proton to give HO_2 which has a characteristic pK_a of 4.5 - 4.9 (20-22), and hydrogen peroxide is a relatively weak acid, releasing a proton with pK_a of 11.7 (19). The second ionization of hydrogen peroxide does not appear feasible to any extent in aqueous solution, and the pK_a given for this step was calculated but not measured (23). Thus in aqueous medium at pH 7, O_2^- would be fully deprotonated and H_2O_2 would be fully protonated.

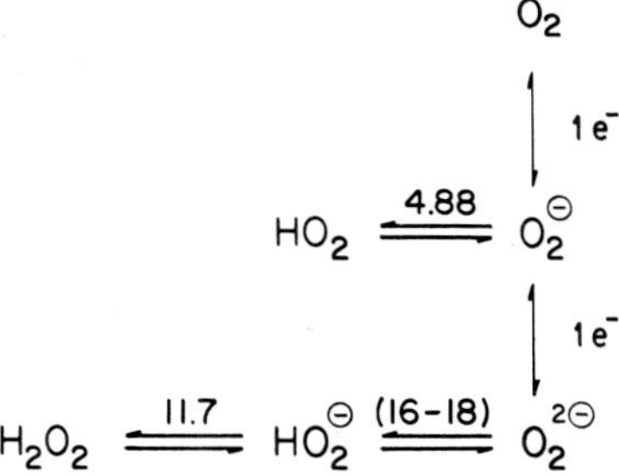

Fig.5. *Acid-base characteristics of dioxygen, superoxide, and hydrogen peroxide. The numbers on the horizontal correspond to pK_a values.*

The energetics of dioxygen reduction to water in aqueous solution have recently been tabulated and interpreted by Philip George (24). His results, incorporating the correction of Sutin (cf. Reference 24), for the redox potentials of the various couples at pH 7 are given in Table II. In what follows, the O_2/O_2^- and O_2^-/H_2O_2 couples will be discussed in some detail and some revised values for their potentials, also presented in Table II, will be considered.

B. *The O_2/O_2^- Couple*

The value reported in reference 24 for this couple is −0.59 V. This value was calculated from the thermodynamic parameters listed in Table 5 of George's article (24) and it agrees quite well with the value calculated earlier by Latimer (25). While this might tend to inspire confidence in the

TABLE II

Oxidation-Reduction Potentials for the Reduction of Oxygen to Water Via Hydrogen Peroxide at pH 7.0[a]

		$E^{o'}$ (volts) George (24)	$E^{o'}$ Revised (see text)
$O_2 + e^-$	$\rightleftarrows O_2^-$	-0.59	<u>-0.33</u>
$O_2^- + e^- + 2H^+$	$\rightleftarrows H_2O_2$	1.12	<u>0.87</u>(0.94)
$H_2O_2 + e^- + H^+$	$\rightleftarrows OH\cdot + H_2O$	0.38	
$OH\cdot + e^- + H^+$	$\rightleftarrows H_2O$	2.33	
$1/2\ O_2 + e^- + H^+$	$\rightleftarrows 1/2\ H_2O_2$	<u>0.27</u>	(0.305)
$1/2\ H_2O_2 + e^- + H^+$	$\rightleftarrows H_2O$	1.35	
$1/4\ O_2 + e^- + H^+$	$\rightleftarrows 1/2\ H_2O$	0.81	

[a]Our estimate of most reliable values of those disputed are underlined

correctness of the number, there has actually been considerable disagreement in the literature concerning the value of this redox potential; some of the proposed values are given in Table III. Until very recently the empirical estimates of the potential of the couple had been based on simple kinetic arguments rather than on any thermodynamic measurement. The value of Latimer was questioned on this basis by Yamazaki and Piette (26) who argued that the value of the potential should be between 0 and -0.3 V.

The potential of the dioxygen/superoxide couple has now been measured under a variety of conditions by different methods. The results are summarized in Table III. It can be seen that the potential depends on the nature of the solvent in which the measurement was made. The average value (of those given in the Table) in aprotic solvents is -0.57 V while in water the value is -0.31 V. This difference, as was pointed out by Chevalet *et al.* (27), is due to a less favorable solvation of the superoxide anion in aprotic solvents. Stated differently, the thermodynamic stability of superoxide increases with the ability of the solvent molecules to solvate the anion. It is expected therefore that the potential required to generate the anion would continue to decrease as some general measure of the solvating capability of the solvent decreased, such as the bulk dielectric constant. Solvation of superoxide ion dramatically affects its chemical reactivity; this will be discussed in some detail below (Section III.E.).

C. *The O_2^-/H_2O_2 Couple*

Given reliable values for the potentials of the couples O_2/O_2^- and O_2/H_2O_2, the value of the O_2^-/H_2O_2 couple can be calculated from equation 4.

$$E^{o'}(O_2^-/H_2O_2) = 2\,E^{o'}(O_2/H_2O_2) - E^{o'}(O_2/O_2^-) \qquad [4]$$

TABLE III

Oxidation-Reduction Potentials of the O_2/O_2^- Couple

Potential, Volts (vs. NHE) O_2/O_2^-	System	References
Calculated		
-0.56		(25)
-0.59		(24)
-0.19		(34)
-0.125 to 0.004		(23)
Measured		
-0.5	DMSO/0.1 M tetramethylammonium chloride	(43)
-0.53	DMSO/0.1 M tetrabutylammonium perchlorate	(44)
-0.49	DMSO/0.1 M tetrabutylammonium perchlorate	(45)
-0.61	DMSO/0.1 M tetraethylammonium perchlorate	(46)
-0.63	DMF/0.1 M tetrabutylammonium perchlorate	(44)
-0.56	DMF/0.1 M tetrabutylammonium perchlorate	(45)
-0.58	acetonitrile/0.1 M tetrabutylammonium perchlorate	(44)
-0.65	pyridine/0.1 M tetrabutylammonium perchlorate	(44)
-0.64	acetone/0.1 M tetrabutylammonium perchlorate	(44)
-0.55	methylene chloride/0.1 tetrabutylammonium perchlorate	(44)
-0.33	pH 7, aqueous, coupled to duroquinone	(28)
-0.31	pH 7, aqueous, coupled to duroquinone	(28,40)
-0.33		(41)
-0.27 to -0.33	pH 7, aqueous, coupled to benzoquinone and cytochrome *c*	(42)
-0.33	pH 7.2, aqueous, coupled to 2.5-dimethylbenzoquinone	(40)
-0.28	pH 12, aqueous solution	(34)
-0.27[a]	1 M sodium hydroxide	(27)

[a]Divisek and Kastening (34) suggest that the number here should be -0.29 V

There seem to be two values for the $E^{o\prime}$ of the O_2/H_2O_2 couple which could be selected. In his monograph entitled "The Electrochemistry of Oxygen", Hoare (175) records values between ∿ 0.25 and 0.28 V, and he suggests a value of 0.27 may be most reliable. Indeed, this is consistent with the known free energy of formation of H_2O_2 in water, -31.47 kcal/mole (24), which also corresponds to 0.27 V for $E^{o\prime}$. The other value which should be considered comes from the assessment of Wood (28) who has recently reviewed some experimental data and suggests the value of $E^{o\prime}$ to be 0.305 ± 0.005 V. This value implies that the free energy of formation of H_2O_2 is 1.8 kcal/mole too positive. This seems unlikely, and it is possible that measured potentials reviewed by Wood (28), contain some systematic contribution amounting to approximately 0.04 V.

Taking a value of -0.33 V for the O_2/O_2^- couple (28) and 0.27 V for the O_2/H_2O_2 couple (24) the potential of the O_2^-/H_2O_2 half-reaction is 0.87 V. The corresponding values based on the evaluation of Wood are placed in parentheses in Table II.

Certain conclusions can be drawn from the data presented in Tables II and III concerning the chemical and biochemical properties of superoxide ion. (a) Superoxide production can clearly be driven by a number of biological reducing agents in water as experience now dictates must be possible. (b) It is unlikely, however, that these same reducing agents will form O_2^- in a lipophilic medium due to the lower redox potential. (c) Superoxide does not appear to be as strong an oxidant as was implied by the value 1.12 V of George (24) for the O_2^-/H_2O_2 couple.

D. *pH Dependence of the Redox Couples*

As a visual aid to the reader, the information in Fig.5 and Table II is combined in Fig.6 to show the redox potentials of the various couples over the 0-14 pH range.

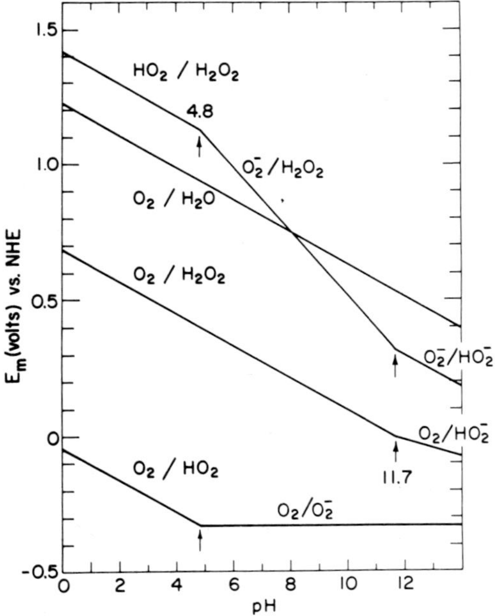

Fig.6. *Reduction potentials for the indicated couples as a function of pH. Values were taken from Table II and Fig.5.*

E. *Solvation of Superoxide*

Kebarle and his co-workers (29-31) have carried out a number of studies on the solvation of ions in the gas phase. These workers have measured the thermodynamic parameters for the formation of mono, di, tri, etc. solvate complexes of the halogen anions and of superoxide. Each anion can bind up to three water molecules with considerable affinity and the order of decreasing affinity for any particular state of hydration is $F^- > O_2^- > Cl^- > Br^- > I^-$. Yamdagni *et al.* (31) have suggested that O_2^- may be considered as an anion having a "radius" falling between those of F^- and Cl^- and these authors predicted that the total enthalpy of hydration of O_2^- in water would fall between that of F^- (-113 kcal/mole) and that of Cl^- (-81.3 kcal/mole), suggesting a value of approximately -100 kcal/mole. Indeed Baxendale *et al.* (32) sometime earlier had calculated the enthalpy of hydration of O_2^- based on measurements of the heat of ionization of superoxide and found it to be -94 kcal/mole, substantially more negative than the value of -70 kcal/mole calculated by Evans and Uri (33).

Further evidence for the very strong interaction between water and superoxide comes from a comparison of the redox potentials measured in water with those measured in organic solvents (cf. Table III). The average redox potential in the organic solvents listed is 0.26 V lower than the average value determined in aqueous solution. As Chevalet *et al.* (27) have pointed out, considerable free energy would be released on transferring O_2^- from the organic solvent to water, i.e. approximately 6 kcal/mole.

What relevance do these observations have to the chemistry of superoxide? First let us consider the effect of solvent on the redox potential of the O_2/O_2^- couple. The data suggests that O_2^- in non-solvating solutions is a rather strong reducing agent. The immediate implication is that O_2^- will not be formed in the hydrophobic environments of a cell because of the paucity of biological reducing agents as strong as -0.57 V. Thus superoxide formation should be restricted to a medium which allows adequate solvation, i.e. the aqueous milieu.

The strong solvation of superoxide in water will also influence its nucleophilicity. Since O_2^- has considerable nucleophilic character in non-aqueous systems (Section IV.D.i.), it might also be surmised to be equally nucleophilic in aqueous solution. This cannot be the case and indeed Divisek and Kastening (34) have obtained results which suggest aqueous O_2^- is only a mild nucleophile toward alkyl bromides. A good deal more supporting evidence can be found by considering the analogous case of fluoride ion which in organic solvents offering poor solvation is a powerful nucleophile. Indeed "naked" fluoride is an interesting new reagent in organic chemistry (121). By contrast, fluoride ion in aqueous solution is well known to be a very poor nucleophile (cf. reference 35). Thus one should not expect the chemical behavior of superoxide in an organic solvent to be emulated in an aqueous environment.

IV. CHEMISTRY OF SUPEROXIDE

A. *Methods of Preparation in Aqueous Solution*

Since superoxide is very unstable toward dismutation in protic solvents such as water, special procedures have been designed to generate either a "pulse" or a steady state "flux" of superoxide in aqueous solution. By contrast, superoxide is quite stable in aprotic solvents, and solutions of O_2^-

having concentrations as high as 0.2 M have been prepared and are stable over long periods of time. In this section we will consider methods for producing a "pulse" of O_2^- in aqueous solution and for creating a "flux" of O_2^- by chemical or electrochemical means.

i. *"Pulse" Procedures*

a. Pulse radiolysis. Superoxide can be produced in an oxygentated aqueous solution by the passage of a beam of electrons through the solution. The processes leading to superoxide have been discussed in detail in the monograph by Draganic and Draganic (47) and are presented here only in abbreviated form:

$$H_2O \xrightarrow{\sim\sim} H_2O^* \rightarrow OH\cdot + H^+ + e_{aq} \quad (1)$$

$$e_{aq}^- + O_2 \rightarrow O_2^- \quad (2)$$

The yield of O_2^- can be substantially increased by including formate which reacts with OH·; the product of this reaction goes on to react with O_2 to yield superoxide (reactions 3 and 4). Concentrations of O_2^- as high as 200 μM can thus be obtained within a few microseconds after passage of the

$$OH\cdot + HCO_2^- \rightarrow H_2O + CO_2^{\dot{-}} \quad (3)$$

$$CO_2^{\dot{-}} + O_2 \rightarrow O_2^- + CO_2 \quad (4)$$

electron beam through the solution. This technique has proved most valuable in investigating the chemical properties of the superoxide ion as well as the mechanism of superoxide dismutases.

b. Photolysis of H_2O_2. Superoxide can be produced in aqueous solution by the photolysis of H_2O_2 with ultraviolet radiation (48). The sequence of events which occurs is as follows:

$$H_2O_2 \xrightarrow{h\nu} 2\ OH\cdot \quad (5)$$

$$OH\cdot + H_2O_2 \rightarrow H_2O + HO_2 \quad (6)$$

$$HO_2 \rightarrow O_2^- + H^+ \quad (7)$$

At low pH reaction 7 will not occur (cf. Fig.5). The yield of superoxide in this process is comparable to that obtained using the pulse radiolysis techniques but has the disadvantage that fairly high concentrations of peroxide must be present.

c. Stopped-flow. The pulse radiolytic methods for studying superoxide chemistry have the advantageous feature of introducing O_2^- into aqueous solution in a very short time period. The technique suffers most from its requirement of specialized sources of ionizing radiation which are available at only a few research institutions. A procedure for mixing superoxide stabilized in an aprotic solvent with water in a time period of slightly less than 6 msec has recently been developed in one of our laboratories (49) and some of the results obtained with this system will be presented below.

ii. *Chemical "flux" Methods*

Superoxide can be formed *in situ* by the oxidation of hydrogen peroxide by IO_4^- (18,50) or by Ce(IV) (51) ions. It can also be formed from molecular oxygen and a number of substances having a structure basically derived from quinol. Production of superoxide by systems of the latter type has generally been inferred by using superoxide dismutase as a means of modifying the kinetics of the autoxidation reaction. The specific case of flavin autoxidation has been studied in more detail by Ballou *et al.* (52) who correlated superoxide production and flavin autoxidation by utilizing a combination of rapid freeze-quench, e.p.r., and optical techniques. The time course of this reaction at pH 8.4 and 10.6 is shown in Fig.7. Addition of small amounts of superoxide dismutase drastically lowered the yield of superoxide and provided one of the early unequivocal demonstrations of the catalytically promoted dismutation of superoxide. In spite of the apparent simplicity of flavin autoxidation as a well defined source of superoxide, there are several observations which suggest that the light-mediated reduction of flavin is more complicated than when flavin that is otherwise reduced is used as a source of superoxide (53). One complication was noted by Strickland and Massey (54) who showed that flavin-EDTA mixtures under continuous irradiation with visible light would effect a dioxygen-dependent, superoxide dismutase-inhibitable hydroxylation of p-hydroxybenzoic acid but that superoxide itself would not effect the same hydroxylation when introduced as a solution in an aprotic solvent. This observation was later confirmed by Fee and Hildenbrand (17). Another complexity noted in this system was that of Stoien and Wang (55) who found that flavin may participate in direct photochemical reaction with organic substances such as tryptophan. As a general precaution all such methods for *in situ* production of superoxide should be considered as potentially complicated by side reactions which are unrelated to superoxide production or action.

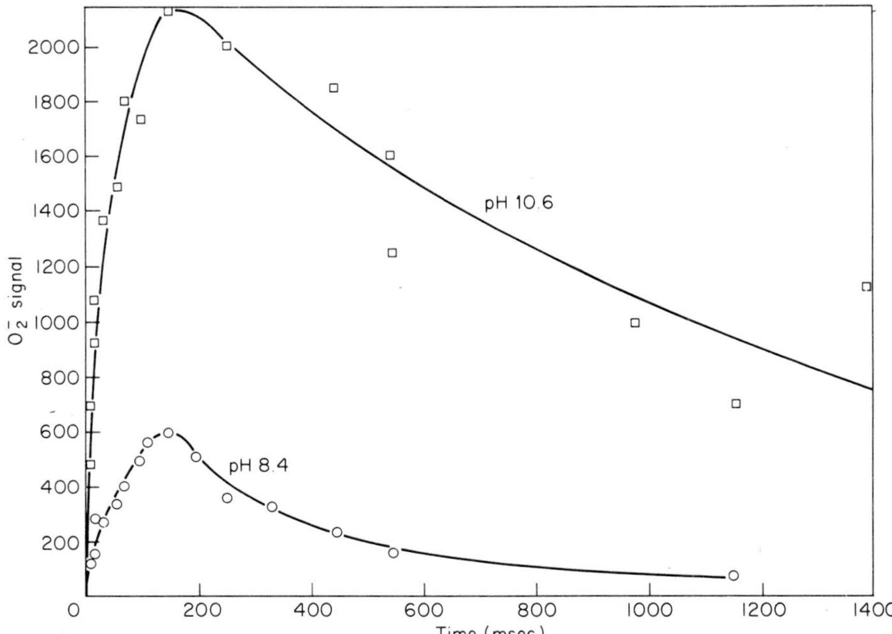

Fig.7. *Formation and decay of superoxide during the reaction of oxygen with fully reduced tetraacetylriboflavin. Taken from Reference 52.*

iii. *Electrochemical Procedures*

Kastening and Kazemifard (56) demonstrated in 1970 that the cathodic reduction of O_2 in aqueous solution can be restricted almost completely to the transfer of one electron to form superoxide. This was also shown independently by Chevalet *et al.* (27) who have offered a most lucid discussion of the rationale for producing superoxide at the dropping mercury electrode without the interfering reduction to hydrogen peroxide. The reduction of dioxygen at a cathode suspended in an aqueous medium usually occurs essentially by a two electron reaction. This reaction can be broken down into its two one-electron steps in the following manner (reactions 8 and 8a).

$$Me(e^-) + O_2 \xrightarrow{\text{slow}} Me + O_2^- \qquad (8)$$

$$Me(e^-)\,H_2O + O_2^- \longrightarrow Me + O_2H^- + OH^- \qquad (8a)$$

($Me(e^-)$ represents electrode surface with available electron. $Me(e^-)\,H_2O$ represents same with a bound water molecule.)

In the presence of a surfactant such as quinoline (27) or triphenylphosphine oxide (56), the second step is retarded since the adsorbed water molecule which is acting as a proton donor in the second step is excluded by a compact, hydrophobic film of surfactant molecules covering the electrode. Thus under these conditions, O_2 will be reduced to superoxide by electron transfer across the hydrophobic film with the second step being blocked, and the superoxide will diffuse away from the electrode, its fate now being dependent only upon its own chemical reactivity. Divisek and Kastening (34) and Chevalet *et al.* (27) using ingenious electrochemical techniques have been able to deduce certain of the properties of the superoxide ion such as the redox potential of the O_2/O_2^- couple, the rate of the dismutation reaction at various pH values, and the chemical reactivity toward certain organic substances, the results of which are considered in other portions of this review.

When superoxide is formed at the electrode as described above, it diffuses away from the electrode and dismutes spontaneously or catalytically to form hydrogen peroxide and dioxygen; the latter can then diffuse back to the electrode to reform superoxide.

The cathodic current is determined by the electrode reaction (reaction 8) and the concentration of O_2 at the electrode surface is determined by the rate of dismutation (reaction 9, C = catalyst) and by the relevant diffusion

$$O_2^- + C \xrightarrow{H^+} 1/2\ O_2 + 1/2\ H_2O_2 + C \qquad (9)$$

rates of O_2^- and O_2. Rigo *et al.* (57-59) have pointed out that the reduction of dioxygen at a cathode in the presence of hydrogen peroxide and catalase is quite analogous to the superoxide system. In this case the electrode reaction is a two-electron reduction (reaction 10) and the concentration of

$$O_2 + 2e^- \xrightarrow{H_2O} H_2O_2 \qquad (10)$$

dioxygen at the electrode is determined by the rate of the catalytic reaction (reaction 11) and by the relevant diffusion rates of peroxide and dioxygen.

$$H_2O_2 + C \rightarrow 1/2\ O_2 + H_2O + C \qquad (11)$$

Koutecky *et al.* (60) have studied this system in great detail and have carried out a rigorous mathematical analysis of the differential expressions which can relate the electrode current to the rate of the catalytic decomposition of hydrogen peroxide. They found that the following simplified expression quite accurately describes the cathodic current (equation 5,

$$\frac{i_\ell}{i_d} = 1 + \frac{k\ C\ t_1}{7.42 + 1.25\ k\ C\ t_1} \qquad [5]$$

where i_ℓ and i_d are the mean limiting current in the presence and absence of catalase, k is the second order rate constant for the catalytic decomposition of hydrogen peroxide, [C] is the concentration of catalase, and t_1 is the drop time of the electrode). Rigo *et al.* (57-59) have applied this equation directly to the superoxide system and have calculated thereby the catalytic dismutation constant for bovine superoxide dismutase finding it to be in agreement with values determined by pulse radiolysis techniques (61).

In general the electrolytic method has many advantages over chemical flux methods but it does suffer from the fact that direct physical observation of O_2^- is not being made. The procedure of Rigo *et al.* (58) would seem to be very valuable for the assay of large numbers of samples for superoxide dismutase activity, but evaluation of potential interfering reactions must yet be carried out.

B. *Preparation of Superoxide in Aprotic Solvents*

Superoxide solutions have been formed in several organic solvents by electrolytic reduction of dioxygen. These include dimethylsulfoxide (62,43, 44,45,46,17), pyridine (62,44), acetone (62,44), dimethylformamide (62,44,17), and methylene chloride (62,44). Sawyer and Roberts (43) have found that solutions of O_2^- in DMSO containing tetraethylammonium perchlorate as supporting electrolyte can have a decay rate less than 3% per hour. Of course the presence of small quantities of free protons will substantially increase the rate of decomposition of superoxide and, depending on the particular system, may also lead to significant decomposition of solvent, possibly involving free radicals (63,17,45,44).

With the discovery that potassium superoxide could be readily dissolved in DMSO (64) and less polar organic solvents (65) by use of crown ethers to complex the cation, superoxide has become a common "bench top" chemical, and this discovery has facilitated research on the chemical behavior of O_2^- in these organic solvents. Although detailed studies of the stabilities of these solutions have not been carried out, it is likely that they are comparable to the solutions prepared electrolytically.

An additional method for obtaining stable solutions of superoxide in aprotic solvents was developed by McElroy and Hashman (66) and has recently been refined by Foote and co-workers (67). The procedure is based on the reaction of KO_2 with the pentahydrate of tetramethylammonium hydroxide in the solid state followed by extraction of the tetramethylammonium superoxide into liquid ammonia. While somewhat more cumbersome than dissolution with crown ethers, it does avoid the possibility of reactions involving the crown ether itself.

C. *Reactions (and Non-Reactions) of Superoxide in Aqueous Solution*

i. *The Dismutation Reaction*

The principal reaction of superoxide in aqueous solution is dismutation. It has been followed spectrophotometrically in pulsed radiolysis and stopped flow experiments, by e.p.r. after rapid freeze-quench, and by electrometric techniques of polarography (27,34,49,56). The reaction is usually a second order process where one superoxide molecule transfers an electron to another superoxide molecule with the generation of excessive charge density being avoided by the involvement of a proton.

This reaction has been studied over a very broad pH range; the pH profile of the second order rate constant is shown in Fig.8.

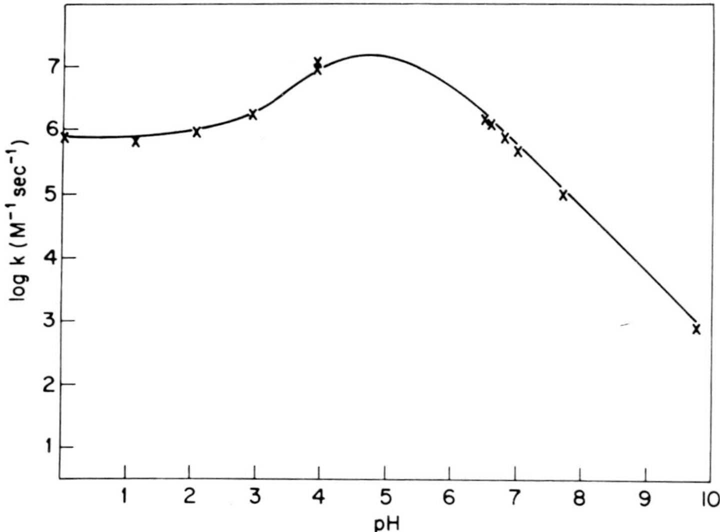

Fig.8. *pH rate profile for the second order dismutation of superoxide. Values were taken from Table 5 of Reference 14 and the solid line was calculated from the information in Table IV using* $pK_{HO_2} = 4.88$

It can be seen that above approximately pH 5 the rate constant decreases by a factor of 10 for each unit increase in pH. The reaction in this region clearly involves the incorporation of a single proton into the transition state.

$$\left.\begin{matrix} O_2^- + HO_2 \\ 2O_2^- + H^+ \end{matrix}\right\} \rightarrow O_2HO_2^- \xrightarrow{\ddagger} \xrightarrow{H^+} O_2 + H_2O_2 \qquad (12)$$

At extremely high pH considerable difficulty is encountered in measuring the rate of the second order reaction (reaction 13), but Divisek and Kastening (34) have reported a value of 6 $M^{-1}sec^{-1}$ at pH $\gtrsim$12. This is approximately

$$O_2^- + O_2^- \xrightarrow{H^+} O_2 + HO_2^-, \qquad (13)$$

the value that would be expected by extrapolating the pH rate profile (Fig.8) in the pH 6-10 range to the higher pH. At pH $\sim$ 4.8, corresponding to pK_a of

HO_2 (20-22), the reaction proceeds with maximum velocity. The rate constant at this pH is that for reaction 14. Below pH 4.8 the rate decreases until a

$$HO_2 + O_2^- \xrightarrow{H^+} O_2 + H_2O_2 \qquad (14)$$

plateau is reached at approximately pH 2 corresponding to reaction 15. The

$$2\ HO_2 \rightarrow H_2O_2 + O_2 \qquad (15)$$

generally accepted rate constants and reported activation energies are listed in Table IV.

TABLE IV

Rates and Activation Energies of the Dismutation Reactions

Reaction	$k(M^{-1}sec^{-1})$	E_a(kcal/mole)	References
$HO_2 + HO_2$	7.6×10^5	4.9 ~ 5.9	14,16,68-70
$O_2^- + HO_2$	1.5×10^7	2.1	14,16,69
$O_2^- + O_2^-$	6		34

Bielski and Saito (68) determined the deuterium (solvent) isotope effect in the pH range 0 to 5 finding a constant value of k_h/k_d of 7 below pH 2 which dropped linearly to approximately 3.5 at pH 5. In 0.8 N H_2SO_4 (D_2SO_4) the isotope effect originates with an increase in the activation energy from 5.8 ± 0.5 to 7.3 ± 0.4 kcal/mole. The authors explained these results in terms of the reaction between two perhydroxyl radicals at low pH and the reaction between superoxide and perhydroxyl at the higher pH values. It would be of some interest to examine the isotope effect at higher pH values since it is known that at least one proton must be involved in the transition state. There has been some discussion regarding the possibility of forming $H_2O_2^+$ at very low pH values (71), but the existence of this species is currently in doubt (14,69). A more complete discussion of the dismutation reaction can be found in references 14 and 16.

There are a number of substances which catalyze the dismutation reaction including the various protein catalysts termed superoxide dismutases which are discussed at some length in this volume. Other substances which catalyze the reaction are aquo- and other coordination complexes of copper (48,72,73, 135,168) and iron EDTA complexes (74,75). It seems quite certain that many other catalysts for this process will be discovered. It is likely that the transition metal catalysts, like the protein catalysts, function by undergoing cyclic oxidation-reduction (equations 16 and 17), but very little is

$$O_2^- + M^{n+} \rightarrow M^{(n-1)+} + O_2 \qquad (16)$$

$$M^{(n-1)+} + O_2^- + 2H^+ \rightarrow M^{n+} + H_2O_2 \qquad (17)$$

known about the mechanisms by which catalysis may occur. It is clear that complexation of Cu^{2+} decreases its efficacy as such a catalyst, but it is not clear why. Studies of the separate reduction and oxidation steps are necessary to ascertain if these reactions are inner or outer sphere, and to assess the effect of different redox potentials and the effect of different coordination geometries. Bielski and co-workers (76) have carried out extensive studies on the catalysis of the reaction by a large variety of substances and have found that almost all organic compounds tested facilitate the reaction to some degree. This was also noted by Divisek and Kastening (34) and may be due to the presence of trace metal impurities. The ability of O_2^- to react both as a reductant and an oxidant forms the basis of this type of catalysis.

The kinetics of the dismutation reaction are not always simple second order; this has been discussed in some detail by Rabani and Nielson (77) and documented by Bielski and Chan (109). Initially there is a second order decay but as the concentration of the superoxide decreases a first order component begins to dominate. This is demonstrated in Fig.9 which was taken from a stopped-flow experiment.

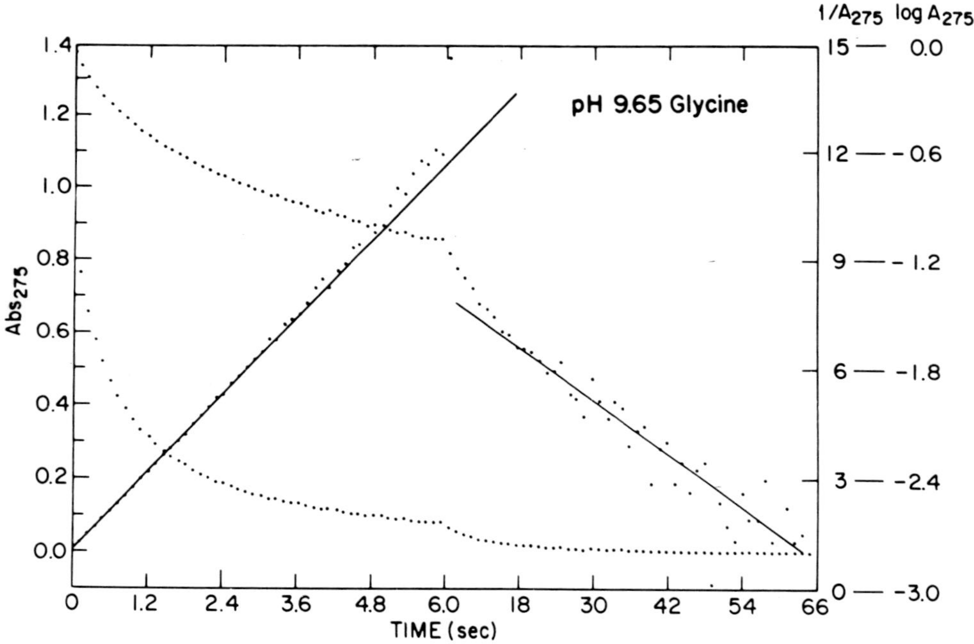

Fig.9. *Demonstration of the first- and second-order components of superoxide decay during a stopped-flow experiment.*

The second order component is evident in the beginning of the reaction where the inverse plot is linear. At the later stages of the reaction the inverse plot is no longer linear but a first order (log) plot becomes linear. The contribution of the first order reaction depends critically on the level of the contaminants in the solution (49.76) and for that reason it is very difficult to prevent its occurrence. One of the consequences of this decomposition pathway is that reducing the concentration of superoxide to very low levels will not confer a corresponding long term stability on that species.

ii. *Superoxide as a One-Electron Reductant*

Superoxide can act as a one-electron reductant toward a variety of substances. In these reactions there seem to be elements of both kinetic and thermodynamic control of the reaction rate. The thermodynamic effects are evident in the work of Patel and Willson (78) who have summarized the rates of the reactions between O_2^- and quinone derivatives as well as the reverse reaction between the radical anion and dioxygen. These along with standard, *two-electron* midpoint potentials are shown in Table V.

TABLE V

Rates of Reactions Between Quinoid Type Compounds and O_2^- and of Reactions Between Semi-Quinoid Type Compounds and Oxygen

(Adapted from Ref. 78)

	$E^{o\prime}$V	O_2^- + S kx10^9 $M^{-1}sec^{-1}$	$S^{\cdot-}$ + O_2 kx10^9 $M^{-1}sec^{-1}$
p-benzoquinone	0.29	1.0	<0.01
2-methylbenzoquinone	0.24	0.76	<0.01
2,3-dimethylbenzoquinone	0.18	0.45	<0.01
2,5-dimethylbenzoquinone	-	0.36	<0.01
2,6-dimethylbenzoquinone	-	0.58	<0.01
Duroquinone	0.05	0.005	0.2 ± 0.5
Vitamin K	-0.06	<0.0002	0.2
2,3-dimethylnaphthoquinone	-0.07	0.004	0.2
Lipoate (ox)	-0.29	-	0.9
NAD^+	-0.325	-	1.9
Dithiothreitol	-0.33	-	2.8
CO_2	-0.42	-	2.4

[a]Two-electron potentials (see text)

It is evident from these data that when the redox potential of the quinone-hydroquinone couple drops below zero volts, the rate of the reaction between the quinone and O_2^- becomes very small. The one-electron Q/QH couple is generally expected to have a lower redox potential than the two-electron Q/QH_2 couple. For example, in the case of duroquinone the former potential has been estimated at -0.25 V (28). Thus the quinones in Table V are presumably arranged in decreasing order of one-electron as well as two-electron oxidizing capability. The data shown in Table V also demonstrate that only those substances having strongly reducing "semiquinone" intermediates are able to form O_2^- by reaction with dioxygen.

Bielski and Gebicki (16) have tabulated several other one electron reduction reactions of superoxide. One reaction that has been used to scavenge O_2^- is that with tetranitromethane which proceeds with a rate

$$O_2^- + C(NO_2)_4 \rightarrow O_2 + C(NO_2)_3^- + NO_2 \quad (18)$$

constant of approximately $2 \times 10^9\ M^{-1}sec^{-1}$ (79). Another widely used scavenging reaction is the reduction of para-nitro bluetetrazolium to form formazan. Rapp *et al.* (147) have indicated that four O_2^- are required for the overall reaction.

The reduction of ferricytochrome *c* by O_2^- is a well known reaction which has been studied by a number of methods. The reaction forms the basis of several assays for the formation of superoxide in biological systems (see Chapter II of this volume). The rate of the reaction was first determined directly by Ballou *et al.* (52) and reported to be $1.6 \times 10^5\ M^{-1}sec^{-1}$ at pH 8.4. Land and Swallow (80) determined the rate at pH 8.5 using pulse radiolysis procedures and found a constant of $1.1 \times 10^5\ M^{-1}sec^{-1}$. Simic *et al.* (81) studied the reaction over the pH range 6 to 10.5 finding that the rate was constant at approximately $1.5 \times 10^6\ M^{-1}sec^{-1}$ below pH 7 and decreased 10-fold for each unit increase in pH above 7. Within the considerable limits of error all reported rates for this reaction are consistent.

Bielski and Gebicki (82) have measured the rates of the reaction between O_2^- and Compound I of horseradish peroxidase over the accessible pH range 3.8 to 8.8. They showed that the overall second order reaction dropped dramatically from approximately pH 4 to 6 and they concluded that the reaction of Compound I with HO_2 had a rate constant of $2.2 \times 10^8\ M^{-1}\ sec^{-1}$ while for O_2^- the rate constant was $1.6 \times 10^6\ M^{-1}sec^{-1}$.

Allen *et al.* (90) have shown that HO_2 is an effective reductant of ferric ion.

iii. *The Formation of Superoxide-Metal Ion Complexes*

There have been several studies of the binding of perhydroxyl radical to metal ions (see Reference 14 for a review). For the most part these have involved the use of the reaction between Ce(IV) ion and hydrogen peroxide to form the perhydroxyl radical under acidic conditions, and observation of the radical complexes has been made by room-temperature e.p.r. flow techniques. It has been established by this method that HO_2 reacts with certain metal ions in their highest oxidation state or with their H_2O_2 complexes to form fairly stable, $K \gtrsim 10^3\ M^{-1}$, metal-perhydroxyl complexes. Competition studies show that the complexation is rapid and reversible and that the exchange of HO_2 among different metal ions probably proceeds via the free perhydroxyl radical. The metal ions which have been examined are Th(IV), U(VI), Ti(VI), Mo(VI), Zr(IV), Hf(IV), and Ce(III). Some of these will react with HO_2 only as their peroxy complexes, Ti(IV) and V(V), while others will react both as the free metal in its highest oxidation state or its peroxy complex, U(VI), Th(IV), and Ce(III). With the exception of the Ce(III)·HO_2 complex, all the metal-perhydroxyl complexes decay very much more slowly than free HO_2 and either first- or second-order decay has been observed.

The e.p.r. spectra of these complexes are characterized by their very narrow linewidths, generally less than 1.5 gauss, compared to the room temperature spectrum of HO_2 which has a linewidth of ca. 27 gauss as shown in Fig.10. As reported by Samuni and Czapski (83) and Shimuzu *et al.* (84), each complex has a slightly different value near g = 2 and most of the spin density remains on the perhydroxyl group. The very narrow linewidth could be interpreted to mean that the proton of the perhydroxyl group is no longer bound to the original radical in the metal complex. The reaction would then be written

$$M^{n+} + HO_2 \rightleftarrows M^{n+} — O_2^- + H^+$$

Little is known about the structures of these complexes.

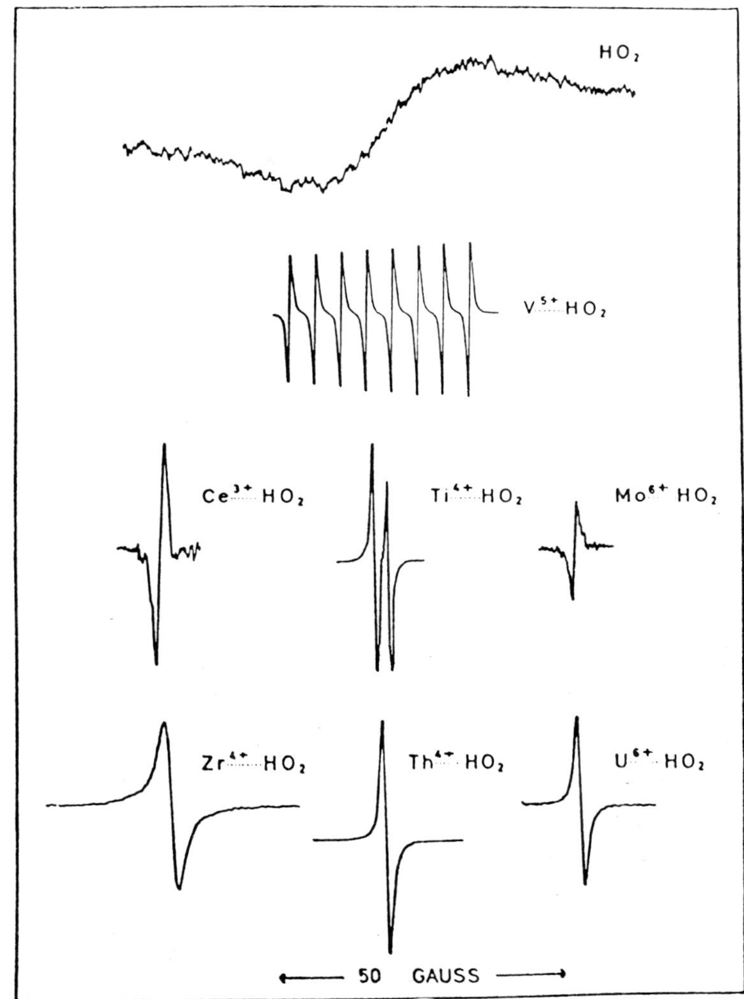

Fig.10. *Room temperature e.p.r. spectra of HO_2 and various metal ion-superoxide complexes. Reference 14.*

It is proper to include the various complexes of Co(III) and superoxide in this section. Either mononuclear or binuclear μ-superoxo bridged complexes are often formed by the reaction between Co(II) complexes and dioxygen in aqueous and non-aqueous solvents (169). The resulting complexes have low temperature e.p.r. spectra suggestive of superoxide ion but the unpaired electron spends some time on the Co as evidenced by the Co(I = 7/2) hyperfine structure (85). Complexes of this type have recently been shown to form by the reaction between O_2^- and Co(III) (86, cf. also Section IV.D.iv).

Recently Bray and his co-workers (13) have demonstrated that superoxide readily forms complexes with the alkaline earth cations, Ca^{2+} and Ba^{2+}, and forms a weaker complex with Na^+ and that these ions drastically reduce the rate of superoxide dismutation. The reactivity of these superoxo-metal complexes toward organic substances deserves consideration.

iv. *The Haber-Weiss Reaction (a Non-Reaction?)*

In 1934 Haber and Weiss (87) proposed that the following reactions were occurring in mixtures of iron salts and hydrogen peroxide in order to account for an unexpectedly large amount of hydrogen peroxide decomposition.

$$Fe^{2+} + H_2O_2 \rightarrow Fe^{3+} + OH\cdot + OH^- \quad (19)$$

$$OH\cdot + H_2O_2 \rightarrow H_2O + O_2^- + H^+ \quad (20)$$

$$O_2^- + H_2O_2 \rightarrow O_2 + OH^- + OH\cdot \quad (21)$$

$$Fe^{2+} + OH\cdot \rightarrow Fe^{3+} + OH^- \quad (22)$$

Reaction 21 is currently termed the Haber-Weiss reaction. The feasibility of this reaction is of great importance because it has been invoked recently in attempts to explain certain effects observed in biochemical and biological systems. This reaction was in fact of considerable concern to chemists at the time of its proposal. Accordingly it was subjected to some rather severe tests over the years and no evidence was found that it is an efficient process, if indeed it occurs at all. Since much of this work seems to have escaped notice in recent years, it will be discussed here in some detail and some recent results from one of our laboratories will be presented.

The Haber-Weiss mechanism for ferrous ion-mediated decomposition predicts that O_2 evolution will increase indefinitely with increasing H_2O_2 concentration due to the reaction between HO_2 and H_2O_2. Barb *et al.* (88), however, found that O_2 evolution reached a limit as peroxide concentration increased. These workers offered additional evidence that the O_2 evolving step also regenerated ferrous ion. With these observations at hand, Barb *et al.* (88) proposed a mechanism for this process which did not involve a reaction between HO_2 and H_2O_2 and which was compatible with their experimental results. In this scheme perhydroxyl radical is produced by reaction 23 which has

$$Fe^{2+} + H_2O_2 \rightarrow Fe^{3+} + OH^- + OH\cdot \quad (19)$$

$$Fe^{2+} + OH\cdot \rightarrow Fe^{3+} + OH^- \quad (22)$$

$$H_2O_2 + OH\cdot \rightarrow H_2O + HO_2 \quad (23)$$

$$Fe^{2+} + HO_2 \rightarrow Fe^{3+} + HO_2^- \quad (24)$$

$$Fe^{3+} + HO_2 \rightarrow Fe^{2+} + O_2 + H^+ \quad (25)$$

subsequently received confirmation; its specific rate was determined by Schwarz (89) to be $4.5 \times 10^7\ M^{-1}sec^{-1}$. It could be argued, however, that this system may not be suitable to test such a subtle point since it is potentially very complicated and there seems to be disagreement as to whether free radicals of the types indicated are actually formed. Indeed, Kremer has proposed an alternative mechanism in which no free radicals are involved. (For a brief discussion cf. Ref. 91.)

Perhaps the most convincing evidence that the Haber-Weiss reaction does not occur was obtained by George (92) who found that potassium superoxide dispersed in a 95% solution of H_2O_2 yielded precisely the amount of dioxygen predicted by simple dismutation of the superoxide. Had significant reduction of H_2O_2 by O_2^- occurred, more than the stoichiometric amount of O_2 would have been evolved. Barb *et al.* (88) have reviewed the other observations made before 1950 which cast serious doubt on this part of the proposed mechanism of Haber and Weiss.

McClune and Fee (49) have recently measured the rate of superoxide dismutation in various buffers with the concentration of H_2O_2 ranging from 0 to 100 mM and found that the decay kinetics are completely independent of peroxide in pyrophosphate and glycine while phosphate and borate buffers show slight enhancements at high peroxide concentration. This was ascribed to subtle effects of H_2O_2 on the trace metal contaminants.

As will be discussed below, there is some suggestive evidence that aprotic solutions of O_2^- can effect the reductive cleavage of alkylhydroperoxides, but it would be misleading to use this as evidence for the analogous reaction with hydrogen peroxide. Indeed all the chemical evidence supports the idea that in aqueous solution superoxide will *not* directly effect the reductive cleavage of H_2O_2.

The reader should be apprised that, while the results of chemical experiments suggest that the Haber-Weiss reaction does not occur, there is a body of biochemical results which suggests it may be an important factor in biological systems (cf. Ref. 177 for an introduction to this literature and Chapter by Cohen in this book for a fuller discussion). In the complex reaction mixtures being studied a product is formed or an effect observed which can be reasonably associated with the known chemistry of OH. (cf. Section V), and inhibition is effected by superoxide dismutase or by catalase. This pattern of inhibition is thought to result from the removal of one of the Haber-Weiss reactants, O_2^- or H_2O_2, effectively stopping the production of OH· by that route. This interpretation is necessarily clouded by a lack of information relevant to the following types of questions: How specific are chemical traps for OH· (cf. Section V)? What role might organic peroxides, ROOH, play? Can the Haber-Weiss reaction be effectively catalyzed by trace metals? Do superoxide dismutases and catalase have absolute specificities for O_2^- and H_2O_2? In the opinion of the authors, simple explanations of these complex phenomena should be regarded with considerable reservations until experimental evidence bearing on these questions becomes available.

v. *Superoxide as an Oxidant*

Superoxide can act both as a one electron reductant and as a one-electron oxidant and the latter is thermodynamically favored as indicated by the redox potentials shown in Table II. Only a few reactions are known in which superoxide acts as the oxidizing agent suggesting that reactions may be kinetically rather than thermodynamically controlled. In their review Bielski and Gebicki (16) list only a few reactions in which strong reductants react with superoxide and these generally occur in acidic solution. Perhaps the most relevant and best studied of these is the reaction between perhydroxyl and ferrous ion. Jayson *et al.* (93) have found that the rate constant for this reaction is 2 x 10^6 $M^{-1}sec^{-1}$ and the immediate product appears to be an outer sphere complex of Fe^{3+} and HO_2^-. The reaction of superoxide with ferrous ion, which may be of some importance in biological systems, has not been studied.

Bielski and Chan (109) have shown that HO_2 and O_2^- oxidize lactate dehydrogenase-bound NADH with rate constants of 2 x 10^6 $M^{-1}sec^{-1}$ and 1 x 10^5 $M^{-1}sec^{-1}$ respectively. This reaction is initiated by superoxide and chain propagated by O_2 and protein bound NAD· radicals.

It would appear that the availability of protons from a substrate as well as its redox potential is important in determining if it will be oxidized by superoxide. Certain hydroquinones can be oxidized by superoxide (94,95 and references therein) depending on their redox potentials. The availability of a proton is presumably important in avoiding even incipient formation of the dianion of peroxide. The oxidation does not occur with hydroquinones that have two-electron redox potentials much above 0.4 volts (95). Thus even with substances of this type, superoxide is not expressing its full thermodynamic capacity as an oxidizing agent.

vi. *The Formation of Singlet Dioxygen from Superoxide (a Non-Reaction?)*

In recent years there have been numerous communications reporting the formation of singlet dioxygen in complex systems undergoing autoxidation (see the 20 citations quoted in note 17 of reference 96). The assignments are based either on the use of chemical trapping agents presumed to be specific for singlet dioxygen or by the observation of weak chemiluminescence presumed to be associated with the $^1\Delta_g$ to $^3\Sigma_g^-$ transition (cf. Fig.3). The effects observed were attenuated by addition of superoxide dismutases and it was therefore concluded that uncatalyzed dismutation of superoxide ions gave excited state singlet dioxygen (reaction 26) whereas the protein-catalyzed

$$2\ O_2^- + 2\ H^+ \rightarrow (^1\Delta_g)\ O_2 + H_2O_2 \qquad (26)$$

reaction gave ground state (triplet) dioxygen. While sufficient energy is released in this reaction (27.6 kcal) to form $^1\Delta_g$ dioxygen, there has been no experimental substantiation for its occurrence. Indeed Poupko and Rosenthal (97) were unable to detect any singlet dioxygen formation from superoxide when using the chemical trapping agents 2,3-dimethyl-2-butene, 2-methyl-2-pentene, and cyclohexene. In another well-defined study, Nilsson and Kearns (98) obtained similar results. Kajiwara and Kearns also observed the weak emission originally reported by Khan (99) upon dissolution of potassium superoxide in DMSO/water or D_2O mixtures but found that the expected increase in chemiluminescence in D_2O (100) was not observed. The authors concluded that singlet dioxygen is not responsible for the weak chemiluminescence and therefore that it is not formed in the above reaction.

In the cases where chemical trapping agents have been used to indicate formation of singlet dioxygen, the conclusions are also in jeopardy. Howard and Mendenhall (101) have shown that at least one of these singlet dioxygen trapping agents, 1,3-diphenylisobenzofuran (DPBF), is very sensitive to autooxidation by ground state dioxygen in the presence of any free radical initiators and they have pointed out the necessity to have carefully controlled experiments to assess this. These comments are particularly relevant to the experiments of Mayeda and Bard who reported that DPBF was converted to *ortho*-dibenzoylbenzene (ODB) when electrochemically generated superoxide was allowed to react with electrochemically generated ferricenium ion (a strong one-electron oxidizing agent) to form ferrocene and dioxygen which was reasoned to be in an excited singlet state (102). The same authors (103) also reported that DPBF was converted to ODB when electrochemically generated superoxide was allowed to dismute in a mixed organic/water solvent. The small amount of singlet dioxygen trapping agent which was modified was further attenuated in the presence of added superoxide dismutase. The control experiment involved dioxygen plus DPBF under the same conditions. The work of Howard and Mendenhall (101) would suggest that this is an inadequate control experiment since autooxidation of the DPBF could be initiated by free radicals present in the experiment.

There are further complications in obtaining an unequivocal answer to the question of whether or not superoxide dismutation can yield singlet dioxygen. Stauff *et al.* (104) have recently reported that a carbonate dependent chemiluminescence can arise from autooxidizing systems and, Hodgson and Fridovich (105) have clearly demonstrated that the chemiluminescence of the xanthine oxidase-xanthine system is directly proportional to the one-half power of the carbonate concentration and is negligible in the absence of carbonate. These observations suggest that considerable caution should be exercised when invoking the presence of singlet dioxygen in even very well defined systems.

vii. *Spin Trapping Reactions for Superoxide and Hydroxyl Radicals*

Harbour *et al.* (106) have reported an extremely efficient system for trapping and identifying both hydroxyl and superoxide radicals. The following reactions yield nitroxide radicals which can be identified by their characteristic e.p.r. spectra.

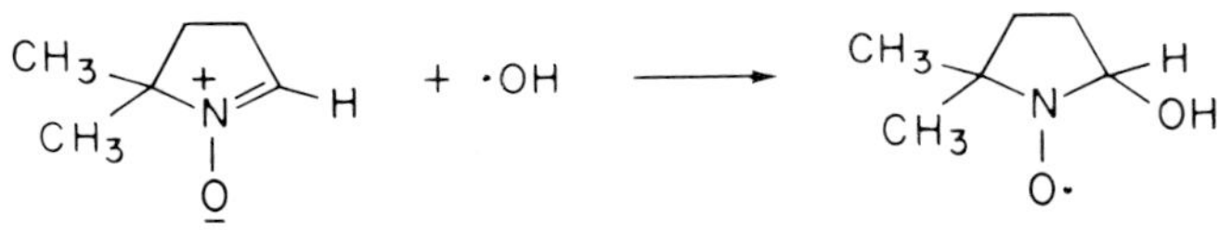

$$(CH_3)_2\text{-pyrroline-}N^{+}\text{-}\underline{O}^{-}\text{(=CH)} + \cdot OH \longrightarrow (CH_3)_2\text{-pyrrolidine-}N\text{-}O\cdot\text{(CH)(OH)} \qquad (27)$$

$$(CH_3)_2\text{-pyrroline-}N^{+}\text{-}\underline{O}^{-}\text{(=CH)} + O_2^{-} \xrightarrow{H_2O} (CH_3)_2\text{-pyrrolidine-}N\text{-}O\cdot\text{(CH)(O-O-H)} \qquad (28)$$

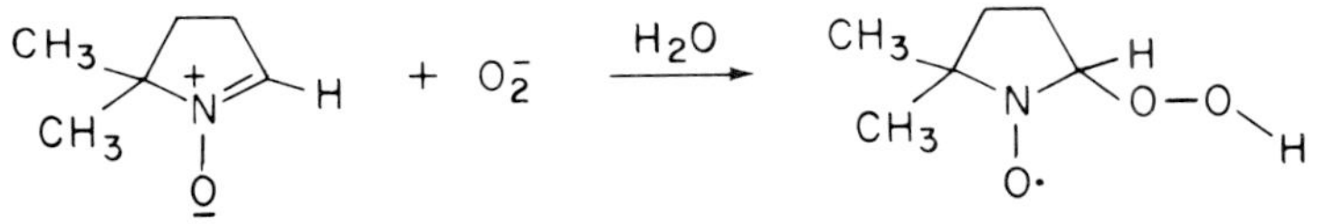

This type of reaction would suggest that superoxide ion is capable of adding to a highly activated double bond although the details of this reaction mechanism are not known. Harbour and Bolton (107) have subsequently used this spin-trapping system to demonstrate unequivocally the formation of superoxide ions in illuminated chloroplasts. This sytem thus seems to be very efficient, but quantitative results concerning the kinetics of the reaction and possible interfering reactions are still needed. The enhanced formation of the adduct in the presence of methylviologen could, for example, result from a reaction between one-electron reduced nitrone and dioxygen. The reaction between MV^+ and O_2^- is probably extremely rapid (108,176).

viii. *Superoxide as a Nucleophile in Aqueous Solution*

As indicated in the following sections superoxide has considerable nucleophilic character when dissolved in aprotic solvents. The only experimental gauge we have to compare its nucleophilicity in organic solvents and in water comes from the work of Divisek and Kastening (34) who have described some of their preliminary results concerning the reaction between electrogenerated superoxide (mercury pool cathode) and organic halides. The experiments were carried out by adding the organic halides to a basic solution of 0.5 to 1 mM superoxide. It was observed that the main effect of the added organic substances was to accelerate the dismutation reaction, but small quantities of organic peroxides could be extracted from the acidified solution after the reaction was complete. Yields: allyl iodide, 10%; allyl bromide or benzyl bromide, 5%; propyl bromide, none. While it is not clear on which reagent the yield was calculated, corresponding amounts of free halide were found after correction for hydrolysis of the organic halides. The authors state that Hg ions may somehow be involved in the reaction and that the yields of organic peroxides are increased if the reaction is carried out in the presence of the mercury pool. It is not clear what role the metal ions are playing in this system and it is thus not possible to determine if superoxide is attacking a free organic halide molecule or if it is attacking, for example, a complex such as R-X...Hg. Taking an even more pessimistic view, it has not been shown that peroxide itself does not attack such a complex.

D. *Reactions of Superoxide in Aprotic Solvents*

Studies of the reactions of superoxide with organic compounds in non-aqueous solvents have been motivated by interest in the biochemistry of superoxide and by the possibility of developing a new reagent for organic synthesis. Elucidating the reactions of superoxide with biological and other organic substrates in *aqueous* media is not an easy task because the competing disproportionation reaction generates hydrogen peroxide, itself often reactive toward many of these substances. The stability of superoxide in aprotic solvents makes investigation of its reactions with organic substrates in such non-aqueous systems considerably more straightforward. It is likely, however, that the reactivity of superoxide will show considerable variation with solvent. The reactivities of small, negatively-charged anions such as fluoride, hydroxide, and chloride are quite different in aprotic *versus* protic solvents where they are involved in hydrogen-bonding interactions with solvent (35). This variation in reactivity with solvent should not be ignored in considering possible biological reactions of superoxide.

In aprotic solvents, superoxide behaves toward organic substrates as a nucleophile, a reducing agent, an oxidizing agent, and a base. It remains to be determined what degree if any superoxide will show reactivity typical of a free radical. The organic reactions of superoxide can be classified on this basis and will be discussed in this order.

i. *Superoxide as a Nucleophile*

The best studied type of reactivity of superoxide toward organic substrates is its nucleophilicity. Nucleophilic reactions of solid sodium and potassium superoxide suspended in diethyl ether, benzene, and tetrahydrofuran (THF) were observed more than 15 years ago (110-112). The reactivity of superoxide could not be assessed on the basis of these experiments, however, because of the lack of solubility of the salts in the non-polar solvents used. In 1970, two reports appeared which suggested the nucleophilic character of electrochemically generated superoxide in dimethylformamide (DMF) and dimethylsulfoxide (DMSO) (113,114). In more recent studies (65,115-119), potassium superoxide solubilized by crown ether in DMSO, benzene, DMF, dimethoxyethane or diethyl ether has been used.

Superoxide dissolved in aprotic solvents appears to have considerable nucleophilic character. A comparison of the reactivity of KO_2 with that of KI was carried out by observing their reaction with 1-bromo-octane in DMSO using 18-crown-6 as a phase transfer catalyst. Under similar conditions, the half-life for the reaction of KI was approximately 20 hours while that of the reaction of KO_2 was less than a minute (115).

In general, the reactivity of superoxide with organic halides and sulfonate esters also appears to be nucleophilic. The effects of substrate structure and the nature of the leaving group on the relative reactivity (benzyl>primary>secondary>tertiary>aryl, I>Br>Cl) are consistent with an S_N2 mechanism (114,115,120). Consistent with this mechanism is the fact that primary halides give predominantly substitution reactions, secondary halides give both substitution with inversion of carbon configuration and elimination, and tertiary halides give predominantly elimination (65,115).

The products obtained in this reaction appear to be solvent dependent. In benzene the principal product after aqueous workup is the dialkyl peroxide (65). The mechanism proposed for the reaction is described by reactions 29-32 (113,65).

$$O_2^- + RX \rightarrow RO_2^\cdot + X^- \quad (29)$$

$$RO_2^\cdot + O_2^- \rightarrow RO_2^- + O_2 \quad (30)$$

$$RO_2^- + RX \rightarrow RO_2R + X^- \quad (31)$$

$$\text{Overall } 2RX + 2O_2^- \rightarrow RO_2R + O_2 + 2X^- \quad (32)$$

When DMSO is present, the reaction product, after aqueous workup, is the alcohol (115,118). It is possible that the alcohol arises from direct reaction of superoxide with the dialkylperoxide. Alternatively, it is known that hydrogen peroxide reacts with DMSO to give dimethylsulfone (122). Moreover, alkylperoxide anions also react with DMSO to give the sulfone and the alkoxide (reaction 33) (123). It is probable therefore that in DMSO the alkyl peroxide anion formed by reactions 29 and 30 reacts with DMSO as in reaction 33 giving alkoxide anion which is converted to alcohol upon addition of water.

$$RO_2^- + (CH_3)_2SO \rightarrow RO^- + (CH_3)_2SO_2 \quad (33)$$

In support of this hypthesis, dimethylsulfone has been isolated from the reaction mixture (124).

Other nucleophilic reactions of superoxide have been studied. Superoxide reacts with carboxylic esters to give the acid and the alcohol after aqueous workup (equation 34) (116). The initial step of this reaction is presumed to

$$RCO_2R' \xrightarrow[\text{benzene}]{KO_2\text{-crown ether}} \xrightarrow[H^+]{H_2O} RCO_2H + R'OH \quad (34)$$

presumed to be addition of O_2^- to the carbonyl carbon. Several observations support this presumption. When the ester of an optically active secondary alcohol reacts with superoxide, the resulting alcohol is produced with 99% retention of configuration (116). Thus the attack of superoxide is not occurring at the alkyl carbon. Moreover the effect of the structure of the leaving group on the relative rate of reaction (R' = phenyl>primary>secondary >tertiary) is consistent with attack on the carbonyl carbon (116).

Superoxide was also found to cleave certain phosphate esters (116) but does not seem to react with simple amides and reacts very slowly with nitriles (116). A kinetic study of the reaction of electrochemically generated superoxide with aryl benzoates in DMF demonstrated that the reaction obeys second order kinetics (125).

Superoxide also reacts with acyl chlorides in a synthetically useful reaction to give diacylperoxides (equation 35) (117). Nucleophilic reactions

$$2\,\mathrm{RC(=O)Cl} + 2\,\mathrm{KO_2} \xrightarrow[\text{benzene}]{\text{crown ether}} \mathrm{RC(=O)OOC(=O)R} + 2\,\mathrm{KCl} + \mathrm{O_2} \qquad (35)$$

of superoxide in aprotic solvents are summarized in Table VI.

TABLE VI

Nucleophilic Reactions of Superoxide in Aprotic Solvents

Substrate	Products[a]	References
alkyl halide alkyl sulfonate ester	alcohol or dialkyl peroxide, olefin	115,118,119
carboxylate ester	acid and alcohol	116,125
phosphate ester	ester and alcohol	116
acyl chloride	diacyl peroxide	117
amide	no reaction	116

[a]After aqueous work up

ii. *Superoxide as a Reducing Agent*

Superoxide acts as a reducing agent in several types of reactions. These include reduction of quinones, nitrosubstituted aromatics, tetranitromethane, and nitrobluetetrazolium dye and the reductive cleavage of alkylhydroperoxides. It is possible that other organic reactions of superoxide involve reduction as a first step, although the overall reaction may not be a reduction.

The most straightforward organic reductions by superoxide are those of several quinones and nitro-substituted aromatics in DMSO solution which give the radical anions (reactions 36 and 37) which have been identified by their

$$\text{p-benzoquinone} + O_2^- \longrightarrow \text{semiquinone radical anion } (^-O\text{–}C_6H_4\text{–}O\cdot) + O_2 \qquad (36)$$

$$C_6H_5NO_2 + O_2^- \longrightarrow [C_6H_5NO_2]^{\cdot -} + O_2 \qquad (37)$$

characteristic e.p.r. spectra (97). Superoxide has also been observed to reduce N-methylacridine to its radical anion in a chemiluminescent reaction (170).

The reductive cleavage of alkylhydroperoxides by superoxide (111,112,67) (reaction 38) is of considerable interest because of its resemblance to the

$$O_2^- + ROOH \rightarrow O_2 + RO\cdot + OH^- \tag{38}$$

postulated Haber-Weiss process, i.e. the reductive cleavage of hydrogen peroxide to give hydroxide and hydroxyl radical (reaction 38, R = H, see section IV.C.iv). Foote has demonstrated that the reaction of *tert*-butyl hydroperoxide with superoxide in acetonitrile gives acetone and *tert*-butyl alcohol (67) (reaction 39). These are the characteristic products derived from the

$$(CH_3)_3COOH \xrightarrow[O_2^-]{CH_3CN} CH_3\overset{O}{C}CH_3 + (CH_3)_3COH \tag{39}$$

tert-butoxy radical (reaction 40 and 41).

$$(CH_3)_3CO\cdot \rightarrow CH_3\overset{\overset{O}{\|}}{C}CH_3 + CH_3\cdot \tag{40}$$

$$(CH_3)_3CO\cdot + RH \rightarrow (CH_3)_3COH + R\cdot \tag{41}$$

Two points should be made about this reaction. The first is that dialkylperoxides react more slowly with O_2^- than do alkylhydroperoxides. For example, di-*tert*-butylperoxide does not react (67) and diethylperoxide reacts much more slowly than the alkylhydroperoxide (124) with superoxide under similar conditions. This may be due to either thermodynamic or kinetic differences between the reactions, but it could also be due to the presence of an acidic hydrogen on the alkylhydroperoxide. Although superoxide is presumably a weaker base than alkoxide anion (see below), primary alcohols have been observed to react rapidly with superoxide to give the alkoxide anion (111,124) (reaction 42). This reaction is accompanied by vigorous evolution

$$ROH + O_2^- \rightleftarrows RO^- + HO_2 \tag{42}$$

$$HO_2 \rightarrow \frac{1}{2} H_2O_2 + \frac{1}{2} O_2 \tag{42a}$$

of O_2 and it is therefore presumed to be driven by the rapid and strongly exothermic disproportionation of HO_2 (reactions 42a). Alkylhydroperoxides are generally more acidic than alcohols and therefore the possibility that superoxide is deprotonating the hydroperoxide (reaction 43) should perhaps be

$$ROOH + O_2^- \rightleftarrows ROO^- + HO_2 \tag{43}$$

considered.

The second point that must be stressed concerning this very interesting nonaqueous reaction is that while it may demonstrate the feasibility of the Haber-Weiss reaction in aqueous solution, it certainly does not suggest that it is probable. There are important differences between this non-aqueous system and the aqueous one, in particular, the longer lifetime of superoxide and its enhanced reducing power in a nonaqueous system (Table III) and the decreased bond strength of the O-O bond in alkyl peroxides compared to H_2O_2 (39), all these factors increasing the probability of the reaction.

iii. *Superoxide as an Oxidant*

Superoxide will oxidize a number of organic compounds but the mechanisms of such reactions have not, in general, been investigated. The compounds that have been most extensively studied are α-keto, α-hydroxy, and α-halo ketones, esters and carboxylic acids in non-aqueous media and catechol derivatives, including epinephrine and tiron, in both aqueous and non-aqueous media.

Reactions of α-keto, α-hydroxy and α-halo ketones, esters and carboxylic acids with superoxide have been carried out in benzene or tetrahydrofuran (167,111,112). The reaction observed is oxidative cleavage of the C-C bond (reaction 44). The mechanism of this reaction remains to be elucidated.

$$\begin{matrix} \text{RC(=O)C(=O)R}' \\ \text{or} \\ \text{RC(=X)C(=O)R}' \end{matrix} + KO_2 \xrightarrow[\text{benzene}]{\text{crown ether}} \xrightarrow{H_3O^+} RCO_2H + R'CO_2H \qquad (44)$$

R' = OH, OR, alkyl, aryl
X = OH, Cl, Br

It has been observed that oxidative cleavage of α-hydroxyl carbonyl compounds does not occur unless there is also an α-hydrogen (167). The resemblance of this reaction to that of certain dioxygenases has been noted (167).

Reactions of superoxide with catechol derivatives have been studied in aqueous and non-aqueous media. In aqueous solution, epinephrine is oxidized to adenochrome by an unknown mechanism (reaction 45) (126) and tiron (1,2-dihydroxybenzene-3,5-disulfonic acid) (127,128) to the *o*-semiquinone radical (reaction 46).

$$\text{epinephrine (HO, HO, C(H)(OH), CH}_2\text{, NH, CH}_3) \xrightarrow{O_2^-} \text{adrenochrome (O, }^-\text{O, H, OH, N}^+\text{, CH}_3) \qquad (45)$$

$$\text{tiron (HO, HO, SO}_3^-\text{, SO}_3^-) \xrightarrow{O_2^-} \text{semiquinone (}^-\text{O, }^\bullet\text{O, SO}_3^-\text{, SO}_3^-) + H_2O_2 \qquad (46)$$

An extensive investigation of the reactions of catechols with KO_2 suspended in CH_3CH or THF has been reported (129). In the case of 9,10-dihydroxyphenanthrene, oxidation to the quinone was observed followed by oxidative cleavage of the C-C bond to give diphenic acid (reactions 47 and 48). Reaction of 3,5-di-*tert*-butylcatechol also gives the orthoquinone as a first product which reacts further to give the products of oxidative cleavage (reactions 49 and 50). The blue color characteristic of the *o*-semiquinone was observed during the reaction of either the catechol or the quinone.

(47)

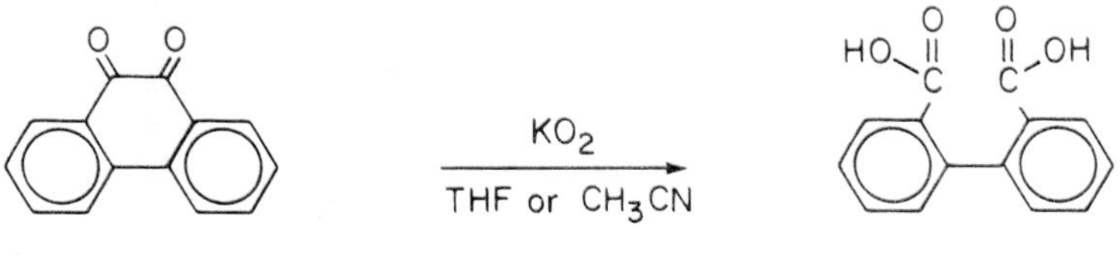

(48)

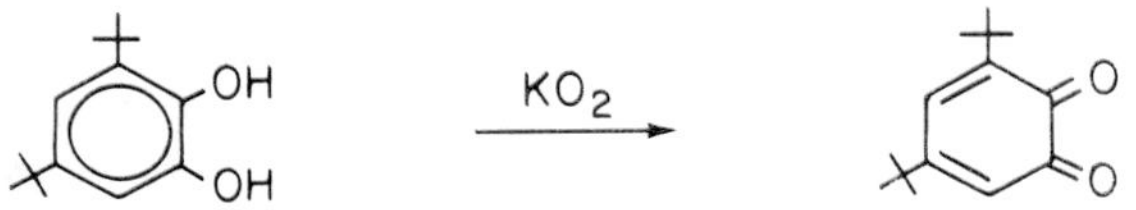

(49)

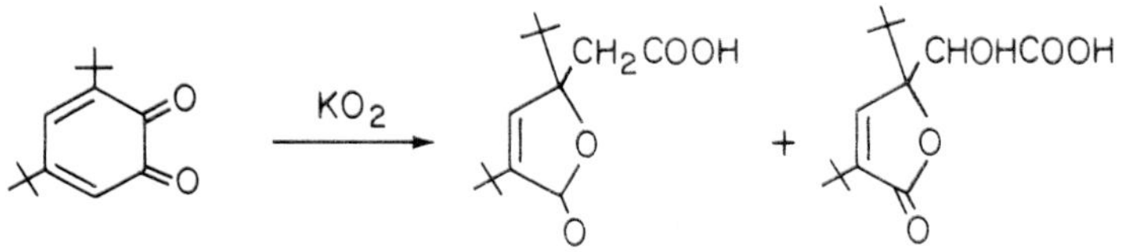

(50)

This has caused Foote to suggest that the initial reaction is the abstraction of a hydrogen atom (reaction 51) in the case of the catechol or a one

(51)

electron reduction (reaction 52) in the case of the quinone. Although

(52)

hydrogen abstraction by O_2^- has been proposed in a large number of reactions (126,127,130), it is difficult to distinguish experimentally from a mechanism where superoxide abstracts a proton to form HO_2 which then oxidizes the anion (reaction 53) or a mechanism whereby O_2^- abstracts a proton to yield HO_2 and the anion, which can react with dioxygen formed by dismutation of HO_2 (reaction 54) or already present in the reaction mixture. Without an estimate of the H-O bond energy of HO_2^-, which does not appear to be readily available, it is not possible to assess the probability of H· abstraction by

O_2^- (131,132).

$$\text{C}_6\text{H}_4(\text{OH})_2 + O_2^- \longrightarrow \text{C}_6\text{H}_4(\text{O}^-)(\text{OH}) + HO_2^\bullet \longrightarrow \text{C}_6\text{H}_4(\text{O}^\bullet)(\text{OH}) + HO_2^- \qquad (53)$$

$$\text{C}_6\text{H}_4(\text{O}^-)(\text{OH}) + O_2 \longrightarrow \text{C}_6\text{H}_4(\text{O}^\cdot)(\text{OH}) + O_2^- \qquad (54)$$

The reactions of catechol derivatives with KO_2 in acetonitrile or THF also generate O_2 and probably H_2O_2; it is not clear how these species may be involved in these reactions (129).

The basicity of superoxide in non-aqueous solvents is difficult to assess. The difficulty in measuring this property is that the acid form, i.e. HO_2, is unstable, rapidly disproportionating to H_2O_2 and O_2. Thus although primary alcohols (111,112,124) and phenols (124) react rapidly to give the alkoxide and phenoxide radicals respectively, this reaction is probably driven by the dismutation of HO_2 and thus cannot be used as a measure of the basicity of superoxide. The basicity of superoxide in the gas phase has been measured and it is found to be a very strong base (173).

The free radical character of superoxide is also difficult to assess. The only free radical reactions that have been proposed to occur are hydrogen atom abstractions from relatively good hydrogen atom donors (126,127,129,130). These compounds are also relatively acidic, however, and the possibility that the observed products arise from proton abstraction followed by oxidation cannot be excluded (134).

A number of other organic reactions of superoxide have been reported (97, 110-113,130,134,136,137) for which it is not clear what category of reactivity of superoxide is being observed. One such reaction should especially be noted. Superoxide appears to react slowly with neat acetonitrile (17,171) and other nitriles (172) but the mechanism is unclear. Abstraction of an α-proton has been proposed as a first step (172). Further studies of all of these reactions should prove interesting.

iv. *Inorganic Reactions of Superoxide*

The inorganic chemistry of superoxide that is known to date is derived mainly from studies of the reactions of O_2^- with transition metal ions or complexes in aqueous or non-aqueous solution. In aqueous solution, reactions involving superoxide or hydroperoxyl radical have been postulated as intermediate steps in reactions of hydrogen peroxide (138) or of dioxygen (139) with transition metal ions.

Reactions of superoxide in non-aqueous solvents are in general more straightforward to study than in aqueous solution, although reactions with solvent are always a complicating possibility. Two types of reactions of superoxide with transition metal complexes have been observed, formation of dioxygen coordination complexes or one-electron reduction. Oxidation reactions have not been observed, presumably because superoxide is a poor oxidant in the absence of protons (140).

Two reports have been published of complex formation by reaction of electrochemically generated superoxide solutions in DMF with transition metal complexes. Vitamin B_{12a}, which is a cobalt (III) complex, reacts with superoxide at $-50^{\circ}C$ to give superoxocobalamin whose presence was demonstrated by its characteristic e.p.r. spectrum (86). This complex had previously been prepared by reaction of B_{12r}(Co(II)) with O_2 (reaction 55). It has also been

$$\underset{B_{12r}}{Co(II)} + O_2 \rightarrow Co(III)O_2^- \leftarrow \underset{B_{12a}}{Co(III)} + O_2^- \qquad (55)$$

reported that superoxide solutions react with iron(III)-protoporphyrin IX dimethylester cation at $-50^{\circ}C$ to give a complex with an absorption spectrum similar to that of oxymyoglobin (141).

One-electron reductions of transition metal complexes have been observed in DMSO using crown ether solubilized KO_2. Both $Cu(II)(phen)^{2+}{}_2$ (64) (phen = 1,10-phenanthroline) and $Mn(III)TPP^+$ (142) (TPP = tetraphenylporphine) are rapidly reduced by superoxide to $Cu(I)(phen)^+{}_2$ and Mn(II)TPP respectively. It is interesting to note that these two products are sensitive to oxidation by O_2, although O_2 is presumably the product of the superoxide reductions (reactions 56 and 57). This illustrates the fact that O_2 is a poor one-

$$Cu(phen)_2^{2+} + O_2^- \rightarrow Cu(phen)_2^+ + O_2 \qquad (56)$$

$$MnTPP^+ + O_2^- \rightarrow MnTPP + O_2 \qquad (57)$$

electron oxidizing agent, especially in aprotic media. In fact oxidation of $Cu(phen)_2^+$ by O_2 requires the presence of protons (64) and oxidation of MnTPP by O_2, which is probably not a one-electron process and which may occur through a binuclear peroxide intermediate, is relatively slow compared to the reduction of $MnTPP^+$ by O_2^- (142).

One other aspect of the inorganic chemistry of superoxide should be mentioned. A number of superoxide complexes of transition metals have been studied which are prepared not from superoxide but by reacting a lower valence transition metal complex with O_2 (reaction 58). Reactions of a large

$$M^{n+} + O_2 \rightarrow M^{(n+1)} + O_2^- \qquad (58)$$

number of cobalt (II) complexes with O_2 have been shown to give cobalt(III)-superoxide complexes (85,143,169, and section IV.C.iii. above) and more recently the reactions of chromium(II) and iron(II) porphyrins with dioxygen have been shown to give chromium(III) (144) and iron(III) (145) superoxide complexes respectively. Since complexes of this type are undoubtedly involved in some metal-containing oxygenase reactions, the reactivity of coordinated superoxide (and peroxide) is of considerable interest, but little is known to date of the reactivity of coordinated superoxide. It has been observed that $Co(III)-O_2^-$ complexes in non-aqueous solvents react with several reducing agents such as *p*-dihydroxybenzene, ascorbic acid, or thiols (146). In the case of *p*-dihydroxybenzene, the semiquinone radical anion is formed in a reaction similar to that observed for free superoxide (129) (see above).

V. COMPARISON OF THE CHEMICAL PROPERTIES OF SPECIES POTENTIALLY DERIVED FROM DIOXYGEN

The species related to dioxygen that are of interest to the biochemist are dioxygen and its activated states, superoxide (and the perhydroxyl radical), hydrogen peroxide, and hydroxyl radical. Much is known about the chemistry of these species and their reactions with a large number of organic and inorganic substrates. From the biochemist's point of view, however, it is perhaps unfortunate that much of the knowledge about the chemistry of these species has been obtained in non-aqueous solvents, at high concentration, or under other conditions which cannot easily be compared to those of biological systems. Nevertheless, it is certainly of interest to compare what we know of the relative reactivities of these species in an attempt to understand the role they may play in biological systems.

Ground State Dioxygen

The organic chemistry of the ground state dioxygen molecule, O_2 ($^3\Sigma_g$), is very different from that of its first excited state O_2('Δ_g). At first glance $^3\Sigma_g$ dioxygen appears to be surprisingly unreactive toward organic substrates especially considering the exothermicity of the overall reactions (148). There is, however, a spin restriction involved in the reaction of triplet dioxygen with a singlet substrate to give singlet products directly and this is at least partly responsible for the lack of reactivity (139,150,148). The spin restriction can be avoided, however, by a free radical mechanism and this type of reaction pathway (149) is in fact generally observed (reactions 59-61). Reaction of $R\cdot(\downarrow)$ with $O_2(\uparrow\uparrow)$ to give $RO_2\cdot(\uparrow)$ (reaction 60) does

$$X\cdot + RH \rightarrow R\cdot + XH \qquad (59)$$

$$R\cdot + O_2 \rightarrow RO_2\cdot \qquad (60)$$

$$RO_2\cdot + RH \rightarrow ROOH + R\cdot \quad \text{etc.} \qquad (61)$$

not suffer from a spin restriction and can be very rapid (152).

Another important factor is that dioxygen is a rather poor one-electron oxidant (Table II). Reaction of ground state (triplet) dioxygen with good one-electron reducing agents such as carbanions (150) and reduced dyes (52) to form superoxide and the organic radical are spin allowed, i.e. $R^-(\downarrow\uparrow)$ + $O_2(\uparrow\uparrow) \rightarrow R\cdot(\uparrow) + O_2^-(\uparrow)$, and proceed rapidly. Most organic substrates, however, are not strong enough reducing agents to show this type of reactivity.

The autooxidation of metal-containing compounds (133) can be divided into two principal categories: the reactions of main group compounds and the reactions of transition metal complexes. The former proceed primarily but not exclusively by initial one-electron reduction of O_2 to O_2^- and concomitant formation of an electron-deficient metal containing intermediate (R_nM + $O_2 \rightarrow R_nM^+ + O_2^-$) whose subsequent reactions can be exceedingly complex (151 and references therein). Although a similar initial step is involved in the autooxidation of certain electron-rich transition metal organometallic compounds (133,174), most transition metal ions and complexes are not strong enough reducing agents to reduce O_2 by one electron (Table II). In fact the mechanism often observed is oxidation through a binuclear peroxide bridged intermediate which bypasses the unfavorable one-electron step (139,152). In some cases, instead of complete oxidation of a transition metal complex, a stable metal-dioxygen complex is isolated (143) as in the case of several

Co(II) complexes (see above). Such species, which are analogous to the oxy forms of hemoglobin and myoglobin, have also been postulated as intermediates in the oxidation of transition metal ions and complexes; analogous dioxygen complexes are assumed to be formed in some oxygenase enzyme mechanisms (153).

"Singlet" Dioxygen

Excited state dioxygen is well known to be highly reactive and demonstrably toxic to living organisms (154). The high reactivity of singlet O_2 presumably results from the lack of a spin restriction and the exothermicity of its reactions with many organic substrates. Most of the chemistry of singlet O_2 has been studied in nonaqueous solvents where it is generated by chemical or photochemical means (155,156). Typically, singlet dioxygen reacts with olefins, dienes, and aromatic hydrocarbons (for example reactions 62-64). In the context of this review, it is interesting to note that

$$>C=C< \quad + \quad (^1\Delta)\,O_2 \longrightarrow -\overset{OOH}{\underset{|}{C}}-C\!\!=\qquad (62)$$

$$\text{cyclohexadiene} \quad + \quad (^1\Delta)\,O_2 \longrightarrow \text{endoperoxide (O–O bridge)}\qquad (63)$$

$$\text{anthracene} \quad + \quad (^1\Delta)\,O_2 \longrightarrow \text{9,10-endoperoxide (O–O bridge)}\qquad (64)$$

superoxide has been observed to quench singlet dioxygen (157), but this is probably not biologically significant, since aqueous concentrations of superoxide would never be large enough to cause a significant amount of quenching (157).

Superoxide

The Chemistry of O_2^- in aqueous solution appears to be quite restricted. It serves as a reducing agent and as a one-electron oxidant toward organic substances. One factor which restricts its ability to act as an oxidant appears to be the availability of a proton to neutralize the charge being transferred from substrate to O_2^-. Thus, O_2^- can act as an oxidant toward certain quinols, ascorbic acid (158) and thiols (116,159) all of which have readily available protons. There is also evidence that superoxide can act as a nucleophile in aqueous solution and may add efficiently to highly activated double bonds (106). It does not, however, add to ordinary olefins (110-112) and is not capable of abstracting H· from carbon atoms in general.

Any reaction of the type described above must effectively compete with the very rapid dismutation reaction which appears to be catalyzed by a variety of materials. Thus when one speaks of O_2^- as being able to react with a particular compound, in order for the reaction to be of consequence, it must proceed with a rate comparable to that of the dismutation.

In aprotic solvents O_2^- is a good nucleophile and a good one-electron reductant but a relatively poor oxidant, again due to the paucity of protons.

It is the opinion of the authors that no chemical reactions have yet been found which would explain how superoxide could act as a "deleterious" or "cytotoxic" species. So far, superoxide itself does not seem to have the necessary reactivity. Moreover, it has not yet been possible in controlled chemical experiments to demonstrate that superoxide can initiate reactions which generate substances of known toxicity such as singlet dioxygen, hydroxyl radical, carbon-based sigma-type free radicals, or organic peroxides.

Hydrogen Peroxide

The chemistry of hydrogen peroxide toward organic substances is that of a powerful oxidant but reactions are often surprisingly slow. Hydrogen peroxide and its anion, HO_2^-, are very powerful nucleophiles (160) and high reactivity can be seen in this limited area. Many of the other reactions of hydrogen peroxide are hydroxyl radical reactions that are initiated by cleavage of the O-O bond by either heat, light, or the presence of transition metals (148). Thus hydrogen peroxide like dioxygen is relatively unreactive in spite of the high exothermicity of its reactions. Hydrogen peroxide reacts with transition metal ions and complexes much more readily than with organic substrates. Here a large number of oxidations have been observed (161,19). Other reactions of hydrogen peroxide that have been studied are Fenton's reagent-type reactions (162) (reaction 65) and catalase-type reactions (163) (reaction 66).

$$Fe^{2+} + H_2O_2 \rightarrow Fe^{3+} + HO^- + HO\cdot \qquad (65)$$

$$2H_2O_2 \xrightarrow{\text{Cu(II), Fe(III), and others}} O_2 + 2H_2O \qquad (66)$$

Hydroxyl Radical

The hydroxyl radical, formed by homolytic or reductive cleavage of hydrogen peroxide, is a powerful one-electron oxidant. Heckner and Landsberg (164) have estimated the potential of the $OH\cdot/OH^-$ couple to be 1.4 ± 0.1 V and other workers (165) have confirmed this value. Clearly most of the energy released in the reduction of O_2 to H_2O comes from the half-reaction $OH\cdot + e^- + H^+ \rightarrow H_2O$, $E^{o\prime} = 2.33$ V.

Mainly because of its extreme oxidizing capability, $OH\cdot$ is highly and indiscriminately reactive. It can remove either a single electron from a substrate or abstract $H\cdot$ to leave a more stable but still reactive free radical (reactions 67 and 68) (47).

$$OH\cdot + RH \rightarrow OH^- + RH^+ \qquad (67)$$

$$OH\cdot + RH \rightarrow H_2O + R\cdot \qquad (68)$$

Most of the reactions of $OH\cdot$ which have been observed proceed with bimolecular rate constants of 10^7 to 10^{10} $M^{-1}sec^{-1}$. The chemical properties of $OH\cdot$ have been discussed at length in Chapter 4 of Reference 47 and many other reactions have been tabulated in Reference 166.

ACKNOWLEDGEMENTS

The authors wish to thank J. San Filippo for helpful discussions,

P.G. Hildenbrand for Fig.4 (Undergraduate Thesis in Chemistry, Rensselaer Polytechnic Institute, 1973), and G.J. McClune for preparing Fig.9. We also acknowledge the expert help of Marion Frost, Janice Short and Cindy Baran in preparing the manuscript. JAF was supported by USPHS grant GM 21519 and acknowledges receipt of a travel stipend from the National Science Foundation. JSV was supported by a grant from the National Science Foundation (CHE 75-14463) and is the recipient of a National Institutes of Health Research Career Development Grant.

REFERENCES

1. McCord, J.M. and Fridovich, I. (1969). *J. Biol. Chem. 244,* 6049-6055.
2. Fridovich, I. (1972). *Accts. Chem. Res. 5,* 321-326.
3. Fridovich, I. (1974). *New Engl. J. Med. 290,* 624-625.
4. McCord, J.M., Keele, B.B. Jr. and Fridovich, I. (1971). *Proc. Nat. Acad. Sci. 68,* 1024-1027.
5. Vol'nov, I.I. (1966). In "Peroxides, Superoxides, and Ozonides of Alkali and Alkaline Earth Metals", Plenum Press, N.Y.
6. Vannerberg, N.G. (1962). *Prog. Inorg. Chem. 4,* 125-197.
7. Hunziker, H.E. and Wendt, H.R. (1976). *J. Chem. Phys. 60,* 4622-4623.
8. Merkel, P.B. and Kearns, D.R. (1972). *J. Am. Chem. Soc. 94,* 7244-7253.
9. Cade, P.E., Bader, R.F.W. and Pelletier, J. (1971). *J. Chem. Phys. 54,* 3517-3533.
10. Neuman, E.W. (1934). *J. Chem. Phys. 2,* 31-33.
11. Känzig, W. and Cohen, M.H. (1959). *Phys. Rev. Letters 3,* 509-510.
12. Lunsford, J.H. (1973). *Catalysis Revs. 8,* 135-157.
13. Bray, R.C. (1976). Personal communication.
14. Czapski, G. (1971). *Ann. Rev. Phys. Chem. 22,* 171-208.
15. Pshezhetskii, S.Y., Kotov, A.G., Milinchuk, V.K., Roginskii, V.A. and Tupikov, V.I. (1974). In "EPR of Free Radicals in Radiation Chemistry", John Wiley and Sons, New York, Chapter III.
16. Bielski, B.H.J. and Gebicki, J.M. (1970). *Advances in Radiation Chemistry 2,* 177-279.
17. Fee. J.A. and Hildenbrand, P.G. (1974). *FEBS Letters 39,* 79-82.
18. Knowles, P.F., Gibson, J.F., Pick, F.M. and Bray, R.C. (1969). *Biochem. J. 111,* 53-58.
19. Ardon, M. (1965). In "Oxygen", W.A. Benjamin, N.Y.
20. Czapski, G. and Dorfman, L.M. (1964). *J. Phys. Chem. 68,* 1169-1177.
21. Weeks, J.L. and Rabani, J. (1966). *J. Phys. Chem. 70,* 2100-2106.
22. Behar, D., Czapski, G., Rabani, J., Dorfman, L.M. and Schwarz, H.A. (1970) *J. Phys. Chem. 74,* 3209-3213.
23. Jacq, J. and Bloch, O. (1970). *Electrochim. Acta 15,* 1945-1966.
24. George, P. (1965). In "Oxidases and Related Redox Systems" (King, T.E., Mason, H.S. and Morrison, M. Eds.), John Wiley and Sons, N.Y. pp 3-36.
25. Latimer, W.M. (1952). In "The Oxidation States of the Elements and Their Potentials in Aqueous Solution", 2nd Ed. Prentice-Hall, Engelwood Cliffs, N.J.
26. Yamazaki, I. and Piette, L.H. (1963). *Biochim. Biophys. Acta 77,* 47-64.
27. Chevalet, J., Rouelle, F., Gierst, L. and Lambert, J.P. (1972). *J. Electroanal. Chem. 39,* 201-216.
28. Wood, P.M. (1974). *FEBS Letters 44,* 22-44.
29. Kebarle, P., Arshadi, M. and Scarborough, J. (1968). *J. Chem. Phys. 49,* 817-822.

30. Arshadi, M. and Kebarle, P. (1970). *J. Phys. Chem. 74,* 1483-1485.
31. Yamdagni, R., Payzant, J.D. and Kebarle, P. (1973). *Can. J. Chem. 51,* 2507-2511.
32. Baxendale, J.H., Ward, M.D. and Wardman, P. (1971). *Trans. Faraday Soc. 67,* 2532-2537.
33. Evans, M.G. and Uri, N. (1949). *Trans. Faraday Soc. 45,* 224-230.
34. Divisek, J. and Kastening, B. (1975). *J. Electroanal. Chem. 65,* 603-621.
35. Parker, A.J. (1967). *Adv. Phys. Org. Chem. 5,* 173-235.
36. Cotton, F.A. and Wilkinson, G. (1972). In "Advanced Inorganic Chemistry" 3rd ed., Wiley, New York, p. 106, 414-417.
37. Abrahams, S.C., Collin, R.L. and Lipscomb, W.N. (1951). *Acta Cryst. 4,* 15-20.
38. Gray, P. (1956). *Trans. Faraday Soc. 52,* 344-353.
39. Benson, S.W. and Shaw, R. (1970). In "Organic Peroxides" (Swern, D. Ed.) Wiley-Interscience, New York, Vol. 1, Chapter 2.
40. Ilan, Y.A., Meisel, D. and Czapski, G. (1974). *Isr. J. Chem. 12,* 891-895.
41. Berdnikov, V.M. and Zhuravleva, O.S. (1972). *Zhur. Fiz. Khim. 46,* 2658-2661. CA. 78:48904 k.
42. Sawada, Y., Iyanagi, T., and Yamazaki, I. (1975). *Biochemistry 14,* 3761-3764.
43. Sawyer, D.T. and Roberts, J.L. Jr. (1966). *J. Electroanal. Chem. 12,* 90-101.
44. Peover, M.E. and White, B.S. (1966). *Electrochim. Acta 11,* 1061-1067.
45. Maricle, D.L. and Hodgson, W.G. (1965). *Anal. Chem. 37,* 1562-1565.
46. Johnson, E.L., Pool, K.H. and Hamm, R.E. (1966). *Anal. Chem. 38,* 183-185.
47. Draganic, I.G. and Draganic, Z.D. (1971). In "The Radiation Chemistry of Water", Academic Press, New York.
48. Baxendale, J.H. (1962). *Radiat. Res. 17,* 312-326.
49. McClune, G.J. and Fee, J.A. (1976). *FEBS Letters 67,* 294-298.
50. Anbar, M. (1961). *J. Am. Chem. Soc. 83,* 2031-2037.
51. Saito, E. and Bielski, B.H.J. (1961). *J. Am. Chem. Soc. 83,* 4467-4468.
52. Ballou, D., Palmer, G. and Massey, V. (1969). *Biochem. Biophys. Res. Commun. 36,* 898-904.
53. Frisell, W.R., Chung, C.W. and MacKenzie, C.G. (1959). *J. Biol. Chem. 234,* 1297-1302.
54. Strickland, S. and Massey, V. (1973). In "Oxidases and Related Redox Systems" (King, T.E., Mason, H.S. and Morrison, M. Eds.), Vol. 1, University Park Press, Baltimore, pp. 189-194.
55. Stoien, J.D. and Wang, R.J. (1974). *Proc. Nat. Acad. Sci. 71,* 3961-3965.
56. Kastening, B. and Kazemifard, G. (1970). *Ber. Bunsenges. Phys. Chem. 74,* 551-556.
57. Rigo, A., Tomat, R. and Rotilio, G. (1974). *J. Electroanal. Chem. and Interfacial Electrochem. 57,* 291-296.
58. Rigo, A., Viglino, P. and Rotilio, G. (1975). *Anal. Biochem. 68,* 1-8.
59. Rigo, A., Viglino, P., Rotilio, G. and Tomat, R. (1975). *FEBS Letters 50,* 86-88.
60. Koutecky, J., Brdicka, R. and Hanus, V. (1953). *Coll. Czechoslov. Chem. Commun. 18,* 611-627.
61. Rotilio, G., Bray, R.C. and Fielden, E.M. (1972). *Biochim. Biophys. Acta 268,* 605-609.
62. Peover, M.E. and White, B.S. (1965). *Chem. Commun.* 183-184.
63. Slough, W. (1965). *Chem. Commun.* 184-185, but see also *ibid*, p. 420.

64. Valentine, J.S. and Curtis, A.B. (1975). *J. Am. Soc. 97*, 224-226.
65. Johnson, R.A. and Nidy, E.G. (1975). *J. Org. Chem. 40*, 1680-1681.
66. McElroy, A.D. and Hashman, J.S. (1964). *Inorg. Chem. 3*, 1798-1799.
67. Peters, J.W. and Foote, C.S. (1976). *J. Am. Chem. Soc. 98*, 873-875.
68. Bielski, B.H.J. and Saito, E. (1971). *J. Phys. Chem. 75*, 2263-2266.
69. Bielski, B.H.J. and Schwarz, H.A. (1968). *J. Phys. Chem. 72*, 3836-3841.
70. Bielski, B.H.J. and Saito, E. (1962). *J. Phys. Chem. 66*, 2266-2268.
71. Bielski, B.H.J. and Allen, A.O. (1967). *Proc. Second Tihany Symposium on Radiation Chemistry*, pp. 81-86.
72. Rabani, J., Klug-Roth, D. and Lilie, J. (1973). *J. Phys. Chem. 77*, 1169-1175.
73. Brigelius, R., Spöttl, R., Bors, W., Lengfelder, E., Saran, M. and Weser, U. (1974). *FEBS Letters 47*, 72-75.
74. Halliwell, B. (1975). *FEBS Letters 56*, 34-38.
75. McClune, G.J., McCluskey, G., Groves, J. and Fee, J.A. (1976). Unpublished results.
76. Bielski, B.H.J. (1976). Personal communication.
77. Rabani, J. and Nielsen, S.O. (1969). *J. Phys. Chem. 73*, 3736-3744.
78. Patel, K.B. and Willson, R.L. (1973). *J. Chem. Soc. Faraday Trans. I. 69*, 814-825.
79. Rabani, J., Mulac, W.A. and Matheson, M.S. (1965). *J. Phys. Chem. 69*, 53-70.
80. Land, E.J. and Swallow, A.J. (1971). *Arch. Biochem. Biophys. 145*, 365-372.
81. Simic, M.G., Taub, I.A., Tocci, J. and Hurwitz, P.A. (1975). *Biochem. Biophys. Res. Commun. 62*, 161-167.
82. Bielski, B.H.J. and Gebicki, J.M. (1974). *Biochim. Biophys. Acta 364*, 233-235.
83. Samuni, A. and Czapski, G. (1970). *Isr. J. Chem. 8*, 563-573.
84. Shimizu, Y., Shiga, T. and Kuwata, K. (1970). *J. Phys. Chem. 74*, 2929-2935.
85. Hoffman, B.M., Diemente, D.L. and Basolo, F. (1970). *J. Am. Chem. Soc. 92*, 61-65.
86. Ellis, J., Pratt, J.M. and Green, M. (1973). *J. Chem. Soc., Chem. Commun.* 781-782.
87. Haber, F. and Weiss, J. (1934). *Proc. Roy. Soc., Ser. A 147*, 332-351.
88. Barb, W.G., Baxendale, J.H., George, P. and Hargrave, K.R. (1951). *Trans. Faraday Soc. 47*, 591-616.
89. Schwarz, H.A. (1962). *J. Phys. Chem. 66*, 255-262.
90. Allen, A.O., Hogan, V.D. and Rothschild, W.G. (1957). *Radiation Res. 7*, 603-608.
91. Kremer, M.L. (1971). *Isr. J. Chem. 9*, 321-327.
92. George, P. (1947). *Disc. Faraday Soc. 2*, 196-205; cf. also discussion presentation by P. George, pp. 219-220.
93. Jayson, G.G., Parsons, B.J. and Swallow, A.J. (1973). *J. Chem. Soc., Faraday Trans. I, 69*, 236-242.
94. Bielski, B.H.J. and Becker, R.R. (1960). *J. Am. Chem. Soc. 82*, 2164-2166.
95. Rao, P.S. and Hayon, E. (1975). *J. Phys. Chem. 79*, 397-402.
96. Smith, L.L. and Kulig, M.J. (1976). *J. Am. Chem. Soc. 98*, 1027-1029.
97. Poupko, R. and Rosenthal, I. (1973). *J. Phys. Chem. 77*, 1722-1724.
98. Nilsson, R. and Kearns, D.R. (1974). *J. Phys. Chem. 78*, 1681-1683.
99. Khan, A.U. (1970). *Science 168*, 476-477.

100. Kajiwara, T. and Kearns, D.R. (1973). *J. Am. Chem. Soc. 95*, 5886-5890.
101. Howard, J.A. and Mendenhall, G.D. (1975). *Can. J. Chem. 53*, 2199-2201.
102. Mayeda, E.A. and Bard, A.J. (1973). *J. Am. Chem. Soc. 95*, 6223-6226.
103. Mayeda, E.A. and Bard, A.J. (1974). *J. Am. Chem. Soc. 96*, 4023-4024.
104. Stauff, J., Sander, U. and Jaeschke, W. (1973). In "Chemiluminescence and Bioluminescence" (Courmier, M.J., Hercules, D.M. and Lee, J. Eds.) Plenum Press, New York, pp. 131-141.
105. Hodgson, E.K. and Fridovich, I. (1976). *Arch. Biochem. Biophys. 172*, 202-205.
106. Harbour, J.R., Chew, V. and Bolton, J.R. (1974). *Can. J. Chem. 52*, 3549-3553.
107. Harbour, J.R. and Bolton, J.R. (1975). *Biochem. Biophys. Res. Commun. 64*, 803-807.
108. Farrington, J.A., Ebart, M., Land, E.J. and Fletcher, K. (1973). *Biochim. Biophys. Acta 314*, 372-381.
109. Bielski, B.H.J. and Chan, P.C. (1976). *J. Biol. Chem. 251*, 3841-3844.
110. Schmidt, M. and Bipp, H. (1960). *Z. Anorg. Allg. Chem. 303*, 190-209.
111. LeBerre, A. and Berguer, Y. (1965). *C.R. Acad. Sci., Paris, 260*, 1995-1998.
112. LeBerre, A. and Berguer, Y. (1966). *Bull. Soc. Chim. Fr.*, 2363-2374.
113. Dietz, R., Forno, A.E.J., Larcombe, B.E. and Peover, M.E. (1970). *J. Chem. Soc. B*, 816-820.
114. Merritt, M.V. and Sawyer, D.T. (1970). *J. Org. Chem. 35*, 2157-2159.
115. San Filippo, J. Jr., Chern, C.I. and Valentine, J.S. (1975). *J. Org. Chem. 40*, 1678-1680.
116. San Filippo, J. Jr., Romano, L.J., Chern, C.I. and Valentine, J.S. (1976). *J. Org. Chem. 41*, 586-588.
117. Johnson, R.A. (1976). *Tetrahedron Lett.*, 331-332.
118. Corey, E.J., Nicolaou, K.C., Shibasaki, M., Machida, Y. and Shiner, C.S. (1975). *Tetrahedron Lett.*, 3183-3186.
119. Corey, E.J., Nicolaou, K.C. and Shibasaki, M. (1975). *J. Chem. Soc., Chem. Commun.*, 658-659.
120. Streitwieser, A. Jr. (1962). In "Solvolytic Displacement Reactions", McGraw-Hill, New York.
121. Liotta, C.L. and Harris, H.P. (1974). *J. Am. Chem. Soc. 96*, 2250-2252.
122. March, J. (1968). In "Advanced Organic Chemistry", McGraw-Hill, New York, p. 887.
123. Howard, J.A. (1973). In "Free Radicals", Vol.11, (J.K. Kochi, Ed.) John Wiley, New York, p. 38.
124. San Filippo, J. Jr. (1976). Personal communication.
125. Magno, F. and Bontempelli, G. (1976). *J. Electroanal. Chem. 68*, 337-344.
126. Misra, H.P. and Fridovich, I. (1972). *J. Biol. Chem. 247*, 3170-3175.
127. Miller, R.W. and Rapp, U. (1973). *J. Biol. Chem. 248*, 6084-6090.
128. Greenstock, C.L. and Miller, R.W. (1975). *Biochim. Biophys. Acta 396*, 11-16.
129. Moro-oka, Y. and Foote, C.S. (1976). *J. Am. Chem. Soc. 98*, 1510-1514.
130. Tezuka, M., Ohkatsu, Y. and Osa, T. (1975). *Bull. Chem. Soc. Jap. 48*, 1471-1474.
131. Pryor, W.A. (1966). In "Free Radicals", McGraw-Hill, New York, pp. 150-155.
132. Swarc, M. (1962). *Chem. Soc. (London), Spec. Publ. 16*, 94.
133. Ingold, K.U. and Roberts, B.P. (1971). In "Free Radical Substitution Reactions", John Wiley and Sons, New York.

134. Dietz, R., Peover, M.E. and Rothbaum, P. (1970). *Chemie-Ing. Techn. 42,* 185-189.
135. Klug-Roth, D. and Rabani, J. (1976). *J. Phys. Chem. 80,* 588-591.
136. Rosenthal, I. and Frimer, A. (1975). *Tetrahedron Lett.*, 3731-3732.
137. Levonowich, P.F., Tannenbaum, H.P. and Dougherty, R.C. (1975). *J. Chem. Soc. Chem. Commun.*, 597-598.
138. Weiss, J. (1952). *Adv. Catal. 4,* 343-365.
139. Taube, H. (1965). *J. Gen. Physiol. 49(2),* 29-50.
140. Goolsby, A.D. and Sawyer, D.T. (1968). *Anal. Chem. 40,* 83-86.
141. Hill, H.A.O., Turner, D.R. and Pellizer, G. (1974). *Biochem. Biophys. Res. Commun. 56,* 739-744.
142. Valentine, J.S. and Quinn, A.E. (1976). *Inorg. Chem.,* in press.
143. Vaska, L. (1976). *Accounts Chem. Res. 9,* 175-183.
144. Reed, C.A. (1976). Personal communication.
145. Collman, J.P. (1976). Personal communication.
146. Abel, E.W., Pratt, J.M., Whelan, R. and Wilkinson, P.J. (1974). *J. Am. Chem. Soc. 96,* 7119-7120.
147. Rapp, U., Adams, W.C. and Miller, R.W. (1973). *Can. J. Biochem. 51,* 158-171.
148. Hamilton, G.A. (1974). In "Molecular Mechanisms of Oxygen Activation" (Hayaishi, O., Ed.), Academic Press, New York, p. 405-451.
149. Uri, N. (1962). In "Advances in Chemistry" Series No.36, pp. 102-112.
150. March, J. (1968). In "Advanced Organic Chemistry", McGraw Hill, New York, pp. 543-545.
151. Panek, E.J. and Whitesides, G.M. (1972). *J. Am. Chem. Soc. 94,* 8768-8775.
152. Fallab, S. (1967). *Angew. Chem. Int. Ed. Engl. 6,* 496-507.
153. Hayaishi, O. (1974). In "Molecular Mechanisms of Oxygen Activation" (Hayaishi, O., Ed.), Academic Press, New York, pp. 1-28.
154. Trozzolo, A.M. (Ed.) (1970). *Ann. N.Y. Acad. Sci. 171, No.1,* pp. 1-302.
155. Foote, C.S. (1968). *Accts. Chem. Res. 1,* 104-110.
156. Kearns, D.R. (1971). *Chem. Rev. 71,* 395-427.
157. Guiraud, H.J. and Foote, C.S. (1976). *J. Am. Chem. Soc. 98,* 1984-1986.
158. Barr, N.F. and King C.G. (1956). *J. Am. Chem. Soc. 78,* 303-305.
159. Swallow, A.J. (1952). *J. Chem. Soc.,* 1334-1339.
160. Jencks, W.P. (1969). In "Catalysis in Chemistry and Enzymology" p. 107-111.
161. Cotton, F.A. and Wilkinson, G. (1972). In "Advanced Inorganic Chemistry" 3rd Ed., John Wiley and Sons, New York, p. 416.
162. Walling, C. and Johnson, R.A. (1975). *J. Am. Chem. Soc. 97,* 363-367.
163. Sigel, H. (1969). *Angew. Chem. Int. Ed. Engl. 8,* 167-177.
164. Hechner, K.H. and Landsber, R. (1965). *Zeit Phys. Chem. 230,* 63-72.
165. Stein, G. (1965). *J. Chem. Phys. 42,* 2986.
166. Neta, P. and Dorfman, L.M. (1968). *Advan. Chem. Ser. 81,* 222-230.
167. San Filippo, J. Jr., Chern, C.I. and Valentine, J.S. (1976). *J. Org. Chem. 41,* 1077-1078.
168. Brigelius, R., Hartmann, H.J., Bors, W., Saran, M., Lengfelder, E. and Weser, U. (1975). *Z. Physiol. Chem. 356,* 739-745.
169. Wilkins, R.G. (1971). *Adv. Chem. Ser. 100,* 111-134.
170. Rosenthal, I. and Bercovici, T. (1973). *J. Chem. Soc., Chem. Commun.* 200-201.
171. Valentine, J.S. and Fallon, P. (1975). Unpublished results.

172. Tezuka, M., Hamada, H., Ohkatsu, Y. and Osa, T. (1976). *Denki Kagaku* *44*, 17-21.
173. Dzidic, I., Carroll, D.I., Stillwell, R.N. and Horning, E.C. (1974). *J. Am. Chem. Soc.* *96*, 5258-5259.
174. Whitesides, G.M., San Filippo, J. Jr., Casey, C.P. and Panek, E.J. (1967). *J. Am. Chem. Soc.* *89*, 5302-5303.
175. Hoare, J.P. (1968). In "The Electrochemistry of Oxygen", Interscience Publishers, New York, p. 64F.
176. Thorneley, R.N.F. (1974). *Biochim. Biophys. Acta* *333*, 487-496.
177. Bors, W., Saran, M., Lengfelder, E., Spöttl, R. and Michel, C. (1974). *Curr. Topics Radiat. Res.* *9*, 247-309.

STUDIES ON THE SUPEROXIDE ION: COMPLEX FORMATION WITH BARIUM AND CALCIUM IONS DETECTED BY e.p.r. SPECTROSCOPY AND KINETICALLY

R.C. BRAY and G.N. MAUTNER

School of Molecular Sciences
University of Sussex
Brighton, Sussex, England

E.M. FIELDEN and C.I. CARLE

Physics Division
Institute of Cancer Research
Sutton, Surrey, England

SUMMARY

The superoxide radical, O_2^-, when present in aqueous solution, may readily be trapped by the rapid freezing procedure of R.C. Bray (*Biochemical Journal 81*, (1961), 189-195) and studied by electron paramagnetic resonance (e.p.r.) spectroscopy at low temperatures. Such work, using the xanthine oxidase system as the source of the radical, yielded unequivocal evidence in the late 1960's for 1-electron reduction of oxygen by a biological system. For physical or biochemical studies, pulse radiolysis may also be used to generate O_2^-, and, potassium superoxide in anhydrous dimethyl sulphoxide (DMSO) is a particularly convenient source of the radical.

O_2^-, trapped by the rapid freezing method in water or in aqueous DMSO, exhibits an axial e.p.r. spectrum, which, owing to fast relaxation, is detectable only at temperatures below about 220°K. Parameters of the signal are reported for a variety of frozen aqueous media. With neutral or alkaline solutions in the presence of alkali metal ions, over a wide range of concentrations and with a variety of anions, the very broad $g_{||}$ feature is at about 2.09. The same spectrum, believed to be due to uncomplexed O_2^-, is observed with 0.1 mM Ba^{++}. However, when the concentration of Ba^{++} ions is increased, two new species are detected. The first, maximally developed at 3 mM Ba^{++}, has a sharper $g_{||}$, shifted to 2.077, while the second, observed with e.g. 250 mM Ba^{++}, is still sharper and has $g_{||}$ 2.065. Enhanced kinetic stability of O_2^- in solution at room temperature with high concentrations of Ba^{++}, implies that existence of the second species, at least, is not confined to the frozen matrix. Ca^{++} produces effects comparable to those with Ba^{++} but with still larger shifts in e.p.r. parameters. The possible nature of the species at low and high metal ion concentrations, as well as the biological implications, are discussed.

INTRODUCTION

Electron paramagnetic resonance (e.p.r.) spectroscopy has played an important part in work on the superoxide anion free radical, O_2^-. Though the radical is relatively unstable in aqueous solution, nevertheless its lifetime is such that it would be expected that it could be trapped by the rapid freezing method of Bray (1961), and so studied in the frozen state by e.p.r. In fact, the radical was almost certainly observed during work using this method on xanthine oxidase reported by Bray *et al.* (1964), though at this stage the superoxide e.p.r. spectrum was confused with that of the flavin semiquinone radical from the enzyme (see Fig.1). Later work by Bray and Knowles (1968) indeed made it clear that some oxygen radical species produced by the enzyme was detectable by e.p.r. However, because of the confusion which existed in the literature at this time about O_2^- in slightly alkaline aqueous solution (Czapski and Dorfman, 1964: "alkaline stabilized" O_2^-), these workers were reluctant to identify their species as superoxide. Further confusion arose since the early work from Fridovich and co-workers (e.g. Greenlee *et al.*, 1962) was interpreted as indicating enzyme-bound rather than free O_2^- although by 1968 McCord and Fridovich had clearly and effectively demonstrated that the radical is liberated into free solution. Subsequent e.p.r. work, including studies with oxygen-17, fully confirmed that the superoxide radical is indeed the radical produced in free solution by 1-electron reduction of oxygen by xanthine oxidase (Bray *et al.*, 1968; Knowles *et al.*, 1969; Nilsson *et al.*, 1969; Bray *et al.*, 1970). These studies furnished what is perhaps the most unequivocal evidence in existence for 1-electron reduction of oxygen by a biological system.

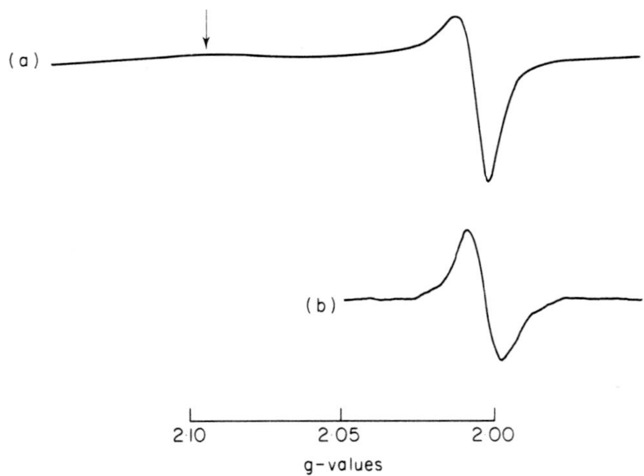

Fig.1. *E.p.r. spectra: (a) of* O_2^- *and (b) of FADH· in desulpho xanthine oxidase.*
The rapid freezing procedure was used and spectra were recorded at about 120°K at 9.1 GHz. The FADH· signal was generated by reducing desulpho xanthine oxidase at pH 8.2 with 0.3 M dithionite for 50 msec. (Bray et al., *1975 and M.J. Barber, unpublished; the desulpho enzyme was preferred to the functional form since there is less interference by Mo(v)).* O_2^- *was obtained by pulse radiolysis of oxygen-saturated sodium formate in dilute aqueous NaOH. The scale of g-values is approximate only. Obviously, the spectra could easily be confused if the* $g_{||}$ *feature of* O_2^- *at g 2.09 (arrow) was not detected.*

The xanthine oxidase system remains a good and useful source of superoxide radicals for biochemical or physical studies. Steady-state concentrations of the radical of up to 5 - 10 μM may be maintained for long periods (B. Almark and R.C. Bray, unpublished). However, a number of other sources of the radical must also be mentioned. Much work has of course depended on pulse radiolysis and Nilsson *et al.* (1969) combined pulse radiolysis with rapid freezing to trap the radical and observe its e.p.r. spectrum. This procedure was later used in work on superoxide dismutase by Fielden *et al.* (1974). Work by Ballou *et al.* (1969) suggested that it is the reduced flavin centre in the xanthine oxidase molecule which reduces the oxygen to superoxide, by showing that the radical is also produced during autoxidation of free flavin molecules. The O_2^- radical may also readily be generated by oxidation of hydrogen peroxide with periodate ions (Knowles *et al.*, 1969) or electrochemically by reduction of oxygen, in organic solvents such as dimethylsulphoxide (DMSO) (Fee and Hildenbrand, 1974). In the present work on the superoxide ion, potassium superoxide dissolved in DMSO was used as a convenient source of the radical for e.p.r. work, while parallel studies were carried out by pulse radiolysis.

Though in frozen aqueous solution the superoxide radical always shows an axial type of e.p.r. spectrum with a relatively broad $g_{||}$ component, of the general type shown in Fig.1a, nevertheless depending on the precise conditions there may be very substanial changes in the form of the spectrum. For instance Knowles *et al.* (1969), reported variations in $g_{||}$ from 2.076 to 2.121, while Pick (1971) reported still larger variations in the spectrum. The present studies arose from an investigation into the origins of these changes in the e.p.r. spectrum of the superoxide radical (see also Carle, 1976).

MATERIALS AND METHODS

Pulse radiolysis, e.p.r. and rapid freezing were carried out as described by Fielden *et al.* (1974). DMSO was distilled from calcium hydride under reduced pressure in an atmosphere of nitrogen, immediately before use. Potassium superoxide (from Ralph H. Emanual Limited, London, HA0 1PY) was rapidly ground in a pestle and mortar before being shaken thoroughly with DMSO. After centrifuging off undissolved solid, the solution, containing up to 6 mM O_2^-, was filled into the syringe of the rapid freezing apparatus. Once in the syringe, such solutions deteriorated at a rate of only about 1% per hour at room temperature. We did not find it necessary to use a crown ether to solubilize KO_2 (Valentine and Curtis, 1975).

In rapid freezing experiments in which KO_2 was the source of O_2^-, 0.2 vol. of KO_2 in DMSO was mixed with 1 vol. of an appropriate aqueous solution in the rapid freezing apparatus and after an appropriate time interval the mixture was frozen rapidly in isopentane, for examination by e.p.r. Thus, in all such work the final medium contained 18% $^W/W$ of DMSO. When pulse radiolysis was the source of O_2^-, the oxygenated solution was passed through a glass irradiation coil at a flow rate such that each element of solution received several pulses of radiation. As in the KO_2 method, the solution was then frozen by the normal rapid freezing technique.

RESULTS

Effect of the Medium on the e.p.r. Parameters of O_2^- in Frozen Aqueous Solution

We examined the e.p.r. spectrum of O_2^-, obtained either from pulse radiolysis or from potassium superoxide, in a large number of different frozen aqueous media. Some typical spectra are reproduced in Fig.2. Clearly the medium has a substantial effect on the spectrum. In particular the position and line width of the $g_{\parallel}$ feature is distinctly variable.

Under most conditions we obtained spectra exemplified by the ones in Fig.2b-e, showing a very broad $g_{\parallel}$ in the region of g = 2.09. Within this group there were only minor, though definite, variations in the spectrum. We obtained spectra like this either from pulse radiolysis (Fig.2b) or by working with potassium superoxide in media containing DMSO at a final concentration of 18% $^W/W$ (Fig.2c-e). Such spectra could be obtained at all pH values within the range 7 to 13, at all ionic strengths within the range of about 0.003 to 1, in the presence of any of the anions ClO_4^-, Cl^-, $P_2O_7^{4-}$, CO_3^{2-}, SO_4^{2-}, acetate or formate and with any of the cations Na^+, K^+, Li^+ or Cs^+.

Much more distinctive spectra were obtained in the presence of barium or calcium ions (Fig.2f and 2g). However, before we discuss these we must first consider the spectrum of Fig.2a. This spectrum, with a somewhat sharpened and shifted $g_{\parallel}$ feature, was obtained only in pulse radiolysis at low ionic strength in the presence of low concentrations of ethanol. It could not be reproduced in the experiments with potassium superoxide in DMSO. Further, it was not obtained in the pulse radiolysis with ethanol if 0.1 M formate was added, whether this was present during irradiation, or whether it was used to dilute the solution between irradiation and freezing. This excludes the possibility that the spectrum of Fig.2a is due to peroxyethanol radical. We are therefore somewhat at a loss to explain why this spectrum is different from those discussed in the previous paragraph and we will not consider this point again in subsequent sections of this paper. However, it seems likely that the spectrum may arise by some specific interaction of the ethanol. Both our $g_{\parallel}$ value of 2.078 and the form of the spectrum is very like that reported by Bennett *et al.* (1968) for O_2^- produced by using sodium and ethanol in the rotating cryostat. These workers obtained $g_{\parallel}$ 2.079. Possibly in our work, ethanol is exerting its influence through effects on water structure (compare work on iron-sulphur proteins by Coffman and Stavens, 1970, and by Cammack *et al.*, 1971).

The effect of barium ions on the O_2^- e.p.r. spectrum is illustrated in Fig.2f and in Fig.3. Remarkable changes of spectrum take place as the concentration of this cation is increased from 0.1 mM to 250 mM. Again, the most notable changes are in the position and width of the $g_{\parallel}$ peak. Clearly, there is evidence for the existence of three distinct species. The broad $g_{\parallel}$ line at g 2.095, observed at low ionic strength (Fig.3a; this spectrum is similar to those in Figs.2b-e), is replaced by a sharper peak, maximally developed at around 3 mM barium, at g 2.077 (Fig.3c) and this in turn is replaced with 250 mM barium, by an even sharper line at g 2.065 (Fig.3e). These effects may be observed either at pH 10 or at pH 7 (compare Figs.3c and 3f).

When calcium ions were used in place of barium ions, similar effects were noted (Figs.2g and 4). However, in this case the three $g_{\parallel}$ peaks were even better separated than those observed with barium.

In our experiments the position of the $g_{\parallel}$ feature (measured at the downward "peak" on the derivative trace) was always near to 2.001, in agreement with Knowles *et al.* (1969).

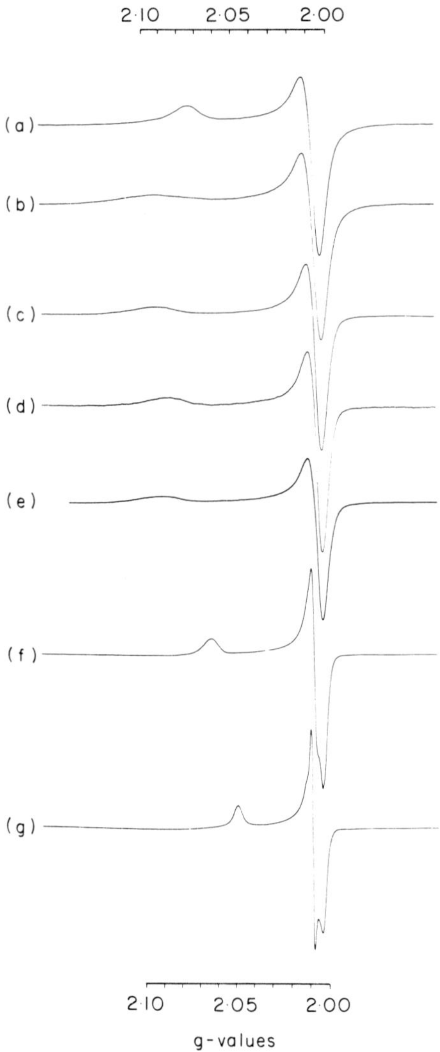

Fig.2. *E.p.r. spectra of the superoxide ion in various media, obtained either by pulse radiolysis or from KO_2*
All samples were prepared by the rapid freezing method. (a) and (b) were from pulse radiolysis, while (c) - (g) were from KO_2 and contained DMSO (final concentration 18% W/W. Final concentrations of cations were as follows, with the main anion indicated in parentheses:
(a) Na^+, about 1 mM (hydroxide or carbonate, with 85 mM ethanol);
(b) Na^+, 100 mM (formate); (c) Na^+, 30 mM (pyrophosphate);
(d) Na^+, 700 mM (chloride); (e) Cs^+, 100 mM (chloride);
(f) Ba^{++}, 250 mM (chloride); (g) Ca^{++}, 250 mM (chloride).
pH values were 10 - 11 in (a) - (c) and (e) and about 7 in (d), (f) and (g). Reaction times ranged from 5 - 200 msec. E.p.r. spectra were recorded at 9.1 GHz and about 110°K, with 1 or 5 Gauss modulation and 10 or 100 mW microwave power. The scale of g-values is approximate only.

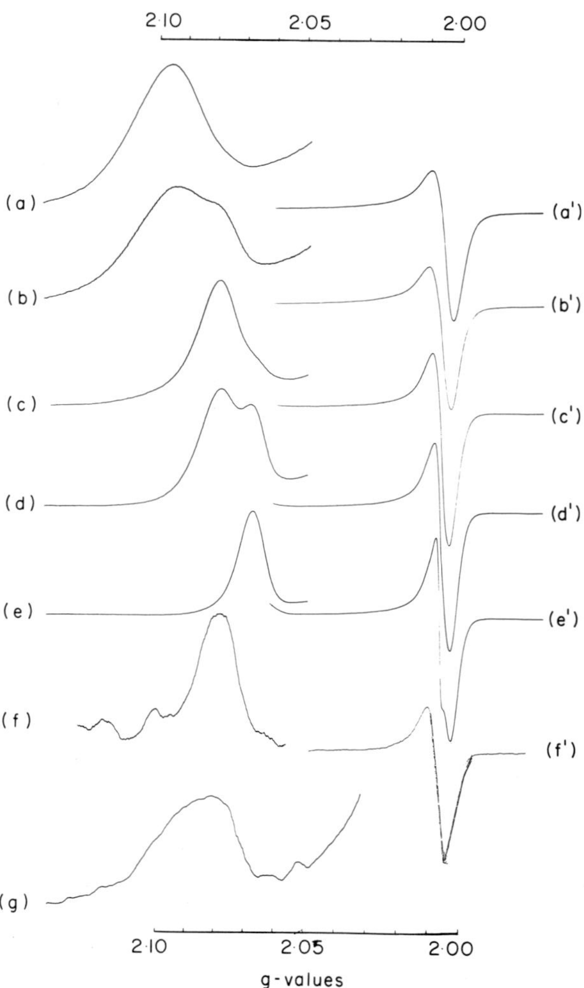

Fig.3. *E.p.r. spectra of the superoxide ion in media containing barium ions at various concentrations.*
The $g_{||}$ and $g_{\perp}$ regions of the spectra are shown at different gain settings. All samples were prepared from KO_2 by the rapid freezing method and contained DMSO (final concentration 18% W/W). The samples contained Ba^{++} (as the hydroxide or chloride) at the following final concentrations: (a), (a^1) 0.1 mM; (b), (b^1) 0.3 mM; (c), (c^1) 3 mM; (d), (d^1) 8 mM; (e), (e^1) 250 mM; (f), (f^1) 3 mM; (g) 3 mM; in addition, Na^+ was present in (g) at 700 mM. pH values were 9.7 to 10.3 in (a) - (e) and about 7.0 (HEPES buffer) in (f), (f^1), (g). Reaction times were 5 - 200 msec. Spectra were recorded at 9.1 GHz and about 110°K, with modulations of 1, 5 and 8 Gauss in (a^1) - (e^1), (a) - (e), (f^1) and (f) - (g) respectively and with microwave powers of 10 mW except in (f) and (g) where 100 mW was used. The scale of g-values is approximate only.

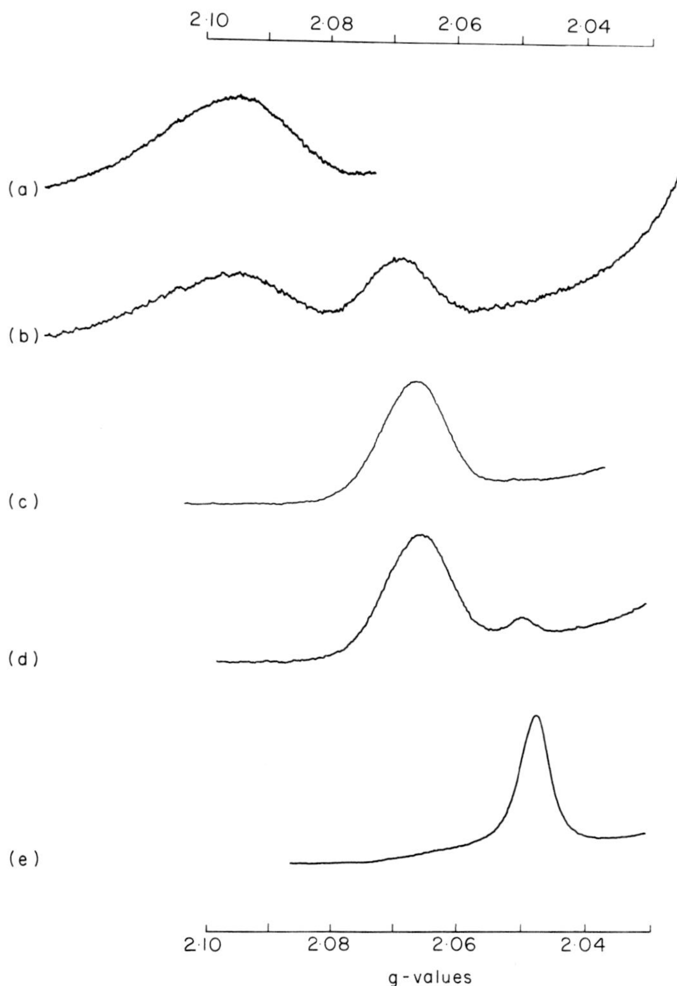

Fig.4. *E.p.r. spectra of the superoxide ion in media containing calcium ions at various concentrations.*
The $g_{||}$ region of the spectrum only is shown. All samples were prepared from KO_2 by the rapid freezing method and contained DMSO (final concentration 18% W/W). The samples contained Ca^{++} (as the hydroxide or chloride) at the following final concentrations: (a) 0.1 mM; (b) 0.3 mM; (c) 3 mM; (d) 8 mM; (e) 250 mM. pH values were 9.1 - 9.8. Spectra were recorded at 9.1 GHz and about 110°K, with 1 Gauss modulation and 10 mW microwave power. The scale of g-values is approximate only.

Effect of Temperature on the e.p.r. Spectrum of O_2^-

Fig.5 illustrates the effect of varying the recording temperature on the e.p.r. spectrum of samples of O_2^- trapped by rapid freezing. The amplitude of the signal, normalised for the Boltzman effect of temperature on signal intensity, is plotted against the absolute temperature. The spectrum is fully sharpened only at temperatures below 40°K, while at higher temperatures it broadens owing to fast relaxation, half maximum broadening being achieved at about 150°K. By about 220°K broadening was so extensive that the spectrum was scarcely detectable. This effect was reversible and on lowering the temperature again, the spectrum again sharpened. The samples illustrated in Fig.5 were frozen at pH 13. With other samples in different media, similar broadening effects were observed. However at lower pH values, irreversible disappearance of the signal, due to decomposition in the frozen state, set in at about 200°K, that is before the fully broadened spectrum had been reached.

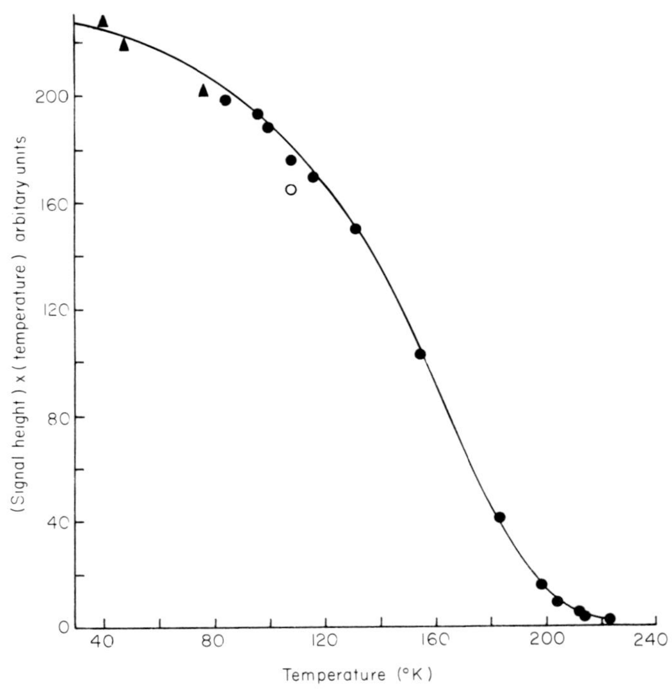

Fig.5. *Effect of varying the recording temperature on the e.p.r. spectrum of* O_2^-.
Amplitude of the $g\perp$ *feature (in arbitrary units) was multiplied by the absolute temperature in order to correct for the Boltzman effect and the product is plotted against the absolute temperature. The point indicated by the* open circle *was obtained by re-cooling after warming to the highest temperature used. A sample of* O_2^- *prepared from* KO_2 *by the rapid freezing method, in a final medium containing 0.1 M NaOH and 18%* $^W/W$ *DMSO was employed.* Triangles *and* circles *correspond to two separate experiments.*

Kinetic and Other Studies on O_2^- *in Solution at Room Temperature*

The changes in the form of the e.p.r. spectrum, as discussed above, in the presence of calcium or barium ions, provide strong evidence for interaction between O_2^- and these cations. However, the possibility has to be considered that this interaction, detected at low temperatures, is peculiar to the frozen aqueous matrix and has no relevance to the situation in solution at room temperature. In fact there has been fairly extensive work on the effect of various solid matrices on the e.p.r. parameters of O_2^-, summarized by Atkins and Symons (1967) and by Knowles *et al.* (1969). We hoped to obtain additional information relating to the existence, or otherwise, of interactions in liquid solution, by studying the kinetics of decay of O_2^- to O_2 and H_2O_2 as a function of the composition of the medium. Decay of the $HO_2^{\bullet}$ radical may readily be studied at room temperature by flow e.p.r. methods (see for example Meisel *et al.*, 1973). However, such methods are not applicable to work on O_2^-, for reasons discussed in the previous sub-section. The two methods available to us for kinetic studies were therefore rapid freezing using e.p.r. detection and conventional pulse radiolysis using optical detection.

Data from a considerable number of kinetic experiments is summarized in Table I and typical decay curves using the two methods are illustrated in Fig.6. In agreement with the literature, second order decay of O_2^- was observed under most conditions by either method. Rapid freezing experiments were carried out with a relatively high concentration of superoxide, while in pulse radiolysis a range of lower concentrations was employed. All our experiments were carried out in dilute HEPES buffer pH 7, with the addition of 0.5 mM EDTA to counter the possible effects of contamination by copper or iron ions. As before 18% $^W/W$ DMSO was present in the rapid freezing experiments. In no case did we observe changes in the form of the e.p.r. signal with time.

TABLE I

Decay Kinetics of O_2^- in Various Media

Pulse radiolysis or rapid freezing e.p.r. was used to follow the decay of O_2^- at about 20 - 25°C. Second order decay kinetics were observed unless otherwise indicated. For all experiments the aqueous medium contained HEPES (10 mM) and EDTA (0.5 mM); where indicated NaCl (0.9 M), $BaCl_2$ (0.3 M) or $CaCl_2$ (0.25 M) was also present; the pH was adjusted to 6.7 to 7.2 with NaOH. To initiate the reaction in the rapid freezing experiments, 0.2 vol. KO_2 in anhydrous DMSO was mixed with the aqueous medium. In pulse radiolysis experiments, unless otherwise noted, 0.1 M Na formate was present as hydroxyl scavenger and measurements were made at 245 nm.

Source of O_2^-	Detection of O_2^-	Approx. initial O_2^- (mM)	Control: rate constant ($M^{-1}sec^{-1}$)	$[t_{\frac{1}{2}}]$ as a fraction of that in control		
				NaCl	$BaCl_2$	$CaCl_2$
KO_2 (in DMSO)	e.p.r.	1	3.9×10^6	1.9	15	-
4 MeV electrons	optical	0.13	1.0×10^6	1.6	5.4	5.2[a]
4 MeV electrons	optical	0.013	-	-	-	3.5[a]
4 MeV electrons	optical	0.006[b]	1.7×10^6	1.3	1.1[c]	-
4 MeV electrons	optical	0.03[d]	-	3.8[a]	-	9.7[a]

[a]Decay apparently 1st order

[b]Without formate

[c]Some deviations from 2nd order kinetics

[d]In the presence of bovine superoxide dismutase (approx. 9 μg/ml)

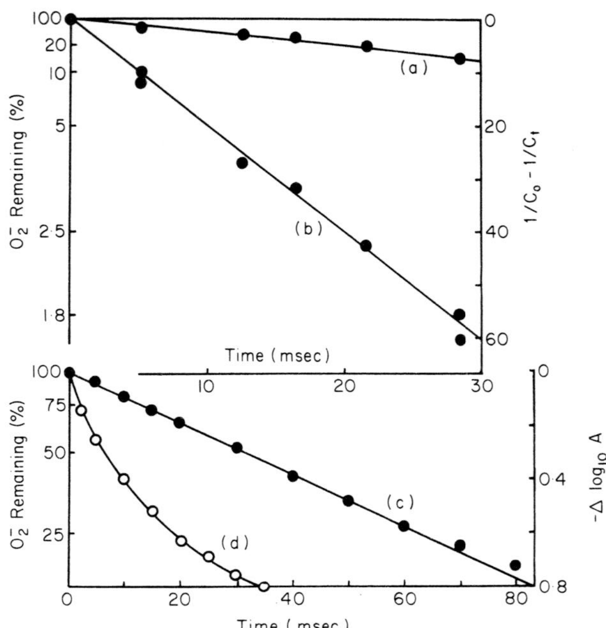

Fig.6. *First and second order decay kinetics for O_2^- in various media. Decay was followed by the rapid freezing method (starting from KO_2 in DMSO) in curves (a) and (b) and by pulse radiolysis at 245 nm in curves (c) and (d). In (a) 0.25 M $BaCl_2$ and in (b) 0.7 M NaCl, respectively (final concentrations) were present and second order decay was found (reciprocal plots); in (c) 0.25 M $CaCl_2$ was present and decay was first order (semi-log plot), whereas in the control, (d), without $CaCl_2$ and at lower ionic strength, second order decay was again observed. The pH (HEPES buffer) was 6.8 - 7.0 and the temperature about 20 - 25°C; other conditions were as in Table I (caption and lines 1 and 2). Initial O_2^- concentrations were 0.8 or 1.1 mM (from KO_2) and 0.13 mM (pulse radiolysis: radiation dose 24 k rad.). A: absorbance at 245 nm; C_o, C_t: O_2^- concentrations (mM) at times 0 and t. Each point in the rapid freezing experiments is the average of triplicate determinations (integrated e.p.r. signal intensity), the point at time 0 corresponding to values obtained on replacing the buffered aqueous solutions by dilute NaOH.*

We first carried out control experiments at low ionic strength (about 0.003 M), then compared these with decay curves obtained at relatively high ionic strengths, which were obtained by the addition of sodium chloride, barium chloride or calcium chloride (Table I). Second order decay rate constants obtained in the controls using pulse radiolysis were (1.4 ± 0.4) x 10^6 $M^{-1}sec^{-1}$. The corresponding value from the rapid freezing experiments with KO_2 was 3.9 x 10^6 $M^{-1}sec^{-1}$. Either of these values agrees adequately with the literature (Behar *et al.*, 1970; Ballou *et al.*, 1969) and the origin of the relatively small discrepancy between the values measured by our two methods is not certain, though warming of the solution on mixing DMSO with water in the rapid freezing experiments with KO_2 might be partly responsible. It seems unlikely that it is due to more specific effects of DMSO, since, at the concentrations employed, this would have relatively little effect on the dielectric constant. Furthermore, in one pulse radiolysis experiment carried out in the presence of DMSO, decay rates seemed to be little affected by the presence of the solvent, though there were complications in relation to hydroxyl radical scavenging.

In all experiments using either experimental method, increasing the ionic strength slowed the reaction rate (Table I). We attribute the slowing by 0.9 M NaCl to a primary kinetic salt effect on the pK of $HO_2^{\cdot}$. The pK would be lowered on increasing the ionic strength. Since, in the neutral and alkaline region decay of O_2^- depends on reaction between O_2^- and $HO_2^{\cdot}$ (Behar *et al.*, 1970), increasing the ionic strength would therefore be expected to slow down the decay in agreement with our findings.

A more marked stabilization of O_2^- than that obtained with sodium chloride was observed when barium chloride or calcium chloride was used to produce a similar increase in the ionic strength (Table I). Decay was in almost all cases slowed several fold compared with that in the presence of sodium chloride. The one exception (Table I) was an experiment at a rather low concentration of O_2^- in the presence of barium ions but even here, though barium had little effect on the decay rate, nevertheless it had some effect in that pure second order decay kinetics were no longer followed in its presence. Calcium had an even more marked effect on the decay kinetics than did barium, in that, apart from the reaction being slowed down, apparently pure first order decay was always observed with calcium, in contrast to the second order decay found in most cases with barium (see Fig.6). Finally, it is noteworthy that calcium ions stabilized O_2^-, not only against spontaneous decay, but also against decay mediated by bovine superoxide dismutase (Table I).

In additional experiments, the concentration of calcium chloride was varied. 100 mM calcium chloride stabilized superoxide to about the same extent as did 250 mM (Table I), but at 10 mM, decay was as in the control.

In the pulse radiolysis work, aside from the kinetic experiments, some small effects of high concentrations of barium chloride on the ultraviolet absorption spectrum of the superoxide radical in solution were also noted. There was little change in the position of the maximum at 245 nm or on the shape of the curve, but the extinction coefficient was decreased by about 25% by 0.3 M $BaCl_2$ (Fig.7).

DISCUSSION

Nature of the Interaction Between O_2^- and Ca^{++} or Ba^{++}

Ion pair formation between superoxide and alkali metal ions is not expected in aqueous solution and our results provide no evidence for this. The changes brought about in the e.p.r. spectrum of the radical by even the highest concentrations of such ions were small and, more particularly, cesium ions, which would be expected to show relatively large hyperfine splittings in an ion pair, did not modify the e.p.r. spectrum (Fig.2e). No doubt the species present under such conditions is free O_2^- (see below).

On the other hand, our results provide clear evidence for interaction of O_2^- with metal ions at high concentrations of calcium or barium chloride. Complex formation between O_2^- and metal ions does not seem to have been reported previously, though it is well known that $HO_2^{\cdot}$ forms metal complexes (e.g. Meisel, 1974). Our experiments with high concentrations of calcium or barium are unequivocal in demonstrating O_2^- - metal interaction in solution. Here, effects were observed not only in the frozen aqueous medium employed for recording the e.p.r. spectra, but also at room temperature, on the u.v. spectrum of superoxide and also kinetically, as demonstrated by two independent methods. In confirmation that these various phenomena arise from a common origin, the threshold concentrations of calcium required for observing the kinetic and the final e.p.r. spectral effects seemed similar to one another. Our data indicate a dissociation constant of perhaps 50 mM for the species formed from O_2^- in the presence of high concentrations of calcium or barium ions. Such a value is not greatly different from the dissociation constant, for instance, for the complex between barium ions and nitrate ions, for which a value of about 120 mM has been reported (Sillén and Martell, 1964).

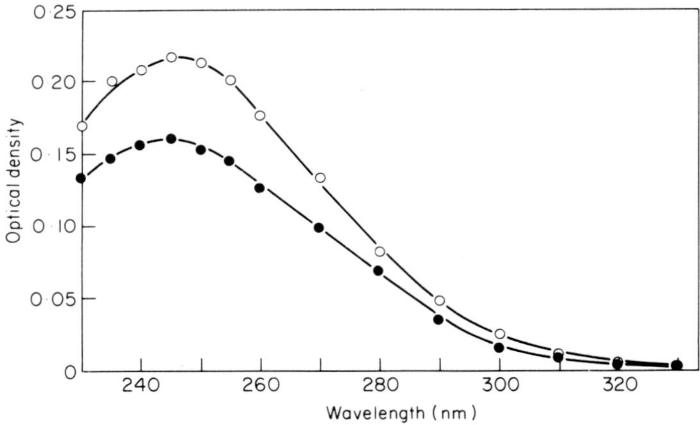

Fig.7. *Ultraviolet absorption spectrum of* O_2^- *in the presence and absence of* $BaCl_2$
Open circles, *alone;* full circles, *with 0.3 M* $BaCl_2$. *Pulse radiolysis (25 k rad; light path 0.7 cm) was used at pH 7.1 (HEPES buffer). Conditions were as in Table I.*

Decay kinetics for the superoxide radical in the presence of calcium and barium ions is clearly complex, as is indicated by the change over from second order to first order kinetics. Similarly, decay of HO_2 complexes is also reported to be complicated (Meisel, 1973).

At lower concentrations of calcium and barium there is no positive kinetic evidence for interaction between superoxide and the metal ions and we have only the effects on the low temperature e.p.r. spectra to consider. Nevertheless, these effects, as shown in Figs.2 to 4, are quite striking. Furthermore, it seems significant that with either metal ion the three species, detected in the $g_{||}$ region as the metal ion concentration is increased from very low to very high values, seem to form a distinct series with progressive decreases in both line width of the peak and in g-value. This leads us to believe that the intermediate species, that is the one present maximally at about 3 mM calcium or barium, is, like the high concentration species, likely not to be confined to the frozen matrix, but to be present also in the solutions, at least near their freezing point (about $-10^{\circ}C$ for 18% $^W/W$ DMSO). There seems to be no reason to suspect that formation of this intermediate species depends in any way on the presence of DMSO in our e.p.r. samples; at the freezing point of such solutions, the dielectric constant will be slightly higher rather than lower than that of water at 0° (P. Douzou, personal communication). Moreover, the same species also seems to have been detected by rapid freezing e.p.r. in the absence of

DMSO when using pulse radiolysis rather than KO_2 as the source of O_2^- (Pick, 1971).

One additional experiment, which is consistent with the view that the species detected with 3 mM barium represents some form of complex between barium ions and superoxide ions is illustrated in Figs.3f and g. Increasing the ionic strength by addition of sodium chloride to the medium shifted the form of the e.p.r. spectrum towards that characteristic of lower barium concentrations (compare Figs.3b and 3g). This may be explained as a primary kinetic salt effect, leading to weakening of the complex between the positively and negatively charged species.

We now consider the actual nature of the various species. Some possibilities are summarized in Table II for barium ions, and no doubt the situation is similar with calcium ions. The species at very low barium concentrations is the same as that in the presence of the alkali metal ions and is no doubt the free (presumably hydrated) superoxide anion radical. Surprisingly tight binding is indicated for the species present at intermediate barium concentrations (around 3 mM). Though further work is required, the most likely candidate for this species may be a solvent-shared or a solvent-separated ion pair (Sharp and Symons, 1972; Eigen and Tamm, 1962). At high concentrations of the metal ion, the change in the e.p.r. spectrum is greater than in the intermediate species. This might be due to the loss of the intervening water molecule, to give a contact species, this change being induced, perhaps, by the binding of a chloride ion (Table II). Alternatively two barium ions (with, or without associated chlorides to reduce the net charge) might be associated with a single O_2^-.

Another triple ion which might be considered at low barium and high O_2^- concentrations would be formed by interaction of one barium ion with two superoxide radicals. However, if this species was present, it would presumably not be detected by e.p.r., whereas it would contribute in pulse radiolysis. The general agreement between kinetic results obtained by the two methods therefore suggests that this species cannot be present in large amounts.

Biological and Analytical Significance of Formation of the Complexes

The tentative conclusion that at a concentration of around 3 mM calcium ions the superoxide radical exists in solution largely as a solvent-shared ion pair with the metal is clearly of biological interest, since the metal is present at such concentrations in most biological fluids (Prosser, 1973). Although formation of such a species does not seem to affect the stability of the superoxide ion, it does of course mean that it would carry a positive, rather than a negative charge in solution, and this could well be important in controlling its diffusion or transport.

In relation to the stabilizing action of the metal ion in the high concentration species, it is interesting that superoxide adsorbed on magnesium oxide surfaces is indefinitely stable (Tench and Holroyd, 1968). Though there seems to be no evidence for a similar situation on calcium oxide, it could, nevertheless, be speculated that superoxide radicals might under some conditions become adsorbed on to bone, with possible serious harmful effects.

In an entirely different context, our finding of specific changes and sharpening in the e.p.r. spectrum of O_2^- induced by calcium or barium ions could well prove useful in analytical work, as an aid to identification of O_2^- in the presence of other free radical species.

TABLE II

Possible Nature of the Three O_2^- Species Detected in the Presence of Ba^{++} by e.p.r.

Though it may be possible that some of the interactions detected between the ions by e.p.r. exist only in the frozen matrix and not in free solution, nevertheless, for the reasons discussed in the text, this is considered unlikely. The most probable species at intermediate Ba^{++} concentrations is shown as a solvent-shared or solvent separated ion pair. The possibility that water molecules may intervene also in some of the other species is not excluded.

Conditions	Species considered most probable	Other possible species
Very low or zero Ba^{++}	O_2^-	-
Intermediate Ba^{++}	Ba^{++} ...O(H)(H).....O_2^-	$Ba^{++}...O_2^-$
High Ba^{++}	$Cl^-... Ba^{++}... O_2^-$	$Ba^{++}...O_2^-$ $Ba^{++}...O_2^-...Ba^{++}$ $Ba^{++}...O_2^-...Ba^{++}...Cl^-$

ACKNOWLEDGEMENTS

We thank Professor M.C.R. Symons for helpful discussions. The work was supported by the Medical Research Council and by the Science Research Council.

REFERENCES

1. Atkins, P.W. and Symons, M.C.R. (1967). In "The Structure of Inorganic Radicals", Elsevier, Amsterdam pp 110-111 and 247-251.
2. Ballou, D., Palmer, G. and Massey, V. (1969). *Biochem. Biophys. Res. Com. 36,* 898-904.
3. Behar, D., Czapski, G., Rabani, J., Dorfman, L.M. and Schwarz, H.A. (1970). *J. Phys. Chem. 74,* 3209-3213.
4. Bennett, J.E., Mile, B. and Thomas, A. (1969). *Trans. Faraday Soc. 64,* 3200-3209.
5. Bray, R.C. (1961). *Biochem. J. 81,* 189-195.
6. Bray, R.C., Barber, M.J., Lowe, D.J., Fox, R. and Cammack, R. (1975). Proc. 10th FEBS Meeting, Paris, Vol.40 "Enzymes and Electron Transport Systems" (ed. Desnuelle, P. and Michelson, A.M.), North Holland Publ. Co. Amsterdam.
7. Bray, R.C. and Knowles, P.F. (1968). *Proc. Roy. Soc. A302,* 351-353.

8. Bray, R.C., Knowles, P.F., Pick, F.M. and Gibson, J.F. (1968). *Hoppe-Seyler's Z. Physiol. Chem. 349,* 1591-1592.
9. Bray, R.C., Palmer, G. and Beinert, H. (1964). *J. Biol. Chem. 239,* 2667-2676.
10. Bray, R.C., Pick, F.M. and Samuel, D. (1970). *Eur. J. Biochem. 15,* 353-355.
11. Cammack, R., Rao, K.K. and Hall, D.O. (1971). *Biochem. Biophys. Res. Com. 44,* 8-14.
12. Carle, C.I. (1976). Ph.D. Thesis, University of London.
13. Coffman, R.E. and Stavens, B.W. (1970). *Biochem. Biophys. Res. Com. 41,* 163-169.
14. Czapski, G. and Dorfman, L.M. (1964). *J. Phys. Chem. 68,* 1169-1176.
15. Eigen, E. and Tamm, K. (1962). *Z. Elektrochem., 66,* 107-121.
16. Fee, J.A. and Hildenbrand, P.G. (1974). *FEBS Lett. 39,* 79-82.
17. Fielden, E.M., Roberts, P.B., Bray, R.C., Lowe, D.J., Mautner, G.N., Rotilio, G. and Calabrese, L. (1974). *Biochem. J. 139,* 49-60.
18. Greenlee, L., Fridovich, I. and Handler, P. (1962). *Biochemistry 1,* 779-783.
19. Knowles, P.F., Gibson, J.F., Pick, F.M. and Bray, R.C. (1969). *Biochem. J. 111,* 53-58.
20. McCord, J.M. and Fridovich, I. (1968). *J. Biol. Chem. 243,* 5753-5760.
21. Meisel, D., Czapski, G. and Samuni, A. (1973). *J. Am. Chem. Soc. 95,* 4148-4153.
22. Meisel, D., Levanon, D. and Czapski, G. (1974). *J. Phys. Chem. 78,* 779-782.
23. Nilsson, R., Pick, F.M., Bray, R.C. and Fielden, M. (1969). *Acta Chem. Scand. 23,* 2554-2556.
24. Pick, F.M. (1971). Ph.D. Thesis, University of London.
25. Prosser, C.L. (1973). In "Comparative Animal Physiology", Saunders and Company, Philadelphia, 3rd Ed., pp 79-110.
26. Sharp, J.H. and Symons, M.C. (1972). In "Ions and Ion Pairs in Organic Reactions" (ed. M. Szwarc), Wiley-Interscience, New York, Vol.*1,* pp 177-262.
27. Sillén, L.G. and Martell, A.E. (1964). In "Stability Constants of Metal Ion Complexes", Chemical Society Special Publications No.17, London.
28. Tench, A.J. and Holroyd, P. (1968). *Chem. Comm.* pp 471-473.
29. Valentine, J.S. and Curtis, A.B. (1975). *J. Am. Chem. Soc. 97,* 224-226.

PRODUCTION OF SUPEROXIDE BY METAL IONS

A.M. MICHELSON

Institut de Biologie Physico-Chimique
Service de Biochimie-Physique
13, rue P. et M. Curie
75005 Paris, France

A mechanism for the autoxidation of ferrous ions in aqueous solution involving production of superoxide ions was first proposed by Weiss (1935).

$$Fe^{2+} + O_2 \rightleftharpoons Fe^{3+} + O_2^{\overline{\cdot}} \quad (1)$$

$$Fe^{2+} + O_2^{\overline{\cdot}} + 2H^+ \longrightarrow Fe^{3+} + H_2O_2 \quad (2)$$

$$Fe^{2+} + H_2O_2 \longrightarrow Fe^{3+} + {}^{-}OH + {}^{\cdot}OH \quad (3)$$

$$Fe^{2+} + {}^{\cdot}OH \longrightarrow Fe^{3+} + {}^{-}OH \quad (4)$$

A more recent paper by the same author (Weiss, 1953) suggests that equation 1 is better represented by formation of an ion pair complex.

$$Fe^{2+} + O_2 \rightleftharpoons Fe^{3+} \ldots O_2^{\overline{\cdot}}$$

The reverse process, leading to restitution of ferrous ions and O_2 was found to be inhibited in the presence of suitable anions such as F^-, $P_2O_7^{4-}$ or ${}^{-}OH$, that is, stabilisation of an $O_2^{\overline{\cdot}}$ containing complex is obtained by formation of ligands with certain anions, or at alkaline pH

$$Fe^{3+} \ldots O_2^{\overline{\cdot}} + F^- \longrightarrow F^- \ldots Fe^{3+} \ldots O_2^{\overline{\cdot}}$$

$$Fe^{3+} \ldots O_2^{\overline{\cdot}} + {}^{-}OH \longrightarrow HO^- \ldots Fe^{3+} \ldots O_2^{\overline{\cdot}}$$

Hydroxylation of benzoic acid by ferrous ions (and O_2) in phosphate buffer pH 7.2 has been described by Nofre *et al.* (1961), presumably via the formation of hydroxyl radicals. Other ligands and chelating agents were studied (as well as other metals of the transition group) and in the case of ferrous ions/EDTA a rather sharp optimum for hydroxylation was observed at pH 7.2 with a 50% decrease at pH 6 or pH 8.2. The optimum ratio of EDTA : Fe^{2+} was found to be equimolar.

Such hydroxylations, also studied by Goscin and Fridovich (1972) are relatively slow and perhaps arise from the action of ferric ions on H_2O_2. This H_2O_2, or at least a large part, is probably derived from the pulse of $O_2^{\overline{\cdot}}$ produced within the first second after injection of ferrous ions into buffer containing a suitable ligand.

Some of the chemical properties of the system Fe^{2+}/O_2/phosphate have been developed (Michelson, 1973). Injection of ferrous ions into aqueous solutions of phosphate gives rise to light emission. The intensity of light emitted is a function of a number of factors including concentration of ferrous ions and of phosphate and the pH. This emission, perhaps mediated by traces of carbonate (see chapter on Superoxide and Chemiluminescence) is feeble, but nevertheless reproducible and can be attributed to the production of superoxide ions formed by reduction of molecular oxygen by ferrous ions and a ligand such as phosphate. Increase in phosphate concentration gives an increased light yield and slightly increased kinetics but for practical purposes a convenient concentration is 0.1 M. The influence of ferrous ion concentration is presented in figure 1. It can be seen that an optimum for activity is obtained at about 3.3×10^{-4} M Fe^{2+} measured either in absence or in presence of luminol. Although the life time of emission due to oxidation of luminol is independent of the concentration of Fe^{2+} (t 1/2 $\sim$ 0.5 sec), excess of ferrous ions (the concentration of dissolved oxygen may be considered to be about 2×10^{-4} M) stabilise the $O_2^{\overline{\cdot}}$ formed and between 10^{-4} M and 10^{-3} M Fe^{2+} a sharp increase in the life time of emission due to $O_2^{\overline{\cdot}}$ (in the absence of luminol) is observed. This stabilisation may be attributed to complex formation of the type $X^- \ldots Fe^{3+} \ldots O_2^{\overline{\cdot}}$ (where X is phosphate), dissociation of this complex, i.e. $X^- \ldots Fe^{3+} \ldots O_2^{\overline{\cdot}} \longrightarrow X^- \ldots\ldots Fe^{3+} + O_2^{\overline{\cdot}}$, being inhibited by an excess of cations.

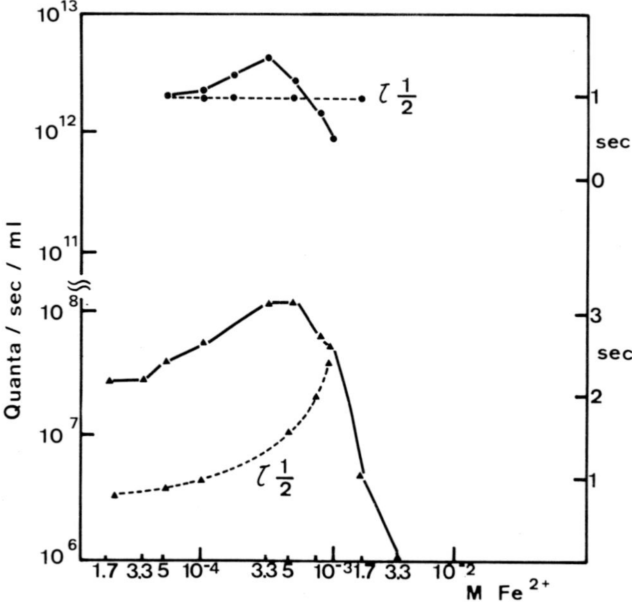

Fig.1. *Influence of concentration of* Fe^{2+} *on Imax of light emission and half peak times (τ 1/2) of the system* Fe^{2+}/O_2*/phosphate at pH 10.45 (lower curves ▲—▲), and of the oxidation of luminol by this system at pH 10.45 (upper curves •—•).*

The effect of pH on light emission by injection of Fe^{2+} into phosphate solutions is shown in figure 2. There is a strong increase in Imax at alkaline pH with a plateau at about pH 10.5. The half-life is short (about 0.4 sec) and is relatively independent of pH. When luminol is present, the effect of increase in pH is quite significant with a 3.5 million fold increase in maximal intensity between pH 7 and pH 11. As in the absence of luminol, light emission reaches a plateau at about pH 10.5. Near neutrality (pH 7 to 8) the oxidation of luminol is relatively slow but increases rapidly with increase in pH (10 fold between pH 7.5 and 9) as shown in figure 2.

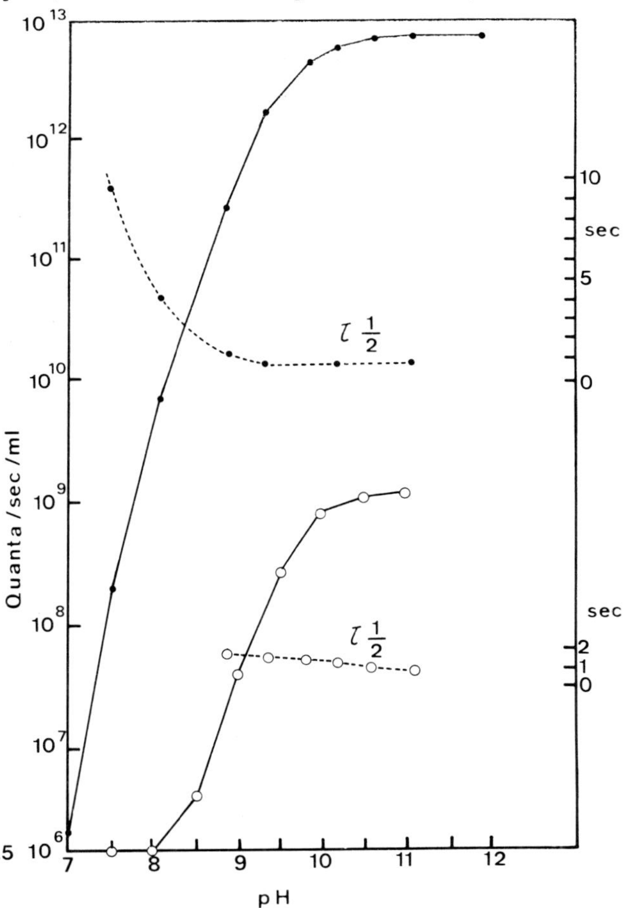

Fig.2. *Variation of Imax of light emission and half peak times (τ 1/2) with pH of the system Fe^{2+}/O_2/phosphate (lower curves X——X) and of the oxidation of luminol (10^{-4} M) by this system (upper curves •------•).*

The short life time of the pulse of $O_2^{\overline{\cdot}}$ produced by the reduction of molecular oxygen with ferrous ions explains the results previously described (Henry and Michelson, 1970) for the oxidation of Pholad luciferin by injection of ferrous ions, in which the order of addition of the reagents plays an important role. Thus injection of ferrous ions into phosphate plus luciferin, or of phosphate into luciferin plus Fe^{2+}, causes light emission whereas injection of luciferin into phosphate plus ferrous ions is inactive.

When ferrous ions are replaced by the same concentration of Ni^{2+} or Co^{2+}

in the system with O_2 and phosphate, oxidation of luminol at pH 10 is some 10,000 fold less effective, and indeed the emission can be attributed to the presence of 0.01% of ferrous ion contamination in the cobalt or nickel salts.

ACTION OF SUPEROXIDE DISMUTASE

Since it is postulated that injection of ferrous ions into phosphate buffer under aerobic conditions produces $O_2^{\cdot-}$ the effects of superoxide dismutase and catalase on light emission of the system alone, and of the chemiluminescence of luminol and Pholad luciferin induced by this system were examined. As shown in Table I, light emission due to decomposition of $O_2^{\cdot-}$ is strongly inhibited by bovine erythrocuprein (Hartz and Deutsch, 1969; McCord and Fridovich, 1969) whereas catalase results in a significant increase in Imax at pH 9.0 and pH 10.0. The inhibition of oxidation of luminol in this system by superoxide dismutase as a function of pH is shown in figure 3 which demonstrates a pH optimum of about pH 9.5 for bovine erythrocuprein. Again the effect of catalase is negligible (Table I). Emission of light due to oxidation of Pholad luciferin by this system is also strongly (82.5%) inhibited at pH 10 with superoxide dismutase (Table I), though not at pH 7. Thus with three different sources of light emission, superoxide dismutase strongly inhibits the reaction which may therefore be attributed to the production of $O_2^{\cdot-}$.

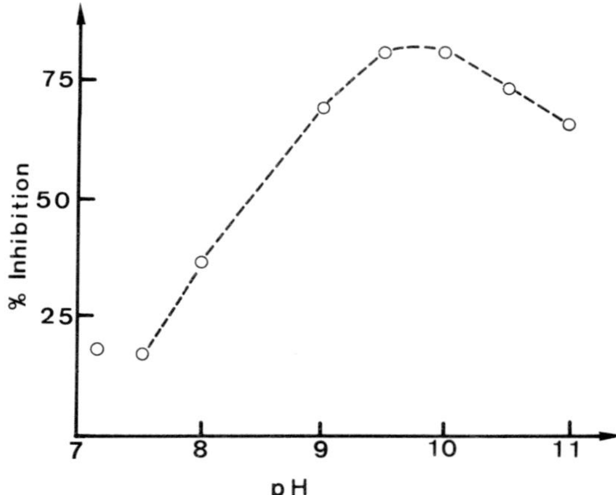

Fig.3. *Inhibition of the oxidation of luminol with Fe^{2+}/O_2/phosphate by superoxide dismutase (erythrocuprein) as a function of pH.*

As mentioned above, Co^{2+} and Ni^{2+} ions in presence of phosphate do not give significant amounts of $O_2^{\cdot-}$. It was of interest to compare these three anions with other ligands. The order of stability of the higher oxidation state of the free ions is $Fe^{3+} > Co^{3+} > Ni^{3+}$, but of greater interest are changes in oxidation-reduction potential for each metal as a consequence of co-ordination with various ligands.

The Pholas dactylus bioluminescent system involves enzymic production of $O_2^{\cdot-}$ radicals (the luciferase is a copper containing glycoprotein). There is a large stimulation of the reaction in presence of dihydroxyfumaric acid (DHF) (Michelson and Isambert, 1972). It was shown that the stimulation occurs as a result of increased production of $O_2^{\cdot-}$, and it is also clear that apart from considerations of redox potential the respective bond dissociation energies for reactions such as $AH^{\cdot} + LH_2 \longrightarrow AH_2 + LH^{\cdot}$ must play an important role.

TABLE I

Effect of Superoxide Dismutase on the Fe^{2+}/O_2 System

Fe^{2+}/O_2/Phosphate Alone	Imax (quanta/sec/ml)	% Inhibition
Control	5.65×10^8	-
+ dismutase	0.70×10^8	87.7
+ catalase	16.30×10^8	-
Plus Pholad Luciferin		
Control	400×10^8	-
+ dismutase	70×10^8	82.5
Plus 10^{-4} M Luminol		
Control	6.48×10^{12}	-
+ dismutase	1.24×10^{12}	81.0
+ catalase	7.96×10^{12}	-

Final concentrations in the cuvette were 3.3×10^{-4} M Fe^{2+} in 0.1 M phosphate pH 10.0 and where necessary 10^{-4} M luminol or 25 µl of a stock solution of Pholad luciferin (0.192 mg of protein/ml) in a final volume of 3 ml. Reaction was initiated by injection of 1 ml of 10^{-3} M Fe^{2+} into 2 ml of the other components in the cuve, which contained where necessary 50 µg of catalase or 60 µg of purified bovine erythrocuprein.

Only feeble oxidative activity is shown by Fe^{2+}, Co^{2+} and Ni^{2+} with ligands such as cyanide or ammonia. However, when dihydroxyfumaric acid is added to such ineffective systems there is very strong stimulation of oxidation of luminol. When DHF is present as ligand, the kinetics of oxidation of luminol by Fe^{2+}/O_2 remain very fast as with phosphate, but with Co^{2+} a continuous reaction occurs which is inhibited by both catalase and superoxide dismutase indicating production of both H_2O_2 and $O_2^{\overline{\cdot}}$ or indirectly of $HO^{\cdot}$ radicals (Fig.4a). With Ni^{2+} and DHF a similar continuous oxidation is seen, after an initial rapid phase (Fig.4b). This reaction is strongly inhibited by superoxide dismutase and it may be concluded that a continuous production of $O_2^{\overline{\cdot}}$ occurs (in contrast with the rapid reaction $Fe^{2+} + O_2 \longrightarrow Fe^{3+} + O_2^{\overline{\cdot}}$). This is probably the consequence of a slow rate of oxidation of Ni^{II} (with two molecules of DHF) to the Ni^{III} form with reduction of O_2 to $O_2^{\overline{\cdot}}$. Since Ni^{III} is generally unstable further reaction by withdrawal of an electron from DHF can occur to give a Ni^{II} complex.

$$DH_2 + -\overset{|}{\underset{|}{Ni^{III}}}- \longrightarrow DH^{\cdot} + H^+ + -\overset{|}{\underset{|}{Ni^{II}}}-$$

Production of $O_2^{\overline{\cdot}}$ would also be obtained by the reaction
$DH^{\cdot} + O_2 \longrightarrow D + H^+ + O_2^{\overline{\cdot}}$

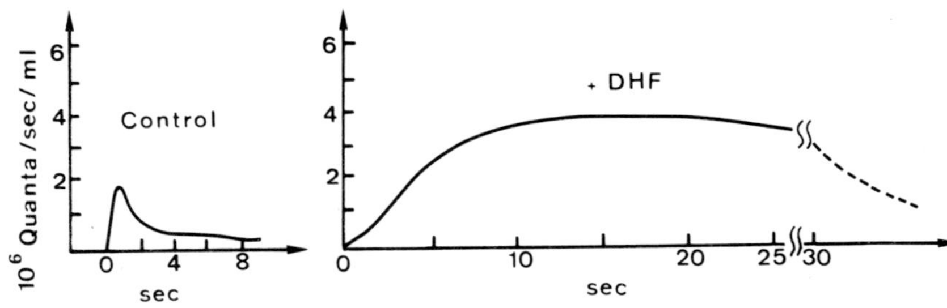

Fig.4a. *Effect of dihydroxyfumaric acid on the oxidation of luminol by* $Co^{2+}/O_2/NH_4Ac$ *pH 9.0.*

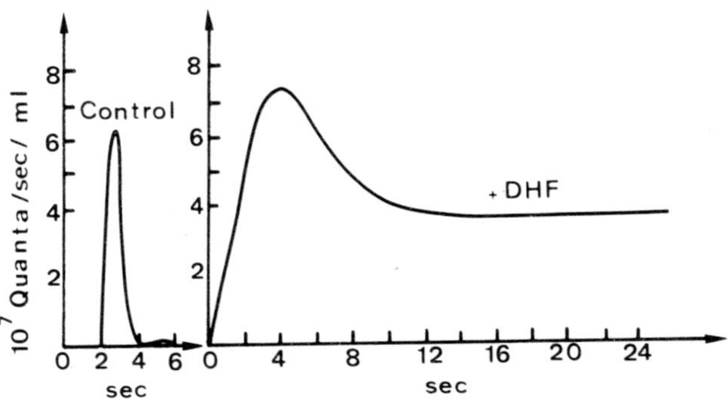

Fig.4b. *Effect of dihydroxyfumaric acid on the oxidation of luminol by* $Ni^{2+}/O_2/NH_4Ac$ *pH 9.8.*

We now consider tetraglycine as ligand for the three metal ions. Fe^{2+}/O_2/tetraglycine is much less effective than the phosphate system since tetraglycine is a poor ligand for Fe^{2+}. With Co^{2+}, tetraglycine gives rise to continuous oxidation of luminol due to production of H_2O_2 and to a lesser extent $O_2^{\overline{\cdot}}$. Addition of DHF can stimulate the system probably by increased production of $O_2^{\overline{\cdot}}$ (Fig.5), since SOD inhibits.

It is known that Ni^{2+} and tetraglycine form a square planar complex (Martin *et al.*, 1960) with the terminal amino group and the three peptide nitrogens co-ordinated, the NH hydrogens being ionized (Freeman, 1966). A recent report has shown that although Ni^{II} is not readily oxidised, activation by peptide co-ordination leads to reaction with molecular oxygen and oxidation of the ligand (Paniago *et al.*, 1971). Maximal rates of O_2 consumption were found to occur at pH 7 to 8 and at pH 8 there was an induction period followed by an autocatalytic reaction. In the present work the optimum pH for oxidation of luminol by Ni^{2+}/O_2/tetraglycine was between pH 9 to 10. An immediate rapid reaction occurs, followed (at pH 9.5) by an induction period after which there is oxidation of luminol with light emission of an oscillatory pattern which can continue for at least 20 min. Superoxide dismutase, but not catalase, inhibits this emission (Fig.6). Oxidation of luminol by Ni^{2+}/O_2/tetraglycine is also stimulated by DHF and this stimulated emission is reduced in presence of superoxide dismutase, as

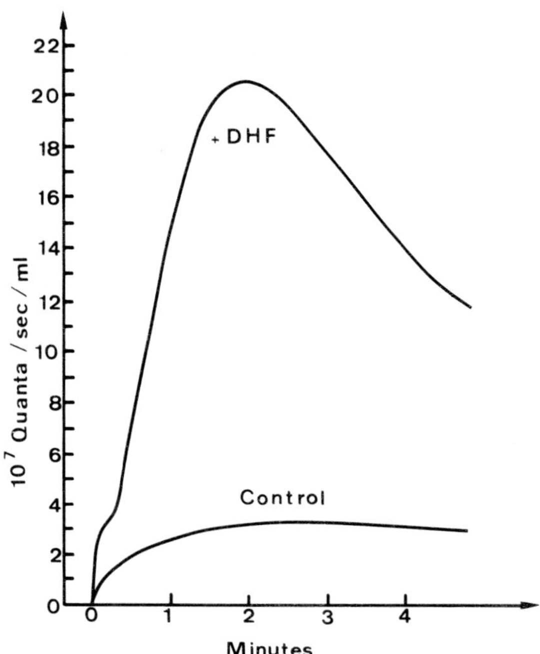

Fig.5. *Effect of 10^{-4} M dihydroxyfumaric acid on the oxidation of luminol (10^{-4} M) by Co^{2+} (3.3×10^{-5} M)/O_2/tetraglycine (3.3×10^{-4} M) at pH 9.8.*

is emission in the absence of DHF. At pH 9.0 - 9.5, DHF not only increases the intensity of the short-lived emission seen immediately on injection of Ni^{2+}, but also gives rise to a continuous emission (Fig.7). This emission is strongly inhibited on injection of superoxide dismutase and can be attributed to a continuous production of $O_2^{\overline{\cdot}}$. At pH 9.5 to 9.8 as we have previously mentioned, various secondary effects are seen in the oxidation of luminol by Ni^{2+}/O_2/tetraglycine. These oscillatory effects on light emission are abolished by DHF which gives rise to a continuous production of light which is again diminished by superoxide dismutase.

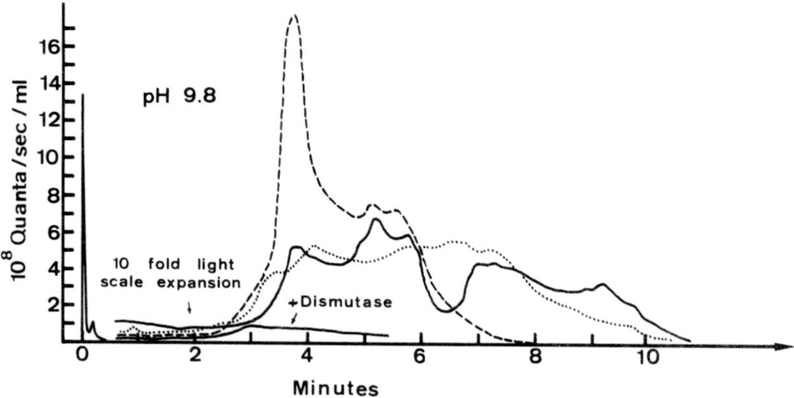

Fig.6. *Oxidation of luminol by Ni^{2+}/O_2/tetraglycine at pH 9.8.*

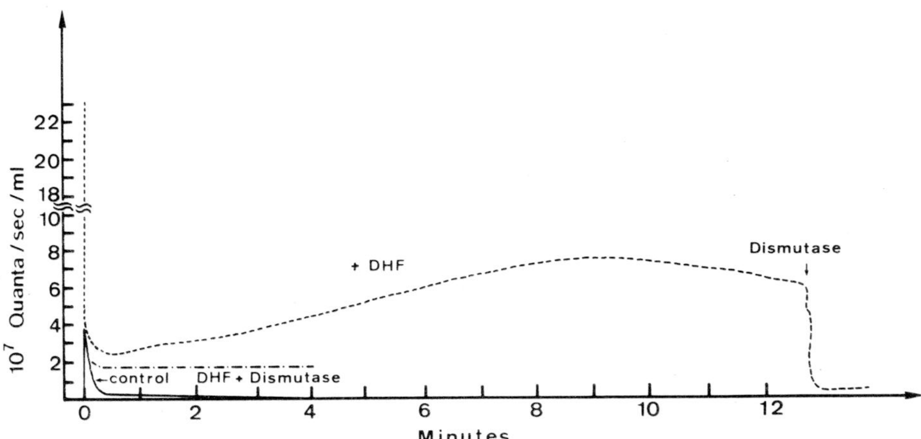

Fig.7. *Effect of dihydroxyfumaric acid on the oxidation of luminol by Ni^{2+}/O_2/tetraglycine at pH 9.0. The cuve contained 2 ml of 5 x 10^{-2} M ammonium acetate adjusted to pH 9.0 with ammonium hydroxide, 30 µl of 10^{-2} M luminol and 100 µl of 10^{-2} M tetraglycine. Reaction was initiated by injection of 1 ml of 10^{-3} M metal ion. Final concentrations were 10^{-4} M luminol, 3.3 x 10^{-4} M tetraglycine and 3.3 x 10^{-4} M metal ion.*

GENERAL MECHANISM

The above results indicate formation of $O_2^{\overline{\cdot}}$ by one electron reduction of molecular oxygen when any of the metal ions Fe^{2+}, Co^{2+} or Ni^{2+} are suitably co-ordinated, presumably by a process $X - Me^{2+} + O_2 \longrightarrow X - Me^{2+} — O_2 \longrightarrow X - Me^{3+} \ldots O_2^{\overline{\cdot}}$. Attack of $O_2^{\overline{\cdot}}$ on luminol (or luciferin) could readily give rise to an intermediate hydroperoxide or other activated form.

$$LH_2 + O_2^{\overline{\cdot}} \longrightarrow LH^{\cdot} + HOO^-$$

$$LH^{\cdot} + O_2^{\overline{\cdot}} \longrightarrow LH - OO^- \text{ or } L + HOO^-$$

followed by degradation of this intermediate with emission of light. In principle such free radical intermediates could give rise to a chain reaction of the type

$$LH + O_2 \longrightarrow LH - OO^{\cdot}$$

$$LH - OO^{\cdot} + LH_2 \longrightarrow LH - OOH + LH^{\cdot}$$

Stimulation by DHF of oxidation by Ni^{2+}/O_2/tetraglycine is probably $X - Me^{3+} + DH_2 \longrightarrow X - Me^{2+} + DH^{\cdot} + H^+$ followed by

$$DH^{\cdot} + O_2 \longrightarrow D + H^+ + O_2^{\overline{\cdot}}$$

as well as by

$$DH^{\cdot} + LH_2 \longrightarrow DH_2 + LH^{\cdot}$$

Evidently the course of such reactions will be governed by relative concentrations of the components and more particularly by their relative bond dissociation energies as well as the respective redox potentials at a given pH.

With respect to the oscillatory light emission and induction period after a rapid flash obtained in the oxidation of luminol by Ni^{2+}/O_2/tetraglycine it is likely that initial production of $O_2^{\overline{\cdot}}$ is followed by a slower process giving rise to a peptide radical.

$$—Ni^{III}— \longrightarrow —Ni^{II}— + \text{peptide radical}$$

either from co-ordinated peptide or from small amounts of free peptide. This can then begin an autocatalytic free radical chain reaction which however is very susceptible to various factors. Indeed, the complete lack of reproducibility in detail (but not in general form) of this system indicates that small amounts of free Ni^{2+} or tetraglycine influence the entire process. In addition it may be noted that such effects are seen only at pH >9.5 and that DHF converts the system into a continuous process of light emission of higher intensity.

In the general area of lipid autoxidation, excited singlet oxygen has been proposed (Rawls and Santen, 1971) as an initiator and indeed, reaction of this species with methyl linoleate is some 10^3 to 10^4 times faster than is the reaction of O_2. The reaction RH + O_2 $\longrightarrow$ R OOH has a very high activation energy, of the order 35 to 65 kcal/mol and it is clear that an intermediate step is necessary (Labuza, 1971). The present work shows that under appropriate conditions all three metal ions can give rise to superoxide radicals, and it is suggested that this is the initiation step in the autoxidation of lipids in various foodstuffs rather than formation of excited singlet oxygen or of metal hydroperoxide complexes which have also been proposed (Uri, 1961). The autoxidation of unsaturated lipids is indeed inhibited by superoxide dismutase.

Finally, such studies are of interest with respect to the adaptation of life from anaerobic geological prebiotic conditions to the aerobic form which is of major importance at present. Oxygen itself is not an efficient oxidising agent whereas one electron reduction to the superoxide ion gives a more reactive species. This reduction, catalysed by metal ions, could well have played an extremely important role in the adaptation and survival of primitive systems, particularly since closely related activities, those of superoxide dismutases, could rapidly control the toxic action of $O_2^{\overline{\cdot}}$ by similar oxidoreduction processes which give rise to O_2 and H_2O_2. It is of interest that the oxidizing activities of metal ions such as Fe^{2+}, Co^{2+} and Ni^{2+} in the presence of O_2 are greatly enhanced by co-ordination with specific ligands including simple oligopeptides, suggesting processes for the development in an evolutionary sense of specific oxidising capacity.

REFERENCES

1. Freeman, H.C. (1966). *Adv. Protein Chem. 22*, 332 (257-424).
2. Goscin, S. and Fridovich, I. (1972). *Arch. Biochem. Biophys. 153*, 778-783.
3. Hartz, J.W. and Deutsch, H.F. (1969). *J. Biol. Chem. 244*, 4565-4572.
4. Henry, J.P. and Michelson, A.M. (1970). *Biochim. Biophys. Acta 205*, 451-458.
5. Labuza, T.P. (1971) *Critical Reviews in Food Technology 2*, 355-397.
6. McCord, J.M. and Fridovich, I. (1969). *J. Biol. Chem. 244*, 6049-6055.
7. Martin, R.B., Chamberlin, M. and Edsall, J.T. (1960). *J. Amer. Chem. Soc. 82*, 495-498.
8. Michelson, A.M. and Isambert, M.F. (1972). *Biochimie 55*, 619-663.

9. Michelson, A.M. (1973). *Biochimie 55*, 465-479; 925-942.
10. Nofre, C., Cier, A. and Lefier, A. (1961). *Bull. Soc. Chim. France*, 530-535.
11. Paniago, E.B., Weatherburn, D.C. and Margerum, D.W. (1971). *Chem. Communications*, 1427-1428.
12. Rawls, H.R. and Van Santen, P.J. (1971). *Ann. N.Y. Acad. Sci. 171*, 135-137.
13. Uri, N. (1961). In "Autoxidation and Antioxidants" Vol.1 p.55 (1-83) (W.O. Lundberg, ed.), J. Wiley and Sons, New York.
14. Weiss, J. (1935). *Naturwissenschaften 23*, 64-69.
15. Weiss, J. (1953). *Experientia 9*, 61-67.

CHEMICAL PRODUCTION OF SUPEROXIDE ANIONS BY REACTION BETWEEN RIBOFLAVIN, OXYGEN, AND REDUCED NICOTINAMIDE ADENINE DINUCLEOTIDE. SPECIFICITY OF TESTS FOR $O_2^{\dot{-}}$

A.M. MICHELSON

Institut de Biologie Physico-Chimique
Service de Biochimie-Physique
13, rue P. et M. Curie
75005 Paris, France

Photoreduction of flavins is often used as a source of production of superoxide ions ($O_2^{\dot{-}}$) during the reoxidation. A hydrogen donor is generally present, such as methionine, EDTA or NADH to facilitate the photoreduction at 365 nm. We describe in this report the production of $O_2^{\dot{-}}$ by mixtures of flavins and NADH in absence of irradiation. Reagents which are commonly used to detect $O_2^{\dot{-}}$ include ferri cytochrome *c* (reduced), tetranitromethane (reduced), nitroblue tetrazolium (reduced) and luminol (oxidized). In view of a number of conflicting observations, we have reinvestigated the specificity of these various diagnostic tests.

RESULTS AND DISCUSSION

Reduction of Fe^{3+} to Fe^{2+} by Photoreduced FMN

We have earlier shown that autoxidation of ferrous ions in phosphate buffer gives rise to considerable amounts of $O_2^{\dot{-}}$ (Michelson, 1973). Thus photosensitized reduction of ferric ions could provide an alternate continuous source of $O_2^{\dot{-}}$. Photoreduction of FMN in presence of Fe^{3+} and bipyridyl gives rise to ferrous ions as measured by complex formation with the bipyridyl (Table I). Although the reaction is inhibited about 35% by 6×10^{-9} M superoxide dismutase (SOD), exactly the same inhibition is obtained in presence of heat denatured SOD, hence $O_2^{\dot{-}}$ is not directly involved and the reduction of Fe^{3+} presumably occurs directly by reaction with $FMNH_2$ (or an intermediate flavin radical, FMNH˙ which would react with protein, native or denatured). Similar reduction of Fe^{3+} in ferritin can be observed via photoreduced FMN and the reduction is more efficient under anaerobic conditions than aerobic, again indicating direct reaction in the absence of oxygen intermediates (Wauters, Crichton and Michelson, 1976).

Replacement of FMN by riboflavin gave similar results except that reduction of Fe^{3+} was some 20 to 30% less efficient.

TABLE I

Reduction of Fe^{3+} by $FMNH_2$

		ΔOD 540 nm	
min	Control	+ 10^{-8} M SOD	+ 10^{-8} M denatured SOD
5	0.089	0.054	0.049
10	0.153	0.101	0.094
15	0.202	0.132	0.126

The cuvette contained 1.9 ml of 0.1 M cacodylate-borate buffer pH 7.0 with additions of 1.0 ml of 3 x 10^{-4} M $FeCl_3$, 0.1 ml of 3 x 10^{-3} M bipyridyl and 5 μl of 10^{-2} M FMN in 10^{-2} M EDTA in a final volume of 3 ml to give final concentrations of 10^{-4} M Fe^{3+} and bipyridyl and 1.67 x 10^{-5} FMN. Irradiation of the cuvette at 365 nm with a B 100 lamp. Formation of the bipyridyl complex of Fe^{2+} was measured at 540 nm.

Reduction of Cytochrome c *and Tetranitromethane by Ferrous Ions*

As shown in Fig.1, ferrous ions are extremely efficient for the reduction of cytochrome *c* and react in a stoichiometric fashion since the same amount of reduction is obtained with 5 x 10^{-6} M Fe^{2+} and 10^{-5} M cytochrome *c* (Fe^{2+} limiting) as with 5 x 10^{-6} M cytochrome *c* and 5 x 10^{-5} M Fe^{2+} (excess). Since ferrous ions can readily give rise to $O_2^{\bar{\cdot}}$ the effect of superoxide dismutase (6 x 10^{-9} M) was examined. No variation was observed and hence direct reduction of cytochrome *c* by ferrous ions occurs in accord with the earlier observations by Zipper *et al.* (1972).

Similarly ferrous ions directly reduce tetranitromethane in absence of intermediate production of $O_2^{\bar{\cdot}}$ as shown in Fig.2.

Hence, in presence of ferrous ions, neither reagent is of value for the detection of $O_2^{\bar{\cdot}}$.

Reduction of Fe^{3+} by NADH/FMN

The reduction of Fe^{3+} by mixtures of NADH and FMN was then examined, using dipyridyl to detect formation of Fe^{2+}. Under the conditions used, no reduction of Fe^{3+} by NADH or by FMN separately occurs. (At acid pH, mixtures of Fe^{3+}, NADH and bipyridyl give rise to a rapid increase in absorption at 540 nm which exceeds the epsilon of ferrous bipyridyl which could be formed.) The results are shown in Fig.3. It can be seen that NADH plus FMN (in the absence of photoreduction) rapidly convert Fe^{3+} to Fe^{2+}. This reduction is not affected by native superoxide dismutase (6 x 10^{-9} M) or by the same quantity of denatured enzyme and hence does not proceed *via* production of $O_2^{\bar{\cdot}}$. Indeed, exactly the same kinetics are observed under anaerobic conditions (Table II) and hence direct reduction is involved. The reaction goes to completion and the same absorption is obtained with an equivalent amount of ferrous dipyridyl (5 x 10^{-4} M gives an absorption of 1.40 at 540 nm) prepared from ferrous sulphate and dipyridyl. Similar results have been reported by Crighton *et al.* (1975a, 1975b) for the reduction of ferritin by FMN/NADH.

It can also be seen that decrease in concentration of FMN does not seriously affect the rate of reduction until less than half the stoichiometric quantity is present. Whereas replacement of the FMN by FAD results in a marked decrease in the kinetics of reduction, and corresponds to the equivalent of less than one tenth the stoichiometric amounts of FMN (Fig.3), riboflavin is more efficient and with equal concentrations of all components the reduction is complete in about twelve minutes at 22°C. The reaction may be written

$$H^+ + FMN + NADH \rightleftharpoons FMNH_2 + NAD^+$$

$$2\ Fe^{3+} + FMNH_2 \longrightarrow FMN + 2\ Fe^{2+} + 2\ H^+$$

$$FMN + NADH + 2\ Fe^{3+} \longrightarrow FMN + NAD^+ + H^+ + 2\ Fe^{2+}$$

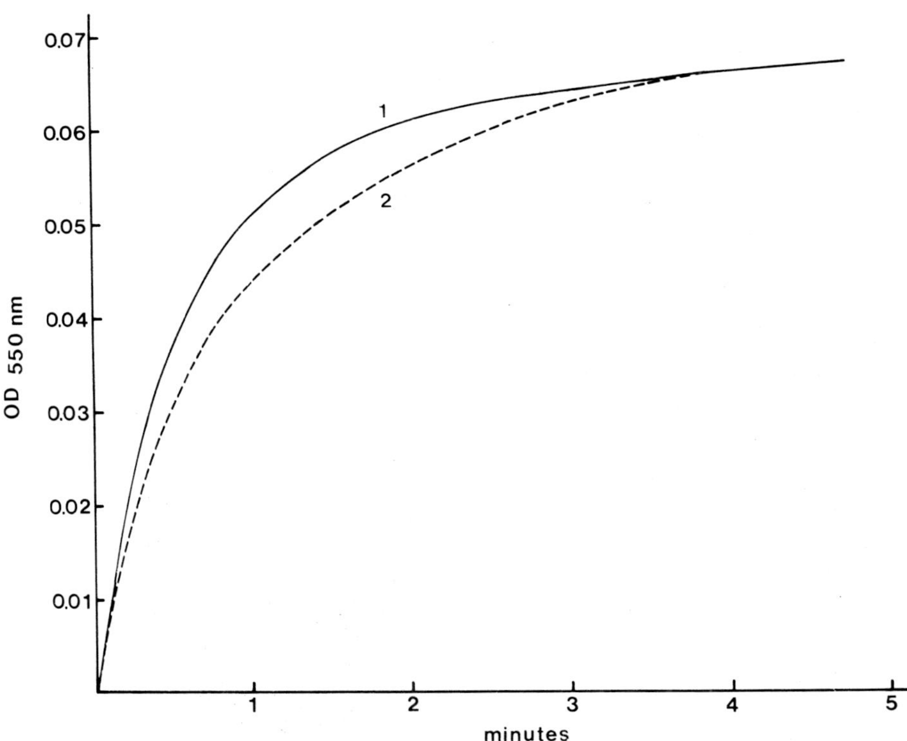

Fig.1. *Reduction of cytochrome* c *by ferrous ions. The cuvette contains ferricytochrome* c *in 2 ml of 0.1 M cacodylate-borate buffer pH 7.0. Reaction is initiated by injection of 1 ml of an aqueous solution of* $FeSO_4$. *Final concentrations: curve 1, 5 x* 10^{-6} *M cytochrome* c *and 5 x* 10^{-5} *M* Fe^{2+}; *curve 2,* 10^{-5} *M cytochrome* c *and 5 x* 10^{-6} *M* Fe^{2+}. *The reaction is followed at 550 nm in a Cary 15.*

TABLE II

Reduction of Fe^{3+} by NADH/flavin

OD at 540 nm

Min	FMN	FMN anaerobic	1/2 FMN	1/3 FMN	1/5 FMN	1/10 FMN	FAD	Riboflavin
1	0.120	0.120	0.102	0.080	0.060	0.030	0.024	0.150
2	0.240	0.230	0.195	0.160	0.120	0.077	0.050	0.292
3	0.340	0.330	0.292	0.245	0.180	0.177	0.075	0.429
5	0.500	0.485	0.462	0.392	0.293	0.207	0.125	0.680
8	0.688	0.622	0.666	0.577	0.458	0.315	0.200	0.973
10	0.786	-	-	-	-	0.391	-	1.070
20	1.150	-	-	-	-	-	-	1.121

Conditions as in Fig.3.

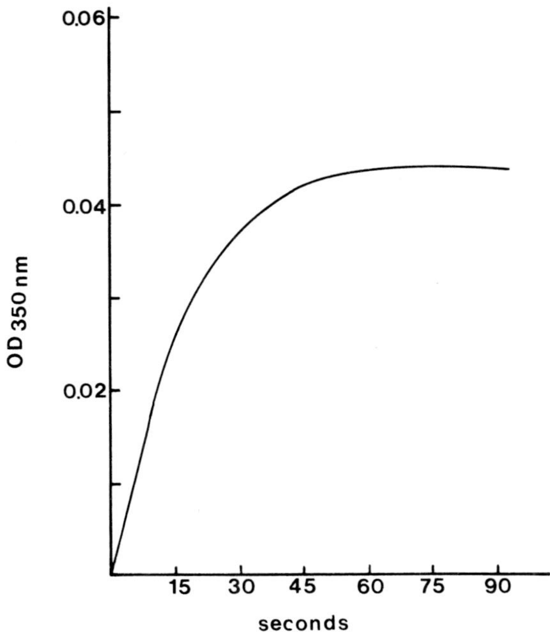

Fig.2. *Reduction of tetranitromethane by ferrous ions. The cuvette contains 1.8 ml of 0.1 M cacodylate-borate buffer pH 7.0 and 0.2 ml of 7.5 x 10^{-5} M tetranitromethane. The reaction is initiated by injection of 1 ml of 1.5 x 10^{-4} M $FeSO_4$ to give a final volume of 3 ml containing 5 x 10^{-6} M tetranitromethane and 5 x 10^{-5} M Fe^{2+}. Reduction is followed at 350 nm. Exactly the same curve is obtained in presence of 10 or 200 units of superoxide dismutase/ml (6 x 10^{-9} M and 1.2 x 10^{-7} M).*

Reduction of Cytochrome c *by FMN/NADH*

Since ferric ions are readily reduced by FMN plus NADH and cytochrome *c* is itself rapidly reduced by ferrous ions it was decided to couple the two systems. Indeed, reduction of cytochrome *c* by mixtures of FMN, NADH and Fe^{3+} occurs rapidly. However, although presence of added ferric ions stimulates reduction (by about 26% for FMN and 10% for riboflavin) as shown in Fig.4, Fe^{3+} is not essential. This is at least partly due to the presence of traces of free ferric ions present in the preparation of cytochrome *c*. The kinetics of reduction of cytochrome *c* by NADH and varying amounts of FMN (in the absence of added Fe^{3+}) are shown in Fig.5. It can be seen (in contrast with the reduction of Fe^{3+} by FMN/NADH) that the reaction is almost a linear function of the concentration of FMN between ratios FMN/NADH of 0.1 to 2, (insert Fig.5). It may be noted that superoxide dismutase (6 x 10^{-9} M) has no effect whatsoever on the kinetics of reduction (Table III) and hence $O_2^{\overline{\cdot}}$ is not involved (in contrast with the reduction of nitroblue tetrazolium by the same system).

When FMN is replaced by an equivalent amount of FAD, the rate of reduction is reduced by about 18% whereas riboflavin is much more efficient than FMN by a factor of 2.2 to 2.5. No reduction of cytochrome *c* by FMN occurs in the absence of NADH, and initiation of the reaction by injection of either FMN or NADH gives identical results. Cytochrome *c* is very slowly reduced by

NADH alone (Table III) but at a rate about 1% of that obtained with a mixture of NADH and riboflavin. Similar results are obtained in presence of Fe^{3+} (5×10^{-5} M) as shown in Fig.6. Again riboflavin is more than twice as efficient as FMN, and doubling the concentration of riboflavin relative to NADH doubles the rate of reduction of cytochrome *c*, while reduction to a ratio of 0.5 diminishes the rate by 50%.

It may be noted that the reaction goes to completion (Fig.4) and gives the same absorption as an equivalent solution of cytochrome *c* reduced by dithionite (1.482 at 550 nm for the preparation used at 10^{-4} M). The linearity of the reduction until completion is striking (Fig.4) and with a ratio of 2 riboflavin: 1 NADH 100% reduction is attained in less than 10 min under the conditions used.

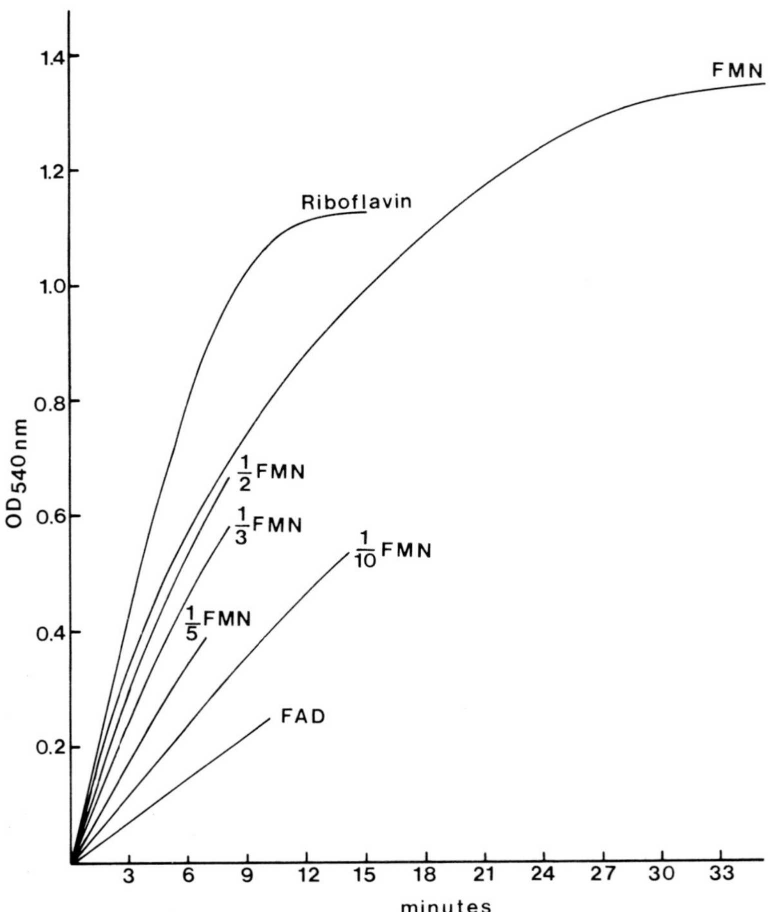

Fig.3. *Reduction of Fe^{3+} by NADH/Flavin. The cuvette contains 1.7 ml 0.1 M cacodylate-borate buffer pH 7.0 and 0.1 ml each of 1.5×10^{-2} M bipyridyl, NADH and $FeCl_3$. Reaction is initiated by injection of 1 ml of 1.5×10^{-3} M FMN to give final concentrations of 5×10^{-4} M for all components. Final pH is 6.75 (due to acidic solution of $FeCl_3$) and the reaction is followed at 540 nm in a Cary 15. (Volume 3 ml). Riboflavin and FAD were also at 5×10^{-4} M when replacing FMN. Variation of concentration of FMN was by dilution of stock solutions to obtain the final concentration by injection of 1 ml.*

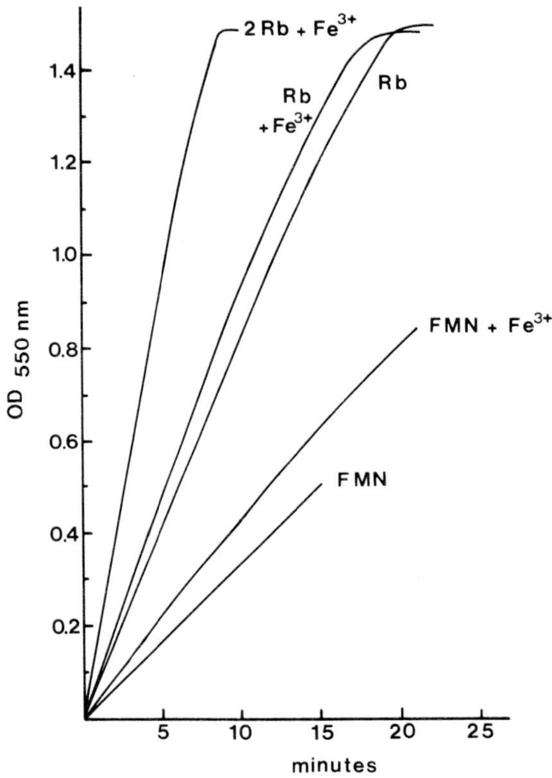

Fig.4. *Reduction of cytochrome* c *by NADH plus FMN or riboflavin in absence or presence of ferric ions. The cuvette contains 1.8 or 1.7 ml of 0.1 M cacodylate-borate buffer pH 7.1, 0.1 ml of 3 x 10^{-3} M cytochrome* c, *0.1 ml of 6 x 10^{-3} M NADH and when used 0.1 ml of 1.5 x 10^{-3} M Fe^{3+}. Reduction was initiated by injection of 1 ml of flavin (in water) to give a final volume of 3 ml and concentrations of 10^{-4} M cytochrome* c, *2 x 10^{-4} M NADH, 2 x 10^{-4} M FMN (or riboflavin) and 5 x 10^{-5} M Fe^{3+} when used. For the reaction with 2 riboflavin 1 ml of a solution at twice the concentration was injected. Reduction was followed at 550 nm in a Cary 15 at 26°C. The blank was a similar solution of cytochrome* c *plus NADH (and Fe^{3+} when used).*

Reduction of cytochrome *c* via photoreduced flavin has been repeatedly described (Merkel and Nicherson, 1954; Vernon, 1959; McCord and Fridovich, 1968; Massey *et al.*, 1969) and more recently by Schmidt and Butler (1976) who indicated an inhibitory effect of superoxide dismutase. However the concentrations of SOD used (6 x 10^{-6} M) were excessively high and may not necessarily imply an intermediate reduction by $O_2^{\cdot-}$ under aerobic conditions. We thus conclude that reduction of cytochrome *c* in the dark by mixtures of flavins and NADH proceeds via the chemical reduction of flavin to intermediates which reduce the cytochrome directly.

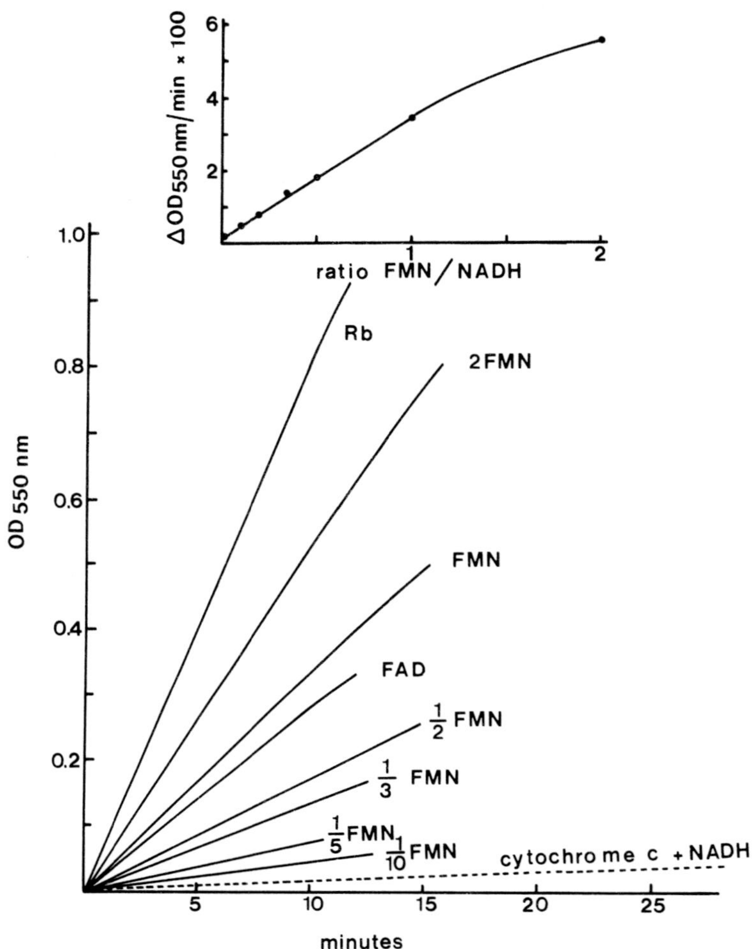

Fig.5. *Influence of the ratio FMN/NADH on the reduction of cytochrome* c. *Comparison of FMN riboflavin and FAD. Conditions as in Fig.4 (absence of* Fe^{3+}*). All measurements were observed against a blank containing NADH and cytochrome* c. *The rate of variation of this blank with time against buffer is also shown. Insert. Variation of rate as a function of the ratio FMN/NADH.*

$$H^+ + FMN + NADH \longrightarrow FMNH_2 + NAD^+$$

$$FMNH_2 + \text{cyt. } c\ (Fe^{3+}) \longrightarrow FMNH^{\cdot} + \text{cyt. } c\ (Fe^{2+}) + H^+$$

$$FMNH^{\cdot} + \text{cyt. } c\ (Fe^{3+}) \longrightarrow FMN + \text{cyt. } c\ (Fe^{2+}) + H^+$$

rather than $$FMNH_2 + 2O_2 \longrightarrow 2O_2^{\cdot-} + FMN + 2\ H^+$$

$$O_2^{\cdot-} + \text{cyt. } c\ (Fe^{3+}) \longrightarrow O_2 + \text{cyt. } c\ (Fe^{2+})$$

As shown by McCord and Fridovich (1970) using the system ferredoxin-TPN reductase, reduction of cytochrome *c* mediated aerobically by FMN (or FAD) is strongly inhibited by SOD, but not under anaerobic conditions, although the rate of reduction of cytochrome *c* is only slightly affected in absence of oxygen. Hence reduced FMN can act directly as a one electron reductant or via production of $O_2^{\overline{\cdot}}$, depending on the conditions. The role of the acceptor of any $O_2^{\overline{\cdot}}$ produced is shown in studies on reduced methylene blue which under aerobic conditions reduces cytochrome *c* directly but oxidizes epinephrine (McCord and Fridovich, 1970) or luminol (Michelson, 1974) via formation of $O_2^{\overline{\cdot}}$.

TABLE III

Reduction of Cytochrome *c* by NADH Plus FMN or Riboflavin

		ΔOD 550 nm per min x 100
1)	FMN	3.42
2)	FMN + 10^{-8} M SOD	3.40
3)	2 FMN	5.49
4)	1/2 FMN	1.80
5)	1/3 FMN	1.33
6)	1/5 FMN	0.78
7)	1/10 FMN	0.48
8)	FAD	2.81
9)	Riboflavin	8.30
	Above rates measured against a blank containing cytochrome *c* and NADH only.	
10)	No flavin (blank against buffer)	0.15
11)	FMN + Fe^{3+}	4.32
12)	1/2 Riboflavin + Fe^{3+}	4.68
13)	Riboflavin + Fe^{3+}	9.18
14)	2 Riboflavin + Fe^{3+}	19.00
	Above rates measured against a blank containing cytochrome *c*, NADH and Fe^{3+} only.	
15)	No flavin (blank against buffer)	0.19

Conditions as in Fig.4.

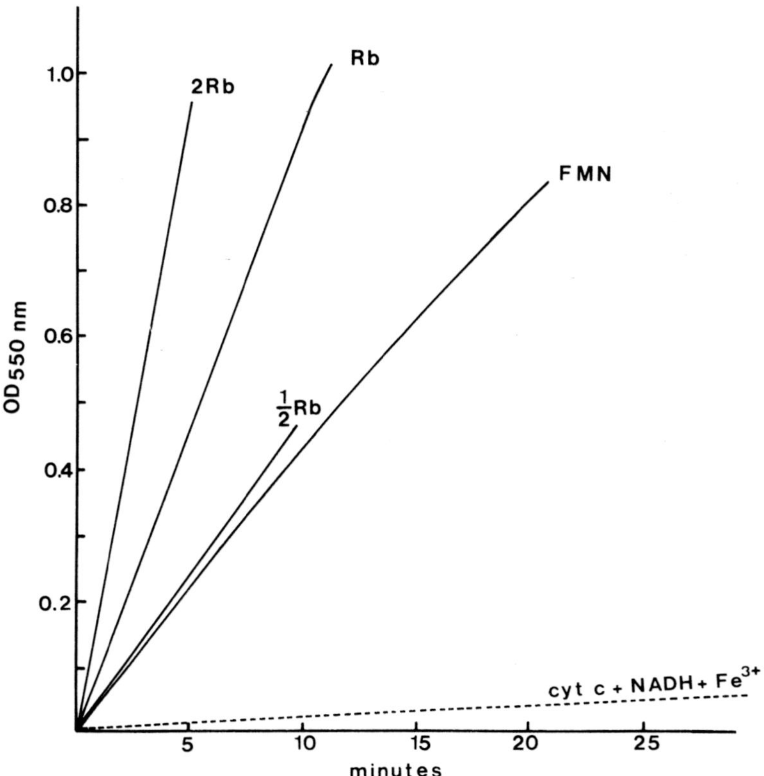

Fig.6. *Influence of the ratio riboflavin/NADH on the reduction of cytochrome* c *in presence of* Fe^{3+}. *Conditions as in Fig.4. Measurements were recorded against a blank containing cytochrome* c, *NADH and* Fe^{3+}. *The rate of variation of this blank is also shown.*

Reduction of Methemoglobin by NADH/Riboflavin

Since cytochrome *c* was readily reduced by the action of NADH and flavins it was of interest to examine the reduction of methemoglobin. As shown in Fig.7 reduction to oxyhemoglobin readily occurs and again riboflavin is about 2 to 3 times more efficient than the equivalent amount of FMN. The peak at 628 nm (0.242 OD at time zero) essentially disappears (OD 628 nm at end = 0.028) with formation of two new λmax at 541 nm (0.873 OD) and 576 nm (0.891 OD at end, 0.280 OD at beginning) characteristic of oxyhemoglobin. In addition, the Soret band of methemoglobin at 406 nm is shifted to 413 nm at the end of the reaction. Although superoxide dismutase (2.4×10^{-8} M) inhibited the blank (reduction of methemoglobin by NADH alone) by about 50%, no inhibition of the complete reaction was observed (as was also the case with reduction of cytochrome *c*) and the methemoglobin is reduced directly by NADH/riboflavin in absence of an $O_2^{\overline{\cdot}}$ intermediate.

Reduction of Nitroblue Tetrazolium

The reduction of nitroblue tetrazolium (also used as indicative of superoxide radicals) was investigated. Whereas ferrous ions show no significant effect, reduction of NBT readily occurs with mixtures of NADH and riboflavin (but not by either component separately) at pH 7.1 (cacodylate-borate buffer). Again FMN is much less efficient than riboflavin (Fig.8a) though in the photoreduction system they are equivalent. Increase in concentration of riboflavin by a factor of two relative to NADH increases the rate of reduction approximately 3.5 fold. At pH 7.8 (phosphate buffer) the reduction of NBT by riboflavin and NADH (2 : 1) is increased almost two fold and is even more rapid at pH 8.1 (Fig.8b). In contrast with the reduction of ferric ions by NADH/flavin, this reaction is strongly inhibited by superoxide dismutase, but not by the denatured enzyme. Reduction of NBT thus occurs via intermediate production of superoxide radicals. The inhibition by SOD is not constant with time but decreases markedly from 75.7% at 5 min, 63.4% at 10 min to 50.8% at 15 min at pH 7.1 with corresponding figures of 61.2%, 46.3% and 37.3% at pH 7.8 with 10^{-8} M SOD. The reason for this is not clear but may be due to direct reduction of NBT by reduced flavin as reported by Beauchamp and Fridovich (1971) as the concentration of oxygen in solution diminishes.

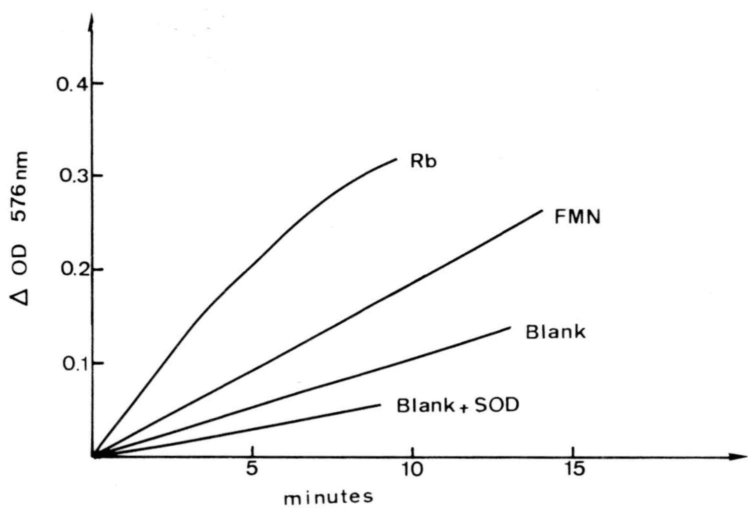

Fig.7. *Reduction of methemoglobin by FMN/Riboflavin and NADH. The cuvette contained 0.2 ml of 6 x 10^{-3} M NADH, 1.6 ml of 0.1 M cacodylate-borate buffer pH 7.0 and methemoglobin to give a final concentration of about 7 x 10^{-5} M. Reaction was initiated by injection of 1 ml of 1.2 x 10^{-3} M riboflavin or FMN and reduction followed by increase in absorption at 576 nm. Final concentrations of flavin and NADH 4 x 10^{-4} M, and the reaction was at 35°C.*

It can therefore be concluded that mixtures of flavin and NADH give rise to $FMNH_2$ which then produces $O_2^{\bar{\cdot}}$. Under certain conditions and with certain acceptors, $O_2^{\bar{\cdot}}$ reduces more efficiently than $FMNH_2$ directly, whereas in other cases the inverse is true. The system is also shown to be surprisingly efficient in the absence of enzymic catalysis or irradiation for the production of reduced flavins.

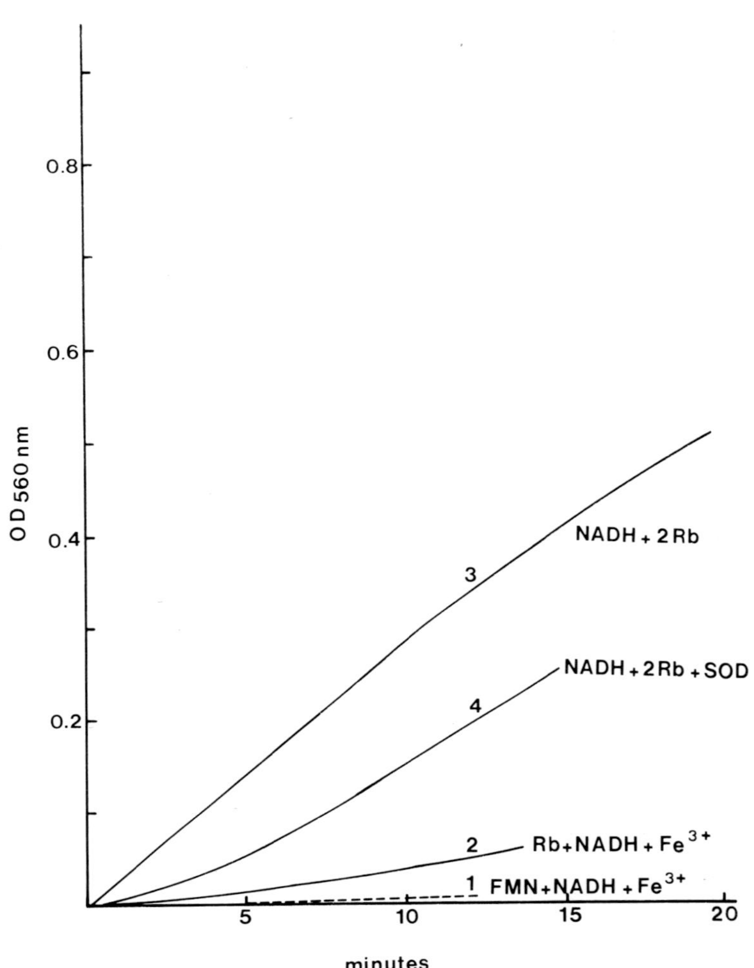

Fig.8a. *Reduction of nitroblue tetrazolium by NADH plus riboflavin. The cuvette contained 1.6 ml of buffer, 0.3 ml of 1.67 x 10^{-3} M NBT, 0.1 ml of 6 x 10^{-3} M NADH, and reaction was initiated by injection of 1 ml of a solution of 6 x 10^{-4} M or 1.2 x 10^{-3} M flavin to give final concentrations of 2 x 10^{-4} M NADH, 1.67 x 10^{-4} M NBT and 2 x 10^{-4} M (or 4 x 10^{-4} M) flavin with a final volume of 3 ml. Final concentration of Fe^{3+} when used was 5 x 10^{-5} M. Curve 1, 2 x 10^{-4} M FMN, 5 x 10^{-5} M Fe^{3+}, 2 x 10^{-4} M NADH, 1.67 x 10^{-4} M NBT, buffer 0.1 M cacodylate-borate pH 7.1; curve 2, as for curve 1 but with 2 x 10^{-4} M riboflavin instead of FMN; curve 3, 4 x 10^{-4} M riboflavin, 2 x 10^{-4} M NADH, 1.67 x 10^{-4} M NBT, buffer 5 x 10^{-2} M phosphate pH 7.8, in absence or in presence of 10^{-8} M heat denatured SOD; curve 4, as for curve 3 but with 10^{-8} M native SOD. The superoxide dismutase used was pure ferri SOD from* Photobacterium leiognathi. *The reduction was followed at 560 nm in a Cary 15, at 20°C.*

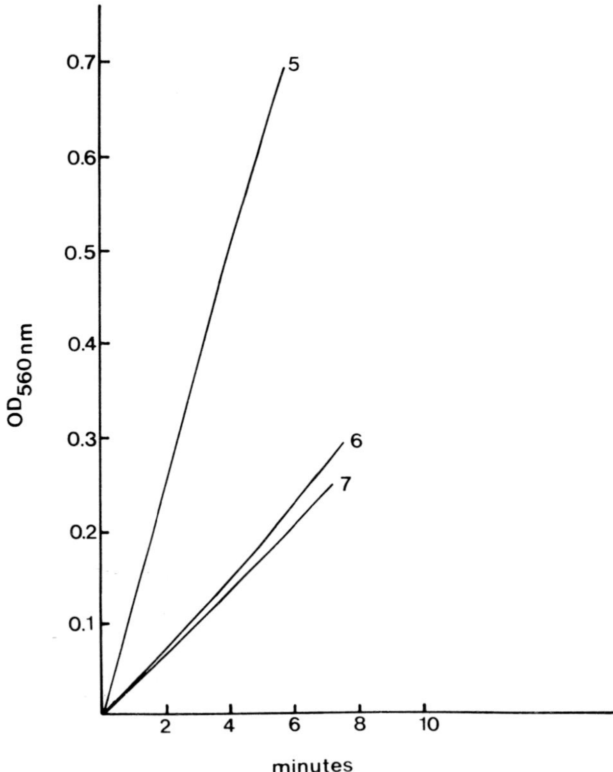

Fig.8b. *Reduction of nitroblue tetrazolium with NADH and FMN/riboflavin. Final concentrations were 4 x 10^{-4} M riboflavin, FMN, and NADH and 1.67 x 10^{-4} M NBT in 0.1 M NaCl 0.05 M phosphate buffer pH 8.1 at 27°C. Curve 5, reaction in absence or in presence of 4 x 10^{-5} M Fe^{3+}, in absence or in presence of 10^{-8} M denatured SOD; curve 6 in presence of 10^{-8} M native SOD; curve 7 with 2 x 10^{-8} M SOD. Injection of NADH.*

Oxidation of Luminol and Pholad Luciferin by FMN/Fe^{3+}/NADH

Since ferrous ions in presence of a suitable ligand such as phosphate rapidly produce superoxide radicals by a one electron reduction of the dissolved molecular oxygen and ferric ions are readily reduced by FMN plus NADH, the oxidation of luminol and of Pholad luciferin by the system NADH/Fe^{3+}/FMN was examined.

At pH 7 (phosphate or borate-cacodylate buffers) no oxidation of luminol as detected by light emission was observed in accord with the pH dependence of luminol oxidation. However, in alkaline solution (pH 9) light emission occurs as shown in Fig.9. The reaction is extremely slow and maximum light emission requires about 30 min. Thereafter oxidation continues with slowly decreasing intensity for at least 10 hours. If FMN is replaced by an equivalent amount of riboflavin a higher rate of oxidation is observed (almost three times the maximum obtained with FMN) and the reaction is complete in about three hours (Fig.9), whereas FAD is about 10 times less efficient than FMN (Table IV). Reduction of the concentration of FMN by half relative to the other components also reduces the rate of oxidation by almost 50%. As shown in Table IV, the oxidation is strongly inhibited by superoxide dismutase (94% for the reaction with FMN and 78% for riboflavin with 10^{-9} M SOD) whereas the same quantity of thermally denatured SOD is practically without effect. Catalase also shows an inhibitory effect but like the strong stimulation caused by horse radish peroxidase this is not necessarily due to formation of H_2O_2 as will be discussed later.

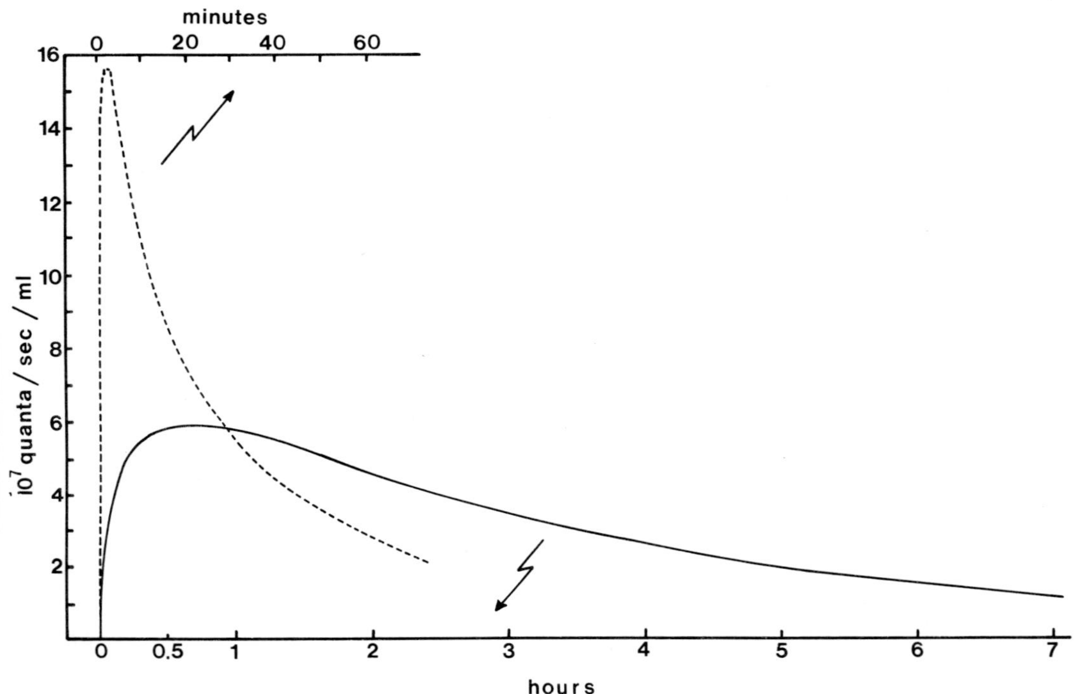

Fig.9. *Oxidation of luminol by NADH, Fe^{3+} and FMN (solid line) or riboflavin (---). The reaction mixture contained 1.8 ml of 0.1 M cacodylate-borate buffer pH 9.05, 0.1 ml of 1.5 x 10^{-2} M Fe^{3+}, 0.1 ml of 1.5 x 10^{-2} M NADH a,d 30 µl of 10^{-2} M luminol. Reaction was initiated by injection of 1 ml of 1.5 x 10^{-3} M FMN or riboflavin. Final concentrations 5 x 10^{-4} M NADH, FMN and Fe^{3+}, 10^{-4} M luminol. The solution of ferric chloride was partially neutralised with NaOH to incipient precipitation to minimise pH changes (final pH 9.05).*

Similar results were obtained with oxidation of Pholad luciferin (Fig.10a) at pH 7.1. Again FAD is less efficient and riboflavin more than doubles the rate relative to FMN. The reaction is less inhibited by superoxide dismutase (due to the higher reaction rate of luciferin with $O_2^{\bar{\cdot}}$ compared with luminol) but denatured SOD is without effect. Catalase also inhibits (to be discussed later) but this inhibition is not strictly additive to that obtained with SOD and to some extent is competitive (Table IV). Again the extremely long period of oxidation with light emission occurring for several hours (Fig.10a) may be noted.

Oxidation of luminol by FMN plus NADH in absence of Fe^{3+} was not detected, presumably due to the low steady state production of $O_2^{\bar{\cdot}}$. Pholad luciferin is much more sensitive for the detection of $O_2^{\bar{\cdot}}$ than is luminol and under

TABLE IV

Oxidation of Luminol and Luciferin by NADH/Fe^{3+}/FMN

	Quanta/sec/ml	
	At Imax	After 10 min Reaction
Luminol. (Conditions as in Fig.9)		
1) Fe^{2+} ↓	1.3×10^{10} (flash)	0
2) FMN + NADH + Fe^{2+} ↓	5.3×10^{9} (flash)	0
3) NADH + FMN ↓	$< 5 \times 10^{5}$	$< 5 \times 10^{5}$
4) Fe^{3+} ↓	2.25×10^{9} (flash)	0
5) NADH + Fe^{3+} ↓	3.9×10^{7} (flash)	0
6) NADH + Fe^{3+} + FMN ↓	5.8×10^{7}	4.5×10^{7}
As 6 plus 10^{-9} M SOD	2.8×10^{6} (- 95%)	2.8×10^{6} (- 94%)
As 6 plus 10^{-9} M denatured SOD	-	4.3×10^{7} (- 4%)
As 6 plus 0.1 μg catalase	-	3.2×10^{7} (- 29%)
As 6 plus 1 μg catalase	-	1.1×10^{7} (- 76%)
As 6 plus 1 μg peroxidase	-	8.2×10^{7} (+ 82%)
7) NADH + Fe^{3+} + 1/2 FMN ↓	-	2.6×10^{7}
8) NADH + Fe^{3+} + FAD ↓	-	4.8×10^{6}
9) NADH + Fe^{3+} + Riboflavin ↓	1.56×10^{8}	1.35×10^{8}
As 9 plus 10^{-9} M SOD	3.45×10^{7} (- 78%)	3.45×10^{7} (- 75%)
As 9 plus 10^{-9} M denatured SOD	1.30×10^{8} (- 17%)	-
As 9 plus 1 μg catalase	1.13×10^{8} (- 28%)	-
As 9 plus 1 μg peroxidase	2.11×10^{8} (+ 35%)	-
Luciferin. (Conditions as in Fig.10a)		
1) NADH + Fe^{3+} + FMN ↓	1.93×10^{8}	1.29×10^{8}
As 1 plus 5×10^{-9} M SOD	-	1.01×10^{8} (- 22%)
As 1 plus 5×10^{-9} M denatured SOD	-	1.31×10^{8} (0%)
2) NADH + Fe^{3+} + FAD ↓	-	7.3×10^{7}
3) NADH + Fe^{3+} + Riboflavin ↓	4.22×10^{8}	
		After 20 min reaction
4) NADH + Fe^{3+} + FMN ↓	1.92×10^{8}	1.59×10^{8}
As 4 plus 5×10^{-8} M SOD	-	9.3×10^{7} (- 41%)
As 4 plus 0.2 μg catalase	-	9.5×10^{7} (- 40%)
As 4 plus 5×10^{-8} M SOD plus 0.2 μg catalase	-	6.2×10^{7} (- 61%)

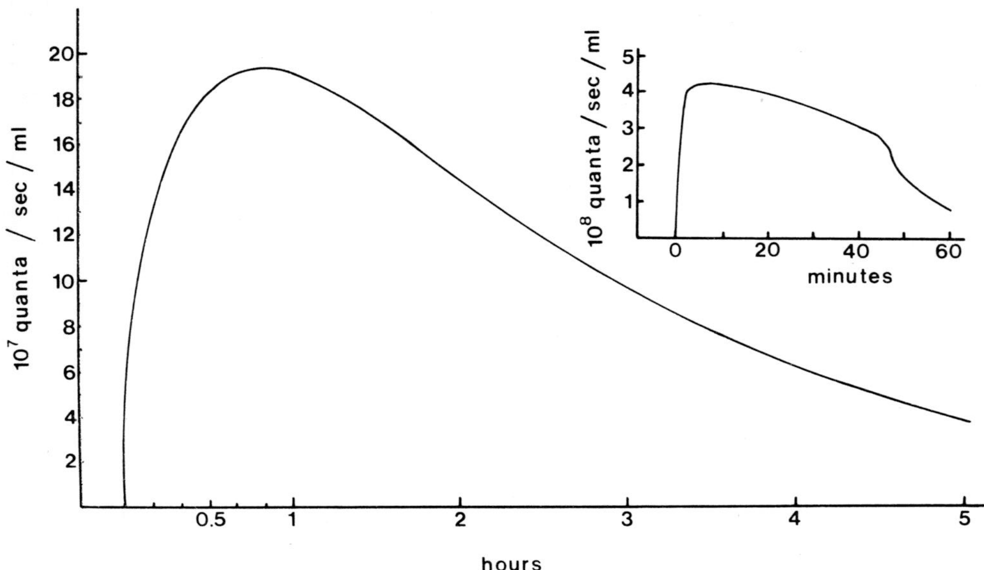

Fig.10a. *Oxidation of Pholad luciferin by NADH, Fe^{3+} and FMN. Conditions as in Fig.8 except that 0.1 M cacodylate-borate buffer pH 7.3 containing 10^{-3} M phosphate was used (final pH 7.1). Luminol was replaced by 20 µl of a solution of Pholad luciferin, corresponding to a final concentration of about 10^{-7} M.*
Insert: Oxidation of Pholad luciferin by NADH, Fe^{3+} and riboflavin. Same conditions as above.

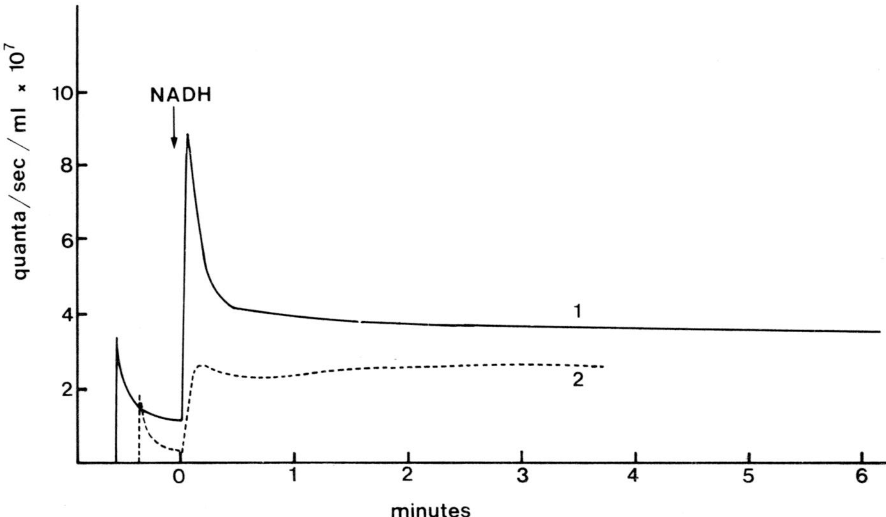

Fig.10b. *Oxidation of Pholad luciferin by FMN/riboflavin and NADH. Final concentrations were 4 x 10^{-4} M riboflavin, FMN and NADH and 10^{-7} M luciferin, in 0.1 M NaCl 0.05 M phosphate buffer pH 8.1. Curve 1, in presence or in absence of 4.5 x 10^{-5} M Fe^{3+}, in absence or in presence of 6 x 10^{-8} M denatured SOD; curve 2 in presence of 6 x 10^{-8} M native SOD. Injection of NADH.*

somewhat different conditions was readily oxidised by FMN/riboflavin and NADH in the absence of ferric ions (Fig.10b) at pH 8.1. Indeed at this pH, presence of Fe^{3+} (4.5×10^{-5} M) in no way changed the kinetics. Again oxidation continues for a long period of time (steady state for at least 15 min). Rather large quantities of SOD inhibit the reaction whereas the same amount of heat denatured SOD is without effect (Fig.10b).

Finally, several minor points may be noted. In the complete system NADH, Fe^{3+}/FMN/luminol the same result is obtained if the final component injected is FMN or NADH. Injection of ferric ions into luminol alone gives a pulse of light emission which is diminished 60 fold in presence of NADH. This is presumably due to decomposition of trace peroxides and is eliminated by pre-incubation of the mixtures containing Fe^{3+} for two minutes before injection of the final component. The system provides a simple chemical model of electron transfer from NADH to FMN to Fe^{3+} to O_2.

Effects of Catalase and Peroxidase on the Dosage of Superoxide Dismutase. Capture of $O_2^{\overline{\cdot}}$ by these enzymes

Various indirect dosages of superoxide dismutase activity have been described. Thus production of a steady state level of $O_2^{\overline{\cdot}}$ by xanthine oxidase-O_2-xanthine (or hypoxanthine) is detected via light emission on oxidation of luminol (Fig.11A). Addition of superoxide dismutase inhibits this steady state light emission (Hodgson and Fridovich, 1973) and indeed the unity of activity is defined in terms of 50% inhibition. It can be seen (Table V) that catalase also inhibits the reaction to some degree though the affinity of catalase for $O_2^{\overline{\cdot}}$ is at least 10 fold less than that of SOD. Oxidation of luminol by the levels of H_2O_2 produced by the xanthine oxidase system (by spontaneous dismutation of $O_2^{\overline{\cdot}}$) gives light at least 100 fold less than that observed, and hence the inhibition due to catalase is not due to destruction of H_2O_2 but rather to removal of $O_2^{\overline{\cdot}}$. This is further confirmed in studies of stimulation of the oxidation by peroxidase.

As is well known, peroxidase forms a complex with $O_2^{\overline{\cdot}}$ and can indeed be used to detect the superoxide radical. The effect of added peroxidase to the above system of oxidation of luminol is striking (Table V) and a 100 fold stimulation of light emission is readily obtained. Again this stimulated oxidation is not due to utilisation of H_2O_2 possibly present since catalase (6 μg) inhibits but slightly (Fig.11C) whereas SOD (1.6 μg) inhibits strongly (about 80%) at pH 9 (Fig.11B). The effects of the two enzymes are not additive in the region of inhibition studied.

Reduction of nitroblue tetrazolium by $O_2^{\overline{\cdot}}$ (produced via photoreduction of riboflavin) is also used as a means of estimating superoxide dismutase activity. This test for the detection of $O_2^{\overline{\cdot}}$ is at least 100 times less sensitive than luminol at pH 9 and in this case μg quantities of catalase have no effect on the interaction of NBT and $O_2^{\overline{\cdot}}$. Peroxidase however is competitive with NBT for $O_2^{\overline{\cdot}}$ and in μg quantities causes an apparent diminution in the level of $O_2^{\overline{\cdot}}$ as detected by reduction of NBT (Table V). Both catalase and peroxidase were free of superoxide dismutase within the limits of detection by gel electrophoresis, but trace contamination of catalase by SOD cannot be excluded.

Similar results have been reported by Kovacs and Matkovics (1975) for the inhibition of reduction of cytochrome *c*, via xanthine oxidase generated $O_2^{\overline{\cdot}}$, by horse radish peroxidase and catalase as well as with the autoxidation of adrenaline and by Sawada and Yamazaki (1973) with respect to horse radish peroxidase.

TABLE V

Effects of Catalase and Peroxidase on the Dosage of Superoxide Dismutase

Luminol Test	Quanta/sec/ml	
	pH 7.8	pH 9.0
1) Control	1.05×10^7	1.60×10^7
2) Plus 0.8 μg SOD	2.50×10^6 (- 76%)	5.00×10^5 (- 97%)
3) Plus 0.6 μg catalase	7.80×10^6 (- 26%)	1.15×10^7 (- 28%)
4) Plus 6.0 μg catalase	4.50×10^6 (- 57%)	6.60×10^6 (- 59%)
5) Plus 10 μg peroxidase	1.44×10^9	1.84×10^9
6) As 5 plus 1.6 μg SOD	1.25×10^9 (- 13%)	3.90×10^8 (- 79%)
7) Plus 6.0 μg peroxidase	1.08×10^9	1.32×10^9
8) As 7 plus 0.6 μg catalase	-	1.34×10^9 (+ 1.5%)
9) As 7 plus 6.0 μg catalase	1.00×10^9 (- 7%)	1.07×10^9 (- 19%)
10) As 7 plus 1.6 μg SOD	7.80×10^8 (- 28%)	2.2×10^8 (- 83%)
11) As 7 plus 6.0 μg catalase and 1.6 μg SOD	-	2.0×10^8 (- 85%)

At pH 7.8 the reaction mixture contains 1.1 ml H_2O, 0.3 ml 1 M phosphate pH 7.8, 0.3 ml of 10^{-3} M EDTA, 0.3 ml of 10^{-3} M luminol and 50 μg of xanthine oxidase. Reaction is initiated by injection of 1 ml of 3×10^{-4} M hypoxanthine to give a final volume of 3 ml.

At pH 9.0 the reaction mixture contains 1.37 ml H_2O, 0.3 ml 1 M glycine pH 9.0, 0.3 ml 10^{-3} M EDTA, 30 μl of 10^{-3} M luminol and 5 μg of xanthine oxidase. Reaction is initiated by injection of 1 ml of 3×10^{-5} M hypoxanthine to give a final volume of 3 ml.

Reduction of Nitrotetrazolium Blue	Optical Density at 560 nm
1) Control	0.245
2) Plus 0.6 μg catalase	0.245
3) Plus 6.0 μg catalase	0.245
4) Plus 0.6 μg peroxidase	0.241
5) Plus 6.0 μg peroxidase	0.164

The reaction mixture (2 ml) contains 1.8×10^{-6} M riboflavin, 9×10^{-3} M methionine, 1.8×10^{-5} M KCN, 1.67×10^{-4} M nitrotetrazolium blue in 4.5×10^{-2} M phosphate pH 7.8.

The cuvette is irradiated at 365 nm for 2 mins with a B 100 A (Mineralight) lamp at a distance of 20 cm and the increase in absorption at 560 nm due to reduction of the NTB by $O_2^{\bar{\cdot}}$ via photoreduced riboflavin plus O_2 is measured.

It may be noted that pure ferri SOD from *P. leiognathi* has a specific activity of 10 units (luminol pH 9) per μg, compared with 50 units/μg pure bovine erythrocuprein.

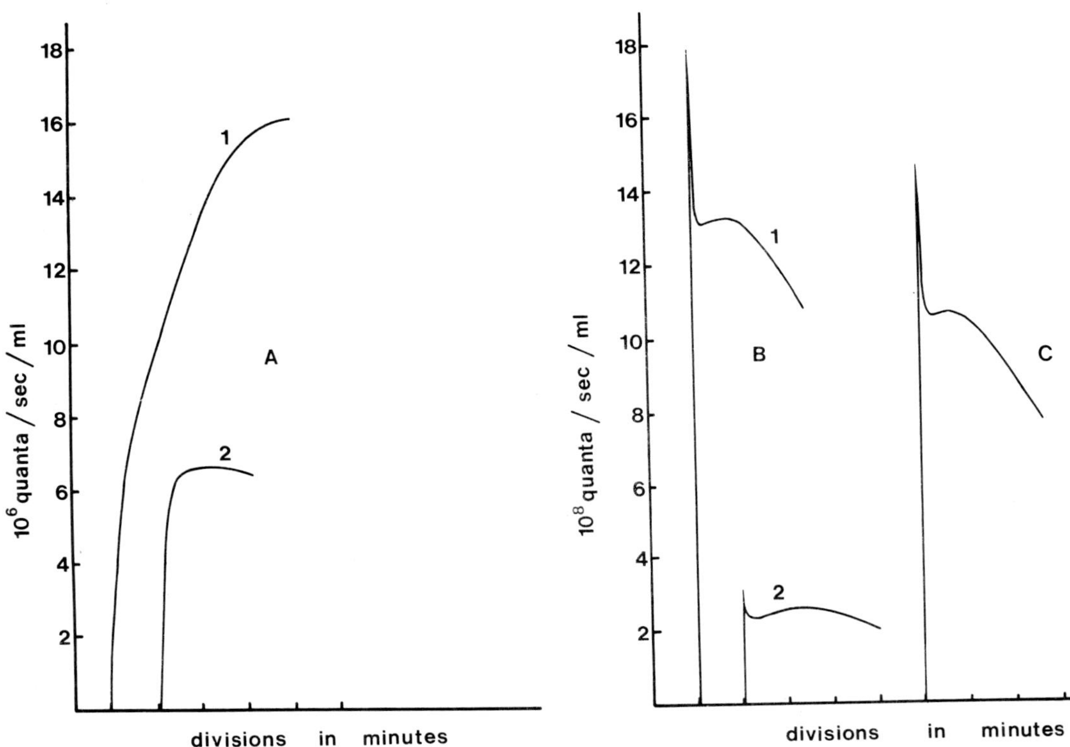

Fig.11. *Effect of catalase and peroxidase on the oxidation of luminol by $O_2^{\bar{\cdot}}$ produced by the system xanthine oxidase-O_2-hypoxanthine at pH 9.0. Conditions as in Table V.*
A. (1) Control: (2) plus 6 μg catalase
B. (1) Control plus 6 μg peroxidase; (2) plus 1.6 μg SOD in addition
C. Control plus 6 μg peroxidase plus 6 μg catalase.

Thus depending on the test used, both catalase and peroxidase can give rise to apparent SOD activity. It is to be noted that peroxidase has a higher affinity for $O_2^{\bar{\cdot}}$ than luminol and the complex thus formed is extremely efficient for the oxidation of luminol, much more so than free $O_2^{\bar{\cdot}}$, possibly since spontaneous dismutation of $O_2^{\bar{\cdot}}$ is in this way considerably diminished.

$$\text{peroxidase } Fe^{3+} \xrightarrow{O_2^{\bar{\cdot}}} \text{Complex III } (Fe^{3+} \ldots O_2^{\bar{\cdot}})$$

$$\text{Complex III} \xrightarrow{LH_2} \text{peroxidase } Fe^{3+} + LH^{\cdot} + HOO^{-}$$

These results provide an alternate explanation for the suggestion of Agro *et al.* (1973) that both catalase and SOD prevent oxidations attributed to singlet oxygen.

Whereas superoxide dismutase appears to be specific for the substrate $O_2^{\bar{\cdot}}$, the substrate is not specific to the enzyme, and indeed apparent SOD activity can arise from the presence of various small molecules apart from peroxidase or catalase in extracts prepared for estimation.

REFERENCES

1. Agrò, A., Finazzi, de Sole, P., Rotilio, G. and Mondovi, B. (1973). *Ital. J. of Biochemistry 22,* 217-231.
2. Beauchamp, C. and Fridovich, I. (1971). *Analyt. Biochem. 44,* 276-287.
3. Crichton, R.R., Roman, F. and Wauters, M. (1975a). *Biochem. Soc. Transactions 3,* 946-948.
4. Crichton, R.R., Wauters, M. and Roman F. (1975b). In "Proteins of Iron Storage and Transport in Biochemistry and Medicine", ed. R.R. Crichton, North Holland Publishing Company, Amsterdam, 287-294.
5. Hodgson, E.K. and Fridovich, I. (1973). *Photochem. Photobiol. 18,* 451-455.
6. Kovacs, K. and Matkovics, B. (1975). *Enzyme 20,* 1-5.
7. McCord, J.M. and Fridovich, I. (1970). *J. Biol. Chem. 245,* 1374-1377.
8. Massey, V., Strickland, S., Mayhew, S.A., Howel, L.G., Engel, P.C., Mathews, R.G., Schuhman, M. and Sullivan, P.A. (1969). *Biochem. Biophys. Res. Communs. 36,* 891-897.
9. Merkel, H.R. and Nickerson, W.J. (1954). *Biochim. Biophys. Acta 14,* 303-311.
10. Michelson, A.M. (1973). *Biochimie 55,* 465-479.
11. Michelson, A.M. (1974). *FEBS Letters 44,* 97-100.
12. Sawada, Y. and Yamazaki, I. (1973). *Biochim. Biophys. Acta 327,* 257-265.
13. Schmidt, W. and Butler, W.L. (1976). *Photochem. Photobiol. 24,* 71-75.
14. Vernon, L.P. (1959). *Biochim. Biophys. Acta 36,* 177-185.
15. Wauters, M., Crichton, R.R. and Michelson, A.M. (1976). Submitted.
16. Zipper, H., Person, P. and O'Connell, D.J. (1972). *Biochemistry 11,* 4686-4691.

HUMAN ERYTHROCYTE SUPEROXIDE DISMUTASE (ERYTHROCUPREIN)

W.H. BANNISTER, A. ANASTASI and J.V. BANNISTER

Department of Physiology and Biochemistry
Royal University of Malta
Msida, Malta

INTRODUCTION

Before the discovery of superoxide dismutase activity by McCord and Fridovich (1969), human superoxide dismutase was isolated and studied as a copper protein of obscure significance which could be obtained from erythrocytes, brain and liver. The protein was termed erythrocuprein, cerebrocuprein or hepatocuprein depending on its origin (Markowitz *et al.*, 1959; Porter and Ainsworth, 1959; Porter *et al.*, 1964). Erythrocuprein was clearly homologous to the protein first isolated from bovine erythrocytes and called haemocuprein by Mann and Keilin (1939). Carrico and Deutsch (1969) demonstrated the identity of erythrocuprein, cerebrocuprein and hepatocuprein and proposed the name cytocuprein for the protein. Cytocuprein was shown to contain 2 g atoms of zinc in addition to 2 g atoms of copper per mole of protein (Carrico and Deutsch, 1970). It is now clear that cytocuprein corresponds to cytosol superoxide dismutase, which is a cyanide-sensitive cuprozinc enzyme in contrast to the cyanide-insensitive manganese superoxide dismutase of mitochondria (Weisiger and Fridovich, 1973). The vast majority of studies on this superoxide dismutase have been made on the enzyme from bovine erythrocytes which has been fully characterized (Steinman *et al.*, 1974; Richardson *et al.*, 1975). There is a good case for studying the homologous enzymes of other mammals. Thus both human and horse superoxide dismutase (Albergoni and Cassini, 1974; Finazzi Agrò *et al.*, 1974) show interesting differences to the bovine enzyme. The purpose of this paper is to describe the properties of human erythrocyte superoxide dismutase as they are known at the present time and to draw attention to points of controversy. Observations on highly purified superoxide dismutase are given. The term erythrocuprein is retained together with erythrocyte superoxide dismutase for the sake of convenience.

PURIFICATION AND PROPERTIES OF ERYTHROCUPREIN

There has been a certain lack of agreement on the properties of human erythrocuprein. To some extent this has depended on the method of preparation of the protein. The starting material for the preparation of erythrocuprein is the fraction of soluble nonhaemoglobin proteins in a lysate (haemolysate) of red blood cells. Two methods have been followed to obtain this fraction. Markowitz *et al.* (1959) and later workers have precipitated haemoglobin with a mixture of ethanol and chloroform and recovered nonhaemoglobin proteins from the remaining supernatant by precipitation with lead acetate. This is the so-called Tsuchihashi procedure (Tsuchihashi, 1923) and was used by Mann and Keilin (1939) in their preparation of bovine erythrocuprein. An alternative to lead acetate precipitation consists in adding solid K_2HPO_4 to the ethanol-chloroform supernatant to obtain two liquid phases. Crude erythrocuprein is recovered from the lighter ethanol-water phase by precipitation with acetone (McCord and Fridovich, 1969). We have found that the nonhaemoglobin proteins may be recovered from the ethanol-chloroform supernatant by direct adsorption to QAE-Sephadex equilibrated with 0.02 M Tris-HCl buffer, pH 7.5. This method gives the same erythrocuprein and protein recovery as the Tsuchihashi procedure with lead acetate precipitation.

Because of the drastic nature of the Tsuchihashi procedure Deutsch and co-workers (Stansell and Deutsch, 1965a; Hartz and Deutsch, 1969, 1972) have preferred to obtain the nonhaemoglobin proteins from a lysate of red blood cells by direct adsorption to DEAE-cellulose at pH 6.5 - 7.25. The ionic strength of the lysate is lowered by dialysis or dilution. In our hands this procedure has invariably led to the poor recovery of erythrocuprein commented upon by Hartz and Deutsch (1969). It is now abundantly clear that the cuprozinc superoxide dismutases, including bovine (McCord and Fridovich, 1969) and human erythrocuprein (Bannister *et al.*, 1973), are remarkably resistant to the Tsuchihashi procedure so that this procedure in any one of its forms remains a convenient starting method for the preparation of these superoxide dismutases.

Markowitz *et al.* (1959) purified human erythrocuprein by adsorption to tricalcium phosphate gel, precipitation with ethanol and precipitation with ammonium sulphate. The protein contained 0.32 - 0.36% copper. Absorption maxima were observed at 265 and 655 nm. The erythrocuprein was immunologically homogeneous (when tested against its antiserum) and identical with the protein in erythrocyte haemolysates so that its immunological reactivity had not been altered. Kimmel *et al.* (1959) obtained immunologically identical material by continuous flow ionophoresis of crude erythrocuprein on a hanging curtain of filter paper. The protein contained 0.34% copper. It showed a minor faster moving component, equivalent to less than 10% of the total protein, in moving boundary electrophoresis. The erythrocuprein was monodisperse in sedimentation velocity experiments and had a sedimentation coefficient, $s^{o}_{20,w}$, of 3.02 S. Kimmel *et al.* (1959) reported the first amino acid analysis of human erythrocuprein. The protein was observed to contain no methionine, 11.3 half-cystine residues, 1.9 tyrosine residues and 0.6 tryptophan residue. Hexosamine, hexose and sialic acid (3.4, 1.8 and 1.6 residues, respectively) were also measured.

After this first work on human erythrocuprein, Nyman (1960) purified the protein by chromatography on DEAE-cellulose. The protein was homogeneous in sedimentation velocity and moving boundary electrophoresis experiments. It was shown to contain 0.38% copper and the visible absorption maximum was located at 670 ± 5 nm. The protein gave the same electron paramagnetic resonance (e.p.r.) spectrum as the erythrocuprein of Markowitz *et al.* (1959) in a parallel investigation (Malmström and Vänngård, 1960). Shields *et al.*

(1961) also purified human erythrocuprein by chromatography on DEAE-cellulose after the Tsuchihashi procedure. They showed that the original erythrocuprein of Markowitz *et al.* (1959) had contained a small amount of a highly antigenic copper-free impurity which was removed by chromatography on DEAE-cellulose.

Stansell and Deutsch (1965a,b) purified human erythrocuprein by chromatography on DEAE-cellulose and gel filtration on Sephadex G-100, followed by rechromatography on DEAE-cellulose and CM-cellulose and crystallization from 45% saturated ammonium sulphate solution in 0.05 M acetate buffer, pH 5.5. The protein contained 0.38% copper. Absorption maxima were determined at 265 and 655 nm. The erythrocuprein was monodisperse in moving boundary electrophoresis and seemed to have an isoelectric point of approximately 4.6. It was also monodisperse in the analytical ultracentrifuge and showed a sedimentation coefficient, $s^{0}_{20,w}$, of 2.77 S. Starch and polyacrylamide gel electrophoresis revealed a minor faster moving component. The presence and abundance of this component were dependent on the nature and ionic strength of the buffer and ageing of the protein. Amino acid analysis indicated no methionine, 5.1 half-cystine residues and virtually no tyrosine. The reported data indicate 0.3 tyrosine residue. Tyrosine and tryptophan were not detected by spectrophotometric analysis according to Bencze and Schmid (1957). This is an anomalous result. The presence of one free sulphydryl group (on the average 0.8) was established. No free N-terminal group was observed. One hexose residue was found but less than 0.1% of either sialic acid, hexosamine or hexuronic acid was detected.

Stansell and Deutsch (1965b) were concerned to show that ethanol-chloroform treatment of erythrocuprein, as in the Tsuchihashi procedure, altered the properties of the protein. They passed their chromatographically purified erythrocuprein through the Tsuchihashi procedure in the presence of 15% haemoglobin. The protein was recovered by the Tsuchihashi steps of Kimmel *et al.* (1959) and purified by chromatography on CM-cellulose (Stansell, 1966). The treated protein appeared to have lost 8% of its copper and showed a 10% decrease in absorbance at 655 nm. Its sedimentation coefficient was about 9% higher and its intrinsic viscosity was 64% lower than the value estimated for the original protein. It was therefore suggested that the axial ratio had decreased. There was an increase in net negative charge and electrophoretic mobility and increased heterogeneity in starch and polyacrylamide gel electrophoresis. The absorbance of the treated protein decreased by 83% at 190 nm. This suggested an increase in helical content. The protein also failed to crystallize.

The exact significance of these findings has not been clear. It is possible that the intrinsic viscosity of untreated erythrocuprein, given as 0.058 $(g/100\ ml)^{-1}$, was overestimated and the intrinsic viscosity of ethanol-chloroform treated erythrocuprein, given as 0.021 $(g/100\ ml)^{-1}$, was underestimated. The intrinsic viscosity of bovine erythrocuprein which had passed through the Tsuchihashi procedure has been given as 0.033 $(g/100\ ml)^{-1}$ (Bannister *et al.*, 1971). In practice the accuracy needed in sedimentation velocity and viscosity measurements to make hydrodynamic predictions is not easily attained. Just the same it seems that a size isomer of human erythrocuprein can appear under certain circumstances. Thus erythrocuprein prepared by Bannister and Wood (1970) behaved as a single component in isoelectric focussing but showed two components in polyacrylamide gel electrophoresis, one minor and faster migrating. The mobility of the bands as a function of gel concentration was of the kind characteristic of size isomers. Similar observations have been made on bovine erythrocuprein (Bannister *et al.*, 1971).

Hartz and Deutsch (1969) purified human erythrocuprein by chromatography with a strategy different from that of Stansell and Deutsch (1965a). The purification steps involved batch adsorption of a red cell lysate to DEAE-cellulose, gel filtration on Sephadex G-75, chromatography on DEAE-cellulose at pH 8.0 followed by chromatography at pH 5.5, and a final step of gel filtration on Sephadex G-75. The purified erythrocuprein contained 0.38% copper. It was monodisperse in the analytical ultracentrifuge and showed a sedimentation coefficient, $s_{20,w}$, of 2.94 S at a protein concentration of 0.21 mg/ml. Sedimentation equilibrium measurements indicated a weight-average molecular weight near 33600. The erythrocuprein was heterogeneous in cellulose acetate or starch gel electrophoresis with at least one faster moving component. Isoelectric focussing showed a major component with an isoelectric point of 4.75 and three minor components with lower isoelectric points. The amino acid composition of the erythrocuprein agreed well with that reported by Stansell and Deutsch (1965b) except that 6.8 half-cystine residues were found. Tyrosine was not detected and no hexose was found. Hartz (1968) noted that the absorption spectrum of the protein exhibited a maximum at 265 nm but the minimum at 248 nm reported by Markowitz *et al.* (1959) and Stansell and Deutsch (1965b) was broadened, shifted to 253 ± 3 nm, and nearly obliterated. He also observed phenylalanine fine structure near 265 nm. The e.p.r. spectrum was similar to that reported by Stansell (1966).

The e.p.r. parameters of various human erythrocuprein preparations have tended to be similar to those reported by Malmström and Vänngård (1960) for the preparations of Markowitz *et al.* (1959) and Nyman (1960) and close to those reported by Rotilio *et al.* (1971) for bovine erythrocuprein. Malmström and Vänngård (1960) found values of 2.063 and 2.265 for g_m and $g_{||}$, respectively, and a value of 0.016 cm^{-1} for $A_{||}$. The e.p.r. spectrum of human erythrocuprein which has been subjected to the Tsuchihashi procedure is very similar to the spectrum of protein prepared entirely by chromatography (Bannister *et al.*, 1973). Stansell (1966) reported a g_m value of 2.079 for his erythrocuprein. However, Carrico and Deutsch (1969) have given a g_m value of 2.063 for similarly prepared erythrocuprein.

It may be presumed that the copper-protein complex in human erythrocuprein is closely similar to the complex in the bovine protein. High resolution ^1H nuclear magnetic resonance studies have shown that the copper in the human protein is liganded to histidine residues as in the bovine protein (Stokes *et al.*, 1973). The human protein shows a higher yield of inactivation with Br_2^- radicals than the bovine protein. This suggests that the active site is more vulnerable in the human protein. Although the basic e.p.r. parameters are similar for the two proteins, the spectrum of the human protein seems to be more axial, indicating a lesser strain imposed on the copper by the native conformation and a weaker copper-histidine interaction (Roberts *et al.*, 1974).

Hartz and Deutsch (1969) observed that gel filtration increased the proportion of the more electronegative variants in their erythrocuprein. Hartz (1968) reported the separation of the main component (Fraction S-1) and the variants by chromatography on DEAE-cellulose at pH 5.5. All the fractions were immunologically similar or identical. One of the variants (Fraction S-5) was compared with Fraction S-1. Both fractions had 0.39% copper and the same amino composition. Eight half-cystine residues were found. Fraction S-1 had a sedimentation coefficient, $s_{20,w}$, of 3.24 S at a protein concentration of 0.27 mg/ml, and Fraction S-5 had a sedimentation coefficient, $s_{20,w}$, of 2.79 S at a protein concentration of 0.29 mg/ml. It was suggested that the erythrocuprein of Fraction S-1 might be more compact and that of Fraction S-5 more extended. Sedimentation equilibrium studies indicated the same molecular weight for the two fractions. Fraction S-5 showed less absorption between 190 and 270 nm and more absorption in the visible region of the spectrum. The two fractions had very similar e.p.r. spectra.

The discovery of zinc in human erythrocuprein by Carrico and Deutsch (1970) has been noted. The protein was shown to contain equimolar levels of copper and zinc. The metals were removed by dialysis against 0.05 M KCN at pH 8.0. The apoprotein probably suffered some modification of antigenic sites. It consisted mainly of a component with a sedimentation coefficient, $s_{20,w}$, of 2.86 S at a protein concentration of 0.8 mg/ml and a molecular weight of 32,900 as determined by equilibrium centrifugation. In starch gel electrophoresis it showed heterogeneity similar to that of the holoprotein and increased anodic mobility. The apoprotein showed a marked decrease in absorption between 250 and 290 nm but no change was noted between 200 and 240 nm. The maximum at 675 nm was absent.

Bannister *et al.* (1972) obtained crude human erythrocuprein by the Tsuchihashi procedure and purified the protein by chromatography on DEAE-Sephadex and gel filtration on Sephadex G-75. Isoelectric focussing showed a single protein with an isoelectric point at pH 4.75. Polyacrylamide gel electrophoresis revealed a small amount of component which moved faster than the main protein. The erythrocuprein sedimented as a single peak in the analytical ultracentrifuge. The sedimentation coefficient, $s^{0}_{20,w}$, was estimated to be 3.45 S. A weight-average molecular weight of 33,500 was found by equilibrium sedimentation. The ultraviolet absorption spectrum resembled that of the erythrocuprein fraction S-1 of Hartz (1968). Removal of copper and zinc with cyanide decreased the ultraviolet absorption with some change in the profile of the spectrum. A definite trough appeared at about 250 nm and a small red-shift of the spectrum was observed. A similar spectrum was observed after acidification of the holoprotein. Difference spectra obtained by subtraction showed a small maximum at 292-294 nm which was attributed to exposure of tryptophan to the solvent in apo and acidified protein. The holo and apo protein showed ultraviolet fluorescence with an uncorrected emission maximum at 342-345 nm. The quantum yield of fluorescence was about three-fold higher in the apoprotein. The fluorescence disappeared on addition of N-bromosuccinimide at pH 4.0 as observed in earlier work (Bannister *et al.*, 1968). It was estimated that N-bromosuccinimide destroyed 2 tryptophan residues per mole of protein.

Bannister *et al.* (1972) described a well-resolved band at 322 nm in the absorption spectrum of human erythrocuprein. The band corresponded to a strong positive contribution (at 325 nm) in the circular dichroism (CD) spectrum. It was attributed to a charge-transfer transition of the cupric ions because of its characteristics and its disappearance in the cyanide-treated (apo) protein. Subsequent work showed that the 320 nm band persisted in apoprotein prepared by treatment with EDTA at pH 3.8 (Calabrese *et al.*, 1975). The band in this apoprotein was destroyed by dithiothreitol, 2-mercaptoethanol, sulphite, borohydride, cyanide, guanidine hydrochloride and sodium dodecylsulphate. It was stable in the pH range 3-12 but disappeared at pH 2 and pH 13. On reaction of the apoprotein with cyanide or borohydride labile sulphur could be measured as thiocyanate or sulphide depending on the reagent. Four gram atoms of labile sulphur per mole of protein were detected. A persulphide group, R-S-SH, was suggested as the most likely structure for the 320 nm chromophore. Although the 320 nm band disappeared at pH 13, cyanolysis gave rise to the same amount of thiocyanate as at lower pH. The formation of trisulphide (R-S-S-S-R') between persulphide and a neighbouring -SH was suggested as a possible explanation of this observation. In comment upon this work it may be noted that human erythrocuprein contains more sulphur than can be accounted for as half-cystine (see below).

SUPEROXIDE DISMUTASE I AND II

Material obtained from human red blood cells by the Tsuchihashi procedure contains two forms of erythrocuprein one of which lacks the 320 nm chromophore. We call these forms erythrocuprein or superoxide dismutase I and II (SOD I and II). The 320 nm chromophore is absent in SOD II. We describe here the purification and properties of SOD I and II.

Crude SOD is obtained by the Tsuchihashi procedure as described by Bannister *et al.* (1971). Chromatography is carried out on QAE-Sephadex the aim being to elute the SOD at the lowest possible ionic strength and the highest possible pH. Tris-HCl (0.02 - 0.03 M; pH 7.5 - 7.0) and cacodylate-HCl (0.02 - 0.03 M; pH 6.5 - 5.0) buffers are used. The pH is decreased in steps of 0.5 pH unit. The SOD from 8 - 10 litres of red blood cells can be monitored visually. Contaminants in front of the SOD are removed by exhaustive elution while the SOD band moves slowly down the column. Contaminants behind the SOD are removed by removing the top of the column when this can be done without disturbing the SOD band. SOD II is prepared from column tops obtained during the chromatography of SOD I. It is eluted at pH 5.5 - 5.0 while SOD I is eluted at pH 6.5 - 6.0. A drawback of the method is the elution of SOD in very dilute form. However, after cleaning a column in front and behind the SOD band, concentrated SOD can be eluted by adding 0.3 M NaCl to the buffer. Each SOD is rechromatographed twice. It is then kept in water at 2 - 4°C or lyophilized. Apoprotein is prepared with EDTA to preserve the 320 nm band (Calabrese *et al.*, 1975). The procedure of Fee (1973) is followed to remove any bound EDTA.

SOD I and SOD II prepared as described here are homogeneous in polyacrylamide gel electrophoresis (Fig.1). The corresponding apoproteins (prepared by dialysis against 10 mM EDTA for 48 h at 2 - 4°C) are also homogeneous (Fig.2). SOD II has a somewhat higher electrophoretic mobility than SOD I. Apo SOD II also moves faster than apo SOD I and each apoprotein appears to move faster than the corresponding holoprotein. Both SOD I and SOD II contain 2 g atoms of copper and 2 g atoms of zinc per mole of protein as determined on the basis of dry weight estimation and metal analysis by atomic absorption spectrophotometry. The apo SOD I described here contained 8% residual copper and 32% residual zinc. The apo SOD II contained 10% residual copper and 27% residual zinc. The residual copper was associated with residual enzyme activity.

Superoxide dismutase activity was measured according to McCord and Fridovich (1969). A rate constant, k_{SOD}, for the enzyme catalysed dismutation,

$$2\ O_2^- + 2\ H^+ \xrightarrow{SOD} O_2 + H_2O_2, \quad (1)$$

was calculated with the assumption of Sawada and Yamazaki (1973):

$$V/V_c - 1 = k_{SOD}\left[SOD\right] / k_c \ \text{cyt}\ c^{3+}, \quad (2)$$

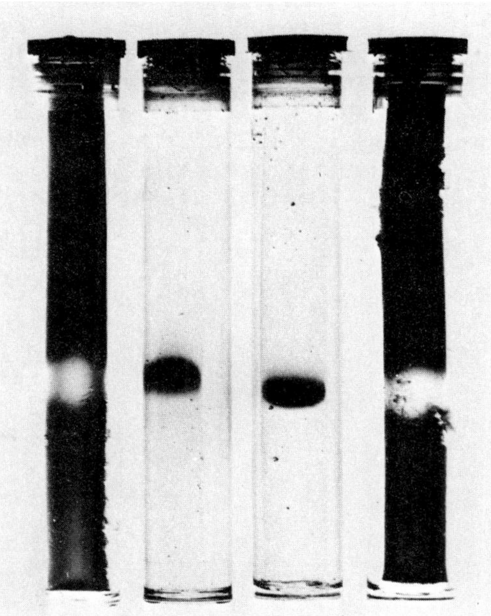

Fig.1. *Polyacrylamide gel electropherograms of human erythrocyte superoxide dismutase. Left to right: SOD I (enzyme stain), SOD I (protein stain), SOD II (protein stain), SOD II (enzyme stain). Conditions: 5% gels; continuous buffer system with 0.05 M glycine-NaOH, pH 9.5; protein stain Coomassie blue; enzyme stain according to Beauchamp and Fridovich (1971).*

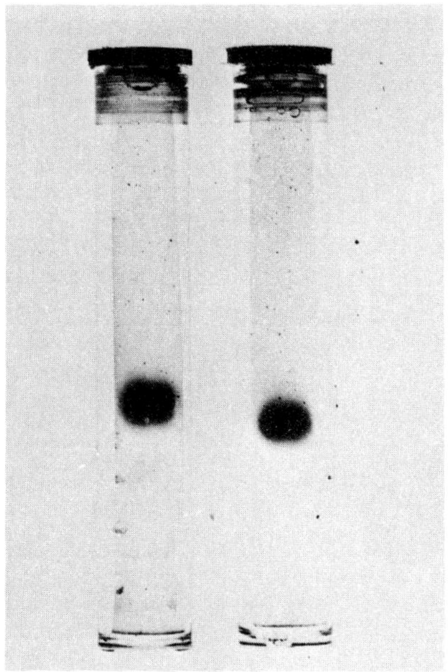

Fig.2. *Polyacrylamide gel electropherograms of human erythrocyte apo superoxide dismutase. Left hand side, apo SOD I; right hand side, apo SOD II. Gels and buffer as in Fig.1; stain Coomassie blue.*

where V is the rate of reduction of ferricytochrome *c* by superoxide anions in the absence of SOD and V_c is the rate of reduction in the presence of SOD. For k_c, the rate constant for the reduction of ferricytochrome *c* by superoxide anions, the value of $2 \times 10^5\ M^{-1}\ sec^{-1}$ obtained in pulse radiolysis studies (Rotilio *et al.*, 1972; Klug *et al.*, 1972; Asada *et al.*, 1974) was used. The linear assumption (2) was valid up to a concentration of 2.5×10^{-9} M for SOD I and SOD II and a concentration of 5.0×10^{-8} M for apo SOD I and apo SOD II (Fig.3). Results obtained are given in Table I.

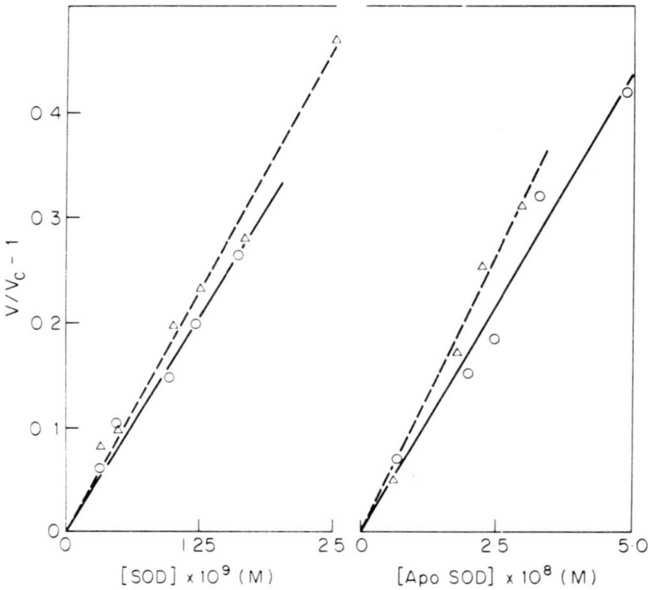

Fig.3. *Effect of human erythrocyte superoxide dismutase on the relative rate of reduction of ferricytochrome* c *by superoxide anions (V/V_c - 1; see text). Assay system according to McCord and Fridovich (1969); ferricytochrome* c *concentration 3.5×10^{-5} M. O——O, SOD I, apo SOD I; Δ-----Δ, SOD II, apo SOD II.*

TABLE I

Enzymatic Activity of Human Erythrocyte Superoxide Dismutase[a]

	Specific activity (Units/mg)	k_{SOD}/k_c cyt c^{3+} (M^{-1})	k_{SOD} ($M^{-1}\ sec^{-1}$)
SOD I	3320	1.67 (± 0.06) x 10^8	1.17 x 10^9
SOD II	3550	1.85 (± 0.05) x 10^8	1.29 x 10^9
Apo SOD I	150	8.73 (± 0.38) x 10^6	6.11 x 10^7
Apo SOD II	140	1.06 (± 0.04) x 10^7	7.42 x 10^7

[a]Assay system according to McCord and Fridovich (1969); computation of k_{SOD} according to Sawada and Yamazaki (1973).

TABLE II

Amino Acid Composition of Human Erythrocyte Superoxide Dismutase

	Kimmel[a]	Stansell[b]	Hartz[c]	Bannister[d]	SOD I[e]	SOD II[e]
	(No. of residues/mole)					
Aspartic acid	41.3	40.2	36.5	38	33.0 ± 2.7	34.2 ± 2.3
Threonine	18.2	16.9	15.8	15	14.8 ± 0.7	14.2 ± 0.1
Serine	19.7	19.8	19.8	19	16.4 ± 0.6	15.0 ± 0.6
Glutamic acid	28.2	28.1	26.3	27	32.1 ± 1.1	30.5 ± 0.6
Proline	11.1	11.3	10.4	11	12 ± 0.3	9.7 ± 1.1
Glycine	45.6	52.2	50.1	60	47.1 ± 0.7	45.3 ± 0.8
Alanine	20.8	21.4	19.8	23	20.7 ± 1.3	20.5 ± 1
Valine	23.0	28.4	27.6	26	27.5 ± 1.5	28.4 ± 1.8
Methionine	0	0	0	0	0	0
Isoleucine	11.0	16.0	16.4	13	17.2 ± 0.4	18.9 ± 1.3
Leucine	17.8	19.7	17.4	18	20.4 ± 0.7	20.8 ± 0.6
Tyrosine	1.9	0.3	0	0	2.6 ± 0.3	2.7 ± 0.3
Phenylalanine	9.9	8.6	7.8	8	8.0 ± 0.2	9.8 ± 0.5
Lysine	20.3	22.9	22.2	20	23.1 ± 1.5	25.3 ± 1
Histidine	14.5	16.5	15.8	14	15.7 ± 0.8	17.4 ± 0.5
Ammonia	24.5	28.0	23.2	-	12.2 ± 1.0	13.2 ± 1.8
Arginine	9.1	8.6	6.8	8	9 ± 0.6	10.3 ± 0.5
Half-cystine	11.3	5.1	6.8	-	7.1 ± 0.3	7.2 ± 0.6
Tryptophan	0.6	0	0	2	2.0 ± 0.0	2.0 ± 0.0

[a]Kimmel *et al.*, (1959)

[b]Stansell and Deutsch (1965b)

[c]Hartz and Deutsch (1969)

[d]Bannister *et al.*, (1972)

[e]Mean ± S.E. of values obtained by duplicate analysis of 24- and 72-hour hydrolysates except for threonine and serine for which 24-hour hydrolysis, and for valine and isoleucine for which 72-hour hydrolysis values are shown. Half-cystine was determined as cysteic acid (Hirs, 1967); 24-hour hydrolysis values are given. Tryptophan was determined by oxidation with N-bromosuccinimide (Spande and Witkop, 1967) in 50% acetic acid. Hexosamine was not detected. A molecular weight of 33500 is assumed.

The apoproteins had about 4% residual activity. The k_{SOD} values for SOD I and SOD II are similar and in good agreement with the value of 1.4×10^9 M^{-1} sec^{-1} at pH 7.5 determined in a previous pulse radiolysis study on human erythrocuprein (Bannister *et al.*, 1973). The k_{SOD} values for the apoproteins were only 5% or so of the values for the holoproteins.

The amino acid composition of SOD I and SOD II is given in Table II. The amino acid determinations are closely similar if not identical and in substantial agreement with previous determinations for human erythrocuprein. The major points of disagreement on the amino acid composition of human erythrocuprein have been the number of half-cystine residues and the presence of tyrosine and tryptophan. The present findings on SOD I and SOD II indicate 7 or 8 half-cystine residues. After correction for 90 - 95% recovery of cysteic acid (Hirs, 1967) the values of Table II are compatible with 8 residues. Hartz (1968) reported this number of residues for his erythrocuprein fractions S-1 and S-5. It is interesting that elementary analysis of human erythrocuprein has indicated 1.05 - 1.13% sulphur (Kimmel *et al.*, 1959; Nyman, 1960) corresponding to 11 - 12 sulphur atoms per mole of protein. These can be accounted for by 7 - 8 half-cystine residues and the four atoms of labile sulphur reported by Calabrese *et al.* (1975). Presumably SOD II has the same total sulphur as SOD I but the 320 nm chromophore has disappeared by a mechanism such as the trisulphide formation suggested by Calabrese *et al.* (1975). This awaits confirmation.

In contradiction to Hartz and Deutsch (1972) we have not been able to demonstrate two free sulphydryl groups in the protein with the Ellman reagent (Ellman, 1959). In 8 M urea sulphydryl groups were not measurable in SOD I or SOD II. Approximately 0.3 and 0.8 of sulphydryl group per mole of protein were detected in apo SOD I and apo SOD II, respectively. This suggests one poorly reactive group. Stansell and Deutsch (1965b) and Hartz and Deutsch (1969) found approximately one group by spectrophotometric titration with p-mercuribenzoate. Hartz and Deutsch (1972) demonstrated two sulphydryl groups by alkylation in 7 M guanidine hydrochloride or 8 M urea and by means of the Ellman reagent in 8 M urea. Six additional half-cystine residues were demonstrated after reduction and alkyaltion in 7 M guanidine hydrochloride. Eight residues of cysteic acid were found after performic acid oxidation. Dissociation of the protein into subunits of approximately 16000 molecular weight was facilitated by alkylation of sulphydryl groups, by removal of the copper and zinc (with 70% formic acid), by reduction and alkylation, or by succinylation. The subunits were demonstrated by detergent-gel electrophoresis and by agarose gel filtration and sedimentation equilibrium in 7M guanidine hydrochloride. After incubation of SOD I or SOD II for 1 h at 45°C in the presence of 4 M urea and 1% sodium dodecylsulphate, with or without 1% (v/v) 2-mercaptoethanol, detergent-gel electrophoresis according to Fairbanks *et al.* (1971) shows approximately 25% conversion into subunits of about 16000 apparent molecular weight. The rest of the protein is seen in forms with approximately 55000 and 170000 apparent molecular weight. These forms may reflect aggregation but anomalous binding of sodium dodecylsulphate giving anomalously low electrophoretic mobilities has to be excluded.

By analogy with bovine erythrocyte superoxide dismutase, the native cuprozinc superoxide dismutases consist of two identical subunits of approximately 16000 molecular weight, which are noncovalently linked. It may be mentioned here that Hartz and Deutsch (1972) have cast doubts on the identity of the subunits in human erythrocyte superoxide dismutase. By peptide mapping they found from 30 to 34 peptides after tryptic digestion of the reduced and alkylated protein. Eight arginine peptides were identified. The total number of peptides tallied with the total number of lysine and arginine residues, and the number of arginine peptides tallied with the number of arginine residues. This suggested nonidentical subunits. After tryptic digestion of superoxide dismutase which had been alkylated with ^{14}C-iodoacetamide, Hartz and Deutsch (1972) found an approximately equal distribution of the radioactivity and S-carboxymethylcysteine between the acid soluble and insoluble fractions of the digest. Two nonidentical radioactive peptides were isolated from these fractions by a combination of ion exchange chromatography and paper electrophoresis in conjunction with chymotryptic digestion. The soluble tryptic peptide was a hexapeptide with a COOH-terminal lysine. From the insoluble fraction a chymotryptic octa- or nonapeptide with a COOH-terminal arginine was recovered. The specific radioactivity of each peptide was approximately half that of the starting labelled protein. This work demonstrates the alkylation of two sulphydryl groups in nonidentical regions of human erythrocyte superoxide dismutase. It does not necessarily imply that the cysteine peptides came from nonidentical subunits. It is also possible that an originally free sulphydryl group and a different unmasked group on identical subunits were alkylated. Guanidine hydrochloride in the reaction mixture might unmask a sulphydryl group by destruction of labile sulphur (Calabrese *et al.*, 1975).

Nonidentity of the subunits is problematic because of the observation of a genetic variant of human cytosol superoxide dismutase and a hybrid enzyme. Cytosol superoxide dismutase (SOD-1) is polymorphic in northern Sweden and northern Finland with genetically controlled electrophoretic variation in starch or polyacrylamide gel electrophoresis. The common phenotype SOD-1 1 shows one major zone and one or two minor anodically faster moving zones. The rare phenotype SOD-1 2 shows a similar isozyme pattern but with an overall slower electrophoretic mobility. The phenotype SOD-1 2-1 shows the SOD-1 1 and SOD-1 2 zones and in addition a hybrid enzyme with intermediate electrophoretic mobility. The observation of this hybrid enzyme implies that the superoxide dismutase is at least a dimer formed through free recombination between two equal polypeptide subunits (Beckman *et al.*, 1973).

To return to the amino acid composition of the protein, Table II indicates 2 or 3 tyrosine residues for SOD I and SOD II. The data are not incompatible with one residue per subunit of about 16000 molecular weight. Of previous analyses only that of Kimmel *et al.* (1959) has shown the presence of tyrosine (2 residues) in human erythrocuprein. Tyrosine was not found by Bannister *et al.* (1972). In the present work we have added a crystal of phenol to protect tyrosine from oxidation during acid hydrolysis. We have also determined tyrosine by spectrophotometric analysis of the apoproteins (Table III). Two-wavelength solutions for tryptophan and tyrosine (Edelhoch, 1967; Donovan, 1969) tended to overestimate the former and underestimate the latter with respect to expected values, and this type of analysis is probably vitiated by residual hyperchromic effect in the apoproteins (see Figs.4 and 5 and below). The tyrosine values based on ionisation are more reliable.

The present finding of 2 tryptophan residues by oxidation with N-bromosuccinimide (Table III) is in agreement with previous work (Bannister *et al.*, 1972). Two residues were determined in SOD I, SOD II and the apoproteins. Both SOD I and SOD II showed ultraviolet fluorescence suggestive of tryptophan. Uncorrected fluorescence spectra were measured by means of an Aminco-Bowman spectrophotofluorimeter. A 3 mm x 3 mm cell was used to decrease inner filter artifacts (Chen *et al.*, 1969). The spectra showed maxima at 235, 282 and 288 nm in excitation and a maximum at 348 nm with a half-width of about 60 nm in emission. The quantum yield of fluorescence measured with respect to the fluorescence of L-tryptophan (Teale and Weber, 1957) was determined to be 0.04 for SOD I, as in previous work on human erythrocuprein (Bannister *et al.*, 1972), and 0.08 for SOD II at pH 6.90. The tryptophan fluorescence is therefore half-quenched in SOD I. At pH 2.25 the quantum yield of fluorescence increased to 0.06 in SOD I and 0.11 in SOD II. It is possible that the quenching of tryptophan fluorescence in SOD I is linked to the presence of the 320 nm chromophore. Thus, on the one hand in human erythrocuprein corresponding to SOD I there is an almost three-fold increase in quantum yield after treatment with cyanide to prepare apoprotein. On the other hand titration and therefore reconstitution of the apoprotein with zinc or copper produces little quenching of the fluorescence (Bannister *et al.*, 1972). Quenching of tryptophan fluorescence in SOD I is therefore related to the 320 nm chromophore, which is destroyed by cyanide, rather than to the metals.

TABLE III

Spectrophotometric Determination of Tryptophan and Tyrosine in Human Erythrocyte Superoxide Dismutase

Conditions	Tryptophan (No. of residues/mole)	Tyrosine (No. of residues/mole)
6.0 M guanidine HCl pH 6.5[a]		
Apo SOD I	3.6	1.7
Apo SOD II	2.6	2.3
Acid pH[b]		
Apo SOD I	3.5	0.2
Apo SOD II	2.7	1.1
6.0 M guanidine HCl pH 12.5[c]		
Apo SOD I	-	1.7
Apo SOD II	-	1.8

[a]According to Edelhoch (1967).

[b]Spectrophotometric analysis as suggested by Donovan (1969) using molar extinction coefficients for tryptophan and tyrosine (280 and 288 nm) given by Mihalyi (1968). The pH was 1.9.

[c]According to Edelhoch (1967) utilising increase in absorbance, with respect to pH 7.0, at 295 and 300 nm due to tyrosine ionisation. Average values for 295 and 300 nm data are given.

Ultraviolet absorption spectra of SOD I, SOD II and the apoproteins are shown in Fig.4 (neutral pH) and Fig.5 (ph 1.9).

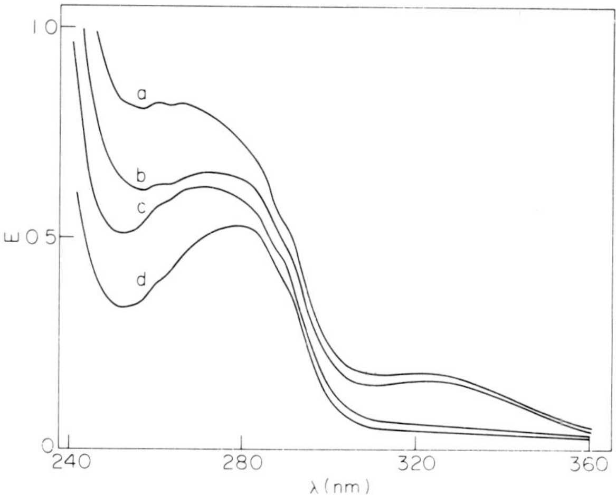

Fig.4. *Ultraviolet absorption spectra of human erythrocyte superoxide dismutase. The spectra were recorded at the following protein concentrations in water: SOD I, 1.93 mg/ml; SOD II, 2 mg/ml; apo SOD I, 1.96 mg/ml; apo SOD II, 1.78 mg/ml. Extinction values are given for 1 mg/ml solutions. a, SOD I; c, SOD II; b, apo SOD I; d, apo SOD II.*

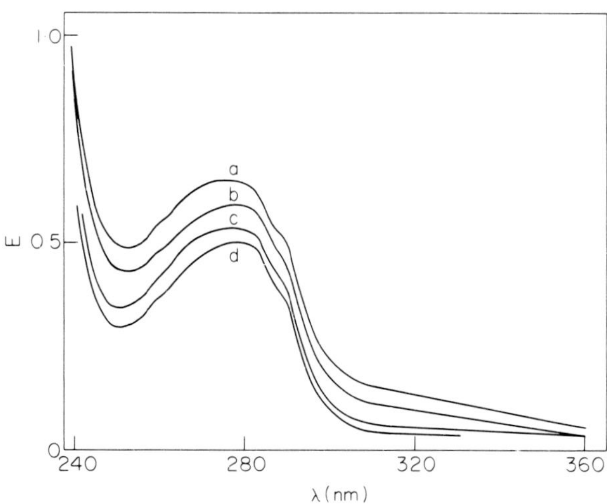

Fig.5. *Ultraviolet absorption spectra of human erythrocyte superoxide dismutase. The spectra were recorded at the following protein concentrations in acidified water (pH 1.9): SOD I, 1.91 mg/ml; SOD II, 1.98 mg/ml; apo SOD I, 1.94 mg/ml; apo SOD II, 1.76 mg/ml. Extinction values as in Fig.4. a, SOD I; c, SOD II; b, apo SOD I; d, apo SOD II.*

The salient features of the spectra are given in Table IV. The spectrum of SOD I is similar to the spectrum of human erythrocuprein reported by Bannister *et al.* (1972) and the spectrum of erythrocuprein fraction S-1 of Hartz (1968). The spectrum of SOD II resembles the spectrum of erythrocuprein fraction S-5 of Hartz (1968). A hyperchromic effect is evident in the spectrum of SOD I with respect to the spectrum of SOD II and in the holoprotein spectra with respect to the corresponding apoprotein spectra. The apoprotein spectra show some normalization of the aromatic absorption, particularly the spectrum of apo SOD II. Considerable normalization of all spectra is evident at acid pH but hyperchromic effects are still present. The apoprotein of SOD II shows the lowest absorbance but the lowest point may not have been reached because of residual metal content. The 320 nm band is not discernible in the spectra of SOD I and apo SOD I at acid pH, but there is considerable residual absorbance in the 320 nm region (Fig.5). A similar pattern of spectra was observed in 6.0 M guanidine hydrochloride (pH 6.5) as at acid pH, but the level of absorbance remained high.

TABLE IV

Salient Features of Ultraviolet Absorption Spectra of Human Erythrocyte Superoxide Dismutase

	Maxima λ_{max}	$E_{1\,cm}^{0.1\%}$	Inflexious λ (nm)	Minima λ (nm)
Neutral pH				
SOD I	266 320-323	0.819 0.184	280 290	-
SOD II	270-272	0.626	280 290	252
Apo SOD I	270-275 322-324	0.656 0.165	280 290	257
Apo SOD II	279-280	0.531	290	252
Acid pH				
SOD I	274-277	0.647	290	252-253
SOD II	276	0.532	290	251
Apo SOD I	278-279	0.591	290	252-254
Apo SOD II	279	0.501	290	251

Various ultraviolet difference spectra are shown in Figs.6 and 7. The spectra were recorded directly by means of a Pye-Unicam SP 700A spectrophotometer. The solution with lower absorbance was placed in the reference beam. Hence the spectra have a positive sense. The nomenclature refers to the reference solution. The apoprotein difference spectra with respect to the corresponding holoproteins (Fig.6a,b) do not throw much light on the nature of the hyperchromism in the holoproteins. The spectra show a band in the 240 nm region and a broad hump in the 260 nm region. The apo SOD II difference spectrum (Fig.6a) shows a shoulder at 290 nm which is much attenuated in the apo SOD I spectrum (Fig.6b). The SOD II difference spectra with respect to SOD I (Fig.6d) and that of apo SOD II with respect to apo SOD I (Fig.6c) also show a band in the region of 240 nm with a long tail to the near ultraviolet. The spectra show the expected band at 320 nm. The molar absorbance at 321 to 324 nm for the SOD II and apo SOD II difference spectra (Fig.6d,c) is 4490 and 4200 M^{-1} cm^{-1}, respectively (with correction for the experimental difference in protein concentration of sample and reference solutions). These values compare with the value of approximately 4600 M^{-1} cm^{-1} estimated for the 320 nm chromophore by Calabrese *et al.* (1975).

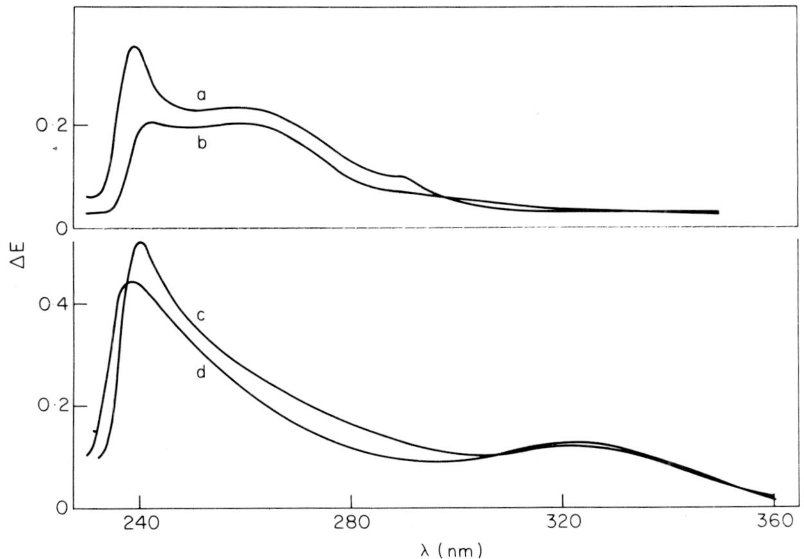

Fig.6. *Human erythrocyte superoxide dismutase difference spectra. Reference solutions apo SOD II (a), apo SOD I (b), apo SOD II (c) and SOD II (d) with sample solutions SOD II, SOD I, apo SOD I and SOD I, respectively. Protein concentrations were those of Fig.4, in water. Extinction values are given for 1 mg/ml solutions (with primary reference to sample solution and without correction for the small disparity in working concentrations of sample and reference solutions).*

The acid difference spectra (Fig.7) show bands at about 240, 285, 293 and 325 nm. Fine structure bands attributable to phenylalanine effects are also evident between 248 and 268 nm. The band at 285 nm may be due to exposure of tyrosine and tryptophan residues. However, the band at 293 nm is probably a tryptophan band or largely so. Where this band is least complicated by the short wavelength tail of 320 nm absorption, i.e. in the apo SOD II difference spectrum (Fig.7d), the molar absorbance (of 1600 M^{-1} cm^{-1}) is that expected for the transfer of one indole chromophore from a protein interior to the solvent (Donovan, 1969). The molar absorbance at 325 nm for the SOD I and apo SOD I acid difference spectra (Fig.7a,c) is 2750 M^{-1} cm^{-1}. Presumably this is an underestimate for the 320 nm chromophore due to residual absorbance at the acid pH.

Ultraviolet CD spectra of SOD I, SOD II and the apoproteins are presented in Figs.8 to 11. Circular dichroism was measured by means of a Jasco J-40C spectropolarimeter. Spectra were recorded in duplicate, averaged and smoothed by 5-point least squares approximation at 1 nm intervals. The ultraviolet CD of SOD I (Fig.8a) is similar to that reported previously for human erythrocuprein (Bannister *et al.*, 1972). The CD has positive contributions at about 230, 255, 285 and 325 nm. Closely similar contributions are seen in the CD spectrum of apo SOD I (Fig.9a). The CD of SOD II (Fig.8b) and the corresponding apoprotein (Fig.9b) is more negative. The 230 nm band is discernible on a negative envelope. The 325 nm band is absent. Sulphite and dithiothreitol destroyed the 325 nm contribution and caused the CD of SOD I to approach that of SOD II (Fig.10). The 230 nm band remained as a peak on a negative envelope. Acidification made the CD more negative in all preparations (Fig.11) and the 230 nm contribution was either considerably attenuated or abolished.

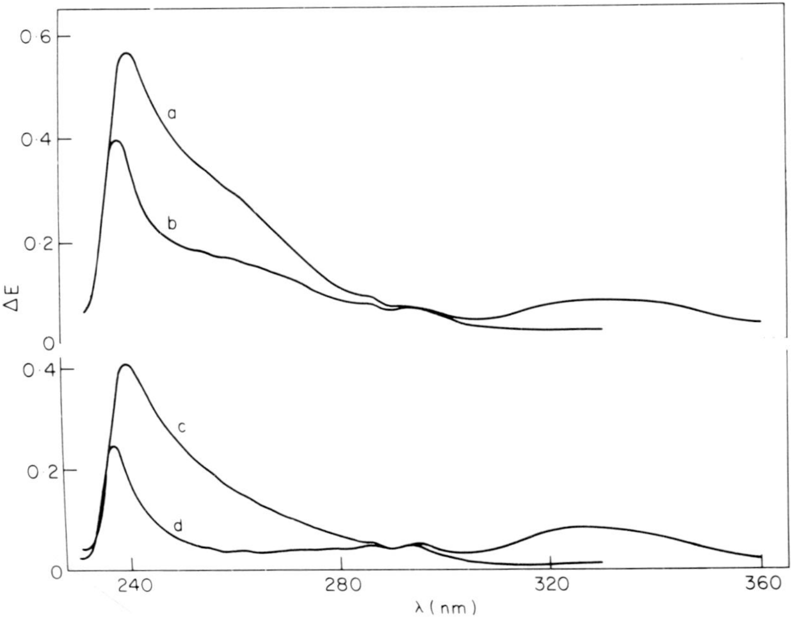

Fig.7. *Human erythrocyte superoxide dismutase acid difference spectra. Protein concentrations were those of Fig.4, in water, for sample, and those of Fig.5, in acidified water (pH 1.9), for reference solutions. Extinction values as in Fig.6. a, SOD I; b, SOD II; c, apo SOD I; d, apo SOD II.*

It is not possible to assign the expected cystine or aromatic contributions to the ultraviolet CD on present evidence. Clearly the ultraviolet envelopes have multiple overlapping effects. The CD is not dominated by metal effects but by the 230 and 325 nm bands, as can be seen by comparing the spectra in Figs.8 to 11. The 325 nm contribution corresponds to the 320 nm band in absorption which has been attributed to labile sulphur (Calabrese *et al.*, 1975). Without implying a chromophoric relationship, it may be noted that there is some correlation between the 230 and 325 nm contributions to the CD. The CD at 230 nm is negative, with a peak or shoulder in this region, with lack or destruction of the 325 nm band. This situation obtains in bovine erythrocyte superoxide dismutase (Wood *et al.*, 1971) which lacks the 320 nm chromophore. It has been thought that the 320 nm band was zinc-

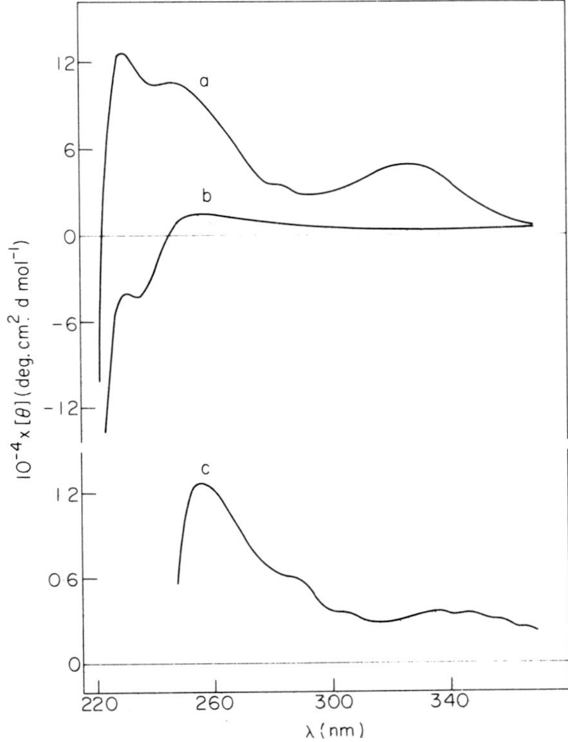

Fig.8. *Ultraviolet circular dichroism spectra of human erythrocyte superoxide dismutase. a, SOD I; b, SOD II; c, to illustrate observed profile of SOD II spectrum. Protein was dissolved in water. Ellipticity refers to molecular ellipticity on a metal subunit basis or 16300 molecular weight.*

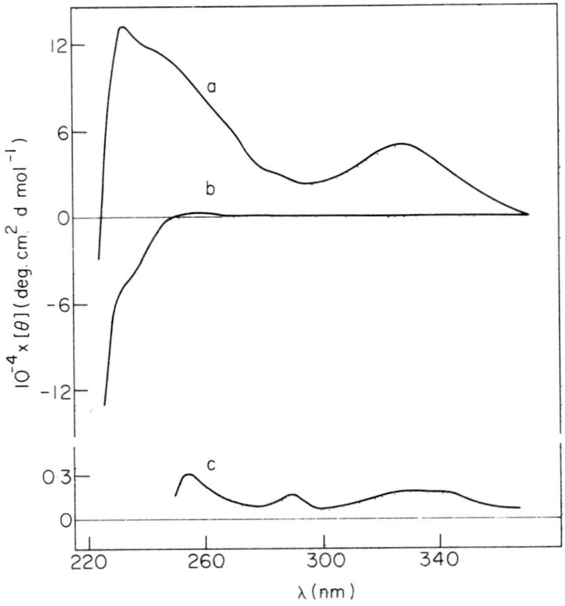

Fig.9. *Ultraviolet circular dichroism spectra of human erythrocyte apo superoxide dismutase. a, apo SOD I; b, apo SOD II; c, to illustrate observed profile of apo SOD II spectrum. Protein was dissolved in water. Ellipticity as in Fig.8.*

dependent because it almost disappeared in apoerythrocuprein prepared with cyanide which contained residual copper but no zinc (Bannister *et al.*, 1972). The spectrum observed was rather similar to that of apo SOD II (Fig.9b) which had about 30% residual zinc. The destructive effect of cyanide on the 320 nm chromophore was not known in this work.

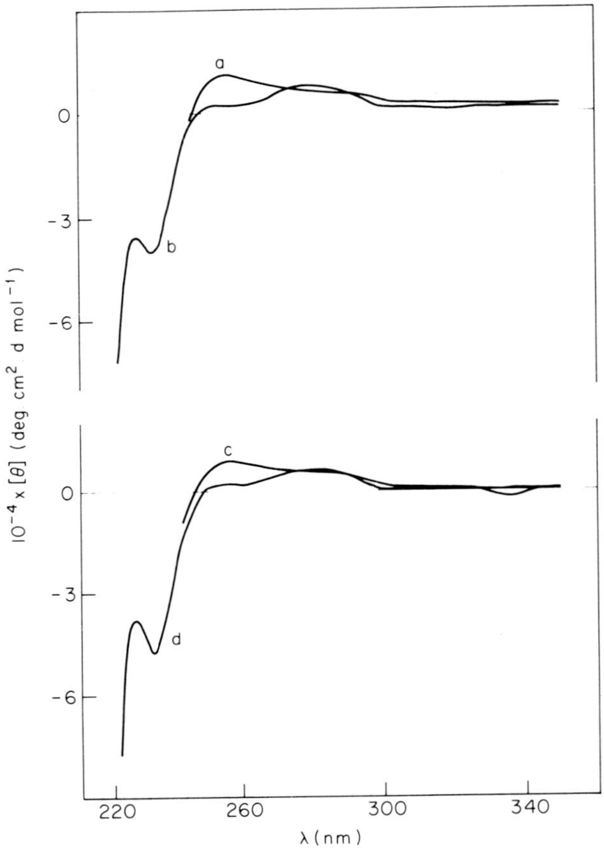

Fig.10. *Ultraviolet circular dichroism spectra of human erythrocyte superoxide dismutase to illustrate the effect of sulphite and dithiothreitol. Protein was dissolved in 0.1 M pyrophosphate buffer, pH 8.3, containing 20 mM sodium sulphite (a,c) or 0.2 mM dithiothreitol (b,d). a,b, SOD I; c,d, SOD II. Ellipticity as in Fig.7.*

Disappearance of the 230 nm contribution at acid pH is associated with loss of β-sheet structure. Far ultraviolet CD spectra are presented in Fig.12. The spectra were fitted with the reference spectra of Chen *et al.* (1975) for helix, β-sheet and unordered structure in order to estimate the percentage of the secondary structures. The rationale of this type of procedure has been given previously (Bannister and Bannister, 1974). The reference spectra gave an excellent fit of the spectrum of bovine erythrocyte superoxide dismutase (Table V) which encourages interpretation of the spectra of the human protein. Table V shows that the human protein, whether SOD I or SOD II in holo or apo form, contains less than 10% helix and about 50% β-sheet structure. The structural composition is therefore closely similar to that

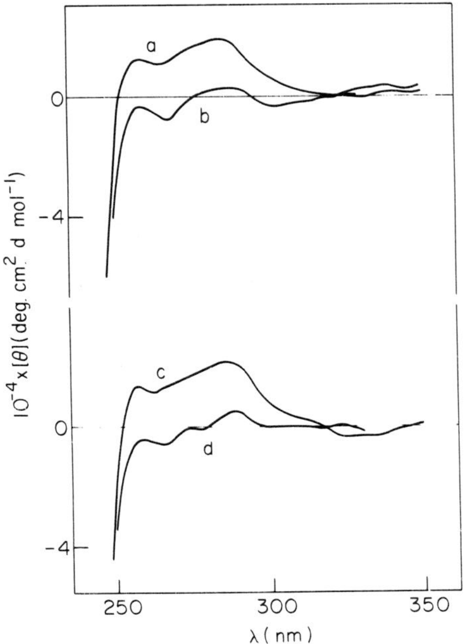

Fig.11. *Ultraviolet circular dichroism spectra of human erythrocyte superoxide dismutase to illustrate the effect of acid pH. Protein was dissolved in acidified water (pH 1.9). a, SOD I; b, SOD II; c, apo SOD I; d, apo SOD II. Ellipticity as in Fig.7.*

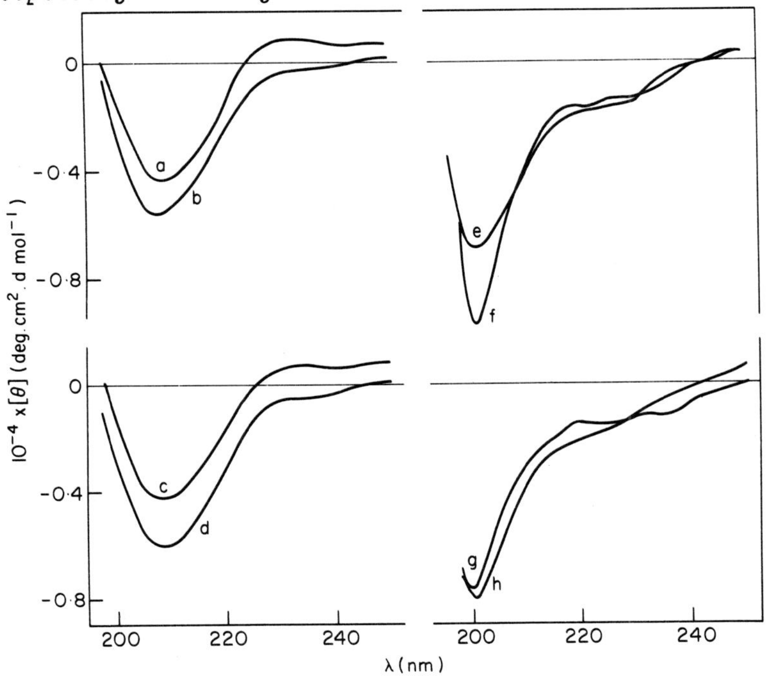

Fig.12. *Far ultraviolet circular dichroism spectra of human erythrocyte superoxide dismutase. Protein was dissolved in water at neutral (a-d) or acid pH (pH 1.9; e-h). a,e, SOD I; b,f, SOD II; c,g, apo SOD I; d,h, apo SOD II. Ellipticity refers to mean residue ellipticity per mean residue weight of 110.*

of bovine erythrocyte superoxide dismutase (Richardson *et al.*, 1975). At acid pH β-sheet structure decreases to about 20% and there is a corresponding increase in unordered structure. Possibly the percentage of helix is little altered.

TABLE V

Structural Composition of Human Erythrocyte Superoxide Dismutase as Determined from Far Ultraviolet Circular Dichroism Spectra[a]

	Helix (%)	β-Sheet (%)	Unordered structure (%)	RMS[b]
Neutral pH				
SOD I	3.6 ± 1.8	58.1 ± 7.3	38.3 ± 2.7	660
SOD II	8.6 ± 1.4	46.6 ± 5.6	44.7 ± 2.1	610
Apo SOD I	4.6 ± 1.9	54.6 ± 7.6	40.7 ± 2.9	660
Apo SOD II	8.6 ± 1.1	47.7 ± 4.5	43.6 ± 1.7	530
Erythrocuprein[c]	4.5 ± 2.2	54.8 ± 8.9	40.7 ± 3.4	650
Bovine SOD[d]	10.1 ± 0.7	44.5 ± 3.8	45.3 ± 1.4	440
Acid pH				
SOD I	11.1 ± 1.2	20.9 ± 4.7	67.9 ± 1.8	460
SOD II	9.0 ± 1.0	16.4 ± 4.1	74.5 ± 1.6	460
Apo SOD I	10.4 ± 1.1	19.4 ± 4.3	70.2 ± 1.6	440
Apo SOD II	9.6 ± 1.2	18.0 ± 4.8	72.4 ± 1.8	420

[a]Unconstrained least squares fitting of reference spectra of Chen *et al.* (1975) at 1 nm intervals between 200 and 225 nm. Values include S.E.

[b]Root mean square of residuals for fitted spectrum in $deg.cm^2.dmol^{-1}$ of mean residue ellipticity.

[c]Fit of human erythrocuprein spectrum of Bannister *et al.* (1972).

[d]Fit of bovine erythrocyte superoxide dismutase spectrum of Wood *et al.* (1971) for purposes of comparison. Bovine erythrocyte SOD has 3 - 8% helix and about 45% β-sheet structure by X-ray diffraction analysis (Richardson *et al.*, 1975).

ACKNOWLEDGEMENTS

W.H.B. thanks the Wellcome Trust for financial support of this work.

REFERENCES

1. Albergoni, V. and Cassini, A. (1974). *Comp. Biochem. Physiol.* *47B*, 767-777.

2. Asada, K., Takahashi, M. and Nagate, M. (1974). *Agric. and Biol. Chem.* *38*, 471-473.
3. Bannister, W.H. and Bannister, J.V. (1974). *Int. J. Biochem.* *5*, 673-677.
4. Bannister, J.V., Bannister, W.H., Bray, R.C., Fielden, E.M., Roberts, P.B. and Rotilio, G. (1973). *FEBS Lett.* *32*, 303-306.
5. Bannister, J., Bannister, W. and Wood, E. (1971). *Eur. J. Biochem.* *18*, 178-186.
6. Bannister, W.H., Dalgleish, D.G., Bannister, J.V. and Wood, E.J. (1972). *Int. J. Biochem.* *3*, 560-568.
7. Bannister, W.H., Salisbury, C.M. and Wood, E.J. (1968). *Biochem. Biophys. Acta* *168*, 392-394.
8. Bannister, W.H. and Wood, E.J. (1970). *Life Sci. II* *9*, 229-233.
9. Beauchamp, C. and Fridovich, I. (1971). *Analyt. Biochem.* *44*, 276-287.
10. Beckman, G., Lundgren, E. and Tärnvik, A. (1973). *Human Hered.* *23*, 338-345.
11. Bencze, W.L. and Schmid, K. (1957). *Analyt. Chem.* *29*, 1193-1196.
12. Calabrese, L., Federici, G., Bannister, W.H., Bannister, J.V., Rotilio, G. and Finazzi-Agro, A. (1975). *Eur. J. Biochem.* *56*, 305-309.
13. Carrico, R.J. and Deutsch, H.F. (1969). *J. Biol. Chem.* *244*, 6087-6093.
14. Carrico, R.J. and Deutsch, H.F. (1970). *J. Biol. Chem.* *245*, 723-727.
15. Chen, R.F., Edelhoch, H. and Steiner, R.F. (1969). In "Physical Principles and Techniques of Protein Chemistry" (S.J. Leach, ed.) part A, pp 171-244. Academic Press, New York and London.
16. Chen, Y.H., Yang, J.T. and Cau, K.H. (1974). *Biochemistry* *13*, 3350-3359.
17. Donovan, J.W. (1969). In "Physical Principles and Techniques of Protein Chemistry" (S.J. Leach, ed.) part A, pp 101-170. Academic Press, New York and London.
18. Edelhoch, H. (1967). *Biochemistry* *6*, 1948-1954.
19. Ellman, G.L. (1959). *Archs Biochem. Biophys.* *82*, 70-77.
20. Fairbanks, G., Steck, T.L. and Wallach, D.F.H. (1971). *Biochemistry* *10*, 2606-2616.
21. Fee, J.A. (1973). *J. Biol. Chem.* *248*, 4229-4234.
22. Finazzi-Agrò, A., Albergoni, V. and Cassini, A. (1974). *FEBS Lett.* *39*, 164-166.
23. Hartz, J.W. (1968). Ph.D. thesis, University of Wisconsin.
24. Hartz, J.W. and Deutsch, H.F. (1969). *J. Biol. Chem.* *244*, 4565-4572.
25. Hartz, J.W. and Deutsch, H.F. (1972). *J. Biol. Chem.* *247*, 7043-7050.
26. Hirs, C.H.W. (1967). *Meth. Enzym.* *11*, 59-62.
27. Kimmel, J.R., Markowitz, H. and Brown, D.M. (1959). *J. Biol. Chem.* *234*, 46-50.
28. Klug, D., Rabani, J. and Fridovich, I. (1972). *J. Biol. Chem.* *247*, 4839-4842.
29. McCord, J.M. and Fridovich, I. (1969). *J. Biol. Chem.* *244*, 6049-6055.
30. Malmström, B.G. and Vänngård, T. (1960). *J. Mol. Biol.* *2*, 118-124.
31. Mann, T. and Keilin, D. (1939). *Proc. R. Soc. B* *126*, 303-315.
32. Markowitz, H., Cartwright, G.E. and Wintrobe, M.M. (1959). *J. Biol. Chem.* *234*, 40-45.
33. Mihalyi, E. (1968). *J. Chem. Engng. Data* *13*, 179-182.
34. Nyman, P.O. (1960). *Biochim. Biophys. Acta* *45*, 387-389.
35. Porter, H. and Ainsworth, S. (1959). *J. Neurochem.* *5*, 91-98.
36. Porter, H., Sweeney, M. and Porter, E.M. (1964). *Archs Biochem. Biophys.* *105*, 319-325.

37. Richardson, J.S., Thomas, K.A., Rubin, B.H. and Richardson, D.C. (1975). *Proc. Nat. Acad. Sci. U.S.A. 72,* 1349-1353.
38. Roberts, P.B., Fielden, E.M., Rotilio, G., Calabrese, L., Bannister, J.V. and Bannister, W.H. (1974). *Radiat. Res. 60,* 441-452.
39. Rotilio, G., Bray, R.C. and Fielden, E.M. (1972). *Biochim. Biophys. Acta 268,* 605-609.
40. Rotilio, G., Finazzi Agrò, A., Calabrese, L., Bossa, F., Guerrieri, P. and Mondovì, B. (1971). *Biochemistry 10,* 616-621.
41. Sawada, Y. and Yamazaki, I. (1973). *Biochim. Biophys. Acta 327,* 257-265.
42. Shields, G.S., Markowitz, H., Klassen, W.H., Cartwright, G.E. and Wintrobe, M.M. (1961). *J. Clin. Invest. 40,* 2007-2015.
43. Spande, T.F. and Witkop, B. (1967). *Meth. Enzym. 11,* 498-506.
44. Stansell, M.J. (1966). Ph.D. thesis, University of Wisconsin.
45. Stansell, M.J. and Deutsch, H.F. (1965a). *J. Biol. Chem. 240,* 4299-4305.
46. Stansell, M.J. and Deutsch, H.F. (1965b). *J. Biol. Chem. 240,* 4306-4311.
47. Steinman, H.M., Naik, V.R., Abernethy, J.L. and Hill, R.L. (1974). *J. Biol. Chem. 249,* 7326-7338.
48. Stokes, A.M., Hill, H.A.O., Bannister, W.H. and Bannister, J.V. (1973). *FEBS Lett. 32,* 119-123.
49. Teale, F.W.J. and Weber, G. (1957). *Biochem. J. 65,* 476-482.
50. Tsuchihashi, M. (1923). *Biochem. Z. 140,* 63-112.
51. Weisiger, R.A. and Fridovich, I. (1973). *J. Biol. Chem. 248,* 3582-3592.
52. Wood, E., Dalgleish, D. and Bannister, W. (1971). *Eur. J. Biochem. 18,* 187-193.

A MANGANESE-CONTAINING SUPEROXIDE DISMUTASE FROM HUMAN LIVER

J.M. McCORD, J.A. BOYLE*, E.D. DAY, Jr.,
L.J. RIZZOLO* and M.L. SALIN†

Departments of Medicine and Biochemistry
Duke University Medical Center
Durham, North Carolina 27710, U.S.A.

Superoxide dismutases have been isolated from numerous sources. Three distinct types or families of superoxide dismutases have been described, each containing a different metallic prosthetic group. The best characterized of these is the copper-containing enzyme present in the cytosol of eukaryotic cells. (Members of this family also contain zinc, which appears to play a structural role rather than a catalytic role.) Copper-containing superoxide dismutases, blue-green in color, have been isolated from mammalian tissues (McCord and Fridovich, 1969), plants (Sawada *et al.*, 1972), chicken liver (Weisiger and Fridovich, 1973a), yeast (Goscin and Fridovich, 1972) and fungi (Misra and Fridovich, 1972). All of these cytoplasmic enzymes have molecular weights near 32,000, are composed of two identical subunits, and contain one atom of copper and one atom of zinc per subunit.

When superoxide dismutase (SOD) was first isolated from a prokaryotic source, *Escherichia coli* B, it was found to be an enzyme of a different color - pink. Although similar in specific activity to the bovine cuprozinc enzyme, this bacterial enzyme contained no copper and no zinc. It had a molecular weight of 40,000, was composed of two identical subunits, and contained slightly more than one atom of manganese per molecule (Keele *et al.*, 1970). The enzyme was subsequently isolated from another prokaryotic source, *Streptococcus mutans*, with similar results (Vance *et al.*, 1972). Shortly thereafter, Weisiger and Fridovich noted that chicken liver mitochondria contained an SOD electrophoretically distinct from the cytoplasmic enzyme. Upon purification from isolated chicken liver mitochondria, this second type of avian enzyme was found to be a manganoenzyme, very similar in many respects to the manganoenzymes isolated from bacteria. Its primary difference was in molecular weight; it was a tetramer of molecular weight 80,000 composed of four identical subunits (Weisiger and Fridovich, 1973a).

*Predoctoral trainees of the United States Public Health Service
†Postdoctoral Fellow of the United States Public Health Service

The first member of the remaining family of superoxide dismutases was also isolated from *E. coli* after polyacrylamide gel electropherograms stained for dismutase activity revealed that a second SOD existed in that organism. Yost and Fridovich (1973) reported that the new enzyme was yellow in color and contained one atom of iron per dimeric molecule. Several other members of the iron-containing family have since been isolated from other bacterial sources.

We have now isolated a manganese-containing SOD from a mammalian source, human liver, and find it to be a protein of molecular weight 85,300, composed of four identical subunits and containing four atoms of manganese per molecule. Contrary to earlier indications, the manganese SOD is not exclusively a mitochondrial enzyme. In several species the manganese enzyme was found to be present in the cytoplasm at concentrations comparable to, or exceeding that of the cuprozinc SOD.

MATERIALS AND METHODS

SOD was assayed as described by McCord and Fridovich (1969). In crude homogenates or in the early stages of purification, manganese SOD could be assayed in the presence of cupro-zinc SOD, the latter being completely inhibited by the addition of 2 mM cyanide to the assay mixture. After polyacrylamide disc gel electrophoresis by the method of Davis (1964), SOD activities were visualized by the activity stain devised by Beauchamp and Fridovich (1971) as modified by Salin and McCord (1974). SDS gel electrophoresis was performed by the method of Weber and Osborn (1969). Samples were prepared by boiling for 1 minute in 1% sodium dodecyl sulfate containing mercaptoethanol. Lysozyme, chymotrypsin, ovalbumin, catalase, and β-galactosidase were used as molecular weight standard to estimate the subunit weight of the purified human manganese enzyme. Molecular weight of the native enzyme was determined by sedimentation equilibrium centrifugation in a Beckman Model E ultracentrifuge equipped with ultraviolet scanning optics. Amino acid composition was generously performed by Dr. Howard Steinman. Manganese was quantitatively determined by atomic absorption spectroscopy using a Perkin-Elmer 107 instrument equipped with a graphite furnace. Protein was determined by the methods of Lowry *et al.*, (1951) and Murphy and Kies (1960).

Human liver was obtained from autopsy at Duke University Medical Center. For comparative purposes, liver was obtained from rat, guinea pig, chicken, and the baboon *Papio ursinus* by necropsy immediately after the death of the animal. Bovine liver was obtained frozen from Swift and Company.

Subcellular Fractionation

Approximately 1 g of liver was washed in deionized water, blotted dry, cut into small pieces and placed in a Potter-Elvehjem tissue homogenizer containing 10 ml of cold 0.25 M sucrose in 20 mM Tris-Cl, pH 7.45. After five strokes, the homogenate was filtered through six layers of cheese cloth and centrifuged at 100 x g for 5 minutes. The pellet, consisting of unbroken cells and connective tissue, was discarded. The supernate, labeled crude homogenate, was centrifuged at 650 x g for 5 minutes to obtain the pelleted nuclear fraction. The resultant supernate was further centrifuged at 10,000 x g for 10 minutes, yielding the mitochondrial fraction. A final spin at 144,000 x g for 1 hour in a Spinco Model L centrifuge yielded a microsomal pellet and a supernate referred to as the cytosolic fraction. All pelleted fractions were resuspended in 20 mM Tris-Cl, pH 7.45, containing 0.2% Triton X-100, and were sonicated to solublize membrane bound enzyme. Cytochrome oxidase (Warton and Tzagoloff, 1955) and glutamic acid dehydrogenase (Strecker, 1955) were assayed as mitochondrial marker enzymes.

RESULTS

Purification of the Enzyme

Human liver (516 g) was homogenized in a blender with 3.5 volumes of cold 50 mM potassium phosphate buffer, pH 7.8, containing 0.1 mM EDTA. The homogenate was centrifuged for 30 minutes at 14,000 x g. The supernate was heated to 65° for 5 minutes in aliquots of 300 ml, then cooled in an ice bath and centrifuged for 5 minutes at 10,000 x g to remove precipitated protein. Ammonium sulfate was added to the supernate to a concentration of 430 g/l, or 65% saturation. This mixture was stirred at room temperature for 90 minutes, then centrifuged for 10 minutes at 12,000 x g. The precipitate was discarded. An additional 107 g/l ammonium sulfate was added to the supernate, bringing it to 80% saturation, and the mixture was again stirred for 90 minutes at room temperature, then centrifuged for 10 minutes at 12,000 x g. The precipitate was dissolved in a minimal amount of 5 mM potassium phosphate buffer, pH 7.8, containing 10 μM EDTA, and exhaustively dialyzed against the same buffer. The sample was then run through a 1.5 x 25 cm column of Whatman DE-52 cellulose ion-exchange resin and eluted with the same buffer. The SOD activity does not bind to the column under these conditions. The sample was concentrated by ultrafiltration on an Amicon PM-10 membrane, and dialyzed against 40 mM potassium acetate buffer, pH 5.5, for 16 hours at 4°. The sample was applied to a 2.5 x 25 cm column of Whatman CM-52 cellulose ion-exchange resin equilibrated in the same buffer, then eluted by a linear gradient of potassium acetate, pH 5.5, ranging from 40 mM to 0.2 M. Fractions containing the activity were pooled, concentrated by ultrafiltration, and applied to a 1.5 x 85 cm column of Bio-Gel A-0.5 m, equilibrated and eluted with 50 mM potassium phosphate buffer, pH 7.8, containing 0.1 mM EDTA. The yield was 52 mg.

The results of a purification procedure are summarized in Table I. After the ammonium sulfate cut, the SOD activity was completely insensitive to cyanide, indicating that beyond that point the manganese SOD had been completely separated from the cuprozinc SOD. The final gel filtration step did not significantly increase the specific activity of the preparation, but effectively removed a very minor contaminant which made a significant contribution to the visible absorption spectrum of the preparation in the region near 420 nm. Because of its intense absorption in this region of the spectrum, the contaminant was most likely a heme-containing protein.

Purity

The specific activity of the final product was 3600 standard units/mg protein. This is in the range of specific activities previously reported for superoxide dismutases isolated from various other sources. Fig.1 shows the results obtained when the purified human manganese SOD was subjected to polyacrylamide disc gel electrophoresis. Gel 1 shows that the native enzyme produced two closely spaced bands when stained for protein, but Gel 2 shows that both these bands possessed SOD activity. Hence no extraneous protein contaminants were visible by this procedure. Gel 3 shows a single band from the SDS-denatured enzyme, electrophoresed according to the method of Weber and Osborn (1969). Again, no contaminants were visible.

TABLE I

Purification of Manganese Superoxide Dismutase from Human Liver

Purification Stage	Volume	Total Protein	Total Units	Specific Activity	Purification Factor	% Yield
	ml	*mg*	$x\ 10^5$	*units/mg*	*-fold*	
Supernate from						
Homogenate	1800	25,650	11.4	44	1.0	100
Heat step	1750	6,672	11.4	169	3.8	99.7
65% $(NH_4)_2SO_4$	1700	2,657	7.3	271	6.2	63.7
80% $(NH_4)_2SO_4$ ppt.						
redissolved	100	1,225	7.6	611	13.9	66.4
After dialysis	134	1,064	5.4	506	11.5	47.7
After DE52 and						
dialysis	110	606	5.1	839	19.1	45.0
After CM52	84	52	1.9	3600	80.8	16.4

Protein concentration, at all stages except the final one, was estimated by the method of Lowry *et al.* (1952). At the final stage, protein was determined by the method of Murphy and Kies (1960), corrected for the anomalous behavior discussed under "Spectral Properties".

Spectral Properties

The optical absorption spectrum of the human manganese SOD shown in Fig.2 is representative of the family of manganese enzymes. The absorption in the near UV is that of a typical protein, reflecting the enzyme's average content of tryptophan and tyrosine and contrasting dramatically with members of the cuprozinc family in this regard. The molar absorptivity at 283.5 nm is 1.97 x 10^5 M^{-1} cm^{-1}. In the visible range, there is a weak maximum at 480 nm with a shoulder at about 600 nm. The enzyme had a molar absorptivity at 480 nm of 2050 M^{-1} cm^{-1}.

The shoulder at 225 nm, probably due to the manganese chromophore, results in an anomalously low ratio of absorbances at 215 and 225 nm of 1.31. The 215/225 ratio for most other proteins, including cuprozinc SOD, is very close to 2.0. Obviously, if this ratio is used to determine the concentration of unknown solutions of manganese SOD by the method of Murphy and Kies (1960), erroneous results will be obtained unless a correction is introduced. This anomaly, discovered by Weisiger and Fridovich (1973a), led to underestimates of enzyme concentration in the earlier studies reported by Keele *et al.* (1970) and Vance *et al.* (1972). For the manganese dismutases a difference in absorbance (A_{215}-A_{225}, 1 cm) of 1.0 indicates a protein concentration of 221 μg/ml.

Amino Acid Composition

Table II shows the amino acid composition of the human manganese SOD. The composition is quite similar to those reported previously for the *E. coli* and chicken enzymes (Keele *et al.*, 1970; Weisiger and Fridovich, 1973a).

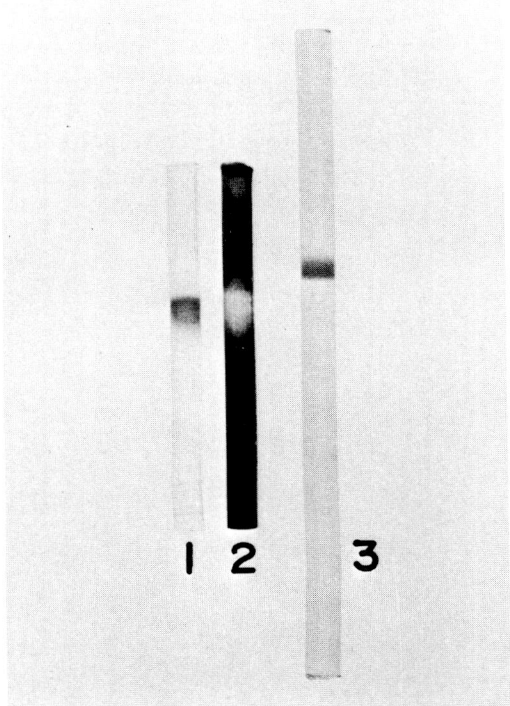

Fig.1. *Polyacrylamide disc gel electrophoresis of human manganese SOD. On Gels 1 and 2, native enzyme was electrophoresed according to Davis (1964). Gel 1 was stained for protein with Amido Black; Gel 2 was stained for SOD activity. The sample applied to Gel 3 was denatured and reduced by boiling in sodium dodecyl sulfate and mercaptoethanol, then electrophoresed according to Weber and Osborn (1969).*

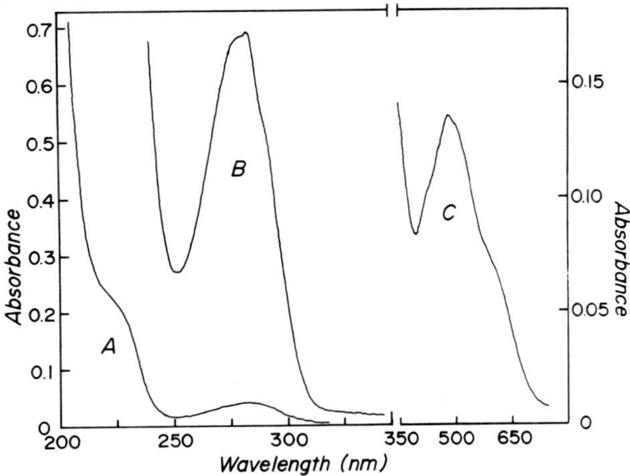

Fig.2. *Absorption spectrum of human liver manganese superoxide dismutase. Enzyme concentrations were as follows: spectrum A, 14.6 μg/ml; spectrum B, 0.293 mg/ml; spectrum C, 5.85 mg/ml. Absorption maxima occur at 283.5 and at 480 nm.*

TABLE II

Amino Acid Composition of Human Liver Manganese Superoxide Dismutase

Amino Acid	Moles/Mole Enzyme	Residues per Subunit (nearest integar)
Lysine	60.0	15
Histidine	34.2	9
Arginine	20.5	5
Aspartic Acid	90.0	23
Threonine	21.4	5
Serine	25.8	6
Glutamic Acid	88.5	22
Proline	40.0	10
Glycine	75.7	19
Alanine	67.2	17
Valine	45.8	11
Methionine	10.0	3
Isoleucine	34.3	9
Leucine	64.3	16
Tyrosine	27.2	7
Phenylalanine	20.1	5
Cysteic Acid	10.0	3
Tryptophan	19.1	5
Total Residues	754.3	190

Molecular Weight

The enzyme was subjected to sedimentation equilibrium analysis at 17,000 RPM. When equilibrium was attained, an ultraviolet scan of the cell was obtained. When the log of the absorbance was plotted versus r^2, a straight line resulted. Partial specific volume was calculated from the amino acid composition according to Cohen and Edsall (1943) and found to be 0.729. Molecular weight was then calculated to be 85,300.

A number of molecular weight standards were prepared in SDS and electrophoresed under conditions identical to Gel 3 in Fig.1. A comparison of relative mobilities indicated the molecular weight of the single band which resulted from the human liver SOD to be 21,500. Hence, the native enzyme appears to be composed of four such subunits.

Manganese Content

Atomic absorbtion analysis, based on spectral parameters, indicated the presence of 3.9 ± 0.5 g atoms of manganese per mole of enzyme, or one atom of manganese per subunit. Ashed samples of human liver were analyzed for manganese content, and found to contain 3.1×10^{-5} moles of manganese per kg liver. This manganese was completely extractable into the soluble fraction of well-homogenized liver. From measurements of cyanide-resistant activity present in crude homogenates of liver, one finds 614 mg manganese SOD per kg liver, or 7.2 μmoles enzyme per kg liver. This amount of enzyme would contain 2.9×10^{-5} moles of manganese, accounting for 94% of the total liver manganese. In contrast, rat liver was found to contain about 3 times the manganese of human liver, but only 5% as manganese SOD. Thus, in the rat, the enzyme accounts for only about 1.5% of the total manganese in the liver.

Relative Abundance of Mn SOD in Different Species

A comparison of the relative amounts of manganese and cuprozinc dismutases in liver homogenates from four species may be seen in Fig.3.

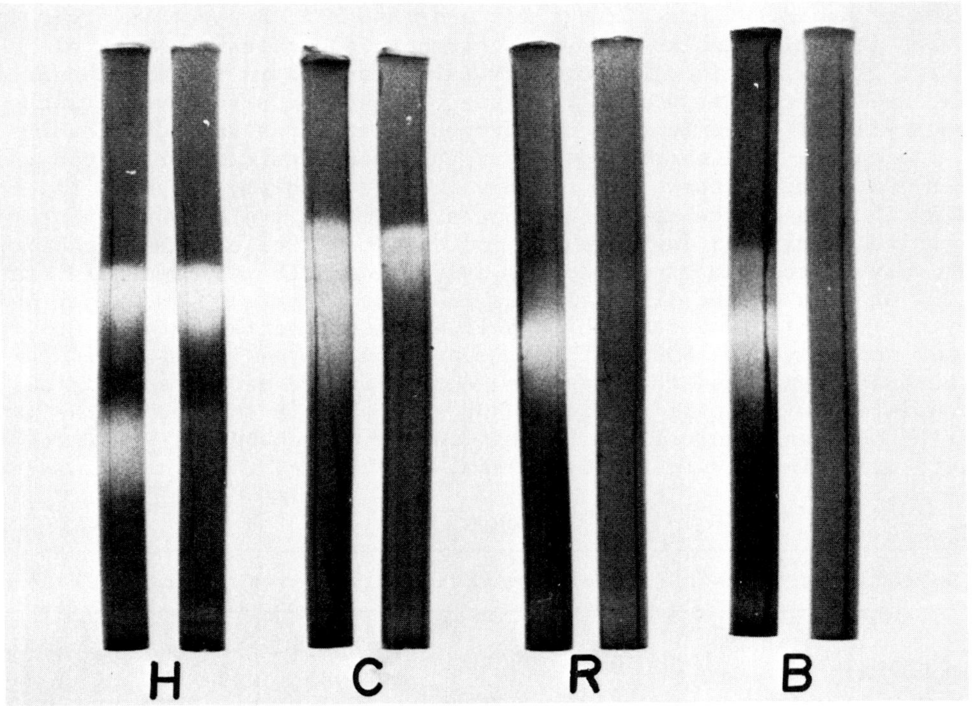

Fig.3. *Polyacrylamide gels stained for superoxide dismutases in liver homogenates of various species. The staining procedure was that of Beauchamp and Fridovich (1971) as modified by Salin and McCord (1974). All gels received equivalent amounts of supernate from homogenized human (H), chicken (C), rat (R), or bovine (B) liver. Paired gels for each species were treated identically except that the gel on the right in each pair was stained in the presence of 2 mM cyanide, eliminating any acromatic zones due to cuprozinc superoxide dismutases.*

All gels received equivalent amounts of liver homogenate and were stained for SOD activity in the presence and absence of cyanide. Cyanide inhibits the cuprozinc SOD, but does not affect the manganese enzymes, thereby permitting an identification of the bands on the gels. The human liver homogenate shows two distinct bands of activity, the slower moving band being the cyanide-insensitive manganese enzyme, and accounting for well over half the total activity. The two types of enzyme are less well-separated in the chicken liver homogenate, but again the cyanide-insensitive manganese SOD accounts for a sizable fraction of the total activity, about 50%. In marked contrast, however, the homogenates of rat and bovine liver show only a cyanide-sensitive cuprozinc SOD. Under optimal conditions, a very faint cyanide-insensitive band may be seen with these species. Guinea pig liver homogenate, not shown, yielded results identical to rat.

Subcellular Distribution of SOD

Because human liver was found to contain such an abundance of manganese SOD, twenty times more than an equivalent weight of rat liver, we wondered whether the enzyme were located exclusively in the mitochondria. Human liver was obtained from autopsy, homogenized, and subjected to differential centrifugation as described under Materials and Methods. Because Weisiger and Fridovich (1973b) found the enzyme located in the soluble matrix of the mitochondria, glutamic acid dehydrogenase, also located in the mitochondrial matrix, was selected as a marker enzyme. When the human liver fractions were assayed, all the glutamic acid dehydrogenase appeared in the cytosolic fraction, indicating that mitochondrial rupture had occurred during the hours between death and autopsy. To circumvent this problem, we chose to examine the distribution of the enzyme in another primate, the baboon *Papio ursinus*. The results summarized in Table III indicate that while the mitochondrial marker enzymes are almost completely contained in the sedimentable fractions, only 30% of each of the dismutases is in these fractions. Weisiger and Fridovich (1973b) found that chicken liver mitochondria contain both cuprozinc SOD and manganese SOD, with the cuprozinc enzyme localized in the intermembrane space and the manganese enzyme in the matrix space. Our results are consistent with their findings, but show that in the baboon the manganese SOD is by no means exclusively located in the mitochondrial matrix. It is, in fact, predominantly in the cytoplasm.

TABLE III

Subcellular Distribution of Superoxide Dismutases in *Papio Ursinus*

Fraction	Cu-Zn SOD	Mn SOD	Glutamate Dehydrogenase	Cytochrome Oxidase
Soluble	70±8%	70±5%	6±9%	11±16%
Sedimentable (144,000 x g)	30±8%	30±5%	94±9%	89±16%
Units/gram	84±18	146±49		

DISCUSSION

The manganese SOD isolated from human liver is quite similar in physical composition and spectral properties to two other manganese dismutases isolated from eukaryotic, although non-mammalian, sources: *Saccharomyces cerevisiae* (Ravindranath and Fridovich, 1975) and chicken liver (Weisiger and Fridovich, 1973a). These three enzymes are each composed of four apparently identical subunits, and have a tetrameric molecular weight near 80,000.

The human enzyme as isolated with a full complement of four manganese atoms per molecule had a molar absorptivity at 480 nm of 2050 M^{-1} cm^{-1}, or about 500 per manganese atom. This is in good agreement with results reported for other manganese-containing dismutases (Keele *et al.*, 1970; Vance *et al.*, 1972; Weisiger and Fridovich, 1973a). The human and yeast enzymes clearly contain one atom of manganese per subunit, but the question of stoichiometry of metal content becomes less clear when one examines other members of the manganese and iron families of enzymes. The chicken enzyme contained 2.3 atoms of manganese per tetramer, easily rationalized by assuming that some manganese was lost during the purification procedure. Certain other members of the families, however, have been found to contain suspiciously close to 1.0 atom of metal per dimer, even though the subunits are apparently identical as judged by the sequence data. If the manganese content of the *E. coli* enzyme reported by Keele *et al.* (1970) is recalculated, for example, correcting for the spectral anomaly in the short UV, the result is about 1.1 atoms of manganese per dimer. The iron SOD from *E. coli* was found to contain 1.04 atoms of iron per dimer (Yost and Fridovich, 1973); the iron SOD from *Plectonema broyanum* was found by Misra and Keele (1975) to contain 0.94 atoms of iron per dimer, although Asada *et al.* (1975) reported 2.0 atoms per dimer for the same enzyme. This is clearly an area that merits further study.

Of the three families of dismutases, it appears that the manganese enzymes are most widely distributed, being found in both prokaryotes and eukaryotes. The cuprozinc enzymes seem to be found exclusively in eukaryotic organisms, although one possible exception has been raised by Puget and Michelson (1974) with the isolation of "bacteriocuprein" from a marine bacterium. The molecule was found to contain one or two atoms of copper and two atoms of zinc per mole, and is composed of two non-identical subunits. Spectrally and immunologically it appears to be quite different from the eukaryotic cuprozinc enzymes. Hence, it may not be genetically related to the eukaryotic enzymes. Sequencing the enzyme may be required to answer the question conclusively.

Three recent independent studies on the subcellular distribution of SOD in rat liver are in excellent agreement with one another, all finding that the manganese enzyme accounts for 7 to 8% of the total activity present, and that it is apparently located exclusively in the matrix of the mitochondria (Tyler, 1975; Peeters-Joris *et al.*, 1975; and Panchenko *et al.*, 1975). Our examination of human liver produced a rather unexpected and surprising observation: human liver contains 20 times the amount of manganese SOD, but less than half the amount of cuprozinc enzyme contained in an equal weight of rat liver. Although the ratio of the two enzymes in primate liver varied considerably from individual to individual, there was on the average twice as much manganese SOD as cuprozinc SOD. In chicken liver, we found roughly equivalent amounts. These findings contrast sharply and dramatically with the situations existing in rat and bovine livers, as is graphically portrayed in Fig.3. The fluctuations in manganese SOD content are particularly intriguing in view of the quite constant concentrations of cuprozinc SOD found from tissue to tissue, and organism to organism.

ACKNOWLEDGEMENTS

This work was supported by U.S. Public Health Service grants AM 17091 and HL 17603.

REFERENCES

1. Asada, K., Yoshikawa, K., Takahashi, M., Maeda, Y. and Enmanji, K. (1975). *J. Biol. Chem.* *250*, 2801-2807.

2. Beauchamp, C. and Fridovich, I. (1971). *Anal. Biochem. 44*, 276-287.
3. Cohen, E.J. and Esall, J. (1943). In "Proteins, Amino Acids and Peptides". Reinhold Publishing Corporation, New York.
4. Davis, B.J. (1964). *Ann. N.Y. Acad. Sci. 121*, 404-427.
5. Goscin, S.A. and Fridovich, I. (1972). *Biochem. Biophys. Acta 289*, 276-283.
6. Keele, B.B., Jr., McCord, J.M. and Fridovich, I. (1970). *J. Biol. Chem. 245*, 6176-6181.
7. Lowry, O.H., Rosebrough, N.J., Farr, A.L. and Randall, R.J. (1951). *J. Biol. Chem. 193*, 265-275.
8. McCord, J.M. and Fridovich, I. (1969). *J. Biol. Chem. 244*, 6049-6055.
9. Misra, H.P. and Fridovich, I. (1972). *J. Biol. Chem. 247*, 3410-3414.
10. Murphy, J.B. and Kies, M.W. (1960). *Biochem. Biophys. Acta 45*, 382-384.
11. Panchenko, L.F., Brusov, O.S., Gerasimov, A.M. and Loktaeva, T.D. (1975). *FEBS Lett. 55*, 84-87.
12. Peeters-Joris, C., Vandevoorde, A. and Baudhuin, P. (1975). *Biochem. J. 150*, 31-39.
13. Puget, K. and Michelson, A.M. (1974). *Biochem. Biophys. Res. Comm. 58*, 830-838.
14. Ravindranath, S.D. and Fridovich, I. (1975). *J. Biol. Chem. 250*, 6107-6112.
15. Salin, M.L. and McCord, J.M. (1974). *J. Clin. Invest. 54*, 1005-1009.
16. Sawada, Y., Ohyama, T. and Yamazaki, I. (1972). *Biochem. Biophys. Acta 268*, 305-312.
17. Strecker, H.J. (1955). In "Methods in Enzymology" (S.P. Colowick and N.O. Kaplan, eds.) Vol.II, Academic Press, New York and London, pp 220-225.
18. Tyler, D.D. (1975). *Biochem. J. 147*, 493-504.
19. Vance, P.G., Keele, B.B., Jr. and Rajagopalan, K.V. (1972). *J. Biol. Chem. 247*, 4782-4786.
20. Weber, K. and Osborn, M. (1969). *J. Biol. Chem. 244*, 4406-4412.
21. Weisiger, R.A. and Fridovich, I. (1973a). *J. Biol. Chem. 248*, 3582-3592.
22. Weisiger, R.A. and Fridovich, I. (1973b). *J. Biol. Chem. 248*, 4793-4796.
23. Wharton, D.C. and Tzagoloff, A. (1967). In "Methods in Enzymology" (R.W. Estabrook and M.E. Pullman, eds.) Vol.X, Academic Press, New York and London, pp 245-250.
24. Yost, F.J. and Fridovich, I. (1973). *J. Biol. Chem. 248*, 4905-4908.

SUPEROXIDE DISMUTASES FROM PROCARYOTE AND EUCARYOTE BIOLUMINESCENT ORGANISMS

K. PUGET, F. LAVELLE and A.M. MICHELSON

Institut de Biologie Physico-Chimique
Service de Biochimie-Physique
13, rue P. et M. Curie
75005 Paris, France

Superoxide radicals were identified as an intermediate in the bioluminescence of the boring mollusc *Pholas dactylus* by Henry *et al.*, (1973). The radicals are produced by luciferase, a copper containing glycoprotein (M.wt. 300,000) during oxidation of luciferin, also a glycoprotein (of M.wt. 35,000). Following this identification, the purification and study of superoxide dismutases from various bioluminescent organisms, both procaryote and eucaryote was undertaken.

PLEUROTUS OLEARIUS

This fungus normally grows on olive trees. It emits a blue green luminescence and can be cultured in aqueous media. Using suitable extraction procedures and purification techniques, including separation on QAE-Sephadex, two distinct superoxide dismutases were isolated (Lavelle *et al.*, 1974; Lavelle and Michelson, 1975). Both enzymes contain two atoms of manganese per molecule and are of similar molecular weight (76,000 and 78,000) with four subunits of 20 000 daltons. Other superoxide dismutases containing iron or copper were not detected.

Both manganese enzymes have the same specific activity but whereas the minor SOD (20%) contains four identical subunits, the major SOD (80%) contains two types of subunit (both of the same molecular weight) and hence has an $\alpha_2\beta_2$ structure. This second enzyme has an absorption band with λmax = 480 nm and $\varepsilon max = 850\ M^{-1}\ cm^{-1}$, characteristic of certain octahedral complexes of "high spin" trivalent manganese. The spectrum is not modified by oxidants but disappears in presence of reductants, without loss of activity. No EPR signal can be detected in the native enzyme at $77^{\circ}K$, in accord with other proteins containing trivalent manganese such as the Mn-SOD from *Escherichia coli* (Keale *et al.*, 1970). The characteristic signal of Mn^{2+} appears after liberation of the free cation by acid treatment of the protein. It thus appears that in the resting state the enzyme contains M^{III} and that during the catalytic cycle, the metal oscillates between Mn^{III} and Mn^{II}

$$Mn^{III} + O_2^{\overline{\cdot}} \longrightarrow Mn^{II} + O_2$$

$$2\ H^+ + Mn^{II} + O_2^{\overline{\cdot}} \longrightarrow Mn^{III} + H_2O_2$$

No immunochemical cross reactions using immunodiffusion and immunoelectrophoresis could be detected between the two manganese SODs using antisera prepared from the major enzyme. This, in addition to the different types of quaternary structures, suggests that the two SODs are isoenzymes rather than two forms synthesised from the same gene. Although the cellular distribution is as yet undefined, it is likely that the major enzyme is cytoplasmic and hence *Pleurotus olearius* is the first example of a eucaryote cell containing a cytoplasmic manganese SOD. It may be noted that human manganese SOD has recently been found to be present in the cytoplasm, as well as the mitochondria (McCord, 1976; McCord *et al.*, this volume).

BIOLUMINESCENT MARINE BACTERIA

Marine bacteria, in particular bioluminescent bacteria, possess a major superoxide dismutase which contains iron and no other metal. Rigorous search for a manganese containing enzyme, earlier regarded as characteristic of procaryotes (Keele *et al.*, 1970) gave negative results for *Photobacterium leiognathi* and *Photobacterium sepia* (Puget and Michelson, 1974a). Two independent classes of ferri SOD exist, both of molecular weight approximately 40500 but which are readily distinguished by immunochemical techniques. Other differences such as slight variations in thermostability, pH stability and isoelectric point are also present. In each case the enzyme consists of two subunits of identical molecular weight 20 000. These two classes characterise "free living" bioluminescent bacteria (*P. sepia*) and symbiont bacteria (*P. leiognathi*) and immunochemical techniques using antibodies to the enzymes provide an easy technique for classification. (A third group of bacteria represented by *Photobacterium fischeri* also exists, since a third Fe SOD, immunologically different from the two others is present in these bacteria.)

It is of interest that the two different Fe-SODs are closely related to the two different luciferases (a "slow" luciferase is characteristic of "free living" bacteria and a "fast luciferase" of symbiotic bacteria) suggesting genetic relationships between luciferase and Fe-SOD (Table I).

As with other SODs, the *P. leiognathi* ferri SOD is very stable and shows only a 50% loss in activity after 40 min at 60°C and in addition is only slowly denatured by 5 M guanidine HCl and is completely active in 50% dimethylsulphoxide at 22°C.

The loss of activity and of iron on photodestruction at 254 nm of tryptophane residues of *P. leiognathi* SOD indicates the possibility that these are involved in the active site and may indeed be the ligands of the metal. This would be compatible with the strong blue shift (25 nm) of tryptophane fluorescence emission in the native enzyme (λmax 335 nm) compared with tryptophane (λmax 360 nm) in the same buffer, giving rise to a red shift and increase in fluorescence on denaturation. Considerable similarity in amino acid composition of the ferri SOD from *P. sepia* and mangano SOD of *E. coli* (Yost and Fridovich, 1973) exists and it would be extremely interesting to have a comparative amino-terminal sequence (Steinman and Hill, 1973). Although the enzymes described here are similar to the iron containing SOD isolated from *E. coli* (Yost and Fridovich, 1973) some differences exist. Thus reduction

TABLE I

	Reaction with antibody to superoxide dismutase from *P. sepia*		Reaction with antibody to superoxide dismutase from *P. leiognathi*	
	Positive (free living)	Negative (symbiotic)	Positive (symbiotic)	Negative (free living)
	P. sepia	*P. leiognathi*	*P. leiognathi*	*P. sepia*
Strain of marine biolumi-nescent bacteria	79 106 1 SH 15 MAV	PP 1 H 4 PPA 1 141 L 1 L 7 153 SH 21 *P. fischeri* 665 H 7 M JP F1S	PP 1 H 4 PPA 1 141 L 1 L 7 153 SH 21 665 H 7 M JP F 1S	79 106 1 SH 15 MAV *P. fischeri*
Luciferase	slow	fast	fast	slow

with dithionite causes loss of activity whereas bleaching of the *E. coli* enzyme occurs without loss of activity. The superoxide dismutase from *E. coli* contains one atom of Fe per molecule of protein and has an ε_{350} of 1675 M^{-1} cm^{-1} compared with a molar extinction coefficient at 350 nm of 2500 for *P. leiognathi* SOD. These values are low compared with ε values for other non hematinic ferri proteins. Although the values of Fe are less than 2 atoms per molecule for the two enzymes from *P. sepia* and *P. leiognathi* other preparations from strain 1 (free living) and strain 153 (symbiont) gave values of 1.92 and 2.04 atoms Fe respectively (Halbwachs, Puget and Michelson, unpublished results). Both the *E. coli* enzyme and the dismutases from bioluminescent bacteria contain iron in the ferric form. Differences in the electron spin resonance spectra nevertheless exist (Henry *et al.*, 1974). The g-values were nearly identical for the two superoxide dismutases: *P. sepia*: 4.80, 3.98 and 3.64; *P. leiognathi*: 4.81, 3.94 and 3.63.

These resonances are indicative of ferric high-spin iron atoms ($^6S_{5/2}$) in a nearly rhombic symmetry and arise from transitions between the two sublevels of the middle Kramers doublet split by Zeeman energy. The symmetry of the iron ion appears to be partially tetragonal and not completely rhombic (70%).

The resonance observed at 77°K was about 100 gauss broad and a unique set of three lines could be described. The two iron atoms appear therefore to be indistinguishable and not coupled. Total oxidation of the iron atoms by addition of ferricyanide did not modify the g-values of the resonance and in most cases did not enhance the signal intensity. However in one preparation of *P. sepia* superoxide dismutase a two-fold enhancement was observed. It seems therefore that the iron atoms are nearly completely oxidized in aerobic solutions: this is not in contradiction to a possible shuttle in the valence state of the two iron atoms in the steady-state. Complete disappearance of the EPR signals occurs on addition of dithionite showing a complete reduction of the iron. Study of the catalytic properties and the mechanism of action of samples of Fe-SOD from *P. leiognathi* containing various quantities of iron between 1 and 2 atoms per mole by the technique of radiolysis by accelerated electrons has shown that two types of iron are present, one gram atom of Fe which is active and a second atom of Fe in variable amounts which is inactive (Lavelle *et al.*, 1977). The rate constant was $k = 1.2 \times 10^8\ M^{-1}\ sec^{-1}$ at pH 10.0 (Rotilio *et al.*, 1976) using a polarographic technique (compared with $k = 1.3 \times 10^8\ M^{-1}\ sec^{-1}$ determined by pulse radiolysis, Lavelle *et al.*, 1976). The Michaelis plot indicates inhibition of the Fe-SOD by O_2 (or $O_2^{\bar{\cdot}}$) at high concentrations of $O_2^{\bar{\cdot}}$ (Rotilio *et al.*, 1976).

Estimation of H_2O_2 produced from $O_2^{\bar{\cdot}}$ (γ irradiation, cobalt) in absence and in presence of Fe-SOD showed no significant differences. Hence the SOD does not change the stoichiometry of the dismutation of $O_2^{\bar{\cdot}}$ and catalyses only the reaction

$$2\ H^+ + 2\ O_2^{\bar{\cdot}} \longrightarrow O_2 + H_2O_2$$

FERRI-SOD FROM BACTERIA (*P. LEIOGNATHI*) GROWN AT DIFFERENT IRON CONCENTRATIONS

Bacteria were cultured as previously described (Puget and Michelson, 1974a) in the same medium to which was added ferrous sulphate to a final concentration of 5×10^{-5} M. For low iron content, the culture medium was treated with 10 g/litre of alumina (Alcoa A 305, Bacteriological grade) at 110°C for 20 mins, cooled and filtered then resterilised at 110°C for 15 mins. Iron content was estimated (bathophenanthroline) to be 4.8×10^{-6} M. Growth of the bacteria was at 28°C in presence of oxygen. The "low iron" medium yielded 75 g wet weight of bacteria while the yield from "high iron" was 109 g, both with 12 l of medium.

In each case the bacteria were lysed and the ferri superoxide dismutase separated and purified (Puget and Michelson, 1974b) as previously described. A solution of the ferri-SOD in 10^{-2} M phosphate pH 7.8 was passed through a column of Chelex 100 equilibrated in the same buffer and iron estimated with an Instrumentation Laboratories atomic absorption spectrophotometer (IL 253). Enzyme from bacteria grown in an iron poor medium contained 1.12 g atoms of iron (average of 4 determinations) per molecule (molecular weight 40 660 ± 2500) whereas that from the iron rich medium 1.15 g atoms of iron (average of 4 determinations). It is thus clear that no difference in iron content is present. No difference in thermal stability of the two enzymes was detected. However, in contrast with earlier preparations containing higher iron content, thermal loss of activity (170 μg of protein per ml in 10^{-2} M phosphate pH 7.8 at 60°C) was first order. Finally, specific activities of a number of preparations of ferri-SODs containing 1.15, 1.25, 1.25, 1.34 and 1.53 g atoms of iron per mole respectively were essentially identical and showed no correlation as has been reported by Yamakura (1976) for the ferri-SOD from *Pseudomonas ovalis*.

We therefore conclude that the physiologically functional enzyme contains one atom of iron which is catalytically active, but that a second atom iron can be bound but has no effect on activity. It is possible that variations in amount of the second iron result from pH differences from one preparation to another during purification. Indeed, dialysis of the enzyme at pH 5.5 (but not at pH 6.1) reduces the iron content to one g atom, with no loss in activity.

ATTEMPTED CROSS-LINKING OF SUBUNITS IN FERRI-SOD (*P. LEIOGNATHI*)

In order to stabilize Fe-SOD, attempts were made to cross-link the two subunits with the bifunctional reagent dimethylsuberimidate which acylates exposed ε-amino groups of lysine residues. Cross-linking limits are about 8.8 Å. Alternatively this reagent could acylate lysine residues to convert the basic amino groups into amidohexyl amidates in addition to cross linking (inter and intra subunit).

A solution of pure ferri-SOD (1 mg) in 1 ml of 0.1 M phosphate pH 9.35 was added to 20 mg of dimethylsuberimidate dihydrochloride previously neutralised with 20 µl of 2 N NaOH (final pH 9.5) and the solution left overnight at 4°C, then dialysed against 10^{-2} M phosphate pH 7.8. The specific activity was decreased by about 20% from the original. A control in which the enzyme was treated identically except for the presence of dimethylsuberimidate showed no loss in activity.

That considerable reaction with external lysine ε-amino groups occurred was shown by the lack of staining by Comassie blue on acrylamide gel electrophoresis of the treated enzyme. However no cross-linking between subunits occurred since electrophoresis of denatured enzyme on acrylamide gels in presence of sodium dodecyl sulphate in both cases gave the same Rf. Internal lysine amino groups were liberated and normal staining by Comassie blue occurred after denaturation of the dimethylsuberimidate treated SOD. Thermal stability of the treated enzyme was identical with that of the native SOD (100 µg protein/ml in 10^{-2} M phosphate pH 7.8 at 60°C) whereas stability with respect to acid pH (pH 4-5) was about 10% less than the untreated enzyme, despite modification of external lysine ε-amino groups.

METAL REPLACEMENTS IN Fe-SOD (*P. LEIOGNATHI*)

In view of the replacement of zinc by cobalt in bovine erythrocuprein (Rotilio *et al.*, 1973) replacement of Fe by treatment of the Fe-SOD apoenzyme with different di- and tri-valent metals was examined (Puget *et al.*, 1976).

Preparation of Apoenzyme

A solution of ferri SOD (34 ml at 18.38 mg of protein/ml; purity 75%) was dialysed at 4°C against 5 x 10^{-2} M acetate buffer pH 3.8 containing 10^{-3} M EDTA and 2 x 10^{-4} M mercaptoethanol until the pH inside the dialysis bag was pH 3.8. The solution was then dialysed against 10^{-2} M phosphate pH 7.8 to neutrality; no activity and no iron remained. The apoenzyme was purified by column chromatography on DEAE-Sephadex A-50 using a gradient of 5 x 10^{-3} M to 1 M phosphate pH 7.8.

Reconstitution

Treatment of the apoenzyme with ferric ions did not lead to recovery of activity. However, treatment with ferrous ions (in a 3 to 10 g atom excess) led to about 50% recovery of the initial activity. A typical experiment was as follows. To 1 ml of the apoenzyme (3.5 mg/ml) in 5 x 10^{-3} M phosphate pH 7.8 was added 0.05 ml of 10^{-2} M ferrous sulphate and the solution kept at 4°C for 6 hours then dialysed to remove excess iron. Dosage indicated 45% of the initial activity.

Replacement by Other Metals

In each case 0.1 ml of 2 x 10^{-2} M metal ion was added to 2 ml of 3.5 x 10^{-4} M apoenzyme in 5 x 10^{-3} M phosphate pH 7.8 (final concentration 10^{-3} M metal) using $FeCl_3$, $CuSO_4$, $Zn(Ac)_2$, $MnCl_2$, $CoCl_2$, $NiCl_2$, $CrCl_2$ (under argon) and $CrCl_3$. The samples were left at 4°C for 48 hours and then tested for activity. No significant recovery of activity was observed (1 unit per mg compared with 10 000 units/mg for pure ferri-SOD, for all cases except manganese reconstitution which has 10 units/mg).

Each sample was then chromatographed on a column of DEAE Sephadex A-50 (15 ml) using a gradient of 5 x 10^{-3} M to 1 M phosphate pH 7.8 (100 ml of each), the protein being eluted at 0.26 to 0.28 M phosphate. Each fraction of the peak of protein was then dialysed against 10^{-2} M phosphate pH 7.8 and the different metals estimated by Atomic Absorption Spectroscopy as well as measurement of the ultraviolet absorption at 280 nm.

As shown in Fig.1 the metal content follows protein concentration in the case of Cu^{2+}, Mn^{2+}, Zn^{2+}, Ni^{2+}, Co^{2+} and Cr^{2+} whereas attempted fixation of Cr^{3+} (as with Fe^{3+}) does not lead to specific attachment to the apoenzyme. Thus the ligands at the site of metal binding appear to be specific for divalent metal ions, though a surprising lack of specificity is shown for the metal. That the same ligand centre is occupied by each metal as was originally filled by iron in the native SOD is strongly indicated by the fact that in all cases (except Cr^{3+}) addition of Fe^{2+} did not lead to reconstitution of enzymic activity, that is, the original ligands are no longer available once they are bound to the other metals. In addition it is clear that once bound, the replacement metal is not displaced by Fe^{2+}.

It may be noted that the apoenzyme, denatured SOD and the various metal replacement proteins all precipitate with the antibody to ferri-SOD.

Apart from correlation of metal content with protein concentration in the peak of proteïn, for each reconstitution from the apoenzyme, the number of gram atoms of metal per mole of protein (M.wt. 40,000) was determined for the tube with the highest protein concentration. Similarly, peak tubes were combined in each case, dialysed against 10^{-2} M phosphate pH 7.8 and the metal/protein ratios redetermined. The results are shown in Table II. Protein concentration was determined by interferometry with a Carl Zeiss (Jena) Interferometer LI 3 and metal content with an Instrumentation Laboratory Inc. Atomic Absorption Spectrophotometer IL 253. The results indicate that replacement with Mn^{2+}, Zn^{2+}, Ni^{2+}, Co^{2+} and Cr^{2+} all give reconstituted metalloprotein containing one gram atom of metal per mole of protein (two identical subunits of M.wt. 20,000) whereas with Cu^{2+} and Fe^{2+} two atoms of metal can be bound.

BACTERIOCUPREIN

A second SOD, bacteriocuprein, was isolated from *P. leiognathi* during purification of the major Fe-SOD (Puget and Michelson, 1974b). No manganese was present in this SOD which has a molecular weight of 33,000 with two different subunits of M.wt. 17,000 and 15,000. Metal analysis showed that the enzyme contained copper and zinc, and thus this SOD resembles the typical

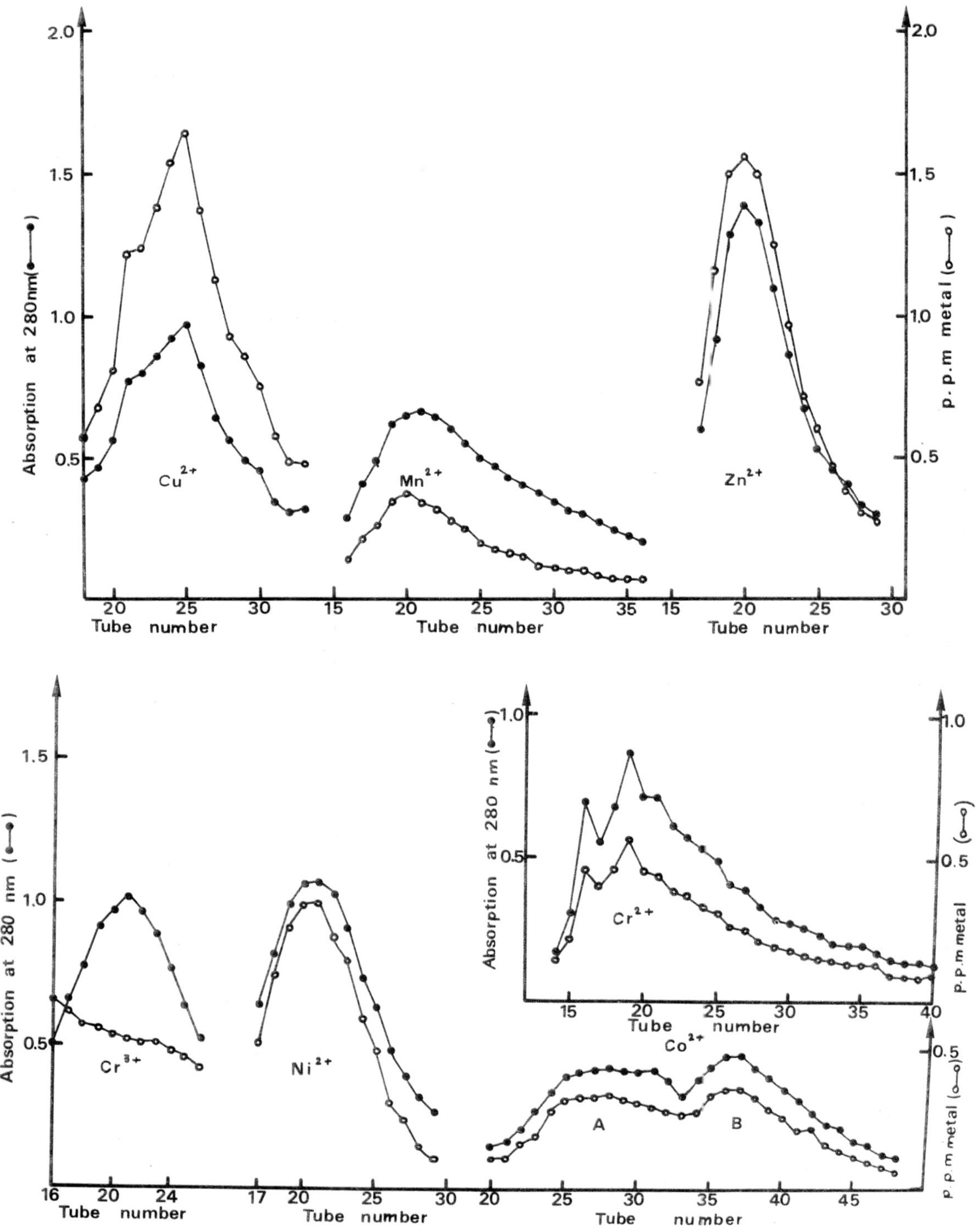

Fig.1. *Fixation of different metals to the apoenzyme of ferri-SOD. Chromatography on DEAE Sephadex A-50; ●——●, absorption at 280 nm; o——o, metal determined by Atomic Absorption Spectroscopy in parts per million.*

eucaryote Cu/Zn superoxide dismutases.

TABLE II

g Atom/Mole Protein

Metal	Peak Tube	Combined Peak	$A_{1\%}^{1\ cm}$ at 280 nm
Cu^{2+}	1.82	1.56	16.8
Mn^{2+}	0.75	0.45	18.4
Zn^{2+}	1.12	1.03	16.2
Ni^{2+}	1.06	0.98	16.8
Co^{2+} (A)	0.94	0.64	17.4
Co^{2+} (B)	0.79	0.53	15.4
Cr^{2+}	0.84	0.64	16.0
Fe^{2+}	-	1.68	-

Absorption at 280 nm as Function of Iron Content

g Atom Fe/Mole Protein	$A_{1\%}^{1\ cm}$ at 280 nm
1.56	21.0
1.48	20.7
1.40	19.8
1.36	18.9

The effect of increased partial pressure of oxygen (approximately 5 fold) was interesting in that preferential synthesis of the iron enzyme is observed though some stimulation of bacteriocuprein also occurs (Table III). This is in contrast with *E. coli* which also contains a ferri superoxide dismutase (different from that of *P. leiognathi*) with a second enzyme containing manganese. In this case the iron enzyme (although periplasmic) apparently does not vary in response to changes in p O_2 during growth whereas the level of the mangano enzyme is markedly dependent and shows a strong increase with increase in p O_2 (Gregory and Fridovich, 1973; Gregory *et al.*, 1973).

The ultraviolet absorption spectrum of bacteriocuprein is quite different from that of bovine erythrocuprein (λmax 258 nm) which contains no tryptophane. Some similarity with the absorption spectrum of human erythrocuprein can be seen.

A number of differences exist between bacteriocuprein and erythrocuprein (bovine or human). Whereas identical subunits are present in the mammalian enzymes, bacteriocuprein has two different subunits. In addition, the bacterial superoxide dismutase is a much more basic protein with an isoelectric point 8.25 compared with 4.95 for bovine (Bannister *et al.*, 1971) and 4.6 for human erythrocuprein (Stansell and Deutsch, 1965), and is readily separated from the erythrocupreins by electrophoresis on acrylamide gels (relative R_{fs}, 0.13 for bacteriocuprein, 0.25 for bovine erythrocuprein and

0.36 for the human enzyme). No immunological cross reactions could be detected among the three superoxide dismutases, despite the very close similarity of molecular weights and the presence of copper and zinc in all cases. The thermal and pH stability of bacteriocuprein resembles the two erythrocupreins while the inhibitory effect of cyanide is similar to that reported (Bannister *et al.*, 1973) for human erythrocuprein (50% inhibition at 1.6 x 10^{-5} M KCN with 2 x 10^{-6} M enzyme) and is 25% at 10^{-6} M KCN, 50% at 2.2 x 10^{-6} M and 75% at 4.8 x 10^{-6} M KCN for 2 x 10^{-11} M bacteriocuprein at pH 9.

TABLE III

Effect of Oxygen on Superoxide Dismutase Activities in *P. Leiognathi*

	Units		Ratio of Units
	Air	Oxygen	O_2/air
Total units	28 220	102 480	3.63
Bacteriocuprein	12 920	28 880	2.23
Ferri SOD	15 300	73 600	4.81
Ratio units of Ferri SOD/cupro SOD	1.18	2.55	

The first preparation of the enzyme contained 1.02 atoms of Cu and 2.5 atoms of Zn, that is half the copper present in a normal erythrocuprein. The band of absorption in the visible region with λmax at 690 nm (εmax, 100) was about half that of bovine or human erythrocuprein and finally the specific activity was also 50% of that of the two erythrocupreins in accord with the presence of one atom of Cu per molecule.

Incubation of bacteriocuprein with 10^{-5} M copper sulphate overnight at 4°C followed by dialysis, had no effect on the specific activity of the enzyme. The K_M for $O_2^{\overline{\cdot}}$ was 1.2 x 10^{-3} M (Rotilio *et al.*, 1976) while the rate constant was k = 4.02 x 10^8 M^{-1} sec^{-1} at pH 10.0. However, this value increased 41% on addition of one equivalent of Cu^{2+} (Rotilio *et al.*, 1976).

A large scale preparation of bacteriocuprein has recently been re-examined. Following separation from the Fe-SOD by chromatography on DEAE-Sephadex, the bacteriocuprein was separated into three peaks of activity by chromatography on carboxymethyl cellulose (CM 52) at pH 7.0. Each of these three components was further purified on QAE-Sephadex following which each of the peaks could be further separated into two active components by gel electrophoresis or electrofocalisation. It is thus clear that considerable microheterogeneity exists in this preparation, probably artifacts due to deamination (or other modifications) during storage of the crude enzyme in concentrated ammonium sulphate solutions for almost one year. (A new purification is at present in progress in which no storage is effected in order to verify this point. Only two major fractions are present but several minor components can also be seen.)

However, each of the components had a specific activity of 50,000 units/mg (luminol units, Puget and Michelson, 1974a) corresponding to the specific activity of pure bovine (or human) erythrocuprein, rather than 25 000 units/mg as obtained for the first preparation. All three major Cu-SODs appear to contain two non-identical subunits and all contain approximately 2 atoms of copper and 2 atoms of zinc (I, 1.77 Cu, 1.93 Zn; II, 2.12 Cu, 2.04 Zn; III, 2.14 Cu, 2.07 Zn). Thus at least in this preparation, both the activity and the copper content correspond to erythrocuprein. The fact that preparations containing two copper atoms (and 2 Zn) are twice as active as those with 1 Cu (and 2 Zn) suggests that both copper atoms are simultaneously and directly involved in catalytic cycles, rather than a single site mechanism (Fielden *et al.*, 1974).

Bacteriocuprein has been crystallized and it would be of interest to compare X-ray crystallographic aspects with bovine erythrocuprein.

It may be noted that activities are expressed in luminol units (Puget and Michelson, 1974) in which the SOD is estimated in terms of inhibition of the chemiluminescence of luminol induced by $O_2^{\bar{\cdot}}$ produced by xanthine oxidase/O_2/hypoxanthine. This technique is extremely sensitive and can be used to estimate 0.2 ng/ml of erythrocuprein with precision (about 5 - 6 femtomoles per ml). For pure bovine or human erythrocuprein the specific activity is 50 000 units/mg (50% inhibition in the test is considered to be equivalent to 0.1 units) corresponding to 3300 cytochrome *c* units (McCord and Fridovich, 1969). Units based on inhibition of reduction of nitrobluetetrazolium by $O_2^{\bar{\cdot}}$ produced by photoreduction of FMN or riboflavin are also used. Very approximately these correspond to the cytochrome *c* units. However, when different dosage techniques are applied to different SODs - Cu, Mn or Fe-SODs, the factors necessary to convert one unit into another become very different and are related to the nature of the SOD. These differences are shown in Table IV. Thus the Fe-SOD (specific activity 10 000 units/mg) is 5 times less active than bovine erythrocuprein in terms of luminol units but is about 70% as active in terms of cytochrome *c* units. It is clear that a standardized unit of SOD activity is necessary, based not on secondary inhibition techniques (since activity is a function of the nature of the molecule, and its affinity for $O_2^{\bar{\cdot}}$, in competition with the SOD for the $O_2^{\bar{\cdot}}$ as well as the nature of the SOD) but rather on pulse radiolysis techniques. These differences are in some measure due to different assay conditions (pH, ionic strength) and will still be present with pulse radiolysis techniques under different conditions. For example, although the rate constant for Cu-SOD is essentially independent of pH, the Fe-SODs and Mn-SODs show variations (Lavelle *et al.*, 1976). An international unit of SOD activity should be defined in terms of direct measurement of disappearance of substrate for comparison with the various indirect techniques.

TABLE IV

	Ratio of Units Luminol/NBT	Luminol/Cytochrome *c*
Bovine erythrocuprein	21	15
Bacteriocuprein	16.7	24.5
Fe-SOD (*P. leiognathi*)	6.7	4.3
Mn-SOD (*P. olearius*)	12.0	-

INHIBITION OF SOD BY DIETHYLDITHIOCARBAMATE

The Cu/Zn SODs can be readily distinguished from Fe-SODs and Mn-SODs since they are inhibited by cyanide. In principle, diethyldithiocarbamate should also show a specificity of inhibition for the different SODs. However, at 5×10^{-5} M diethyldithiocarbamate, using the inhibition by SOD of the reduction by $O_2^{\cdot-}$ of cytochrome *c*, it was found that all of the SODs, whether Cu, Fe or Mn were inhibited 40 - 70%. Thus the reagent shows no specificity with respect to inhibition.

EVOLUTIONARY SIGNIFICANCE OF BACTERIOCUPREIN

The isolation of a Cu-SOD from a procaryote organism raises some interesting problems. It has been postulated that the molecular properties of superoxide dismutase have been rigidly preserved (McCord *et al.*, 1971; Goscin and Fridovich, 1972; Misra and Fridovich, 1972) during evolution of eucaryotes and that all eucaryotes should contain SOD with properties similar to bovine erythrocuprein, whereas all procaryotes should contain the distinct manganese containing enzyme characteristic of *E. coli* (Keele *et al.*, 1970). Indeed such a hypothesis has been cited in discussions (Cohen, 1973) on the origins of organelles of eucaryotes cells, particularly since mitochondrial SOD in an as yet limited number of species (Weisiger and Fridovich, 1973) is a manganese enzyme similar to that isolated from *E. coli.*

The isolation of a copper and zinc containing SOD, bacteriocuprein, from bacteria (*P. leiognathi*) suggests that with respect to SOD activities there may well have been a monophyletic evolutionary process from procaryote to eucaryote, which began at the moment of transition from anaerobic to aerobic life conditions. Of perhaps equal significance is the absence of a manganoenzyme in these bacteria - no trace could be detected in *P. leiognathi*, though this does not exclude the possible presence of a mangano SOD in other marine bacteria. At least in certain branches of aerobic evolution the SODs of certain eucaryotes and procaryotes may have a common origin, rather than independent lines of evolution of these two classes of cells after passage to aerobic life. However, as pointed out by Fridovich and McCord (private communication) another possibility may well exist, with somewhat far reaching implications. Since *P. leiognathi* is a symbiotic bacterium, found in the Pony fish, a natural genetic engineering process may well have occurred in which the eucaryotic gene for Cu-SOD proper to the fish itself was transferred to the bacteria, which would then lose a less efficient manganese enzyme. We are presently engaged in a survey of the presence or absence of Cu-SOD in the two main classes of bioluminescent marine bacteria - free living and symbiotic - which already correspond to two different luciferases and two immunologically different Fe-SODs, to determine whether the copper enzyme is specific to *P. leiognathi*, to symbiotic marine bacteria in general or to both free living and symbiotic bioluminescent marine bacteria. In addition, a preparation of the Pony fish Cu-SOD will be undertaken, to study possible immunochemical similarities (or identity) with the bacterial enzyme, in order to verify what is possibly the first case of natural genetic engineering involving transfer of a eucaryote gene to a procaryote.

REFERENCES

1. Bannister, J.V., Bannister, W.H. and Wood, E. (1971). *Eur. J. Biochem.* *18*, 178-186.

2. Bannister, J.V., Bannister, W.H., Bray, R.C., Fielden, E.M., Roberts, P.B. and Rotilio, G. (1973). *FEBS Letters 32*, 303-306.
3. Cohen, S.S. (1973). *American Scientist 61*, 437-445.
4. Fielden, E.M., Roberts, P.B., Bray, R.C., Lowe, D.J., Mautner, G.N., Rotilio, G. and Calabrese, L. (1974). *Biochem. J. 139*, 49-60.
5. Goscin, S.A. and Fridovich, I. (1972). *Biochim. Biophys. Acta 289*, 276-283.
6. Gregory, E.M. and Fridovich, I. (1973). *J. Bacteriol. 114*, 1193-1197.
7. Gregory, E.M., Yost, F.J. and Fridovich, I. (1973). *J. Bacteriol. 115*, 987-991.
8. Henry, J.P., Isambert, M.F. and Michelson, A.M. (1973). *Biochimie 55*, 83-93.
9. Henry, Y.A., Puget, K. and Michelson, A.M. (1974). *Fed. Proc. Abstracts 33*, 1321.
10. Keele, B.B., Jr., McCord, J.M. and Fridovich, I. (1970). *J. Biol. Chem. 245*, 6176-6181.
11. Lavelle, F., Durosay, P. and Michelson, A.M. (1974). *Biochimie 56*, 451-458.
12. Lavelle, F. and Michelson, A.M. (1975). *Biochimie 57*, 375-381.
13. Lavelle, F., McAdam, M., Fielden, E.M., Roberts, P.B., Puget, K. and Michelson, A.M. (1977). *Biochem. J. 161*, 3-11.
14. McCord, J.M. (1976). In "Iron and Copper Proteins" Yasunobu, K.T., Mower, H. and Hayaishi, O. (eds.) Plenum Press, New York, pp 540-550.
15. McCord, J.M. and Fridovich, I. (1969). *J. Biol. Chem. 244*, 6049-6055.
16. McCord, J.M., Keele, B.B. and Fridovich, I. (1971). *Proc. Natl. Acad. Sci.* U.S.A. *68*, 1024-1027.
17. Misra, H.P. and Fridovich, I. (1972). *J. Biol. Chem. 247*, 3410-3414.
18. Puget, K. and Michelson, A.M. (1974a). *Biochimie 56*, 1255-1267.
19. Puget, K. and Michelson, A.M. (1974b). *Biochem. Biophys. Research Communs. 58*, 830-838.
20. Puget, K., Durosay, P. and Michelson, A.M. (1976) unpublished results.
21. Rotilio, G., Calabrese, L. and Coleman, J.E. (1973). *J. Biol. Chem. 248*, 3855-3859.
22. Rotilio, G., Puget, K. and Michelson, A.M. (1976) unpublished results.
23. Stansell, M.J. and Deutsch, H.F. (1965). *J. Biol. Chem. 240*, 4306-4311.
24. Steinman, H.M. and Hill, R.L. (1973). *Proc. Natl. Acad. Sci.* U.S.A. *70*, 3725-3729.
25. Weisiger, R.A. and Fridovich, I. (1973). *J. Biol. Chem. 248*, 3582-3592.
26. Yamakura, F. (1976). *Biochem. Biophys. Acta 422*, 250-294.
27. Yost, F.J. and Fridovich, I. (1973). *J. Biol. Chem. 248*, 4905-4908.

SUPEROXIDE DISMUTASE FROM *BACILLUS STEAROTHERMOPHILUS*

J. I. HARRIS

MRC Laboratory of Molecular Biology
Hills Road, Cambridge CB2 2QH, England

Superoxide dismutase is widely distributed among oxygen metabolising organisms (for review see Fridovich, 1975). The enzyme is a metalloprotein and appears to occur in two evolutionarily distinct forms. Copper/zinc-containing enzymes, blue proteins found mainly in the cytoplasm of eukaryotic organisms, are dimeric molecules comprising two chemically identical subunits containing one atom each of copper and of zinc. Removal of the two metals with chelating agents causes total inactivation but activity is largely restored by replacement of copper alone, and fully restored by replacement of both metals (Beem *et al.*, 1974). The amino acid sequence and three-dimensional structure of the Cu/Zn enzyme from bovine erythrocytes has recently been elucidated (Richardson *et al.*, 1975) and the protein ligands to the two metals in each of the subunits have been identified.

Manganese or iron-containing superoxide dismutases found principally in procaryotic organisms and in mitochondria are either dimers or tetramers and are also composed of identical subunit polypeptide chains. The two classes of enzyme are clearly different both in structure and in properties and are believed to be of independent origin (Steinman and Hill, 1973; Bridgen *et al.*, 1975). In order to establish structural and functional relationships between the two classes it is necessary to determine the complete amino acid sequence and three-dimensional structure of a manganese or iron-containing enzyme and with this aim in view the successful crystallisation and preliminary X-ray diffraction study of the reddish-purple Mn-eznyme from *B. stearothermophilus* is now reported. We also describe the preparation and properties of a stable inactive Mn-free apoprotein and the reconstitution from it of the fully active Mn-enzyme. This appears to be indistinguishable from the native enzyme and it is now hoped that the availability of stable forms of both the apo and holoprotein will aid in investigations of the mode of binding of the metal and of its role in catalysis.

PREPARATION OF ENZYME

B. stearothermophilus cells were grown, disrupted and extracted as previously described (Atkinson *et al.*, 1973). Partially purified enzyme fractions were prepared at the Microbiological Research Establishment, Porton, England, by a procedure similar to that described previously for comparable fractions from *E. coli* (Bruton *et al.*, 1975). Pure Mn-superoxide dismutase was prepared from the 'PFK rich' fraction (cf. Hengartner and Harris, 1975) by gel-filtration on G-75 of the fraction that did not absorb onto AMP-Sepharose. The homogeneity of the enzyme preparation was examined by polyacrylamide gel-electrophoresis in the absence and presence of 1% SDS. As shown in Fig.1A,B, it gave a single protein band in both systems. When the 'native' gel was stained for enzyme activity (Fig.1A) the activity stain coincided exactly with the protein stain. The subunit molecular weight was estimated to be 20,000 (cf. Bridgen *et al.*, 1975) and the pure native enzyme contains one atom of manganese per dimer (cf. Table I).

TABLE I

Reconstitution of Mn Superoxide Dismutase

Enzyme		Protein concentration*	$A_{280}/_{480}$ nm	Activity†	Mn‡
Native		9 mg/ml	37	3.9×10^8	0.9
Apo		9 mg/ml	200	$< 2 \times 10^6$	<0.01
Reconstituted	(1)	2 mg/ml	39	3.5×10^8	0.8
	(2)	3 mg/ml	38	3.7×10^8	

*Based on $E_{280\ nm}^{1\%} = 13.2$

†Specific activities were determined by pulse radiolysis assay (Rotilio *et al.*, 1972) and are expressed as second order rate constants ($M^{-1}\ S^{-1}$) for the turnover of O_2 under conditions of exponential decay.

‡Mn contents were determined by neutron activation analysis and are expressed as g atoms Mn per mole (40,000 g) enzyme measured by gel-filtrating on a calibrated column of Sephadex G-100.

CRYSTALLISATION AND PRELIMINARY X-RAY DIFFRACTION STUDY
(Bridgen, Harris and Kolb, 1976)

Crystals were obtained from a solution containing 10 mg/ml of enzyme in 50 mM NaCl/10 mM Tris-HCl, pH 8.0 by dialysis (in 60 µl vials) at 45° against 2 M NaH_2PO_4, pH 8.0. Larger crystals of the same general shape were obtained by dialysis against 35% $(NH_4)_2SO_4$ in 33 mM Tris-HCl, pH 8.0 at 4°. After several weeks clusters of large pink plates with dimensions up to 2 mm x 2 mm x 0.1 mm were obtained and these crystals were used for subsequent analysis.

Individual crystals were mounted in capillaries and 8° precession photographs were taken (Fig.2). From these data it was established that the crystals were monoclinic, the space group was $P2_1$ and the unit cell dimensions were a = 50 Å, b = 70 Å and c = 69 Å with $\beta^* = 68^\circ 55'$. The crystals diffracted strongly to 2 Å and resistance to radiation damage appeared to be good.

The unit cell dimensions correspond to a volume of 2.27×10^5 $Å^3$ which, if the asymmetric unit is the dimer, gives a value for V_m of 2.8 $Å^3$ per dalton, approximately in the centre of the normal range of values (Mathews, 1968). If the asymmetric unit contained either a monomer or two dimers, the corresponding values for V_m would be well outside the range normally found in other protein crystals.

This crystal form appears suitable for a complete structure determination of the Mn superoxide dismutase from *B. stearothermophilus*. Such a study would also be aided by the stability of the enzyme towards denaturation and the ability to remove the manganese atom reversibly.

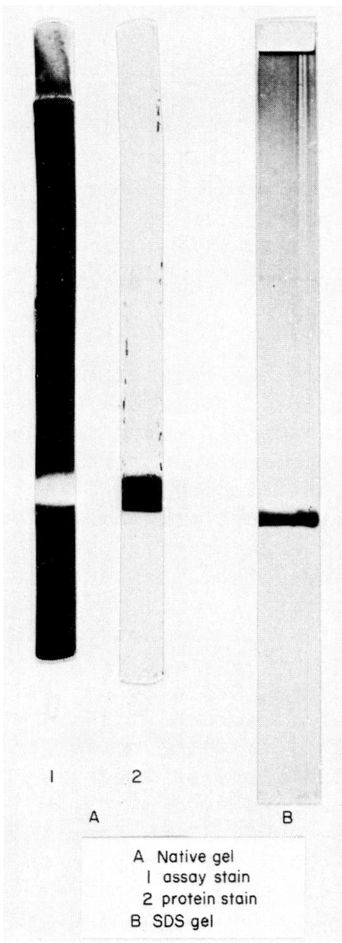

Fig.1. *Polyacrylamide gel-electrophoresis of* B. stearothermophilus *superoxide dismutase. A, 'native' 7.5% gel at pH 8.4 stained (a) with Coomassie blue (5 μg protein); (b) for enzyme activity (0.5 μg protein) by a photochemical method (Beauchamp and Fridovich, 1971). B, SDS gel. The sample (5 μg) was pre-heated in 1% SDS at 100° for 10 min and run on a 12.5% slab gel.*

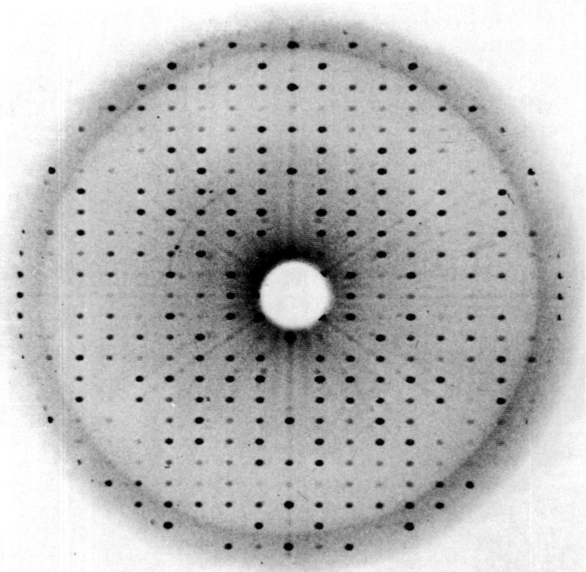

Fig.2. *X-ray diffraction using Cu Kα radiation of superoxide dismutase from* B. stearothermophilus. *8° precession photograph of hko projection showing systematic absences along b*. Some unfiltered Cu Kβ radiation is still present.*

PREPARATION OF Mn-FREE APOPROTEIN (Brock, Harris and Sato, 1976)

Manganese (and iron) containing dismutases contain firmly bound metal and, in contrast to the copper/zinc enzymes, the Mn-containing enzyme from *B. stearothermophilus* is stable in EDTA, even in the presence of 8 M urea at neutral pH. Attempts to prepare a Mn (or Fe) free apoenzyme for independent study have hitherto been frustrated by the instability of the putative apo-enzyme under conditions that ensure complete removal of the metal from the native enzymes (cf. Keele, McCord and Fridovich, 1970; Puget and Michelson, 1974). More recently, however, Ose and Fridovich (1976) have reported the reversible exchange of manganese from *E. coli* superoxide dismutase. The apo-protein as such was, however, again found to be unstable in solution and reconstitution of active Mn-enzyme could only be achieved *in situ* by addition of excess $MnCl_2$ in the presence of the metal chelating agent, 8-hydroxy-quinoline, and 2 M guanidinium hydrochloride; i.e. under conditions that allow the metal to be released from the native enzyme.

Enzymes from thermophilic bacteria are known to be more stable than their counterparts in mesophiles. Thus from *B. stearothermophilus* it has been possible to prepare a stable crystalline apo-(NAD-free) glyceraldehyde 3-phosphate dehydrogenase suitable for X-ray crystallographic study (Suzuki and Harris, 1971) as well as a stable but inactive metal-free class II aldolase from which active metalloenzyme could be reconstituted by addition of the appropriate divalent metal (Hill *et al.*, 1976).

To remove manganese from the *B. stearothermophilus* dismutase it was found necessary to expose the enzyme to 8 M urea/EDTA at acid pH. Under these conditions the reddish-purple colour disappears with concomitant loss of manganese and of enzyme activity. These conditions form the basis of a method for the preparation of a stable apoprotein. In a typical experiment a 9 mg/ml solution of pure Mn-superoxide dismutase in 50 M Tris-HCl pH 7.5 was dialysed for up to 12 h at 4^{0} against 8 M urea/10 mM EDTA adjusted to pH 3.7 with HCl or acetic acid. The resulting colourless solution was next dialysed for up to 12 h against 8 M urea/10 mM EDTA/50 mM Tris-HCl, pH 7.5 followed by dialysis against the same buffer with 0.1 mM EDTA but lacking urea. The slightly turbid solution was centrifuged (if the dialysis against 8 M urea at neutral pH is omitted most of the protein has precipitated at this stage). The soluble protein (approximately 9 mg/ml) contained less than 0.01 atom Mn per mole, did not absorb at 478 - 480 nm and was completely inactive when assayed for superoxide dismutase activity by two different methods (McCord and Fridovich, 1969; Rotilio *et al.*, 1972), Table I. The apoprotein was indistinguishable from the native enzyme when it was analysed by acrylamide gel-electrophoresis at pH 8.9 (Fig.3), or by gel-filtration on Sephadex G-75, showing that the dimeric structure is maintained in the absence of the metal (cf. Sato and Harris, 1976).

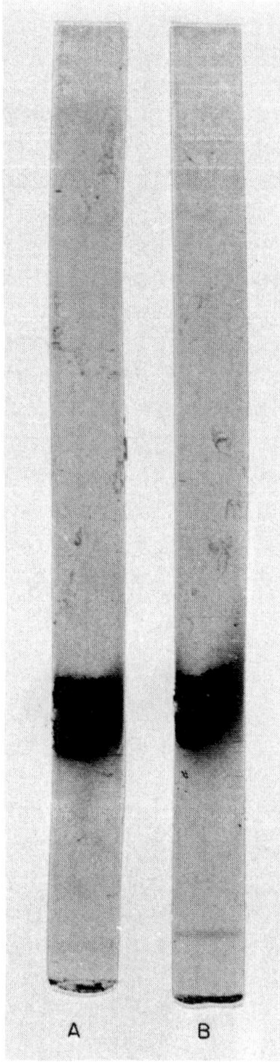

Fig.3. *Polyacrylamide gel-electrophoresis of* B. stearothermophilus *superoxide dismutase. 'Native' 7.5% gel at pH 8.4 stained with Coomassie blue. A, native Mn-enzyme. B, apoprotein.*

RECONSTITUTION OF ACTIVE Mn-ENZYME

Reconstitution of active Mn-enzyme does not occur by addition of $McCl_2$ to the apoprotein in buffer at neutral pH. Total reconstitution of fully active superoxide dismutase was however found to occur after the apoprotein had been exposed to excess $MnCl_2$ in the presence of 8 M urea at pH 3.7. In a typical experiment apoprotein (5 mg/ml) in 50 mM Tris-HCl/0.1 mM EDTA pH 7.5 was dialysed successively for periods of up to 12 h at 4° against:

(a) 8 M urea/10 mM $MnCl_2$ adjusted to pH 3.7 with 1% acetic acid;
(b) 8 M urea/50 mM Tris-HCl/10 mM $McCl_2$, pH 7.5;
(c) 50 mM Tris-HCl/10 mM $MnCl_2$, pH 7.5.

The resulting reddish-purple solution was next dialysed against the same Tris-HCl buffer containing mM EDTA; or alternatively the protein solution was gel-filtered on Sephadex G-50 in the same buffer, in order to remove excess $MnCl_2$.

The extent of reconstitution was assessed by measuring the optical density at 480 nm, manganese content, and regain of superoxide dismutase activity. As shown in Table I the reconstituted enzyme had the same A_{280}/A_{480} nm ratio and content of Mn as the native Mn-enzyme. It was also fully active when assayed by the pulse radiolysis method (Rotilio *et al.*, 1972).

These experiments show that the manganese is essential for superoxide dismutase activity. The results of neutron activation analysis of both native and reconstituted enzymes indicate the presence of only one atom of manganese per dimer. Unlike the copper and zinc in the bovine erythrocyte enzyme (Beem *et al.*, 1974) the manganese atom is very firmly bound to the protein. It appears to be shared between the two subunits and is released only under conditions that cause dissociation of the dimeric structure. The apoprotein forms a stable dimer in the absence of manganese and dissociation of the dimer appears to be an essential step in the reconstitution of the active Mn-enzyme.

Similar results have been obtained (Sato and Harris, 1976) with superoxide dismutase isolated from a more extreme thermophile, *Thermus aquaticus*. This, like the mitochondrial enzyme (Weisiger and Fridovich, 1973), is tetrameric and contains two atoms of manganese per tetramer. Formation of a stable tetrameric apoprotein and reconstitution of active manganese enzyme occur under conditions similar to those now described for *B. stearothermophilus* superoxide dismutase.

These are the first examples of the preparation of stable apoproteins and of reconstituted fully active Mn-enzymes from Mn containing superoxide dismutases. It has not so far proved possible to prepare a stable apoprotein from any mesophile superoxide dismutase although reversible metal exchange has been shown to occur *in situ* with the dimeric *E. coli* Mn-enzyme (Ose and Fridovich, 1976). The colour and NMR properties of the *E. coli* enzyme indicate that the constituent manganese is trivalent in the resting state (Villafranca *et al.*, 1974). Therefore as reconstitution of the *B. stearothermophilus* Mn enzyme was achieved by addition of Mn^{2+} to the apoprotein it must be assumed that the added Mn^{2+} is oxidised to Mn^{3+} during the reconstitution process.

The availability of stable crystals of *B. stearothermophilus*-Mn enzyme suitable for a detailed X-ray crystallographic analysis, and the ability to prepare stable apoprotein and to reconstitute the fully active Mn enzyme, opens up new possibilities for the study of the three-dimensional structure of the metalloenzyme and of the role of the metal in catalysis.

REFERENCES

1. Atkinson, A., Bruton, C.J. and Jakes, R. (1973). *Abstr. FEBS Meet. Dublin No. 71.*
2. Beauchamp, C. and Fridovich, I. (1971). *Anal. Biochem. 44*, 276.
3. Beem, K.M. Rich, W.E. and Rajagopalan, K.V. (1974). *J. Biol. Chem. 249*, 7298-7305.
4. Bridgen, J., Harris, J.I. and Northrop, F. (1975). *FEBS Lett. 49*, 392-395.
5. Bridgen, J., Harris, J.I. and Kolb, E. (1976). *J. Mol. Biol. 105*, 333-335.
6. Brock, C., Harris, J.I. and Sato, S. (1976). *J. Mol. Biol.*, in press.
7. Bruton, C.J., Jakes, R. and Atkinson, A. (1975). *Eur. J. Biochem. 59*, 327-333.
8. Fridovich, I. (1975). *Ann. Rev. Biochem. 44*, 147-157.
9. Hengartner, H. and Harris, J.I. (1975). *FEBS Lett. 55*, 282-285.
10. Hill, H.A.O., Lobb, R.R., Sharp, S.L., Stokes, A.M. Harris, J.I. and Jack, R.S. (1976). *Biochem. J. 153*, 551-560.
11. Keele, B.B., McCord, J.M. and Fridovich, I. (1970). *J. Biol. Chem. 245*, 6176-6181.
12. McCord, J.M. and Fridovich, I. (1969). *J. Biol. Chem. 244*, 6049-6055.
13. Mathews, B.W. (1968). *J. Mol. Biol. 33*, 491-497.
14. Ose, D.E. and Fridovich, I. (1976). *J. Biol. Chem. 251*, 1217-1218.
15. Puget, K. and Michelson, A.M. (1974). *Biochemie 56*, 1255-1267.
16. Richardson, J.S., Thomas, K.A., Rubin, B.H. and Richardson, D.C. (1975). *Proc. Nat. Acad. Sci. U.S.A. 72*, 1349-1353.
17. Rotilio, G., Bray, R.C. and Fielden, E.M. (1972). *Biochim. Biophys. Acta 268*, 605-609.
18. Sato, S. and Harris, J.I. (1976). *Eur. J. Biochem.*, in press.
19. Steinman, H.M. and Hill, R.L. (1973). *Proc. Nat. Acad. Sci. U.S.A. 70*, 3725-3729.
20. Suzuki, I. and Harris, J.I. (1971). *FEBS Lett. 13*, 217-220.
21. Weisiger, R.A. and Fridovich, I. (1973). *J. Biol. Chem. 248*, 3582-3592.

SIGNIFICANCE OF SUPEROXIDE DISMUTASE AND CATALASE ACTIVITIES IN THE STRICT ANAEROBES, SULFATE REDUCING BACTERIA

C.E. HATCHIKIAN and J. LE GALL

Laboratoire de Chimie Bacteriénne, C.N.R.S.
13274 Marseille Cedex 2, France

G.R. BELL

Department of Biochemistry
The University of Georgia
Athens, Ga. 30601, U.S.A.

INTRODUCTION

Sulfate reducing bacteria are unique in the microbial world: although they are strict anaerobes, they possess a "respiratory" type of metabolism in which oxygenated sulfur compounds are reduced by specific reductases into hydrogen sulfide. Their electron transfer chain from the energetic substrates to the terminal reductases is complex and implies quinones, flavoproteins, heme and non-heme iron-proteins. Oxidative phosphorylations have been shown to occur during electron transfer (Peck, 1966). The sulfate reducing bacteria have been divided into two genuses, namely *Desulfotomaculum* and *Desulfovibrio* (Postgate and Campbell, 1966). This study will be related to the latter genus exclusively.

The reports of Hata *et al.* (1964), who isolated sulfate reducing bacteria from the superficial waters of Hiroshima Bay and of Novozhilova and Berezina (1968), who found these bacteria in the sediments of a turbulent, highly aerated lake in a Russian semi-desert, are indications that strict anaerobes may possess a protecting system against temporary exposures to oxygen. Indeed, catalase was found in these organisms (Bell and Le Gall, 1971) and a recent publication by Hewitt and Morris (1975) shows that superoxide dismutase activity is also found in sulfate reducing bacteria and in some clostridial species.

Since several types of catalase and superoxide dismutase have been found in aerobic species and because of the speculations that are made concerning bacterial evolution in connexion with these types of proteins, it was of interest to purify and characterize the two enzymes from a strict anaerobe.

OCCURRENCE OF CATALASE AND SUPEROXIDE DISMUTASE IN *DESULFOVIBRIO*

The results of the survey of *Desulfovibrio* species for catalase and superoxide dismutase activities are shown in Table I. The distribution of these enzymes seems to be erratic throughout the genus *Desulfovibrio*. Catalase activity was detected in extracts of *D. desulfuricans*, strain Norway 4, *D. gigas* and *D. vulgaris*. The specific activity of catalase in extracts of the latter two organisms was about 80 μmoles min^{-1} mg^{-1}. Those organisms which showed no detectable catalase activity were *D. salexigens* and *D. desulfuricans*, strain Essex 6. The sulfate-reducers containing catalase activity also exhibited superoxide dismutase activity. The specific activity of superoxide dismutase in extracts of *D. gigas* and *D. vulgaris* was about 1 unit mg^{-1}; the level of activity was higher in *D. desulfuricans* strain Norway 4 (3.8 units mg^{-1}). Superoxide dismutase activity was not detected in sulfate-reducers lacking catalase activity as *D. salexigens* and *D. desulfuricans* strain Essex 6.

TABLE I

Survey of the Presence of Catalase and Superoxide Dismutase Activities in *Desulfovibrio*

Organism	NCIB[a] number	Catalase[b]	Superoxide Dismutase[c]
Desulfovibrio vulgaris	8,303	+	+
Desulfovibrio gigas	9,332	+	+
Desulfovibrio desulfuricans			
strain Norway 4	8,310	+	+
strain Essex 6	8,307	-	-
Desulfovibrio salexigens	8,403	-	-

[a]NCIB: National Collection of Industrial Bacteria

[b]Catalase was assayed by the spectrophotometric method of Beers and Sizer (1953). Crude cell extracts were used in the assays; protein concentration reached 2 mg protein by assay in the case where activity was not detected.

[c]Superoxide dismutase was assayed by the spectrophotometric method of McCord and Fridovich (1969). Dialysed crude extracts were used in the assays; protein concentration was 2.5 mg protein by assay in the case where activity was not detected.

CATALASE

Purification

Catalase was purified from *D. vulgaris* strain Hildenborough (NCIB 8303). The activity was measured by a modification of the spectrophotometric method of Beers and Sizer (1953) as described by Luck (1963).

The purification scheme was organized to allow the separation of cytochromes, flavodoxin, rubredoxin, hydrogenase and sulfite reductases which are of interest for other studies. A 630 fold purification was obtained from steps involving chromatography on silica gel, alumina gel, DEAE-cellulose and gel filtration. The purified protein was homogenous after gel electrophoresis with a specific activity of 50,000 μmoles H_2O_2 min^{-1} mg^{-1}. At pH 8.3 or 8.9, the catalase migrated as a green band toward the cathode and, after staining with buffalo black, a heavy, well defined protein band was observed at the same position on the gel.

Properties

Determination of the Molecular Weight. The molecular weight of the purified catalase was determined by low speed sedimentation equilibrium experiments. At 0.35 mg ml^{-1} an apparent weight average molecular weight of 230,500 was calculated assuming a partial specific volume of 0.73. The linearity of the dependence of the natural logarithm of protein concentration on the square of the distance from the center of rotation was indicative of a homogeneous preparation.

The molecular weight of catalase from *D. vulgaris* was also estimated from the molecular weights of its subunits as determined by sodium dodecyl sulfate (SDS) polyacrylamide gel electrophoresis. Treatment of the native enzyme with 1.0% SDS and 1% 2-mercaptoethanol (BME) for 6 hours at 37°C followed by dialysis against 0.1% SDS and BME for 24 hours gives two bands in the SDS-polyacrylamide system. The major band corresponded to a molecular weight of 60,000 and the minor band to 120,000. It is apparent that the native catalase (MW 240,000) consists of four euqal-sized subunits.

Absorption Spectrum. The ultraviolet and visible absorption spectrum of *D. vulgaris* catalase is shown in Fig.1. Absorption maxima are at 278, 405, 500, 535, 589 and 623 nm. A calculation of the molar extinction coefficient at 405 nm (based on a molecular weight of 232,000) yields a value of 20.5 x 10^4 M^{-1} cm^{-1}. This spectrum is typical of those reported in the literature for other heme-type catalases which exhibit maxima at 404-406, 500-505, 535-540 and 624-626 nm (Deisseroth and Dounce, 1970). *D. vulgaris* catalase is not reducible in the visible region by sodium dithionite, a property also typical of other catalases. However, the peak at 589 nm has not been previously observed with any catalase, and its presence suggested that a further investigation of the type of heme in this enzyme should be made. It should be pointed out that the hemoprotein nature of *D. vulgaris* catalase could not be established until the latter stages of purification because of the presence of desulfoviridin, the dissimilatory sulfite reductase in these organisms.

Pyridine Hemochromogen Spectrum. Figure 2 shows the absorption spectrum of the alkaline pyridine hemochromogen of *D. vulgaris* catalase (2.6 x 10^{-6} M). This spectrum is representative of pyridine hemochromogen spectra of other protoheme IX hemoproteins (Chance and Herbert, 1950). The heme iron concentration calculated from this data is 4.6 x 10^{-6} M. Using this value the number of hemes per molecule of catalase (MW 232,000) was calculated to be 1.8.

Ferric Complexes of D. Vulgaris *Catalase with Various Ligands.* Table II lists the absorption maxima of the complexes of ferric *D. vulgaris* catalase with water (natural ligand), cyanide, azide or sulfide as the ligand occupying the 6th coordination position. Each of these derivatives was prepared according to Herbert and Pinsent (1948) by adding an excess of the sodium salt of each anion to a sample of enzyme. The maxima for these complexes with catalase from bovine liver are listed as a comparison (Deisseroth and Dounce, 1970).

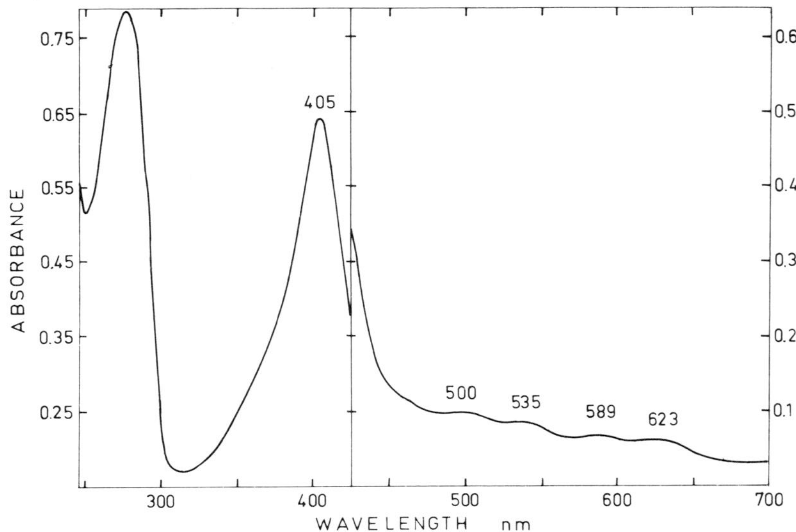

Fig.1. *Absorption spectrum of* D. vulgaris *catalase. The protein concentration was 0.74 mg/ml; a 1 cm light path cuvette was used.*

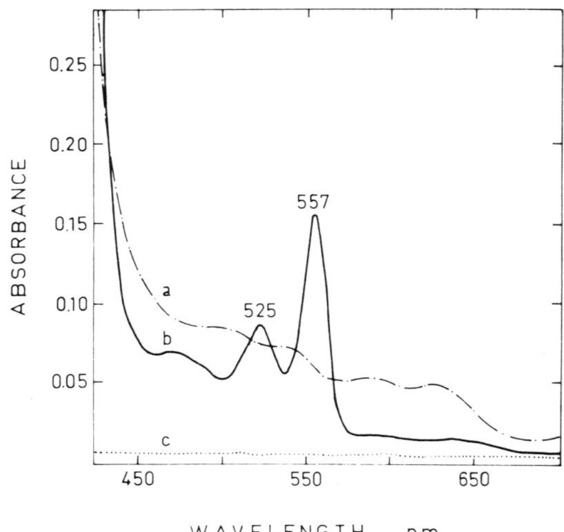

Fig.2. *Pyridine hemochromogen spectrum of* D. vulgaris *catalase. Curve a: absorption spectrum of 2.6 x* 10^{-6} *M catalase in 50 mM potassium phosphate buffer pH 7.6. Curve b: pyridine hemochromogen spectrum from reaction mixture of 2.6 x* 10^{-6} *M catalase, 100 mM NaOH, 100 mM pyridine and a few crystals of dithionite. Curve c: reaction mixture in (b) minus enzyme.*

Comparative Properties of Catalase from D. Vulgaris *and from Other Organisms.* Catalase activity has been demonstrated in several strains of sulfate-reducing bacteria. The purification of *Desulfovibrio vulgaris* catalase to homogeneity represents the first time this enzyme has been purified from a

TABLE II

Absorption Maxima of Complexes of Catalase with Various Ligands

Source	Visible Absorption Maxima of Complex (nm) — Ligand at Sixth Coordination Position of Heme Iron			
	Water	Cyanide	Azide	Sulfide
D. vulgaris	623,589,535,500,405	585,552,424	623,589,535,500,405	625,589,540,517,418
Bovine liver*	630,---,535,500,405	585,555,425	619,---,535,500,411	640,587,548,---,**

*Values for bovine liver catalase are from Deisseroth and Dounce, 1970.

**Not reported

strict anaerobe.

In almost every respect the catalase from *D. vulgaris* is typical of the catalases purified from a number of aerobic cells. The enzymes have strikingly similar molecular weights, subunit structures, and absorption spectra (Table III).

TABLE III

Physicochemical Properties of Catalases from *D. Vulgaris* and Other Sources

Source	Molecular Weight	Specific Activity	Molar Extinction Coeff. of Soret Band (x 10^{-4})	Hemes Per Molecule
*D. vulgaris**	232,000	50,000	20.5	1.8
Bovine liver	240,000	36,000	29.0	2
Equine liver	240,000	45,000	34.0	3
Equine erythrocyte	240,000	61,000	40.2	4
M. lysodeikticus	232,000	99,000	41.0	4
Rps. spheroides	232,000	120,000	40.5	4

*Values for *D. vulgaris* catalase are from this work; all others were given by Deisseroth and Dounce, 1970.

Protoheme IX is the prosthetic group of all heme-type catalases. Bonnischsen (1948) assayed the activity of catalases prepared from a number of sources which contained from 1.5 to 4 hemes per molecule. When plotted these data indicated zero activity at a heme-value of one, which suggested that at least two hemes are required for catalytic activity. The approximate 2-fold difference, then, between the activity of *D. vulgaris* catalase and other bacterial catalases (e.g. *M. lysodeikticus*) is not surprising in that the former catalase has two and the latter four hemes per molecule. As can be seen in Table II there is close agreement in the absorption maxima of the cyanide-catalase complexes of the enzymes from *D. vulgaris* and bovine liver. It is noteworthy that the unique peak at 589 nm observed in the absorption spectrum of *D. vulgaris* catalase persists in the azide-complex of this enzyme and constitutes the only major differences with the azide complex of the bovine liver catalase. Because the sulfide-derivative of the mammalian catalase shows a peak at 587 nm and because large amounts of H_2S are produced from dissimilatory sulfate reduction in *D. vulgaris*, it seemed possible that the peak at 589 nm in the spectrum of pure *D. vulgaris* catalase could be a result of some fraction of the molecules of enzyme having the 6th coordination position of the heme iron atom occupied by sulfide rather than water. However, no increase in absorbance at 589 nm was observed when Na_2S was added to a sample of *D. vulgaris* catalase. Also different peak shifts are observed in the sulfide-catalase spectra of *D. vulgaris* and bovine liver. Nicholls (1961) found that when catalase is exposed to hydrogen sulfide in the presence of hydrogen peroxide, an inactive derivative of catalase and sulfur is formed. These complexes were postulated to be between sulfide and the porphyrin ring of catalase, and three structures were proposed:

1) a methene-bridge thioketone; 2) a methene-bridge thiol and 3) a methene-bridge or pyrrole-ring cyclic thioether. Structure 1 is most probable because thiol reagents had no effect and because a strong Soret absorption of the complex indicating that the conjugated ring structure was present. The presence of such adducts in *D. vulgaris* catalase seems possible.

THE SUPEROXIDE DISMUTASE

Purification

Superoxide dismutase was purified from *D. desulfuricans* strain Norway 4 (NCIB 8310) since this strain exhibited a level of activity about 4-fold higher than that of *D. vulgaris* which was the organism chosen for the purification of catalase.

The crude extract prepared from a wet weight of bacteria slightly higher than 4 kg contained 1.1 x 10^6 units of superoxide dismutase activity. The first steps of purification allowed separation of the electron carriers. Cytochrome c_3 was removed by an amberlite batch process followed by silica gel batch treatment and acidic proteins containing ferredoxin and rubredoxin were separated by DEAE-cellulose batch treatment. The by-product obtained after extraction of cytochrome c_3 and acidic electron carriers was used for the purification of superoxide dismutase. It contained 2.8 x 10^5 units of superoxide dismutase activity, i.e. 26% of the activity present in the crude extract. A 540 fold purification was obtained from steps involving ammonium sulfate fractionation, chromatography on DEAE-cellulose, hydroxyapatite and gel filtration. This procedure of purification produced a superoxide dismutase exhibiting a specific activity of 2,060 units mg^{-1} as measured by the method of McCord and Fridovich (1969).

The purified dismutase was concentrated and stored frozen at - 20°. Polyacryalamide gel electrophoresis (40 μg protein per gel) of the purified enzyme at pH 8.9 and the location of the superoxide dismutase activity on the gel indicated that the preparation was homogeneous.

Since the purification yield was only 11%, it is not possible to be certain that superoxide dismutase activity in these bacteria is due to one or two proteins although the evidence suggests that only the Fe-protein is present, as will be seen later.

Properties

Molecular Weight. The molecular weight of the purified dismutase was determined by gel filtration with a Sephadex G-75 column. From the data obtained, the molecular weight of the native superoxide dismutase was estimated to be 43,000 daltons. The molecular weight of superoxide dismutase was also estimated from the molecular weight of its subunits. The purified dismutase was subjected to sodium dodecyl sulfate gel electrophoresis, in the presence and absence of 2-mercaptoethanol (Weber and Osborn, 1969). In these conditions, subunits of *D. desulfuricans* (Norway 4) superoxide dismutase gave a single band and comparisons of its mobility to that of molecular weight standards yielded a subunit weight of 21,500. Thus, the native enzyme exists as a dimer of two subunits of equal size and these subunits are not joined by interchain disulfide bonds.

Absorption Spectrum. Figure 3 presents the optical absorption spectrum of the purified superoxide dismutase. In the ultraviolet region, besides the typical peak at 280 nm seen with most proteins, shoulders were observed at 290 nm and 260 nm. The molar extinction coefficient at 280 nm was 8.6 x 10^4 M^{-1} cm^{-1}. The visible absorption spectrum exhibited a weak absorption from 320 to 600 nm with a shoulder around 350 nm and a small absorption band at about 550 nm. The molar extinction coefficient at the 350 nm shoulder was 2,580 M^{-1} cm^{-1}. The visible and ultraviolet spectra were comparable to the spectra of iron-superoxide dismutases isolated from *E. coli* (Yost and Fridovich, 1973), *P. leiognathi* (Puget and Michelson, 1974) and the blue-green algae *P. boryanum* (Asada *et al.*, 1975).

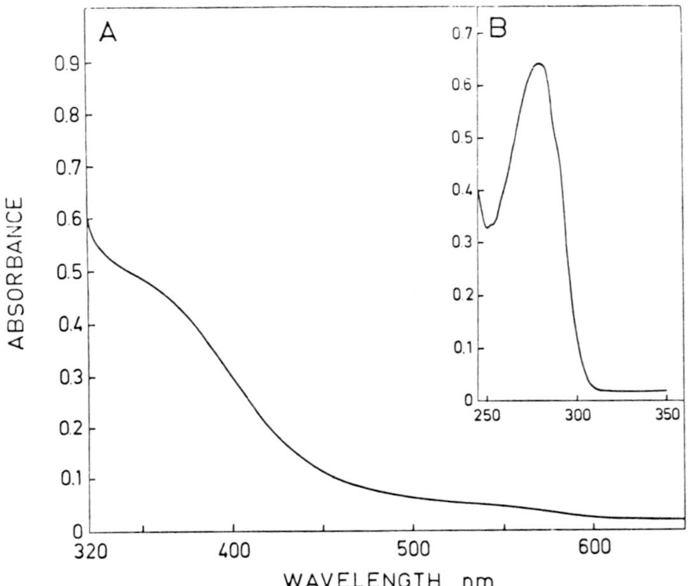

Fig.3. *Absorption spectra of* D. desulfuricans *(Norway 4) superoxide dismutase. A 1 cm light path cuvette was used. A, the visible absorption spectrum of the enzyme at 8 mg/ml in 10 mM potassium phosphate buffer pH 7.6. B, the ultraviolet absorption spectrum of the enzyme at 0.32 mg/ml in the same buffer.*

Metal Analysis. Atomic absorption spectroscopy indicated the presence of 1.6 atom of iron per molecule (average of five estimations with different preparations). The lowest value obtained was 1.35 atom of iron and the highest 1.87. Assays for manganese and copper indicated that these metals were not present.

Amino Acid Analysis. Samples of the enzyme were hydrolyzed, under vacuum, in 6N HCl at 110° for 24 hours. After removal of the HCl, under vacuum, the residues were dissolved in 0.2 N sodium citrate pH 2.2 and subjected to amino acid analysis. The amino acid composition of *D. desulfuricans* (Norway 4) superoxide dismutase is presented in Table IV. It is compared with that of the iron-containing superoxide dismutases from *E. coli*, *P. sepia* and *P. boryanum*. Considerable similarity exists with these dismutases, especially with respect to the high amount of aromatic amino-acids. However, the superoxide dismutase of *Desulfovibrio* appears to have a slightly higher amount of lysine, aspartic and glutamic acids and cysteine.

TABLE IV

Comparative Amino Acid Composition of Iron-containing Superoxide Dismutase

Residues/subunit	*D. desulfuricans* Norway	*P. sepia*[1]	*E. coli*[2]	*P. boryanum*[3]
Lysine	16	10	10	12
Histidine	5	6	6	5
Arginine	4	2	4	1.5
Tryptophane[4]	6	6	4	5
Aspartic acid	27	20	22	23
Threonine	12	13	13	10
Serine	7	8	10	9
Glutamic acid	18	15	16	16
Proline	8	9	9	10
Glycine	19	13	16	15
Alanine	22	20	26	24
Cystine (Half)[5]	2.0		1	1
Valine	8	8	11	10
Methionine[5]	2	3	0	1.5
Isoleucine	7	7	8	2.5
Leucine	15	14	14	17
Tyrosine	8	6	6	5
Phenylalanine	12	10	10	12

[1]From Puget and Michelson (1974)
[2]From Yost and Fridovich (1973)
[3]From Asada *et al.* (1975)
[4]Based on ultraviolet absorbance (Edelhoch, 1967)
[5]Cysteine and methionine were determined as cysteic acid and methionine sulfone after performate oxidation of the protein (Hirs, 1967)

We thank Dr. M. Bruschi for the amino acid analysis.

Comparative Physico-chemical Properties of Iron-containing Superoxide Dismutases. The physico-chemical properties of *D. desulfuricans* (Norway 4) superoxide dismutase and those of Fe-enzymes isolated from aerobic micro-organisms are reported in Table V. These proteins have similar molecular weights, subunit structure, absorption spectra and amino acid compositions. It appears with respect to iron content that, except for *E. coli* dismutase which contains 1 atom per molecule, the other Fe superoxide dismutases contain 2 atoms per molecule. The visible absorption spectrum of *D. desulfuricans* (Norway 4) superoxide dismutase which is comparable to the spectra of the other iron-containing enzymes is suggestive of the presence of Fe^{3+}. As far as the specific activities of these Fe-proteins are concerned, only those of *D. desulfuricans* and *E. coli* which were measured by the same method (McCord and Fridovich, 1969) can be compared. From the data reported here it appears that superoxide dismutase from *D. desulfuricans* (Norway 4) exhibits physico-chemical properties similar to those of ferri-superoxide dismutases isolated from aerobic cells.

TABLE V

Comparative Properties of Iron-containing Superoxide Dismutases from Bacteria

Properties	*D. desulfuricans*	*P. sepia*[1]	*E. coli*[2]	*P. boryanum*[3]
Molecular Weight	43,000	40,660	38,700	41,700
Subunit M. Wt.	21,500	20,000	17,800	21,000
Specific Activity	2,060	5,806	2,470	7,657
ε_{280} (M^{-1} cm^{-1})	86,000	78,880	55,000	66,900
ε_{350} (M^{-1} cm^{-1})	2,580	2,500	1,675	2,860
Iron (atom/mole)	1.6	1.6	1	2
Tryptophane (residues/mole)	12	12	8	10

[1]From Puget and Michelson (1974)

[2]From Yost and Fridovich (1973)

[3]From Asada *et al.* (1975)

DISCUSSION

The mechanism of oxygen toxicity towards certain microorganisms has been a puzzle for decades. It was early suggested by McLeod and Gordan (1923) that oxygen itself was not toxic to anaerobes but rather the divalently reduced derivative, hydrogen peroxide. It was stated that catalase, which catalyzes the conversion of hydrogen peroxide to oxygen and water was only present in aerobic bacteria. However, subsequent investigations revealed that this distinction between aerobes and anaerobes was not unequivocal. Prévot and Thouvenot (1952) reported that several strict anaerobes were catalase-positive. Conversely, McCord *et al.* (1971) noted that many organisms capable of aerobic growth do not contain catalase. The most recent theory of obligate-anaerobiosis is based on the detrimental effects of superoxide anion reported by the same authors. This highly reactive free radical O_2^- has been demonstrated to be generated on reduction of oxygen by reduced clostridial ferredoxin (Misra and Fridovich, 1971a), reduced flavins and flavoproteins (Massey *et al.*, 1969), hydroquinones (Misra and Fridovich, 1971b) and a number of metallo-flavoenzymes (Greenlee *et al.*, 1962; McCord and Fridovich, 1968; Rotilio *et al.*, 1970).

The apparent ubiquity of superoxide dismutase in all oxygen-metabolizing organisms led McCord *et al.* (1971) to suggest that the prime physiological function of the enzyme is to protect the oxygen-metabolizer against the potentially detrimental effects of the superoxide free radical and that the level of this enzyme, not catalase, in an organism is the basis on which the organism can be classified as an aerobe, strict anaerobe, or aerotolerant anaerobe. The authors surveyed the distribution of catalase and superoxide dismutase in various microorganisms and concluded that strict anaerobes exhibited no superoxide dismutase activity and, generally, no catalase activity. Aerotolerant anaerobes, which do not use molecular oxygen as a terminal electron acceptor but which can metabolize to a limited extent, were reported to be devoid of catalase activity but did have superoxide dismutase activity. All aerobes which reduce oxygen via cytochrome systems, contained both activities. Later, Gregory and Fridovich (1973a) confirmed this theory by demonstrating that superoxide dismutase was induced by oxygen in both *S. faecalis* and *E. coli* whereas catalase was not detectable in *S. faecalis* and was unaffected by oxygen in *E. coli*. These results and those obtained with *B. subtilis* (Gregory and Fridovich, 1973b) under the same conditions are consistent with the view that superoxide dismutase is an important component of the defenses of these organisms against the toxicity of oxygen, whereas the catalases are of secondary importance.

The presence of superoxide dismutase and catalase activities in some sulfate reducers appears to be in conflict with the theory of McCord *et al.* (1971). The sulfate-reducing bacteria include a large number of microorganisms exhibiting a strictly anaerobic mode of growth based on reduction of sulfate as terminal electron acceptor. Lactate, which is the energetic substrate generally used by these bacteria, is oxidized to acetate *via* the pyruvate phosphoroclastic system whereas sulfate is reduced to sulfide by electrons produced by the metabolism of lactate. As already noted a number of electron carriers present in these bacteria insures the coupling between oxidation of lactate and reduction of sulfate. These electron carriers include c-type cytochromes (Postgate, 1956; Le Gall *et al.*, 1965; Ambler *et al.*, 1971), b-type cytochrome (Jones, 1971; Hatchikian and Le Gall, 1972), ferredoxin (Le Gall and Dragoni, 1966), flavodoxin (Le Gall and Hatchikian,

1967), rubredoxin (Le Gall and Dragoni, 1966; Bruschi and Le Gall, 1972) and menaquinone (Maroc *et al.*, 1970; Weber *et al.*, 1970). Most of these electron carriers as well as non heme iron proteins such as hydrogenase and adenylyl sulfate reductase (Bramlett, 1973) present in these organisms are potential generators of superoxide anion.

The results reported here allow one to conclude that the catalase of the strict anaerobe sulfate-reducing bacteria is a typical catalase and that their superoxide dismutase is similar to the ferri-superoxide dismutase present in several aerobic microorganisms. The problem of the presence of these two enzymes in such strict anaerobes is posed. Two explanations are possible: one is that, as suggested by Lumsden and Hall (1975), due to the Urey effect (that is to say the photolysis of water that may have taken place under UV radiation before the establishment of the protective ozone layer in the atmosphere) some molecular oxygen was produced in the primitive oceans, against which primitive strict anaerobes had to be protected; the second is that superoxide dismutase and catalase activities in strict anaerobes is a more recent acquisition allowing the survival of the strict anaerobes which are able to synthetize these enzymes, when entering in contact with atmospheric oxygen by accident.

In view of the possible primordial appearance of superoxide dismutase it is to be noted that the ferri- and mangano-superoxide dismutases of *E. coli* have been shown, by sequence homology, to be related to each other and to the mitochondrial mangano-enzyme (Steinman and Hill, 1973). A high degree of sequence homology was also reported between *B. stearothermophilus* Mn-enzyme and the dismutases listed above (Bridgen *et al.*, 1975). The close relationship between the ferri- and mangano-superoxide dismutases suggests that they have evolved from a common ancestral protein. If this is the case it then could be important from an evolutionary point of view to determine the type of enzyme and the amino acid sequences of superoxide dismutase present in primitive microorganisms. The dissimilatory reduction of sulfate is a very primitive process since isotopic distribution of sulfur showed that it started from 2.5 to 3 billion years ago (Ault and Kulp, 1959). The presence of an iron-containing superoxide dismutase both in the sulfate-reducing anaerobe *D. desulfuricans* strain Norway 4 and the photosynthetic anaerobe *Chromatium* (Asada *et al.*, 1975) is in favour of the primitivity of this type of superoxide dismutase as compared with Mn- and Cu-enzymes. Thus, comparisons of amino acid sequences of these proteins may prove to be of interest in the establishment of phylogenetic relationships between these important groups of primitive organisms.

However, the somewhat erratic presence of the enzyme in both sulfate reducing bacteria and clostridia (Hewitt and Morris, 1975) should be kept in mind. This could be an indication that the information for the biosynthesis of the protein comes from a plasmid. If this is the case, the presence of superoxide dismutase could be a very recent acquisition by anaerobes and, if so, its structure cannot be related to the history of the bacteria.

Part of this work is included in the dissertation of G.R. Bell (Studies on electron transfer systems in Desulfovibrio: purification and characterization of hydrogenase, catalase and superoxide dismutase; Ph.D. Thesis, The University of Georgia, Athens, Georgia, U.S.A., 1973).

ACKNOWLEDGEMENTS

We are indebted to Dr. M. Scandellari and R. Bourrelli (from the Laboratoire de Chimie Bactérienne) and to the Fermentation Plant of the University of Georgia for growing the bacteria. We wish also to thank Mrs. N. Forget and G. Bovier-Lapierre for skilful technical assistance.

REFERENCES

1. Ambler, R.P., Bruschi, M. and Le Gall, J. (1971). In "Recent Advances in Microbiology", International Congress of Microbiology, Mexico, p. 25-34.
2. Asada, K., Kanematsu, S., Takahashi, M. and Kono, Y. (1975). In "Symposium on Fe- and Cu-proteins", Honolulu, Hawai, in press.
3. Ault and Kulp (1959). *Geochim. Cosmochim. Acta 16*, 201-235.
4. Beers, R.F. and Sizer, I.W. (1953). *Science 117*, 710-712.
5. Bell, G.R. and Le Gall, J. (1971). *Bacteriological Proceedings*, p. 263.
6. Bonnichsen, R. (1948). *Acta Chem. Scand. 2*, 561-573.
7. Bramlett, R.N. (1973). Studies on the mechanism of adenylyl sulfate reductase from the sulfate reducing bacterium *Desulfovibrio vulgaris*. Ph.D. Thesis, University of Georgia.
8. Bridgen, J., Harris, J.I. and Northrop, F. (1975). *FEBS Letters 49*, 392-395.
9. Bruschi, M. and Le Gall, J. (1972). *Biochim. Biophys. Acta 263*, 279-282.
10. Chance, B. and Herbert, D. (1950). *Biochem. J. 46*, 402-413.
11. Deisseroth, A. and Dounce, A.L. (1970). *Phys. Rev. 50*, 319-375.
12. Edelhoch, H. (1967). *Biochemistry 6*, 1948-1954.
13. Greenlee, L., Fridovich, I. and Handler, P. (1962). *Biochem.*, 779-783.
14. Gregory, E.M. and Fridovich, I. (1973a). *J. Bacteriol 114*, 543-548.
15. Gregory, E.M. and Fridovich, I. (1973b). *J. Bacteriol 114*, 1193-1197.
16. Hata, Y., Kadota, H., Miyoshi, H. and Kimata, M. (1964). In "Second International Conference on Water Pollution Research", p. 287.
17. Hatchikian, E.C. and Le Gall, J. (1972). *Biochim. Biophys. Acta 267*, 479-484.
18. Herbert, D. and Pinsent, J. (1948). *Biochem. J. 43*, 193-203.
19. Hewit, J. and Morris, J.G. (1975). *FEBS Letters, 50*, 315-318.
20. Hirs, C.H.W. (1967). In "Methods in Enzymology" (Colowick, S.P., ed.) Vol.2, pp. 59-62, Academic Press, New York.
21. Jones, H.E. (1971). *Arch. Mikrobiol. 80*, 78-85.
22. Le Gall, J. and Dragoni, N. (1966). *Biochem. Biophys. Res. Comm. 23*, 145-149.
23. Le Gall, J. and Hatchikian, E.C. (1967). *C.R. Acad. Sc. Paris, 264*, 2580-2583.
24. Le Gall, J., Mazza, G. and Dragoni, N. (1965). *Biochim. Biophys. Acta 99*, 385-387.
25. Luck, H. (1963). "Catalase", in "Methods of Enzymological Activity" (H.O. Bergmeyer, ed.), Academic Press, New York, p. 885.
26. Lumsden, J. and Hall, D.O. (1975). *Nature 257*, 670-672.
27. Maroc, J., Azerad, R., Kamen, M.D. and Le Gall, J. (1970). *Biochim. Biophys. Acta 197*, 87-89.
28. Massey, V., Strickland, S., Mayhew, S.G., Howell, L.G., Engel, P.C., Matthews, R.G., Schuman, M. and Sullivan, P.A. (1969). *Biochem. Biophys. Res. Comm. 36*, 891-897.
29. McCord, J.M. and Fridovich, I. (1968). *J. Biol. Chem. 243*, 5753-5760.
30. McCord, J.M. and Fridovich, I. (1969). *J. Biol. Chem. 244*, 6049-6055.

31. McCord, J.M., Keele, B.B., Jr. and Fridovich, I. (1971). *Proc. Nat. Acad. Sci.*, U.S.A., *68*, 1024-1027.
32. McLeod, J.W. and Gordan, J. (1923). *J. Pathol. Bacteriol. 26*, 332-343.
33. Misra, H.P. and Fridovich, I. (1971a). *J. Biol. Chem. 246*, 6886-6890.
34. Misra, H.P. and Fridovich, I. (1971b). *J. Biol. Chem. 247*, 188-192.
35. Nicholls, P. (1961). *Biochem. J. 81*, 374-383.
36. Novozhilova, M.I. and Berezina, F.S. (1968). *Mikrobiologiya 37*, 534.
37. Peck, H.D., Jr. (1966). *Biochem. Biophys. Res. Comm. 22*, 112-118.
38. Postgate, J.R. (1956). *J. Gen. Microbiol. 14*, 545-572.
39. Postgate, J.R. and Campbell, L.L. (1966). *Bact. Rev. 30*, 732-738.
40. Prevot, A.R. and Thouvenot, H. (1952). *Ann. Inst. Pasteur*, *83*, 443-448.
41. Puget, K. and Michelson, A.M. (1974). *Biochimie 56*, 1255-1267.
42. Rotilio, G., Calabrese, L., Finazzi Agro, A. and Mondovi, B. (1970). *Biochim. Biophys. Acta 198*, 618-619.
43. Steinman, H.M. and Hill, R.L. (1973). *Proc. Nat. Acad. Sci.*, U.S.A., *70*, 3725-3729.
44. Weber, M.M., Matschiner, J.T. and Peck, H.D. (1970). *Biochem. Biophys. Res. Comm. 38*, 197-204.
45. Weber, K. and Osborn, M. (1969). *J. Biol. Chem. 244*, 4406-4412.
46. Yost, F.J. and Fridovich, I. (1973). *J. Biol. Chem. 248*, 4905-4908.

STRUCTURE-FUNCTION RELATIONSHIPS IN SUPEROXIDE DISMUTASES

J.A. FEE

Biophysics Research Division
and Department of Biological Chemistry
The University of Michigan
Ann Arbor, MI 49109, U.S.A.

INTRODUCTION

Superoxide dismutases constitute a rather disparate class of metalloproteins having in common the feature that they are effective catalysts of the reaction

$$2\ O_2^- + 2H^+ \rightarrow H_2O_2 + O_2. \qquad [1]$$

Arguments have been made that this is the physiological function of these proteins (1). The purpose of this review is to discuss the physical and chemical measurements relevant to the metal binding sites of the recently determined structure of bovine copper-zinc superoxide dismutase and to offer a molecular mechanism for the catalysis which is consistent with both the structure of the enzyme and the kinetics of the reaction. In addition some recent results from our laboratory concerning the metal binding sites of other superoxide dismutases will be briefly discussed.

THE Zn/Cu PROTEIN OF BOVINE ERYTHROCYTES

Composition and Structure

This protein has a molecular weight of 31,500 daltons, binds two atoms each of Zn and Cu, and consists of two identical polypeptide subunits. The amino acid sequence has been determined by Hill and his co-workers (2-4), and it is presented in Fig.1 which also indicates the residues involved in the metal binding site as deduced from the X-ray structural investigations of the Richardsons.

The structural work has now been extended to 3A resolution (5,6) which has allowed location of the metal binding sites and revealed some interesting structural features of the protein itself such as the "beta-barrel" discussed elsewhere in this volume by Richardson.

The features of the metal binding site which are of interest to us are indicated in Fig.2.

```
 1                                        10
Ac-Ala-Thr-Lys-Ala-Val-Cmc-Val-Leu-Lys-Gly-Asp-

                                    20
Gly-Pro-Val-Gln-Gly-Thr-Ile-His-Phe-Glu-Ala-Lys-

                            30
Gly-Asp-Thr-Val-Val-Val-Thr-Gly-Ser-Ile-Thr-Gly-

                40                  Cu      Cu
Leu-Thr-Glu-Gly-Asp-His-Gly-Phe-His-Val-His-Gln-

        50
Phe-Gly-Asp-Asn-Thr-Gln-Gly-Cmc-Thr-Ser-Ala-Gly-

60  Cu                                  70
Pro-His-Phe-Asn-Pro-Leu-Ser-Lys-Lys-His-Gly-Gly-
    Zn                              Zn
                                    80
Pro-Lys-Asp-Glu-Glu-Arg-His-Val-Gly-Asp-Leu-Gly-
                        Zn          Zn
                        90
Asn-Val-Thr-Ala-Asp-Lys-Asn-Gly-Val-Ala-Ile-Val-

                100
Asp-Ile-Val-Asp-Pro-Leu-Ile-Ser-Leu-Ser-Gly-Glu-

            110                         Cu
Tyr-Ser-Ile-Ile-Gly-Arg-Thr-Met-Val-Val-His-Glu-

120                                     130
Lys-Pro-Asp-Asp-Leu-Gly-Arg-Gly-Gly-Asn-Glu-Glu-

                                    140
Ser-Thr-Lys-Thr-Gly-Asn-Ala-Gly-Ser-Arg-Leu-Ala-

                            150
Cmc-Gly-Val-Ile-Gly-Ile-Ala-Lys
```

Fig.1. *Amino acid sequence of bovine erythrocyte superoxide dismutase. The residues associated with copper have a Cu above the residue abbreviation and those associated with zinc have Zn below the residue abbreviation. Note that histidine-61 is bound to both zinc and copper. Taken from references 3 and 6.*

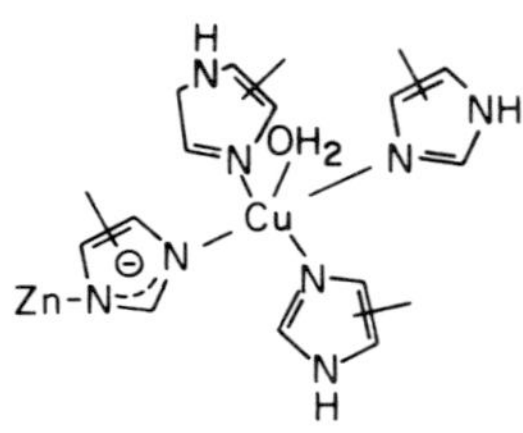

Fig.2. *Metal binding sites of copper and zinc on superoxide dismutase. The zinc is in an approximately tetrahedral arrangement of the side chains of His-61, His-69, His-78, and Asp-81; while the copper resides in an approximate plane with His-44, His-46, His-61 and His-118. The water molecule is not observed by X-ray diffraction, but its presence is deduced from other measurements.*

Spectral Properties of the Cu^{2+}

The Cu^{2+} ion has rather unique spectral properties when they are considered in detail. Bovine erythrocuprein exhibits the optical and CD spectra shown in Fig.3 and the EPR spectrum shown in Fig.4 which is characterized by a rhombic g-tensor having the principal values: 2.265, 2.11, and 2.03 (7,8).

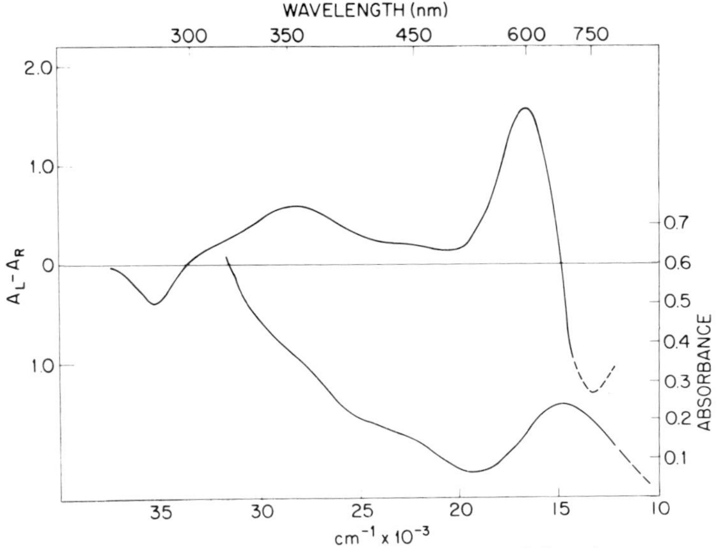

Fig.3. *Optical and circular dichroism spectra of bovine superoxide dismutase. The concentration of copper was approximately 1.8 mM.*

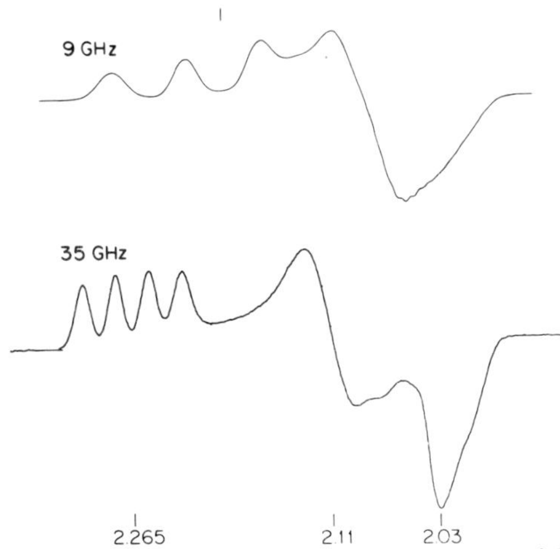

Fig.4. *X- and K-band EPR spectra of bovine superoxide dismutase.*

The information in Figs.3 and 4 can be combined under the assumption of a distorted or rhombic-tetragonal symmetry (9) to yield the approximate energy level diagram for the d-orbitals of the Cu^{2+} shown in Fig.5. This is consistent with all the d-orbitals being affected differently by the ligand field. It is certain that the orbital having the highest energy is $d_{x^2-y^2}$ (10) and the ordering of the levels as indicated in Fig.5 is such as to account approximately for the principal g-values of the EPR spectrum.

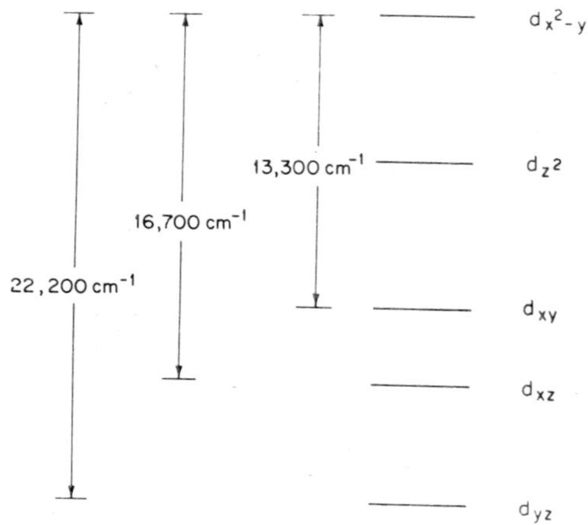

Fig.5. *d-Orbital levels in the copper of superoxide dismutase.*

The tentative assignments are as follows: The near u.v. absorption is too intense to be readily classified as a d-d transition, and it is therefore considered to be a ligand to metal charge transfer transition. The shoulder at 450 nm (22,000 cm^{-1}) corresponds to the energy difference between the d_{yz} and d_{x2-y2} orbitals, the band at 600 nm (16,700 cm^{-1}) is the transition between the d_{xz} and d_{x2-y2} levels, and the long wavelength band, 750 nm (13,300 cm^{-1}) correlates with the d_{xy}-d_{x2-y2} energy difference. There are no other bands out to 2500 nm (11) although the $d_{z2} \rightarrow d_{x2-y2}$ transition may be found at even longer wavelengths (9). These assignments are consistent with the g-values using the following abbreviated expressions (12-14):

$$g_x = g_e - [2\lambda a_x^2/(E_{yz} - E_{x2-y2})] \quad [2]$$

$$g_y = g_e - [2\lambda a_y^2/(E_{xz} - E_{x2-y2})] \quad [3]$$

$$g_z = g_e - [2\lambda a_z^2/(E_{xy} - E_{x2-y2})] \quad [4]$$

where λ is the spin orbit coupling constant, -828 cm^{-1}, g_e is the g-value for the free electron, a_x, a_y, and a_z are combinations of coefficients occasionally referred to as orbital reduction parameters. Since the unpaired electron spends some time on the liganding nitrogen atoms, evident as superhyperfine structure in the $g\perp$ region of the 9 GH_z spectrum of Fig.4, molecular orbital theory must be applied (14) before any other information can be gleaned from the EPR spectrum. This theory has been applied to a number of small Cu complexes, and given a complete set of experimental data such as the principal values and molecular orientation of the hyperfine coupling tensor, g-tensor, and the coupling constants to the nitrogen atoms it should be possible to apply this theory to obtain a fairly quantitative description of the bonding in the complex. These data have not yet been collected, and it is hoped that the single crystal EPR study presently being carried out in the author's laboratory will yield such information.

The EPR spectrum of the native protein is a very sensitive probe for changes in the environment of the Cu^{2+}, and has been used extensively in probing the catalytic site of the Zn/Cu superoxide dismutase in a wide variety of experiments.

Evidence for a Ligand at the Metal Binding Site Having a pKa > 9

Oxidation-Reduction Properties of the Cu. The first evidence for an unusual ligand associated with the metal binding center came from studies of the mid-point potential of the Cu^{2+}/Cu^{+} couple as a function of pH (15)[1]. In this work a ferro/ferricyanide redox buffer was coupled with the Cu^{2+}/Cu^{+} system, and the mid-point potential of the Cu-couple was determined from Nernst type plots at pH values ranging from 4 to 8.7. Above pH 5.5 it was found that the mid-point potential decreased linearly to pH 8.7 with a slope of -0.06 V/pH (Fig.6).

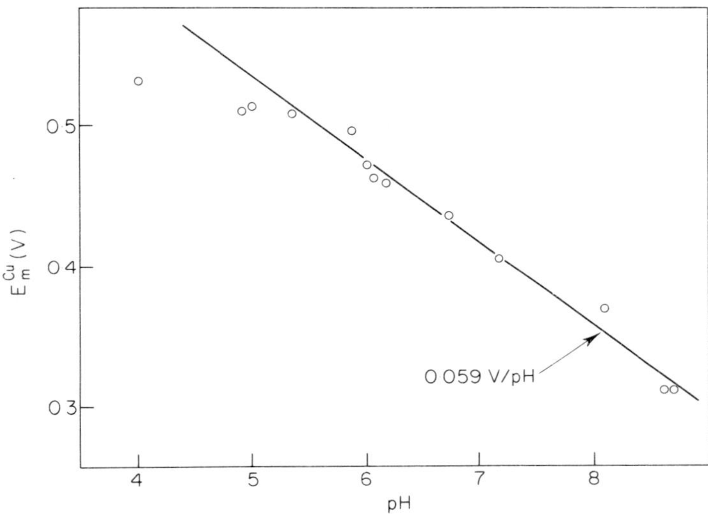

Fig.6. *The mid-point potential of the Cu^{2+}/Cu^{+} couple of superoxide dismutase. Taken from Reference 15.*

This behavior is consistent with a half-reaction of the Cu couple

$$E\ Cu^{2+} + e^{-} + H^{+} \rightleftarrows H^{+}\ E\ Cu^{+}$$

where the proton is bound to some basic group on the reduced protein. That one proton was actually bound with each electron was determined independently (15).

Since Bray and co-workers (16) have shown that the protein is only partially reduced during catalysis the relationship of the "thermodynamic" reduced state as determined in the redox studies to the "kinetic" reduced state is not clear.

Before offering a rationalization of this observation we will explore the possible reason why E_m becomes independent of pH below 5.5.

The Low pH Transition. When solutions of holo-protein are acidified into the pH range 4-3 subtle alterations occur in the spectral properties of the Cu^{2+}. These have been examined in some detail (17) and it was shown that a transition involving the Cu^{2+} occurs between pH 4 and 3 which is entirely reversible, and which does not involve unfolding of the protein or dissociation of the subunits. The changes which have been observed are: a small decrease in 680 nm absorbance, loss of the 450 nm shoulder, and appearance of a new band at 330 nm. The circular dichroism in the 600 to 700 nm region is abolished, while a weak negative band appears at approximately 380 nm and a new band at 310 nm. There are also changes in the EPR spectrum. $A_{\parallel}$ increased from approximately 130 gauss to approximately 150 gauss. Although a 35 GH_z EPR spectrum has not been recorded it is apparent that some of the rhombic character characteristic of the native protein is lost. The process is quite slow on adjusting the pH downward and is rapid in the opposite direction. Recent work has shown that the pH at which the onset of this transition occurs is probably closer to 5 than 4.

Rotilio *et al.* (18) were actually the first to observe the low pH transition but they supposed that it occurred at a lower pH and stated ".. the typical EPR signal of copper of the native bovine enzyme is maintained through the full range of pH 3 - 10". Below pH 3 more drastic changes occur at the copper binding site (17) which are not of interest here.

Interpretation of the pH Dependence of E_m and the Low pH Transition. These two phenomena are thought to be related by the following equilibria

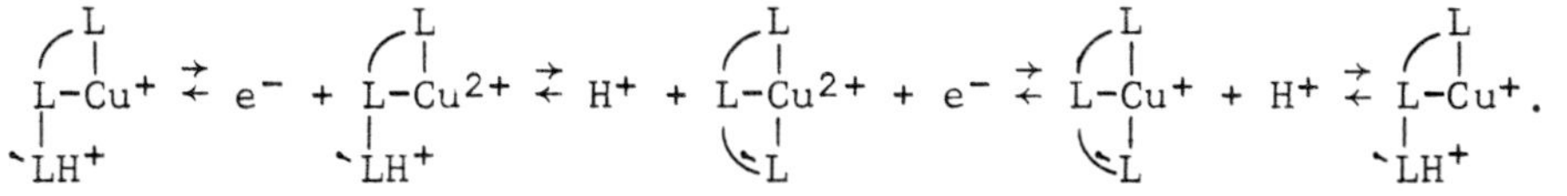

$$\text{L-Cu}^{+}(\text{L})(\text{LH}^{+}) \rightleftarrows e^{-} + \text{L-Cu}^{2+}(\text{L})(\text{LH}^{+}) \rightleftarrows H^{+} + \text{L-Cu}^{2+}(\text{L})(\text{L}) + e^{-} \rightleftarrows \text{L-Cu}^{+}(\text{L})(\text{L}) + H^{+} \rightleftarrows \text{L-Cu}^{+}(\text{L})(\text{LH}^{+}). \quad 5$$

In order to account for the linear decrease of E_m up to pH 8.7, the apparent pK_a of ˋL must be substantially greater than 8.7. The deviation of E_m from the expected linear relation below pH 5.5 can be explained in terms of the protonation of the high pK ligand prior to reduction, and the low pH transition of the holo-protein is simply the protonation of L´.

These concepts are now formulated in terms of the specific structural changes indicated in Fig.7. A similar scheme has been proposed by Hodgson and Fridovich (19).

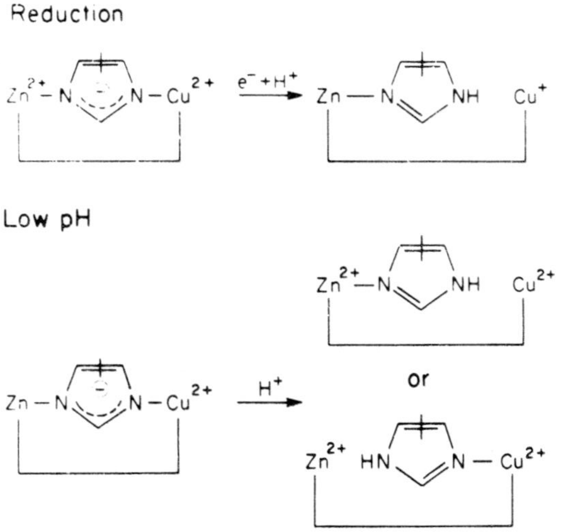

Reduction

Zn^{2+}–N(imidazolate)N–Cu^{2+} $\xrightarrow{e^{-}+H^{+}}$ Zn–N(imidazole)NH Cu^{+}

Low pH

Zn–N(imidazolate)N–Cu^{2+} $\xrightarrow{H^{+}}$ Zn^{2+}–N(imidazole)NH Cu^{2+} or Zn^{2+} HN(imidazole)N–Cu^{2+}

Fig.7. *Projected structural changes at the metal binding sites of superoxide dismutase upon reduction (upper) and at low pH (lower).*

Here we have associated the high pK ligand with the bridging imidazolato anion. This is justified in terms of the structure and the known pK_a values for free, $\geq$14 (20), and metal bound, 11.7 (21,22), imidazole. The indications in Fig.6 regarding which metal imidazole coordinate bond is being broken are on less firm ground. Fee and Ward (23) have argued that the Cu^+ has a coordination position open to solvent and that this is necessary for catalysis. Blumberg and his co-workers (24) have found that only the Cu has altered X-ray absorption properties upon reduction while the Zn ion is largely unaffected. On these grounds then, we argue that reduction leads to the breaking of the Cu-N bond, but a choice cannot be made for the low pH transition.

Evidence for a Bridging Ligand Between the Two Metal Ions

Metal Ion Substitutions. A number of metal ion derivatives of bovine superoxide dismutase have been prepared in which each metal site is occupied by a particular ion. These are indicated in the somewhat optimistic representation of Fig.8. In this discussion we will have concern for only a few of the derivatives which have been prepared, with the purpose of providing evidence for a direct bridging ligand between the two metal atoms. All the derivatives can be prepared by first removing the metal ions from the holo-protein followed by "reconstitution" according to specific procedures which involve recombining metal ions with the apoprotein.

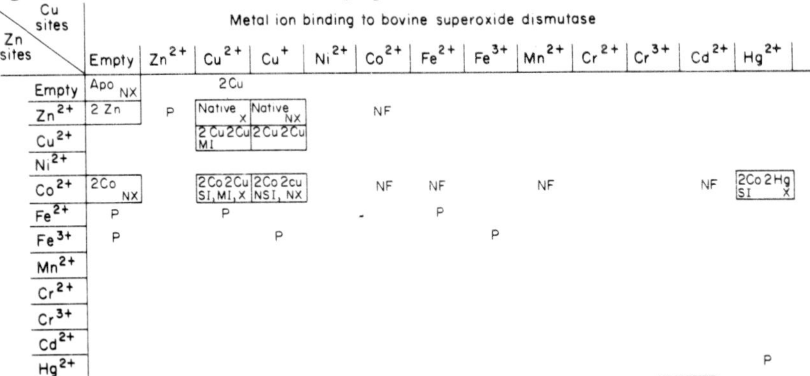

Fig.8. *Metal binding matrix for superoxide dismutase.* ▭ *prepared and characterized; P, preliminary experiments suggest formation; NF, non forming; SI, spectral dependence of one metal on the other; MI, magnetic interaction observed; X, crystals form; NX, no crystals.*

$2Co^{2+}2\ Cu^{2+}$. This derivative was first prepared by Calabrese *et al.* (25) who treated the protein with 0.5 M $CoCl_2$ for an extended period of time and noted that Zn^{2+} was displaced and there was less EPR detectable Cu^{2+} than total Cu present. This suggested some sort of magnetic interaction between the Co^{2+} and the Cu^{2+} ions. A different approach was taken in the author's laboratory which involved treatment of apoprotein with stoichiometric quantities of Co^{2+} or Co^{2+} plus Cu^{2+}. The following were subsequently reported (26): optical, circular dichroic, EPR and crystallographic (27) properties. The optical spectra of 2 Co^{2+} protein and the reduced i.e., 2 $Co^{2+}2\ Cu^{+}$-protein were identical whereas the optical absorption spectrum due to Co^{2+} was distinctly different in the oxidized derivative (Fig.9). Further, a comparison of the spectral properties of Cu^{2+} in native and $2Co^{2+}2Cu^{2+}$ proteins showed small but distinct differences. Taken together these observations suggested that the environments of the two ions were interacting in some way. The close proximity of the two metal ions in the $2Co^{2+}2Cu^{2+}$ derivative was demonstrated

by examining the EPR properties. In this respect, the $2Co^{2+}2Cu^{2+}$ protein has a very broad and weak signal in the g4 region observable at temperatures less than 77°K and a small amount of adventitious Cu^{2+} evident in the g2 region. Upon reduction of the Cu^{2+} with dithionite a new signal appeared in the region of g4 which arises from Co^{2+} and is identical to that of the $2Co^{2+}$-derivative (Fig.10). These results, later confirmed by Rotilio *et al.* (28), were consistent with the two metal ions being in very close proximity. It was subsequently shown by magnetic susceptibility measurements (Fig.11) that there was orbital overlap between the metals which resulted in their being antiferromagnetically coupled (27). This can only occur if some ligand acts to bridge the metals (29).

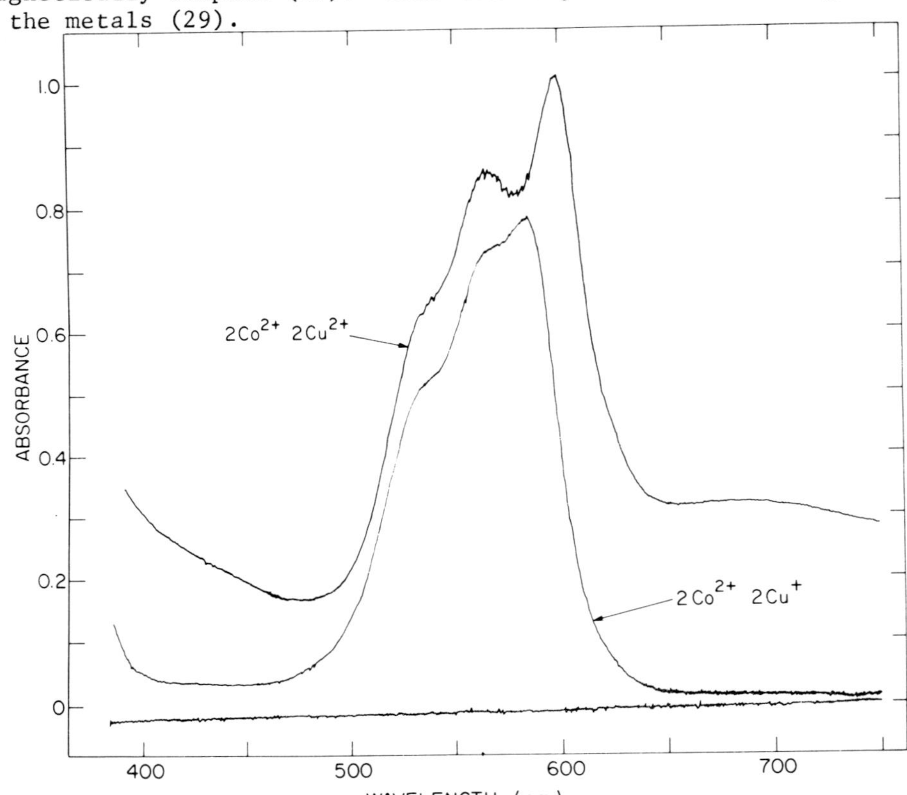

Fig.9. *Optical absorption spectra of 2 Co^{2+} $2Cu^{2+}$- and 2 Co^{2+} $2Cu^{+}$ derivatives of superoxide dismutase. Reference 27.*

The $2Cu^{2+}2Cu^{2+}$-Derivative. Further evidence for a bridging ligand derives from studies with the $4Cu^{2+}$ protein. This was found in 1973 (30) by the observation of an unusual EPR signal in solutions of apoprotein treated with an excess of Cu^{2+}. Subsequently, the derivative was prepared by direct titration of the apoprotein with Cu^{2+} as shown in Fig.12, and it was found that the unusual EPR spectrum disappeared on lowering the temperature below 10°K (Fig. 13) and therefore probably derived from the excited triplet state of a spin-coupled pair of Cu^{2+} ions. By fitting the temperature dependence of the EPR spectrum with an appropriate partition function it was shown that the triplet state was approximately 50 cm^{-1} above the ground state singlet (31). This again is good evidence for ligand mediated overlap between the Zn^{2+} and Cu^{2+} binding sites provided, of course, the additional Cu^{2+} is occupying the Zn^{2+} binding site.

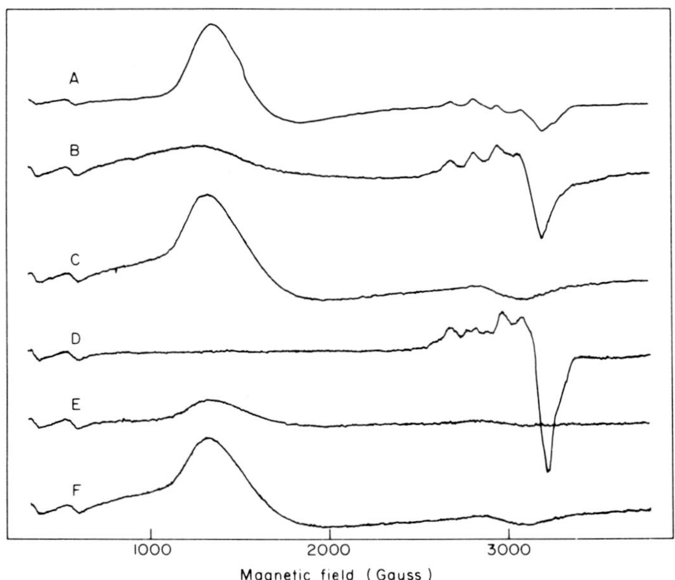

Fig.10. *Low temperature (20°K) EPR spectra of SOD derivatives. A, 2 Co^{2+}; B, 2 Co^{2+} 2 Cu^{2+}; C, 2 Co^{2+} 2 Cu^{+} (reduction with dithionite). At 50°K; D, 2 Co^{2+} 2 Cu^{2+}; E, 2 Co^{2+} 2 Cu^{+} (reduction with dithionite) and at 34°K, F, 2 Co^{2+} 2 Cu^{+}.*

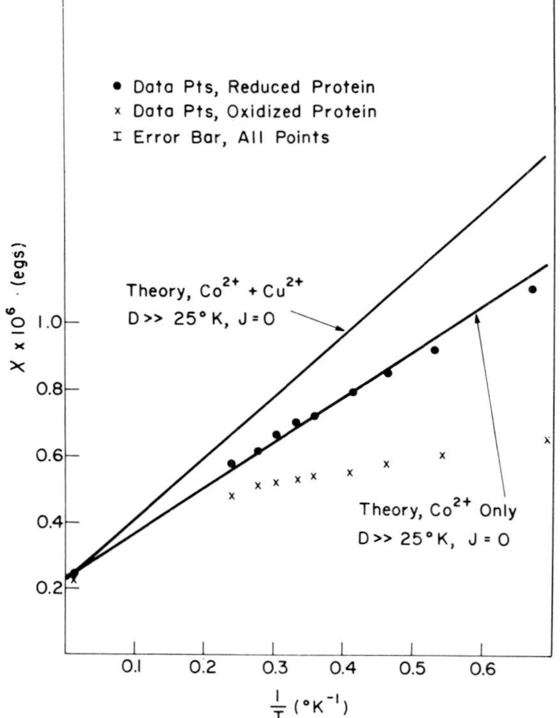

Fig.11. *Magnetic susceptibility of 2 Co^{2+} 2 Cu^{2+-} and 2 Co^{2+} 2 Cu^{+} derivatives in the temperature range 1.4 - 77°K. Taken from Reference 27.*

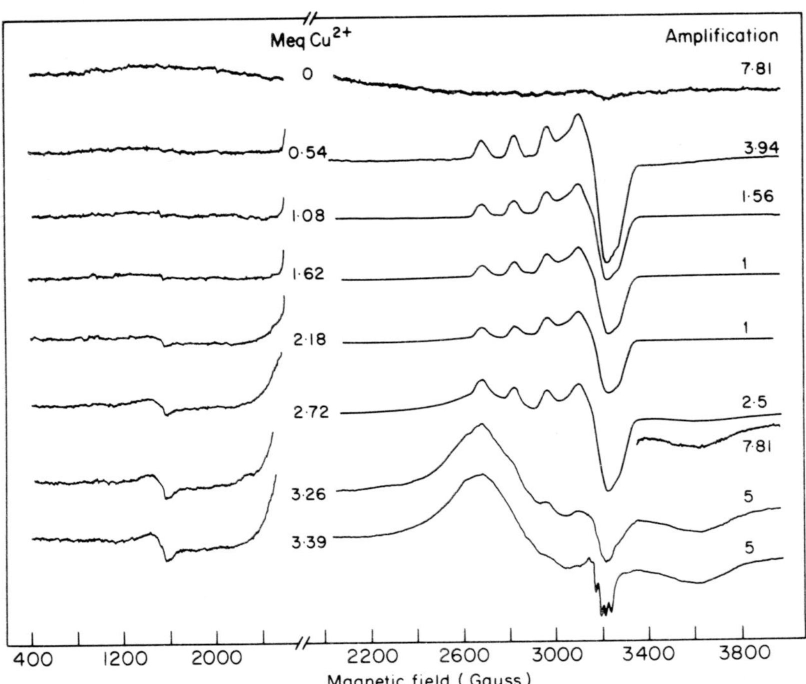

Fig.12. *EPR titration of apo-superoxide dismutase with cupric ion. Taken from Reference 31.*

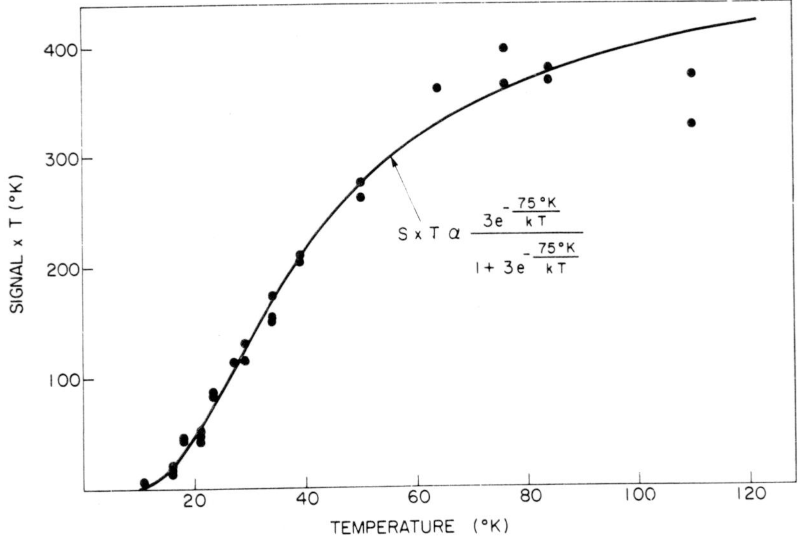

Fig.13. *Temperature dependence of the EPR spectrum of 2 Cu^{2+} 2 Cu^{2+}-derivative of superoxide dismutase. Taken from Reference 31.*

^{67}Zn Substitution. It was shown that ^{67}Zn substitution did not result in any apparent broadening of the EPR spectrum of the Zn/Cu protein (30), which may indicate a very low spin density on the Zn^{2+}, possibly because of largely ionic bonding (32).

Interpretation of the Magnetic Interaction in the Above Derivatives. In the case of the $2Co^{2+}2Cu^{2+}$-protein there seems to be a very strong antiferromagnetic interaction between Co^{2+} and Cu^{2+}. The magnetic susceptibility results (27) did allow us to set a lower limit, >> 5 cm^{-1}. Rigo *et al.* (33) have suggested on the basis of a water proton relaxation study that the magnitude of the coupling is greater at room temperature, i.e., $\gtrsim$ 210 cm^{-1}. The coupling constant in the $4Cu^{2+}$ derivative is lower, 50 cm^{-1} (31). The levels involved in magnetic coupling are shown in Fig.14. It should be emphasized that the coupling constant is extremely sensitive to minor structural perturbations (34).

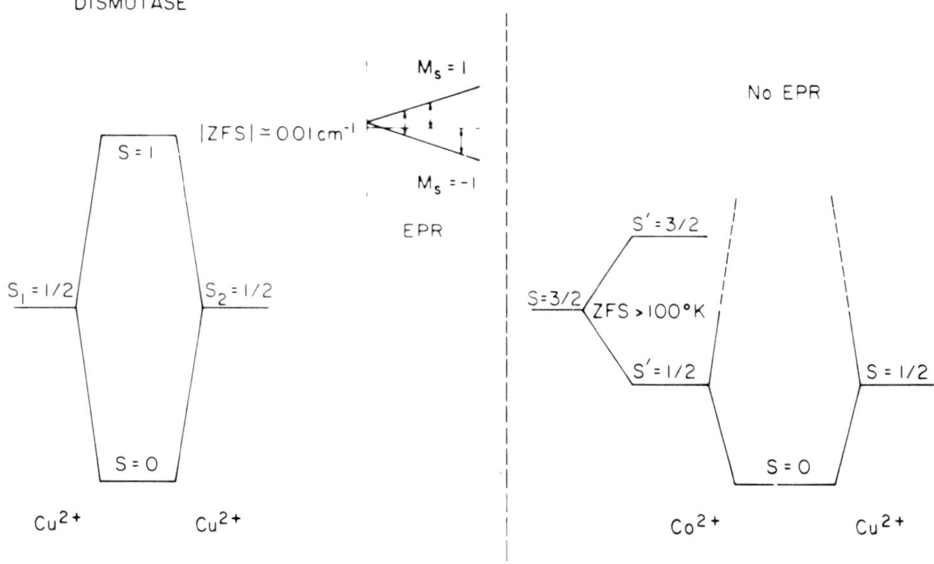

Fig.14. *Magnetic interactions in 2 Cu^{2+} 2 Cu^{2+}- and 2 Co^{2+} 2 Cu^{2+}-derivatives of superoxide dismutase.*

The possibility should be raised here that the $2Co^{2+}2Cu^{2+}$- and $2Cu^{2+}2Cu^{2+}$ derivatives do not represent true metal substitution derivatives. While the $2Co^{2+}2Cu^{2+}$ protein emulates the native protein in almost every way, and monoclinic crystals of native and $2Co2Cu^{2+}$ derivatives have identical space group dimensions (27), slightly different reflection intensities evident in the precession photographs (35) raise the possibility that a significant conformational change has occurred upon substituting Co^{2+} for Zn^{2+}. It would thus be useful to carry out a difference-Fourier analysis of these diffractions. At present, however, it is reasonable to assign to histidine-61 the role both as bridging ligand as evidenced by the magnetic measurements and as the high pK ligand as evidenced in the redox and low pH behaviors.

Evidence for Nitrogenous Ligands to the Cu^{2+}

There are a number of observations which suggest that nitrogen atoms are coordinated to the Cu^{2+}. These involve EPR, NMR and chemical modification.

The EPR measurements have been made on both native protein (36) and its cyanide complex which exhibits additional ligand hyperfine structure in the

EPR spectrum (30,36-38). A typical spectrum showing only the low-field ^{63}Cu-hyperfine line is presented in Fig.15. The smooth line in Fig.15 was calculated assuming interaction of the unpaired electron with three equivalent ^{14}N atoms. Hafner and Coleman (39) have found that a single cyanide binds to the Cu^{2+} and by using $^{13}C^{14}N^-$ they found that the EPR spectrum was split by 45 gauss by the ^{13}C nucleus. The only conclusion which can be drawn from these observations is that each Cu^{2+} is bonded to at least three identical N-atoms and one cyanide ion.

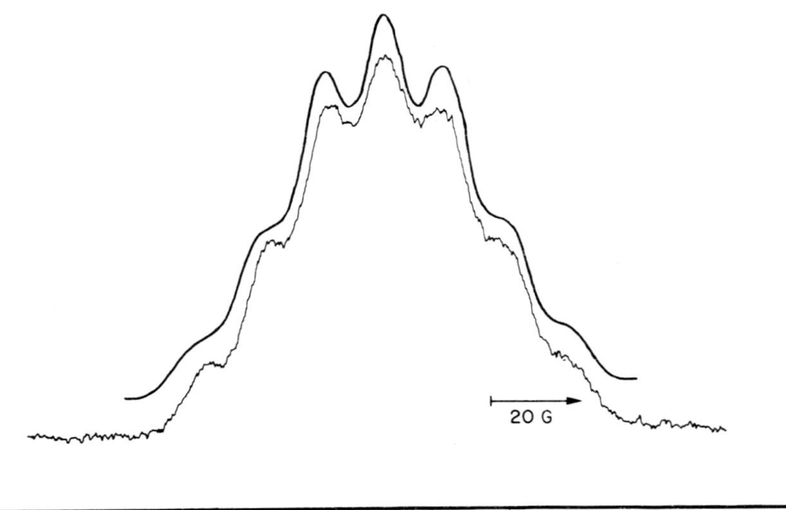

Fig.15. *Experimental and computed superhyperfine pattern on the $M_I = -3/2$ hyperfine line of Cu^{2+}-cyano superoxide dismutase. The computed line was based on three equivalent ^{14}N atoms and a line-width of 12 gauss.*

The photo oxidation studies of Forman *et al.* (40) indicated that histidine was available for reaction in the apoprotein but not in the holo-protein. Studies on the inactivation of the protein with Br_2^- and other radicals (41) also suggested the presence of histidine residues near the metal binding sites. The reaction of the apoprotein with diazotized sulfanilamide (40) indicated the presence of approximately 4 histidine residues while none was reactive toward this reagent in the holo-protein. Bray and his co-workers (42) demonstrated that one histidine residue per subunit was destroyed upon inactivation with H_2O_2 and that the Cu^{2+} binding site was thereby modified.

High resolution NMR studies by Stokes *et al.* (43) have suggested that approximately six histidine residues are in proximity to the Cu^{2+} but not all are necessarily bonded to the copper.

Thus, there existed considerable evidence for histidine binding to the Cu^{2+} prior to the determination of the structure by X-ray crystallographic techniques. There now seems to be some conflict as to how many nitrogen atoms are actually bound to Cu^{2+}. On one hand the EPR experiments are only consistent with three while the combined diffraction and amino acid sequence results would appear to require four nitrogen atoms bound to Cu^{2+}. At present there does not seem to be a solution to this problem but it could be rationalized in different ways. First it is important to understand the origin of the ligand hyperfine interaction in a Cu^{2+} complex of this type.

Since the unpaired electron resides in the d_{x2-y2} orbital it will only have a strong interaction with atoms which effectively overlap with this orbital. In practice this means that only N-atoms of in-plane ligands will yield hyperfine interaction of the magnitude observed while a N-atom interacting with the d_{z2} orbital, for example, would not be resolved. It should also be pointed out that cyanide is a very strong ligand to Cu^{2+} and invariably occupies an in-plane coordination position (cf. Ref.44 for an excellent discussion of dynamic Jahn-Teller inversions in Cu^{2+} complexes). It is possible then that when cyanide coordinates to the Cu^{2+} either one of the histidine ligands is displaced from the Cu^{2+} or shifts from an in-plane position to an axial position. The bridging ligand does not appear to be broken (28), and unpublished results from the author's laboratory show that ^{-}CN, not HCN, is bound at pH 9 indicating that the bridging ligand is intact.

Evidence for Open Coordination Positions on the Metals of Oxidized and Reduced Superoxide Dismutase

The importance of open coordination positions for catalysis of the dismutation reaction has been emphazied for some time (23,45,46) and a complete discussion of their possible role in catalysis will be given below.

The Oxidized Protein. Gaber *et al.* (46) presented evidence for at least one, rapidly exchanging, water molecule in the first coordination sphere of superoxide dismutase bound Cu^{2+}, and it was shown (37) that this water molecule was probably displaced by cyanide (Fig.16) and by azide upon their binding to the Cu^{2+}.

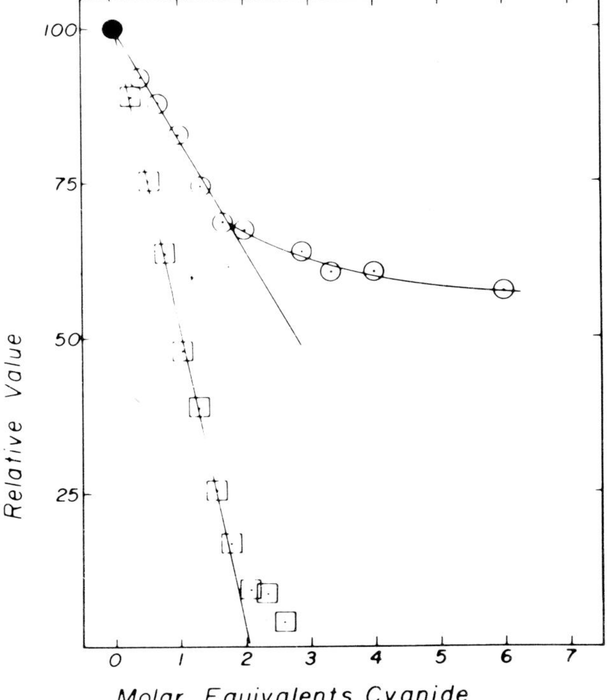

Fig.16. *Titration of oxidized superoxide dismutase with cyanide:* ⊙, *spectral changes;* ⊡ , *water relaxivity. From Reference 37.*

The strong relaxivity of this Cu^{2+} toward water protons has been confirmed by others (33,37). However, it should be pointed out that the claim of Rotilio *et al.* (38) to have observed the water molecule by exchanging the protein into D_2O and recording a modified EPR spectrum is ill-founded. The altered EPR spectrum clearly arose from a slight denaturation during the exchange process, and it was pointed out later (17) that the EPR spectrum was *identical* in D_2O and H_2O.

Further evidence of the accessibility of the Cu^{2+} to solute molecules comes from anion binding studies. In the oxidized protein, at least, the following anions have been shown to bind to the Cu: CN^- (8,37-39), N_3^- (8,37), SCN^-, OCN^- (37). Contrary to what was claimed in an earlier report (37), as was later pointed out by the present author (26), Zn^{2+} does not appear to be accessible to anions, and this has been emphasized in other publications (28, 48).

The Reduced Protein. The only evidence for an open coordination position on the reduced protein comes from a study of the effect of the protein on the relaxation of the ^{35}Cl nucleus of chloride ions (23). The reduced protein has a molar relaxivity which is approximately twice that of apoprotein or of the $2Zn^{2+}$ protein. As shown in Fig.17 the relaxivity is lowered to that of the apoprotein by addition of a small molar excess of KCN. It is assumed that the higher relaxivity of the reduced protein is due to Cl^- binding to a metal ion where it is subject to a larger electric field gradient than when free in water and that CN^- displaces Cl^- from this position. It was found that the relaxivity increased with decreasing temperature, indicating rapid exchange of the bound Cl^- ions (49). We presume that this coordination position is on Cu^+ and that it is the site of catalysis.

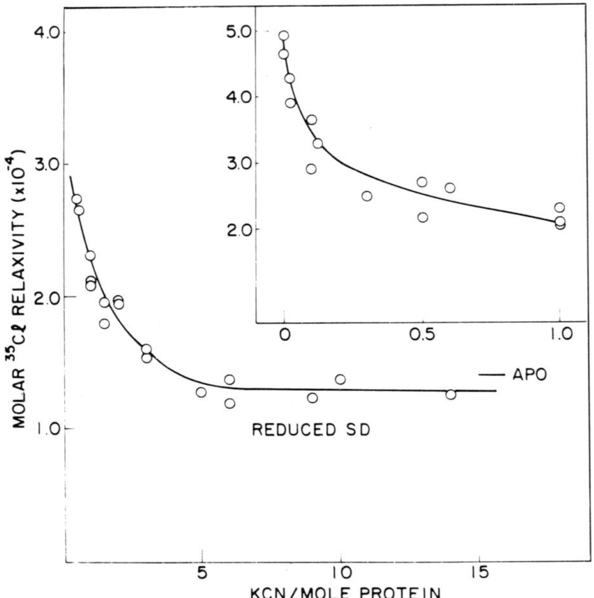

Fig.17. *Titration of reduced superoxide dismutase with cyanide; O, ^{35}Cl relaxivity. High relaxivity at low cyanide to protein ratio is due to a trace of adventitious Cu^{2+}. Reference 23.*

Anions as Inhibitors. A number of studies have shown that CN^- is an inhibitor of the dismutase activity and azide also inhibits, but less strongly (19). Inhibition probably results from blocking the access of O_2^- to the Cu at some state of the catalytic reaction. Rigo *et al.* (50) have reported the inhibition constant for CN^- to be 1.8×10^{-6} M and for N_3^- 1.43×10^{-2} M. At pH values between 10 and 12 activity is reversibly lost (50-52), and Rotilio *et al.* (38) have suggested that this is due to the ionization of the bound water molecule, $Cu^{2+}\ OH_2 \rightleftarrows Cu^{2+}OH^- + H^+$

The Importance of Open Coordination Position for Catalysis. It has been shown by Bray and co-workers (16) and Rabani and co-workers (53) that the Cu undergoes a cyclic reduction and oxidation during catalysis.

$$E\ Cu^{2+} + O_2^- \rightarrow E\ Cu\ + O_2 \qquad [6]$$

$$E\ Cu^+ + O_2^- + 2H^+ \rightarrow E\ Cu^{2+} + H_2O_2 \qquad [7]$$

It is reasonable to ask what the elementary steps are in these reactions. Fee and Gaber (45) suggested that inner sphere electron transfer would occur in the first step, and proposed the following scheme

$$E\ Cu^{2+}\ OH_2 + O_2^- \rightleftarrows E\ Cu^{2+}\ OH_2 \,..\, O_2^- \xrightarrow{k_{cat}} E\ Cu^{2+}\ O_2\ + H_2O \xrightarrow{fast}$$

$$E\ Cu^+ + O_2 + H_2O \qquad [8]$$

based on the well known ligand substitution process formulated by Eigen and Tamm (54). Is such a mechanism reasonable? This question is approachable by experiment. It is clear from the above discussion that the Cu^{2+} does indeed bind a water molecule, and it exchanges with a time constant, $10^{-8} \lesssim \tau_M <<$ 4×10^{-6} sec (46). Assuming the above mechanism

$$k_{cat} = \frac{K_{os}}{\tau_M} \qquad [9]$$

where K_{os} is the formation constant of the outer-sphere complex indicated in brackets. Setting $k_{cat} = 1 \times 10^9\ M^{-1}sec^{-1}$ (16,53) K_{os} must lie between $4 \times 10^3\ M^{-1}$ and $10\ M^{-1}$. The former value seems excessively high and an exchange time of 4 μsec appears to be quite long for a Cu^{2+} bound water molecule. In general, single ligands bound to Cu^{2+} have been found to exchange much more rapidly (55). However, Rigo *et al.* (56) have presented some data which suggests that saturation occurs in the range of 0.1 - 0.5 mM O_2^-, but the quantitative aspects of this work bear further scrutiny (57).

Is such a mechanism necessary? This could also be examined experimentally by measuring the rate of reduction of Cu^{2+}-anion complexes by O_2^-. Since O_2^- can act as an effective one-electron reducing agent, reduction by O_2^- of complexes in which the water molecule is replaced by another ligand, at rates comparable to the catalytic rate, would suggest that inner sphere electron transfer is not necessary. These experiments have not been done.

Turning now to the second step in which the Cu^+ is acting as a reducing agent toward an O_2^- molecule, it is clear that simple electron transfer *cannot* be occurring.

$$E\ Cu^+ + O_2^- \rightarrow E\ Cu^{2+} + O_2^{2-} \qquad [10]$$

This type of process is energetically unfeasible and it appears to occur only at highly polarized electrodes (58).

Chevalet *et al.* (58) have presented an elegant discussion of the reduction of molecular oxygen to H_2O_2 at a polarized cathode, and Fee and Ward (23) have pointed out the analogy of their mechanism to that of catalyzed dismutation. According to Chevalet *et al.* (58) the following mechanism occurs at a mercury cathode under normal conditions

$$M(e) + O_2 \xrightarrow{\text{fast}} M\ O_2^-$$

$$M(e)\text{-}(H_2O)\text{-}O_2^- \longrightarrow M + H\ O_2^- + OH^-.$$

The first step produces superoxide bound to the electrode surface. The second step requires an electrode bound water molecule which acts as a *proton donor* to the electrode bound superoxide. Transfer of the electron occurs only to an O_2^- interacting with a proton so as to avoid forming the very unstable O_2^{2-}. This mechanism can be altered (58,59). In the presence of a surfactant which can form a hydrophobic film over the electrode and yet allow electron transfer from the electrode to external dioxygen, the superoxide forms but is not bound to the electrode. The second step is thus blocked because the O_2^- can freely diffuse away from the electrode. This of course, is the basis of the polarographic assay procedure introduced by Rigo *et al.* (50).

The protein catalyst must be acting in a similar fashion to avoid formation of the dianion, and Fee and Ward (22) have proposed the following mechanism for the second step of catalysis

$$E.\ Cu^+\ OH_2 + O_2^- \rightleftarrows E\ Cu^+\ OH_2..O_2^- \rightleftarrows E\ Cu^+\ O_2^-...H\text{-}O\text{-}H \rightarrow$$

$$E\ Cu^{2+}\text{-}O\text{-}O\text{-}H + OH^- \xrightarrow{H_2O} E\ Cu^{2+}\ OH_2 + H_2O_2 \qquad [11]$$

in which O_2^- bonds to the open coordination position on the Cu^+, again by an Eigen-Tamm mechanism, and interacts with the proton of a water molecule. Electron transfer then occurs to form the peroxy anion coordinated to Cu^{2+} which is rapidly released. A mechanism in which the bridging ligand acts as the proton donor has been described (19).

It is possible that any sequestering agent which maintains an open coordination position on a redox active metal will form an effective catalyst of superoxide dismutation. It may further be speculated that catalysis of superoxide dismutation by erythrocuprein could be a secondary reaction unrelated to some as yet undiscovered physiological function.

THE IRON CONTAINING PROTEIN FROM *E. coli B*

The iron containing superoxide dismutase from *E. coli* was discovered by Yost and Fridovich (60) who reported that it had a molecular weight of 38,700 daltons, consisted of two identical subunits, and contained a single ferric atom per mole. Similar proteins were subsequently isolated from other microbial sources (61-63), and it was reported that these proteins contained from 1.4 to 2 g-atoms Fe per mole of protein.

Our recent work (64) on this protein will be briefly summarized. Our purified protein contains 2 g-atoms Fe per mole of protein. Cyanide does not associate with the ferric iron to a detectable degree, while both fluoride and azide have substantial affinity for the ferric ion. Optical and EPR titrations showed that each ferric ion was able to bind *two* fluoride ions. The first having an association constant of 520 M^{-1} and the second approximately 25 M^{-1}. Measurements of the superoxide dismutase activity in the presence of fluoride indicated an inhibition constant of 30 M^{-1}.

At room temperature only one azide ion binds per ferric ion to give a distinct spectral change, and assay by a number of procedures including our newly developed stopped-flow procedure showed that azide even at very high concentrations did not inhibit superoxide dismutase activity. Upon freezing to liquid nitrogen temperatures solutions of the iron protein in the presence of excess azide turned brilliant pink in a totally reversible process which we reasoned was due to the binding of an additional azide to the iron, and this was supported by EPR studies.

Room temperature optical titrations of iron superoxide dismutase with fluoride and azide are shown in Fig.18 with theoretical fits to a model in which two fluoride ions bind but only one azide.

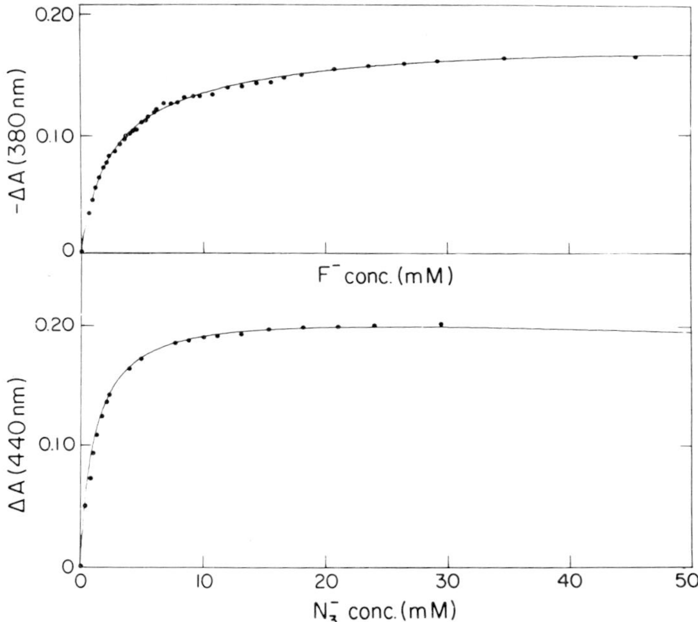

Fig.18. *Optical titrations of* E. coli B *iron superoxide dismutase with fluoride and azide. Solid lines are "best fit" to a model involving two* F^- *binding sites per iron and one* N_3^- *binding site per iron. Reference 64.*

Our present working model is that two coordination positions are open to solvent on each iron atom but that only one of these is essential for the dismutase activity.

We have recently obtained high quality crystals of this superoxide dismutase as shown in Fig.19, and in collaboration with Prof. M.L. Ludwig and her associates we are proceeding with a crystal structure determination.

THE MANGANESE CONTAINING PROTEIN FROM *E. coli B.*

This protein was discovered by Keele *et al.* (65) who reported a molecular weight of 40,000 with two identical subunits. The manganese content has varied from two to one depending on the type of measurement used for protein concentration (66). Keele *et al.* (65) postulated that the pink color of the resting protein was due to the presence of manganic, Mn(III), ion. Again, our recent work (67) will only be summarized in brief. We find a manganese

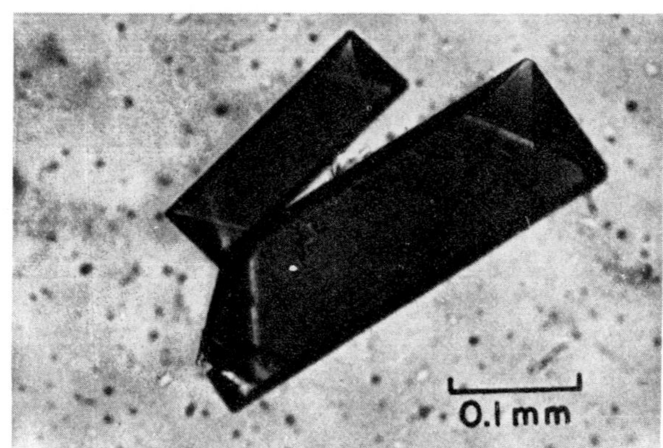

Fig.19. *Single crystals of iron superoxide dismutase from* E. coli B. *Taken by T.O. Slykhouse.*

content of 0.7 - 1.8 Mn per mole based on Lowry determination of protein concentration. The resting form of the protein does not exhibit any EPR spectrum even at very low temperatures, and its measured magnetic susceptibility is consistent with Mn(III). By contrast, the colorless protein obtained by treatment with sodium dithionite has a rhombic g 4.3 EPR signal arising from Mn(II), and the magnetic susceptibility is consistent with Mn(II) as well. The reduced form of the protein is quite unstable which probably accounts for the variable manganese contents, and the fact that we have not yet been able to develop a preparative procedure suitable for large scale purification of this protein.

The results described above provide the first unequivocal evidence for the existence of Mn(III) in a biological system, and confirm the suggestions of other workers (65,68,69).

FOOTNOTE

[1]The reference to Klug *et al.* (1972) on p. 4898 of this article is incorrect; apologies to the authors.

ACKNOWLEDGEMENTS

The author would like to thank the following collaborators who have been so helpful over the past several years: Bruce Gaber, Ray Ward, Seymour Koenig, Rodney Brown, Bill Phillips and Tom Moss. Special thanks go to Martha Ludwig for her help with the crystallographic experiments, to Bob Lieberman for Fig.15, and to Janice Short and Marion Frost for excellent help in preparing the manuscript. Support for this work was from USPHS grants GM 18869 and GM 21519. A travel stipend from the National Science Foundation is also acknowledged.

REFERENCES

1. McCord, J.M., Keele, B.B. Jr., and Fridovich, I. (1971). *Proc. Nat. Acad. Sci. 68,* 1024-1027.
2. Evans, H.J., Steinman, H.M. and Hill, R.L. (1974). *J. Biol. Chem. 249,* 7315-7325.
3. Steinman, H.M., Naik, V.R., Abernethy, J.L. and Hill, R.L. (1974). *J.*

Biol. Chem. *249,* 7326-7337.
4. Abernethy, J.L., Steinman, H.M. and Hill, R.L. (1974). *J. Biol. Chem.* *249,* 7339-7347.
5. Richardson, J.S., Thomas, K.A., Rubin, B.H. and Richardson, D.C. (1975). *Proc. Nat. Acad. Sci.* *72,* 1349-1353.
6. Richardson, J.S., Thomas, K.A. and Richardson, D.C. (1975). *Biochem. Biophys. Res. Commun.* *63,* 986-992.
7. Fee, J.A. and Gaber, B.P. (1971). *Fed. Proc.* *30,* Abstract 1409.
8. Rotilio, G., Morpurgo, L., Giovagnoli, C., Calabrese, L. and Mondovi, B. (1972). *Biochemistry* *11,* 2187-2191.
9. Hathaway, B.J. (1971). In "Essays in Chemistry" (Bradley, J.N., Gillard, R.D. and Hudson, R.F. eds.), Academic Press, London, Vol.2, pp 61-92.
10. Vänngård, T. (1972). In "Biological Applications of Electron Spin Resonance" (Bolton, J.R., Borg, D.C. and Schwarz, H.M. eds.), Wiley-Interscience, pp 411-447.
11. Fee, J.A., unpublished results.
12. Maki, A.H. and McGarvey, B.R. (1958). *J. Chem. Phys.* *29,* 31-34.
13. Maki, A.H. and McGarvey, B.R. (1958). *J. Chem. Phys.* *29,* 35-38.
14. Fujimoto, M. and Janecka, J. (1971). *J. Chem. Phys.* *55,* 1152-1156.
15. Fee, J.A. and Dicorleto, P.D. (1973). *Biochemistry* *12,* 4893-4899.
16. Fielden, E.M., Roberts, P.B., Bray, R.C., Lowe, D.J., Mautner, G.N. Rotilio, G. and Calabrese, L. (1974). *Biochem. J.* *139,* 49-60.
17. Fee, J.A. and Phillips, W.D. (1975). *Biochim. Biophys. Acta* *412,* 26-38.
18. Rotilio, G., Calabrese, L., Bossa, F., Barra, D., Finazzi-Agro, A. and Mondovi, B. (1972). *Biochemistry* *11,* 2182-2187.
19. Hodgson, E.K. and Fridovich, I. (1975). *Biochemistry* *14,* 5294-5299.
20. Yagil, G. (1967). *Tetrahedron* *23,* 2855-2861.
21. Morris, J.P. and Martin, R.B. (1970). *J. Am. Chem. Soc.* *92,* 1543-1546.
22. Martin, R.B. (1974). *Proc. Nat. Acad. Sci.* *71,* 4346-4347.
23. Fee, J.A. and Ward, R.L. (1976). *Biochem. Biophys. Res. Commun.*, in press.
24. Blumberg, W., personal communication.
25. Calabrese, L., Rotilio, G. and Mondovi, B. (1972). *Biochim. Biophys. Acta* *263,* 827-829.
26. Fee, J.A. (1973). *J. Biol. Chem.* *248,* 4229-4234.
27. Moss, T.H. and Fee, J.A. (1975). *Biochem. Biophys. Res. Commun.* *66,* 799-808.
28. Rotilio, G., Calabrese, L., Mondovi, B. and Blumberg, W.E. (1974). *J. Biol. Chem.* *249,* 3157-3160.
29. Martin, R.L. (1968). In "New Pathways in Inorganic Chemistry" (Ebsworth, E.A.V., Maddock, A.G. and Sharpe, A.G. eds.), Cambridge University Press, Cambridge, pp 175-231.
30. Fee, J.A. (1973). *Biochim. Biophys. Acta* *295,* 107-116.
31. Fee, J.A. and Briggs, R.G. (1975). *Biochim. Biophys. Acta* *400,* 439-450.
32. Ahrland, S. (1973). *Structure and Bonding, Vol.15,* 167-188.
33. Rigo, A., Terenzi, M., Franconi, C., Mondovi, B., Calabrese, L. and Rotilio, G. (1974). *FEBS Letters* *39,* 154-156.
34. Hay, P.J., Thibeault, J.C. and Hoffman, R. (1975). *J. Am. Chem. Soc.* *97,* 4884-4899.
35. Fee, J.A. and Ludwig, M.L., unpublished observations.
36. Beem, K.M., Rich, W.E. and Rajagopalan, K.V. (1974). *J. Biol. Chem.* *249,* 7298-7305.
37. Fee, J.A. and Gaber, B.P. (1972). *J. Biol. Chem.* *247,* 60-65.

38. Rotilio, G., Finazzi-Agro, A., Calabrese, L., Bossa, F., Guerrieri, P. and Mondovi, B. (1971). *Biochemistry 10*, 616-621.
39. Haffner, P.H. and Coleman, J.E. (1973). *J. Biol. Chem. 248*, 6626-6629.
40. Forman, H.J., Evans, H.J., Hill, R.L. and Fridovich, I. (1973). *Biochemistry 12*, 823-827.
41. Roberts, P.B., Fielden, E.M., Rotilio, G., Calabrese, L., Bannister, J.V. and Bannister, W.H. (1974). *Radiat. Res. 60*, 441-452.
42. Bray, R.C., Cockle, S.A., Fielden, E.M., Roberts, P.B., Rotilio, G. and Calabrese, L. (1974). *Biochem. J. 139*, 43-48.
43. Stokes, A.M., Hill, H.A.O., Bannister, W.H. and Bannister, J.V. (1973). *FEBS Letters 32*, 119-123.
44. Moss, D.B., Lin, C.-T. and Rohrabacher, D.B. (1973). *J. Am. Chem. Soc. 95*, 5179-5185.
45. Fee, J.A. and Gaber, B.P. (1973). In "Oxidases and Related Systems" (King, T.E., Mason, H.S. and Morrison, M. eds.), University Park Press, Baltimore, pp 77-82.
46. Gaber, B.P., Brown, R.D., Koenig, S.H. and Fee, J.A. (1972). *Biochim. Biophys. Acta 271*, 1-5.
47. Boden, N., Holmes, M.C. and Knowles, P.F. (1974). *Biochim. Biophys. Res. Commun. 57*, 845-848.
48. Rotilio, G., Calabrese, L. and Coleman, J.E. (1973). *J. Biol. Chem. 248*, 3855-3859.
49. Ward, R.L. and Culp, M.L. (1972). *Arch. Biophys. 150*, 436-439.
50. Rigo, A., Viglino, P. and Rotilio, G. (1975). *Anal. Biochem. 68*, 1-8.
51. Roberts, P.G., Fielden, E.M., Rotilio, G., Calabrese, L., Bannister, J.V. and Bannister, W.H. (1974). *Radiat. Res. 60*, 441-452.
52. McClune, G.J. and Fee, J.A., submitted for publication.
53. Klug-Roth, D., Fridovich, I. and Rabani, J. (1973). *J. Am. Chem. Soc. 95*, 2786-2790.
54. Eigen, M. and Tamm, K. (1962). *Z. Electrochem. 66*, 93-106.
55. Eigen, M. and Wilkins, R.G. (1965). *Adv. Chem. Series No.49*, American Chemical Society, Washington, pp 55-67.
56. Rigo, A., Viglino, P. and Rotilio, G. (1975). *Biochem. Biophys. Res. Commun. 63*, 1013-1018.
57. Koutecky, J., Brdicka, R. and Hanus, V. (1953). *Coll. Czech. Chem. Commun. 18*, 611-627.
58. Chevalet, J., Rouelle, F. Gierst, L. and Lambert, J.P. (1972). *J. Electroanal. Chem. 39*, 201-216.
59. Kastening, B. and Karemifard, G. (1970). *Ber. Bunsenges, Phys. Chem. 74*, 551-556.
60. Yost, F.J. Jr. and Fridovich, I. (1973). *J. Biol. Chem. 248*, 4905-4908.
61. Asada, K., Yoshikawa, K., Takahashi, M.-A., Maeda, Y. and Enmanji, K. (1975). *J. Biol. Chem. 250*, 3801-3807.
62. Puget, K. and Michelson, A.M. ~~(1974)~~. ~~*Biochemi 56*~~, ~~1255-1267~~.
63. Yamakura, G. (1976). *Biochim. Biophys. Acta 422*, 280-294.
64. Slykhouse, T.O. and Fee, J.A. (1976). *J. Biol. Chem.*, in press.
65. Keele, B.B. Jr., McCord, J.M. and Fridovich, I. (1970). *J. Biol. Chem. 245*, 6176-6181.
66. Weisiger, R.A. and Fridovich, I. (1973). *J. Biol. Chem. 248*, 3582-3592.
67. Fee, J.A., Shapiro, E. and Moss, T.H., submitted for publication.
68. Lavelle, F., Durosay, P. and Michelson, A.M. (1974). *Biochimie 56*, 451-458.
69. Lavelle, F. and Michelson, A.M. (1975). *Biochimie 57*, 375-381.

THE COBALT DERIVATIVES OF BOVINE SUPEROXIDE DISMUTASE

G. ROTILIO and L. CALABRESE

Institutes of Biological Chemistry
University of Camerino (Camerino, Italy)
and Rome (Rome, Italy)

and C.N.R. Center for Molecular Biology
Rome, Italy

The cobalt derivatives of Cu, Zn superoxide dismutase have been prepared. Because of its optical and magnetic properties the cobaltous ion (Lindskog, 1970) is a useful probe of the geometry of zinc sites and of protein conformational changes, provided the metal substitution does not alter the native structure and activity of the enzyme.

This paper will review some properties of the cobalt derivatives of bovine superoxide dismutase: the Co(II), Cu(II) protein; the Co(II), Cu(I) protein; and the Co(II), copper-free protein. The matter will be divided in two sections, one dealing with the basic characterization of the derivatives, methods of preparation, extent of metal substitution, enzyme activity, optical and magnetic properties, and the other one concerning the sensitivity of the cobalt chromophore to various treatments.

BASIC PROPERTIES

The Co(II), Cu(II) protein has been prepared by two procedures. One method (Calabrese *et al.*, 1972) is an exchange dialysis of the native enzyme against 0.5% Co(II) in 0.1 M acetate buffer, pH 5.4, at 4°C, followed by exhaustive dialysis against water. Under these conditions the copper site is not affected, as shown by copper analysis with atomic absorption and by the integrity of its typical 680 nm absorption band. The zinc is replaced by cobalt ~~to a variable extent, though never~~ 100%, depending on the time of dialysis. The activity of the derivative is identical to that of the native enzyme, independent of the extent of the metal substitution. In the other method (Fee, 1973), 2 Co(II) and 2 Cu(II) are subsequently added to the apoprotein, thus giving a complete replacement of zinc by cobalt. This derivative is about half as active as the native enzyme.

The Co(II), Cu(I) protein is obtained by treating the Co(II), Cu(II) protein with various reductants, such as ferrocyanide (Rotilio *et al.*, 1973a, 1974) and dithionite (Fee, 1973), as the Co(II) is insensitive to any reducing agent and the reduced protein is stable to air for a long time at pH 5-6. The Co(II), copper-free derivative is obtained either by simple addition of 2 Co(II) to the apoprotein (Fee, 1973) or by exchange dialysis (Calabrese *et al.*, 1976) of the zinc-containing, copper-free protein (Rotilio *et al.*, 1972). The usual yields of the latter method are approximately 1 Co/protein after 24 hours dialysis and a maximum of 1.5 after 48 hours. The dialysis time necessary to reach the same extent of replacement as in the copper-containing protein is halved in the copper-free protein.

An important question is whether the two metals are in the native zinc and copper sites in these derivatives. Beside the enzyme activity, the spectral properties have to be considered. Figs.1 and 2 report some relevant optical and EPR spectra of the preparations of Calabrese *et al.* (1972, 1976). The identity of the U.V. spectra (Fig.1), which are very sensitive to alterations of the concentration, valence and symmetry of the copper site (Rotilio *et al.*, 1972), is a further evidence that copper is in the same site in the native and in the cobalt substituted protein. The visible spectra (Fig.1) show the intense absorption with a maximum near 600 nm due to the Co(II). The typical band of the Cu(II) is unaltered in the oxidized protein, though the EPR spectra (Fig.2) show the disappearance of a portion of the copper EPR signal corresponding to the amount of substituted cobalt and the absence of any cobalt EPR even at very low temperature. If the copper is absent or reduced an EPR signal of Co(II) near g = 4 is observed. These results suggested a magnetic coupling between cobalt and copper due to the close proximity of the sites (Rotilio *et al.*, 1974). X-ray crystallography (Richardson *et al.*, 1975) has shown that the zinc and copper sites are joined to each other by a bridging imidazole (His 61) and thus supports the contention that all the cobalt of this preparation is in the zinc site. However this may not be true in samples where the cobalt substitution exceeds the 1:1 Co/protein ratio. In Fig.3 the extent of cobalt incorporation and the decrease of the EPR-detectable copper are plotted against the dialysis time. It appears that the two parameters correlate fairly well up to approximately 1 Co per protein, while above this value the EPR detectable copper levels out. The increase of the cobalt content is paralleled by the increase of the cobalt absorption band. The samples with more than 1 Co/protein usually show the cobalt EPR signal, its intensity ranging from 10% to 30% of the total signal observed after copper reduction. It is evident that the cobalt exceeding 50% substitution is not coupled to the copper to the same extent, though it has the same optical and EPR spectra. In Fee's preparation 30% of the cobalt is observed by EPR, and the copper optical band is modified with respect to the native protein (Fee, 1973). More recently Moss and Fee (1975) have reported that this preparation contains only 50% of the metals in the form of Co(II)-Cu(II) pairs. Therefore it seems that the exchange method gives the cleanest Co(II), Cu(II) derivatives, provided that 50% replacement is not exceeded. It can be also concluded, from the identity of the optical and EPR spectra (Figs.2 and 3), of the Co(II), Cu(I) and Co(II), copper-free proteins, that the exchange method, with the above mentioned limitation, also places all the cobalt in the zinc site in the copper-free derivative.

The nature of the cobalt-copper magnetic interaction has been investigated more deeply by EPR, NMR and magnetic susceptibility. Temperature dependence of the EPR spectra between 1.4 and 80 K (Rotilio *et al.*, 1974) led to the suggestion that the coupling was antiferromagnetic with an exchange constant $>100\ cm^{-1}$. This was apparently confirmed by the absence of the enhancement of the water proton relaxation rate in solutions of the

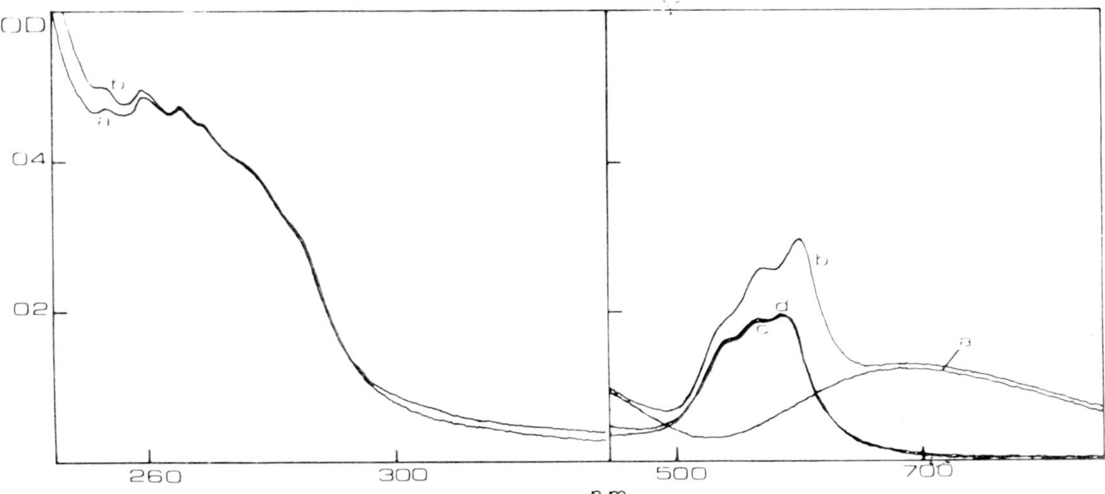

Fig.1. *U.V. (left) and visible (right) spectra of cobalt derivatives of bovine superoxide dismutase. a) Zn(II), Cu(II) enzyme; b) Co(II), Cu(II) enzyme; c) Co(II), Cu(I) enzyme, as obtained by addition of excess ferrocyanide; d) Co(II), copper-free enzyme. The protein concentration was 5×10^{-5} M for the U.V. spectra and 4.5×10^{-4} M for the visible spectra. The extent of substitution was 1.1 Co/protein.*

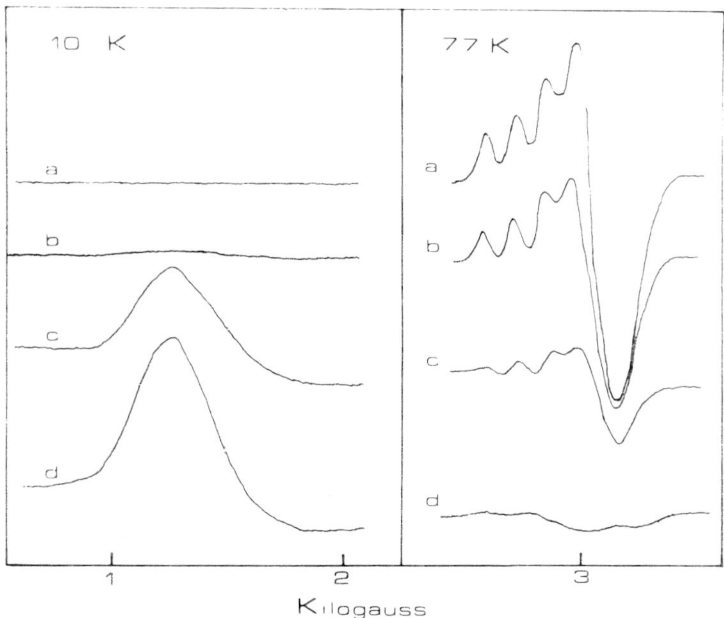

Fig.2. *EPR spectra of cobalt derivatives of bovine superoxide dismutase. a) 5×10^{-4} M Zn(II), Cu(II) enzyme; b) 5×10^{-4} M Co(II), Cu(II) enzyme with approximately 0.8 Co/protein; c) the same as b), half reduced by ferrocyanide; d) 5×10^{-4} M Co(II), copper-free enzyme. The cobalt and the copper signals of the spectrum were recorded at different temperatures to avoid copper saturation.*

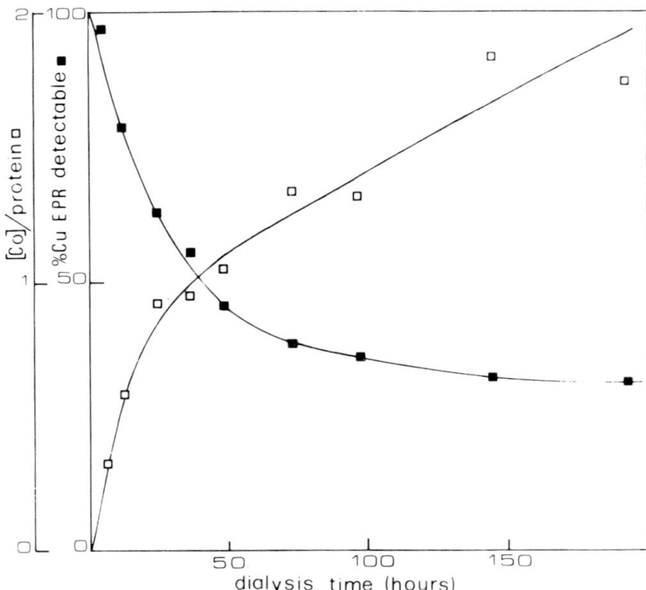

Fig.3. *Dependence of the extent of cobalt substitution and of the EPR-detectable copper on the dialysis time in the procedure of Calabrese* et al. *(1972). Cobalt was determined by atomic absorption.*

Co(II), Cu(II) protein (Rigo *et al.*, 1974). Magnetic susceptibility measurements (Moss and Fee, 1975) have proved the existence of antiferromagnetic coupling but were unable to measure the coupling constant, because of the magnetic heterogeneity of the preparation. Magnetic susceptibility measurements are in progress in Rome with the other type of preparation and seem to indicate a much lower constant than that previously suggested (Rotilio *et al.*, 1974).

Beside the existence of copper-zinc clusters, the other important information given by the spectral properties of the cobalt derivatives concerns the geometry of the zinc site. Optical (Calabrese *et al.*, 1972), EPR (Rotilio *et al.*, 1974) and MCD (Rotilio *et al.*, 1973b) data indicate a tetrahedral geometry for the cobalt chromophore. Proton relaxation rates (Rigo *et al.*, 1974) and EPR measurements show that the cobalt site is not exposed to the solvent. Both facts are confirmed by the X-ray data (Richardson *et al.*, 1975). The geometry of the site and its exclusion from solvent access account for the structural role attributed to the zinc (Rotilio *et al.*, 1972). This role can now be more clearly outlined in the light of the closer association of copper and zinc. The cluster type of structure which is now recognized appears to determine the particular symmetry and redox activity of the active site, as will be discussed in greater detail in the next section.

SENSITIVITY OF THE COBALT CHROMOPHORE TO VARIOUS TREATMENTS

The spectra of the cobalt chromophore are modified under various experimental conditions and the different patterns of reaction observed, especially in relation to the presence and the valence state of the copper, give information on the structural and functional role of the imidazole group bridging the two metals. The various reactions can be grouped according to their different dependence on the conditions of the copper site.

Reaction With OH^- And CN^- (Calabrese *et al.*, 1976)

With these reagents the copper reacts reversibly at its water binding position and the intermetal bridge is not affected. The cobalt reacts very slowly with an irreversible process only in the copper-free derivative.

Reaction With H_2O_2

This reaction does not occur in the absence of copper (Calabrese *et al.*, 1976). In the Co(II), Cu(II) protein copper is reduced, then slowly re-oxidizes. In this process denaturation of the copper site occurs, as indicated by EPR (Rotilio *et al.*, 1973a) and activity measurements (Bray *et al.*, 1974). A histidine is destroyed in this reaction (Bray *et al.*, 1974). Also the cobalt chromophore is irreversibly modified during the reoxidation of copper (Calabrese *et al.*, 1976), and the destruction of one of the surrounding histidines (Richardson *et al.*, 1975) can account for the observed changes.

Reaction With H^+

At acid pH (<pH 5) the cobalt chromophore undergoes the same optical and EPR changes either in the presence or in the absence of copper (Calabrese *et al.*, 1976). Moreover, acid titration of the EPR-detectable copper in the Co(II), Cu(II) protein brings about a reversible uncoupling of the cobalt and copper spin systems with an apparent pK = 3.5 (Calabrese *et al.*, 1975). Protonation of the imidazole nitrogen facing the cobalt could account for this magnetic uncoupling and for the far greater alterations of the cobalt chromophore in comparison with those of the copper. In turn, the detachment from the zinc site changes the copper EPR line shape to axial and renders the copper site non-reducible by ferrocyanide (Morpurgo *et al.*, 1976).

Reaction With Reductants

Reducing agents (including hydrated electrons) have no effect on the cobalt spectrum in the copper-free protein. On the other hand reduction of copper brings about a characteristic shift of the maximum of the cobalt optical absorption to shorter wavelengths, which renders the spectrum of the Co(II), Cu(I) protein identical with that of the Co(II), copper-free protein (Fig.1). This suggests that copper is released from the bridging imidazole upon reduction through protonation of the copper-facing imidazole nitrogen. Also, the pH dependence of the superoxide dismutase-ferrocyanide redox equilibrium (Fee and Di Corleto, 1973) indicates that a ligand with pK >9 is protonated and released from the copper coordination sphere when reduction of copper occurs and, in fact the pK of the second nitrogen of a metal bound imidazole is >9 (Sundberg and Martin, 1974).

On the basis of these results and of the X-ray data, which indicate penta-coordination of the copper in the oxidized protein (Richardson *et al.*, 1975), it has been proposed (Morpurgo *et al.*, 1976; Rotilio *et al.*, 1976) that the copper site changes between penta-coordinated and tetra-coordinated (possibly tetrahedral) during catalysis. This would account for the extremely fast rates of the catalytic redox cycle, the two coordination geometries suited for the two valence states being achieved without much displacement of the ligand atoms and the protonation of the peroxide anion produced in the reoxidation step being favoured by the parallel deprotonation of the imidazole nitrogen facing copper. A scheme incorporating these features has been proposed (Hodgson and Fridovich, 1975).

REFERENCES

1. Calabrese, L., Rotilio, G. and Mondovi, B. (1972). *Biochim. Biophys. Acta, 263*, 827-829.
2. Calabrese, L., Cocco, D., Morpurgo, L., Mondovi, B. and Rotilio, G. (1975). *FEBS Letters, 59*, 29-31.
3. Calabrese, L., Cocco, D., Morpurgo, L., Mondovi, B. and Rotilio, G. (1976). *Eur. J. Biochem., 64*, 465-470.
4. Fee, J.A. (1973). *J. Biol. Chem. 248*, 4229-4234.
5. Fee, J.A. and Di Corleto, P.E. (1973). *Biochemistry, 12*, 4893-4899.
6. Hodgson, E.K. and Fridovich, I. (1975). *Biochemistry, 14*, 5294-5299.
7. Lindskog, S. (1970). *Struct. Bonding 9*, 153-196.
8. Morpurgo, L., Mavelli, I., Calabrese, L., Finazzi Agrò, A. and Rotilio, G. (1976). *Biochem. Biophys. Res. Commun. 70*, 607-614.
9. Moss, T.H. and Fee, J.A. (1975). *Biochem. Biophys. Res. Commun. 66*, 799-808.
10. Richardon, J.A., Thomas, K.A., Rubin, B.M. and Richardson, A.C. (1975). *Proc. Natl. Acad. Sci. U.S.A. 72*, 1349-1363.
11. Rigo, A., Terenzi, M., Franconi, C., Mondovi, B., Calabrese, L. and Rotilio, G. (1974). *FEBS Letters, 39*, 154-156.
12. Rotilio, G., Calabrese, L., Bossa, F., Barra, D., Finazzi Agrò, A. and Mondovi, B. (1972). *Biochemistry 11*, 2182-2187.
13. Rotilio, G., Morpurgo, L., Calabrese, L. and Mondovi, B. (1973a). *Biochim. Biophys. Acta 302*, 229-235.
14. Rotilio, G., Calabrese, L. and Coleman, J.E. (1973b). *J. Biol. Chem. 248*, 3853-3859.
15. Rotilio, G., Calabrese, L., Mondovi, B. and Blumberg, W.E. (1975). *J. Biol. Chem. 249*, 3157-3160.
16. Rotilio, G., Morpurgo, L., Calabrese, L., Finazzi Agrò, A. and Mondovi, B. (1976). In "Metal-Ligand Interactions in Organic and Bio-chemistry" (B. Pullman, ed.) in press, Reidel, Dordrecht, Holland.
17. Sundberg, R.T. and Martin, R.B. (1974). *Chem. Rev. 74*, 471-517.

CHEMICAL MODIFICATION STUDIES ON A FERRI-SUPEROXIDE DISMUTASE FROM MARINE BACTERIA

B. VANOPDENBOSCH and R.R. CRICHTON

Unité de Biochimie
Université Catholique de Louvain
1348 Louvain-la-Neuve, Belgium

K. PUGET

Institut de Biologie Physico-Chimique
13, rue P. et M. Curie
75005 Paris, France

INTRODUCTION

In the study of the mechanism of enzyme action a number of techniques have been employed to identify amino acid residues involved in the catalytic site. The use of chemical modification of specific amino acid residues has proved to be of great value in these studies (Means and Feeney, 1971; Glazer *et al.*, 1975; Sigman and Mooser, 1975). Although more recently techniques such as X-ray crystallography, magnetic resonance and kinetics have somewhat reduced the importance of chemical modification in the elucidation of enzymic mechanisms, this approach still provides in many cases the best evidence for identification of amino acid residues involved in catalysis.

Having already used these techniques to establish the role of specific amino acid residues both in the catalytic site and in subunit-subunit interactions in apoferritin from horse spleen (Bryce and Crichton, 1973; Wetz and Crichton, 1976) and more recently from a number of other mammalian sources (Limet, 1976; Ysenbaert, 1976), we were intrigued by the possibility of extending our chemical modification studies to superoxide dismutase. Thus, when Dr. Michelson offered to supply us with the ferri-SOD from the bioluminescent marine bacterium *Photobacterium leiognathi*, we were delighted. As will become clear from the brief discussion on mechanism at the end of this article, there may well be parallels between apoferritin, which oxidises iron by the activation of molecular oxygen, and the ferri-SOD which alternately oxidises and reduces superoxide, no doubt using iron as electron donor and acceptor.

CHARACTERIZATION AND ASSAY OF THE ENZYME

The ferri-SOD from *P. leiognathi*, was prepared as described by Puget and Michelson (1974). The enzyme has a molecular weight (estimated from the sedimentation constant) of 40,660, an isoelectric point of 4.4 and contains 1.6 atoms of iron per molecule (Puget and Michelson, 1974). The preparation on which we carried out our experiments gave a single band by polyacrylamide gel electrophoresis in sodium dodecyl sulphate corresponding to a molecular weight of 18,960, in good agreement with the value of 20,000 reported by Puget and Michelson (1974). It appears that the active form of the enzyme is a dimer of molecular weight 38,000 - 40,000 containing one iron atom per subunit. On polyacrylamide gel electrophoresis in Tris-asparagine buffer, pH 7.3 two bands were observed which we presume to be different aggregated forms of the enzyme, either monomer/dimer or dimer/tetramer.

The amino acid composition of the enzyme is given in Table I both as residues per molecule (38,000) and as residues per subunit. There is good agreement with the published values of Puget and Michelson (1974). The protein has a high content of non-polar amino acids (51%). On the basis of its isoelectric point and its content of basic amino acids we calculate that of the 38 glx and asx residues no more than 13 - 14 are present as acids and the rest are in the amide form.

TABLE I

Amino Acid Composition - SOD *(P. Leognathi)*

Amino Acid	Residues/Molecule	Residues/Subunit
Asx	37.2 ± 1.4	18.6
Thr	19.8 ± 1.2	9.9
Ser	17.4 ± 0.8	8.7
Glx	37.7 ± 1.1	18.9
Pro	13.9 ± 0.5	7.0
Gly	24.6 ± 0.9	12.3
Ala	37.2 ± 1.5	18.6
Val	20.7 ± 0.7	10.4
Met	1.5 ± 0.4	0.8
Ile	12.7 ± 0.5	6.4
Leu	28.3 ± 1.5	14.2
Tyr	12.6 ± 0.5	6.3
Phe	18.5 ± 0.5	9.8
His	10.7 ± 0.4	5.4
Lys	17.1 ± 0.7	8.6
Arg	4.9 ± 0.2	2.5
Trp	11.5 ± 0.7	5.8

Amino acid analysis was carried out after 16 h hydrolysis. The results given are the mean ± standard error of 18 analyses. Tryptophan was determined separately by the method of Edelhoch (1967). Cysteine was not determined.

The activity of the SOD at pH 7.8 in 10 mM potassium phosphate buffer was determined using xanthine oxidase/hypoxanthine to generate superoxide and cytochrome *c* reduction as the detection system. One unit of SOD activity in the system (which gave 50% inhibition of cytochrome *c* reduction) corresponded to 0.45 μg of protein (specific activity, 2220 units/mg). The apoenzyme was prepared by dialysis at pH 3.8 in sodium acetate buffer (50 mM) containing 1 mM EDTA and 0.2 mM 2-mercaptoethanol followed by dialysis against potassium phosphate buffer, pH 7.8 (10 mM and then 5 mM). Reconstitution of the enzyme was by incubation with ferrous sulphate followed by dialysis to remove the excess of iron (A.M. Michelson, personal communication). The reconstituted SOD had one unit of activity per 1.58 μg of protein (35% recovery of activity).

CHEMICAL MODIFICATION CONDITIONS

We decided to modify histidine and tryptophan residues and carboxyl groups both in the native holoenzyme and in the apoenzyme. The reagents used were chosen for their specificity and for the ease with which we would determine the number of amino acid residues modified.

Histidine was modified using ethoxyformic anhydride (diethyl pyrocarbonate) and diazonium-1-H-tetrazole. Diethyl pyrocarbonate reacts with both amino groups and imidazole functions but not with sulphydryls and rarely with indoles (Means and Feeney, 1971). The number of histidine residues modified was determined by measuring the increase in absorbance at 230 nm (Melchior and Fahrney, 1970). Diazonium tetrazole is an electrophilic reagent which is quite specific for histidine and tyrosine residues (Horinishi *et al.*, 1974; Sokolovsky and Vallee, 1966). The number of histidine residues modified was determined spectrophotometrically (Takenaka *et al.*, 1969). The extent of tyrosine modification was also determined spectrophotometrically. The alkylation of tryptophan at pH 5.5 by hydroxy-nitrobenzyl-bromide (Barman and Koshland, 1967) can be readily measured at 410 nm and is quite specific for the indole nucleus. Finally the carbodiimide promoted conversion of protein carboxyl groups by glycinamide provides a convenient and specific method for selective modification of carboxyl groups (Hoare and Koshland, 1967). Full details of the precise conditions used are reported elsewhere (Vanopdenbosch, 1976; Vanopdenbosch and Crichton, in preparation).

AMINO ACID RESIDUES INVOLVED IN THE CATALYTIC SITE

Table II summarises the results of the chemical modification experiments. With diethyl pyrocarbonate and diazonium tetrazole we find 2 histidine residues modified per subunit (of a total of 5 residues/subunit). The diazonium tetrazole-modified protein is virtually completely inactive in O_2^- dismutation (Fig.1). The same number of histidine residues were modified in the apoenzyme and reconstitution of the modified apo-SOD with iron did not lead to recovery of activity (Fig.1). Assuming that we have not changed the protein conformation during the modification this suggests that at least one histidine residue per subunit is involved in the catalytic site of the enzyme.

Modification of 2.6 tryptophan residues per subunit also results in almost complete loss of activity, and the alkylated apoenzyme remains inactive after reconstitution (Fig.1). We may also conclude that at least one tryptophan residue per subunit is involved in the catalytic site. Involvement of tryptophan in the active site was previously adduced by Puget and Michelson (1974) who observed a loss of activity of the ferri-SOD on

TABLE II

Number of Amino Acid Residues Modified in Ferri-SOD

The chemical modifications were carried out as described in the text. The number of residues modified in the holoenzymes and in the apoenzyme per subunit (molecular weight 19,000) is given.

Reagent	Number of Residues Modified in the ferri-SOD	in apo-SOD
Diazonium-1-H-tetrazole	1.85	1.90
Diethylpyro-carbonate	2.3	2.35
Hydroxy-nitrobenzyl-bromide	2.6	3.45
Glycinamide	5.35	4.95

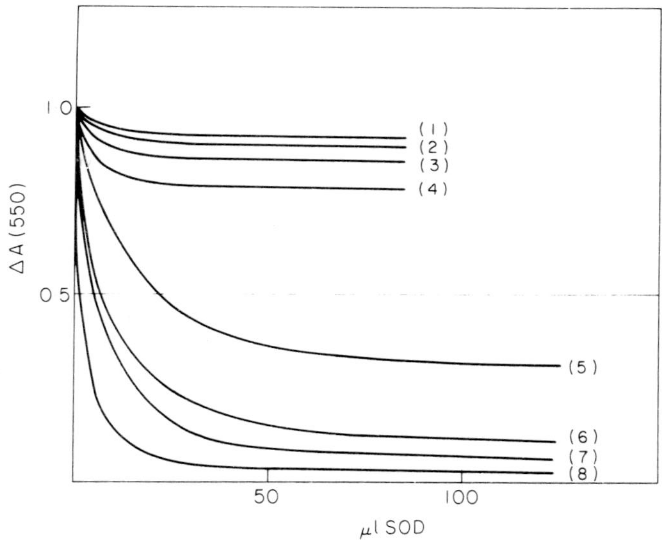

Fig.1. *Dismutase activity of ferri-SOD and of modified dismutases. The activity (measured with the xanthine oxidase/hypoxanthine/cytochrome* c *system) is followed by measuring the change in absorption at 550 nm in function of the volume of SOD solution added to the incubation mixture. The various preparations used were as follows:*

1. *ferri-SOD modified by diazonium tetrazole*
2. *apo-SOD modified by diazonium tetrazole and then reconstituted with iron.*
3. *ferri-SOD modified by hydroxy-nitrobenzyl-bromide*
4. *apo-SOD modified by hydroxy-nitrobenzyl-bromide and then reconstituted with iron*
5. *ferri-SOD modified by glycineamide*
6. *apo-SOD modified by glycineamide and then reconstituted with iron*
7. *ferri-SOD unmodified*
8. *apo-SOD unmodified, reconstituted with iron.*

photo-oxidation which proceeded even faster than the loss in tryptophan fluorescence.

In contrast amidation of 5 carboxyl groups per subunit out of a total of 13 - 14 has only a slight effect on the SOD activity (Fig.1). The carboxyl-modified apoenzyme is also active after reconstitution (Fig.1). Thus either carboxylate functions are not involved in the active site or else are buried in the interior of the active site.

AMINO ACID RESIDUES INVOLVED IN IRON BINDING

The rationale involved in these experiments was the following. If an amino acid residue is involved in metal binding we would expect that in the holoenzyme the residue would not be modified whereas in the apoenzyme the amino acid would be accessible to modification (although this need not necessarily be the case if the metal binding site became buried as a result of a conformational change in the apoenzyme). Thus if we find more residues of a particular amino acid residue modified in the apoenzyme than in the holoenzyme we may attribute a role in metal binding to that amino acid. Inspection of Table II shows that the only case for which this is true is for tryptophan, where we find one additional residue alkylated in the apoenzyme than in the holoenzyme. This is in good agreement with the results of Puget and Michelson (1974) who found that in parallel with the loss of activity and of tryptophan fluorescence on irradiation of the enzyme the iron was totally lost from the enzyme. The blue shift in the tryptophan fluorescence emission maximum of the enzyme observed by these authors is also compatible with a role for tryptophan as a ligand for the iron.

In the case of the Cu-Zn SOD from bovine erythrocytes (Richardson *et al.*, 1975) it has been established that the Cu has four histidine ligands in a somewhat distorted square planar configuration and the Zn has three histidines and one aspartate in an approximately tetrahedral arrangement. We cannot exclude that histidine residues are involved in metal binding as in catalytic activity, but a more definitive conclusion must await further studies.

SPECULATIONS CONCERNING MECHANISM

In the course of our detailed studies on the mechanism of iron oxidation catalysed by apoferritin (Bryce and Crichton, 1973; Crighton *et al.*, 1975; Wetz and Crichton, 1976; Wauters *et al.*, 1976; Crichton and Roman, 1976) we have been able to establish that the activation of dioxygen is most probably achieved by the formation of an end-on complex in which the dioxygen is coordinated to two iron atoms. This formal superoxo complex is transformed into a peroxo complex (Fig.2) which is subsequently hydrolysed to form ferric oxyhydroxide: the ferric oxyhydroxide then migrates to the interior of the protein shell.

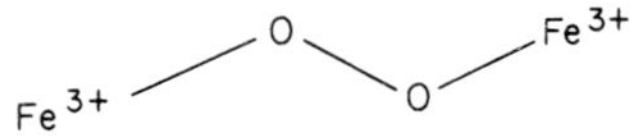

$$Fe^{3+}-O-O-Fe^{3+}$$

Fig.2. *The active site peroxo complex of apoferritin; the peroxide is complexed end-on by two ferric iron atoms.*

In ferri-SOD the two iron atoms are either sufficiently close to fix the superoxide in an end-on superoxo complex (Fig.3) with the O_2^- coordinated to both iron atoms, or else in an end-on superoxo complex with one iron atom

(Fig.4). There is at present no evidence from EPR that the iron atoms were coupled nor could they be distinguished (Puget and Michelson, 1974). More recent pulse radiolysis and EPR studies show that only one of the two iron atoms is active, in the sense that it undergoes valence changes during the catalytic cycle (Lavelle *et al.*, 1977).

Fig.3. *Superoxo complex with the superoxide complexed by two iron atoms.*

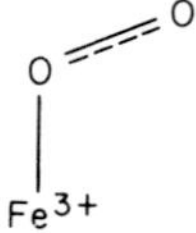

Fig.4. *Superoxo complex with the superoxide complexed by one iron atom.*

By comparison with the apoferritin system (Fig.2) we propose that, in the event that the two iron atoms of the ferri-SOD are sufficiently close to fix superoxide in a complex of the type shown in Fig.3, a reaction sequence of the type illustrated in Scheme 1 could operate. In agreement with the EPR and pulse radiolysis data (Lavelle *et al.*, 1976) only one of the two iron atoms undergoes alternate reduction and oxidation, while the other serves to anchor the superoxide anion in place and to activate the oxygen-oxygen bond. It seems inherently unlikely that O_2^- reacts directly with amino acid residues in the protein. However, once the superoxide is bound by the iron atoms the essential histidine residue (or residues) could assist in electron withdrawal and addition as well as in liberation of the peroxide.

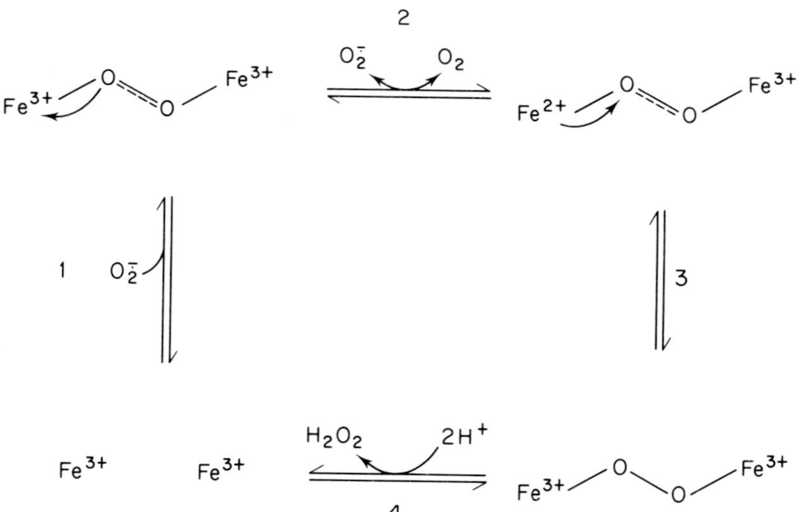

Scheme 1. *Proposed reaction sequence for ferri-SOD. The first superoxide molecule (1) is bound between the two ferric iron atoms of the active centre and (2) is oxidised with liberation of dioxygen and reduction of one iron atom. The second superoxide molecule binds to the ferro-enzyme and (3) is reduced to form a peroxo complex; the 'active' iron atom is reoxidised. In the final step (4) the peroxo complex is cleaved to liberate hydrogen peroxide and regenerate the enzyme active site.*

The present model has the additional advantage of attributing a function to the second "inert" metal atom that one so often finds in superoxide dismutases, which instead of serving no function becomes an essential ligand for the substrate. By analogy we propose that in the Cu-Zn SOD the inactive Zn might serve essentially the same function (Fig.5). It should be noted that from the X-ray structure we know that the Cu and Zn atoms are about 6Å apart (Richardson *et al.*, 1975). For peroxo complexes metal-oxygen distances of 1.9 - 2.0Å have been reported (Vaska, 1976). The oxygen-oxygen distances in O_2^- and $O_2^=$ are known to be 1.33Å and 1.49Å respectively (Vaska, 1976). These values are certainly not incompatible with the model proposed in Fig.5.

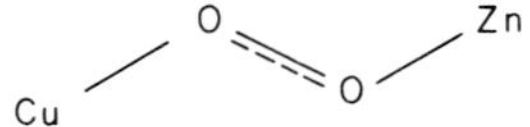

Fig.5. *Superoxo complex with the superoxide complexed by one copper and one zinc atom.*

ACKNOWLEDGEMENTS

We are most grateful to Mick Michelson for many valuable discussions. We also thank Mrs. Francine Brouwers who carried out the amino acid analyses.

REFERENCES

1. Barman, T.E. and Koshland, D.E. (1967). *J. Biol. Chem. 242,* 5771-5776.
2. Bryce, C.F.A. and Crichton, R.R. (1973). *Biochem. J. 133,* 301-309.
3. Crichton, R.R., Wauters, M. and Roman, F. (1975). In "Proteins of Iron Storage and Transport in Biochemistry and Medicine" (R.R. Crichton, ed.) pp 287-294, North Holland, Amsterdam.
4. Crichton, R.R. and Roman, F. (1976) in preparation.
5. Edelhoch, H. (1967). *Biochemistry 6,* 1948-1954.
6. Glazer, A.N., Delange, R.J. and Sigman, D.S. (1975). In "Chemical Modification of Proteins", North Holland, Amsterdam.
7. Hoare, D.G. and Koshland, D.E. (1967). *J. Biol. Chem. 242,* 2447-2453.
8. Horinishi, H., Hachimori, Y., Kurihara, K. and Shibata, K. (1964). *Biochem. Biophys. Acta 86,* 477-489.
9. Lavelle, F., McAdam, M.E., Fielden, E.M., Roberts, P.B., Puget, K. and Michelson, A.M. (1977). *Biochem. J. 161,* 3-11.
10. Limet, J. (1976). *Mémoire de Licence, Université de Louvain.*
11. Means, G.E. and Feeney, R.E. (1971). In "Chemical Modification of Proteins", Holden-Day, San Francisco.
12. Melchior, W.B. and Fahrney, D. (1970). *Biochemistry 9,* 251-258.
13. Puget, K. and Michelson, A.M. (1974). *Biochimie 56,* 1255-1267.
14. Richardson, J.S., Thomas, K.A., Rubin, B.H. and Richardson, D.C. (1975). *Proc. Natl. Acad. Sci.* (U.S.A.) *72,* 1349-1353.
15. Sigman, D.S. and Mooser, G. (1975). *Ann. Rev. Biochem. 44,* 889-931.
16. Sokolovsky, M. and Vallee, B.L. (1966). *Biochemistry 5,* 3574-3581.
17. Takenaka, A., Zusuki, T., Takenaka, O., Horinishi, H. and Shibata, K. (1969). *Biochem. Biophys. Acta 194,* 293-300.
18. Vaska, I. (1976). *Acc. Chem. Res. 9,* 175-183.
19. Vanopdenbosch, B. (1976). *Mémoire de Licence, Université de Louvain.*
20. Wauters, M., Crichton, R.R. and Michelson, A.M. (1976). Submitted for publication.
21. Wetz, K. and Crichton, R.R. (1976). *Eur. J. Biochm. 61,* 545-550.
22. Ysenbaert, D. (1976). *Mémoire de Licence, Université de Louvain.*

INTERACTION BETWEEN SUBUNITS IN Cu,Zn SUPEROXIDE DISMUTASES

G. ROTILIO

Institute of Biological Chemistry
University of Camerino, Camerino, Italy
and C.N.R. Center for Molecular Biology, Rome, Italy

A. RIGO and P. VIGLINO

Institute of Physical Chemistry
University of Venice, Venice, Italy

L. CALABRESE

Institute of Biological Chemistry, University of Rome
and C.N.R. Center for Molecular Biology, Rome, Italy

INTRODUCTION

Pulse radiolysis work on bovine superoxide dismutase (Rotilio *et al.*, 1972; Klug *et al.*, 1972; Klug-Roth *et al.*, 1973; Fielden *et al.*, 1974) has shown that (1) the copper of the enzyme undergoes reduction and reoxidation with equal rates during catalysis and (2) the steady state value of cupric copper after reaction with excess superoxide is more than 50% and can be accounted for by only one of the two copper sites functioning in catalysis (Fielden *et al.*, 1974). The latter contention conflicts with the equivalence of the two sites which results from EPR (Fielden *et al.*, 1974), sequence (Steinman *et al.*, 1974) and X-ray (Richardson *et al.*, 1975) evidence. Therefore the observed steady state level has to be interpreted either assuming a heterogeneity of the enzyme preparations which would very often give half functional and half non functional molecules with identical EPR spectra, or in terms of anticooperative interaction between the two sites, one of which would become unreactive when the other is reacting with the substrate. The latter alternative implies intersubunit interaction in these enzymes which in fact are always dimeric proteins.

The data presented in the following paper, though preliminary, suggest the involvement of interactions between the subunits of copper-zinc superoxide dismutases under different experimental conditions and thus support the hypothesis of a more general occurrence of this phenomenon.

COPPER DISTRIBUTION BETWEEN SUBUNITS IN SAMPLES OF COPPER-FREE PROTEIN RECOMBINED WITH LESS THAN STOICHIOMETRIC AMOUNTS OF COPPER

Samples of copper-free protein which give the best results in terms of metal content and reconstitution ability can be prepared by dialysis against 0.05 M CN^- in 0.1 M phosphate buffer pH 6.0 for 8 - 10 hours after reduction of the copper with excess ferrocyanide. Readdition of stoichiometric Cu^{2+} to this derivative leads to full recovery of all the properties of the native protein. Disc gel electrophoresis of samples recombined at equilibrium with less than stoichiometric copper gives three bands (Fig.1).

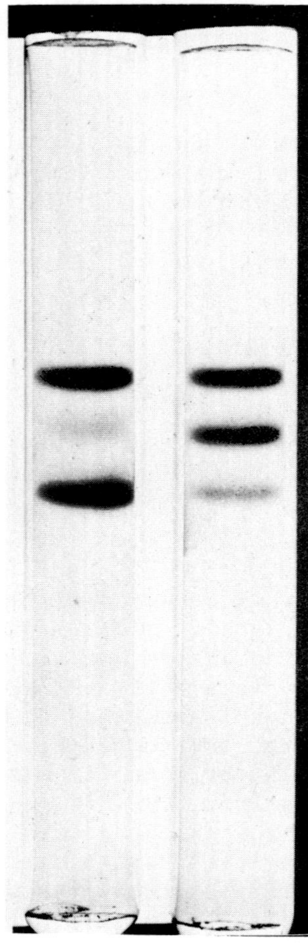

Fig.1. *Disc gel electrophoresis of bovine superoxide dismutase. On the left: a mixture of holoprotein (top) and copper-free protein (bottom). On the right: copper-free protein recombined with approximately 1.4* Cu^{2+}*/protein. 50 μg of protein were applied to each gel.*

One has the same mobility as the holo protein, another that of the copper-free protein, while the band with intermediate mobility is likely to be the protein carrying a single copper per dimer. Table I reports the relative concentrations of the three bands, as evaluated by densitometric analysis of electropherograms of samples recombined with increasing amounts of copper. It appears that the distribution of the three species is not statistic and that the binding of the second copper by the dimer is favoured with respect to the binding of the first one. If the two copper sites are initially equivalent in the metal binding process, this result suggests that binding at one site brings about conformational changes which facilitate copper binding by the other subunit.

TABLE I

Copper Distribution in the Recombination of Copper-Free Bovine Superoxide Dismutase

Cu^{2+} added per mole of protein	N_0	N_1	N_2
0.11	90 (92)	9 (8)	1 (0)
0.66	52 (46)	30 (46)	18 (8)
0.78	46 (36)	30 (49)	24 (15)
1.20	26 (17)	28 (49)	46 (34)
1.44	17 (7)	22 (40)	61 (53)
1.65	10 (4)	15 (30)	75 (66)

N_0, N_1 and N_2 indicate the fractions of molecules containing 0, 1 and 2 copper atoms respectively. The recombination was carried out at pH 8 (sodium borate, $\mu = 0.02$). The figures in parentheses indicate the N_0, N_1, N_2 values expected for a statistical distribution.

KINETICS OF RECOMBINATION WITH COPPER OF THE COPPER-FREE PROTEIN

Enzyme activity and u.v. absorption at 260 nm are recovered after addition of copper to the copper-free protein with parallel first order processes which are strongly pH dependent, being much faster at pH 8 than at pH 5. The measured activation energies were 20 Kcal/mole at pH 5 and 2 Kcal/mole at pH 8. They were determined by polarographic measurements (Rigo *et al.*, 1975) of the rate of the activity recovery as a function of temperature in the range 9 - 37°C. The EPR spectrum shows parallel changes from a non specific type of copper signal to the native line shape. All these data indicate that binding of copper is followed by a relatively slow rearrangement step leading to the reconstitution of the active site according to the following scheme:

$$Cu^{2+}(H_2O)_6 + P \underset{}{\overset{a}{\rightleftharpoons}} (P_i - Cu^{2+})_a \overset{b}{\rightleftharpoons} (P_i - Cu^{2+})^* \overset{c}{\rightleftharpoons} (P_i - Cu^{2+})$$
$$(P_i - Cu^{2+})_a \rightleftharpoons (P_j - Cu^{2+})_a$$

where P is the copper-free protein, $(P_{i,j}-Cu^{2+})_a$ indicates a series of different copper-protein complexes with equivalent energy, $(P_i - Cu^{2+})^*$ represents the copper already coordinated to the active site ligands in a complex which still lacks the active configuration and $(P_i - Cu^{2+})$ is the final state possessing the native configuration and activity. Step b controls the distribution of copper between subunits in the recombination process, while step c accounts for the first order recovery of enzyme activity. The rate constant of the latter process was found to decrease as the copper concentration was increased up to the stoichiometric Cu^{2+}/protein ratio. Table II shows the data obtained at low pH, but the same phenomenon was observed at any pH between pH 5 and 10.5.

A likely explanation is that step c becomes slower when the other site in the dimer is already occupied by copper.

TABLE II

Kinetics of Recombination of Copper-Free Bovine Superoxide Dismutase

Exp.	$[Cu^{2+}]/[protein]$	$k(h^{-1})$
1	0 - 1.8	3.0
2	0 - 0.6	9.2
3	0.6 - 1.2	5.1
4	1.2 - 1.8	3.5

Stoichiometric Cu^{2+} was added to the copper-free protein at pH 5.9 either in a single step (exp.1) or in multiple steps (exps.2-4). The second column indicated the copper added in terms of a concentration interval, between that of the copper bound to the protein before the addition and the final value after the addition. The superoxide dismutase activity was measured according to Rigo *et al.* (1975).

SPECTRAL AND ENZYMIC PROPERTIES OF THE PROTEIN RECOMBINED WITH INCREASING AMOUNTS OF COPPER

Figs.2, 3 and 4 show that most properties of the copper site of superoxide dismutase - i.e. EPR spectrum, enhancement of relaxivity of the water protons, u.v. absorption at 260 nm and enzyme activity of samples recombined at pH 5.2 - are linear functions of the copper content of recombined samples of copper-free protein, with sharp changes of slope around $[Cu^{2+}]/[protein] = 2$. This confirms the spectroscopic equivalence of the two copper sites and demonstrates that with this method of preparation no spurious species are observed in the recombination, at variance with other preparations (Fee *et al.*, 1973). The activity versus $[Cu^{2+}]/[protein]$ plot is however non linear for samples recombined at higher pH (Fig.4) and can be accounted for by assuming that the first bound copper is more active than the second copper. A more precise evaluation of the relative specific activity of the copper ions in protein molecules containing one and two Cu^{2+} respectively was obtained by determining the actual concentration of the different types of molecules by disc gel electrophoresis. For each distribution the observed activity constant is given by the following equation.

$$k_{obs} = k_{Cu_1} N_1 + 2 k_{Cu_2} N_2$$

where N_1 and N_2 are the fractions of molecules with one and two Cu^{2+} respectively and k_{Cu_1} and k_{Cu_2} are the specific rate constants for the copper in the molecules with one and two copper atoms respectively. Table III reports the data relative to a recombination experiment at pH 8 and shows that the mean value of the k_{Cu_1}/k_{Cu_2} ratio is about two. This result supports the hypothesis that the reaction of one copper site with the substrate prevents the site on the other subunit from catalytic function.

TABLE III

Relative Specific Rate Constants for the Molecules Containing 1 and 2 Copper Atoms in the Recombination of Copper-Free Bovine Superoxide Dismutase

$[Cu^{2+}]/[protein]$	N_0	N_1	N_2	k_{obs} (normalized)	k_{Cu_1}/k_{Cu_2}
0.08	0.92	0.08	-	0.09	2.16
0.79	0.45	0.31	0.24	0.53	1.87
1.19	0.26	0.29	0.45	0.73	1.93
1.44	0.17	0.22	0.61	0.82	1.95

The recombination was carried out at pH 8 (sodium borate, μ = 0.02). The N_0, N_1 and N_2 values were evaluated by disc gel electrophoresis. The superoxide dismutase activity constant was determined according to Rigo *et al.* (1974).

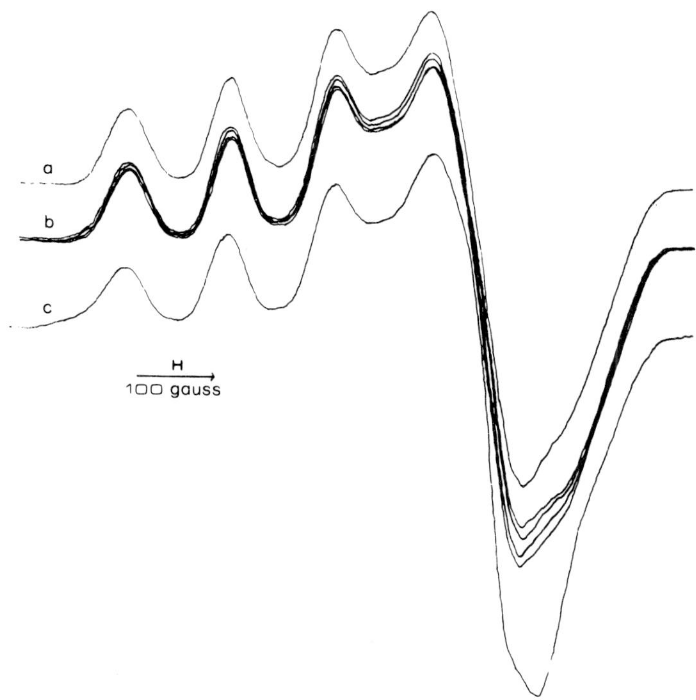

Fig.2. X-band EPR spectra of bovine superoxide dismutase. Spectrum a) is the native holoprotein. Spectra under b) and c) are copper-free protein samples recombined with 0.4, 0.8, 1.2, 1.6 (b) and 1.9 (c) Cu^{2+} per protein. The protein concentration was 5 x 10^{-4} M, the pH was 7.6 (borate, μ = 0.02), the temperature -170°C. The amplification was adjusted to give approximately the same amplitude.

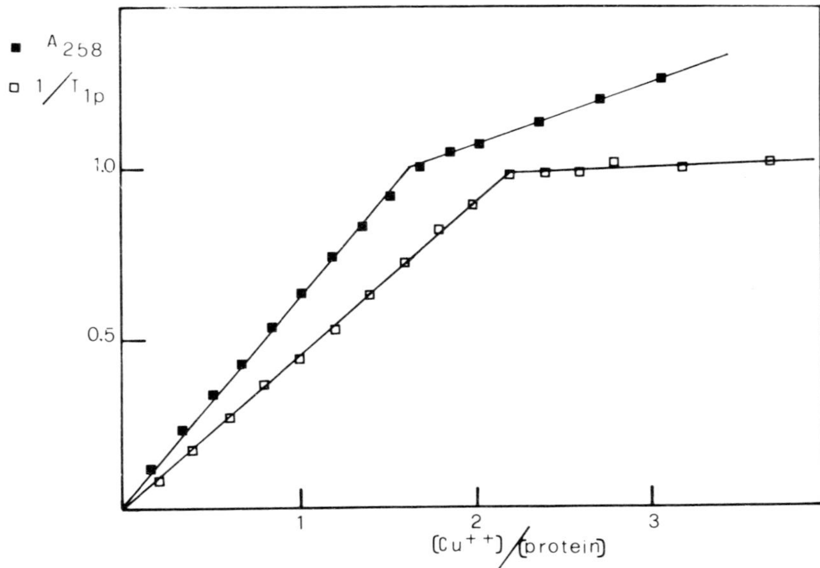

Fig.3. *Normalized values of the paramagnetic contribution to the water protons spin-lattice relaxation rates ($1/T_1$) and of the absorbance increments at 258 nm (A_{258}) as functions of the Cu^{2+} added to copper-free bovine superoxide dismutase. Borate (μ = 0.02), pH 7.6. T_1 was measured by the 90°- τ -90° pulse sequence on a pulsed NMR apparatus working at 16.0 MHz.*

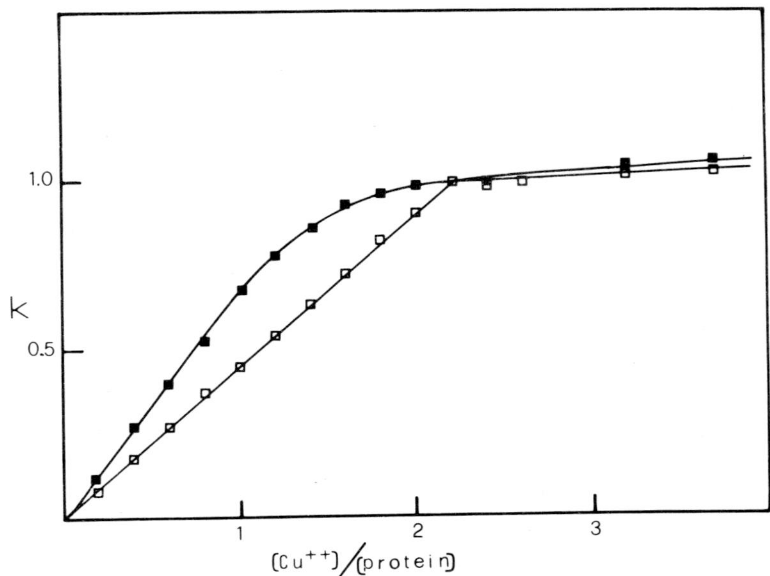

Fig.4. *Normalized activity constants as functions of the Cu^{2+} added to copper-free bovine superoxide dismutase. □ : pH 5.2; ■ : pH 7.6. The activity constant was measured polarographically according to Rigo* et al. *(1974).*

THE WHEAT GERM SUPEROXIDE DISMUTASE SYSTEM

A complementary evidence for the existence of intersubunit interactions in Cu,Zn superoxide dismutase will derive from the isolation of a dimer containing only one copper or of an active monomer. It has already been shown (Marmocchi *et al.*, 1974) that one of the two cytoplasmic isozymes of wheat germ (Beauchamp and Fridovich, 1973) is split by SDS into inactive monomers which recover the enzymic activity if the detergent is removed. Fig.5 shows that the EPR spectrum of the inactive monomer (spectrum b) has different line shape with respect to the native enzyme (spectrum a) but substantially the same hyperfine splitting. This indicates that the copper is not released from its site by the SDS treatment and accounts for the good recovery of the native line shape after removal of the detergent (spectrum c). A kinetic analysis of the recovery process shows that 50% of activity is regained in a first order process lasting about 30 min at pH 8 and 25°C. The full original activity is reached only after days in a second first order step. The relative concentration of monomeric and dimeric forms in "recovering" samples can be evaluated by mobility analysis in SDS disc gel electrophoresis after stopping the recovery process at various times with re-addition of SDS under controlled conditions. The enzyme was found to be mostly in the monomeric form when the first phase of activity recovery was completed and the EPR spectrum was already substantially of the native type. Reassociation into dimers occurred in the slow phase, when further changes of the EPR spectrum toward the native line shape were also observed. These results suggest that refolding of monomers is the first step observed after SDS removal, and this creates partially active copper sites, probably because the fully active configuration can be reached only in the dimeric structure.

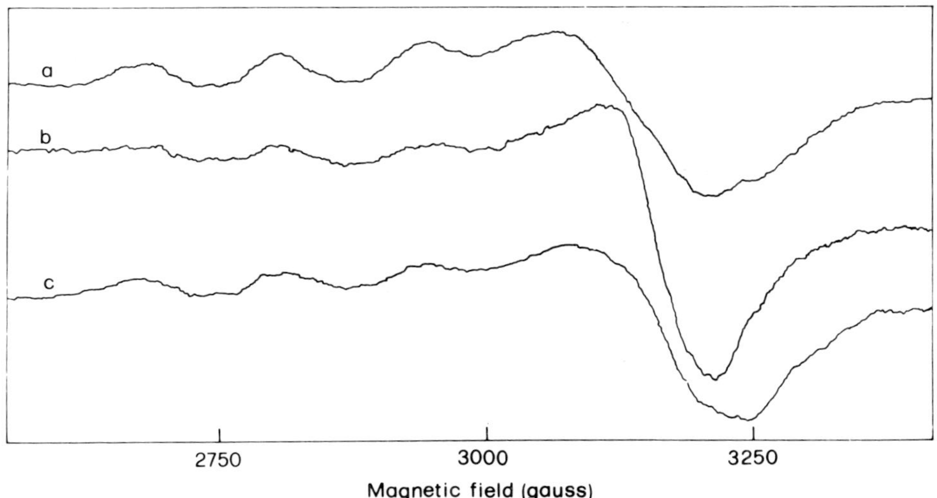

Fig.5. *X-band EPR spectra of superoxide dismutase from wheat germ. 1 x 10^{-4} M native protein (a) was incubated 48 hours in 4% SDS at 38°C (b). SDS was then removed by addition of KCl and the spectrum (c) taken after 20 hours standing at room temperature. 0.1 M Na phosphate buffer pH 7.4. Temperature: - 150°C.*

REFERENCES

1. Beauchamp, C.O. and Fridovich, I. (1973). *Biochim. Biophys. Acta 317*, 50-64.
2. Fee, J.A., Natter, R. and Baker, G.S.T. (1973). *Biochim. Biophys. Acta 295*, 96-106.
3. Fielden, P.M., Roberts, P.B., Bray, R.C., Lowe, D.J., Mautner, G.R. Rotilio, G. and Calabrese, L. (1974). *Biochem. J. 139*, 49-60.
4. Klug, D., Rabani, J. and Fridovich, I. (1972). *J. Biol. Chem. 247*, 4839-4842.
5. Klug-Roth, D., Fridovich, I. and Rabani, J. (1973). *J. Am. Chem. Soc. 95*, 2786-2790.
6. Marmocchi, F., Venardi, G., Caulini, G. and Rotilio, G. (1974). *FEBS Letters 44*, 337-339.
7. Richardson, J.A., Thomas, K.A., Rubin, B.M. and Richardson, A.C. (1975). *Proc. Natl. Acad. Sci.* U.S.A. *72*, 1349-1363.
8. Rigo, A., Viglino, P. and Rotilio, G. (1975). *Anal. Biochem. 68*, 1-8.
9. Rotilio, G., Bray, R.C. and Fielden, E.M. (1972). *Biochim. Biophys. Acta 268*, 605-609.
10. Steinman, H.M., Naik, V.R., Abernethy, J.L. and Hill, R.L. (1974). *J. Biol. Chem. 249*, 7326-7338.

DO ALL THE COPPER ATOMS IN BOVINE SUPEROXIDE DISMUTASE FUNCTION IN CATALYSIS?

S.A. COCKLE* and R.C. BRAY

School of Molecular Sciences
University of Sussex
Brighton, Sussex, Gt. Britain

Fielden *et al.* (1974) presented evidence that during steady state turnover of superoxide by bovine superoxide dismutase, the amount of Cu(II) present was not exactly 50% of the total copper as would have been expected. Instead, after correcting for some copper which was difficult to reduce chemically, they found about 75% (or perhaps slightly less) of the metal was in the Cu(II) state in the steady state when starting from the oxidized enzyme, whereas the corresponding figure when approaching the steady state from the reduced enzyme, was only about 25% (or perhaps slightly less). They concluded that probably only half of the active sites in the enzyme were functioning in catalysis. The evidence from X-ray crystallography is, however, that the enzyme possesses two identical active sites. The object of the present communication is to draw attention to another possible explanation of the data of Fielden *et al.* (1974) and to report very briefly on additional but as yet incomplete experiments.

As in the earlier work, changes in the Cu(II) content on putting the enzyme into the steady state were followed by e.p.r., using rapid freezing. However, by using potassium superoxide dissolved in DMSO as the source of O_2^-, in preference to pulse radiolysis, we were able to work at higher enzyme concentrations. Results were found to depend on the sample of enzyme which was used but in all cases the "gap" between the amount of Cu(II) in the steady states approached from the oxidized and from the reduced sites, respectively, was less than the 50% reported previously. In one case using carefully prepared enzyme and with a good excess of superoxide, the gap was only 16%. A parameter which can perhaps be measured more precisely than can the "gap" (and which in any case is not complicated by the presence or absence of Cu(II) which is difficult to reduce), is the decrease in the Cu(II) content on going to the steady state from the oxidized enzyme.

*Present address: Department of Biochemistry, McMaster University Health Sciences Centre, Hamilton, Ontario, Canada.

Whereas, Fielden *et al.* (1974) reported a decrease of from 18 to 27%, we now find higher values of from 24 to 41%. We therefore have serious reservations about interpreting the results in terms of only one of the two active sites of the enzyme molecule functioning at any one time in catalysis. We wonder instead whether micro-inhomogeneity of the enzyme preparations might not be the cause of the phenomenon. Thus, preparations might be contaminated with variable amounts of a non-functional modification of the enzyme having quite similar e.p.r. properties to the functional form. If this were the case the enzyme would be analagous in this respect to xanthine oxidase, which is normally contaminated with a non-functional modification (Bray, 1975). If the extent of reduction of copper in the enzyme in the steady state could be used to estimate the percentage of the active sites which are in the functional state, then the percent functionality of the present enzyme samples would range from 48 to 82. However, this suggestion would obviously need to be confirmed by careful specific activity measurements and unfortunately at the present state of this work, this has not been done. The nature of the postulated non-functional form of the enzyme would clearly need to be investigated.

Apart from studying the intensity of the Cu(II) e.p.r. signal, we also examined changes in its form. There were some small changes on reducing the enzyme by superoxide, however these are probably not relevant to the possible presence of non-functional enzyme molecules in the preparations, since the changes were similar with all the samples we examined. Furthermore, the spectrum did not seem to depend on whether we started from oxidized or from reduced enzyme. The changes might be related to the medium which was used for the rapid freezing experiments and we have not as yet excluded the possibility that they were due to pH increases caused by the presence of potassium superoxide at the rather low buffer concentrations employed.

REFERENCES

1. Bray, R.C. (1975). In "The Enzymes", ed. Boyer, P.D. Third Edition, Academic Press Vol. XII, pp. 299-419.
2. Fielden, E.M., Roberts, P.B., Bray, R.C., Lowe, D.J., Mautner, G.N., Rotilio, G. and Calabrese, L. (1974). *Biochem. J. 139*, 49-60.

THE THREE-DIMENSIONAL STRUCTURE OF Cu,Zn SUPEROXIDE DISMUTASE

D.C. RICHARDSON

Department of Biochemistry
Duke University Medical Center
Durham, North Carolina 27710, U.S.A.

Of the two major classes of superoxide dismutases, no information is yet available on the three-dimensional structure of any of the Mn or Fe superoxide dismutases, although X-ray crystallographic work is in progress on the Mn enzyme from *Escherichia coli* (Beem *et al.*, 1976). However, the three-dimensional structure of one of the Cu,Zn superoxide dismutases (from bovine erythrocytes) has been determined to 3A resolution (Richardson *et al.*, 1975a, b).

Bovine Cu,Zn superoxide dismutase has two 151-residue subunits of identical amino acid sequence (Steinman *et al.*, 1974), each with one copper and one zinc. The crystals used in the X-ray work have two molecules (4 subunits) per asymmetric unit (Richardson *et al.*, 1972). The two subunits within each dimer molecule are related by a local twofold axis, but the contact between the two dimers is asymmetrical, involving a non-180° rotation plus a translation (Thomas *et al.*, 1972). There is no indication that the dimer-dimer association is significant in solution. In the crystal, at 3A resolution, differences in backbone conformation between the 4 different subunits appear to be quite minor, and most if not all differences in side chain position are due to nonequivalent crystal packing environments. Therefore it seems likely that the two subunits of a Cu^{++},Zn^{++} superoxide dismutase dimer in solution would be exactly, or very closely, equivalent in conformation. Averaged α-carbon and metal coordinates are available in Richardson *et al.* (1975b), from the Brookhaven Data Bank, and in the Global Atlas of Macromolecular Structures on Microfiche edited by Richard Feldmann at the National Institutes of Health. The most up-to-date set of coordinates will always be available from the latter two sources.

Figure 1 is a simplified schematic drawing of the bovine Cu,Zn superoxide dismutase subunit, and Figure 2 is a stereo photograph of a wire model of the α-carbon backbone. The dominant feature is a large cylindrical barrel made up of 8 extended chains of entirely antiparallel β structure; it contains about 75 residues (50% of the backbone). The interior of the β barrel is packed with hydrophobic side chains. The barrel is slightly flattened in cross section: the average distance between main chains is 12 or 13A in the shorter direction and 16 or 17A in the longer direction. The half of the barrel toward the outside of the subunit is quite regular; the 4 chains on that side have a twist of only about 10° per residue and are within hydrogen-bonding distance of one another for as much as 9 or 10 residues. The less regular, more twisted, half of the β barrel is almost entirely internal to the subunit; that side contributes 4 ligands to the metals and one end to the disulfide bridge. In Figure 3 the β structure is diagrammed as if spread out flat and viewed from the outside of the barrel. Residues shown zigzagging downwards on the heavy lines point toward the interior of the barrel.

Fig.1. *Simplified schematic drawing of the structure of the bovine Cu,Zn superoxide dismutase subunit.*

The rest of the structure of the subunit is made up of two long loops of non-repetitive structure. Each of the loops begins and ends in two adjacent chains on the less regular side of the β barrel. The first such loop (residues 48-79) has two distinct halves. The first half, the "disulfide loop", has its end held back against the barrel by the disulfide bridge and participates extensively in the contact between subunits. The second, quite hydrophilic, half makes a figure eight shape and contributes 3 ligands to the zinc. The second loop (residues 119-141), also very hydrophilic, can be seen at the far right side in Figures 1 and 2. Although it extends quite far out into solution from the rest of the molecule and does not make particularly close packing contacts in the crystal, it still appears quite well ordered in the electron density map.

The most certain piece of α-helix is in the external loop and is 1 or $1\frac{1}{2}$ turns long. There are two other places in the subunit where there may be a single helical-type hydrogen bond, but at this resolution it is unclear whether the residues involved are really in α-helical conformation. The helix content, therefore, is low: between 3% and 8%

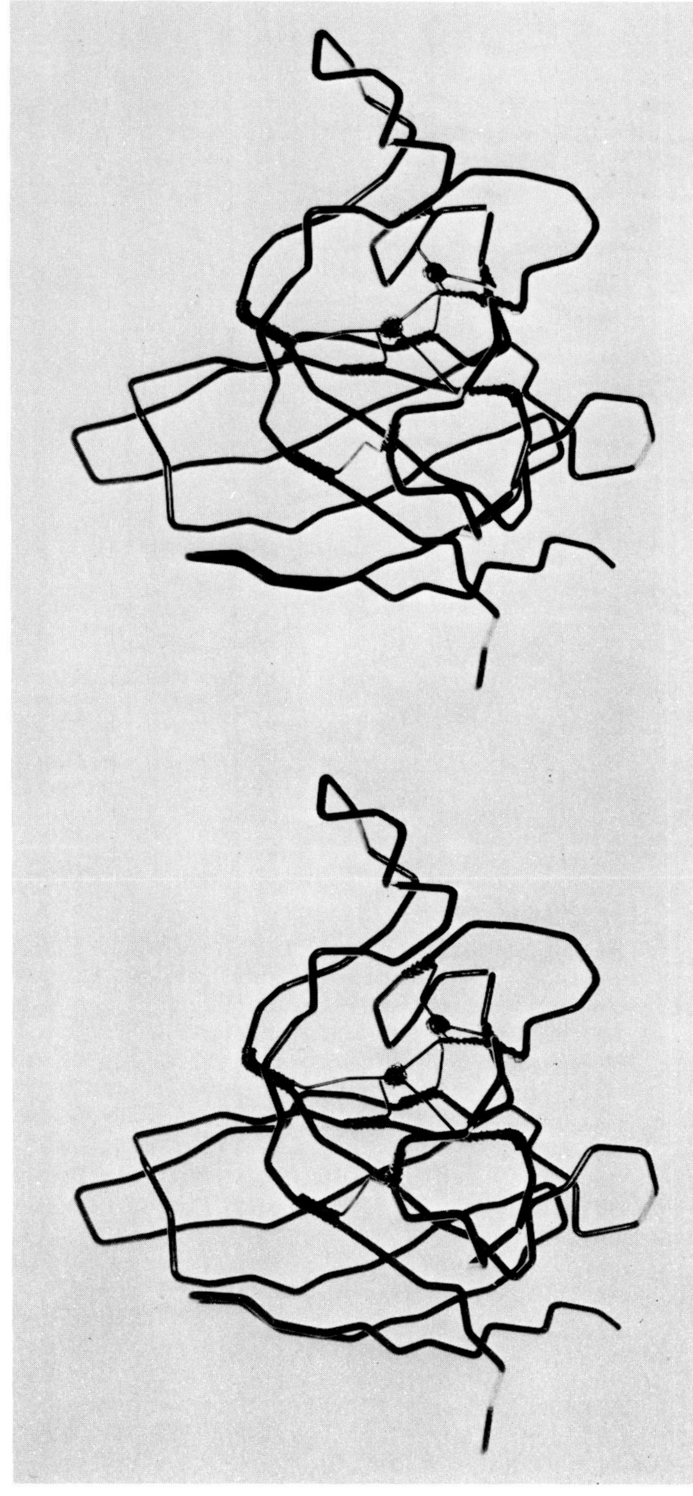

Fig.2. *Stereo photograph of a wire model of the polypeptide backbone of the Cu,Zn superoxide dismutase subunit, with the metals and their ligands and the disulfide bridge indicated. The view of the Cu (left-hand ball) is from its solvent accessible side.*

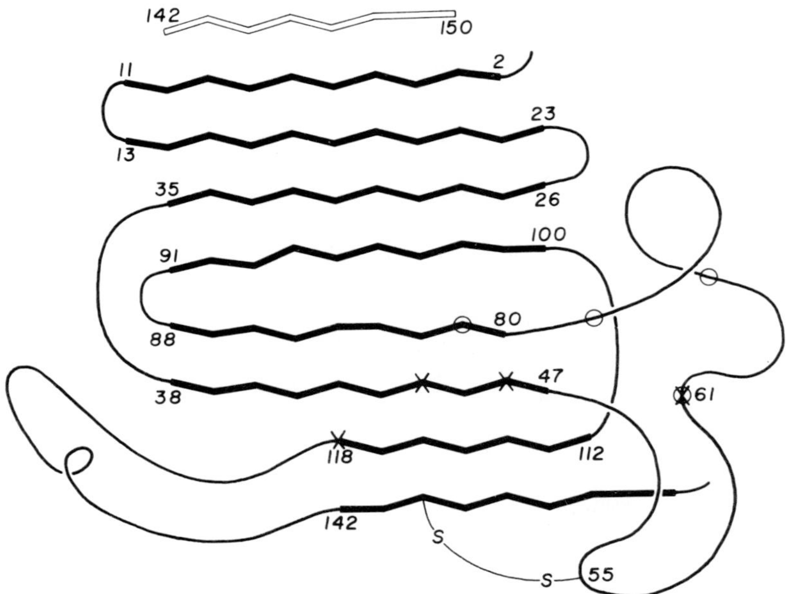

Fig.3. *Diagram showing the 8 strands of the Cu,Zn superoxide dismutase β sheet and their topological connectivity. The barrel is spread flat and shown from the outside, and the C-terminal strand is repeated at the top to show its position relative to the N-terminal strand. Those residues which probably participate in the β sheet hydrogen bonding are shown in heavy zigzag lines; however, there are not hydrogen bonds between all residues shown opposite one another. When the line zigzags downward, the corresponding side group in the structure points toward the interior of the barrel. Circles mark the Zn ligands and Xs the Cu ligands.*

The relationship of the two subunits of a dimer molecule is shown in Figure 4. The contact area across the local twofold axis is broad and closely fitted. It involves part of the outside surface of the β barrel, the last few residues at the C terminal, and the disulfide loop. It consists principally of hydrophobic side chain interactions: there are at most two main chain and perhaps two or three side chain hydrogen bonds between the subunits, while there are 12 to 14 hydrophobic side groups from each subunit that are apparently in Van der Waals contact. Val 146 and Ile 111 both interact with their counterparts across the local two fold axis, and at the closest approach between main chains, Gly 49 and Gly 112 of one subunit are opposite Gly 148 of the other subunit.

The two Cu atoms on opposite subunits within the superoxide dismutase dimer are approximately 34A apart, ruling out the possibility of direct interaction between the two active sites. The Cu and Zn on a single subunit are only about 6A apart; between them is the imidazole ring of His 61, to which both metals appear liganded. The protein ligands to the Zn are His 61, His 69, His 78, and Asp 81, in approximately tetrahedral arrangement. In addition to His 61, the Cu is liganded by His 44, His 46, and His 118, and has one side relatively open to solvent access.

Detailed model-to-map fitting of the 3A resolution electron density map

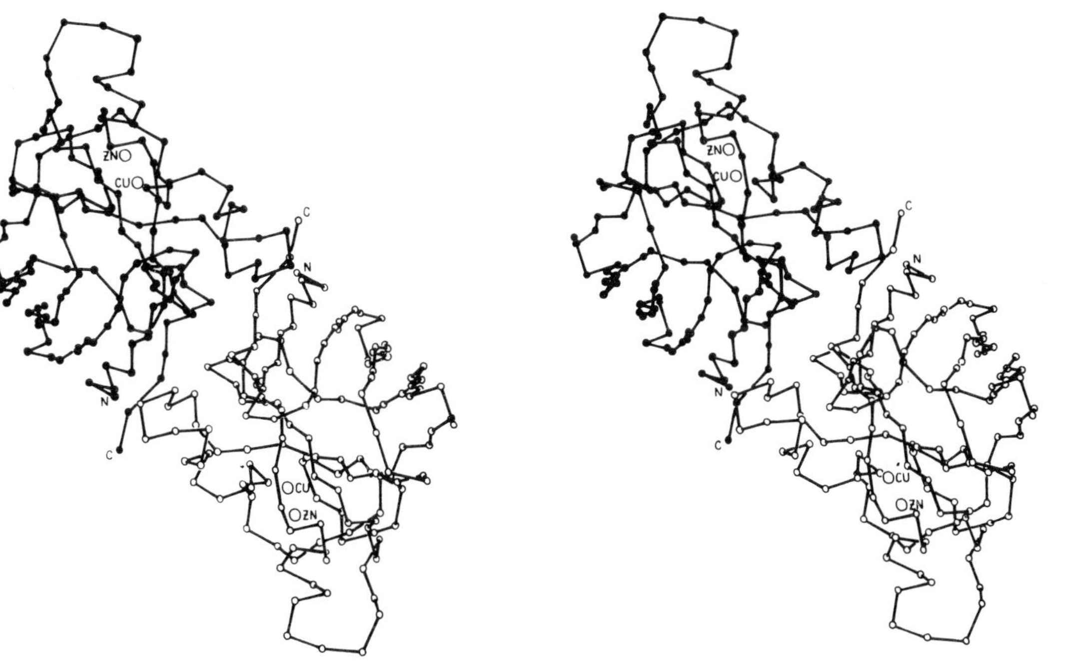

Fig.4. *Stereo drawing of a bovine Cu,Zn superoxide dismutase dimer molecule, viewed down the local twofold axis of the subunit contact. The α-carbons are shown as solid circles for one subunit and open circles for the other.*

on an interactive computer graphics system has made it clear that the ring of His 61 is not directly on a line between the Cu and Zn, although it is close enough to interact with both of them. In general, the conformation of the fitted models agrees extremely well among the four subunits, although there is inaccuracy in such features as the orientation of the histidine rings, since this map is only at 3A resolution. There is some variation in the appearance of the electron density for the bridging histidine, although these differences are more likely to be due to phasing inaccuracies in the map than to a genuine conformational difference. In order to fit into the density in each of the four subunits, the position of the imidazole ring of His 61 must be somewhat above the line between the two metals and considerably above the plane of the other three copper ligands. However, since the map does not show the orientation of the ring except indirectly through constraints imposed by the backbone position, the bridging imidazole ring can either be fit with a symmetrical bending of the bonds to both metals, or it can be fit in an asymmetrical position where the proximal nitrogen has a straight bond to the zinc and the distal nitrogen sits above the copper. The latter possibility is illustrated in Figure 5.

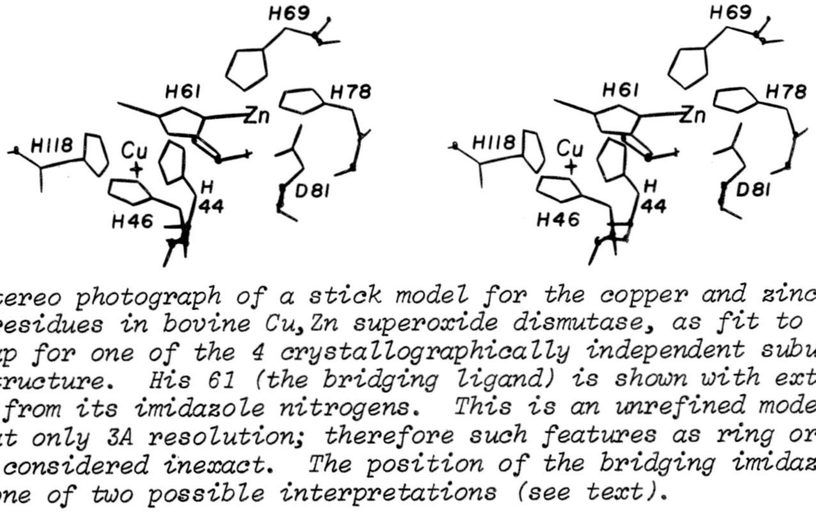

Fig.5. *Stereo photograph of a stick model for the copper and zinc and their 7 ligand residues in bovine Cu, Zn superoxide dismutase, as fit to the electron density map for one of the 4 crystallographically independent subunits in the crystal structure. His 61 (the bridging ligand) is shown with extra "bonds" extending from its imidazole nitrogens. This is an unrefined model fitting to a map at only 3A resolution; therefore such features as ring orientations should be considered inexact. The position of the bridging imidazole ring is shown in one of two possible interpretations (see text).*

Figure 5 shows a stick model for the copper and zinc and their 7 ligand residues, as fitted to the electron density map for one of the four subunits. The His 61 imidazole is shown with extra "bonds" extending from the ring nitrogens. The one from the proximal nitrogen ends at the zinc position (metal positions can be located by the high local electron density maxima, as well as by inference from the geometry of their surrounding ligands), so that His 61 is fit as a normal histidine ligand to a tetrahedrally coordinated zinc. The nitrogens of His 44, His 46, and His 118 lie approximately at three corners of a square plane centered around the copper, with the fourth corner unoccupied, and this fitting places the distal nitrogen of His 61 somewhat above that square plane (on the side accessible to solvent) in a position intermediate between planar and axial.

One of the surprising features of the Cu,Zn superoxide dismutase structure is its close resemblance to the basic folding pattern of an immunoglobulin domain, which consists of a 7-stranded barrel (or sandwich) of β structure. If the N-terminal strand in the superoxide dismutase β barrel is ignored, the rest of the topological connectivity is identical for these two functionally different proteins, and the two long loops of the superoxide dismutase structure come at places equivalent to hypervariable region loops in the immunoglobulin structure. There is no resemblance in disulfide position or in subunit or domain contact geometry, however, and no detectable sequence homology. This comparison is discussed in detail elsewhere (Richardson *et al.*, 1976). A matching and least-squares fit of α-carbon coordinates was done between various pairs of these structures. The match obtained between superoxide dismutase and any one of the immunoglobulin domains is slightly less close than the match between immunoglobulin constant and variable domains. An analysis was made of the probability that two topologies of this sort would match by chance. Including allowance for the empirically estimated probabilities of occurrence for individual connection types, the overall probability of a chance match was estimated as about 1/3000. The most plausible explanation for the resemblance between the two proteins is some sort of evolutionary relationship, but it is also possible that the folding of β sheets is limited by enough additional constraints to make a chance match reasonably likely.

REFERENCES

1. Beem, K.M., Richardson, J.S. and Richardson, D.C. (1976). *J. Mol. Biol. 105*, in press.
2. Richardson, D.C., Bier, C.J. and Richardson, J.S. (1972). *J. Biol. Chem. 247*, 6368-6369.
3. Richardson, J.S., Thomas, K.A., Rubin, B.H. and Richardson, D.C. (1975a). *Proc. Nat. Acad. Sci. U.S.A. 72*, 1349-1353.
4. Richardson, J.S., Thomas, K.A. and Richardson, D.C. (1975b). *Biochem. Biophys. Res. Commun. 63*, 986-992.
5. Richardson, J.S., Richardson, D.C., Thomas, K.A., Silverton, E.W. and Davies, D.R. (1976). *J. Mol. Biol. 102*, 221-235.
6. Steinman, H.M., Naik, V.R., Abernethy, J.L. and Hill, R.L. (1974). *J. Biol. Chem. 249*, 7326-7338.
7. Thomas, K.A., Rubin, B.H., Bier, C.J., Richardson, J.S. and Richardson, D.C. (1974). *J. Biol. Chem. 249*, 5677-5683.

AMINO ACID SEQUENCE HOMOLOGIES AMONG SUPEROXIDE DISMUTASES

J.I. HARRIS

MRC Laboratory of Molecular Biology
University Postgraduate Medical School
Hills Road, Cambridge, CB2 2QH, England

H.M. STEINMAN*

Department of Biochemistry
Duke University Medical Center
Durham, North Carolina 27710, U.S.A.

Free radicals, once viewed as esoteric molecules fascinating solely to electron beam chemists, are now recognised to be ubiquitous physiological agents exercising predominant roles both in normal metabolism and in certain disease states. The discovery of superoxide dismutase (McCord and Fridovich, 1968, 1969) fostered this biological *volte face* by stimulating study of the chemical and biological properties of the superoxide ion itself, as well as investigations of the structure, function and evolution of the enzyme protein responsible for its dismutation to hydrogen peroxide and molecular oxygen (Fridovich, 1974, 1975).

Superoxide dismutase has attracted the attention of protein chemists for a variety of reasons: these cupro-zinc, mangano, or ferri metalloenzymes are abundant in oxygen-metabolising organisms, and the subunit polypeptide chains of each are apparently chemically identical and of a size (about 150 - 200 residues) which is ideal for direct study by contemporary methods of sequence analysis. Moreover, representatives of each metalloenzyme class have been crystallized, so that cooperative sequence and X-ray analyses can be implemented to determine three-dimensional molecular structures and structure-function relationships. (For recent reviews, see Fridovich, 1974, 1975).

Beyond the amino acid sequence and three-dimensional structure of cupro-zinc dismutase from bovine erythrocytes (Richardson *et al.*, 1975; Steinman *et al.*, 1974), little information is available about the extent of sequence homology within the cupro-zinc class of enzymes. Manganese- and iron-containing enzymes on the other hand are amenable to automated methods of amino-terminal sequence analysis (cupro-zinc enzymes having blocked amino-terminal residues) and by this means have been shown to constitute a highly homologous class of proteins (Steinman and Hill, 1973; Bridgen *et al.*, 1975). Moreover, comparisons of amino-terminal sequences of superoxide dismutases of

*Current address: Department of Biochemistry, Albert Einstein College of Medicine, 1300 Morris Park Avenue, Bronx, New York 10461, U.S.A.

prokaryotic, eurkaryotic, and mitochondrial origin have been interpreted to show:

(1) that the cupro-zinc class and the manganese-iron class are of independent evolutionary origins, and

(2) that mitochondria have evolved from prokaryotes by endosymbiosis.

We now report recent progress, in the form of hitherto unpublished sequence data from two laboratories, on a wider range of manganese and iron-containing superoxide dismutases from prokaryotic and eukaryotic sources, including photosynthetic organisms. While we are aware of the pitfalls inherent in definitively tracing evolutionary relationships from limited sequence data, particularly when sequences are not correlated with three-dimensional structure, we believe that these sequence data are nevertheless eminently pertinent to the evolution of oxygen-metabolising organisms.

Amino-terminal sequences of eleven manganese or iron superoxide dismutases (seven with Mn, four with Fe) have been investigated by automated analysis with a Beckman Sequenator. The *Spirulina platensis* (Fe) and *Rhodopseudomonas spheroides* (Mn) enzymes were obtained from Drs. J. Lumsden and D.O. Hall and the *Photobacterium leiognathi* (Fe) enzyme from Drs. K. Puget and A.M. Michelson (Lumsden and Hall, 1976; Puget and Michelson, 1974). The *B. stearothermophilus* (Mn) and *T. aquaticus* (Mn) enzymes were purified according to published procedures (Bridgen *et al.*, 1975; Sato and Harris, 1976). The amino-terminal sequences of these five enzymes were determined in Cambridge (Bridgen *et al.*, 1975; Sato and Harris, 1976; and unpublished results of J.I. Harris and F. Northrop*). The *plectonema boryanum* (Fe), *E. coli* (Fe), *E. coli* (Mn), *Saccharomyces cerevisiae* (Mn), chicken liver mitochondrial (Mn), and human liver (Mn) enzymes were obtained from members of the laboratories of Dr. Irwin Fridovich and Dr. Joe McCord (Misra and Keele, 1975; Yost and Fridovich, 1973; Keele *et al.*, 1970; Ravindranath and Fridovich, 1975; Weisiger and Fridovich, 1973; McCord *et al.*, 1977) and were sequenced in Durham (Steinman and Hill, 1973; Steinman, 1975-76). The *E. coli* (Mn) sequence beyond residue 24 was established from a peptide overlapping the amino terminal region (Steinman, 1975-76).

These 11 amino-terminal sequences are presented in Fig.1, with that of bovine erythrocyte cupro-zinc dismutase (Steinman *et al.*, 1974). Sequences (1) and (2) are from blue-green alga dismutases, (3) - (8) from bacterial dismutases, (9) - (11) from eukaryotic mangano dismutases, and (12) from the bovine cupro-zinc dismutase. Compared with otherwise homologous sequences of aerobic respirers, sequences (1), (2) and (8), from photosynthetic prokaryotes, are out of phase, by insertion of one residue, Met-19. Similarly, homology is maximized with the *T. aquaticus* sequence, (7), by aligning residue 3 with residue 1 of the other sequences.

Considerable sequence homology among prokaryotic and eukaryotic dismutases of the manganese-iron class is apparent on inspection of Fig.1. Thus, nine out of 35 amino-terminal residues (26%) are identical in the eleven sequences, excluding the unidentified residues in (2), (4), (9), and (11), which might either increase or decrease this percentage. The replacements at most sites are compatible with single base changes in nucleotide codons. This degree of sequence identity among manganese-iron enzymes is comparable to that found among eukaryotic cytochromes *c*, where 31% of the residues are invariant (Smith, 1970). If the comparison were restricted to mangano

*See also Lumsden *et al.* (1977). This volume pp 437-450.

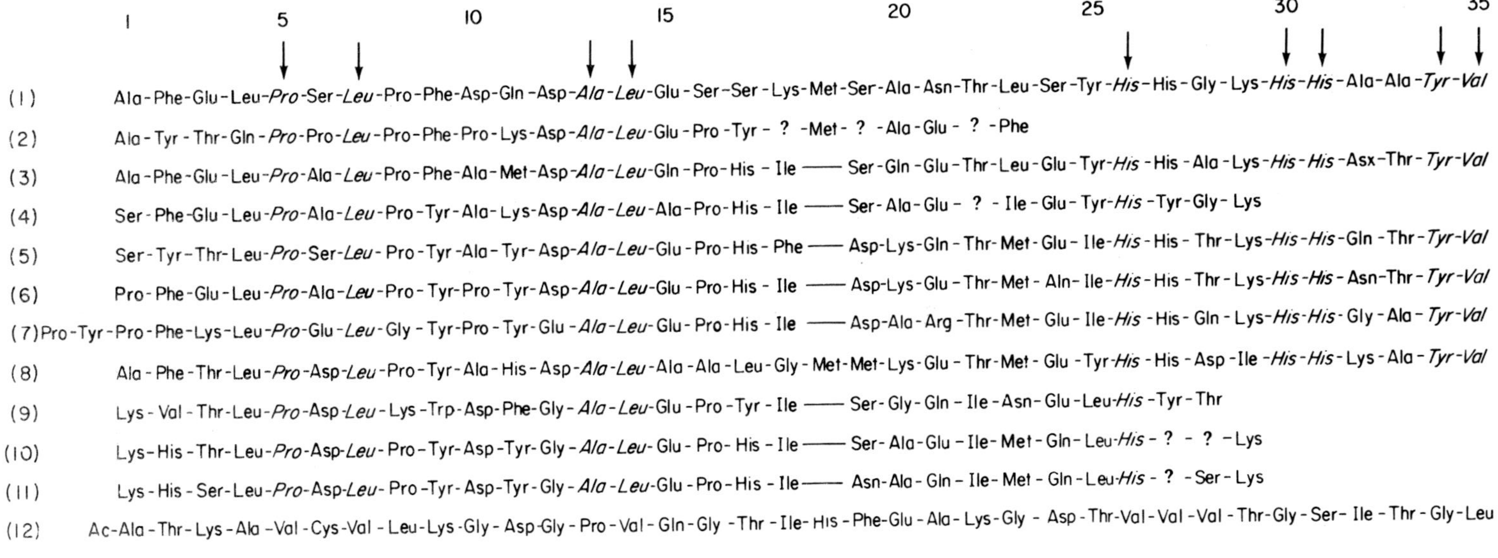

(1) *Spirulina platensis* (Fe)
(2) *Plectonema boryanum* (Fe)
(3) *Photobacterium leiognathi* (Fe)
(4) *Escherichia coli* (Fe)
(5) *Escherichia coli* (Mn)
(6) *Bacillus stearothermophilus* (Mn)
(7) *Thermus aquaticus* (Mn)
(8) *Rhodopseudomonas spheroides* (Mn)
(9) *Saccharomyces cerevisiae* mitochondrial (Mn)
(10) Chicken liver mitochondrial (Mn)
(11) Human liver mitochondrial (Mn)
(12) Bovine erythrocyte (Cu-Zn)

The methionine at residue 19, in sequences (1), (2) and (8), and the two amino terminal residues in sequence (7) have been excluded from the consecutive numbering shown on top. A "?" indicates that a residue is present but could not be identified; a "—" indicates that no residue is present and that a gap has been introduced to maximize homology. The arrows above indicate the residues (italicized) which are invariant in all Mn and Fe dimutases.

Fig.1. *Amino terminal sequences of superoxide dismutases.*

enzymes from three different bacterial sources (*E. coli* (5) and two thermophiles (6) and (7)), 21 (60%) of the amino terminal 35 residues are identical. This value may be compared with a sequence identity of 52% in cytochromes *c*-551 from three species of *Pseudomonas* (Ambler and Wynn, 1973) and 44% in the amino-terminal 50 residues of five bacterial tryptophan synthetase α-chains (Li and Yanofsky, 1972).

None of the 9 invariant residues in the dismutase sequences is glycine, an amino acid which occupies 8 of the 35 invariant sites (23%) in eukaryotic cytochromes *c*. Glycine residues are crucial for the numerous β-turns in bovine erythrocyte dismutase (Richardson *et al.*, 1975). The relative scarcity of the amino acid may be important to the secondary structure of the amino terminal regions of mangano and ferri dismutases.

Based upon these limited data the rate of evolutionary divergence among manganese-iron superoxide dismutases appears to be relatively slow, consistent with their essential physiological role, wherein even slight decreases in catalytic effectiveness might constitute a lethal mutation. The true significance of the invariant residues will, however, emerge only when their structural and functional roles in the dismutase molecule are ascertained.

The present superoxide dismutase data may be evaluated by comparison with the known sequence and three-dimensional data for another essential multimeric enzyme, triosephosphate isomerase. Among vertebrates, triosephosphate isomerase is a slowly evolving protein: the frequency of substitution is only two amino acids per 100 residues per 100 million years (Kolb *et al.*, 1974), and the sequences of the polypeptide subunits from chicken muscle and from the thermophile, *B. stearothermophilus*, are identical in about 100 of the 250 residues (40%) (Artavanis and Harris, 1976). Moreover, the invariant residues in these sequences are known to be important for catalytic activity, or involved in interdomain or intersubunit contacts, in the three-dimensional structure (Banner *et al.*, 1975). Comparison of the amino-terminal sequences of manganese dismutases from the *same* two species, chicken (10) and *B. stearothermophilus* (6), shows that 15 of the amino terminal 26 residues (about 60%) are conserved, a degree of conservation appreciably higher than found for triosephosphate isomerase. This observation assumes that comparisons based upon 20% of the residues, from the amino terminus of the polypeptide chain, are representative of the entire subunit of some 185 - 200 residues, an assumption which, however, is unlikely to be valid. Indeed, the amino-terminal 20% of the triosephosphate isomerases have a sequence identity of only 25%, a value considerably lower than that based upon the whole subunit chain. These considerations illustrate the dangers in over-interpretation of the limited sequence data presently available for superoxide dismutases.

The significance of sequence comparisons between the eleven mangano or ferri enzymes and the single representative of the cupro-zinc class must be evaluated with cognizance that the lengths of the two types of subunits differ by approximately 40 residues. The absence of significant homology, evident between the amino-terminal sequences of mangano or ferri and cupro-zinc sequences listed in Fig.1, is also apparent upon comparison (Bridgen *et al.*, 1975) of the amino terminal 60 residues of the *B. stearothermophilus* mangano enzyme with the entire bovine erythrocyte sequence, consistent with the disparity of enzymological properties of the two dismutase classes. Divergent evolution from a common ancestor appears to be ruled out by these sequence data. Distinguishing between convergent or parallel evolutionary mechanisms will require comparisons of complete primary structures as well as of three-dimensional structures of exemplary proteins within each class of dismutase. Further sequence analysis will also be necessary with the cupro-zinc dismutases, to ascertain if they are as homologous among themselves as the manganese-iron class of enzymes have proved to be.

The striking sequence similarities between mitochondrial and prokaryotic dismutases support the endosymbiotic hypothesis of mitochondrial origin. Essential to arguments for endosymbiosis are certain inferences about the evolutionary origin of dismutases, which distinguish them from other enzymes that also occur in both cytoplasm and mitochondria. In so far as superoxide dismutase functions as a defense against oxygen toxicity, it would not have been essential for survival until oxygen levels in the earth's atmosphere reached a crucial level. The accumulation of oxygen on earth has been attributed largely to photosynthesis, initially by photosynthetic blue-green algae, a prokaryotic species which is believed to have evolved shortly before the prokaryote and eukaryote cell lines diverged. If the superoxide dismutase molecule were to have evolved after the earth's atmosphere became oxidizing, then it is likely to have initially appeared after the blue-green algae and thus after divergence of prokaryotes and eukaryotes. The sequence homology among prokaryotic dismutases and eukaryotic mitochondrial dismutases could not then be attributed to mitochondria or protomitochondria which existed in a common ancestor of prokaryotes and eukaryotes, but would be consistent with the hypothesis that eukaryotic mitochondria evolved from prokaryotes by intracellular symbiosis. The endosymbiotic explanation of mitochondrial origin is, however, a broad concept which impinges upon many aspects of biology and biochemistry, and it must be stressed that the sequence similarities in dismutases are here presented merely as evidence in support of the endosymbiosis theory, and not as definitive proof. Amino acid sequence and three-dimensional structural studies of manganese and iron enzymes, now in progress, promise to provide more conclusive evidence about the evolutionary origin of mitochondria and about the structural, functional and evolutionary relationships between the different classes of superoxide dismutase.

ACKNOWLEDGEMENT

H.S. expresses his deep appreciation to Dr. Robert L. Hill for use of Beckman Sequencer and Gas Chromatograph facilities while H.S. was a member of the Department of Biochemistry at Duke University.

REFERENCES

1. Ambler, R.P. and Wynn, M. (1973). *Biochem. J. 131,* 485.
2. Artavanis, S. and Harris, J.I. (1976). *Eur. J. Biochem.,* in press.
3. Banner, D.W., Bloomer, A.C., Petsko, G.A., Phillips, D.C., Pogson, C.I., Wilson, I.A., Corran, P.H., Furth, A.J., Milman, J.D., Offord, R.E., Priddle, J.D. and Waley, S.G. (1975). *Nature 255,* 609.
4. Bridgen, J., Harris, J.I. and Northrop, F. (1975). *FEBS Letters 49,* 392.
5. Fridovich, I. (1974). *Adv. Enzymol. 41,* 35.
6. Fridovich, I. (1975). *Ann. Rev. Biochem. 44,* 147.
7. Keele, B.B., Jr., McCord, J.M. and Fridovich, I. (1970). *J. Biol. Chem. 245,* 6176.
8. Kolb, E., Harris, J.I. and Bridgen, J. (1974). *Biochem. J. 137,* 185.
9. Li, S.-L. and Yanofsky, C. (1972). *J. Biol. Chem. 247,* 1031.
10. Lumsden, J. and Hall, D.O. (1976). *Biochim. Biophys. Acta 438,* 380.

11. McCord, J.M., Boyle, J.A., Day, E.D., Jr., Rizzolo, L.J. and Salin, M.L. (1977). In "Superoxide and Superoxide Dismutase" (Michelson, A.M., McCord, J.M. and Fridovich, I., eds.), Academic Press, London (this volume), p. 129-138.
12. McCord, J.M. and Fridovich, I. (1968). *J. Biol. Chem. 243,* 5753.
13. McCord, J.M. and Fridovich, I. (1969). *J. Biol. Chem. 249,* 6049.
14. Misra, H.P. and Keele, B.B., Jr. (1975). *Biochim. Biophys. Acta 379,* 418.
15. Puget, K. and Michelson, A.M. (1974). *Biochem. Biophys. Res. Commun. 58,* 380.
16. Ravindranath, S.D. and Fridovich, I. (1975). *J. Biol. Chem. 250,* 6107.
17. Richardson, J.S., Thomas, K.A., Rubin, B.H. and Richardson, D.C. (1975). *Proc. Natl. Acad. Sci. (U.S.A.) 72,* 1349.
18. Sato, I. and Harris, J.I. (1976). *Eur. J. Biochem.,* in press.
19. Smith, E.L. (1970). In "The Enzymes", Third Edition, Vol. 1 (Boyer, P.D., ed.), Academic Press, New York, p. 267.
20. Steinman, H.M. (1975-76) unpublished data.
21. Steinman, H.M. and Hill, R.L. (1973). *Proc. Natl. Acad. Sci. (U.S.A.) 70,* 3725.
22. Steinman, H.M., Naik, V.R., Abernathy, J.L. and Hill, R.L. (1974). *J. Biol. Chem. 249,* 7326.
23. Weisiger, R.A. and Fridovich, I. (1973). *J. Biol. Chem. 248,* 3582.
24. Yost, F.J., Jr. and Fridovich, I. (1973). *J. Biol. Chem. 248,* 4905.

THE ROLE OF THE SUPEROXIDE DISMUTASES IN PULMONARY OXYGEN TOXICITY

J.D. CRAPO

Department of Medicine
Duke University Medical Center
Durham, North Carolina 27710, U.S.A.

INTRODUCTION

Most mammals die after prolonged exposures to pure oxygen atmospheres. After an initial asymptomatic period there is the gradual onset of irritation to the respiratory system, and then the animal experiences progressive dyspnea. Death usually follows shortly after the onset of dyspnea and is due to pulmonary edema and profound hypoxemia. Pathologic evaluation of the lungs of these animals reveals diffuse damage to the capillary endothelium. The endothelial cells are markedly swollen and many are dead and degenerating. This is thought to be the insult which leads to the terminal interstitial and alveolar edema. The alveolar epithelium is usually intact, but the epithelial cells are swollen, Alveolar Type II cells have proliferated, and the alveolar surface is commonly lined with a "hyaline membrane" (Weibel, 1971).

The mechanisms of the toxic reactions of oxygen are not completely understood, but a number of theories have been proposed. Evidence from both *in vivo* and *in vitro* experiments demonstrate that increased oxygen tensions produce alterations in cellular metabolism (Haugaard, 1965). A number of enzymes containing sulfhydryl groups are inactivated by oxygen and the sulfhydryl groups of several smaller compounds such as reduced glutathione, lipoic acid, and coenzyme A are oxidized. These phenomena have been proposed as important components of oxygen toxicity (Haugaard, 1968). Oxygen may also damage cell membranes by lipid peroxidation, which can have significant effects on metabolic processes at both the cellular and subcellular levels. One toxic product of oxygen metabolism is hydrogen peroxide. Catalase and peroxidase have been thought to be important in protecting against oxygen toxicity (Callow, 1923).

Gerschman (1964) was the first to propose that oxygen toxicity may be related to the formation of free radicals with subsequent chain reactions leading to uncontrolled destructive oxidations. Gerschman noted that the damage caused to tissue by ionizing radiation and that caused by increased pressures of oxygen were similar. Many agents protect against damage from both sources. Antioxidants such as alpha-tocopherol act as nonspecific scavengers of free radicals, and have been found to reduce the toxic effects of oxygen. Gerschman further proposed that our normal atmosphere of 20% oxygen is potentially toxic and that animals had to develop defenses in order to survive in and utilize the oxygen in our atmosphere. She suggested that these mechanisms of protection from harmful oxidations were limited, and that they could be overwhelmed by exposure to radiation or to high tensions of oxygen.

Before the discovery of superoxide dismutase (SOD) in 1969, adequate tools were not available to study the free radical reactions associated with oxidative metabolism, and no data were available to indicate which free radicals were important in the genesis of oxygen toxicity. Recent work with the superoxide free radical, the hydroxyl radical and singlet oxygen have suggested that these species may be the agents that initiate the major toxic reactions of oxygen. High oxygen tensions have been shown to lead to an increased production of superoxide free radicals by several common biological reactions (Fridovich, 1974). Hydroxyl radicals ($OH^{\cdot}$) can be produced from the reaction of superoxide and its dismutation product H_2O_2 by the Haber and Weiss reaction. It has been recently proposed that this reaction also liberates electronically excited singlet oxygen:

$$H_2O_2 + O_2^- \longrightarrow OH^- + OH\cdot + O_2^* \qquad \text{(Kellogg, 1975)}$$

McCord and Fridovich (1969), after isolating and purifying SOD, proposed that it played a pivotal role in protecting the organism against the deleterious actions of the superoxide radical. They tested this thesis by relating the SOD activity of micro-organisms to the capacity of those organisms to tolerate oxygen. They found a significant correlation between the ability of micro-organisms to grow in and utilize oxygen and their content of SOD (McCord *et al.*, 1971). Further evidence to support this theory was found when large increases in SOD activity were induced in *E. coli B* and *Streptococcus faecalis* exposed to hyperbaric oxygen. These "induced" bacteria were then found to be more resistant to the deleterious effects of an extremely high atmosphere of oxygen (46 atmospheres) than were bacteria grown under anaerobic conditions (Gregory and Fridovich, 1973a,b). *E. coli B* was shown to have a distinct ferri-superoxide dismutase in the periplasmic space. When this enzyme was depleted by growth in iron-deficient media, the cells were rendered hypersensitive to an exogenous source of O_2^- (Gregory *et al.*, 1973).

This paper will review work which extends the initial observations on the role of superoxide dismutase in oxygen toxicity from micro-organisms to mammalian pulmonary oxygen toxicity.

INDUCTION OF SUPEROXIDE DISMUTASE IN OXYGEN TOLERANT RATS

Adult rats, when exposed to 100% oxygen uniformly die after 60-72 hours of continuous exposure. If those same animals are first exposed to 85% oxygen for 5-7 days, they become "tolerant", and can then be placed in 100% oxygen and will survive for prolonged periods. These oxygen-adapted rats provided an excellent model in which to test the theory that SOD is an essential agent in protecting against the toxic effects of oxygen. During the initial sublethal exposure to oxygen, SOD might be induced and subsequently could enable the animal to survive in a pure oxygen atmosphere. A series of large male rats were exposed to 85% oxygen for 1, 2, 3, 4, 5 and 7 days (Crapo and Tierney, 1974). At each of these time intervals, a portion of each group were killed and their lungs were assayed for SOD activity. The remainder of the animals in each group were placed in 100% oxygen to evaluate whether "tolerance" had occurred (Fig.1). Total pulmonary SOD activity was found to increase about 50%. The change in SOD activity occurred between the third and the fifth days of the 85% oxygen exposure, which correlates well with the time when tolerance develops. After being transferred to 100% oxygen the animals all survived at least 24-72 hours. Of those animals which survived 72 hours in 100% oxygen, only one died subsequently, although many of them were exposed for prolonged periods of up to 21 days.

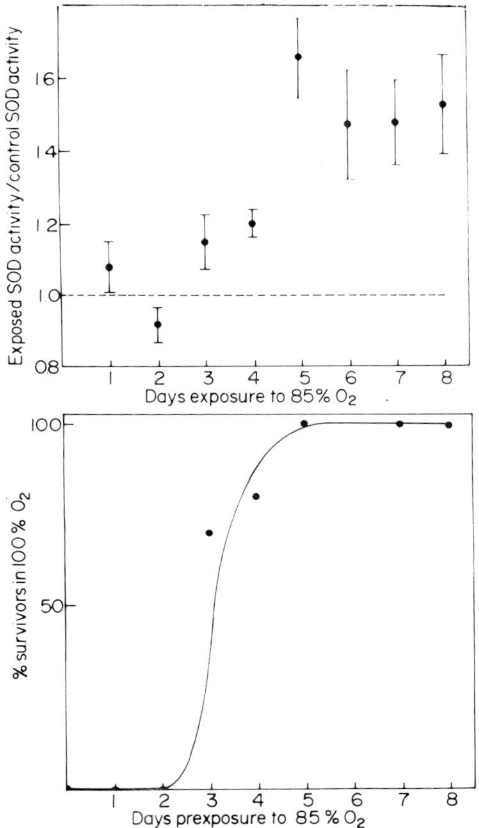

Fig.1. *Time course of changes in pulmonary SOD activity and development of tolerance during 85% oxygen exposure. Top figure is a composite of several individual experiments and is graphed as composite mean ± SD for each time period. Each individual experiment showed a significant increase in pulmonary SOD levels among exposed animals by 4th day that persisted throughout day 8. In each instance beginning with day 4, $P < 0.05$ and often $P < 0.01$ $n = 10$ for bottom figure.*
Published in: *Amer. J. Physiol. 226,* 1404 (1974)

The duration of oxygen tolerance was also tested by exposing a large group of rats to 85% oxygen for 7 days, and then placing them in air for variable periods of time. The total SOD content in the lung was found to remain elevated for the first 7-14 days and then to decrease gradually over a 30-day interval (Fig.2). Other animals from the same groups were tested for tolerance and were found to retain their ability to survive in 100% oxygen for 14 days, and then to gradually lose this over the remainder of the 30-day period. The close correlation between the time course for the development and subsequent loss of oxygen tolerance and the time course for the increase and subsequent decrease in SOD activity suggests that there may be a cause and effect relationship.

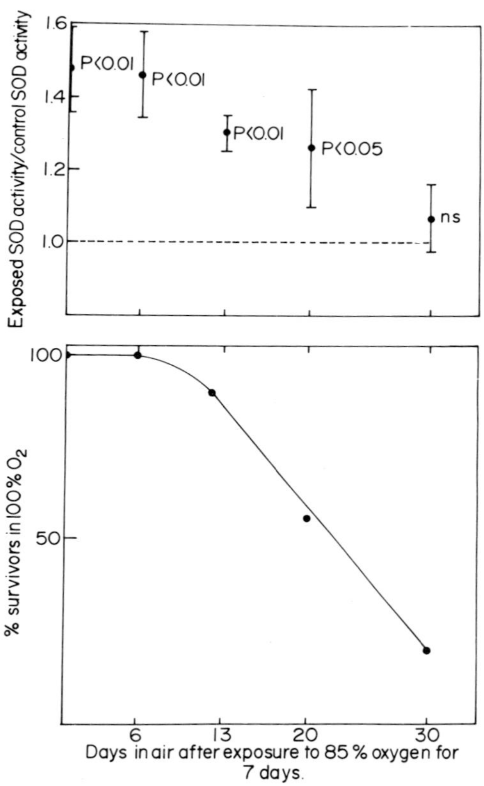

Fig.2. *Duration of increased pulmonary SOD activity after return to air compared to duration of the tolerant state. In top figure, all data are means ± SD with n = 6. n = 10 for bottom figure.*
Published in: *Amer. J. Physiol. 226,* 1404 (1974).

To evaluate the species specificity of the oxygen tolerant state, guinea pigs, hamsters and mice were also evaluated. Fourteen guinea pigs were exposed to 85% oxygen, and after 7 days, ten had died and pulmonary SOD activity was found to be unchanged in the survivors (Table I). Fourteen hamsters and 16 mice were exposed to 85% oxygen for 7 days, and then ten from each species were further exposed to 100% oxygen and the remainder used to determine activity of pulmonary SOD (Table I). Only the rat demonstrated both an increase in SOD and tolerance to pure oxygen. Guinea pigs and hamsters showed no change in SOD activity and did not develop tolerance. Mice showed a small increment in SOD activity and were not tolerant to 100% oxygen (Crapo and Tierney, 1974).

TABLE I

Pulmonary Superoxide Dismutase Activity in Different Species After 7-day Exposure to 85% Oxygen

	Lung Weight	SOD/Lung, U	Percent Increase	Percent Survival in 100% O_2 (7 days)
Rat				
Control	1.20 ± .03	777 ± 48	46	0
Exposed	1.74 ± .08	1,130 ± 87		100
			(P <0.01)	
Guinea Pig				
Control	2.61 ± .15	1,142 ± 64	9	Died in
Exposed	5.18 ± .3	1,247 ± 89		85% O_2
			(NS)	
Hamster				
Control	.588 ± .01	250 ± 3	0	0
Exposed	.562 ± .03	249 ± 13		0
			(NS)	
Mice				
Control	.17 ± .01	122 ± 3	18	0
Exposed	.21 ± .03	145 ± 3		0
			(P <0.01)	

Lung weight and SOD/lung values are means ± SE. NS = P <0.10 in two-sided test.
Published in: *Amer. J. Physiol. 226,* 1405 (1974).

PURIFICATION OF RAT SOD AND ANTIBODY TITRATION ASSAYS OF PULMONARY SOD

The initial observation demonstrating that rats made tolerant to oxygen acquire a simultaneous 50% increase in the activity of pulmonary SOD raised a number of further questions. One is the need to define whether that increase in SOD activity was due to an increase in the specific activity of the enzyme such as might be caused by a structural change in the enzyme or the loss of an inhibitor, or whether there was an increase in the absolute number of enzyme molecules. This question was studied by first purifying the rat cuprozinc SOD (Crapo and McCord, 1976) using a modification of the procedure originally reported by McCord and Fridovich (1969). The enzyme was found to be very similar to previous cuprozinc enzymes isolated from mammalian species. It has a molecular weight of 31,000, and the amino acid composition, the content of copper and zinc, and the ultraviolet and light absorption spectra were all similar to previously isolated cuprozinc dismutases. The purified rat enzyme was injected into rabbits using Freund's incomplete adjuvant to produce a rabbit antisera specific for rat cuprozinc SOD.

Six rats were exposed to 85% oxygen for 7 days, after which their lungs were removed, homogenized and asayed for SOD using an antibody titration assay specific for the rat cuprozinc dismutase, (Crapo and McCord, 1976).

The results are shown in Table II. They demonstrate a 40-50% increase in superoxide dismutase which can be demonstrated both by activity assays as previously noted, and by the antibody titration technique. One can use the purified cuprozinc SOD as a standard and convert the activity assay to a direct measurement of enzyme concentration if one makes the following two assumptions: (1) The mitochondrial dismutase, if present, contributes negligible activity at pH 10, and (2) the specific activity of the cuprozinc SOD did not change during either the oxygen exposure or during its purification. This was done, and the results are shown in the second column of Table II. These results suggest that the above assumptions are valid. All of the SOD activity present when the cytochrome *c* assay is used at pH 10.0 can be accounted for by the amount of cuprozinc SOD found by the antibody titration assay.

TABLE II

Pulmonary SOD in Oxygen Adapted Animals

	Spectrophotometric Assay		Antibody Titer
	Units/lung	μg/lung†	μg/lung
Controls	2419 ± 55	126 ± 2.7	132 ± 6.9
85% Oxygen Treated	3590 ± 172*	187 ± 8.9*	186 ± 4.5*

n = 6. All results are ± S.E.
* P <0.01 when compared to the control value
† Calculated assuming that in the rat only the Cu-Zn SOD is active when assays are performed at pH of 10.0 (0.052 μg Cu-Zn SOD = 1 unit in the pH 10 assay)
Published in: *Amer. J. Physiol. 231,* (in press) 1976.

Antibody titration assays measure enzyme protein rather than enzyme activity. The above use of this assay technique demonstrates that the 50% increase in SOD activity in oxygen tolerant rats is due to an increase in the total amount of cuprozinc SOD present and not to a change in its specific activity. The purification of rat cuprozinc SOD also led to the finding that SOD assays performed at high pH measure primarily, if not exclusively, the cuprozinc SOD in the rat. This was confirmed by the close correlation between the content of SOD determined by antibody titrations and that determined by activity assays using purified cuprozinc SOD as a standard. These findings do not rule out a concurrent change in another dismutase such as the manganese SOD.

ATTEMPTS TO MODIFY PULMONARY OXYGEN TOXICITY BY EXOGENOUSLY ADMINISTERED SUPEROXIDE DISMUTASE

Since SOD is an unusually stable and poorly antigenic enzyme, it is reasonable to postulate that administering it to the lungs of oxygen exposed animals could modify the toxic effects of oxygen. Its substrate, the superoxide free radical, is produced during the biochemical reduction of oxygen which primarily occurs in the intracellular space. If that is the only space in which superoxide is produced in significant amounts, there would be no need for protection by SOD other than inside the cell. A protein such as SOD cannot be expected to cross cell membranes and it is essential when one considers the exogenous administration of such a large molecule to consider the possibility that it may not be delivered to the site where its activity is needed. There is some evidence that superoxide could be produced in the extracellular space. Activated polymorphonuclear leukocytes actively secrete this free radical in large amounts (Salin and McCord, 1974). Further, when cellular degeneration begins to occur, the release of large numbers of enzymes and flavoproteins into the extracellular space may lead to the production of superoxide in that space during the late stages of oxygen toxicity. The intracellular space is protected from the toxic effects of superoxide by the activity of endogenous SOD, but the extracellular space has no such defense mechanism. These facts suggest that the production of small amounts of superoxide radicals in the extracellular space could be responsible for a major part of the toxicity of high oxygen tensions.

Administering SOD may be effective in reducing inflammation in the extracellular space. This is suggested by (1) the demonstration that extracellular superoxide is produced in inflamed joints by white cells, and in *in vitro* experiments at least part of its toxic reactions can be blocked by adding SOD (Salin and McCord, 1974); (2) bacteria exposed to an extracellular flux of superoxide are readily protected by the administration of SOD (Gregory and Fridovich, 1973b); and (3) the toxic effects of paraquat, a herbicide which is thought to auto-oxidize to produce the superoxide free radical, has been modified in rats by the exogenous administration of superoxide dismutase (Autor, 1974).

To test the effects of SOD on oxygen toxicity, the bovine cuprozinc SOD was administered to oxygen toxic rats (Crapo *et al.*, 1976). Initially this was done using subcutaneous injections. The clearance of the enzyme was so rapid that significant levels in the lung could not be maintained. Subsequently an aerosol was used to administer SOD directly and continuously to the lung. The aerosolized dismutase was maintained in the lung in amounts approximately equal to the endogenous cuprozinc SOD in rat lung. However, adequate exposure to aerosolized superoxide dismutase failed to modify either the time course or the cumulative toxicity of 100% oxygen (Fig.3).

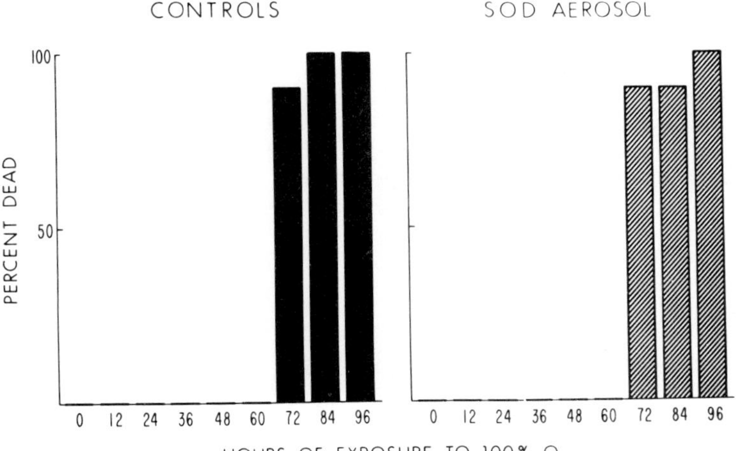

Fig.3. *Effect of aerosolized superoxide dismutase on the mortality of rats exposed to 100% oxygen. n = 10 for each group.*

This failure of exogenous superoxide dismutase to provide protection does not rule out a role for the endogenous enzyme as an essential protective agent in oxygen toxicity, but it does indicate that the superoxide dismutases are primarily needed within the intracellular space. These results would suggest that oxygen toxicity is not caused by production of large amounts of the superoxide free radical in the extracellular space, and if we are to modify pharmacologically the toxic effects of pure atmospheres of oxygen, agents must be found which can cross the cell membrane. We need a small molecule analog of superoxide dismutase which can reach the intracellular space where it can be expected to increase the essential defenses against oxygen's toxic effects.

ACKNOWLEDGEMENTS

This work was supported in part by National Institute of Health Grant HL 17603.

REFERENCES

1. Autor, A.P. (1974). *Life Sci. 14,* 1309-1312.
2. Callow, A.B. (1923). *J. Pathol. Bacteriol. 26,* 320-325.
3. Crapo, J.D. and McCord, J.M. (1976). *Am. J. Physiol. 321,* (Oct. 1976 - in press).
4. Crapo, J.D. and Tierney, D.F. (1974). *Am. J. Physiol. 226,* 1401-1407.
5. Crapo, J.D., DeLong, D.M., Hasler, G.R. and Drew, R.T. (1976). *Am. Rev. Resp. Disease,* submitted.
6. Fridovich, I. (1974). *Adv. Enzymol. Relat. Areas Mol. Biol. 41,* 35-97.
7. Gerschman, R. (1964). In "Oxygen in the Animal Organism", Macmillan, New York, p. 475.
8. Gregory, E.M. and Fridovich, I. (1973a). *J. Bacteriol. 114,* 543-548.
9. Gregory, E.M. and Fridovich, I. (1973b). *J. Bacteriol. 114,* 1193-1197.
10. Gregory, E.M., Yost, F.J. and Fridovich, I. (1973). *J. Bacteriol. 115,* 987-991.
11. Haugaard, N. (1965). *Ann. N.Y. Acad. Sci. 117,* 736-744.
12. Haugaard, N. (1968). *Physiol. Rev. 48,* 311-373.
13. Kellogg, E.W. III and Fridovich, I. (1975). *J. Biol. Chem. 250,* 8812-8817.
14. McCord, J.M. (1974). *Science 185,* 529-531.
15. McCord, J.M. and Fridovich, I. (1969). *J. Biol. Chem. 244,* 6049-6055.
16. McCord, J.M., Keele, Jr. B.B. and Fridovich, I. (1971). *Proc. Natl. Acad. Sci. U.S.A. 68,* 1024-1027.
17. Salin, M. and McCord, J.M. (1974). *J. Clin. Invest. 54,* 1005-1009.
18. Weibel, E.R. (1971). *Arch. Internal Med. 128,* 54-56.

GENERATION OF SUPEROXIDE RADICAL BY HEME PROTEINS AND ITS BIOLOGICAL RELEVANCE

G. ROTILIO

Institute of Biological Chemistry
University of Camerino, Camerino, Italy
and C.N.R. Center for Molecular Biology, Rome, Italy

E. FIORETTI and G. FALCIONI

Laboratory of Molecular Biology
University of Camerino, Camerino, Italy

M. BRUNORI

Institute of Chemistry, Faculty of Medicine
University of Rome, Rome, Italy
and C.N.R. Center for Molecular Biology, Rome, Italy

Generation of O_2^- from endogenous sources inside erythrocytes and granulocytes has frequently been reported in the recent literature as a crucial biochemical event in the intracellular production of reactive species which may produce lesions of the red cell membrane, the granulocyte membrane, or the bacterial membrane (during phagocytosis). In the erythrocyte this may offer a clue to the molecular understanding of phenomena such as hemolytic events or shortening of the cell's life span; in the granulocyte during phagocytosis superoxide may be a source of chemical weapons, such as hydroxyl radical and H_2O_2, to kill bacteria. The essential experimental support for these ideas is represented by the detection of O_2^- formation through the combined use of chromogenic scavengers and superoxide dismutase in the process of autoxidation of unstable hemoglobins (Misra and Fridovich, 1972; Wever *et al.*, 1973; Brunori *et al.*, 1975) or during phagocytosis in intact granulocytes (Babior *et al.*, 1973). Our knowledge concerning the role of O_2^- generation in erythrocytes and granulocytes *in vivo* is still far from complete and newly acquired information may be complementary for the two systems. Thus, under some conditions, hemoglobin, the major protein component of the red cell, can generate O_2^- but it is unknown (i) if O_2^- is actually generated *in vivo* in the erythrocyte, (ii) what is the target of its potentially damaging action and (iii) what is the chemical nature of the damage. On the other hand, it is demonstrated that granulocytes produce O_2^- *in vivo* and that O_2^- contributes to the bactericidal action of the cell (Johnston *et al.*, 1975), but knowledge of the biochemical source and the mechanism of production of superoxide is still incomplete.

In view of these general considerations additional experimental work should be directed towards the understanding of the still obscure aspects of the role of O_2^- in the two systems. The purpose of the work presented below is to obtain experimental data relevant to this problem.

Concerning the erythrocyte, we have concentrated our attention on an *in vivo* situation which may be reproduced in its essential biochemical features *in vitro*. This situation is represented by the red blood cell disease known as *Thalassemia major*, where an excess of either α or β hemoglobin chains is produced because of inherited defects of synthesis. Under these conditions, autoxidation of the excess α (or β) chains has been recognized to occur inside the cell leading to formation of hemichromes which form aggregates known as Heinz-bodies (Rachmilewitz *et al.*, 1969). Moreover, in this disease oxidative alterations of the red cell membrane have been clearly demonstrated (Stocks *et al.*, 1972; Kahane and Rachmilewitz, 1976). Demonstration of the generation of the O_2^- during *in vitro* autoxidation of the isolated hemoglobin chains would contribute to bridging the gap in the understanding of the molecular mechanisms of membrane alteration in thalassemia and, in turn, would provide a more solid experimental frame to envisage an *in vivo* role for O_2^-. Cooxidation of adrenaline with formation of the coloured product adrenochrome, and inhibition of this process by superoxide dismutase (McCord and Fridovich, 1969; Misra and Fridovich, 1972) were used to reveal O_2^- production during autoxidation of oxy hemoglobin chains.

A number of events is associated with autoxidation of isolated hemoglobin chains according to the following overall scheme (Brunori *et al.*, 1975):

$$\text{Oxy chains} \xrightarrow{a} \text{met chains} \xrightarrow{b} \text{hemichromes} \xrightarrow{c} \text{precipitate}$$

At pH 5.5, step *c* is much slower than step *a*; therefore autoxidation can be followed spectrophotometrically at this lower pH without interference from precipitation. Table I reports the initial rate of autoxidation of the oxy chains from hemoglobin A under various conditions. Several features stand out: (i) a remarkable inhibition of the rate of autoxidation by EDTA; (ii) a smaller, though still significant, inhibition by superoxide dismutase and catalase; (iii) an increase in the rate of oxidation in the presence of adrenaline, an effect inhibited by both enzymes; (iv) a far greater oxidizing effect of adrenochrome, which however is insensitive to the addition of superoxide dismutase. Fig.1 shows that co-oxidation of adrenaline in the presence of the α chains is strongly inhibited by superoxide dismutase.

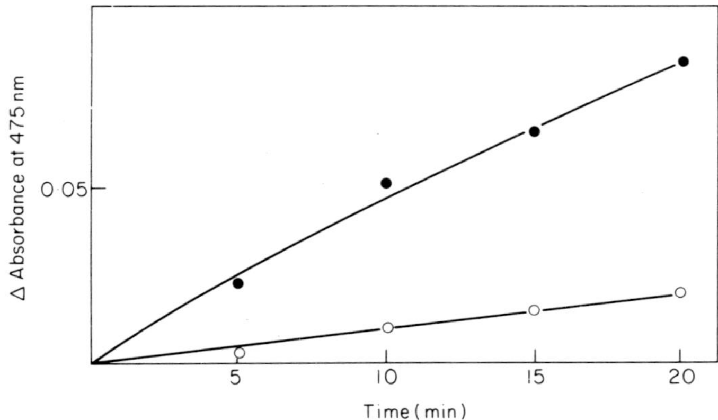

Fig.1. *Effects of superoxide dismutase on the co-oxidation of adrenaline by oxy hemoglobin chains.* ● : *α oxy chains + adrenaline;* ○ : *α oxy chains + adrenaline + superoxide dismutase. 0.4 M acetate buffer pH 5.5; temperature 37°C; protein concentration: 3 - 5 mg/ml; adrenaline concentration: 5 mM; superoxide dismutase concentration: 1.5 μM.*

TABLE I

Autoxidation of Oxy Human Hemoglobin Chains in Various Conditions

Incubation Mixture	$\Delta E/min \times 10^3$
A : 3×10^{-5}M oxy chain + 1×10^{-4}M EDTA	1.8
B : 3×10^{-5}M oxy chain + 1×10^{-4}M EDTA + 1.5 µM superoxide dismutase	1.4
C : 3×10^{-5}M oxy chain + 1×10^{-4}M EDTA + 0.2 µM catalase	1.3
D : 3×10^{-5}M oxy chain + 1×10^{-3}M EDTA	0.7
E : A + 5×10^{-4} adrenaline	3.8
F : E + 1.5 µM superoxide dismutase	2.1
G : E + 0.2 µM catalase	2.8
H : D + 5×10^{-4}M adrenaline	3.6
I : A + 5×10^{-4}M adrenochrome	9.0
J : I + 1.5 µM superoxide dismutase	9.0
K : I + 0.2 µM catalase	4.5

0.4 M acetate buffer pH 5.5. Temperature: 37°C. Wavelength of observation: 576 nm.

The inhibitory effect of EDTA on the rate of autoxidation of hemoglobin chains suggests that this process is catalyzed by metal ions, such as Cu^{++} or Fe^{3+}, similar to what was reported by Rifkind (1974) with hemoglobin. Increased concentrations of ions have been reported in thalassemic red cells (Pollack and Rachmilewitz, 1973). It appears also that O_2^- is produced in this autoxidation and contributes somewhat, probably through the products of its spontaneous dismutation (H_2O_2 and singlet oxygen, Finazzi Agro *et al.*, 1972) to the oxidation of the chains. This could account for the small inhibitory effect by superoxide dismutase and catalase. A mechanism for autoxidation, which accounts for the role of metal ions, could be as follows:

$$Hb\ Fe^{3+}\ O_2^- + Me^{n+1} \longrightarrow Hb\ Fe^{3+} + O_2 + Me^n$$

$$Me^n + O_2 \longrightarrow Me^{n+1} + O_2^-$$

or

$$Hb\ Fe^{3+}\ O_2^- \xrightarrow{Me^{n+1}} Hb\ Fe^{3+} + O_2^-$$

Then:

$$O_2^- + O_2^- \xrightarrow{2H+} H_2O_2 + O_2$$

The effect of adrenaline, on the other hand, seems to be unaffected by metals, since it is insensitive to addition of EDTA, while it is strongly inhibited by superoxide dismutase. It is likely that the species responsible for the effect of adrenaline is adrenochrome, which produces a large increase in the rate of autoxidation, insensitive to superoxide dismutase. Thus the effect of superoxide dismutase on the oxidation induced by adrenaline may be understood in terms of inhibition by this enzyme of adrenochrome formation. This is known to proceed through a free radical mechanism whose initiation step could be, in this case, the reaction between adrenaline and oxy hemoglobin chains with formation of superoxide acting as a chain propagator. Since oxidation of adrenaline to adrenochrome involves the loss of four hydrogens, and two more hydrogens are lost in the further oxidation of adrenochrome (Harrison, 1961), we may suggest the following sequence of reactions, where adrenochrome is RH_2 and the final oxidation product is R:

$$RH_6 + Fe^{3+}\,O_2^- + H \longrightarrow RH_5^{\bullet} + Fe^{3+} + H_2O_2 \quad (a)$$

$$RH_5^{\bullet} + O_2 \longrightarrow RH_4 + O_2^- + H^+ \quad (b)$$

$$RH_4 + O_2^- + H^+ \longrightarrow RH_3^{\bullet} + H_2O_2 \quad (c)$$

$$RH_3^{\bullet} + O_2 \longrightarrow RH_2 + O_2^- + H^+ \quad (d)$$

$$RH_2 + Fe^{3+}\,O_2^- + H^+ \longrightarrow RH^{\bullet} + Fe^{3+} + H_2O_2 \quad (e)$$

$$RH^{\bullet} + O_2 \longrightarrow R + O_2^- + H^+ \quad (f)$$

This would account for (1) the insensitivity to EDTA, (2) the inhibition by superoxide dismutase of the effect of adrenaline, if this is due to formation of adrenochrome, and (3) the insensitivity to superoxide dismutase of the effect of adrenochrome on the oxy chains.

It can be concluded from these experiments that autoxidation of isolated α (or β) chains of human Hb in the presence of catalytic amounts of transition metal ions and of reducing organic molecules gives rise to O_2^- as *one* of the O_2 radicals. Such conditions are very likely to occur in the thalassemic cell.

Similar to hemoglobin for the red cell, myeloperoxidase represents a major protein component of the granules of granulocytes, which are liberated in the vacuoles during phagocytosis. Since ferrous peroxidase forms an oxygen adduct (Wittemberg *et al.*, 1967) which decays to ferric peroxidase (Phelps *et al.*, 1974), the demonstration that O_2^- is produced during decay of oxyperoxidase would be in favour of a role for myeloperoxidase as a potential source of the radical in phagocytosis.

A series of experiments analogous to those with Hb chains was designed to follow the decay of the oxygenated adduct of horse radish peroxidase, which is very similar to that of myeloperoxidase (Wittemberg *et al.*, 1967; Odajima and Yamazaki, 1972). Oxyperoxidase was prepared by photo dissociation of the CO derivative in the presence of O_2, and its decay to the ferric form was followed spectrophotometrically at pH 7.4 and 10°C, where half time of the decay is of the order of several minutes (Phelps *et al.*, 1974; Rotilio *et al.*, 1975). Catalase or superoxide dismutase had no effect on the half time of decay, which was accelerated by adrenaline, as in the case of oxy hemoglobin chains. Liberation of O_2^- was followed through the formation of adrenochrome at 462 nm, a wavelength isosobestic for oxy and ferric peroxidase. It was unequivocally revealed by the increase of absorption at this wavelength and inhibition of this process by superoxide dismutase and catalase (Rotilio *et al.*, 1975). The data on the effects of the two enzymes are shown in Table II, and indicate that O_2^- contributes to adrenochrome formation much more than H_2O_2.

TABLE II

Inhibition by Superoxide Dismutase and Catalase of the Co-oxidation of Adrenaline by Oxyperoxidase

Enzyme Concentration	Per cent inhibition
0.1 μM catalase	2
0.1 μM superoxide dismutase[1]	8
0.3 μM catalase	10
0.3 μM superoxide dismutase[1]	42
1.5 μM catalase	15
1.5 μM superoxide dismutase[1]	50
2.4 μM catalase	16
2.4 μM superoxide dismutase[1]	49

Peroxidase 25 μM - Adrenaline 2 mM. 0.1 potassium phosphate buffer, pH 7.4. Temperature: 10°C.

[1]The effect of superoxide dismutase was tested in the presence of saturating amounts of catalase (i.e. 1.5 μM) and the percentage inhibition calculated by reference to the experiments containing saturating amounts of catalase and no superoxide dismutase.

These results merely show that oxyperoxidase, if formed, may be a source of O_2^- in the granulocyte. Additional work is needed to understand the relationships between the myeloperoxidase system and the system responsible for the respiratory burst of phagocytosis, which involves cyanide-insensitive utilization of NADPH (Zatti and Rossi, 1966) and is associated with O_2^- generation (Patriarca *et al.*, 1975).

REFERENCES

1. Babior, M.B., Kipnes, R.S. and Curnette, J.I. (1973). *J. Clin. Invest. 52*, 741-743.
2. Brunori, M., Falcioni, G., Fioretti, E., Giardina, B. and Rotilio, G. (1975). *Eur. J. Biochem. 53*, 99-104.
3. Finazzi Agrò, A., Giovagnoli, C., De Sole, P., Calabrese, L., Rotilio, G. and Mondovi, B. (1972). *FEBS Letters 21*, 183-185.
4. Harrison, H.V. (1963). *Arch. Biochem. Biophys. 101*, 116-130.
5. Johnston, R.B., Keele, B.B., Misra, H.P., Lehmeyer, J.E., Webb, S.L., Baehner, R.L. and Rajagopalan, K.V. (1975). *J. Clin. Invest. 55*, 1357-1372.
6. Kahane, I. and Rachmilewitz, E.A. (1976). *Isr. J. Med. Sci. 12*, 11-12.
7. McCord, J.M. and Fridovich, I. (1969). *J. Biol. Chem. 244*, 6049-6055.
8. Misra, H.P. and Fridovich, I. (1972). *J. Biol. Chem. 247*, 6960-6962.
9. Odajima, T. and Yamazaki, I. (1972). *Biochim. Biophys. Acta 284*, 355-359.
10. Patriarca, P., Dri, P., Kakinuma, K., Tedesco, F. and Rossi, F. (1975). *Biochim. Biophys. Acta 385*, 380-386.

11. Phelps, C., Giacometti, G., Brunori, M. and Antonini, E. (1974). *Biochem. J. 141*, 265-272.
12. Polliak, A., and Rachmilewitz, E.A. (1973). *Br. J. Haematol. 24*, 319.
13. Rachmilewitz, E.A., Peisach, J., Bradley, T.B. and Blumberg, W.E. (1969). *Nature* (Lond.) *222*, 248-250.
14. Rifkind, J.M. (1974). *Biochemistry 13*, 2475-2481.
15. Rotilio, G., Falcioni, G., Fioretti, E. and Brunori, M. (1975). *Biochem. J. 145*, 405-407.
16. Stocks, J., Offerman, E.L., Modell, C.B. and Dormandy, T.L. (1972). *Br. J. Haematol. 23*, 713-724.
17. Wever, R., Oudega, B. and Van Gelder, B.F. (1973). *Biochim. Biophys. Acta 302*, 475-478.
18. Wittenberg, J.B., Noble, R.W., Wittemberg, B.A., Antonini, E., Brunori, M. and Wyman, J. (1967). *J. Biol. Chem. 242*, 626-634.
19. Zatti, M. and Rossi, F. (1966). *Experimentia 22*, 758-759.

TOXICITY OF SUPEROXIDE RADICAL ANIONS

A.M. MICHELSON

Institut de Biologie Physico-Chimique
Service de Biochimie-Physique
13, rue P. et M. Curie
75005 Paris, France

Superoxide is both an oxidant and a reductant and hence can modify a variety of biologically important molecules. Apart from oxidation of substances such as adrenaline, or reduction of cytochrome *c* (and other heme derivatives), the chemical effects of $O_2^{\bar{\cdot}}$ can be amplified by initiation of free radical chain reactions. We have studied the effects of $O_2^{\bar{\cdot}}$ producing systems on various biological macromolecules, viruses, bacteria and eucaryote cells, and the protection afforded by superoxide dismutase (Lavelle *et al.*, 1973), as shown in the following examples.

Protein - ribonuclease. Photoreduction of FMN was used to produce $O_2^{\bar{\cdot}}$ in presence of ribonuclease. The enzyme rapidly lost activity but was to a large extent protected when SOD was added.

	% Original activity		
Number of Photo-reduction cycles	Control	Plus dismutase denatured at pH 3	Plus 4 x 10^{-8} M superoxide dismutase
0	100	100	100
3	67	56	89
6	45	45	89
9	22	22	67

Lipoprotein - yeast lysine tRNA-ligase. Spontaneous production of $O_2^{\bar{\cdot}}$. This ligase is a lipoprotein complex for which the enzymic activity is a function not only of the presence but also of the integrity of the lipid moiety. Autoxidation of the lipid is extremely rapid (probably due to trace metals which react with dissolved oxygen) and results in loss of enzymic activity. This loss of activity is greatly diminished when native SOD is added (but not denatured SOD). Reduced glutathione showed no protective action while ascorbate accelerated loss of activity.

Days	Control	% Activity + 0.1% ascorbate	Plus 5 x 10^{-8} M superoxide dismutase
0	100	100	100
2	81	31	97
4	62	19	96
7	35	2.8	72
9	27	1.2	67

Nucleoprotein - bacteriophage R 17. Photoreduction of FMN as source of $O_2^{\cdot-}$.

Number of cycles of photoreduction	Infective particles/ml Control (plus denatured dismutase)	Plus 10^{-8} M SOD
0	4.1 x 10^8	4.1 x 10^8
3	4.9 x 10^7	7.5 x 10^7
6	5.0 x 10^6	1.5 x 10^7
9	5.0 x 10^5	2.0 x 10^6

Xanthine oxidase/O_2/hypoxanthine as source of $O_2^{\cdot-}$

Number of enzymic cycles	Infective particles/ml Control	Plus superoxide dismutase
0	5.0 x 10^8	5.0 x 10^8
1	3.5 x 10^8	5.0 x 10^8
2	2.0 x 10^8	3.0 x 10^8

Bacteria - *P. leiognathi*. Photoreduction of FMN at 365 nm as source of $O_2^{\cdot-}$. (See also Gregory and Fridovich, 1973).

Time of irradiation in min	Viable Cells Control	Plus dismutase
0	1 x 10^7	1 x 10^7
30	4 x 10^6	7 x 10^6
60	5 x 10^5	3.9 x 10^6
120	8 x 10^3	6 x 10^4

Protection of Bacteria Against Ultraviolet Irradiation

A significant protection by superoxide dismutase was observed against ultraviolet irradiation (365 nm) in the absence of exogenous FMN. Presumably such irradiation can lead to superoxide radicals by indirect means such as photosensitization.

Thus $O_2^{\cdot}$ can cause extensive damage to various biological structures such as nucleic acids, proteins and lipids and hence provoke lethal alterations when the internal concentration becomes too high to be controlled by the endogenous intracellular enzyme or when $O_2^{\cdot}$ is generated extracellularly. Although the above examples demonstrate a role for $O_2^{\cdot}$, the effects may not always be due to a direct action of this radical, but rather to secondary formation of other radicals such as HO· or $CO_3^{\cdot}$. Nevertheless SOD protects, whether at a primary level, or by interrupting a chain of events which leads to species more reactive than $O_2^{\cdot}$.

MAMMALIAN CELLS

The relatively long life time of superoxide radicals compared with diffusion rates renders possible the active participation of an exogenous enzyme which destroys $O_2^{\cdot}$. Thus external protection becomes feasible and indeed plays a major role when the volume of the medium is compared with the total volume of the cells. Mammalian cells possess a further advantage in that not only can toxic effects be followed, but effects on cell growth, morphology and differentiation are also readily visible. This is particularly true for myoblast cells (Michelson and Buckingham, 1974).

It has been reported that certain antioxidants such as butylated hydroxytoluene (BHT) were efficient in diminishing chromosome breakage caused by carcinogenic hydrocarbons (Schamberger *et al.*, 1973). We therefore examined the effects on mammalian cells in culture of hydrocarbons alone, hydrocarbons activated by systems producing superoxide radicals, and $O_2^{\cdot}$ alone, as well as the protection afforded by SOD. In general we have used a simple photochemical system for production of $O_2^{\cdot}$ but extension to the use of γ rays was also studied, since it has been shown that the major chemical species produced by high energy irradiation under aerobic conditions is indeed production of superoxide radicals (Beher *et al*, 1970). It was also of interest to see whether genetic effects could be observed in mammalian cells treated with $O_2^{\cdot}$ since we have shown (Michelson and Brody, 1974) the strong mutational effects on T 4 bacteriophage by systems producing $O_2^{\cdot}$. Indeed, a probable cause of some spontaneous mutations is formation of superoxide radicals either within the cell or in the medium itself (c/f accumulation of mutants in T 4 on storage (Drake, 1966) for several years).

Calf primary myoblast cultures undergo cell division (0 - 42 h), followed by a stationary phase during which the cells become aligned. This precedes the onset of cell fusion (52 h). Initially two to three nuclei fusions are observed, followed by the development of long multinucleate fusions and the appearance of the muscle contractile apparatus. The cultures underwent treatment at 24 h, during exponential cell growth; observations were made at 12 h intervals thereafter. Striking stimulation or amplification by $O_2^{\cdot}$ (via photoreduction of FMN) of the effects of the carcinogen, methylcholanthrene was observed. A proportion of the cells rapidly assume an abnormal hyperplasmic morphology with a very large cell nucleus and extended cytoplasm (Fig.1a,b). There is some cell death, cell division is inhibited, and very few fusions occur. This type of effect is seen to a lesser degree on treatment with flavin mononucleotide and irradiation alone, and is due essentially to the action of superoxide radicals. The addition of superoxide dismutase increases cell viability and the extent of cell division and fusion (Fig.1c). Toxicity is a function of the quantity of $O_2^{\cdot}$ generated and it is the accumulative dose of irradiation (i.e. of $O_2^{\cdot}$) which is important over various time periods and intensities.

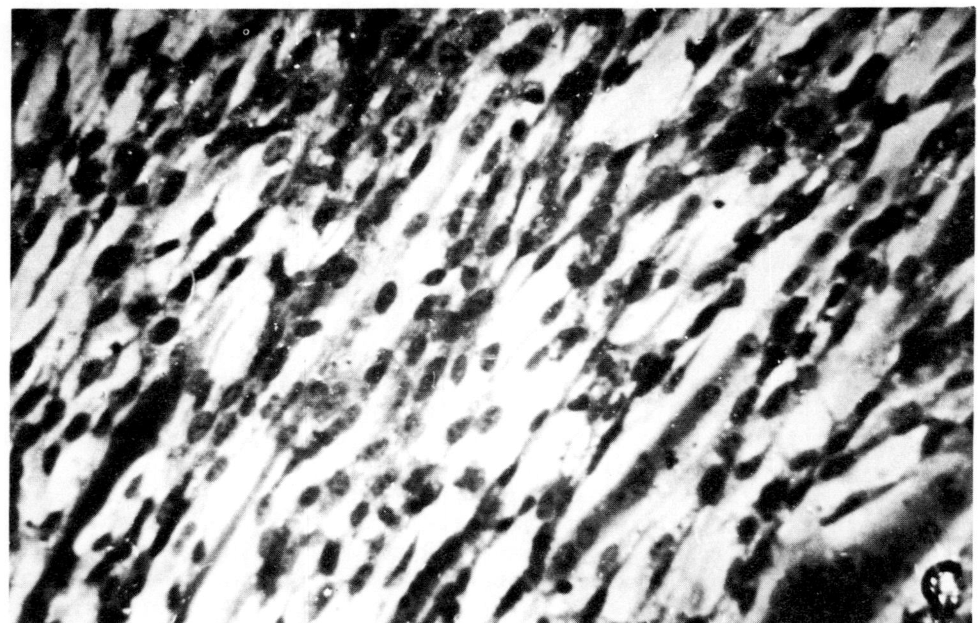

Fig.1a. *Morphological effects on myoblasts; untreated cells.*

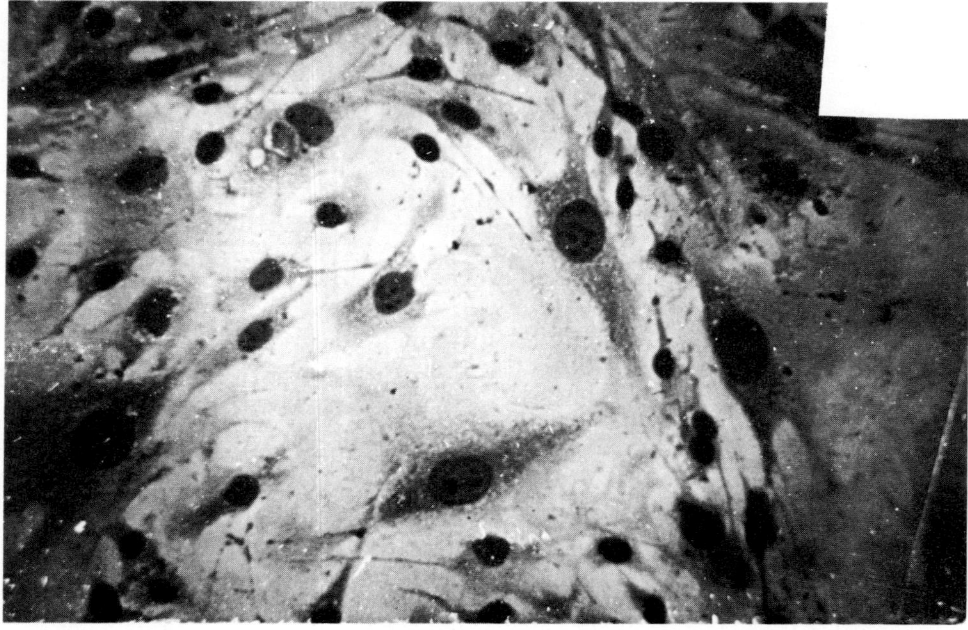

Fig.1b. *Morphological effects on myoblasts; cells irradiated in the presence of methylcholanthrene and FMN.*

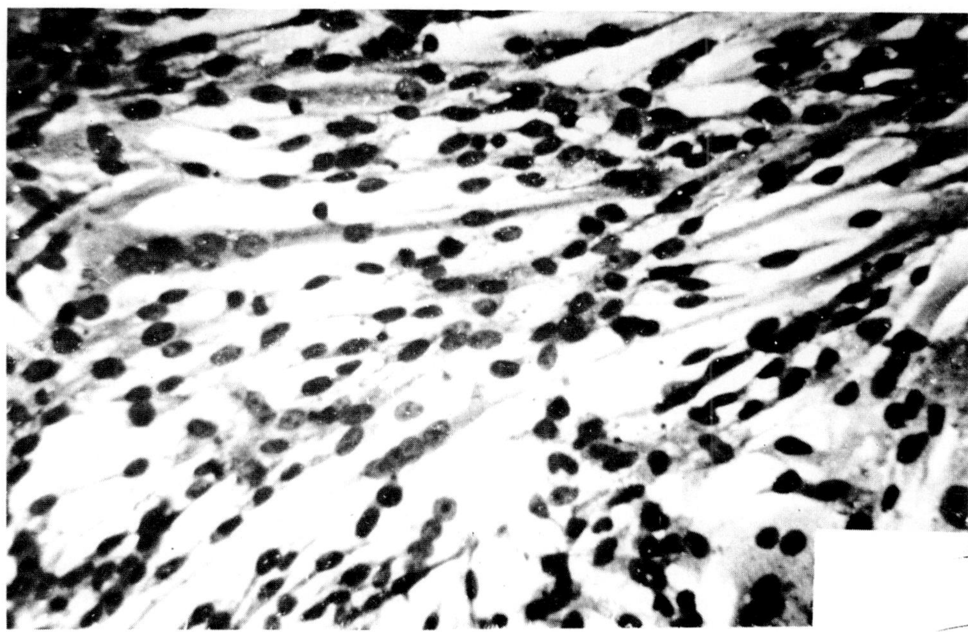

Fig.1c. *Morphological effects on myoblasts; as in Fig.1b, but in presence of SOD.*

The toxic effects of irradiation on cultures to which hydrocarbon and flavin had been added at different times prior to and during myoblast differentiation were examined. During the period of cell division, messenger RNA species, specifying differentiated proteins, notably myosin, are transcribed, but are unstable. In the stationary period preceding fusion these messenger RNA molecules are stabilised, and used in the synthesis of contractile proteins during the subsequent formation of long fusions. Transcription is therefore less necessary for differentiation, once the stable messenger RNA species are present (Buckingham *et al.*, 1974). In fact, myoblast differentiation was severely affected on treatment during the period of cell division. If treatment occurred during the stationary period, fusion although initiated, proceeded abnormally, whereas if cells were treated once fusion had begun no effects could be detected. This suggests that the primary toxic effect of superoxide radicals alone or via hydrocarbons is at the nuclear level affecting DNA synthesis and RNA transcription. We have shown that *in vitro*, $O_2^{\overline{\cdot}}$ producing systems are efficient for the covalent fixation of hydrocarbons to DNA.

The effects of γ-irradiation and the protection afforded by SOD were also examined. With increasing doses of irradiation, changes in cell viability and morphology were essentially similar to those observed with the $O_2^{\overline{\cdot}}$ generating system. Superoxide dismutase greatly diminished the toxic effects. Irradiation in the stationary phase, just prior to fusion had no effect on fusion at low doses; morphological abnormalities were induced at higher doses. Contaminating mononucleate cells were much more sensitive and were eliminated. It may thus be concluded that the primary chemical event leading to biological destruction by γ irradiation under aerobic conditions is in fact production of superoxide radicals, which act at the nuclear level, and that SOD provides at least a partial protection of living cells against γ rays (and presumably other high energy irradiation). Even at very high doses of γ rays (10,000 rads) SOD reduced cell death significantly.

The results provide a convincing demonstration that the effects of carcinogenic hydrocarbons on mammalian cells are greatly stimulated and amplified in presence of a system producing $O_2^{\overline{\cdot}}$. It may even be provisionally concluded that such effects are normally mediated by endogenous production of $O_2^{\overline{\cdot}}$ within the cell. Indeed the amplification is so powerful that hydrocarbons such as anthracene (not normally regarded as carcinogenic) show most of the effects of benzpyrene or methylcholanthrene in presence of $O_2^{\overline{\cdot}}$. It is clear that the primary target is the nucleus in dividing cells; after fusion, the cells become much more resistant. Of more general interest, is the effect of superoxide radicals alone in the absence of specific carcinogens. Again SOD protects against the toxic effects expressed by this radical, which as in the case of hydrocarbons mediated effects, occur principally (if not entirely, at the doses used) on dividing cells. Mutational effects can be seen and indeed a significant number of cells with an abnormal morphology can be observed, the incidence of which is markedly reduced by SOD. Inhibition of differentiation is also striking and again is reduced in presence of exogenous SOD. This raises the possibility that superoxide radicals and SOD may play a role in the initiation and protection against certain malignant processes. Apart from direct action of $O_2^{\overline{\cdot}}$ on DNA, an alternate mechanism would involve phenotypic modifications producing defective DNA repair enzymes (for example oxidation of tryptophane residues in the protein) or DNA polymerases, which could lead to defective enzymes liable to produce more errors in base pairing during repair or replication, thus indirectly causing genetic modifications.

Biological protection against such possibilities is afforded by superoxide dismutases which often have a highly localised distribution. Apart from the usual distribution of Cu-SOD in the cytoplasm and Mn-SOD in the mitochondria (not always followed in all eucaryote cells) there is evidence for a superoxide dismutase in rat liver nuclei, different in R_F (gel electrophoresis) from the cytoplasmic enzyme (Michelson and Isambert, 1975) while two typical Cu-SODs are present in human semen. The mammalian cell thus provides suitable protection for possible modifications of DNA as a result of the action of $O_2^{\overline{\cdot}}$. Of interest also is the observation that during development of the fruit fly *Ceratitis capitata* no changes in superoxide dismutase levels occur during the first three stages, eggs $\longrightarrow$ larvae $\longrightarrow$ pupae, but emergence and development of the adult fly is accompanied by a large increase in mitochondrial SOD per gm of material, with a much smaller relative increase in the cytoplasmic enzyme (Fernandez-Sousa and Michelson, 1976). This increase parallels the 50 fold increase in the volume of mitochondria which occurs in the 24 hours following emergence of the fly, and is related to the high consumption of oxygen during flight (21 litres O_2 per gm dry weight per hour).

DAMAGE DUE TO OTHER ACTIVATED OXYGEN SPECIES: PHOTOHEMOLYSIS OF ERYTHROCYTES

These studies began with the hypothesis that lysis of erythrocyte membranes as a general phenomenon is essentially, if not entirely, due to peroxidation of unsaturated lipids by superoxide radicals, and that by suitable techniques the localisation of such unsaturated lipids in the interior or exterior of the membrane could be readily determined by simple lysis studies. The concept thus gives a major role to the large amounts of superoxide dismutase (∿ 460 μg/g hemoglobin or 260 000 molecules erythrocuprein per red blood cell) present in the erythrocyte as a protective system against production of $O_2^{\overline{\cdot}}$ by autoxidation of hemoglobin, particularly by autoxidation of single chains such as in cases of Thalassemia (see Rotilio *et al.*, this volume, preceding chapter). This hypothesis was particularly attractive since superoxide dismutase efficiently protects certain lipoproteins against autoxidation and loss of enzymic activity (Lavelle *et al.*, 1973). However, the results completely invalidate the hypothesis and indicate that hydroxyl radicals and excited singlet oxygen play a much more important role in hemolysis.

Two techniques were used; the first was direct irradiation at 254 nm while the second involved treatment with a system producing $O_2^{\overline{\cdot}}$, photoreduction of riboflavin at 365 nm in presence of NADH as hydrogen donor (Michelson and Durosay, 1976).

IRRADIATION AT 254 nm

As earlier noted by Roshchupkin *et al.* (1975) a very important oxygen effect is observed (Fig.2) and it is clear that at some stage activated oxygen species are involved.

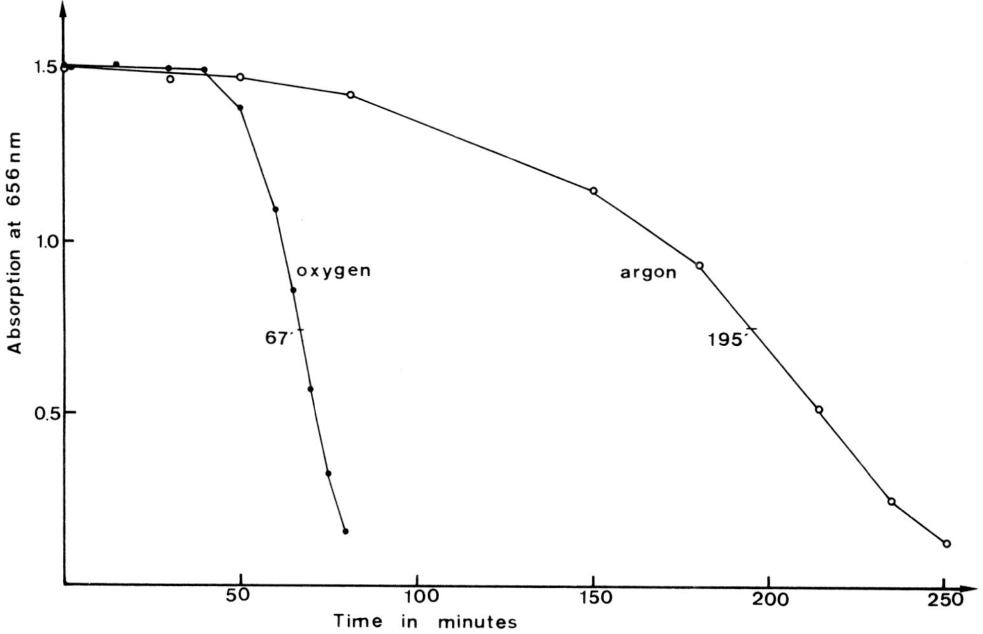

Fig.2. *Oxygen effect on photolysis of erythrocytes.*

It could reasonably be supposed that irradiation of hemoglobin in the erythrocytes at 254 nm in presence of O_2 would generate photosensitized $O_2^{\overline{\cdot}}$, and that at least to some extent (given the life time of $O_2^{\overline{\cdot}}$ and diffusion rates) exogenous superoxide dismutase would afford some protection (c/f. Michelson and Buckingham, 1974). However, absolutely no effect is observed. Conversely one can readily destroy the natural internal defense since erythrocyte superoxide dismutase is entirely cytoplasmic and uniquely as erythrocuprein and hence is 100% inhibited by KCN. No loss of protection (stimulation of lysis) was observed, and indeed an important protective effect is visible. Thus $O_2^{\overline{\cdot}}$ can be eliminated as a major oxygen species involved in photohemolysis at 254 nm. In confirmation of this, when erythrocytes from Trisomy 21 patients, containing 50% more superoxide dismutase per g of hemoglobin were compared with normal erythrocytes the stability, rather than being increased, was slightly diminished.

Excited singlet oxygen is an alternate possibility. However, photohemolysis in D_2O rather than H_2O at 254 nm instead of being accelerated, as would be expected due to the 10 fold increase in lifetime of $^1O_2$ in D_2O (Merkel and Kearns, 1971, 1973; Merkel *et al.*, 1972), is in fact significantly slower (Fig.3A). Thus $^1O_2$ can be eliminated as a major oxygen species involved in photohemolysis at 254 nm. Another species to be considered is H_2O_2. Addition of 10^{-2} M H_2O_2 to human erythrocytes did not lead to lysis and was rapidly destroyed by the endogenous catalase (about 4 mg/g hemoglobin). The levels of glutathione peroxidase (perhaps the primary protection against H_2O_2 at least in the membrane) have no effect on membrane stability even in the case of one subject with essentially zero glutathione peroxidase. Hence H_2O_2 can likewise be eliminated.

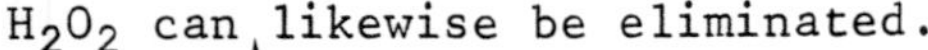

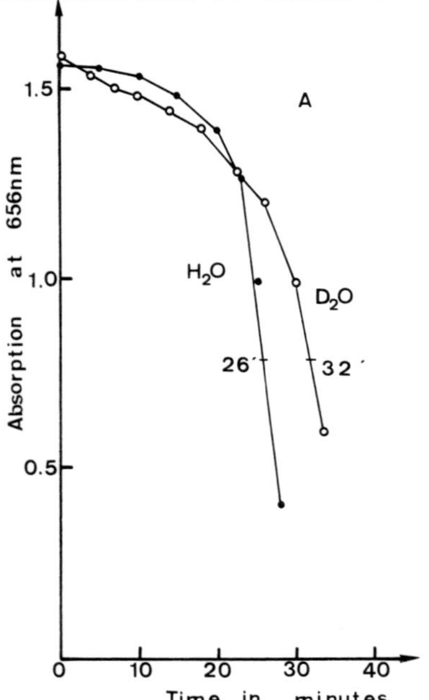

Fig.3A. *Photohemolysis at 254 nm of erythrocytes in water and in deuterium oxide at 10°C (standard isotonic buffer).*

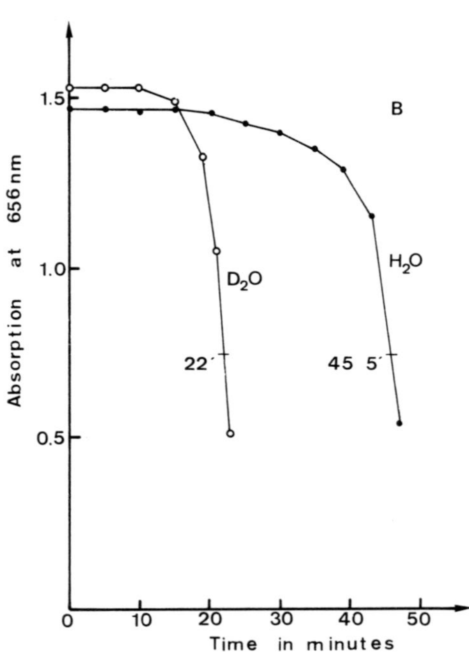

Fig.3B. *Hemolysis of erythrocytes by irradiation of riboflavin and NADH at 365 nm in water and in deuterium oxide at 20°C.*

Nevertheless, exogenous catalase (2 x 10^{-7} M) does protect human erythrocytes against photolysis at 254 nm as shown in Fig.4. Superoxide dismutase does not increase this protection, and when carbonate is present in the medium, hemolysis is stimulated. Hodgson and Fridovich (1976) have suggested the role of the carbonate anion radical $CO_3^{\overline{\cdot}}$ which can be formed by the reaction:

$$HO^{\cdot} + CO_3^{2-} \longrightarrow HO^- + CO_3^{\overline{\cdot}}$$

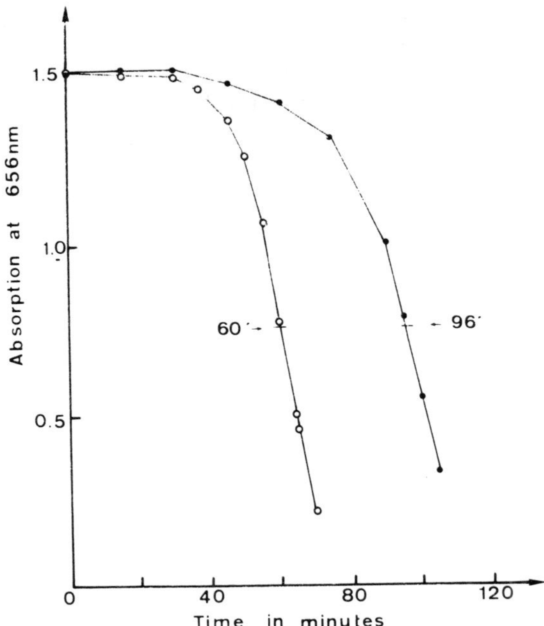

Fig.4. *Effect of catalase (2 x 10^{-7} M) on the lysis of human erythrocytes by irradiation at 254 nm. Conditions as in Fig.1. Open circles, control; solid points, in presence of catalase.*

The lifetime of HO˙ radicals is extremely short compared with $CO_3^{\dot{-}}$ radicals. An explanation which covers all the experimental data is that the primary chemical event causing lysis of erythrocyte membranes by irradiation of red blood cells at 254 nm under the conditions used is hemoglobin photosensitized production of hydroxyl radicals. These can interact with catalase (interconversion of compound I and compound II) which thus affords efficient protection by acting as a radical trap rather than as a catalyst for the reaction $2\ H_2O_2 \longrightarrow 2\ H_2O + O_2$. Indeed, the H_2O_2 is now necessary to reverse the interconversion of compounds I and II. Similarly, in presence of carbonate ions, formation of $CO_3^{\dot{-}}$ leads to increased attack of unsaturated membrane lipids by oxidative carboxylation or lipid radical formation.

Photolysis of erythrocyte ghosts at 254 nm, in accord with the concepts outlined above is less efficient and they show greatly increased resistance, but again catalase offers a marked protection despite the fact that the ghosts are quite resistant to 10^{-2} M H_2O_2. In this case exogenous bovine erythrocuprein even accelerates photodestruction of the membranes. A possible explanation may lie in the peroxidative activity of erythrocuprein in presence of H_2O_2 which has been reported by Hodgson and Fridovich (1975). Alternatively, loss of hydroxyl radicals by the reaction $O_2^{\dot{-}} + HO^{\cdot} \longrightarrow O_2 + HO^-$ could well be prevented by the action of superoxide dismutase.

Photoreduction of riboflavin at 365 nm in the presence of a hydrogen donor (NADH) as a source of $O_2^{\dot{-}}$ as applied to the hemolysis of human erythrocytes is shown in Fig.5. Inhibition of the cytoplasmic superoxide dismutase by KCN showed no influence whatsoever on the lysis, and presence of exogenous erythrocuprein resulted in a very slight acceleration rather than protection. As earlier mentioned, H_2O_2 in a concentration greatly exceeding that produced in the system has no effect either. Nevertheless, exogenous catalase confers a very important protection (Fig.5). On the other hand, presence of carbonate buffer accelerates lysis significantly due to production of $CO_3^{\cdot-}$ or formate radicals ($CO_2^{\cdot-}$) obtained via reduction of CO_2 by $O_2^{\cdot-}$

$$HCO_3^- + O_2^{\cdot-} \longrightarrow HO^- + CO_2^{\cdot-} + O_2$$

Both $CO_3^{\cdot-}$ and $CO_2^{\cdot-}$ could subsequently react with unsaturated lipids.

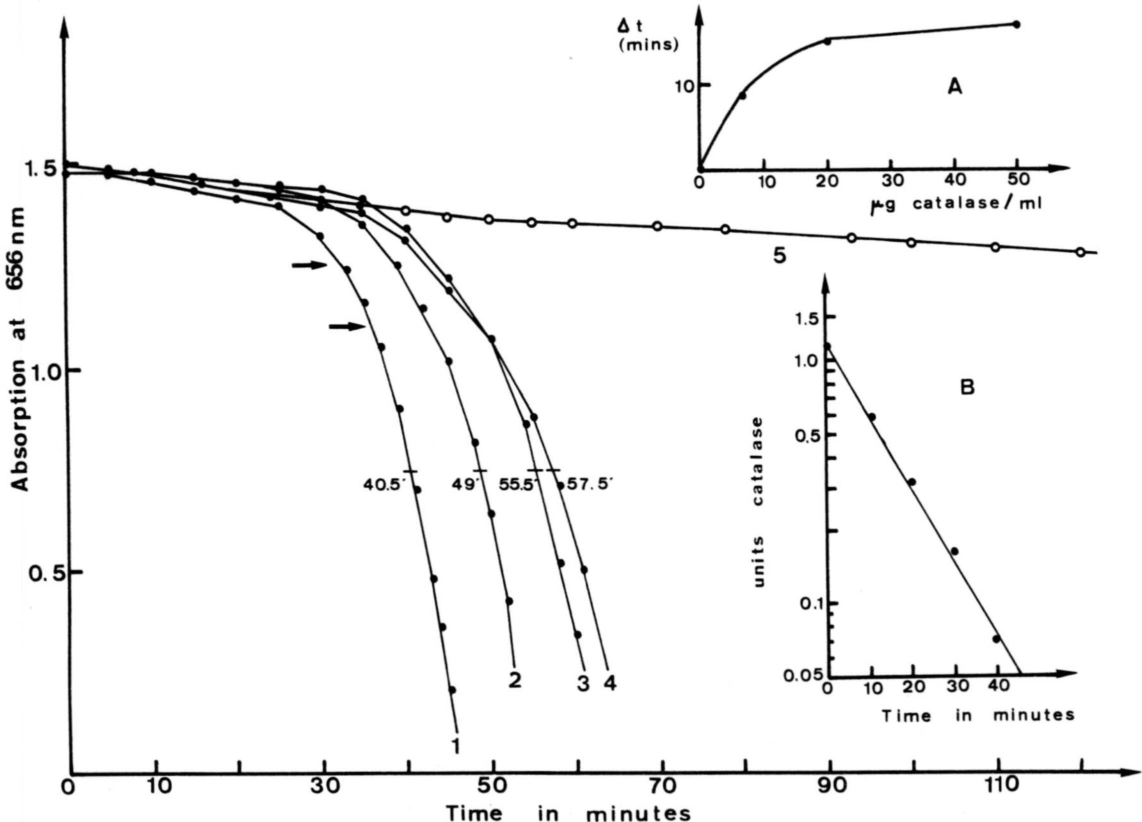

Fig.5. *Lysis of erythrocytes by riboflavin/NADH and h ν 365 nm.*
1) Erythrocytes alone or in presence of 10^{-3} *M KCN*
2) In presence of 6.7 μg catalase/ml (2.7 x 10^{-8} *M)*
3) Plus 20 μg catalase/ml (8 x 10^{-8} *M)*
4) Plus 50 μg catalase/ml (2 x 10^{-7} *M);*
5) Irradiation at 365 nm in absence of riboflavin and NADH.
Insert A. Increase in time for 50% hemolysis as a function of concentration of added catalase. Insert B. Loss in activity of catalase (100 μg/ml, 4 x 10^{-7} *M) on treatment with riboflavin/NADH and hν 365 nm in absence of erythrocytes.*

Thus despite the fact that $O_2^{\cdot-}$ is a major oxygen species produced in this system, these radicals are not involved in membrane lysis. The same arguments as used above can be applied to indicate that hydroxyl radicals in fact play a significant role. However, in this system, hemolysis is very markedly stimulated in D_2O compared with H_2O (Fig.3B). Thus the riboflavin photosensitized system, in addition to producing $O_2^{\overline{\cdot}}$ (and hence HO$^{\cdot}$ via reaction of $O_2^{\overline{\cdot}}$ with H_2O_2) is an excellent method for the production of excited singlet oxygen $^1O_2^*$, and it is this species (together with HO$^{\cdot}$ radicals) which is responsible for membrane lysis. The origin of $^1O_2$ by irradiation of riboflavin at 365 nm is perhaps due to direct energy transfer to dissolved oxygen, rather than dismutation of $O_2^{\overline{\cdot}}$ (see also Hodgson and Fridovich, 1976). Production of singlet oxygen by irradiation of riboflavin has been described by Foote (1968).

REFERENCES

1. Behar, D., Czapski, G., Rabani, J., Dorfman, L.M. and Schwarz, H.A. (1970). *J. Phys. Chem. 74,* 3209-3213.
2. Buckingham, M.E., Caput, D., Cohen, A., Whalen, R.G. and Gros, F. (1974). *Proc. Natl. Acad. Sci. U.S.A. 71,* 1466-1470.
3. Drake, J.W. (1966). *Proc. Natl. Acad. Sci. U.S.A. 55,* 738-743.
4. Fernandez-Sousa, J. and Michelson, A.M. (1976). *Biochem. Biophys. Research Communs.,* in press.
5. Foote, C.S. (1968). *Accounts Chem. Res. 1,* 104-110.
6. Gregory, E.M. and Fridovich, I. (1973). *J. Bacteriol. 114,* 543-548; 1193-1197.
7. Hodgson, E.K. and Fridovich, I. (1975). *Biochemistry 14,* 5299-5303.
8. Hodgson, E.K. and Fridovich, I. (1976). *Arch. Biochem. Biophys. 172,* 202-205.
9. Lavelle, F., Michelson, A.M. and Dimitrijevic, L. (1973). *Biochem. Biophys. Research Commun. 55,* 350-357.
10. Merkel, P.B. and Kearns, D.R. (1971). *Chem. Phys. Letters 12,* 120.
11. Merkel, P.B., Nilsson, R. and Kearns, D.R. (1972). *J. Am. Chem. Soc. 94,* 1030.
12. Merkel, P.B. and Kearns, D.R. (1973). *J. Am. Chem. Soc. 95,* 5336.
13. Michelson, A.M. and Brody, E. (1974). Unpublished results.
14. Michelson, A.M. and Buckingham, M.E. (1974). *Biochem. Biophys. Research Comm. 58,* 1079-1086.
15. Michelson, A.M. and Durosay, P. (1976). *Photochem. Photobiol.,* in press.
16. Michelson, A.M. and Isambert, M.F. (1975). Unpublished work.
17. Roshchupkin, D.I., Pelenitsyn, A.M., Potapenko, A. Ya., Talitsky, V.V. and Vladimirov, Yu. A. (1975). *Photochem. and Photobiol. 21,* 63-69.
18. Shamberger, R.J., Baughman, F.F., Kalchert, S.L., Willis, C.E. and Hoffman, G.C. (1973). *Proc. Natl. Acad. Sci. U.S.A. 70,* 1461-1463.

FREE RADICALS IN LEUKOCYTE METABOLISM AND INFLAMMATION

M.L. SALIN* and J.M. McCORD

Departments of Medicine and Biochemistry
Duke University Medical Center
Durham, North Carolina 27710, U.S.A.

Superoxide is a fairly reactive and potentially cytotoxic free radical anion. A great amount of evidence now supports a reaction first proposed by Haber and Weiss in 1934 in which superoxide and hydrogen peroxide react to produce the hydroxyl free radical, a very powerful oxidizing species. The reaction is probably catalyzed by trace metals as follows:

$$O_2^- + H_2O_2 \xrightarrow{Fe^{++}} O_2 + OH^- + OH^\cdot .$$

Since Beauchamp and Fridovich (1970) invoked this reaction to account for the xanthine oxidase dependent production of ethylene from methional, many other phenomena have been reported which are apparently due to the same reaction. The systems all contain a source of superoxide (and hence a source of hydrogen peroxide as well), and the various phenomena which are monitored can all be inhibited by the addition of superoxide dismutase (SOD) or catalase, either of which removes a reactant and thus prevents the reaction from taking place, or by such compounds as ethanol, mannitol, or benzoate, which serve to scavenge the hydroxyl radical. Among the phenomena thought to be due to hydroxyl radical formation are hydroxylation reactions (Goscin and Fridovich, 1972; Halliwell and Ahluwalia, 1976), damage to DNA (Lown *et al.*, 1976; Cone *et al.*, 1976), depolymerization of structural polysaccharides (McCord, 1974), and bactericidal action (Babior *et al.*, 1975; Klebanoff, 1974; Johnston *et al.*, 1975).

In 1971 a positive identification was made between the enzyme isolated as SOD by McCord and Fridovich (1969) and a protein isolated several years earlier by Huber and co-workers on the basis of its anti-inflammatory properties (Carlson *et al.*, 1973). This anti-inflammatory protein had been

*Postdoctoral Fellow of the United States Public Health Service.

This work was supported by U.S. Public Health Service Grants AM 17091 and HL 17603.

named "orgotein". A relationship between SOD activity and anti-inflammatory activity became obvious with the observation by Babior *et al.* (1973) that phagocytosing polymorphonuclear leukocytes (PMN), effector cells of the acute inflammatory response, release large amounts of superoxide into the medium in which the stimulated cells are suspended. Babior and others have gone on to establish that this radical production represents a significant bactericidal mechanism in leukocyte metabolism. Our own concern has been with an apparently inadvertant and unfortunate side-effect of the process - the release of a sizable fraction of the O_2^- into the medium. We have established that exposure to superoxide could bring about considerable damage to biological components such as structural polysaccharides via hydroxyl radical generation, and have further demonstrated that mammalian extracellular fluids are sufficiently free of SOD and catalase activities to permit OH· production to occur (McCord, 1974). A comparison of the concentrations of SOD in several extracellular fluids with those of two representative tissues, liver and erythrocytes, is made in Table I. Although the amounts of SOD in the fluids are not negligible, they are nevertheless insufficient to effectively scavenge such superoxide as might be produced by phagocytosing or otherwise stimulated PMN.

TABLE I

Superoxide Dismutase Concentrations in Various Tissues and Fluids

Erythrocytes	100 mg/l
Liver	500 mg/kg
Blood plasma	0.7 mg/l
Synovial fluid	1.0 mg/l
Cerebrospinal fluid	0.23 mg/l
Aqueous humor	0.9 mg/l

We have now studied some of the features and characteristics of superoxide production by PMN, and have utilized the leukocyte itself as a model to examine the cytotoxicity of radical production and the potential role of superoxide and hydroxyl radical in the inflammatory process.

MATERIALS AND METHODS

Isolation of Cells

All procedures employed plastic or siliconized glass containers. Venous blood, drawn into heparinized syringes from human volunteers, was diluted fourfold with 0.9% saline. Following the procedure of Boyum (1967), the monocytes and lymphocytes were isolated in a Ficoll-Hypaque gradient. The pellet containing the erythrocytes and PMN was brought to three times the original volume with 0.9% saline containing 3% dextran (Sigma Chemical Co., avg. mol. wt. 500,000). After 20 min of sedimentation under gravity, the supernatant was withdrawn and centrifuged at 120 x g for 10 min. The pellet, containing PMN and some contaminating erythrocytes, was subjected to hypotonic stress by the addition of 10 ml of 0.2% saline. Thirty seconds later, 10 ml of 1.6% saline was added and the suspension centrifuged at 120 x g for

10 min. This procedure was repeated until the pellet was visibly free of erythrocytes. The leukocyte pellet was then suspended in 0.5 ml of Ringer-phosphate buffer, pH 7.4, and examined microscopically. Cell viability judged by exclusion of the dye trypan blue (Fallon *et al.*, 1962) was greater than 98%. The leukocytes were greater than 98% PMN and there was less than one platelet per four PMN.

Quantitative NBT Test

The quantitative nitroblue tetrazolium (NBT) test was performed in siliconized 10 x 75 mm test tubes. Each tube contained 5 x 10^6 PMN and, where indicated, 10^8 *Escherichia coli* in 1 ml of Ringers-phosphate buffer containing 1 mM NBT, 1% glucose, and 5% autologous serum. SOD was present as indicated. Reaction mixtures were incubated in a water bath at 37°. At intervals, aliquots were removed and 200 - 300 cells were microscopically examined. Cells containing the dark blue formazan particles were scored as NBT-positive cells. On occasion, formazan was extracted with pyridine and quantitatively determined by the method of Baehner and Nathan (1968).

Cell Viability Studies

PMN and bacteria were incubated in the medium described above, but without NBT. Long term incubations (longer than three hours) contained in addition 50 U/ml penicillin (Sigma) and 50 μg/ml streptomycin sulfate (Sigma) to retard extraneous bacterial growth. Other additions were made as indicated. At various time intervals, aliquots were withdrawn, mixed with a 1% solution of trypan blue, and examined microscopically. The percentage of cells not stained by the dye is represented as "% viable cells". In certain experiments, a significant number of dead cells lysed. In these cases, the % viable cells was based on the original cell count.

Preparation of Polylysyl-SOD

SOD was isolated from bovine liver by the method of McCord and Fridovich (1969). Polylysyl-SOD (PLSOD) was prepared by incubating 12 mg of purified bovine liver SOD with 5 mM 1-ethyl-3(3-dimethylamino-propyl)carbodiimide and 4 mM polylysine (avg. mol. wt. 2,000, Sigma) for 12 min at 23°. The incubation mixture (vol. 1 ml) was maintained at a pH of 4.5 by the addition of 0.5 M HCl as the reaction proceeded. The sample was then chromatographed on a Sephadex G-25 column (1.5 x 21 cm) equilibrated with 50 mM sodium phosphate buffer, pH 7.0, to separate the unreacted carbodiimide and polylysine from the derivitized enzyme. The product was characterized by agarose gel electrophoresis (Analytical Chemists, Inc.) performed for 10 min at 200 V in Tris-glycine buffer, pH 8.5. The gels were stained for SOD activity by the method of Beauchamp and Fridovich (1971) as modified by Salin and McCord (1974), and revealed that the PLSOD migrated toward the cathode as a heterogeneous group of bands. The native SOD migrates toward the anode as a single band under the same conditions. Quantitative determination of the specific activity of the PLSOD was made using a modification (1 ml total volume) of the standard assay described by McCord and Fridovich (1969), with mathematical conversion to standard units. The product sustained a 55% loss of activity as a result of the treatment. Amino acid analysis, performed by Dr. Howard Steinman, indicated that PLSOD contained an additional 40 - 50 lysine residues per molecule. This corresponds to the covalent attachment of about 3 polylysine chains per molecule, with approximately 15 lysine residues per chain.

Phagocytic Kill of Bacteria by PMN

The effects of native SOD and PLSOD on the bactericidal activity of PMN were observed by incubating 10^6 PMN with 3 x 10^6 *E. coli* in 1 ml of Ringer-phosphate buffer containing 1% glucose and 5% autologous serum, with the indicated concentrations of SOD or PLSOD. At intervals of 4 min, aliquots were removed and immediately diluted 100-fold into 0.2% Triton X-100 to lyse the PMN and release the bacteria, thereby interrupting the killing process. Cell counts were then performed by plating appropriate dilutions of these suspensions on trypticasesoy agar plates, incubating 24 hr at 37°, and counting colonies.

Polyacrylamine Disc Gel Electrophoresis

The procedure of Davis (1964) was used with 10% acrylamide resolving gels. The gels were stained for SOD activity as described above. Human liver was obtained from autopsy at Duke University Medical Center for comparison with leukocytes with regard to SOD content. Approximately 2 g of liver was homogenized at 4° in 40 ml of Ringer-phosphate buffer and centrifuged for 10 min at 5,000 x g. The supernatant was used for electrophoresis.

Miscellaneous Materials

Mannitol, catalase, and bovine serum albumin were products of Sigma Chemical Co. Endotoxin from *Salmonella typhosa* was obtained from Difco Laboratories, Detroit, Michigan. A lymphokine preparation was kindly provided by Mr. Linville Meadows.

RESULTS

SOD Content of PMN

Reports that PMN do not contain cytoplasmic SOD (Beckman *et al.*, 1973) and that phagocytosing leukocytes produce superoxide (Babior *et al.*, 1973) appeared in the literature at the same time. It seemed reasonable that the absence of a cytoplasmic scavenging system in the leukocyte would enhance the survival of the superoxide radical produced by the aerobic metabolism within the cell. A subsequent paper by DeChatelet *et al.* (1974) reported that human peripheral PMN contained only a cyanide-insensitive SOD (presumably the manganese enzyme) which they believed to be located in the cytosol. We examined once more the SOD content of the PMN in an attempt to determine the amount and type(s) of enzyme contained in this cell type. The results of activity-stained disc gel electrophoresis of sonicated human PMN are shown in Fig.1, with homogenized human liver for comparison. The cyanide-sensitive (cuprozinc) and cyanide-insensitive (manganese) SOD bands are electrophoretically identical for the two preparations. Identical results were obtained with human placenta, lung, and pancreas homogenates. We conclude that PMN contain intracellular superoxide dismutases which are apparently identical to the enzymes found in other tissues examined (Salin and McCord, 1974).

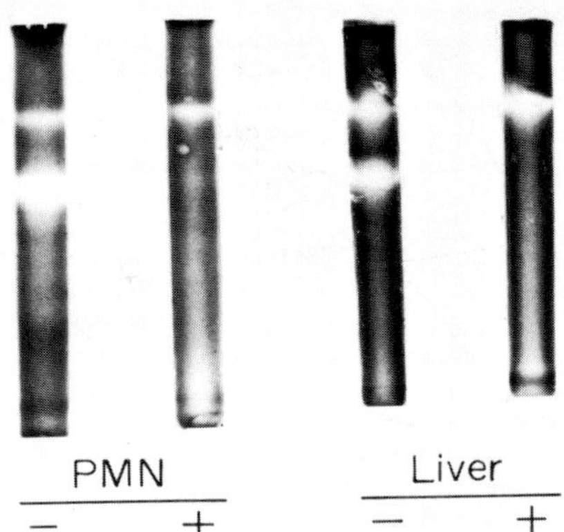

Fig.1. *Polyacrylamide gel electrophoresis stained for SOD activity. Techniques were performed as described under Materials and Methods. Gels labeled PMN received 30 µl of a sonicated suspension of human PMN containing 10^8 cells/ml. Gels labeled Liver received 5 µl of the human liver preparation described in Materials and Methods. The gels designated + were stained in the presence of 1 mM potassium cyanide.*

Superoxide Release by SOD-Containing PMN

Since it appeared that our isolated PMN contained a significant amount of SOD, it was critical to determine whether these cells were also capable of releasing superoxide into the medium. Superoxide production, measured by the rate of SOD-inhibitable cytochrome *c* reduction by phagocytosing PMN, occurred at approximately 1 nanomole/min/10^7 cells. This rate was comparable to that recorded by Curnutte and Babior (1974) and 15 times that originally observed (Babior *et al.*, 1973). The finding that leukocytes which contain cytoplasmic SOD are capable of releasing O_2^- into the medium suggests that the site of formation of the free radical ion is not within the cell but rather resides on the outer surface of the cell membrane.

Superoxide-Mediated NBT Reduction

It is well known that activated leukocytes reduce nitroblue tetrazolium to the insoluble dark blue formazan. This property of the PMN has had wide clinical usage (Baehner and Nathan, 1968). Since it had been demonstrated that activated PMN produce superoxide (Babior *et al.*, 1973), and since superoxide had been shown to reduce NBT efficiently (Beauchamp and Fridovich, 1971), it followed that at least part of the reduction of NBT by leukocytes might be due to superoxide released into the medium. The susceptibility of NBT reduction by phagocytosing PMN to inhibition by SOD is shown in Fig.2. A significant amount of NBT reduction in resting cells was apparent and showed little or no response to the presence of SOD. This is probably due to a variety of dehydrogenases which are capable of reducing NBT directly. However, in cells which were phagocytosing *E. coli*, there was a two- to four-fold increase in the rate of dye reduction. Most of this increase could be inhibited by the addition of SOD. On occasion, NBT reduction was determined quantitatively by pyridine extraction of the incubation mixtures, in addition to counting NBT positive cells. These two methods were in excellent agreement

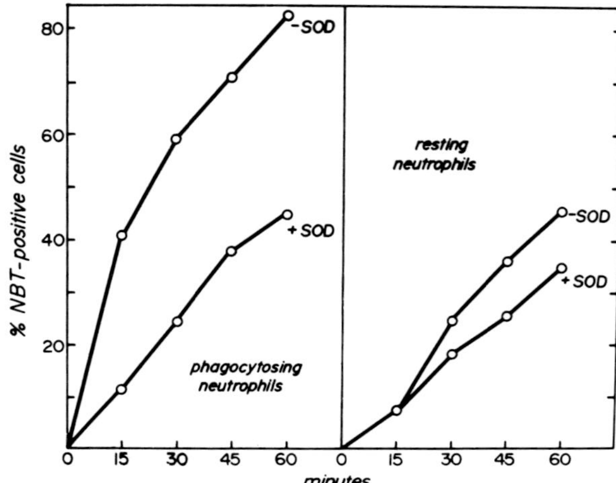

Fig.2. *Effect of SOD on NBT reduction by phagocytosing and resting PMN. Incubations contained 5 x 10⁶ PMN, with (phagocytosing) or without (resting)* E. coli *as described under Materials and Methods. Those labeled + SOD contained 500 μg of SOD.*

Since a variety of non-particulate stimuli can cause increased rate of NBT reduction by PMN, it seemed important to determine whether these stimuli induced superoxide production. Fig.3 illustrates the effects of two soluble factors, endotoxin and lymphokine, on superoxide production by isolated PMN. Endotoxin is a soluble lipopolysaccharide preparation from *Salmonella typhosa*. Lymphokine represents one or more soluble protein factors released from lymphocytes which have been stimulated by a non-specific mitogen (such as concanavalin A) or with an antigen to which they have been sensitized (Snyderman and Altman, 1973). Lymphokines serve as chemotactic factors to other leukocytes. While the non-specific NBT reduction showed a modest increase in the presence of all three stimuli, the superoxide-mediated NBT reduction showed a dramatic increase in each case. We conclude that phagocytosis of particles is not the only circumstance which can induce superoxide production by the PMN.

We have also observed superoxide production by leukocytes other than PMN. A mixed population of monocytes and lymphocytes was isolated using a Ficoll-Hypaque gradient. A differential count of Wright stained cells indicated it contained 11% monocytes and 89% lymphocytes. The cell preparation was tested for its ability to reduce NBT under the same conditions used for PMN. When cells were examined microscopically, essentially none showed the characteristic inclusion of purple formazan particles. Visual inspection of the incubation mixture, however, showed that the tube with the activated cells without SOD was distinctly more purple than those to which SOD had been added. This was confirmed by quantitative pyridine extraction of the formazan. The cells produced superoxide at a rate of 3 nanomoles/min/10^7 cells. It appears that stimulated monocytes, and perhaps lymphocytes also, release superoxide into the external medium, thereby reducing NBT, but these cells do not phagocytose the formazan particles as readily as do PMN.

Protection of Phagocytosing PMN by SOD

The release of significant amounts of superoxide into the medium by activated PMN is perhaps an inadvertant side-effect of the bactericidal process. It is conceivable that the release of radicals into the medium would not only cause the destruction of bacteria but may also result in damage to surrounding tissue or extracellular components. One would expect that the most serious and perhaps lethal radical-induced damage would be incurred by the leukocyte itself. To examine this possibility, we monitored the viability of resting and activated PMN in the presence and absence of SOD. Fig.4 shows that resting PMN (incubated in the absence of *E. coli*) maintained full ability to exclude the dye trypan blue for the first 18 hours of incubation.

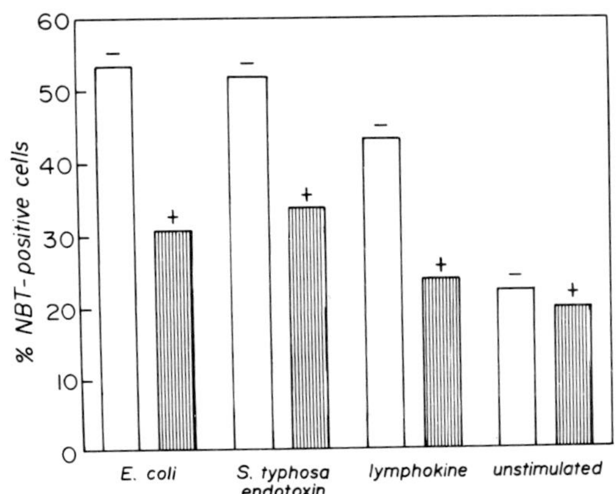

Fig.3. *Effects of several stimulants on superoxide production by PMN. Incubations were performed as described under Materials and Methods, with the following additions:* E. coli, *10^8 cells;* S. typhosa *endotoxin, 400 μg; lymphokine, a ten-fold dilution of supernate from 10^6 concanavalin A-stimulated lymphocytes per ml. SOD was present where indicated (+) at 450 μg/ml.*

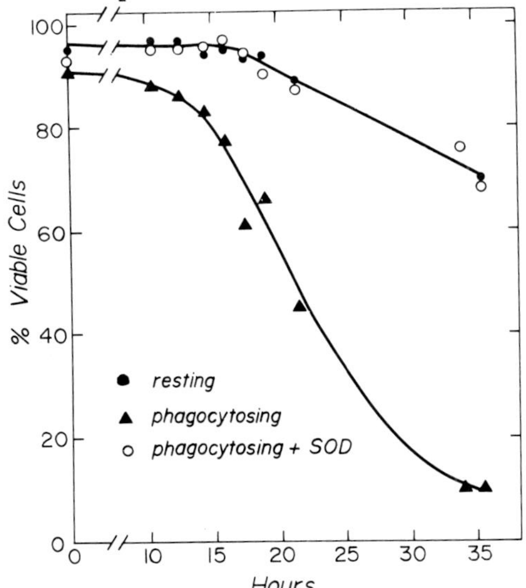

Fig.4. *Protection of phagocytosing PMN by SOD. Incubations were performed as described under Materials and Methods. Those designated "phagocytosing" contained 3 x 10^7 bacteria. Where indicated, SOD was added to a concentration of 300 μg/ml. At intervals, samples were withdrawn, mixed with 1% trypan blue in 0.9% saline, and the cells were examined microscopically.*

Thereafter, the percentage of viable cells slowly decreased, reaching a value of 70% after 35 hours. Phagocytosing cells showed a slow loss of viability during the initial 15 hours of incubation, but after that time proceeded to deteriorate rapidly with 90% of the cells dead after 35 hours. An incubation mixture containing phagocytosing cells plus SOD at a concentration of 300 μg/ml behaved much like resting cells. The presence of SOD at comparable concentrations had no effect upon the process of phagocytosis nor on the bactericidal action by PMN. The data in Fig.5 show the effect of varying concentrations of SOD on the ability to maintain the viability of activated PMN during 40 hours of incubation with bacteria at 37°. The maximal protective effect was attained at 200 μg of SOD/ml.

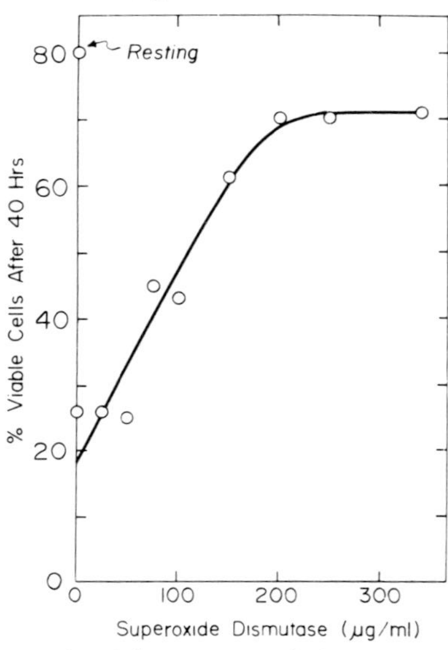

Fig.5. *Effect of SOD concentration on maintenance of PMN viability. Conditions and procedures were as described in Fig.4, except that SOD concentration was varied as indicated, and all incubations were examined only after 40 hours. All cells were phagocytosing except the point labeled "Resting".*

After prolonged incubation, the pH of the reaction mixtures changed less than one unit, and this was not affected by the presence of protectants. The nature of the superoxide-induced death was further investigated by incubating phagocytosing PMN with catalase or mannitol. If superoxide *per se* were responsible for the death of the phagocytosing PMN, neither catalase nor mannitol would protect the leukocytes. If, however, superoxide exerted its toxicity by reacting with H_2O_2 to produce OH·, catalase and mannitol should protect as well as SOD. Fig.6 shows that SOD, catalase, and mannitol were equally effective at maintaining the viability of phagocytosing PMN. Bovine serum albumin, however, failed to protect the leukocytes, thereby ruling out any non-specific protein effect, as does the fact that each incubation mixture contained 5% autologous serum. We therefore conclude that the agent causing the premature death of the cells is the hydroxyl radical produced secondarily from the superoxide radical. Superoxide *per se*, which was not scavenged in those reaction mixtures containing catalase or mannitol, did not cause the premature death of PMN. Likewise, hydrogen peroxide *per se*, which was not scavenged in those mixtures containing SOD or mannitol, did not contribute to the death of the cell. Only when O_2^- and H_2O_2 are present simultaneously is hydroxyl radical formation possible, and only under these conditions is there a premature loss of PMN viability.

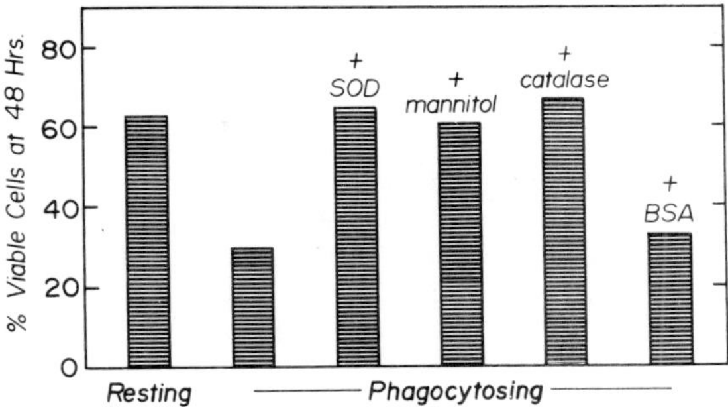

Fig.6. *Protection of phagocytosing PMN by SOD, mannitol, and catalase. Incubation mixtures were as described in Fig.4 with the following additions where indicated: SOD, 250 μg/ml; mannitol, 1 mM; catalase 250 μg/ml; bovine serum albumin (BSA), 250 μg/ml. Viability was determined after 48 hours.*

The time course of cell death varied for cells obtained from different donors, but the maximal rate of cell death was always seen to take place between 12 and 30 hours after phagocytic challenge. This delayed death was not anticipated since the metabolic burst of superoxide production was shown to occur for 20 - 30 min after the onset of phagocytosis, and then subside to the level of a resting cell (Babior *et al.*, 1973). It appeared that the leukocytes were "mortally wounded" by the initial burst of free radical production but that death did not result from this damage until several hours later. To test this hypothesis, PMN were induced to phagocytose bacteria in the presence and absence of SOD. After 1.5 hours, when O_2^- production had ceased, samples were gently centrifuged and resuspended with and without SOD for the remainder of the observation period. The results seen in Fig.7 reveal that premature cell death can be prevented only if SOD is present during that part of the incubation in which the superoxide production takes place, and that the addition of the enzyme after that initial exposure to superoxide resulted in no protection from a premature loss of viability.

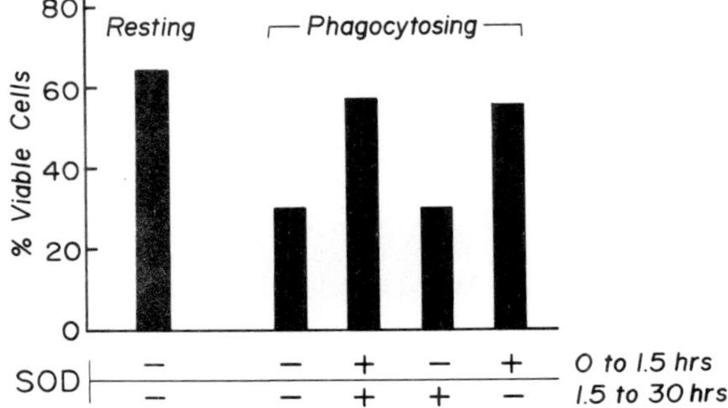

Fig.7. *Time-course of protection of phagocytosing PMN by SOD. Conditions were as described in Fig.4. After allowing 1.5 hours for the completion of phagocytic activity, the cells were centrifuged and resuspended in fresh medium, then incubated an additional 28.5 hours. SOD was present, when and where indicated (+), at 300 μg/ml. The percentage of viable cells at the end of the incubation was based on the number of viable cells present at* t = *0.*

Studies on Polylysyl-SOD

In the long term cell viability studies, the amount of SOD needed to achieve maximal protection of activated PMN was about 200 μg/ml (see Fig.5). This amount of enzyme is at least two orders of magnitude greater than that needed to inhibit superoxide-mediated reactions such as the spectrophotometric assays. It appeared possible that the enzyme in free solution was not effectively encountering its substrate, presumably being produced on the outer surface of the cell membrane. A possible explanation for this phenomenon might be an electrostatic repulsion between the negatively charged SOD and the negatively charged outer cell membrane. Perhaps if a more positively charged SOD could be produced it would not experience the electrostatic repulsion between enzyme and membrane, with the result being that a lower SOD concentration would be needed to protect activated PMN from a premature loss of cell viability.

Fig.8 compares the abilities of positively-charged PLSOD and native SOD to maintain PMN viability through 40 hours of incubation after phagocytosis of bacteria. The PLSOD worked most effectively at a concentration approximately one order of magnitude lower than the optimal concentration for the native SOD. At higher concentrations of PLSOD, however, a precipitous decline

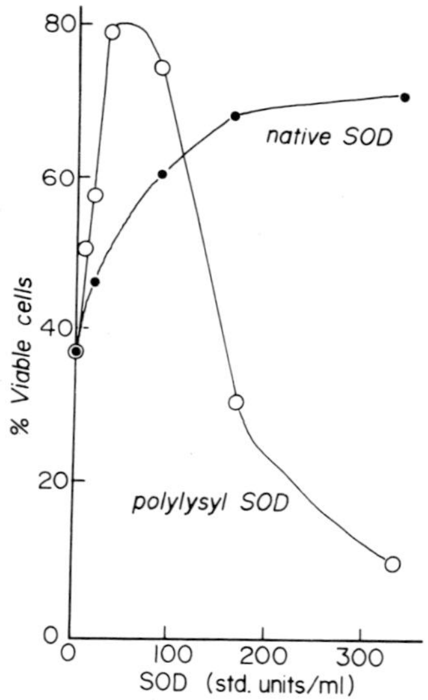

Fig.8. *Effect of various concentrations of PLSOD and SOD on maintenance of viability of phagocytosing PMN. Incubations were as described under Materials and Methods, except that PLSOD or SOD concentrations were varied as indicated and incubations were examined only after 40 hours.*

in cell viability was observed, apparently due to toxic effects of the PLSOD *per se*. Since there is no evidence that native SOD is toxic to PMN at much higher concentrations, we suspected that the toxicity of PLSOD might be due to the polylysine moiety of the molecule. To test this hypothesis, non-phagocytosing PMN were incubated with various concentrations of PLSOD and with equivalent amounts of free polylysine. The results are seen in Fig.9. Both free polylysine and PLSOD were toxic to resting PMN when the polylysine or equivalent polylysine concentration exceeded about 5 μM. We conclude that the deleterious effects of high PLSOD concentrations on PMN viability must be due to the toxic effect of the polylysine upon the cell.

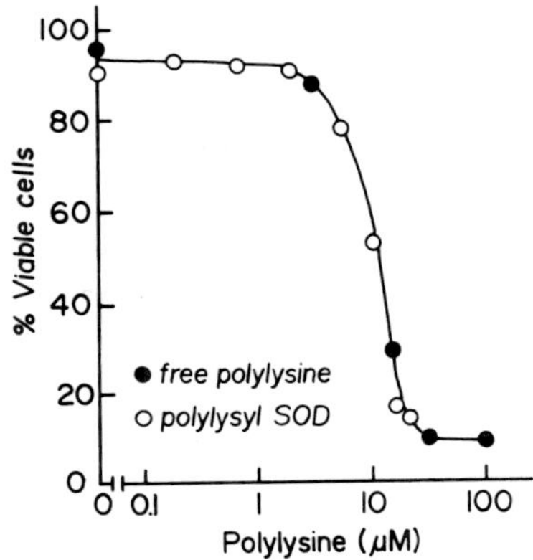

Fig.9. *Effect of polylysine and PLSOD concentrations on viability of resting PMN. Resting cells were incubated as described under Materials and Methods. As indicated, various concentrations of polylysine or the equivalent in PLSOD were added. After three hours incubation at 37°, cell viability was determined by trypan blue exclusion.*

Because PLSOD would be less restricted in approaching the cell membrane, and because it might even bind to the negatively charged surfaces of the PMN or the bacteria, it seemed possible that it might find its way into the phagocytic vacuole much more easily than native SOD, and hence might interfere with the killing process. Johnston *et al.* (1975) demonstrated that native SOD interfered with the killing process only if it were attached to latex particles which were phagocytosed along with the bacteria. We examined the bactericidal efficiency of PMN on *E. coli* in the presence of PLSOD and native SOD, and the results are shown in Fig.10. The concentrations of SOD and PLSOD selected were those necessary to provide maximal protection to the PMN. Native SOD, at 730 U/ml, failed to alter the killing efficiency of the PMN; PLSOD, at only 18 U/ml, produced a marked protection of the bacteria during the early minutes of the killing process. After the initial delay, however, the killing process began and was ultimately successful.

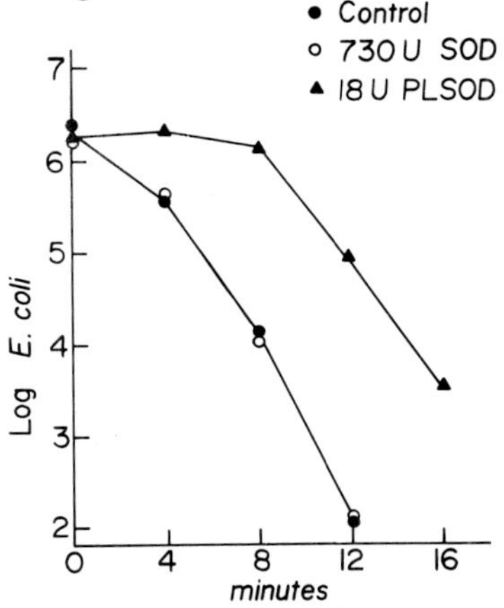

Fig.10. *Effects of PLSOD and native SOD on bactericidal action by PMN.* 10^6 *PMN were incubated with 3 x* 10^6 E. coli *as described under Materials and Methods. At the indicated times, aliquots were removed and the PMN were lysed by detergent treatment to release the bacteria and interrupt the killing process. Plate counts revealed the numbers of surviving bacteria.*

DISCUSSION

While PMN may contain quantitatively less SOD than most other cell types, it appears that they do contain significant amounts of both types of SOD. This makes it very unlikely that the superoxide detectable in the medium in which phagocytosing PMN are suspended is produced *inside* the cell. Rather, we have proposed that superoxide production occurs on the outer surface of the cell's plasma membrane (Salin and McCord, 1974; McCord and Salin, 1975). Up to 80% of the maximal amount of superoxide which could be produced by a suspension of leukocytes (based on oxygen consumption) is detectable in the medium (Drath and Karnovsky, 1975; Roos *et al.*, 1976). This fact supports the hypothesis, as does the recent findings by Briggs *et al.* (1975) that NBT reduction appears to take place on the outer membrane surface, as visualized by electron microscopy.

If superoxide is generated by the PMN for the purpose of killing ingested bacteria, it would seem that the end could be best achieved by generating the radical only on those portions of the membrane which would invaginate to line the phagocytic vacuoles. Apparently, though, the radical is generated on regions of the membrane which remain in contact with the external milieu, such that most of it escapes its intended purpose and instead simply serves to provide a chemical insult to surrounding healthy tissue or to the PMN itself. If our concept of the radical's purpose is correct, certain other responses of the system seem inappropriate. As shown in Fig.4, superoxide production may be triggered by stimuli which do not necessarily involve bacteria. A recent study by Goldstein *et al.* (1975) confirms this point and expands the list of leukocyte stimulators to include heat aggregated IgG and complement components.

Superoxide production is not unique to the PMN. We observed copious production of the radical by a mixed preparation of monocytes and lymphocytes. Drath and Karnovsky (1975) observed equally abundant superoxide production from stimulated alveolar and peritoneal macrophage. Thus, superoxide production may be a significant component of many, if not all types of cell-mediated inflammation. For example, sufficiently large numbers of activated PMN are present in a typically inflamed rheumatoid arthritis joint that superoxide-mediated depolymerization of hyaluronic acid is quantitatively feasible as the *in vivo* mechanism of synovial fluid degradation (McCord, 1974)

The clinically observable anti-inflammatory activity of SOD (Edsmyr *et al.*, 1976; Marberger *et al.*, 1975; Lund-Olesen and Menander, 1974) may be due in part to its protection of extracellular components and nearby tissues from damage by the radical, and in part to its protection of the PMN against premature death and lysis. It is clear that the death of leukocytes and the subsequent release of hydrolytic enzymes and chemotactic factors play major roles in the perpetuation of the inflammatory cycle (Hirsch, 1974).

The data reported herein and elsewhere (McCord, 1974) support the notion that the superoxide radical *per se* may not be so detrimental a species in a physiological context. Through its reaction with H_2O_2, however, it is apparently capable of generating secondary radicals which are capable of causing damage of grave consequence to physiological systems. Because extracellular fluids are deficient in SOD and catalase, such a reaction appears to be of physiological relevance.

REFERENCES

1. Babior, B.M., Curnutte, J.T. and Kipnes, R.S. (1975). *J. Lab. Clin. Med.* *85*, 235-244.
2. Babior, B.M., Kipnes, R.S. and Curnutte, J.T. (1973). *J. Clin. Invest.* *52*, 741-744.
3. Baehner, R.L. and Nathan, D.G. (1968). *New Engl. J. Med.* *278*, 971-976.
4. Beauchamp, C.O. and Fridovich, I. (1970). *J. Biol. Chem.* *245*, 4641-4646.
5. Beauchamp, C.O. and Fridovich, I. (1971). *Anal. Biochem.* *44*, 276-287.
6. Beckman, G., Lundgren, E. and Tarnvik, A. (1973). *Hum. Hered.* *23*, 338-345.
7. Boyum, A. (1967). *Scand. J. Clin. Lab. Invest.* *21*, (Suppl. 97) 77-89.
8. Briggs, R.T., Drath, D.B., Karnovsky, M.L. and Karnovsky, M.J. (1975). *J. Cell Biol.* *67*, 566-586.
9. Carson, S., Vogin, E.E., Huber, W. and Schulte, T.L. (1973). *Toxicol. Appl. Pharmacol.* *26*, 184-202.
10. Cone, R., Hasan, S.K., Lown, J.W. and Morgan, A.R. (1976). *Can. J. Biochem.* 219-223.
11. Curnutte, J.T. and Babior, B.M. (1974). *J. Clin. Invest.* *53*, 1662-1672.
12. Davis, B.J. (1964). *Ann. N.Y. Acad. Sci.* *121*, 404-427.
13. DeChatelet, L.R., McCall, C.E., McPhail, L.C. and Johnston, R.B. Jr., (1974). *J. Clin. Invest.* *53*, 1197-1201.
14. Drath, D.B. and Karnovsky, M.L. (1975). *J. Exptl. Med.* *141*, 257-262.
15. Edsmyr, F., Huber, W. and Menander, K.B. (1976). *Curr. Therap. Res.* *19*, 198-211.
16. Fallon, H.J., Frei, III, E., Davidson, J.D., Trier, J.S. and Burk, D. (1962). *J. Lab. Clin. Med.* *59*, 779-791.
17. Goldstein, I.M., Roos, D., Kaplan, H.B. and Weissmann, G. (1975). *J. Clin. Invest.* *56*, 1155-1163.
18. Goscin, S. and Fridovich, I. (1972). *Arch. Biochem. Biophys.* *153*, 778-783.
19. Halliwell, B. and Ahluwalia, S. (1976). *Biochem. J.* *153*, 513-518.
20. Hirsch, J.G. (1974). In "The Inflammatory Process" (B.W. Zweifach, L. Grant and R.T. McCluskey, eds.), Vol.1, pp 411-443, Academic Press, New York and London.
21. Johnston, R.B. Jr., Keele, B.B. Jr., Misra, H.P., Webb, L.S., Lehmeyer, J.E. and Rajagopalan, K.V. (1975). In "The Phagocytic Cell in Host Resistance" (A. Bellanti and D.H. Dayton, eds.) pp. 61-75, Raven Press, New York.
22. Klebanoff, S.J. (1974). *J. Biol. Chem.* *249*, 3724-3728.
23. Lown, J.W., Begleiter, A., Johnson, D. and Morgan, A.R. (1976). *Can. J. Biochem.* *54*, 110-119.
24. Lund-Olesen, K. and Menander, K.B. (1974). *Curr. Ther. Res. Clin. Exp.* *16*, 706-717.
25. Marberger, H., Huber, W., Bartsch, G., Schulte, T and Swoboda, P. (1974). *Int. Urol. Nephrol.* *6*, 61-74.
26. McCord, J.M. (1974). *Science* *185*, 529-531.
27. McCord, J.M. and Fridovich, I. (1969). *J. Biol. Chem.* *244*, 6049-6055.
28. McCord, J.M. and Salin, M.L. (1975). In "Erythrocyte Structure and Function" (G.J. Brewer, ed.) pp. 731-746, Alan R. Liss, New York.
29. Roos, D., Homan-Müller, J.W.T. and Weening, R.S. (1976). *Biochem. Biophys. Res. Comm.* *68*, 43-50.
30. Salin, M.L. and McCord, J.M. (1974). *J. Clin. Invest.* *54*, 1005-1009.

31. Snyderman, R. and Altman, L. (1973). In "Annual Review of Allergy" (C.A. Frazier, ed.) pp. 377-387, Med. Exam. Publ. Co., Flushing, New York.

RECENT STUDIES ON OXYGEN METABOLISM IN HUMAN NEUTROPHILS: SUPEROXIDE AND CHEMILUMINESCENCE

B.M. BABIOR

Hematology Service
New England Medical Center Hospital
and the Department of Medicine
Tufts University School of Medicine
Boston, Massachusetts 02111, U.S.A.

The discovery of superoxide dismutase by McCord and Fridovich (1) has revolutionized the field of oxygen metabolism. Among other things, it has provided a tool for investigating the role of O_2^- in biological processes. With the aid of this enzyme, it has been possible to show that O_2^-, originally thought to be only a harmful byproduct to be disposed of as quickly and effectively as possible, can be put to good use by living systems. The use discussed here is its role in biological defense, where it is of central importance in the oxygen dependent microbicidal mechanisms of phagocytes.

O_2^- AND NEUTROPHIL OXYGEN METABOLISM

The neutrophil is a phagocyte and as such is a member of a class of cells whose function is to ingest and degrade foreign substances. While most phagocytes tend to be scavengers, taking up any foreign matter which comes their way, neutrophils specialize in the destruction of invading microorganisms. Neutrophils are formed in the bone marrow, circulate for a brief time in the blood, then leave the blood stream to migrate to sites of bacterial invasion, where they perform their task (2).

The importance of oxygen in the destruction of microorganisms by these cells has been recognized since 1933, when Baldridge and Gerard found an increase in oxygen uptake by neutrophils upon exposure to the microorganism *Sarcina lutea* (3). It was originally thought that the increase in oxygen uptake was necessary to provide the energy for phagocytosis (i.e. ingestion of the bacteria). However, when Sbarra and Karnovsky showed in 1959 that phagocytosis occurred anaerobically (4), it became apparent that oxygen uptake served another purpose. It was soon shown that the increase in oxygen consumption associated with phagocytosis represented the activation of a metabolic pathway by which oxidizing agents were produced which participated directly in

Supported by USPHS Grant No. A1-11827 and USAMRDC Contract No. 17-74-C-4055.

bacterial killing by these cells. The various manifestations of this metabolic pathway have been termed the "respiratory burst".

The first oxidizing agent shown to be formed was H_2O_2 (5). This compound, demonstrated to act in collaboration with myeloperoxidase and Cl^- (6) as an antimicrobial agent of unusual potency, was produced by the enzyme-catalyzed reduction of oxygen by a pyridine nucleotide. The fact that the rise in oxygen consumption was accompanied by an increase in the oxidation of glucose via the hexosemonophosphate shunt (4), a pathway which generates NADPH from NADP, led Iyer and Quastel (7) to postulate that H_2O_2 was produced by the action of an NADPH oxidase:

$$H^+ + O_2 + NADPH \rightarrow H_2O_2 + NADP$$

and that the increase in hexosemonophosphate shunt activity served to regenerate the NADPH consumed in this reaction. Others, however, proposed that NADH rather than NADPH was the true electron donor (8), invoking secondary reactions to account for the increase in NADPH production represented by the rise in hexosemonophosphate shunt activity (9-11). This debate has been reviewed by Karnovsky (12).

O_2^- was first shown to be produced by neutrophils in 1973 (13-18). The production of O_2^- by these cells is similar in its characteristics to the increases in oxygen uptake, H_2O_2 production and hexosemonophosphate shunt activity which comprise the respiratory burst. Both oxygen uptake and O_2^- production show a time course characterized by an early lag and a late decline (Fig.1) (4,14,19), and a response to metabolic inhibitors characterized by a marked sensitivity to sulfhydryl reagents but a lack of sensitivity to CN^-, N_3^- and other heme enzyme inhibitors (Table I) (4,5,20,31). Moreover, all these phenomena - the increases in oxygen uptake, O_2^- production, H_2O_2 production and hexosemonophosphate shunt activity - are absent in neutrophils from patients with chronic granulomatous disease (CGD) (12,15,16, 22-24).

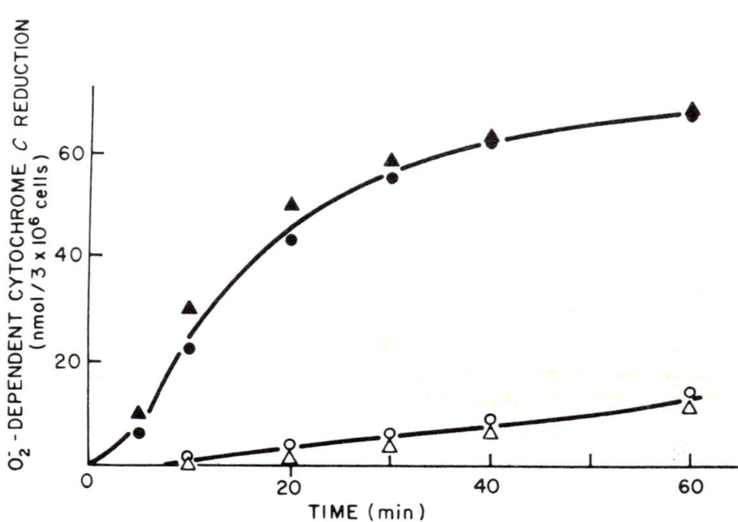

Fig.1. *O_2^- production by human neutrophils as a function of time. o, resting; ●, activated. For experimental details see Ref. 14.*

TABLE I

Effect of Metabolic Inhibitors on Stimulated Oxygen Uptake and O_2^- Production by Neutrophils

Agent	Rate (% of Control)	
	Stimulated Oxygen Uptake	O_2^- Production
Cyanide	59	92
Antimycin A	103	93
2,4-Dinitrophenol	80	105
Sulfhydryl reagents	16	7

For experimental details, see Ref. 4 (oxygen uptake) and 20 (O_2^- production).

This condition, an inherited defect in phagocyte metabolism which leads to recurrent and severe bacterial infections in affected individuals, is characterized by the failure of the cells to manifest the respiratory burst in response to appropriate stimuli. The fact that CGD cells fail to produce O_2^- (Table II) is the strongest evidence that O_2^- production is a property of the metabolic pathway responsible for the respiratory burst.

TABLE II

O_2^- Production by CGD Neutrophils

Subject	O_2^- Production (nmol/3 x 10^6 cells/30 min)
Normal (5)	26.1 ± 3.1 SD
Case 1	0.9
Case 2	0.9

For experimental details, see Ref. 24.

The importance of O_2^- in this pathway is indicated by work showing that at least 70% of the oxygen consumed by stimulated neutrophils can be accounted for as O_2^- (15), and that at least 70% of the H_2O_2 produced by stimulated neutrophils is derived from O_2^- (25). There seems to be general agreement that the characteristic reaction of the respiratory burst - the reaction which is activated when neutrophils are exposed to bacteria - is the one-electron reduction of oxygen to O_2^-.

PROPERTIES OF THE O_2^--FORMING ENZYME OF NEUTROPHILS (SUPEROXIDE SYNTHETASE)

The search for the enzyme responsible for the respiratory burst began shortly after the discovery that H_2O_2 was released into the medium by activated neutrophils. Within a few years, the search had narrowed down to the question of which of two proposed candidates was the enzyme of interest (12.26). One candidate was an NADH oxidase, a flavoprotein present in the cytosol of guinea pig and human neutrophils. The other was an NADPH oxidase, found in the 27,000 x g pellet from homogenates of activated neutrophils, an enzyme thought until quite recently to require manganese. These enzymes were generally assayed by measuring reduced pyridine nucleotide-dependent oxygen uptake. Both activities had been reported to be diminished (but not absent) in CGD neutrophils (27,28); with the NADH oxidase, this finding was reported from one laboratory, but could not be confirmed in two others (22,27,28). In general, the evidence was confusing, and it was difficult on the basis of the oxygen uptake data to decide which if either of these two activities was the enzyme responsible for the respiratory burst.

Very recently, it was discovered that manganese was introducing a major artifact in the NADPH oxidase assay system (29-31). This artifact was expressed as a marked acceleration in oxygen consumption in the presence of the ion, and resulted from the involvement of manganese in a free radical chain reaction in which the other propagating species were O_2^- and the NADP radical. With the omission of Mn^{++} from the assay system, it became possible to show that O_2^-, the initial product of the respiratory burst, was generated in large quantities when NADPH was incubated with the 27,000 x g pellet from homogenates of activated neutrophils (i.e. the "NADPH oxidase") (32). O_2^- was not generated by particles from resting neutrophils, nor was it produced by particles from activated* neutrophils obtained from patients with CGD (Table III).

TABLE III

O_2^- Production by Neutrophil Particles

Particles	O_2^- Production (nmol/min/mg protein)
Normal resting (4)	0.5 ± 0.3 SE
Normal activated (4)	18.3 ± 2.9
CGD activated (3)	0

The figure in parentheses indicates the number of separate particle preparations used in the experiment. For experimental details, see Ref. 32.

These results strongly suggest that the enzyme whose activation results in the sequence of events known as the respiratory burst is a superoxide-forming enzyme which is located in the particulate fraction of neutrophil homogenates. This enzyme is probably identical to the NADPH oxidase which had been studied in connection with the respiratory burst since 1961. Henceforward this enzyme will be termed "superoxide synthetase".

We have initiated a detailed study of the properties of the superoxide

*Neutrophils were activated by exposure to serum-treated yeast cell wall particles (zymosan), a procedure known to initiate the respiratory burst in intact cells.

synthetase from human neutrophils. The first studies were concerned with the kinetic properties of the enzyme and the nature of the reaction which it catalyzed. Studies of substrate specificity (Table IV) revealed that the only reducing agents which the enzyme could use were NADH and NADPH.

TABLE IV

Substrate Specificity of Superoxide Synthetase

Reducing Agent	O_2^- Production (nmol/min/mg protein)
NADPH	5.3
NADH	1.5
Glutathione	< 0.05
Ascorbate	< 0.05
Lactate	< 0.05
Glyceraldehyde-3-phosphate	< 0.05
Glucose-6-phosphate	< 0.05
6-Phosphogluconate	< 0.05

For experimental details, see Ref. 44. The concentration was 0.1 mM for each reducing agent.

With NADPH as the reducing agent, the enzyme consumed 0.55 mole of pyridine nucleotide per mole of O_2^- produced, a result consistent with the following reaction stoichiometry:

$$NADPH + 2O_2 \rightarrow NADP^+ + 2O_2^- + H^+$$

The enzyme was not inhibited by metal chelators or by inhibitors of heme proteins (Table V); the latter finding was particularly important inasmuch as insensitivity of CN^- is one of the hallmarks of the respiratory burst, and it would be expected that an enzyme identified as the mediator of the burst should be similarly insensitive to CN^-.

A comparison of O_2^- production by neutrophil particles with oxygen uptake by an equivalent number of intact cells (both preparations being activated in a similar way) showed that with NADPH as the reducing agent, the activity of the particles accounted for about 30 - 40% of the oxygen taken up by intact cells. Though not perfect, the agreement between the values obtained with whole cells and particles is regarded as satisfactory, especially in view of the likelihood that further investigation would disclose that the conditions used for the assay of O_2^- production by particles were not optimum. A comprehensive study of these conditions has yet to be undertaken.

As can be inferred from the foregoing discussion regarding the two candidate oxidases, a question which has occupied the attention of investigators for a number of years has to do with which of the two pyridine nucleotides is the reducing agent used by intact neutrophils during the respiratory burst (12). Arguments concerning physiological substrates frequently turn on the question of Michaelis constants, the substrate with

TABLE V

Effect of Inhibitors on Superoxide Synthetase

Inhibitor	O_2^- Production (nmol/min/mg protein)
None	6.1
CN^-	6.0
N_3^-	11.0
EDTA	8.8
1,10-Phenanthroline	11.6

For experimental details see Ref. 44. Inhibitor concentrations were: CN^-, 1 mM; N_3^-, 0.5 mM; EDTA, 0.25 mM; 1,10-phenanthroline, 0.25 mM.

the lower K_m being favored as the "real" substrate of the enzyme under debate. Measurements of K_m with the superoxide synthetase clearly favored NADPH as the physiological substrate. These measurements, made at several pH values, showed that the K_m for NADH was approximately 0.7 mM at pH 6.0 as well as pH 7.5, while the K_m for NADPH varied strongly with pH (Fig.2), being lowest at pH 6.0, the pH of the phagocytic vesicle in human neutrophils, and highest at pH 7.5.

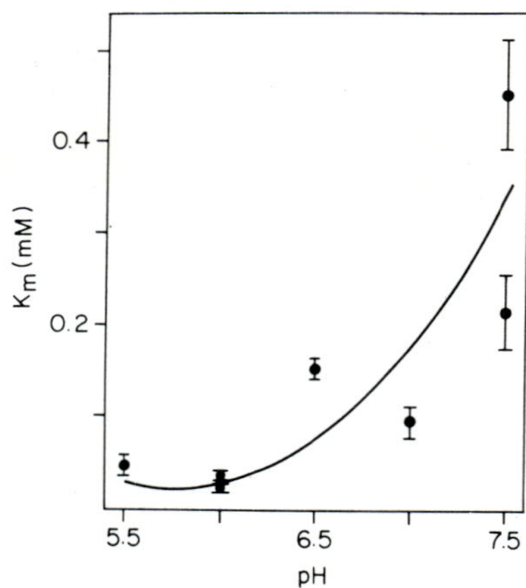

Fig.2. *K_m of NADPH as a function of pH. For experimental details, see Ref. 44.*

In all cases the K_m for NADPH was lower than the K_m for NADH, suggesting the former as the physiological reducing agent. This conclusion is supported by measurements of pyridine nucleotide concentrations performed by Rossi and associates (33), who showed that in the neutrophil the concentration of NADPH exceeds the concentration of NADH by a factor of 5, and that after activation there is a significant fall in the concentration of NADPH but not of NADH.

Attempts to release superoxide synthetase from the particle in soluble form have so far been unsuccessful, but experiments conducted with this goal in mind using the non-ionic detergent Triton X-100 have revealed the enzyme to be a flavoprotein. The evidence for this conclusion is presented in Table VI.

TABLE VI

Flavin Requirement of Superoxide Synthetase

Conditions	O_2^- Production (nmol/min/mg protein)
Control	27.4
Triton + FAD	32.6
Triton + FMN	5.6
Triton + Riboflavin	3.2
Triton + ADP	1.0
Triton + AMP	2.5

Particles were prepared and incubations conducted as described in Ref. 44. The concentration of NADPH was 0.1 mM, and of particle protein 0.1 mg/ml. Triton X-100 and cofactors were present as indicated, at concentrations of 0.045% and 40 uM respectively.

It is seen that the synthesis of O_2^- was virtually abolished when the assay was conducted in the presence of 0.045% Triton-X-100 unless FAD (40 μM) was present in the assay mixture as well. In this capacity, FAD could not be replaced by FMN, riboflavin, AMP or ADP. Superoxide synthetase thus appears to be a member of an already well-recognized class of flavoproteins, including xanthine oxidase (34) and dihydroorotate dehydrogenase (35), which are able to reduce oxygen by one electron. Superoxide synthetase, however, differs from other members of the class in that with this enzyme, O_2^- is generated not as a byproduct of another reaction, but for a specific purpose: to participate in the destruction of pathogenic microorganisms.

CHEMILUMINESCENCE

A manifestation of neutrophil function which has attracted considerable interest recently is the emission of light by cells engaged in phagocytosis (36-40). This phenomenon, first reported by Allen *et al.* in 1972 (36), has been found to be inhibited by superoxide dismutase, catalase and N_3^- (the latter an inhibitor of myeloperoxidase), and is missing in cells from patients with CGD. These observations implicate the products of oxygen metabolism formed during the respiratory burst as mediators of chemiluminescence in neutrophils. Allen *et al.* interpreted this luminescence as an indication that neutrophils form singlet oxygen during the respiratory burst (36). This interpretation, which has been widely accepted, is based on the existence of certain chemiluminescent reactions known to involve singlet oxygen. The reactions cited by Allen *et al.* were the relaxation of singlet oxygen to triplet oxygen, which leads to the emission of a photon, and the formation of dioxetanes by the addition of singlet oxygen across carbon-carbon double bonds, a reaction which leads to light emission because of the spontaneous cleavage of the dioxetane into two carbonyl compounds, one of which is in an electronically excited state and emits a photon as it decays to the ground state (41).

$$^1O_2 + \rangle C{=}C\langle \rightarrow \rangle \overset{O-O}{C-C} \langle \rightarrow \rangle C{=}O + \rangle C{=}O^*$$

The source of singlet oxygen was postulated to be O_2^-, which reacts with itself to form H_2O_2 and can subsequently react with the H_2O_2 to generate singlet oxygen plus hydroxyl radical (42).

We recently obtained a spectrum of the light emitted by neutrophils which have taken up opsonized zymosan. This spectrum, shown in Fig.3, shows that light is emitted by phagocytosing granulocytes over a broad wavelength region beginning between 400 and 450 nm and extending as far into the red as could be detected by the phototube employed in the measurements.

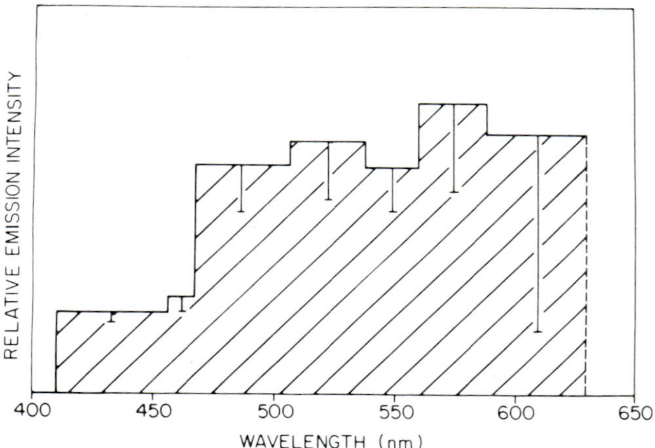

Fig.3. *The spectrum of the light emitted by phagocytosing granulocytes, corrected for photomultiplier tube sensitivity. For experimental details, see Ref. 45.*

This spectrum indicates that very little of the emitted light can be attributed to the relaxation of singlet to triplet oxygen, because the spectrum of light emitted in that transition consists of several sharp bands, the bluest of which is centered at 634 nm, well into the red end of the visible spectrum (43). Further evidence that light emission by neutrophils is not due to the relaxation of singlet to triplet oxygen is provided by the observation that neutrophils stimulated by F^- to produce O_2^- and H_2O_2 in the absence of particles do not luminesce (Table VII). This is the case even though the conditions necessary for the formation of singlet oxygen - namely, the simultaneous presence in a reaction mixture of O_2^- and H_2O_2 - are met when neutrophils are stimulated by F^-. Nonetheless, as mentioned above, neutrophils engaged in the phagocytosis of zymosan luminesce actively, and this luminescence is not quenched by subsequent exposure of the neutrophils to F^- (Table VII).

TABLE VII

Luminescence of F^--Activated and Zymosan-Activated Neutrophils

Preparation	Light Emission (counts/30 s)
F^--activated cells	100
Zymosan-activated cells	204100
Diluted into F^--containing medium	26300
Diluted into F^--free medium	25800

For experimental details, see Ref. 45

Luminescence is also seen when zymosan particles are exposed to an artificial O_2^--generating system (xanthine oxidase plus purine) (Table VIII); the spectrum of the light emitted under these conditions is similar to that of the light emitted by neutrophils which have taken up zymosan. The light produced by zymosan incubated with the artificial O_2^--generating system is quenched by superoxide dismutase, but not by catalase or mannitol, indicating that O_2^- alone is sufficient to elicit light from the particles.

TABLE VIII

Light Emission by Zymosan Plus Xanthine Oxidase-Purine

Conditions	Light Emission (counts/30 s)	
Complete mixture	51400	22200
Omit purine	9500	
Omit zymosan	4400	
Plus dismutase		4400
Plus catalase		22800
Plus mannitol		23000

For experimental details, see Ref. 45.

We interpret these observations to indicate that chemiluminescence by phagocytosing neutrophils results from the oxidation of constituents of the particle by the oxidizing agents generated by the stimulated cells. The identity of the oxidizing agent or agents responsible for the luminescence cannot be established from the data at hand, although the results with the artificial system show that O_2^- alone is capable of reacting with zymosan to produce light of the correct color. In particular, there is no way to ascertain whether or not singlet oxygen is participating in neutrophil chemiluminescence. Even though the evidence excludes the $^1O_2 \rightarrow {}^3O_2$ reaction as a major contributor to light emission by neutrophils, it is possible that singlet oxygen may be reacting with particle components to form chemiluminescent products. On the other hand, these products may originate solely from reactions involving O_2^- and H_2O_2, with no participation of singlet oxygen whatsoever in chemiluminescence.

SUMMARY

The enzyme responsible for the manifestations of the respiratory burst in human neutrophils is superoxide synthetase, a particulate O_2^--forming enzyme detectable in homogenates from activated but not resting neutrophils. This enzyme uses NADPH as the physiological substrate, and catalyzes the following reaction:

$$NADPH + 2O_2 \rightarrow NADP^+ + 2O_2^- + H^+$$

Other oxidizing agents generated by activated neutrophils are formed in secondary reactions involving O_2^-. The chemiluminescence observed in neutrophils engaged in phagocytosis results from the interaction of these oxidizing agents with the constituents of the particles being taken up by the cells.

REFERENCES

1. McCord, J.M. and Fridovich, I. (1969). *J. Biol. Chem. 244,* 6049.
2. Boggs, D.R. (1967). *Semin. Hematol. 4,* 359.
3. Baldridge, C.W. and Gerard, R.W. (1933). *Am. J. Physiol. 103,* 235.
4. Sbarra, A.J. and Karnovsky, M.L. (1959). *J. Biol. Chem. 234,* 1355.
5. Iyer, G.Y.N., Islam, M.F. and Quastel, J.H. (1961). *Nature 192,* 535.
6. Klebanoff, S.J. and Hamon, C.B. (1972). *J. Reticuloendothel. Soc. 12,* 170.
7. Iyer, G.Y.N. and Quastel, J.H. (1963). *Canad. J. Biochem. Physiol. 41,* 427.
8. Evans, W.H. and Karnovsky, M.L. (1961). *J. Biol. Chem. 236,* PC30.
9. Kaplan, N.O., Colowick, S.P. and Neufeld, E.F. (1953). *J. Biol. Chem. 205,* 1.
10. Noseworthy, J. and Karnovsky, M.L. (1972). *Enzyme 13,* 110.
11. Reed, P.W. (1969). *J. Biol. Chem. 244,* 2459.
12. Karnovsky, M.L. (1973). *Fed. Proc. 32,* 1527.
13. Babior, B.M., Kipnes, R.S. and Curnutte, J.T. (1973). *J. Clin. Invest. 52,* 741.
14. Curnutte, J.T. and Babior, B.M. (1974). *J. Clin. Invest. 53,* 1662.
15. Weening, R.S., Wever, R. and Roos, D. (1975). *J. Lab. Clin. Med. 85,* 245.
16. Johnston, R.B., Keele, B.B., Misra, H.P., Lehmeyer, J.E., Webb, L.S., Baehner, R.L. and Rajagopalan, K.V. (1975). *J. Clin. Invest. 55,* 1357.
17. Baehner, R.L., Murrmann, S.K., Davis, J. and Johnson, R.B. (1975). *J. Clin. Invest. 56,* 571.
18. Goldstein, I.M., Roos, D., Kaplam, H.B. and Weissmann, G. (1975). *J. Clin. Invest. 56,* 1155.
19. Malawista, S.E. and Bodel, P.T. (1967). *J. Clin. Invest. 46,* 786.
20. Curnutte, J.T. and Babior, B.M. (1975). *Blood 45,* 851.
21. Patriarca, P., Cramer, R., Tedesco, F. and Kakinuma, K. (1975). *Biochim. Biophys. 385,* 387.
22. Holmes, B., Page, A.R. and Good, R.A. (1967). *J. Clin. Invest. 46,* 1422.
23. Baehner, R.L. and Nathan, D.G. (1968). *New Engl. J. Med. 278,* 971.

24. Curnutte, J.T., Whitten, D.M. and Babior, B.M. (1974). *New Engl. J. Med. 290,* 593.
25. Root, R.K. and Metcalf, J. (1975). *Clin. Res. 23,* 311A.
26. Patriarca, P., Cramer, R., Monocalvo, S., Rossi, F. and Romeo, D. (1971). *Arch. Biochem. Biophys. 145,* 255.
27. Baehner, R.L. and Karnovsky, M.L. (1968). *Science 162,* 1277.
28. Hohn, D.C. and Lehrer, R.I. (1975). *J. Clin. Invest. 55,* 707.
29. Curnutte, J.T., Karnovsky, M.L. and Babior, B.M. (1975). *Clin. Res. 23,* 271A.
30. Patriarca, P., Dri, P., Kakinuma, K., Tedesco, F. and Rossi, F. (1975). *Biochim. Biophys. Acta 385,* 380.
31. Curnutte, J.T., Karnovsky, M.L. and Babior, B.M. (1976). *J. Clin. Invest. 57,* 1059.
32. Curnutte, J.T., Kipnes, R.S. and Babior, B.M. (1975). *New Engl. J. Med. 293,* 628.
33. Selvaraj, R.J. and Sbarra, A.J. (1967). *Biochim. Biophys. Acta 141,* 243.
34. Olson, J.S., Ballou, D.P., Palmer, G. and Massey, V. (1974). *J. Biol. Chem. 249,* 4350.
35. Forman, H.J. and Kennedy, J. (1975). *J. Biol. Chem. 250,* 4322.
36. Allen, R.C., Stjernholm, R.J. and Steele, R.H. (1972). *Biochem. Biophys. Res. Comm. 47,* 679.
37. Webb, L.S., Keele, B.B. Jr. and Johnston, R.B. Jr. (1974). *Infect. Immun. 9,* 1051.
38. Allen, R.C., Yevich, S.J., Orth, R.W. and Steele, R.H. (1974). *Biochem. Biophys. Res. Comm. 60,* 909.
39. Klebanoff, S.J. and Rosen, H.S. (1976). *J. Clin. Invest. 58,* 50.
40. Stjernholm, R.L., Allen, R.C., Steele, R.H., Waring, W.W. and Harris, J.A. (1973). *Infect. Immun. 7,* 313.
41. McCapra, F. (1968). *Chem. Comm.*
42. Kellogg, E.W. and Fridovich, I. (1975). *J. Biol. Chem. 250,* 8812.
43. Browne, R.J. and Ogryzlo, E.A. (1964). *Proc. Chem. Soc.*
44. Babior, B.M., Curnutte, J.T. and McMurrich, B.J. (1976). *J. Clin. Invest. 58.*
45. Cheson, B.D., Christensen, R.L., Sperling, R., Kohler, B.E. and Babior, B.M. (1976). *J. Clin. Invest. 58,* 789.

SUPEROXIDE AND CHEMILUMINESCENCE

J.P. HENRY and A.M. MICHELSON

Institut de Biologie Physico-Chimique
Service de Biochimie-Physique
13, rue P. et M. Curie
75005 Paris, France

INTRODUCTION

Light emission is often used as a means of detection of superoxide anions. Such techniques are relatively simple and can be extremely sensitive. In addition, information is obtained with respect to kinetics at a given instant as well as continuous presentation of the totality of the reaction taking place, thus avoiding the need for multiple estimations.

Two approaches have been used with respect to $O_2^{\cdot-}$. In the first place, dismutation of the radical in aqueous solution gives rise to a low level of light emission which can be followed; alternatively the $O_2^{\cdot-}$ is used to produce a chemiluminescence of much greater intensity by oxidation of suitable detectors such as luminol or Pholad luciferin.

Luminescence Due to Dismutation of $O_2^{\cdot-}$

It is now well established that oxidation of hypoxanthine, xanthine or acetaldehyde by molecular oxygen catalysed by xanthine oxidase produces superoxide radicals (Knowles *et al.*, 1969). Stauff (Stauff *et al.*, 1963; Stauff and Wolf, 1964) and later Arneson (Arneson, 1970) showed that a low level of light emission was associated with the enzymic reaction and that with acetaldehyde as substrate this luminescence was more intense. The emission is related to production of $O_2^{\cdot-}$ and can indeed be observed with other systems producing the radical under conditions in which dismutation is possible such as oxidation of H_2O_2 by periodate, reduction of molecular oxygen by ferrous ions and a suitable ligand, or reoxidation of reduced flavins (Michelson, 1973). More directly, when solutions of $O_2^{\cdot-}$ in aprotic solvents (where dismutation is impossible) are injected into aqueous solutions, light is emitted (Bader and Kuwana, 1965; Legg and Hercules, 1969; Stauff *et al.*, 1973; Michelson, 1973).

While this chemiluminescence is always associated with dismutation of $O_2^{\cdot-}$, it is not specific to this reaction since both singlet oxygen species and hydroxyl radicals can give rise to a similar luminescence. Decomposition of 1-phospha-2,8,9-trioxa-adamantane ozonide (Boda *et al.*, 1974) is an example of $^1O_2$ luminescence whereas the luminescence of Fenton's reagent is probably due to hydroxyl radicals. It remains to be shown whether all these reactions produce light via the same emitter, and if so, the nature of the emitter.

Characteristics. Light emission by the xanthine oxidase system has been studied by Arneson (1970) and Hodgson and Fridovich (1976). Both $O_2^{\cdot-}$ and H_2O_2 are intermediates in the reaction since luminescence is inhibited both by superoxide dismutase and by catalase. (With respect to the latter enzyme some caution must be exercised since catalase may be capable of acting as a hydroxyl radical trap and could also react with $O_2^{\cdot-}$.)

The autocatalytic aspect of the kinetics of light emission (Imax is reached after several minutes only) is probably explicable by the slow accumulation of H_2O_2 due to spontaneous dismutation of $O_2^{\cdot-}$, produced in this system at a very low steady-state concentration ($\sim 10^{-10}$ M). Emission is also inhibited by ethanol and formate ions, both of which react with $HO^{\cdot}$. Hodgson and Fridovich (1976) extending observations by Stauff and collaborators (Stauff and Bergman, 1972; Stauff *et al.*, 1973) have demonstrated that light emission is greatly stimulated in presence of carbonate and bicarbonate anions and have postulated that *luminescence is entirely dependent on the presence of these ions,* since extrapolation of emission versus carbonate concentration resulted in zero luminescence at zero carbonate ion concentration.

Light emission due to pulse dismutation of $O_2^{\cdot-}$ is more difficult to study. Intensity of the luminescence as a function of pH has been studied by Michelson (1973) using the systems periodate/H_2O_2; $FMNH_2/O_2$; Fe^{2+}/O_2/phosphate and injection of electrolytically prepared $O_2^{\cdot-}$ (in an aprotic solvent) into aqueous solution. In the last case the effect of pH between 7 and 10 is not significant. In all cases the emission was inhibited by superoxide dismutase whereas catalase was not efficient, perhaps because of the rapid nature of the reaction (not steady-state dismutation) or that H_2O_2 is not directly involved. As mentioned earlier Stauff *et al.* (1973) showed that presence of bicarbonate anions greatly stimulated light emission.

Spectral characteristics of emission in the above cases have not been defined due to the low level of luminescence and the rapidity of the reactions. Stauff *et al.* (1973) using solutions of $O_2^{\cdot-}$ in hexamethylphosphoramide showed that the main emission was centered at 630 nm, but in aqueous solution the band was at 450 - 550 nm, without further precision. Using the single photon counter technique (Vigny and Duquesne, 1974) we have obtained a well defined emission spectrum for the system acetaldehyde/O_2/xanthine oxidase in the presence of carbonate ions (Vigny, Puget, Henry and Michelson, 1976) as shown in Fig.1. In confirmation of the results of Stauff *et al.*, no emission was observed in the region of 630 nm, the peak being centered at 435 nm, that is, more blue than earlier indicated emission characteristics.

Mechanism. For the moment the mechanism of light emission is subject to controversy and no definite conclusions can as yet be drawn. The main problem is the nature of the emitter. For about ten years, following the suggestion of Stauff *et al.* (1963) together with studies by Kahn and Kasha (1966), singlet oxygen has been generally accepted as playing this role. Thus a dimer of oxygen in the state 2 ($^1\Delta g$) on returning to the stable form ($^3\Sigma g^-$) would produce a calculated emission at 633 nm whereas a dimer of the form ($^1\Delta g$ + $^1\Sigma g^+$) would emit at 478 nm. In accord with this hypothesis Kahn

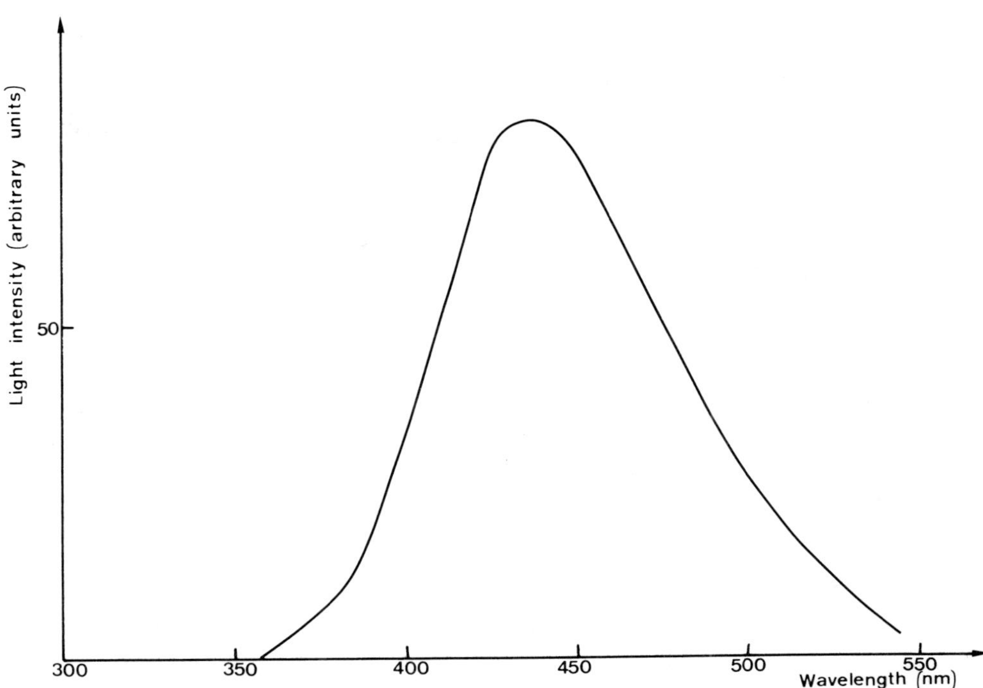

Fig.1. *Emission spectrum of the system xanthine oxidase-acetaldehyde - O_2 - carbonate ions.*

and Kasha (1964) obtained luminescence at 633 nm and 703 nm from the reaction NaOCl + H_2O_2 while reagents producing singlet oxygen during their decomposition such as 1-phospha-2,8,9-trioxa-adamantane ozonide also give rise to a chemiluminescence (Goda *et al.*, 1974). Thus dismutation of $O_2^{\cdot-}$ could be related to production of excited singlet oxygen by one of the following reactions

$$2\ O_2^{\cdot-} + 2\ H^+ \longrightarrow H_2O_2 + {}^1O_2^* \qquad (1)$$

$$O_2^{\cdot-} + HO^{\cdot} \longrightarrow HO^- + {}^1O_2^* \qquad (2)$$

$$O_2^{\cdot-} + H_2O_2 \longrightarrow HO^- + HO^{\cdot} + {}^1O_2^* \qquad (3)$$

All three reactions possess sufficient energy to populate the $^1\Delta g$ state (22 kcal) while reaction 2 (and perhaps 1, Mayeda and Bard, 1974) can theoretically reach the $^1\Sigma g^+$ state (38 kcal). At various times all three reactions have been postulated, the first two to explain chemiluminescent reactions (Goda *et al.*, 1974; Arneson, 1970). In particular, reaction 1 has been proposed for rapid emissions when a pulse of $O_2^{\cdot-}$ is produced (Stauff *et al.*, 1963; Michelson, 1973; Goda *et al.*, 1974). This hypothesis receives experimental support in the case of production of $O_2^{\cdot-}$ by the action of solvated electrons on O_2 in hexamethyl phosphoramide since Stauff *et al.* (1973) have observed an emission at 630 nm corresponding to the transition 2 $(^1\Delta g) \longrightarrow$ 2 $(^3\Sigma g^-)$.

Reaction 2 has been postulated by Arneson (1970) for the emission due to production of $O_2^{\cdot-}$ by the action of xanthine oxidase on acetaldehyde. Light emission requires presence of H_2O_2 as shown by the autocatalytic nature of the reaction, stimulation by added H_2O_2 and inhibition by catalase. Production of $HO^{\cdot}$, necessary for reaction 2 is a result of a postulated Haber-Weiss reaction.

$$O_2^{\overline{\cdot}} + H_2O_2 \longrightarrow HO^- + HO^{\cdot} + O_2 \qquad (4)$$

Inhibition by superoxide dismutase of light emission due to a postulated reaction 1 is explained by loss of energy in presence of the enzyme to give triplet ground state oxygen directly. This concept receives some support from results obtained by Mayeda and Bard (1974) and Schaap *et al.* (1974) who indicate that $^1O_2$ is not a substrate for SOD and is not quenched catalytically by this enzyme.

However, singlet oxygen is now considered highly unlikely as emitter for luminescence due to O_2^- production in aqueous solutions for a number of reasons. Thus production of $HO^{\cdot}$ radicals in the xanthine oxidase reaction must be at an extremely low concentration and will be limited by the very low rate of formation compared with the rapid rate of loss in presence of H_2O_2 by the reaction

$$H^{\cdot} + H_2O_2 \longrightarrow H_2O + O_2^{\overline{\cdot}} + H^+$$

$$(k = 4.5 \times 10^7\ M^{-1}sec^{-1})\ \text{(Schwarz, 1962)}$$

$$(k = 2.25 \times 10^7\ M^{-1}sec^{-1})\ \text{(Thomas, 1965)}$$

Further, spectroscopic studies are not in accord with the hypothesis. Apart from $O_2^{\overline{\cdot}}$ in hexamethyl phosphoramide, the emissions are generally described as green, implying an oxygen dimer ($^1\Delta g + {}^1\Sigma g^+$) of which the existence, while well established in the gas phase, has yet to be demonstrated in aqueous solution, and indeed for the reaction $NaOCl + H_2O_2$, light emission was considered to be due to gas bubbles (Khan and Kasha, 1966). In addition, emission of xanthine oxidase/acetaldehyde in the presence of carbonate has λmax at 435 nm which is not compatible with emission of excited oxygen. The role of carbonate ions must also be considered. Recently Hodgson and Fridovich (1976) have shown that light emission by xanthine oxidase/acetaldehyde shows an absolute dependence on the presence of carbonate ions, and this observation excludes singlet oxygen emission. An explanation was proposed by Stauff *et al.* (1973) and by Hodgson and Fridovich (1976) that formation of carbonate/bicarbonate radicals $CO_3^{\overline{\cdot}}$ and $HCO_3^{\cdot}$ and their dimerisation are necessary steps, being created by the following reactions

$$O_2^{\overline{\cdot}} + H_2O_2 \longrightarrow HO^- + HO^{\cdot} + O_2 \qquad (4)$$

$$HO^{\cdot} + CO_3^{2-} \longrightarrow HO^- + CO_3^{\overline{\cdot}} \qquad (5)$$

$$2\ CO_3^{\overline{\cdot}} \longrightarrow X \longrightarrow h\gamma + \text{products} \qquad (6)$$

It may be noted that the Haber-Weiss reaction (4) is again invoked, though some catalytic effect (perhaps of trace metals such as Fe^{2+}/Fe^{3+}) is clearly necessary to avoid the objections mentioned earlier. A possible alternate mechanism, given the very large excess of carbonate ions could be

$$2\ H^+ + O_2^{\overline{\cdot}} + CO_3^{2-} \longrightarrow H_2O_2 + CO_3^{\overline{\cdot}} \qquad (7)$$

or

$$H^+ + O_2^{\overline{\cdot}} + HCO_3^- \longrightarrow H_2O_2 + CO_3^{\overline{\cdot}}$$

However, oxidation of carbonate ions by $O_2^{\overline{\cdot}}$ may well be more than counterbalanced by the reaction $CO_3^{\cdot-} + O_2^{\cdot-} \quad CO_3^{2-} + O_2$ and a more likely possibility is reduction of CO_2 by $O_2^{\cdot-}$ to give a formate radical (Michelson and Durosay, 1976).

$$CO_2 + O_2^{\cdot-} \longrightarrow O_2 + CO_2^{\cdot-} \qquad (8)$$

or

$$HCO_3^- + O_2^{\cdot-} \longrightarrow HO^- + CO_2^{\cdot-} + O_2$$

Light emission would then involve formate radical combination possibly via dimerisation to give oxalic acid

$$2\ \dot{C}(=O)O^- \longrightarrow {}^-O-C(=O)-C(=O)-O^-$$

rather than dimerisation of the carbonate or bicarbonate radical anions to give a dicarboxylperoxide

$$2\ HO-C(=O)-O^{\cdot} \longrightarrow O=C(OH)-O-O-C(OH)=O$$

followed by decomposition to CO_2 and H_2O_2 as suggested by Stauff *et al.* (1973).

Such schemes explain the effects of superoxide dismutase, catalase and carbonate anions. It is nevertheless clear that the reactions are complex and the various possible roles of $O_2^{\cdot-}$, $HO^{\cdot}$, $CO_3^{\cdot-}$ and $CO_2^{\cdot-}$ remain to be clarified. It may be concluded that presence of a feeble light emission cannot be considered as indicative of $O_2^{\cdot-}$ as has sometimes been reported (Sligar *et al.* 1974). Between $O_2^{\cdot-}$ and the true emitter lie various intermediates, any one of which in the chain can ultimately give rise to luminescence. It would be of interest to determine whether the luminescence observed on treatment of erythrocuprein with H_2O_2 (Hodgson and Fridovich, 1975) in carbonate buffer, is not in fact related to the presence of carbonate ions. Similarly, the effect of carbonate ions on luminescence associated with phagocytosis (Webb *et al.*, 1974) is worth studying. If stimulation occurs it should be possible to obtain spectral characteristics of the emission.

Chemiluminescence Due to $O_2^{\cdot-}$ Acting on a Suitable Reagent

Since $O_2^{\cdot-}$ does not appear to give rise to light emission directly, but via an unknown emitter, it is perhaps more useful to use the $O_2^{\cdot-}$ to oxidise a suitable chemiluminescent product, since this can also greatly increase the intensity of the emission. Such compounds are frequently postulated to pass *via* a peroxide intermediate, readily formed by the action of $O_2^{\cdot-}$, though apart from the case of lophine (White *et al.*, 1965; Dufraisse *et al.*, 1957; Hayaishi and Maeda, 1968) such intermediates have not been isolated.

Luminol has often been used in this connection but a more sensitive reagent is the luciferin (a glycoprotein, M. wt. 34 000) of the boring clam *Pholas dactylus* (Henry *et al.*, 1970; 1973; 1975), with the added advantage that alkaline pH is not necessary for efficient light emission, as is the case with luminol. Thus xanthine oxidase plus xanthine or acetaldehyde induces light emission from luminol (Hodgson and Fridovich, 1973) and from *Pholas* luciferin (Henry and Michelson, 1970; 1973) due to the $O_2^{\cdot-}$ produced.

Both reactions are inhibited by SOD but luciferin requires larger concentrations (about 10 fold) of enzyme for the same inhibition due to the greater affinity for $O_2^{\cdot -}$ compared with luminol. Oxidation of luminol is also inhibited by ethanol and formate, both considered to react with $HO^{\cdot}$ radicals (Hodgson and Fridovich, 1973) while carbonate ions stimulate luminescence about 35 fold (Puget and Michelson, 1976). It may be noted that peroxycarbonates react with luminol with efficient light production and White and Brundrett (1973) have shown that H_2O_2 in sodium carbonate solution gives rise to chemiluminescence of luminol on acidification.

Another enzymic source of steady state production of $O_2^{\cdot -}$ is the reaction of horse radish peroxidase with dihydroxyfumaric acid. This system has been shown to oxidize luminol with light emission (Nilsson, 1969) and also pholad luciferin (Henry *et al.*, 1973).

Oxidation of luminol by $O_2^{\cdot -}$ produced electrolytically by reoxidation of reduced flavins, by IO_4^-/H_2O_2 and by Fe^{2+}/O_2/phosphate has also been described (Michelson, 1973) and in all cases superoxide dismutase inhibits the luminescence whereas catalase (under these conditions of pulse production of $O_2^{\cdot -}$) has a relatively small effect, if any. Similar studies have been reported for Pholad luciferin (Henry and Michelson, 1970; 1973).

Mechanism. Although a luminol peroxide intermediate has never been isolated, present day views of the chemiluminescence of luminol invoke formation of compound I which degrades to the emitter II.

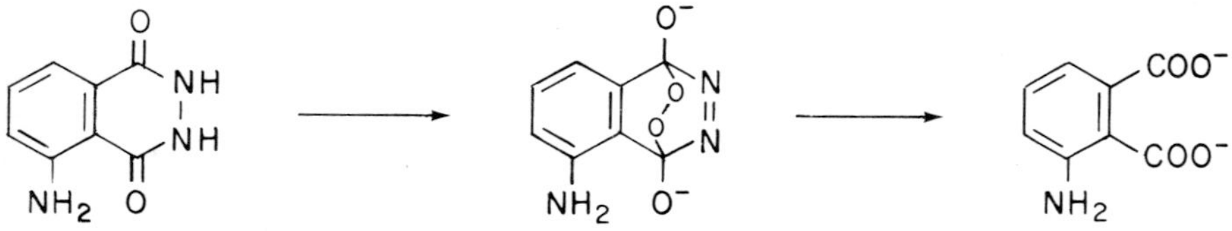

I II

Initiation of the reaction requires ionisation of luminol. In aprotic solutions the dianion must be formed which then reacts with molecular oxygen, whereas in aqueous solution the monoanion is sufficient and formation of the peroxide occurs by a free radical mechanism. Thus reaction with pulses of $O_2^{\cdot -}$ can be represented as

$$H^+ + LH_2 + O_2^{\cdot -} \longrightarrow LH^{\cdot} + H_2O_2$$

$$LH^{\cdot} + O_2^{\cdot -} \longrightarrow LOOH$$

where LH_2 is luminol and LOOH the peroxide intermediate (ionisation of the luminol molecule is neglected in this presentation). Such reactions naturally would be inhibited by SOD but not by catalase. In the case of xanthine oxidase induced chemiluminescence of luminol, the sequence is more complex as mentioned earlier, and production of hydroxyl radicals by the Haber-Weiss reaction is invoked (Hodgson and Fridovich, 1973).

$$O_2^{\cdot -} + H_2O_2 \longrightarrow HO^{\cdot} + O_2$$

followed by $HO^{\cdot} + LH_2 \longrightarrow LH^{\cdot} + H_2O$

For the effect of carbonate on the oxidation of luminol the reactions

$$O_2^{\overline{\cdot}} + CO_3^{2-} + 2\,H^+ \longrightarrow H_2O_2 + CO_3^{\overline{\cdot}}$$

$$LH_2 + CO_3^{\overline{\cdot}} \longrightarrow LH^{\cdot} + H^+ + CO_3^{2-}$$

have been proposed (Puget and Michelson, 1976), to which can be added, if hydroxyl radicals are present,

$$HO^{\cdot} + CO_3^{2-} \longrightarrow HO^- + CO_3^{\overline{\cdot}}$$

However, as discussed earlier, oxidation of carbonate ions by $O_2^{\overline{\cdot}}$ is unlikely, and, in order to avoid the inconveniences of a Haber-Weiss reaction (unless catalysed) a more feasible scheme is reduction by $O_2^{\overline{\cdot}}$ to the formate radical which then oxidises luminol

$$O_2^{\overline{\cdot}} + HCO_3^- \longrightarrow O_2 + HO^- + CO_2^{\overline{\cdot}}$$

$$CO_2^{\overline{\cdot}} + LH_2 \longrightarrow HCOO^- + LH^{\cdot}$$

It is to be noted that in the case of light emission from xanthine oxidase plus acetaldehyde alone, the intensity (i.e. Vmax) as well as total light emitted in a given time is proportional to the square of carbonate concentration, (Hodgson and Fridovich, 1976), whereas when luminol is present (system xanthine oxidase/hypoxanthine) light intensity is proportional to the square root of carbonate anion concentration (Puget and Michelson, 1976).

Thus induced chemiluminescence of luminol (or other reagents) cannot be taken as direct evidence of the simple presence of $O_2^{\overline{\cdot}}$. Apart from oxidation by the free radicals mentioned ($O_2^{\overline{\cdot}}$, $HO^{\cdot}$, $CO_3^{\overline{\cdot}}$, $CO_2^{\overline{\cdot}}$), a variety of reagents cause luminescent oxidation of luminol by simple electron transfer, and in particular electrolytic oxidation which presumably proceeds via

$$LH_2 \longrightarrow LH^{\cdot} + H^+ +$$

$$LH^{\cdot} + O_2 \longrightarrow LHOO^{\cdot}$$

$$LHOO^{\cdot} + LH_2 \longrightarrow LHOOH + LH^{\cdot}$$

and

$$LH^{\cdot} + O_2 \longrightarrow L + H^+ + O_2^{\overline{\cdot}}$$

$$LH^{\cdot} + O_2^{\overline{\cdot}} \longrightarrow LHOO^-$$

The last reaction is necessarily sensitive to SOD and inhibition by this enzyme has been demonstrated by Hodgson and Fridovich (1973) for oxidation of luminol by ferricyanide, hypochlorite and persulphate.

Although these techniques can be used with precaution to detect $O_2^{\overline{\cdot}}$, particularly when the effects of both native and denatured SOD are explored, as already pointed out by Hodgson and Fridovich (1975) superoxide dismutase can also catalyse peroxidation under certain conditions and would thus stimulate rather than inhibit. This last effect could also occur by inhibition of the loss of hydroxyl radicals (more oxidant than $O_2^{\overline{\cdot}}$) by the reaction

$$O_2^{\overline{\cdot}} + HO^{\cdot} \longrightarrow HO^- + O_2$$

Of particular interest is the effect of carbonate ions which merits further studies in biological systems. The presence of carbonate ions has already been shown to stimulate photohemolysis (at 254 nm) of erythrocytes as well as the riboflavin sensitized (365 nm) hemolysis (Michelson and Durosay, 1976).

REFERENCES

1. Arneson, R.M. (1970). *Arch. Biochem. Biophys. 136*, 352-360.
2. Bader, J.M. and Kuwana, I. (1965). *J. Electroanal. Chem. 10*, 104.
3. Dufraisse, C., Etienne, A. and Martel, J. (1957). *Compt. Rend. 244*, 970.
4. Goda, K., Kimura, T., Thayer, A.L., Kees, K. and Schaap, A.P. (1974). *Biochem. Biophys. Res. Comm. 58*, 660-666.
5. Hayaishi, T. and Maeda, K. (1962). *Chem. Soc. Japan 35*, 2057.
6. Henry, J.P. and Michelson, A.M. (1970). *Biochem. Biophys. Acta 205*, 451-458.
7. Henry, J.P., Isambert, M.F. and Michelson, A.M. (1970). *Biochem. Biophys. Acta 205*, 437-450.
8. Henry, J.P. and Michelson, A.M. (1973). *Biochimie 55*, 75-81.
9. Henry, J.P., Isambert, M.F. and Michelson, A.M. (1973). *Biochimie 55*, 83-93.
10. Henry, J.P., Monny, C. and Michelson, A.M. (1975). *Biochemistry 14*, 3458-3466.
11. Hodgson, E.K. and Fridovich, I. (1973). *Photochem. Photobiol. 18*, 451-455.
12. Hodgson, E.K. and Fridovich, I. (1975). *Biochemistry 14*, 5299.
13. Hodgson, E.K. and Fridovich, I. (1976). *Arch. Biochem. Biophys. 172*, 202-205.
14. Khan, A.U. and Kasha, M. (1964). *Nature 204*, 241-243.
15. Khan, A.U. and Kasha, M. (1966). *J. Am. Chem. Soc. 88*, 1574-1576.
16. Knowles, P.F., Gibson, J.F., Pick, F.M. and Bray, R.C. (1969). *Biochem. J. 111*, 53.
17. Legg, K.D. and Hercules, D.M. (1969). *J. Am. Chem. Soc. 91*, 1902.
18. Mayeda, E.A. and Bard, A.J. (1974). *J. Am. Chem. Soc. 96*, 4023-4024.
19. Michelson, A.M. (1973). *Biochimie 55*, 4651479.
20. Michelson, A.M. and Durosay, P. (1976). *Photochem. Photobiol.*, in press.
21. Nilsson, R. (1969). *Biochim. Biophys. Acta 184*, 237-251.
22. Puget, K. and Michelson, A.M. (1976). *Photochem. Photobiol.*, in press.
23. Schaap, A.P., Thayer, A.L., Faler, G.R. Goda, K. and Kimura, T. (1974). *J. Am. Chem. Soc. 96*, 4024-4025.
24. Schwarz, H.A. (1962). *J. Phys. Chem. 66*, 255.
25. Sligar, S.G., Lipscomb, J.D., Debrunner, P.G. and Gunsalus, I.C. (1974). *Biochem. Biophys. Res. Comm. 61*, 290-296.
26. Stauff, J., Schmidkunz, H. and Hartmann, G. (1963). *Nature 198*, 281-282.
27. Stauff, J. and Wolf, H. (1964). *Z. Naturforsch. 19B*, 87-96.
28. Stauff, J. and Bergmann, (1972). *Z. Physik. Chem. 78*, 263.
29. Stauff, J., Sander, U. and Jaeschke, W. (1973). In "Chemiluminescence and Bioluminescence" (Cormier, M.J., Hercules, D.M. and Lee, J. eds.) 131-140, Plenum Press, New York.
30. Thomas, J.K. (1965). *Trans. Faraday Soc. 61*, 702.
31. Vigny, P. and Duquesne, M. (1974). *Photochem. Photobiol. 20*, 15-25.
32. Vigny, P., Puget, K., Henry, J.P. and Michelson, A.M. (1976). Unpublished results.
33. Webb, L.S., Keele, B.B. and Johnston, R.B. (1974). *Infect. Immun. 9*, 1051-1056.
34. White, E.H. and Harding, M.J.C. (1964). *J. Am. Chem. Soc. 86*, 5685.
35. White, E.H. and Brundrett, R.B. (1973). In "Chemiluminescence and Bioluminescence", op. cit., 231-241.

THE INVOLVEMENT OF OXYGEN METABOLITES FROM PHAGOCYTIC CELLS IN BACTERICIDAL ACTIVITY AND INFLAMMATION

R.B. JOHNSTON, Jr. and J.E. LEHMEYER

Departments of Pediatrics and Microbiology
and the Comprehensive Cancer Center
The University of Alabama Medical Center
Birmingham, Alabama 35294, U.S.A.

The work of Karnovsky and co-workers in 1959 began what has developed into a large body of evidence indicating that phagocytizing neutrophils undergo a cyanide-insensitive "respiratory burst" consisting of oxygen consumption from the surrounding medium, generation of hydrogen peroxide (H_2O_2), activation of the hexose monophosphate (HMP) shunt, and reduction of nitroblue tetrazolium (NBT) dye (Karnovsky, 1968; Klebanoff, 1975a). The work of Klebanoff, supported by that of Sbarra and colleagues (1972) related this respiratory burst to the principal biologic function of these cells, the killing of microorganisms, by demonstrating that H_2O_2, peroxidase from azurophilic granules (myeloperoxidase, MPO) and halide ions constitute a potent microbicidal mechanism (Sbarra *et al.*, 1972; Klebanoff, 1975a). Exploration of oxidative metabolism in neutrophils from children with chronic granulomatous disease, in which the cells do not undergo a phagocytosis-associated burst of oxidative metabolism and cannot kill most ingested bacteria, led to a substantiation of the general thesis that the consumption and metabolism of oxygen by phagocytizing cells is essential for their bactericidal capacity.

When evidence began to accumulate through the work of Fridovich, with McCord and others, that superoxide anion ($O_2^{\bar{\cdot}}$) might be generated by cells undergoing aerobic metabolism (reviewed by Fridovich, 1972), the possibility arose that this oxygen by-product could play some part - at least indirectly - in oxygen-dependent microbicidal activity. Babior and colleagues subsequently showed that phagocytizing neutrophils elaborate $O_2^{\bar{\cdot}}$ (Babior *et al.*, 1973), and we presented evidence suggesting that neutrophil $O_2^{\bar{\cdot}}$ and H_2O_2 interact to form hydroxyl radical (·OH) and that this potent oxidant might be an important part of the bactericidal mechanism (Johnston *et al.*, 1973). The generation of luminescence by phagocytizing neutrophils led Allen *et al.* (1972) to hypothesize the formation during ingestion of a fourth oxygen metabolite, the excited singlet state ($^1O_2$). A possible role for this species or its derivative(s) in bacterial killing has been proposed (Allen *et al.*, 1972; Webb *et al.*, 1974; Johnston *et al.*, 1975; Rosen and Klebanoff, 1976).

We have attempted to summarize here work relating cellular generation of $O_2^{\bar{\cdot}}$, ·OH, and singlet oxygen to phagocytic bactericidal activity and chemiluminescence and, potentially, to the tissue damage of inflammation.

PHAGOCYTIC BACTERICIDAL ACTIVITY

Considering the additional reactivity inherent in the oxygen molecule after its reduction to $O_2^{\bar{\cdot}}$, we examined the possibility that this radical might play some role in the killing of phagocytized bacteria. A standard technique was employed in which viable bacteria are mixed with 2.5 x 10^6 neutrophils purified from human blood in a buffer mixture containing 10% human serum with complement activity preserved (Johnston and Baehner, 1970). Samples are taken with micropipets immediately upon addition of the leukocytes and at 30 min intervals for 120 min thereafter. Dilutions are made in distilled water to lyse leukocytes, and pour plates are made with agar for the counting of colonies the next day.

On adding purified bovine copper-zinc superoxide dismutase (SOD) in concentrations of 25 to 200 μg/ml, there was a slight inhibition of phagocytic bactericidal activity, as shown for one experiment in Fig.1A. The results were not striking, but with each attempt to increase the effect by modifying the concentration of enzyme (30 - 200 μg/ml, 1 - 6 x 10^{-6} M), the bacteria used (*S. aureus*, *S. viridans*, and *E. coli*), or the bacteria to phagocyte ratio, some inhibition was always observed. This led us to suspect that the inhibitory effect was real but that we were limited by our ability to introduce the SOD into the phagocytic vacuole where this effect might be most demonstrable. Accordingly, we attempted to bind the enzyme to latex particles which might be taken into the phagocytic vacuole along with the ingested bacteria. (Adherence of enzyme activity to the particles, even after a saline wash, could be demonstrated.) The inhibitory effects of SOD with added latex were much more striking (Fig.1B), and existed with all three bacteria employed (Johnston *et al.*, 1975). Electron micrographs confirmed the fact that latex and bacteria appeared very commonly within the same phagocytic vacuole. There was no significant inhibition by equal concentrations of bovine serum albumin (BSA), heat-denatured SOD, or apoenzyme. The uptake of radio-labeled bacteria in the same reaction mixture was not different in the presence of SOD, BSA, or inactive SOD.

Thus it appeared that $O_2^{\bar{\cdot}}$ plays some role in the killing of ingested bacteria. What, then, about the large body of convincing published data implicating H_2O_2 in this process? The role of H_2O_2 in our system was explored using catalase in the same manner in which SOD had been employed. There was slight but consistent inhibition of phagocytic killing by catalase alone (Fig.2A), which was much improved when the catalase was bound to latex particles (Fig.2B). Again, ingestion was not different with catalase than with BSA, which was used as control since heat-denatured catalase aggregated markedly; the effect was the same with *E. coli*, *S. aureus* and *S. viridans*.

The apparent requirement for both $O_2^{\bar{\cdot}}$ and H_2O_2 made attractive the possibility that these species were interacting to generate the potent oxidant ·OH. The precedent for such a possibility had been established by Haber and Weiss (1934), who hypothesized the reaction $O_2^{\bar{\cdot}} + H_2O_2 \rightarrow \cdot OH + OH^- + O_2$. Accordingly, benzoate, a known scavenger of ·OH, was tested for its ability to inhibit phagocytic killing of the three test bacteria, and definite inhibition was noted (Fig.3). Results were not improved by the presence of latex particles, presumably because the benzoate was not well bound to the latex;

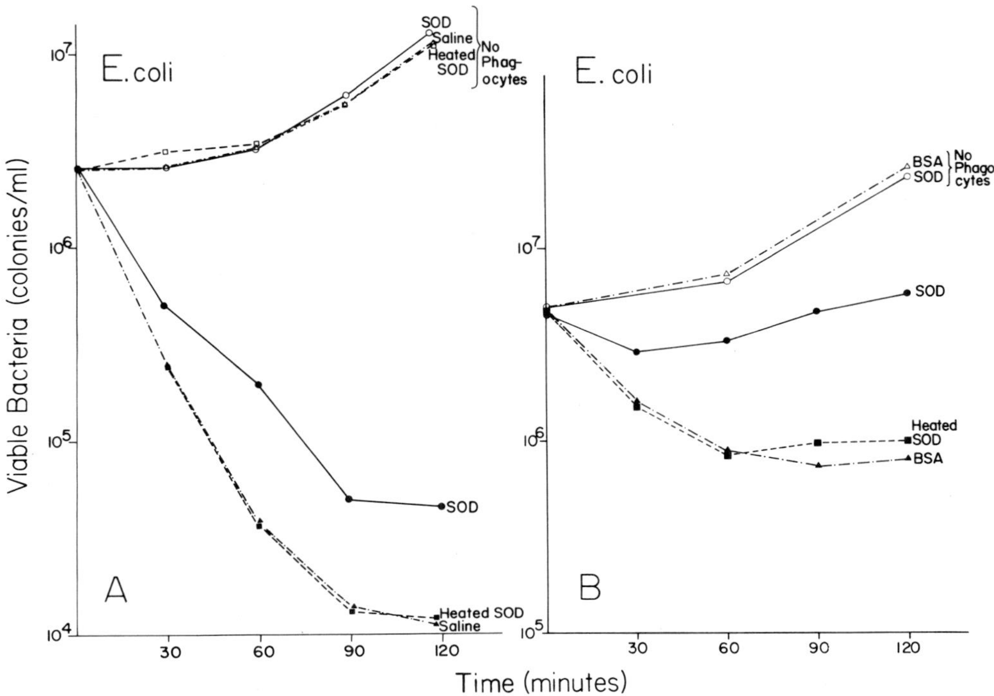

Fig.1. *Inhibition of phagocytic bactericidal activity by SOD in the absence (A) and presence (B) of latex particles. The number of viable bacteria in the 1-ml reaction mixture, determined as colony-forming units, is plotted as a function of the time at which the mixture was sampled. In A, the enzyme was present at a concentration of 100 μg/ml; in B, at 200 μg/ml. Saline or an equal concentration of heat-denatured SOD or BSA were used as controls. 2.5 x 10⁶ phagocytes were employed in all bactericidal assays. Illustrated experiments were selected as representative by rank ordering differences in SOD and control values from 12 experiments without latex and 23 with latex using* E. coli, S. aureus, *or* S. viridans. *(Modified from Johnston* et al., *1975, with permission of the publishers.)*

particles were therefore omitted from these experiments. The results achieved compare favorably with those obtained when SOD or catalase was used in the absence of latex. Similar inhibition was obtained with mannitol (Johnston *et al.*, 1975). Ethanol was not useful as an ·OH scavenger in that it affected ingestion slightly at the point at which it inhibited bacterial killing.

The inhibition of phagocytic bactericidal activity by catalase was reported in 1967 (McRipley and Sbarra; Klebanoff) and more recently (Mandell, 1975). Others have confirmed the inhibition of phagocytic bactericidal activity by SOD (Sagone *et al.*, 1976; Salin and McCord, 1977). Evidence for the killing of bacteria by ·OH in a photochemical $O_2^{\bar{\cdot}}$ generating system has been presented (Gregory and Fridovich, 1974), and it has been shown that bacteria rich in periplasmic, iron-containing SOD and cytoplasmic catalase are more resistant to the killing effects of whole blood (Yost and Fridovich, 1974). On the basis of these studies and our results, we feel it likely - but not conclusively proven - that hydroxyl radical is at least one of the means by which phagocytic cells kill their captured prey.

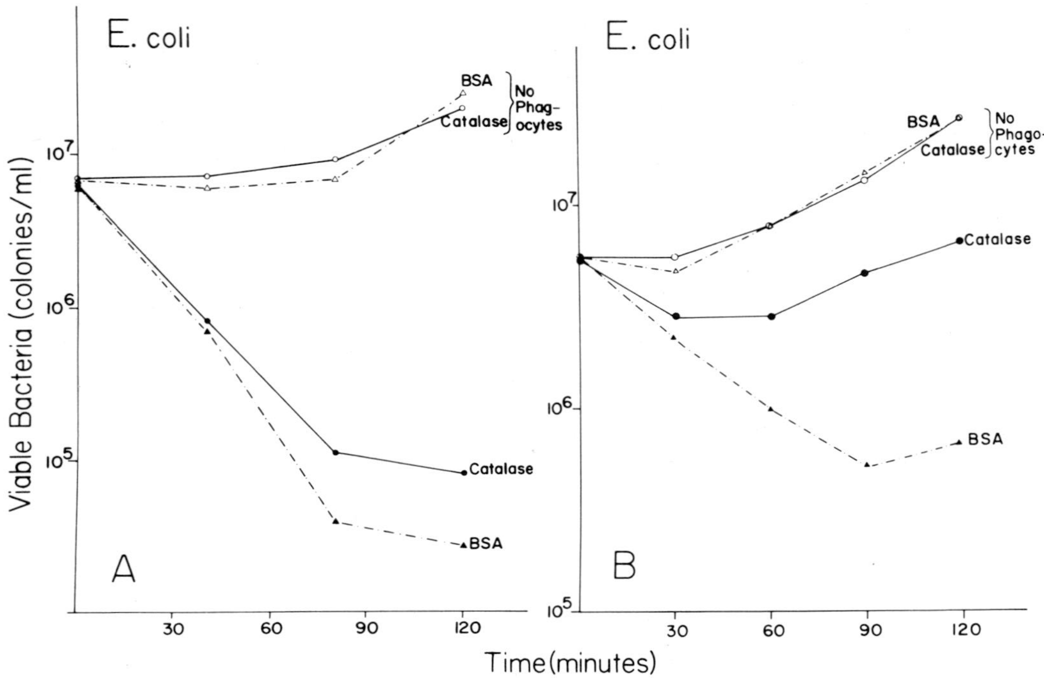

Fig.2. *Inhibition of phagocytic bactericidal activity by catalase, 200 μg/ml, in the absence (A) and presence (B) of latex particles. An equal concentration of BSA was used as control. These results are representative of the effect achieved in seven experiments without and 15 with latex using* E. coli*,* S. aureus*, or* S. viridans.

PHAGOCYTOSIS-ASSOCIATED OXIDATIVE METABOLISM

In interpreting the effects of SOD on phagocytic bactericidal activity, it seemed important to determine the influence this enzyme might have on the phagocytosis-associated respiratory burst. Accordingly, oxidative metabolism of human neutrophils was studied during ingestion of latex particles coated with SOD or an equal concentration of BSA, apo SOD or other proteins as controls (Baehner *et al.*, 1975). The rate and extent of uptake of particles was the same whether they were coated with SOD or BSA. Results are summarized in Table I.

Increased concentrations of SOD would be expected to drive the $O_2^{\overline{\cdot}}$ dismutation reaction to the right, with the generation of H_2O_2 and the release of 1 mole of oxygen for every 2 moles of oxygen consumed. The observed increase in the *initial rate* of oxygen consumption is compatible with this stoichiometry, assuming that H_2O_2 is not immediately returned to H_2O and $\frac{1}{2}O_2$ in the system. The technique used here to measure H_2O_2 production accounts for only 3 - 15% of the consumed oxygen as H_2O_2, which is a gross underestimate according to more recently developed methods (Roos *et al.*, 1977). A significant increase in H_2O_2 production in the presence of SOD was shown, however. SOD induced a marked activation of the HMP shunt, presumably through activation of the glutathione peroxidase pathway by increased amounts of H_2O_2. The elimination of HMP shunt activation by the further addition of catalase supports this interpretation. These results favor the interaction of the glutathione peroxidase pathway with H_2O_2 as an explanation for shunt activation during phagocytosis and as a means of protecting the cell against excess of H_2O_2.

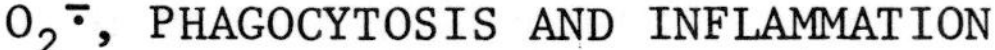

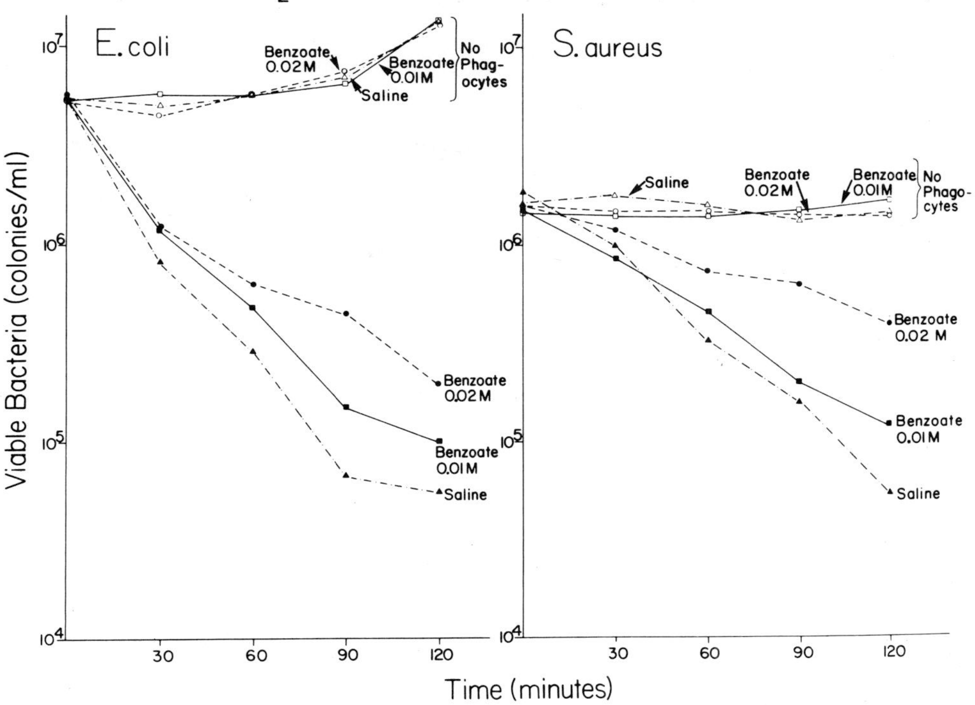

Fig.3. *Inhibition of phagocytic bactericidal activity by sodium benzoate compared to 0.15 M NaCl. Latex particles were not present. The results that are plotted are representative of those achieved in six experiments with* E. coli, *three with staphylococci, and one with streptococci. The presence of latex particles in several of these experiments had no influence on the inhibition achieved by benzoate. The osmolalities of the final reaction mixtures containing the saline and benzoate were approximately equal.*

TABLE I

Effects of Superoxide Dismutase on Phagocytosis-Associated Oxidative Metabolism

Metabolic Activity	Effect of SOD[a]
Oxygen consumption	69% increase
H_2O_2 production (^{14}C-formate oxidation)	38% increase
HMP shunt activation (1-^{14}C-glucose oxidation)	3-fold increase[b]
Iodination of ingested particles	2-fold increase
NBT reduction	62% decrease

a) Results represent a comparison of activity during ingestion of latex coated with SOD, 33 - 100 μg/ml (1 - 3 μM), or an equal protein concentration of BSA, apo SOD, hemoglobin or thrombin. The means of eight experiments performed in duplicate are shown, except for NBT reduction which represents five experiments. All of the effects are statistically significant ($P < 0.001$, *t* test, for each). (Baehner *et al.*, 1975).
b) This increase was removed by the further addition of catalase.

Iodination of ingested zymosan particles was increased two-fold in the presence of SOD, confirming the dependence of this activity on H_2O_2 but - in view of the inhibitory effect of SOD on pagocytic bactericidal activity - separating this phenomenon from bacterial killing. The decrease in NBT reduction by SOD implicates $O_2^{\overline{\cdot}}$ as an important source of this manifestation of the respiratory burst.

PHAGOCYTOSIS-ASSOCIATED CHEMILUMINESCENCE

The possibility that oxygen consumed and further changed during phagocytosis might assume additional molecular forms has been raised by the observation that ingestion is associated with a burst of luminescence (Allen *et al.*, 1972). The reaction occurs in an aqueous medium, can be measured in a liquid scintillation spectrometer in the tritium region, and needs no amplification by fluorescing materials or "antennae" such as luminol (Allen *et al.*, 1972; Webb *et al.*, 1974; Johnston *et al.*, 1975).

The involvement of $O_2^{\overline{\cdot}}$ in this luminescence, at least as an initial carrier of energy, is suggested by the pronounced inhibition of luminescence by SOD, as demonstrated in Fig.4 with human neutrophils (Webb *et al.*, 1974).

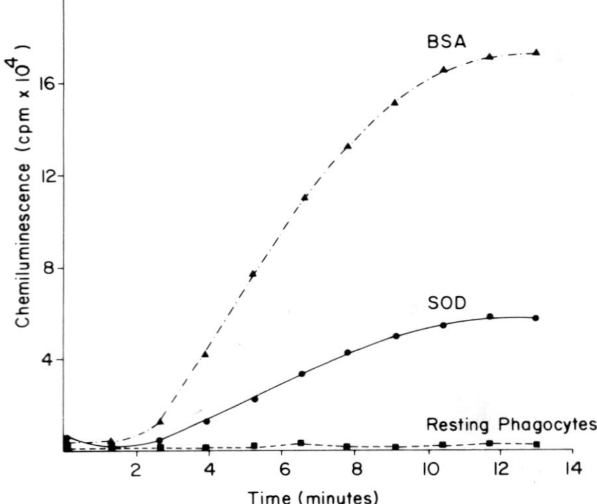

Fig.4. *Inhibition of the rate and extent of phagocytosis-associated chemiluminescence, recorded in counts per min (cpm), by SOD at a concentration of 200 μg/ml.*

As shown here, the technique permits kinetic analysis, and inhibition by SOD of both rate and peak luminescence to about 25% of control is evident. Similar inhibition by SOD has been shown with human monocytes (Johnston *et al.*, 1976). Much less - but consistent - inhibition is achieved by catalase or benzoate (Fig.5), suggesting that H_2O_2 or ·OH or both might be involved in this phenomenon (Webb *et al.*, 1974). The reaction is oxygen-dependent in that bubbling nitrogen into the cell mixture before adding particles to begin the reaction decreased the extent of chemiluminescence by a mean of 55% from that achieved by a control into which air was bubbled, whereas, bubbling oxygen into the cell mixture increased the peak value by 10% ($196 \pm 58 \times 10^3$ cpm, mean ± SD, for air; $89 \pm 38 \times 10^3$ cpm for nitrogen; $215 \pm 41 \times 10^3$ cpm for oxygen; n = 4).

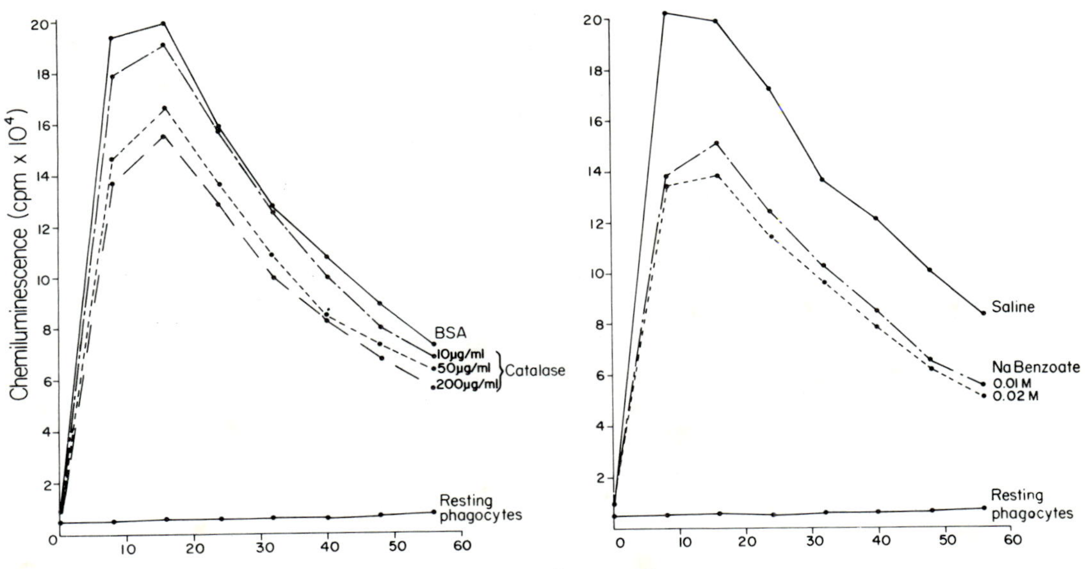

Fig.5. *Inhibition of phagocytic chemiluminescence by catalase (left) and benzoate (right). The BSA concentration was 200 μg/ml. Conductivity of the benzoate solutions was made equal to that of the saline control by the addition of NaCl. Inhibition by catalase was 27% at a concentration of 200 μg/ml; inhibition by benzoate was 31% at 0.02 M.*

The involvement of the enzyme MPO in phagocytic chemiluminescence has been indicated by the recent work of Rosen and Klebanoff (1976), who showed a reduction in peak chemiluminescence by MPO-deficient cells of about 50%. The luminescence of normal neutrophils was inhibited to a similar extent by azide, which inhibits MPO. In addition, the chemical system MPO + H_2O_2 + Cl^- generates chemiluminescence (Allen, 1975; Rosen and Klebanoff, 1976), and this can be inhibited by azide (Rosen and Klebanoff, 1976). (The direct interaction of azide with $^1O_2$ (Hasty *et al.*, 1972) must be considered in interpreting these experiments, but the fact that azide only slightly inhibited the luminescence of MPO-deficient cells (Rosen and Klebanoff, 1976) argues against the significance of a direct quenching, at least in the phagocytosis system.)

The evidence for involvement of $O_2^{\bar{\cdot}}$, MPO, and H_2O_2 in phagocytosis-associated chemiluminescence and the possible mechanisms by which the luminescence might be generated are summarized in Table II. Since about three-fourths of the chemiluminescence can be eliminated by SOD, it seems reasonable to implicate $O_2^{\bar{\cdot}}$ in the phenomenon. In considering the luminescence of a chemical system, Greenlee *et al.* (1962) suggested that direct interaction between $O_2^{\bar{\cdot}}$ and an excitable substrate might occur to form an excited intermediate which would decay and emit light. Such a mechanism seems plausible in phagocytic cells, perhaps through the generation of epoxides, for example (Hamman and Seliger, 1976; Cormier, 1977). The presence of multiple broad bands of absorbance on spectral analysis of phagocytosis-associated luminescence (Babior, 1977) agrees with this concept.

The generation of oxygen in the excited singlet state through spontaneous dismutation of $O_2^{\bar{\cdot}}$, originally proposed as the event responsible for phagocytosis-associated chemiluminescence (Allen *et al.*, 1972; Webb *et al.*, 1974), now appears less likely, based on the following observations: (i) Oxidation of diphenylfuran to dibenzoylethylene, believed to be specific for $^1O_2$, was not shown by an $O_2^{\bar{\cdot}}$-generating system (Nilsson and Kearns, 1974; King *et al.*, 1975); and (ii) Luminescence by the $O_2^{\bar{\cdot}}$-generating xanthine-xanthine oxidase system appears to require the presence of carbonate ions, and the reaction shown in Table II, A2b has been proposed as that responsible for the emission of light by this sytem (Hodgson and Fridovich, 1976). The relatively weak inhibition of phagocytic chemiluminescence by catalase and benzoate does not favor this latter mechanism as a major source of luminescence in phagocytic cells but does support its having some role in this process. Kellogg and Fridovich (1975) have published findings which suggest that $^1O_2$ may be a direct product of the Haber-Weiss reaction (Table II, A2c), and thermodynamic considerations predict that such a reaction is indeed feasible ($\Delta G_0' = -3.5$ kcal) (Koppenol, 1976).

Singlet oxygen might still explain the luminescence of the MPO system, since the oxidation of diphenylfuran to dibenzoylethylene can be shown with this system using chemical reagents (Rosen and Klebanoff, 1976). The system could also act directly on excitable substrate(s).

A role for $^1O_2$ in the killing of phagocytized bacteria has been suggested by the finding that *Sarcina lutea* containing carotenoid, a scavenger of $^1O_2$, were protected against phagocytic bactericidal activity compared to a colorless mutant (Krinsky, 1974). In addition, the $^1O_2$ quencher, DABCO {1,4-diazobycyclo-(2,2,2,) octane}, has been reported to inhibit the antimicrobial effect of purified MPO, H_2O_2 and halide (Klebanoff, 1975b).

OXYGEN METABOLITES AND PHAGOCYTIC BACTERICIDAL ACTIVITY: SYNTHESIS AND HYPOTHESIS

A schematic summary of chemical reactions involved in the conversion of molecular oxygen to metabolites capable of killing phagocytized microorganisms, based on the observations cited above, is shown in Fig.6.

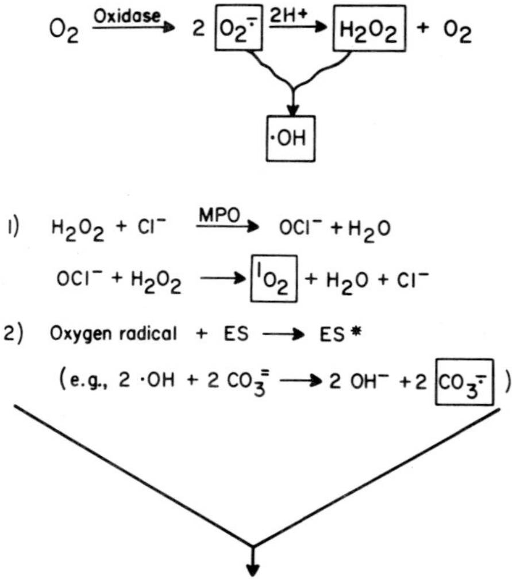

Fig.6. *Summary of the oxidative metabolic reactions that may be involved in the killing of ingested microorganisms by phagocytic cells.*

TABLE II

Phagocytosis-Associated Chemiluminescence

A. DEPENDENT ON $O_2^{\bar{\cdot}}$

1. Evidence

(a) Inhibition (75%) of phagocytic chemiluminescence by SOD

(b) Inhibition of luminescence from xanthine oxidase-xanthine reaction by SOD

2. Possible Mechanisms

(a) Direct effect of $O_2^{\bar{\cdot}}$ on excitable substrate(s):

$$O_2^{\bar{\cdot}} + ES \longrightarrow ES^* + \text{products}$$

(b) Generation of excited carbonate radicals:

$$O_2^{\bar{\cdot}} + H_2O_2 \longrightarrow OH^- + O_2 + \cdot OH$$

$$\cdot OH + CO_3^- \longrightarrow OH^- + CO_3^{\bar{\cdot}}$$

(c) Generation of $^1O_2$ from Haber-Weiss reaction:

$$O_2^{\bar{\cdot}} + H_2O_2 \longrightarrow OH^- + \cdot OH + {}^1O_2$$

(Chemiluminescence from $^1O_2$ relaxation or $^1O_2$ excitation of ES, or both)

B. DEPENDENT ON MPO OR H_2O_2 OR BOTH

1. Evidence

(a) Deficiency of phagocytic chemiluminescence in MPO-deficient PMNs

(b) Inhibition of phagocytic chemiluminescence by azide, which inhibits MPO

(c) Generation of luminescence by MPO + H_2O_2 + Cl^-, and inhibition of this by azide

(d) Inhibition (25%) of phagocytic chemiluminescence by catalase

2. Possible Mechanisms

(a) Generation of $^1O_2$:

$$H_2O_2 + Cl^- \xrightarrow{MPO} H_2O + OCl^-$$

$$OCl^- + H_2O_2 \longrightarrow H_2O + Cl^- + {}^1O_2$$

(b) Direct oxidation of ES:

$$H_2O_2 + ES \xrightarrow{MPO} ES^* + \text{products}$$

Species capable of inflicting damage on ingested microbes are enclosed for emphasis. The "oxidase" shown here is believed to use NADPH or NADH, or both, as electron donor (Karnovsky, 1973; Patriarca *et al.*, 1974).

A summary of how these reactions might occur in a phagocytic cell is shown in Fig.7. The diagram is based upon current understanding. However, several of its aspects are unproven, and it is presented primarily as a theoretical model, open to modification.

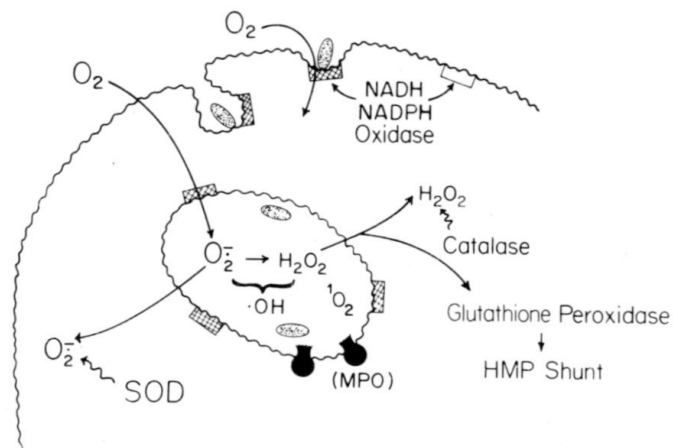

Fig.7. *Schematic diagram, in part hypothetical, of the conversion of oxygen by a phagocytic cell to microbicidal species. The stippled spheres represent bacteria, and the solid circles represent granules rupturing into the phagocytic vacuole.*

Activity begins upon contact of the microorganism, after its reaction with serum antibody and complement, and the cell plasma membrane (Briggs *et al.*, 1975; reviewed in Johnston and Lehmeyer, 1976). The physical placement of the oxidative, "respiratory burst" enzyme at the plasma membrane is currently hypothetical. Wherever this enzyme is actually located, such contact or similar surface perturbation certainly activates it to catalyze the reduction of oxygen to $O_2^{\bar{\cdot}}$. An enzyme associated with plasma membrane would eventually be similarly associated with the phagocytic vacuole and, thus, located between the cell's cytoplasm and captured microorganisms. In this location the enzymatic generation of toxic species would be ideally placed to permit their diffusion into and concentration in the vacuole. Although $O_2^{\bar{\cdot}}$ and H_2O_2 might also diffuse into the cell, exposing intracellular constituents to possible damage, the protective enzymes SOD, catalase, and glutathione peroxidase are present in the cytoplasm. These proteins should not enter the phagocytic vacuole (this has been proven only with catalase) and thus would protect the phagocyte but not its ingested prey. Lysosomal degranulation would introduce MPO into the vacuole, which would enhance the overall oxidative effectiveness of the oxygen metabolites by the mechanisms described above or, perhaps, by others not yet understood. Cationic proteins and hydrolytic enzymes from granules could also contribute to the killing and, certainly digestion of phagocytized microorganisms.

In summary, what should be emphasized here is that reactivity bestowed enzymatically on oxygen, beginning with its reduction to $O_2^{\bar{\cdot}}$ is essential for optimal phagocytic microbicidal activity.

PHAGOCYTIC OXYGEN METABOLITES AND INFLAMMATION

A common feature of inflammation is the local accumulation of polymorphonuclear phagocytes. When the inflammation is due to infection, the phagocytes play an essential role in body defense. These cells could also cause a major part of the tissue injury that occurs with inflammation, and dependence upon phagocytes for development of the inflammatory manifestations of several diseases has been described (reviewed in Johnston and Lehmeyer, 1976). The occurrence of neutrophil lysosomal enzymes and cationic proteins in the urine of animals with experimental nephritis suggests that phagocyte-mediated tissue injury may result from lysosomal degranulation (Hawkins and Cochrane, 1968). However, electron micrographs of the neutrophils that line the site of renal injury have not shown granular depletion (Cochrane *et al.*, 1965), and the possibility exists that other phagocyte factors may play a role in inflammatory tissue damage.

When inflammation has an immunologic origin (as it does, for example in so-called autoimmune diseases like rheumatoid arthritis), antigen, antibody, and complement components, as well as phagocytes, can be demonstrated at the inflammatory site. The *in vitro* model for studying immunologic tissue injury that has been widely employed utilizes immune complexes or aggregated immunoglobulin fixed to tissue such as cartilage or collagen or to the surface of a micropore filter (Henson, 1971; reviewed in Johnston and Lehmeyer, 1976). In this model phagocytes adhere to the noningestable surface and discharge their lysosomal contents at or near the site of attachment. We have used this model, schematized in Fig.8, to explore the possibility that contact of the cell surface with "immune" particles might also induce the release of toxic oxygen species into the surrounding environment.

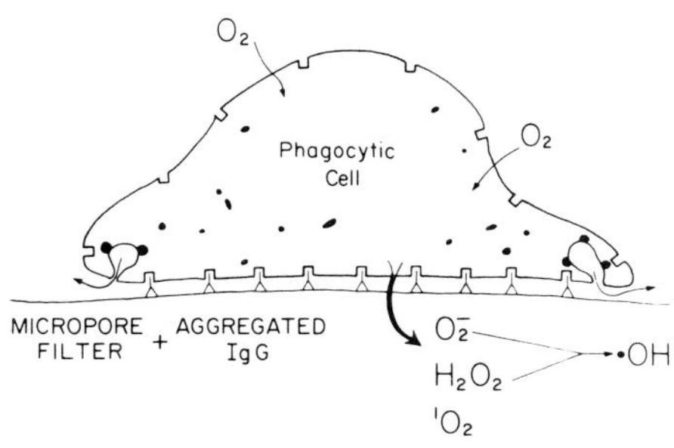

Fig.8. *Schematic diagram of* in vitro *model used to study the role of phagocytic cells in the pathogenesis of immune complex disease. The cell is stimulated by contact of its plasma membrane with fixed aggregated IgG to undergo degranulation and to release toxic oxygen metabolites.*

In the model, purified IgG that has been aggregated by gentle heating is attached to micropore filters by incubation at 37°C. The filters are washed vigorously, and the IgG that remains with the filter is tightly bound. Neutrophils, monocytes or lymphocytes, purified from human blood by centrifugation through a Ficoll-Hypaque mixture, were incubated with the IgG-coated filters in petri dishes at 37°C for 60 min for $O_2^{\cdot-}$ or H_2O_2 determinations, or in scintillation vials for chemiluminescence. $O_2^{\cdot-}$ was determined by cytochrome *c* reduction and H_2O_2 by oxidation of scopoletin (Johnston and Lehmeyer, 1976; Johnston *et al.*, 1976).

The nonphagocytosable surface induced a vigorous release of $O_2^{\overline{\cdot}}$ from both neutrophils and monocytes, as shown in Fig.9. Lymphocytes had no activity in the assay. Chemiluminescence was also induced in both neutrophils and monocytes (Fig.10), although the difference in the activities of the two cell types was considerably greater in this system. The generation of H_2O_2 was studied only in neutrophils: in six experiments done in triplicate, mean H_2O_2 generation was 29.8 nmol/10^6 cells over the 60 min of incubation, greater than a 10-fold rise from levels achieved in the absence of filters.

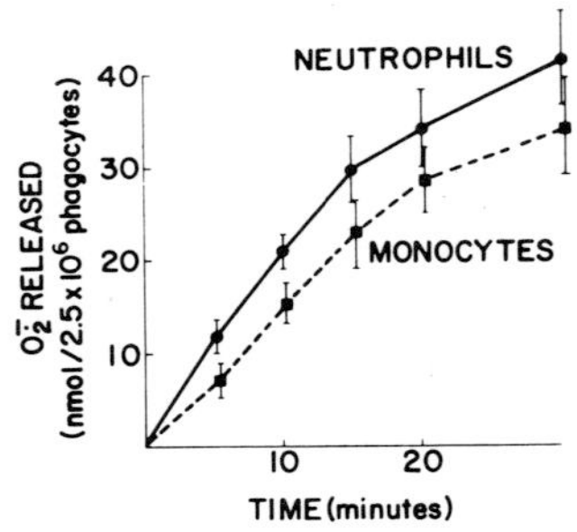

Fig.9. *Kinetics of $O_2^{\overline{\cdot}}$ release from phagocytic cells in contact with IgG-coated micropore filters. 2.5 x 10^6 phagocytes were present in the reaction mixtures. Mean values are plotted for four experiments comparing neutrophils and monocytes from the same individuals; the bars indicate the SEM.*

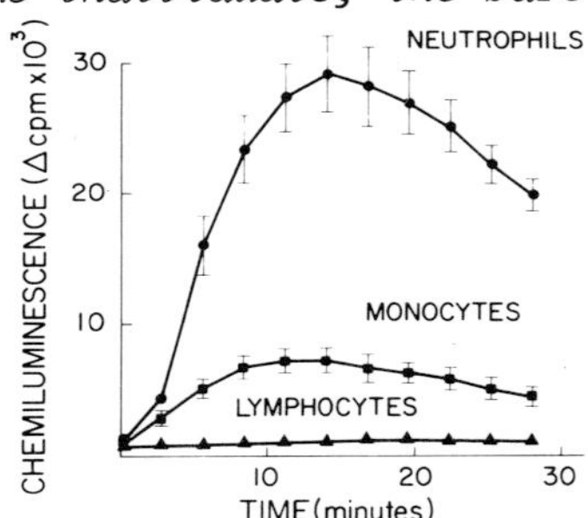

Fig.10. *Generation of chemiluminescence by 5 x 10^6 neutrophils, monocytes or lymphocytes in contact with aggregated IgG bound to micropore filters. Means of averaged duplicate values ± the SEM are plotted for seven experiments. A value of 2000 cpm, representing the mean background for empty vials, has been subtracted. In the absence of IgG-coated filters, values for neutrophils, monocytes, or lymphocytes were not different from those shown for lymphocytes in contact with the filter-IgG. (From Johnston* et al.*, 1976, with permission of the publishers.)*

Neutrophils from three patients with chronic granulomatous disease incubated with IgG-coated filters released less than 1 nmol of $O_2^{\overline{\cdot}}/2.5 \times 10^6$ cells in 60 min, and monocytes from two of these generated similarly insignificant amounts of $O_2^{\overline{\cdot}}$. Neutrophils from two patients generated no significant chemiluminescence; and neutrophils from one generated <3% of the control value of H_2O_2.

Contact of neutrophils or monocytes with the surface-active agent phorbol myristate acetate (PMA), the toxic ingredient of croton oil, also stimulates chemiluminescence and the vigorous release of $O_2^{\overline{\cdot}}$ (DeChatelet *et al.*, 1976;

Johnston *et al.*, 1976). The kinetics of $O_2^{\overline{\cdot}}$ generation by neutrophils and monocytes stimulated by contact with PMA are compared to those on contact with and phagocytosis of opsonized zymosan in Fig.11. The rate of $O_2^{\overline{\cdot}}$ release with PMA was considerably faster than that with zymosan phagocytosis, although there was a consistent lag period on stimulation of either cell type by PMA that could be influenced by PMA concentration. Such a pronounced delay in achieving optimal rates did not exist with opsonized zymosan. The significantly slower rate of $O_2^{\overline{\cdot}}$ generation by monocytes is apparent with either stimulus. At higher concentrations of PMA (330 ng/ml), however, the rates with monocytes were approximately 69% those of neutrophils (mean of two experiments, each comparing results with 2.5 x 10^6 and 5.0 x 10^6 cells).

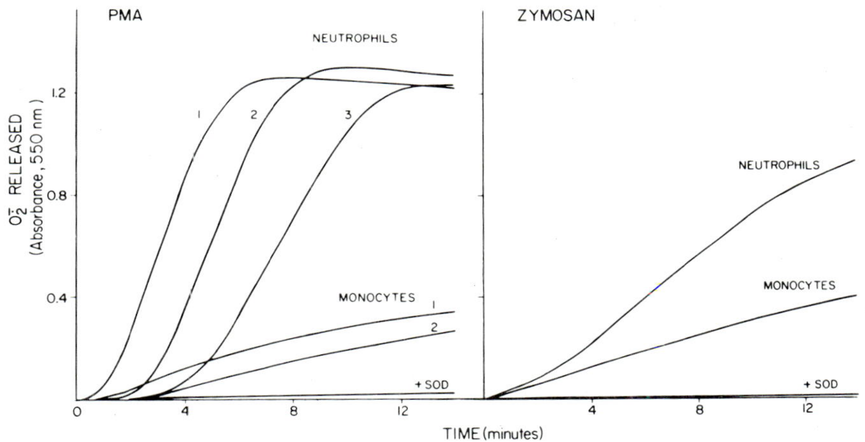

Fig.11. *Kinetics of $O_2^{\overline{\cdot}}$ generation by 2.5 x 10^6 phagocytic cells on contact with PMA at concentrations of 67 (1), 33 (2), and 17 (3) ng/ml (left panel) or with opsonized zymosan (right). Cells and 80 μM cytochrome* c *were prewarmed to 37°C, and the reaction was begun (time 0) by adding the stimulant. Reaction vol was 1.5 ml. SOD, 33 μg/ml, eliminated cytochrome* c *reduction by either cell type (plotted as "+ SOD" here with 67 ng/ml PMA and neutrophils in the left panel, and zymosan and neutrophils in the right panel). Actual tracings of the recording spectrophotometer are represented after superimposition at the same starting point for comparison. The illustrated experiment is representative of three done in duplicate.*

It appears, therefore, that phagocytes capable of undergoing normal oxidative metabolism can be stimulated to do so by surface perturbation, including contact with fixed immune aggregates. If this model for immune complex disease is a valid one, then toxic oxygen metabolites released from neutrophils and monocytes must be considered, along with phagocyte lysosomal hydrolases, as potential mediators of the tissue damage of inflammation. In support of this concept, it has been shown that SOD or catalase can prevent depolymerization of synovial fluid by ·OH derived from $O_2^{\overline{\cdot}}$ and H_2O_2 (McCord, 1974), that injected SOD ameliorates inflammation (Menander-Huber and Huber, 1977) and that neutrophils from patients receiving anti-inflammatory corticosteroid therapy may have decreased capacity to reduce NBT dye (Chretien and Garagusi, 1972), a reaction that depends upon the conversion of oxygen to $O_2^{\overline{\cdot}}$. It should be pointed out, on the other hand, that phagocytic cells accumulate around tumors (Evans, 1975), and these same mechanisms might contribute to the capacity of phagocytes to participate in the rejection of cancer.

CONCLUSION

Among mammalian cells, blood neutrophils and monocytes are particularly adept at converting molecular oxygen to highly reactive metabolites. We have tried to summarize here evidence that this conversion is essential to the principal biological function of these cells, defense against infection. As a byproduct of this capability, however, toxic species may be spilled into the surrounding tissues, and this has important implications for the pathogenesis of certain diseases. The potential significance of information gained regarding the mechanism by which these cells convert oxygen, the location and products of this mechanism, and the effects of these products on microorganisms and tissues should be apparent.

ACKNOWLEDGEMENTS

The helpful comments of Dr. S.J. Klebanoff regarding chemiluminescence and the assistance of Ms. Johanna Griffin with Fig.8 are gratefully acknowledged. This work was supported by USPH grants AI 10286, CA 13148, and CA 16673 from the National Institutes of Health. The paper was written while R.B.J. was a Josiah Macy, Jr. Foundation Faculty Scholar at Rockefeller University, New York.

REFERENCES

1. Allen, R.C. (1975). *Biochem. Biophys. Res. Commun. 63,* 675-683.
2. Allen, R.C., Stjernholm, R.L. and Steele, R.H. (1972). *Biochem. Biophys. Res. Commun. 47,* 679-684.
3. Babior, B.M. (1977). This volume.
4. Babior, B.M. Kipnes, R.S. and Curnutte, J.T. (1973). *J. Clin. Invest. 52,* 741-744.
5. Baehner, R.L., Murrmann, S.K., Davis, J. and Johnston, R.B. Jr. (1975). *J. Clin. Invest. 56,* 571-576.
6. Briggs, R.T., Karnovsky, M.L. and Karnovsky, M.J. (1975). *J. Cell Biol. 64,* 254-260.
7. Chretien, J.H. and Garagusi, V.F. (1972). *J. Reticuloendothel. Soc. 11,* 358-367.
8. Cochrane, C.G., Unanue, E.R. and Dixon, F.J. (1965). *J. Exptl. Med. 122,* 99-116.
9. Cormier, M.J. (1977). This volume.
10. De Chatelet, L.R., Shirley, P.S. and Johnston, R.B. Jr. (1976). *Blood 47,* 545-554.
11. Evans, R. (1975). *Ann. N.Y. Acad. Sci. 256,* 275-287.
12. Fridovich, I. (1972). *Accounts Chem. Res. 5,* 321-326.
13. Greenlee, L., Fridovich, I. and Handler, P. (1962). *Biochemistry 1,* 779-783.
14. Gregory, E.M. and Fridovich, I. (1974). *J. Bacteriol. 117,* 166-169.
15. Haber, F. and Weiss, J. (1934). *Proc. Roy. Soc. Lond. A. Math. Phys. Sci. 147,* 332-351.
16. Hamman, J.P. and Seliger, H.H. (1976). *Biochem. Biophys. Res. Commun. 70,* 675-680.
17. Hasty, N., Merkel, P.B., Radlick, P. and Kearns, D.R. (1972). *Tetrahed. Lett. 1,* 49-52.
18. Hawkins, D. and Cochrane, C.G. (1968). *Immunology 14,* 665-681.
19. Henson, P.M. (1971). *J. Immunol. 107,* 1535-1546.

20. Hodgson, E.K. and Fridovich, I. (1976). *Arch. Biochem. Biophys. 172,* 202-205.
21. Johnston, R.B. Jr. and Baehner, R.L. (1970). *Blood 35,* 350-355.
22. Johnston, R.B. Jr., Keele, B., Webb, L., Kessler, D. and Rajagopalan, K.V. (1973). *J. Clin. Invest. 52,* 44a.
23. Johnston, R.B. Jr., Keele, B.B. Jr., Misra, H.P., Lehmeyer, J.E., Webb, L.S., Baehner, R.L. and Rajagopalan, K.V. (1975). *J. Clin. Invest. 55,* 1357-1372.
24. Johnston, R.B. Jr. and Lehmeyer, J.E. (1976). *J. Clin. Invest. 57,* 836-841.
25. Johnston, R.B. Jr., Lehmeyer, J.E. and Guthrie, L.A. (1976). *J. Exptl. Med. 143,* 1551-1556.
26. Karnovsky, M.L. (1968). *Semin. Hematol. 5,* 156-165.
27. Karnovsky, M.L. (1973). *Fed. Proc. 32,* 1527-1533.
28. Kellogg, E.W. III and Fridovich, I. (1975). *J. Biol. Chem. 250,* 8812-8817.
29. King, M.M., Lai, E.K. and McCay, P.B. (1975). *J. Biol. Chem. 250,* 6496-6502.
30. Klebanoff, S.J. (1967). *J. Exptl. Med. 126,* 1063-1078.
31. Klebanoff, S.J. (1975a). *Semin. Hematol. 12,* 117-142.
32. Klebanoff, S.J. (1975b). In "The Phagocytic Cell in Host Resistance" (J.A. Bellanti and D.H. Dayton, eds.) pp 45-56, Raven Press, New York.
33. Koppenol, W.H. (1976). *Nature 262,* 420-421.
34. Krinsky, N.I. (1974). *Science* (Wash. D.C.) *186,* 363-365.
35. Mandell, G.L. (1975). *J. Clin. Invest. 55,* 561-566.
36. McCord, J.M. (1974). *Science* (Wash. D.C.) *185,* 529-531.
37. McRipley, R.J. and Sbarra, A.J. (1967). *J. Bacteriol. 94,* 1417-1424.
38. Menander-Huber, K.B. and Huber, W. (1977). This volume.
39. Nilsson, R. and Kearns, D.R. (1974). *J. Phys. Chem. 78,* 1681-1683.
40. Patriarca, P., Basford, R.E., Cramer, R., Dri, P. and Rossi, F. (1974). *Biochim. Biophys. Acta 362,* 221-232.
41. Roos, D., Van Schaik, M.L.J., Weening, R.S. and Wever, R. (1977). This volume.
42. Rosen, H. and Klebanoff, S.J. (1976). *J. Clin. Invest. 58,* 50-60.
43. Sagone, A.L. Jr., King, G.W. and Metz, E.N. (1976). *J. Clin. Invest. 57,* 1352-1358.
44. Salin, M.L. and McCord, J.M. (1977). This volume.
45. Sbarra, A.J., Paul, B.B., Jacobs, A.A., Strauss, R.R. and Mitchell, G.W. Jr. (1972). *J. Reticuloendothel. Soc. 12,* 109-126.
46. Webb, L.S., Keele, B.B. Jr. and Johnston, R.B. Jr. (1974). *Infect. Immun. 9,* 1051-1056.
47. Yost, F.J. Jr. and Fridovich, I. (1974). *Arch. Biochem. Biophys. 161,* 395-401.

SUPEROXIDE GENERATION IN RELATION TO OTHER OXIDATIVE REACTIONS IN HUMAN POLYMORPHONUCLEAR LEUKOCYTES

D. ROOS

Central Laboratory of the
Netherlands Red Cross Blood Transfusion Service
P.O. Box 9190, Amsterdam, The Netherlands

M.L.J. van SCHAIK and R.S. WEENING

Pediatric Clinic, Binnen Gasthuis
University of Amsterdam, Amsterdam, The Netherlands

R. WEVER

Laboratory of Biochemistry, B.C.P. Jansen Instituut
University of Amsterdam, Amsterdam, The Netherlands

SUMMARY

Generation of superoxide and hydrogen peroxide by human blood polymorphonuclear leukocytes was determined under optimal conditions. With IgG-coated latex as stimulating particles 65 - 85% of the consumed oxygen could be recovered as superoxide, and 55 - 75% as hydrogen peroxide. With serum-treated zymosan these values were slightly higher: 50 - 90% and 65 - 90% respectively. With the techniques used it was impossible to determine O_2 consumption, O_2^- generation, and H_2O_2 production simultaneously, since in the detection method of O_2^-, based on the superoxide-mediated reduction of cytochrome *c*, oxygen is formed and H_2O_2 formation is inhibited. The high recoveries of superoxide could only be achieved when other O_2^--consuming reactions were competatively inhibited by 150 μM cytochrome *c*. Additional evidence for such competition was obtained by the finding that oxidative injury to polymorphonuclear leukocytes with a deficiency of glutathione reductase could be prevented by the addition of cytochrome *c*.

Addition of superoxide dismutase to polymorphonuclear leukocytes had little effect on O_2 consumption and H_2O_2 production, indicating that dismutation of O_2^- to H_2O_2 and O_2 was completed by either intracellular superoxide dismutase or by spontaneous dismutation. From experiments with cytochalasin-B treated cells and with soluble activators, such as ionophores and phorbol myristate acetate, it can be deduced that superoxide formation takes place at the level of the plasma membrane. A model is presented which was designed according to our findings and literature data. This model explains both the availability of NADPH for intraphagosomal O_2^- production and the high recoveries of O_2^- and H_2O_2 in the extracellular environment.

INTRODUCTION

The oxidative metabolism of polymorphonuclear leukocytes (PMN) is stimulated during phagocytosis: the oxygen consumption (Baldridge and Gerard, 1933) and the hydrogen peroxide production (Iyer *et al.*, 1961) are dramatically increased, while superoxide generation has been demonstrated (Babior *et al.*, 1973) and singlet oxygen production (Allen *et al.*, 1972) has been inferred. It is generally accepted, although based on weak experimental evidence that molecular oxygen is first reduced to superoxide anion radicals before being converted into hydrogen peroxide. Possibly, singlet oxygen is produced in the myeloperoxidase-catalyzed reaction between hydrogen peroxide and chloride (Allen and Stoele, 1973). Each of these oxygen products may have bactericidal properties (Klebanoff, 1975; Johnston *et al.*, 1975).

The intracellular organization of this sytem is still unknown. Several groups of investigators have presented evidence that the primary enzyme of oxygen reduction may be an NADPH oxidase, with most of the activity localized in the granular fraction (Hohn and Lehrer, 1975; Curnutte *et al.*, 1975; Patriarca *et al.*, 1973; De Chatelet *et al.*, 1975). This indicates that this enzyme may be of lysosomal origin, which seems logical in view of the fact that the microbicidal products are needed in the phagosome. On the other hand, if this NADPH oxidase were indeed localized in the granules, it is unclear how NADPH, produced in the cytosol, might reach the NADPH oxidase across the phagosomal membrane. Another uncertain factor is the role of superoxide dismutase (SOD). Salin and McCord (1974) and Johnston *et al.* (1975) have demonstrated the existence of this enzyme in the cytosol of PMN, where it is probably present as a protectant against oxidative injury. Whether this enzyme is also needed for H_2O_2 production from O_2^-, as an alternative to the spontaneous dismutation of O_2^-, is unknown.

In this communication we present evidence that superoxide is indeed a major intermediate in the reduction of molecular oxygen to hydrogen peroxide, that superoxide dismutase does not play an important role in the formation of H_2O_2, and that the primary oxygen-reducing system may be located in the plasma membrane.

MATERIAL AND METHODS

Purified PMN (>95%) were obtained by Ficoll-Isopaque centrifugation of fresh, defibrinated human blood, followed by specific lysis of erythrocytes with isotonic NH_4Cl solution. The cells were suspended in phosphate (10 mM) - buffered 140 mM NaCl, containing 0.6 mM $CaCl_2$, 1.0 mM $MgCl_2$, 5.5 mM glucose, and 0.5% (w/v) human albumin, pH 7.4. Serum-treated zymosan (STZ) (Roos *et al.*, 1976a) and IgG-coated latex (IgG-latex) (Homan-Müller *et al.*, 1975) were prepared as described earlier. PMN were incubated at 37°C in a shaking waterbath (180 cycles/min) with 0.5 mg STZ/ml or 3 x 10^{10} IgG-latex particles/ml.

Oxygen Consumption was measured polarographically in suspensions of PMN with a Clark-type oxygen electrode (Weening *et al.*, 1974).

Superoxide Generation was determined by the superoxide dismutase - inhibitable reduction of cytochrome *c* (Babior *et al.*, 1973; Weening *et al.*, 1975). In brief, PMN (<1 x 10^6/ml) were incubated during 15 min at 37°C with 150 μM horse heart ferricytochrome *c*, Type III (Sigma Chemical Co.) in the presence and absence of 0.4 μM SOD (isolated from bovine blood, McCord and Fridovich, 1969). The amount of reduced cytochrome *c* in the cell-free supernatants was determined spectrophotometrically at 550 nm. The values found in the presence of SOD were subtracted from those found in the absence of SOD.

Hydrogen Peroxide production was measured in cell-free supernatants containing 2 mM NaN_3, with a method based on the oxidation of leucodiacetyl-2,7-dichlorofluorescein by H_2O_2 in the presence of peroxidase to a fluorescent compound (Homan-Müller *et al.*, 1975).

RESULTS AND DISCUSSION

Optimal Test Conditions

In order to determine the stoichiometry of the oxidative reactions in PMN, the test conditions for the oxygen consumption (Weening *et al.*, 1974), the superoxide generation (Weening *et al.*, 1975) and the hydrogen peroxide production (Homan-Müller *et al.*, 1975) were optimized. These reactions can be stimulated by non-opsonized particles in the presence of fresh serum (Weening *et al.*, 1974; 1975), as well as by pre-opsonized particles, such as IgG-latex (Homan-Müller *et al.*, 1975) or STZ (Roos *et al.*, 1976). When superoxide production has to be measured, however, pre-opsonized particles are to be preferred, since some human sera *per se* reduce cytochrome *c* (unpublished results). This precaution is even more stringent when H_2O_2 determinations are carried out: in all human sera tests, H_2O_2 was degraded (Homan-Müller *et al.*, 1975). Even in the absence of serum either KCN (2 mM) or NaN_3 (2 mM) should be added to the incubation mixtures to ensure stability of the H_2O_2 samples. When omitted, H_2O_2 is probably degraded by myeloperoxidase released from the PMN during phagocytosis (Homan-Müller *et al.*, 1975). When conditions for O_2^- generation and H_2O_2 production are compared, azide is to be preferred since cyanide slowly reacts with cytochrome *c* (Horecker and Stannard, 1948).

The amount of cytochrome *c* reduced by phagocytosing PMN is highly dependent on the concentration of ferricytochrome *c* added to the incubation mixtures. Fig.1 shows that at about 150 μM ferricytochrome *c* a plateau in the reduction was reached: apparently, at that concentration, cytochrome *c* competes effectively for O_2^- with other superoxide-consuming reactions.

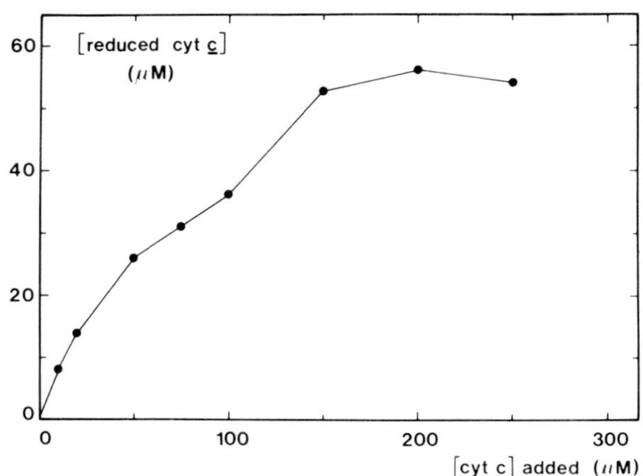

Fig.1. *Increase in ferricytochrome* c *reduction with increasing cytochrome* c *concentrations. PMN (1.0 x 10^6/ml) were pre-incubated for 10 min at 37°C before addition of indicated amounts of cytochrome* c. *After 15 min the incubations were terminated by chilling in ice and centrifugation (2 min, 0°C, 8500 x g). Reduction of cytochrome* c *in the cell-free supernatants was measured spectrophotometrically at 550 nm after appropriate dilution.*

Quantitative Aspects

Using IgG-latex particles to stimulate the oxidative metabolism of human PMN, we recovered 65 - 85% of the consumed oxygen as superoxide (at 150 μM cytochrome *c* added), and 55 - 75% as hydrogen peroxide (Table I). With serum-treated zymosan, which contains both IgG and C3 at its surface (determined with fluorescent antisera), higher metabolic values were found. The recoveries for superoxide and hydrogen peroxide were 50 - 90% and 65 - 90%, respectively. Thus, this indicates that a very large proportion - if not all - of the oxygen, consumed in the respiratory burst during phagocytosis, is converted into hydrogen peroxide. Moreover, superoxide must be an important product of oxygen reduction and not a side-product of minor quantitative importance. Therefore, taking into account the superoxide dismutation reaction, the most likely reaction sequence is:

$$O_2 \longrightarrow O_2^- \longrightarrow H_2O_2$$

TABLE I

Effect of Different Particles on the Oxidative Metabolism of PMN

Parameter	IgG-latex response*	(n)	STZ response*	(n)
Oxygen consumption	18 - 30	10	36 - 58	10
Superoxide generation	16 - 20	4	22 - 46	6
Hydrogen peroxide production	13 - 18	10	31 - 43	8

*in nmoles/10^6 PMN x 15 min (range)
All determinations were carried out in the presence of 2 mM NaN_3

Interfering Conditions

It must be emphasized that the three metabolic parameters have not been, and could not be, measured under identical conditions. The addition of cytochrome *c* interferes with the determination of both the stimulated oxygen consumption (75 - 80% decrease) and the stimulated hydrogen peroxide production (70 - 80% inhibition). Probably, some oxygen is preferentially formed and - among other things - some H_2O_2 formation is prevented:

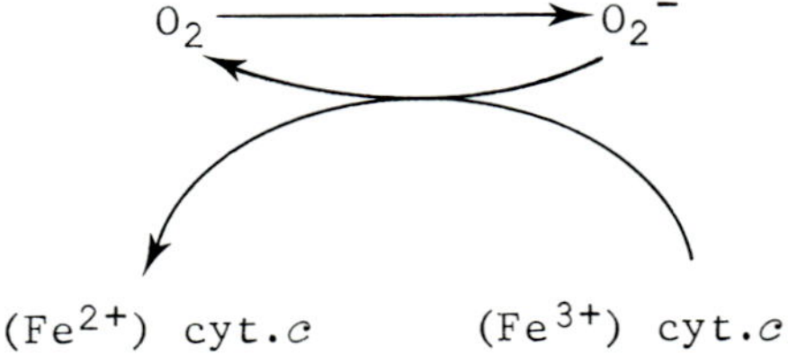

This inhibition of H_2O_2 formation has been interpreted by Root and Metcalf (1976) as an indication of the reaction sequence $O_2 \rightarrow O_2^- \rightarrow H_2O_2$. Although we agree with this conclusion, the data should be considered with caution, since we found that H_2O_2 itself can be broken down by cytochrome *c* (Fig.2), in a sodium azide-insensitive reaction. Hence, the inhibiting effect of cytochrome *c* on the H_2O_2 production by PMN can also be explained by the degradation of H_2O_2 induced by cytochrome *c*, and no conclusions about the reaction sequence can be drawn from this observation.

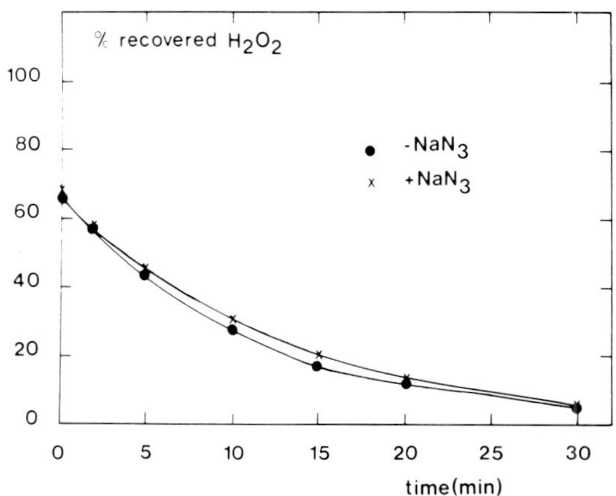

Fig.2. *Breakdown of hydrogen peroxide by cytochrome* c. *H_2O_2 (15 μM) in incubation medium (see Materials and Methods) was preincubated for 5 min at 37°C before addition of cytochrome* c *(90 μM) and NaN_3 (2 mM). At the times indicated samples (100 μl) were taken, chilled in ice, diluted 31-fold with ice-cold distilled water, and assayed for H_2O_2. The results are expressed as percentage of the values found without cytochrome* c.

When studying the PMN from members of a family with a glutathione reductase deficiency (<15% of controls) in all blood cells (Loos *et al.*, 1976) we found that all parameters of the oxidative metabolism except for superoxide production showed only a very brief period (5 - 10 min) of normal activity, thereafter stopping abruptly. The superoxide generation in these cells, as measured by cytochrome *c* reduction, was completely normal up to 30 min of incubation with STZ. Of course, it is impossible to generate O_2^- without consumption of O_2. We tentatively concluded that the cytochrome *c* used for the O_2^- assay protected these cells from oxidative injury by superoxide, hydrogen peroxide, or other toxic oxygen products, by taking away the O_2^- and preventing the subsequent formation of these toxic metabolites. This example conveys a warning about the comparison of results obtained under different conditions.

Role of SOD in H_2O_2 Formation

Cyanide-sensitive superoxide dismutase has been demonstrated in the cytosol of PMN (Johnston *et al.*, 1975; Salin and McCord, 1974), and it has been assumed that this enzyme protects the cells against the toxic effects of O_2^-. The role of SOD in the *formation* of hydrogen peroxide is less clear, however. Addition of SOD to bacteria-phagocytosing PMN has been shown to inhibit the rate of bacterial killing (Johnston *et al.*, 1975). In these experiments, SOD was brought into the phagosomes of the PMN as a coating of latex particles. Under normal conditions, however, SOD is only present in the cytosol and does not enter the phagosomes. Therefore, this enzyme probably plays no important role in H_2O_2 formation. We have measured the influence of SOD, added as the free protein to the incubations, on the stimulated oxygen consumption and H_2O_2 production, to test the role of this enzyme in the formation of bactericidal products.

Addition of 0.4 μM SOD was found to give practically complete inhibition of cytochrome *c* reduction, but only small effects on the oxygen consumption (5 - 15% inhibition) and hydrogen peroxide production (20 - 30% stimulation). These results, summarized in Table II, indicate that the dismutation of O_2^- into O_2 and H_2O_2 is already practically complete in the absence of external and intra-phagosomal SOD.

TABLE II

Effect of Different Additions on the Oxidative Metabolism of PMN

Parameter	Additions			
	SOD (0.4 μM)	cyt. *c* (150 μM)	NaN_3 (2 mM)	KCN (2 mM)
Oxygen consumption	-	--	+	+
Superoxide generation	--	++	±	*)
Hydrogen peroxide production	+	--	++	++

PMN were stimulated with either IgG-latex or STZ
\+ = <50% stimulation over control values
++ = >50% stimulation over control values
\- = <50% inhibition as compared to control values
-- = >50% inhibition as compared to control values
± = no effect
*) = no determination possible due to inhibition of cytochrome *c* by KCN (Horecker and Stannard, 1948).

When we tried to measure the effect of SOD inhibition by KCN, we found that O_2^- could not be determined, due to the blocking of cytochrome *c* by KCN (Horecker and Standard, 1948). Oxygen consumption was stimulated 15 - 30%, probably due to inhibition of catalase, thus preventing breakdown of H_2O_2 in O_2 and H_2O. Hydrogen peroxide production could not be measured without addition of either KCN or NaN_3, probably because of its rapid degradation by cyanide- and azide-sensitive enzymes. If H_2O_2 production were dependent on SOD activity however, inhibition by cyanide might be expected to inhibit H_2O_2 formation. This was not found. Thus, spontaneous dismutation results in almost total conversion of O_2^- into H_2O_2 and O_2.

Protection Against Oxidative Injury

The conclusion from these experiments is that SOD only functions in protecting PMN against oxidative injury from O_2^- or subsequently formed products. Protection against hydrogen peroxide injury may be given by catalase or glutathione peroxidase, both enzymes being present in the cytosol of PMN (Stossel *et al.*, 1971; Karnovsky *et al.*, 1971). Catalase, however, is a very inefficient enzyme at low H_2O_2 concentrations (Cohen and Hochstein, 1963), hence the glutathione system is probably much more active in maintaining a low H_2O_2 concentration in the cytosol. The importance of this system may be derived from the studies of Holmes *et al.* (1970) on glutathione peroxidase deficient PMN, as well as from our own studies on glutathione reductase deficient PMN (Loos *et al.*, 1976), as mentioned above.

From the fact that we found glutathione reductase deficient PMN to be injured by oxidative stress, an important conclusion may be drawn: the glutathione system (glutathione peroxidase coupled to glutathione reductase) fulfills an essential role in the detoxification of oxidative products, probably more important than catalase which was presumably present in normal amounts in these cells.

Localization of the O_2^--Generating System

From our experiments with cytochalasin B-treated PMN (Weening *et al.*, 1974; Roos *et al.*, 1976b), in which the oxidative metabolism was found to be stimulated in the absence of phagocytosis, as well as from similar results obtained with soluble activators of these reactions, such as ionophores (Schell-Frederick, 1974; Romeo *et al.*, 1975; Goldstein *et al.*, 1975), complement component C5a (Goldstein *et al.*, 1975), and phorbol myristate acetate (Goldstein *et al.*, 1975; Repine *et al.*, 1974; DeChatelet *et al.*, 1976) it may be concluded that the triggering of this process takes place at the cell surface. The actual nature of the oxygen reducing system remains to be clarified, however. Patriarca *et al.* (1973), Hohn and Lehrer (1975), Curnutte *et al.* (1975), and DeChatelet *et al.* (1975) have presented evidence that a NADPH oxidase may be the primary enzyme in this reaction. The main activity of this enzyme was found in the granular fraction of PMN, indicating that the NADPH oxidase might be of lysosomal origin or membrane-associated. Moreover, Johnston *et al.* (1975) found most of the NBT reducing activity in the same fraction, although Baehner (1975) localized this activity in microsomes or plasma membranes.

A lysosomal localization leaves some unsolved questions, however. First, how does NADPH, produced in the cytosol, enter the phagosomes when O_2^- is generated inside these structures during the attack on ingested microorganisms? Second, how does this localization explain the high recoveries of superoxide and hydrogen peroxide in the extracellular environment?

Since granular fractions are as a rule heavily contaminated with plasma membranes, we (Roos *et al.*, 1976a,b; Goldstein *et al.*, 1975) and others (Johnston *et al.*, 1975; Salin and McCord, 1974; Baehner, 1975) think that an alternative localization of the O_2^--generating system is possible. If this system were localized in the plasma membrane, these questions would be resolved. First, if the NADPH oxidase were a membrane spanning enzyme, as shown in Fig.3, the reducing agent NADPH might come from the cell interior and the O_2 from the cell exterior, releasing $NADP^+$ inside, but O_2^- and H^+ on the outside of the cells. Moreover, in case of phagosome formation, NADPH does not need to pass through the phagosomal membrane, since the phagosome is surrounded by a piece of *inverted* plasma membrane (see Fig.3). This mechanism would result in *directed* release of O_2^- to the inside of the phagosome, in the direction of the ingested micro-organisms. Second, this localization also explains why such high recoveries of O_2^- and H_2O_2 in the cell-free supernatants are found: triggering the oxidative metabolism of the cells results in formation of O_2^- and H_2O_2 also on the *outside* of the cells. Thus, these products do not pass through the cytosol (which contains SOD) to reach the extracellular environment. This mechanism also explains the enhanced detection of O_2^- from cytochalasin B-treated cells (Weening *et al.*, 1974): when interiorization of the plasma membrane is prevented, all O_2^- will be released at the cell surface into the medium, thus facilitating its reaction with cytochrome *c*.

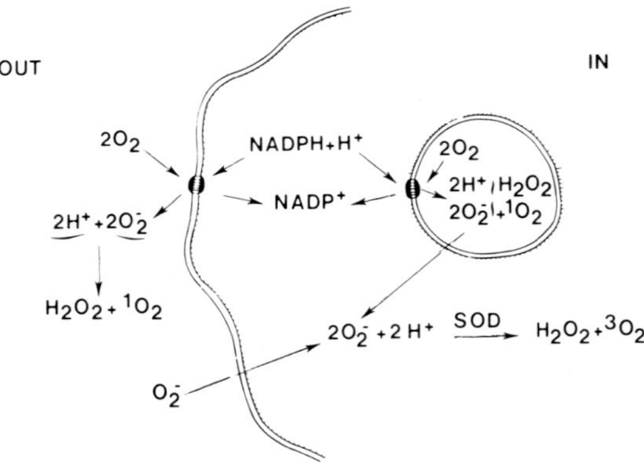

Fig.3. *A model for the localization of the superoxide-generating system in human polymorphonuclear leukocytes. Attachment of particles to, or perturbation of, the cell membrane activates a NADPH oxidase. This enzyme is localized in the plasma membrane and accessible to substrates from both sides of the membrane: from the cell interior for NADPH and from outside for oxygen. As a result, superoxide anions are released into the cell-surrounding medium, or, after phagocytosis, into the phagosomes. From superoxide H_2O_2 and ${}^1O_2$ may be formed, or it may be detoxified in the cytosol by superoxide dismutase. Catalase and glutathione peroxidase protect the cytosol against oxidative injury by H_2O_2.*

The triggering of the oxidative metabolism by particle attachment or membrane perturbation might be thought of as a conformational change in the NADPH oxidase itself or in its direct membrane environment, resulting in the activation of the O_2^- generating system. This would explain the production of superoxide in cell homogenates (Johnston *et al.*, 1975; Curnutte *et al.*, 1975; Babior *et al.*, 1975) as well as the fact that the cyclic AMP-cyclic GMP system is apparently not involved in the triggering of this process in whole cells (Schell-Frederick and Sande, 1974; Babior *et al.*, 1974). A conformational change in the cell membrane might also result in triggering other PMN functions, such as phagocytosis and lysosomal enzyme release. The mechanism of these other phagocytic cell functions differs from the O_2^- production, however, since phagocytosis has been shown to be energy-dependent (Stossel, 1975), and the lysosomal release to be controlled by the cGMP-cAMP system (Zurier *et al.*, 1974; Smith and Ignarro, 1975).

Finally the question arises as to which part of the PMN membrane is triggered into O_2^- production. Does attachment of a particle activate *all* NADPH oxidase complexes in the plasma membrane, or only those at the site of attachment? From our studies it appears that activation of the whole membrane seems probable since so much O_2^- is generated into the extra-cellular medium. However, this may have been caused by the high particle: cell ratios we use (IgG-latex: 10,000; STZ: 30). Nathan and co-workers (1969), on the other hand, found most of the formazan precipitation in the phagosomes (closed and unclosed), around the attached particles, which points to a localized activation of this system. Clearly, this question should be studied at the single cell level. Why such large quantities of superoxide are liberated into the medium is not clear: it may be used for extracellular bacterial killing, but it may also be a reflection of the relative evolutionary newness of the immune response system (McCord and Salin, 1975).

Until more data are available, the above described model of O_2^--generation may be found useful in understanding the oxidative microbicidal mechanism of polymorphonuclear leukocytes.

ACKNOWLEDGEMENTS

Thanks are due to Dr. J.A. Loos for stimulating discussions. This investigation was in part supported by financial aid from the Netherlands Organization for the Advancement of Pure Research (Z.W.O.), The Hague, The Netherlands (grant no. 91-5).

REFERENCES

1. Allen, R.C., Stjernholm, R.L. and Steel, R.H. (1972). *Biochem. Biophys. Res. Commun. 47,* 679-684.
2. Allen, R.C. and Steel, R.H. (1973). *Fed. Proc. 32,* 478a.
3. Babior, B.M., Kipnes, R.S. and Curnutte, J.T. (1973). *J. Clin. Invest. 52,* 741-744.
4. Babior, B.M., Curnutte, J.T. and Kipnes, R.S. (1974). *J. Clin. Invest. 53,* 3a.
5. Babior, B.M., Curnutte, J.T. and Kipnes, R.S. (1975). *J. Clin. Invest. 56*, 1035-1042.
6. Baehner, R.L. (1975). *J. Lab. Clin. Med. 86,* 785-792.
7. Baldridge, C.W. and Gerard, R.W. (1933). *Am. J. Physiol. 103,* 235-236.
8. Cohen, G. and Hochstein, P. (1963). *Biochemistry 2,* 1420-1428.
9. Curnutte, J.T., Kipnes, R.S. and Babior, B.M. (1975). *New Engl. J. Med. 293,* 628-638.
10. DeChatelet, L.R., McPhail, L.C., Mullikin, D. and McCall, C.E. (1975). *J. Clin. Invest. 55,* 714-721.
11. DeChatelet, L.R., Shirley, P.S. and Johnston, R.B. Jr. (1976). *Blood 47,* 545-554.
12. Goldstein, I., Roos, D., Kaplan, H. and Weissmann, G. (1975). *J. Clin. Invest. 56,* 1155-1163.
13. Hohn, D.C. and Lehrer, R.I. (1975). *J. Clin. Invest. 55,* 707-713.
14. Holmes, B., Park, B.H., Malawista, S.E., Quie, P.G., Nelson, D.L. and Good, R.A. (1970). *New Engl. J. Med. 283,* 217-221.
15. Homan-Müller, J.W.T., Weening, R.S. and Roos, D. (1975). *J. Lab. Clin. Med. 85,* 198-207.
16. Horecker, B.L. and Stannard, J.N. (1948). *J. Biol. Chem. 172,* 589-598.
17. Iyer, G.Y.N., Islam, D.M.F. and Quastel, J.H. (1961). *Nature 192,* 535-541.
18. Johnston, R.B. Jr., Keele, B.B., Misra, H.P., Lehmeijer, J.E., Webb, L.S., Baehner, R.L. and Rajagopolan, K.V. (1975). *J. Clin. Invest. 55,* 1357-1372.
19. Karnovsky, M.L., Baehner, R.L., Githens, S., Simmons, S. and Glass, E.A. (1971). In "Immunopathology of Inflammation" (Forscher, B.K. and Houck, J.C., eds.), Exerpta Medica, Amsterdam, p 121.
20. Klebanoff, S.J. (1975). In "The Phagocytic Cell in Host Resistance" (Bellanti, J.A. and Dayton, D.H., eds.), Raven Press, New York, pp 45-59.
21. Loos, J.A., Roos, D., Weening, R.S. and Houwerzijl, J. (1976). *Blood,* in press.
22. McCord, J.M. and Fridovich, I. (1969). *J. Biol. Chem. 244,* 6049-6055.

23. McCord, J.M. and Salin, M.L. (1975). In "Erythrocyte Structure and Function" (Brewer, G.J., ed.), A.R. Liss Inc., New York, pp 731-746.
24. Nathan, D.G., Baehner, R.L. and Weaver, D.K. (1969). *J. Clin. Invest. 48,* 1895-1904.
25. Patriarca, P., Cramer, R., Dri, P., Fant, L., Basford, R.E. and Rossi, F. (1973). *Biochem. Biophys. Res. Commun. 53,* 830-837.
26. Repine, J.E., White, J.G., Clawson, C.C. and Holmes, B.M. (1974). *J. Lab. Clin. Med. 83,* 911-920.
27. Romeo, D., Zabucchi, G., Miani, N. and Rossi, F. (1975). *Nature 253,* 542-544.
28. Roos, D., Homan-Müller, J.W.T. and Weening, R.S. (1976a). *Biochem. Biophys. Res. Commun. 68,* 43-50.
29. Roos, D., Goldstein, I. and Weissmann, G. (1976b). In "Leukocyte Membrane Determinants Regulating Immune Reactivity" (Eijsvoogel, V.P., Roos, D. and Zeijlemaker, W.P., eds.) Academic Press, New York, pp 633-639.
30. Root, R.K. and Metcalf, J. (1976). *Clin. Res. 24,* 352A.
31. Salin, M.L. and McCord, J.M. (1974). *J. Clin. Invest. 54,* 1005-1009.
32. Schell-Frederick, E. (1974). *FEBS Letters 48,* 37-40.
33. Schell-Frederick, E. and Van Sande, J. (1974). *J. Reticuloendothel. Soc. 15,* 139-148.
34. Smith, R.J. and Ignarro, L.J. (1975). *Proc. Nat. Acad. Sci. U.S.A. 72,* 108-112.
35. Stossel, T.P., Pollard, T.D., Mason, R.J. and Vaughn, M. (1971). *J. Clin. Invest. 50,* 1745-1757.
36. Stossel, T.P. (1975). *Semin. Hematol. 12,* 83-116.
37. Weening, R.S., Roos, D. and Loos, J.A. (1974). *J. Lab. Clin. Med. 83,* 570-576.
38. Weening, R.S., Wever, R. and Roos, D. (1975). *J. Lab. Clin. Med. 85,* 245-252.
39. Zurier, R.B., Weissmann, G., Hoffstein, S., Kammerman, S. and Hsin Hsiung Tai, (1974). *J. Clin. Invest. 53,* 297-309.

IN DEFENSE OF HABER-WEISS*

G. COHEN

Department of Neurology
Mount Sinai School of Medicine
New York, New York 10029, U.S.A.

During the course of discussion at this meeting, the Haber-Weiss reaction became the subject of debate. Several investigators invoked the reaction as a source of hydroxyl radicals in biologic or *in vitro* systems. Others objected that insufficient evidence existed to support the reaction. The purpose of this comment is to review briefly the background and evidence in favor of Haber-Weiss.

In 1934, Haber and Weiss proposed the following two reactions (originally the Haber-Willstatter reactions) as part of their description of events in the decomposition of H_2O_2 by iron salts:

$$\cdot OH + H_2O_2 \longrightarrow O_2^{\bar{\cdot}} + H_2O + H^+ \qquad (1)$$

$$O_2^{\bar{\cdot}} + H_2O_2 \longrightarrow \cdot OH + O_2 + OH^- \qquad (2)$$

The reaction at issue at the meeting was reaction 2 in which superoxide and H_2O_2 gave rise to hydroxyl radicals. In their publication, Haber and Weiss designated the superoxide radical in its protonated form ($HO_2\cdot$); at or near neutral pH, however, the superoxide radical would be present as its conjugate base ($O_2^{\bar{\cdot}}$). Reactions 1 and 2 constituted a chain radical mechanism for the decomposition of H_2O_2 with production of oxygen, after an initiating step (e.g., Fe^{2+} plus H_2O_2 yields hydroxyl radicals). One of the strengths of the Haber-Weiss kinetic analysis was that it correctly predicted a burst in oxygen evolution at certain concentration ratios of Fe^{2+}, Fe^{3+} and H_2O_2.

Although the importance of intermediary superoxide and hydroxyl radicals in the overall chain mechanism for decomposition of H_2O_2 has achieved a level of general acceptance, reaction 2 has not been unequivocally demonstrated and it has engendered debate. More recent studies in relatively pure systems (Barb *et al.*, 1949; Walling, 1975) have stressed the importance of iron, either free or in various chelated forms, as part of the chain mechanism. I wish to concern myself with events in biologic systems and will adopt the point of view that the role for iron or other heavy metals, either added or present as contaminants, represents a detail in the overall reaction described by equation 2. That is to say, I wish to stress the validity of the concept that the simultaneous presence of superoxide and H_2O_2 results in the generation of hydroxyl radicals, without giving regard to intervening mechanisms.

In 1970, Beauchamp and Fridovich published the results of their studies on the generation of ethylene from methional (reaction 3) during the xanthine-xanthine oxidase reaction. Ethylene production was inhibited by SOD or catalase

$$\cdot OH + CH_3\text{-}S\text{-}CH_2CH_2\text{-}CHO \longrightarrow CH_2{=}CH_2 + \text{other products} \quad (3)$$

$$\cdot OH + CH_3\text{-}S\text{-}CH_2CH_2\text{-}CO\text{-}COOH \longrightarrow CH_2{=}CH_2 + \text{other products} \quad (4)$$

(cf. reaction 2) or by hydroxyl radical scavengers such as benzoate, formate or ethanol. A lag in ethylene production was eliminated by adding H_2O_2 at zero time, but H_2O_2 alone (xanthine and xanthine oxidase omitted) was ineffective; that is, ethylene production proceeded when superoxide radicals were generated by xanthine oxidase in the presence of H_2O_2. These results were consistent with the production of both superoxide and H_2O_2 during the xanthine oxidase reaction, and with reactions 2 and 3. Recently, reaction 3 was verified in a pulse radiolysis study (Bors *et al.*, 1976), although the authors expressed reservations about complexities in the reaction mechanism.

Several experimental predictions were made and subsequently substantiated (Beauchamp and Fridovich, 1970). First, in the absence of methional, the yield of H_2O_2 was less than theoretical, as would be expected from a decomposition of H_2O_2; but the yield was increased by incorporation of SOD into the reaction medium (cf. reaction 2). Second, in accord with the known oxidant properties of hydroxyl radicals, an oxidation of ferrocytochrome *c* was noted; the phenomenon was inhibited by SOD, catalase or ethanol. Third, when the superoxide-mediated reduction of ferricytochrome *c* was studied, a plateau was reached before the reduction had gone to completion. The plateau indicated a dynamic balance between the reducing action of superoxide radicals and the oxidant properties of hydroxyl radicals; the position of equilibrium was altered predictably by the addition of ethanol or catalase without affecting the rate of oxidation of xanthine. Beauchamp and Fridovich concluded that hydroxyl radicals were produced via reaction 2 during the xanthine-xanthine oxidase reaction. They suggested that the methional reaction (reaction 3), coupled with hydroxyl radical scavengers, might serve to detect hydroxyl radicals in other systems.

Subsequent support for the Haber-Weiss reaction evolved from observations with several substances that reacted spontaneously with oxygen to generate superoxide and H_2O_2. These substances included 6-hydroxydopamine, 6-aminodopamine and dialuric acid, which were of special interest because they induced specific lesions in experimental animals. Studies of ethylene production *in vitro* were conducted with either methional or 2-keto-4-thiobutyric acid (reaction 4), at three pH's (6.5, 7.4 and 8.4), generally in the presence of the metal chelating agent, EDTA (Cohen and Heikkila, 1974; Cohen *et al.*, 1976; Heikkila *et al.*, 1976). Results were qualitatively similar to those reported by Beauchamp and Fridovich for the xanthine oxidase reaction. First, SOD or catalase was inhibitory; second, hydroxyl radical scavengers were inhibitory; third, H_2O_2 alone was ineffective, but H_2O_2 added just prior to 6-aminodopamine spurred ethylene formation during the subsequent autoxidation reaction. In the 6-aminodopamine autoxidation system, good correlations were observed between suppression of ethylene and known rate constants for the reactions of added substances with hydroxyl radicals. Specifically, benzoate was superior to ethanol; in a homologous series of primary alcohols (methanol, ethanol, propanol, butanol) suppression of ethylene increased with increasing chain length; and thiourea was a powerful suppressant, while urea, an agent known to react poorly with hydroxyl radicals, was without effect. Control studies showed that the hydroxyl radical scavengers did not alter the rate of formation of quinoidal products from 6-aminodopamine and that suppression of ethylene could not be attributed to removal of either superoxide or H_2O_2. These experimental observations were readily accommodated by reaction 2, followed by reaction 3 or 4.

The following conclusions emerge: when superoxide radicals and H_2O_2 are present simultaneously, something else is generated. This "something else" is a strong oxidant and possesses the property of interacting with methional to form ethylene. It can be intercepted by hydroxyl radical scavengers in rough concordance with their respective rate constants for reaction with hydroxyl radicals. Thus, the emergent species appears to be identical with the hydroxyl radical.

Other interpretations of the data may be possible. For example, semiquinone intermediates or other products may be capable of interacting with methional to form ethylene. But, the reactive agents would have to exhibit similar properties, mechanisms dependent upon both superoxide and H_2O_2 would be required for maintaining their levels, and the species derived from 6-aminodopamine would have to be imbued with similar reactivities as the hydroxyl radical. Moreover, a reactive organic species would have to be postulated for the xanthine-xanthine oxidase reaction. A simpler interpretation is to invoke reaction 2.

The concept of Haber-Weiss has provided a predictive tool for studies with animals. Thus, hydroxyl radical scavengers were tested as antagonists to the cytotoxic actions of dialurate, 6-aminodopamine and 6-hydroxydopamine in mice (see article by Cohen and Heikkila in this volume). In studies with alloxan (the oxidized form of dialuric acid), prior injection of thiourea blocked both the destruction of the beta cells of the pancreas and the consequent development of a diabetic state, but urea (not a scavenger of hydroxyl radicals) was not protective (Heikkila *et al.*, 1976). Similarly, primary aliphatic alcohols were antidiabetic agents with the longer chain alcohols (butanol, propanol) exhibiting greater efficacy than ethanol or methanol. In studies with 6-hydroxydopamine, injections of n-butanol or ethanol provided some protection against the destruction of sympathetic nerve terminals, while a complex thiourea derivative which was known to penetrate into sympathetic nerves was a very effective antagonist (Cohen *et al.*, 1976). Although this latter agent had never been tested previously as a hydroxyl radical scavenger, it proved to be an effective suppressant of ethylene production *in vitro*; it also antagonized the cytotoxic actions of both 6-aminodopamine and alloxan in experimental animals. Thus, a link between *in vitro* observations and *in vivo* studies has been established.

The Haber-Weiss reaction represents an economical explanation for a variety of experimental observations. It had predictive value in both chemical and biologic experiments. Unless strong evidence to the contrary is presented, Haber-Weiss remains a useful concept.

REFERENCES

1. Barb, W.G., Baxendale, J.H., George, P. and Hargrave, K.R. (1949). *Nature 163*, 692-694.
2. Beauchamp, C. and Fridovich, I. (1970). *J. Biol. Chem. 245*, 4641-4646.
3. Bors, W., Lengfelder, E., Saran, M., Fuchs, C. and Michel, C. (1976). *Biochem. Biophys. Res. Commun. 70*, 81-86.
4. Cohen, G. and Heikkila, R.E. (1974). *J. Biol. Chem. 249*, 2447-2452.
5. Cohen, G., Heikkila, R.E., Allis, B., Cabbat, F., Dembiec, D., MacNamee, D., Mytilineou, C. and Winston, B. (1970). *J. Pharmacol. Exp. Therap.*, in press.
6. Haber, F. and Weiss, J. (1934). *Proc. Roy. Soc. A 147*, 332-351.
7. Heikkila, R.E., Winston, B., Cohen, G. and Barden, H. (1976). *Biochem. Pharmacol. 25*, 1085-1092.
8. Walling, C. (1975). *Accts. Chem. Res. 8*, 125-131.

*EDITORS' FOOTNOTE

Since 1970 various reports have described systems which were inhibited by superoxide dismutase *or* by catalase *or* by such scavengers of OH˙ as alcohols or benzoate (1 - 18). The Haber-Weiss reaction was invoked in all of these cases as a consistent explanation for the data. However, as described elsewhere in this book, there is no chemical evidence that O_2^- directly reduces H_2O_2 to OH^- + OH˙. Nevertheless, biochemically it could do so by way of an iron-containing catalyst. Iron salts or iron-containing proteins were impurities and/or components of those reaction mixtures whose behavior called forth the postulation and indeed iron in many forms is certainly plentiful in biological systems. We conclude that the Haber-Weiss reaction, or at least an iron-catalysed or other variant thereof, remains the most acceptable explanation for the data described in references 1 - 18. The arguments in favor of this view are presented in this chapter.

REFERENCES

1. Beauchamp, C. and Fridovich, I. (1970). *J. Biol. Chem. 245*, 4641-46.
2. Goscin, S.A. and Fridovich, I. (1972). *Arch. Biochem. Biophys. 153*, 778-783.
3. Cohen, G. and Heikkila, R.E. (1974). *J. Biol. Chem. 249*, 2447-2452.
4. Gregory, E.M. and Fridovich, I. (1974). *J. Bacteriol. 112*, 166-169.
5. Okita, R., Bidlack, W.R. and Hochstein, B. (1974). *Fed. Proc. 33*, 1256 Abs.
6. Elstner, E.F. and Konz, J.R. (1974). *FEBS Letters 45*, 18-21.
7. Kellog, E.W. III and Fridovich, I. (1975). *J. Biol. Chem. 250*, 8812-8817.
8. Van Hemmen, J.J. and Meuling, W.J.A. (1975). *Biochim. Biophys. Acta 402*, 133-141.
9. Babior, B., Curnutte, J.T. and Kipnes, R.S. (1975). *J. Lab. Clin. Med. 85*, 235-244.
10. Weiss, S.J. (1976). *Fed. Proc. 35*, Abstr. = 203.
11. Hodgson, E.K. and Fridovich, I. (1976). *Arch. Biochem. Biophys. 172*, 202-205.
12. Halliwell, B. and Ahluwalia, S. (1976). *Biochem. J. 153*, 513-518.
13. Lown, J.W., Begleiter, A., Johnston, D. and Morgan, A.R. (1976). *Can. J. Biochem. 54*, 110-119.

14. Cone, R., Hasan, S.K., Lown, J.W. and Morgan, A.R. (1976). *Can. J. Biochem.* *54*, 219-223.
15. Heikkila, R.E., Winston, B., Cohen, G. and Barden, H. (1972). *Biochem. Pharm.* *25*, 1085-1092.
16. Kon, S. and Schwimmer, S. (1976). *Fed. Proc.* *35*, 1553 Abs.
17. Miryal, J.J., Ackerman, R.S., Blumer, J.L. and Freeman, L.S. (1976). *J. Biol. Chem.* *251*, 3436-3441.
18. Konz, J.R. and Elstner, E.F. (1976). *FEBS Letters* *66*, 8-11.

SUPEROXIDE RADICALS IN MITOCHONDRIA

L. FLOHE and G. LOSCHEN

Research Laboratories of Chemie Grunenthal GmbH
519 Stolberg, W. Germany

A. AZZI and Ch. RICHTER

Instituto di Patologia Generale
Universita di Padova
Padova, Italy

INTRODUCTION

In cells of higher organisms the organelle primarily concerned with oxygen utilization is the mitochondrion. Typically, this organelle catalyses the reduction of molecular oxygen to water with concomitant conservation of chemical energy in form of ATP. The question whether or not this tetravalent reduction is the exclusive way of mitochondrial oxygen utilization did not attract much interest until recently. This situation changed when it was established that two enzymes present in the mitochondrial matrix reacted with partially reduced oxygen species: the manganese containing protein superoxide dismutase (SOD), dismutating the monovalently reduced oxygen O_2^- into O_2 and H_2O_2 (1,2), and the selenoprotein glutathione peroxidase (GSH peroxidase), reacting with the bivalently reduced oxygen species H_2O_2 and other hydroperoxides (3,4). Thus the mitochondrial matrix space contains the enzymatic equipment to eliminate the products of univalent and bivalent oxygen reduction according to equations 1 and 2:

$$2\ O_2^- + 2\ H^+ \xrightarrow{\text{SOD}} H_2O_2 + O_2 \qquad (1)$$

$$2\ GSH + H_2O_2 \xrightarrow{\text{GSH peroxidase}} 2\ H_2O + GSSG \qquad (2)$$

In the following sections, we shall demonstrate a putative role of both enzymes in mitochondria.

EVIDENCE FOR INCOMPLETE OXYGEN REDUCTION IN MITOCHONDRIA

For a long time, mitochondrial H_2O_2 production was only known from the atypical respiratory equipment of the parasite worm ascaris lumbricoides (5). In the late sixties, Jensen (6) and Hinkel *et al.* (7) found some evidence that submitochondrial particles of mammalian origin also produced hydrogen peroxide. Finally, Chance and Oshino (8), Loschen *et al.* (9-11), and Boveris *et al.* (12,13) demonstrated formation of hydrogen peroxide in intact, tightly coupled mitochondria from rat, pigeon and beef heart, which obviously was linked to the respiratory chain.

Thus mitochondria completed the list of cell organelles shown to generate hydrogen peroxide. The rates and conditions under which the various cell organelles produce H_2O_2 are listed in Table I.

The results presented in this table clearly indicate that H_2O_2 is a common product of oxygen reduction in cell metabolism. The general question, however, remains whether H_2O_2 is formed directly in a bivalent reductive step or, indirectly, as a product resulting from the dismutation of primarily formed superoxide anions. The answer to this question is relevant not only to the mechanism of mitochondrial H_2O_2 formation, but also to possible pathophysiological hazards, since H_2O_2 as such seems to be a product of minor toxicity, while O_2^- and the multiple products resulting from its interaction with H_2O_2 and other molecules represent a whole battery of possibly harmful agents. Experimental evidence will be presented to show that mitochondrial H_2O_2 is indeed formed largely, if not exclusively, from superoxide radicals.

SUPEROXIDE PRODUCTION BY SUBMITOCHONDRIAL PARTICLES

In order to prove formation of superoxide radicals by any subcellular particle, either a specific indicator reaction inside the organelle should be available or the superoxide radicals have to be determined after diffusion from the organelle into the surrounding medium. Neither of these requirements for detection of superoxide radicals in intact mitochondria could be met as yet. The usual way to detect O_2^- is to suppress some superoxide dependent reactions by addition of superoxide dismutase. For two reasons, this was not feasible in our case: a) superoxide dismutase added to a suspension of mitochondria cannot penetrate the outer membrane to reach the putative site of superoxide formation, the inner mitochondrial membrane, and b) the endogenous mitochondrial superoxide dismutase will prevent superoxide radicals from diffusing into the incubation medium. These difficulties, unfortunately, ruled out simultaneous investigation of superoxide and peroxide production by intact mitochondria.

However, submitochondrial particles (SMP's), obtained by sonic disruption of intact mitochondria, are membrane vesicles which are turned inside-out and therefore can be expected to be largely freed of the two matrix enzymes SOD and GSH peroxidase. Thus, at least some conditions known to result in mitochondrial H_2O_2 formation could be screened for simultaneous superoxide production.

Like uncoupled mitochondria (9), SMP's produce H_2O_2 only in the presence of antimycin and a suitable substrate (succinate, malate, α-keto-glutarate, etc.) (14). As shown in Fig.1, the rate of H_2O_2 formation in antimycin-inhibited SMP's is roughly doubled after addition of superoxide dismutase.

Taking into account the stoichiometry of the superoxide dismutase reaction (eq.1), we may calculate from these data (Fig.1) a rate of superoxide formation of at least 0.2 nmoles per mg protein/minute. This clearly exceeds the basic rate of H_2O_2 formation in the absence of exogenous superoxide dismutase. Moreover, under the same experimental conditions, O_2^- dismutates spontaneously with an apparent rate constant of about 10^6 $M^{-1}sec^{-1}$ (15). Considering this rapid reaction, it might be assumed that the basic rate of H_2O_2 formation should also be predominantly a consequence of O_2^- dismutation, either occurring spontaneously or catalyzed by traces of mitochondrial

TABLE I

Rates of H_2O_2 Formation By Subcellular Fractions

Cell Organelle (source)	Rate of H_2O_2 Formation nmoles/mg prot./min.	Conditions	Ref.
Peroxisomes (rat liver)	0.9	endogenous substrates	Chance and Oshimo (8)
Peroxisomes (rat liver)	8.6 - 16.4	uric acid supplemented	Boveris *et al.* (12)
Microsomes (rat liver)	1.7	NADH supplemented	Boveris *et al.* (12)
Mitochondria (pigeon heart)	0.7 - 0.8	succinate, state 4	Boveris *et al.* (13)
Mitochondria (rat heart)	0.5	succinate, state 4	Loschen *et al.* (9-11)
SMP's (beef heart)	0.2	succinate, antimycin	Loschen *et al.* (14)
Algeae (Anacystes nidulans)	400 nmoles/mg Chloroph./min.	succinate, illumination	Patterson and Myers (31)
Chloroplasts (spinach, peas)	not determined	ferridoxin or dithionite or illumination	Telfer *et al.* (32)

superoxide dismutase retained in the SMP fraction. An upper limit for O_2^- formation in the SMP system at pH 7 may therefore be estimated to be about 0.4 nmoles per mg protein per minute.

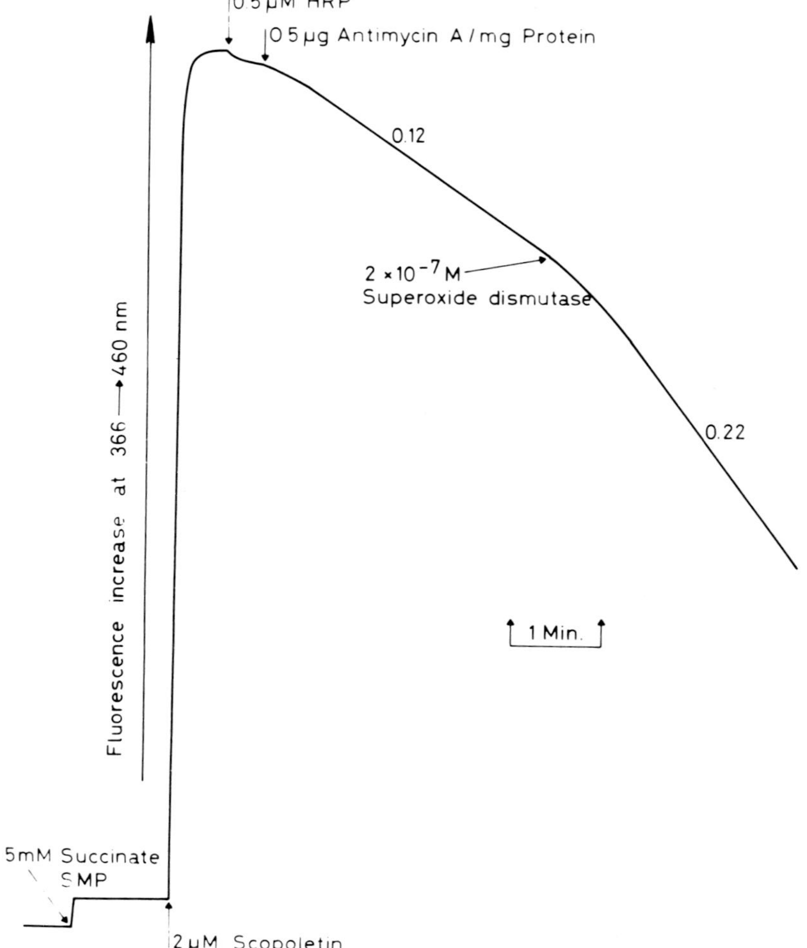

Fig.1. *Stimulation of the rate of mitochondrial H_2O_2 formation by SOD. Mitochondrial membrane fragments obtained by sonic disruption of intact beef heart mitochondria (SMP), in the presence of 2 mM EDTA, were washed three times to remove the mitochondrial superoxide dismutase. H_2O_2 formation was detected by aid of the fluorescent H_2O_2 indicator scopoletin, coupled to horse-radish peroxidase (HRP). The decrease of the scopoletin fluorescence reflects directly the rate of H_2O_2 formation, which is indicated by the numbers on the trace in nmoles H_2O_2 per mg protein and minute.*

Even higher rates of O_2^- formation are suggested when the oxidation of epinephrine is taken as an indicator of superoxide radicals (16). In the presence of excess catalase, the oxidation of epinephrine to adrenochrome may be due to superoxide radicals and can be taken as a reliable indication of O_2^- formation, if it is suppressed by addition of SOD. As can be seen from Fig.2, the rate of adrenochrome formation catalyzed by SMP's, supplemented with succinate and antimycin, decreases by a factor of about 10 upon addition of SOD.

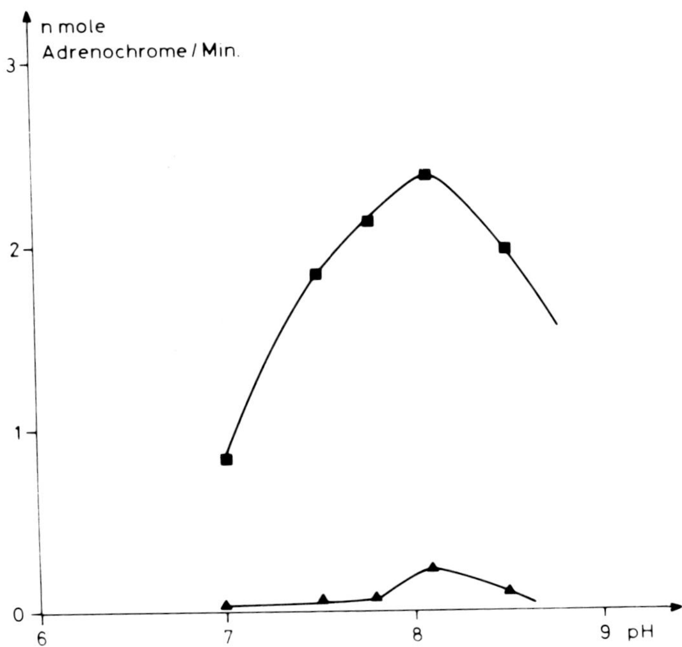

Fig.2. *Inhibition of the rate of adrenochrome formation by SOD in SMP's at various pH. Autoxidation of epinephrine to adrenochrome was induced in SMP's (0.5 mg prot./ml) by the addition of succinate (2 mM) and antimycin (0.5 μg/mg protein) in a reaction medium consisting of 0.25 M sucrose, 0.5 μM catalase, and 50 mM HEPES buffer of various pH-values, as indicated in the figure. The rate of adrenochrome formation was monitored with a dual wavelength spectrophotometer at 485 minus 575 nm and calculated by using the difference of the extinction coefficients at this wavelength pair of 3,260 $M^{-1}cm^{-1}$. The top trace indicates the rate of adrenochrome formation in the absence of SOD, the bottom trace the rate in the presence of 2 x 10^{-7} M SOD.*

A further qualitative proof for the involvement of O_2^- in mitochondrial H_2O_2 formation may be derived from the data presented in Fig.3. Conditions known to induce H_2O_2 formation in SMP's also show a marked luminol chemiluminescence. This luminescence, which by itself is already suggestive of the presence of O_2^- (17), can be completely suppressed by SOD.

The experiments cited so far indicate, by three independent analytical approaches, that O_2^- is formed in SMP's under experimental conditions identical to those in which H_2O_2 is generated, i.e. when exposed to antimycin in the presence of succinate.

Another set of experiments (Fig.3 and 4) reveals that cyanide inhibits both mitochondrial H_2O_2 formation and O_2^- production in SMP's. By means of the GSH peroxidase system as an analytical tool for H_2O_2 detection in the presence of cyanide, it could be shown that 5 mM CH^- completely inhibited the succinate-induced H_2O_2 formation of tightly coupled rat heart mitochondria. Similarly, cyanide suppressed the succinate- and antimycin-induced luminol chemoluminescence (Fig.3) and the adrenochrome formation (18) in SMP's.

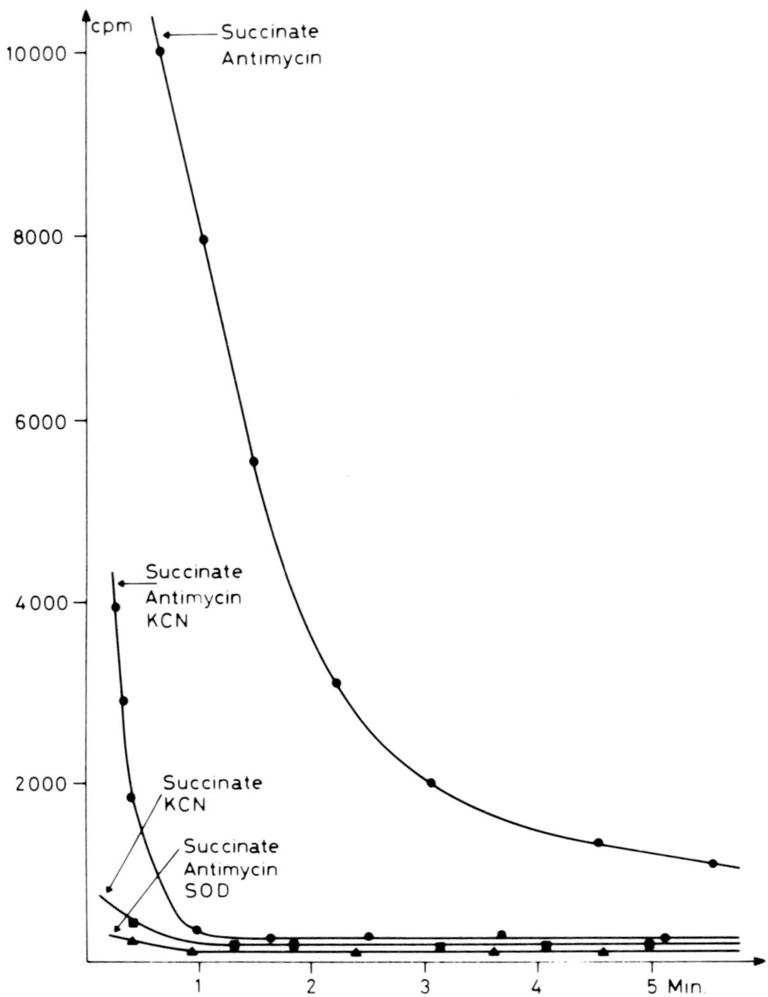

Fig.3. *Antimycin-induced luminol chemiluminescence in SMP's and its inhibition by either* CN^- *or SOD. SMP's (0.5 mg prot./ml) were incubated in a glass vial of a scintillation counter (tritium channel) in 0.25 M sucrose, 50 mM HEPES buffer pH 8.1, 1 mM luminol, and 0.5 μM catalase. Additions were made as indicated in the figure: 3 mM succinate, 0.5 μg antimycin/mg protein, 2 mM* CN^-*, and* 2×10^{-7} *M SOD. The luminol chemiluminescence was printed out every 10 sec as counts per minute.*

One set of experiments appeared to conflict with the assumption that O_2^- is always the precursor of mitochondrial H_2O_2. The rate of H_2O_2 formation in antimycin treated SMP's is increased by SOD at pH values below 7.3, while it is decreased at higher pH-values (14). However, the luminol chemiluminescence and adrenochrome formation can be suppressed by exogenous SOD over the whole pH-range. These discrepancies in the results were initially considered by Loschen *et al.* (14) as an indication of an additional substrate for the SOD formed at high pH-values in the SMP-system. The recent observations of Hodgson and Fridovich (19,20) offer an alternative explanation. In the test system in which O_2^- is indicated by H_2O_2 production upon addition of SOD (see Fig.1), the effect may be reversed at high pH-values by the peroxidase-like action of SOD. Since the other two test systems contain sufficient catalase to remove H_2O_2 from the equilibrium, no peroxidations should be observed in these cases.

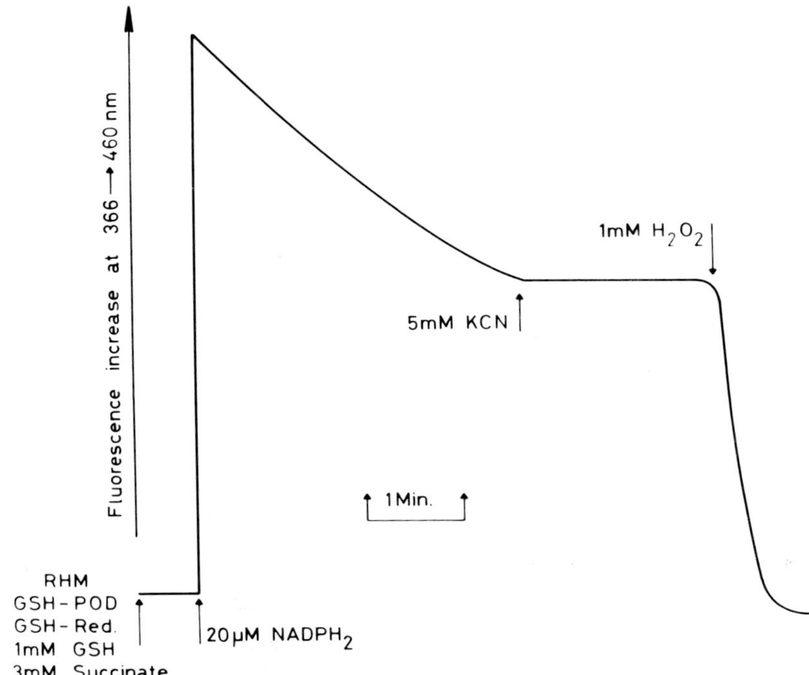

Fig.4. *Inhibition of mitochondrial H_2O_2 formation by CN^-. Formation of H_2O_2 was induced in tightly coupled rat heart mitochondria (respiratory control of 3.5, protein concentration of 1 mg/ml) with succinate in state 4 and monitored by the CN^--insensitive H_2O_2 test: glutathione (1 mM), glutathione peroxidase (1 U/ml), and glutathione reductase (0.5 U/ml) as regenerating system with $NADPH_2$ as electron donor. The CN^- insensitivity of the H_2O_2 detecting system is demonstrated by addition of H_2O_2. Oxidation of $NADPH_2$, indicating H_2O_2 formation, was monitored by fluorescence decrease in an Eppendorf fluorometer.*

In conclusion, O_2^- radicals can be detected in SMP's under the conditions known to result in H_2O_2 generation. Most probably, the formation of H_2O_2 in the respiratory chain of mitochondria is mediated by initially formed superoxide radicals.

THE SITE OF SUPEROXIDE MEDIATED HYDROGEN PEROXIDE FORMATION IN MITOCHONDRIA

In view of the above statement, it may be justified to consider briefly the metabolic conditions resulting in mitochondrial H_2O_2 formation via O_2^- radicals. Most of the pertinent experiments were performed by Loschen and his co-workers and have already been published (9-11,18,21).

The main conclusions derived from different experimental approaches were as follows:

a) Mitochondrial H_2O_2 is a product of the respiratory chain (9).
b) Mitochondrial H_2O_2 is not due to an incomplete reduction of molecular

oxygen by the terminal cytochrome oxidase (10,11).

c) Cytochrome b_{566} or a component of the zero potential pool in equilibrium with this redox carrier is the most likely candidate responsible for H_2O_2 formation (10,11).

d) Cytochrome b_{566} binds ligands isosteric with oxygen (CN^-, CO) and may thus be considered as autoxidizable (18).

It is certainly beyond the scope of this article to describe all the experiments yielding these conclusions, but it may be worthwile to summarize the main arguments.

All respiratory inhibitors blocking electron flow on the substrate site of the b-cytochromes suppress H_2O_2 formation, while antimycin, blocking electron flow on the oxygen site of the b-type cytochromes, enhances peroxide production in uncoupled intact mitochondria and in SMP's (see Fig.5).

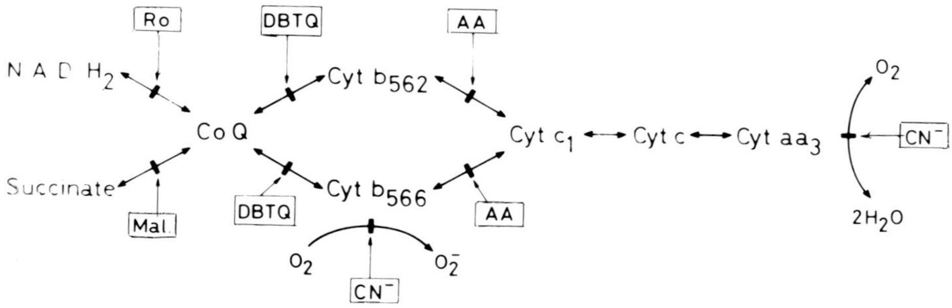

Fig.5. *Schematic presentation of the mitochondrial respiratory chain and the presumed inhibition sites of agents interfering with mitochondrial electron transport to the two sites of oxygen reduction. RO = rotenone, Mal. = malonate, DBTQ = dibromothymoquinone, AA = antimycin A.*

Mitochondrial H_2O_2 formation is further abolished by extraction of ubiquinone (13), but not by cytochrome *c* depletion (25). These observations clearly indicate that a respiratory component of the zero potential pool around the b-type cytochromes has to be reduced in order to react with molecular oxygen.

The isopotential pool with an apparent mid-point potential of about zero, unfortunately, comprises at least seven components: two b-type cytochromes, two flavoproteins, two iron-sulfur proteins and ubiquinone (22). One of these components, however, cytochrome b_{566}, shows some exceptional characteristics in that its apparent redox potential depends on the state of coupling and energization of the mitochondria (22,23). The redox state of cytochrome b_{566} can thus be manipulated selectively. Always and exclusively when cytochrome b_{566} is reduced, mitochondria or SMP's produce H_2O_2. In tightly coupled mitochondria the redox state of cytochrome b_{566} drastically changes with the phosphate potential. In the presence of ADP (state 3 according to ref. 24), cytochrome b_{566} is completely oxidized, but gets reduced in the presence of ATP (state 3 $\longrightarrow$ state 4 transition). Correspondingly, the H_2O_2 production in state 3 is essentially zero and reaches highest values in state 4 (9-11). In substrate-supplied, uncoupled mitochondria and in SMP's antimycin induces a selective reduction of cytochrome b_{566} and a concomitant H_2O_2 formation (11,21). When the redox state of the b-type cytochromes was manipulated by changing the succinate/fumarate ratio of the medium, mitochondrial H_2O_2 formation paralleled the degree of reduction of cytochrome b_{566}, but not of cytochrome b_{562} (21). In cytochrome *c*-depleted mitochondria, all respiratory components on the substrate side of cytochrome *c* are reduced except cytochrome b_{566}. In this system, addition of antimycin again results both in reduction of cytochrome b_{566} and in H_2O_2 formation (25).

The straightforward explanation of these findings is that cytochrome b_{566}, in the reduced state, readily reacts with oxygen. This assumption, of course, conflicts with the dogma that the cytochrome oxidase is the only respiratory chain component which can bind and reduce oxygen. Loschen (18), however, recorded CO difference spectra of SMP's at liquid nitrogen temperatures under conditions characterized by more or less selective cytochrome b_{566} reduction and compared them with the well known carbon monoxide difference spectra of SMP's completely reduced by dithionite. The CO difference spectra of the antimycin-treated SMP's showed peaks at 409, 520 and 554 nm, which could not be attributed to cytochrome aa_3 (Fig.6a and 6b).

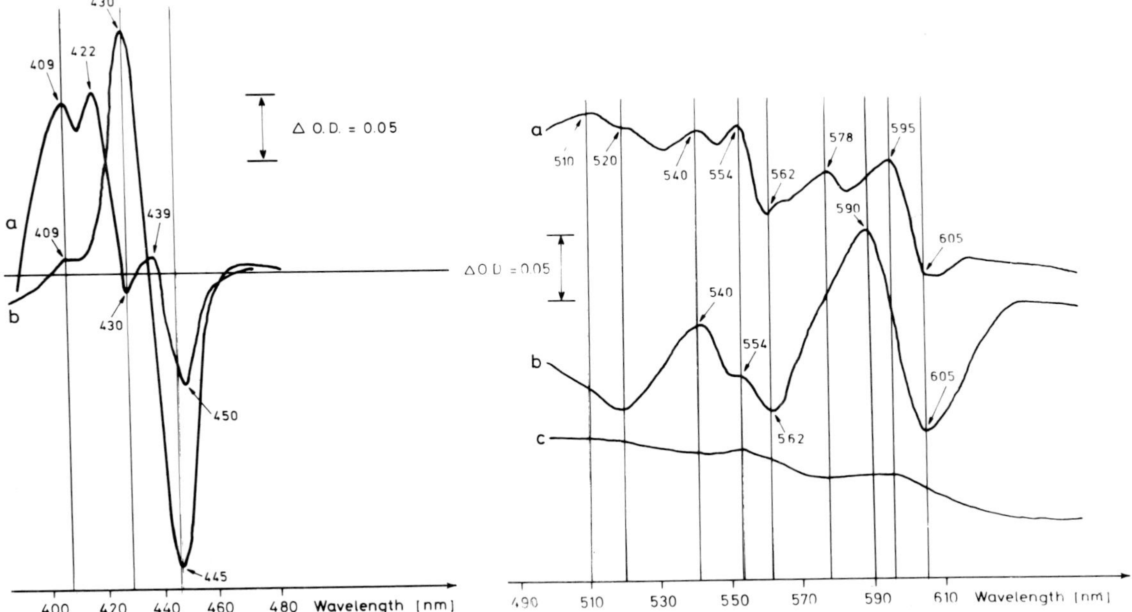

Fig.6. *CO difference spectra of SMP's at liquid nitrogen and room temperatures in the region of the Soret bands (6A) and the α- and β-bands (6B). Curves a of both figures represent the CO difference spectra of SMP's (4 mg prot./ml), recorded at liquid nitrogen temperature in the presence of 3 mM succinate and 5 μg antimycin/mg protein. Curves b of both figures represent the CO difference spectra of SMP's at room temperature reduced by dithionite. Curve c is a CO difference spectrum of SMP's obtained at liquid nitrogen temperature of a fully oxidized respiratory chain. The difference spectra were recorded in a Johnson Foundation split beam and an Aminco-Chance split beam spectrophotometer.*

This experiment proves that cytochrome aa_3 is not the only carbon monoxide-binding cytochrome of mitochondria. The newly observed peaks, however, may characterize the CO derivative of cytochrome b_{566}. In this context, we may recall that cyanide, another typical heme ligand, which is isosteric with oxygen, blocks mitochondrial H_2O_2 and superoxide formation. Cyanide, of course, binds to cytochrome aa_3, but since it blocks H_2O_2 formation completely, which definitely occurs on the substrate side of cytochrome *c*, it must also bind to an additional component, most probably to cytochrome b_{566}. Thus available evidence strongly suggests that cytochrome b_{566}, in the reduced state, binds CN^-, CO and O_2. Since this cytochrome is a monovalent redox catalyst, the primary product of its reaction with oxygen will be the superoxide radical.

POSSIBLE HAZARDS ARISING FROM SUPEROXIDE RADICALS IN MITOCHONDRIA

From the considerations outlined above, we have to face the possibility that H_2O_2 and superoxide radicals occur simultaneously in mitochondria. This situation favours the formation of further harmful radicals or high energy oxygen species and certainly requires an effective enzymatic system to protect the mitochondrial structures from unspecific oxidative damage (19,20,26, 27). Since mitochondria seem to be devoid of catalase, the protective system may consist of mitochondrial superoxide dismutase and glutathione peroxidase in connection with glutathione reductase and isocitrate dehydrogenase supplying NADPH.

Some model experiments reveal the consequences in a biological system insufficiently protected against highly reactive oxygen species. Highly purified inner mitochondrial membranes, when exposed to GSH as autoxidizable agent, are slowly decomposed with concomitant formation of malondialdehyde, a typical product of peroxidized lipids. This membrane peroxidation can be delayed by 10 hours and inhibited by about 50%, when 7×10^{-8} M superoxide dismutase is added to the incubation medium (26,27). This finding is in line with other observations characterizing superoxide dismutase as an efficient protector of biological membranes and the superoxide radical as a reactive metabolite of oxygen with powerful direct or indirect damaging capacity (28, 29).

Due to the presence of endogenous SOD, the role of SOD in intact mitochondria is not easily assessed. In general, agents or systems which prevent lipid peroxidation of mitochondrial membranes also protect intact mitochondria from GSH-dependent high amplitude swelling and loss of contractibility. This could be demonstrated, e.g. for α-tocopherol and the GSH peroxidase system (30), but surprisingly not for exogenous superoxide dismutase (27). Most probably, the SOD added to the medium does not contribute a significant increment to the overall SOD content of the system in order to provide an additional protection against lipid peroxidation mediated by O_2^-. On the other hand, addition of SOD to mitochondria which were simply aged at room temperature protected against the gradual decrease of the respiratory control coefficient, while such an effect could not be achieved by the GSH-peroxidase system (27). This observation suggests that the sites responsible for energetic coupling of respiration represent the most sensitive targets of superoxide radicals.

SUMMARY AND CONCLUSIONS

Under special conditions, the formation of superoxide radicals and H_2O_2 can be detected in SMP's. The rate of superoxide production in this system is higher than that of basic H_2O_2 production. Taking into account the spontaneous rate of O_2^- dismutation, it is reasonable to assume that H_2O_2 formation in SMP's is predominantly. if not exclusively, mediated by O_2^-. In functionally intact mitochondria, H_2O_2 is generated when a component of the zero potential pool gets reduced. Cytochrome b_{566} meets the requirements for the autoxidizable redox catalyst forming H_2O_2. However, since cytochrome b_{566} is a monovalent redox catalyst, the primary product of this oxygen reduction should be O_2^-. In intact mitochondria superoxide radicals are rapidly dimutated by the manganese-containing type of SOD of the mitochondrial matrix. This enzyme is therefore present in mitochondria to protect the mitochondrial membranes from oxidative breakdown. In this function, it is supported by GSH-peroxidase, which further metabolizes the product of O_2^- dismutation. In isolated mitochondria (Table I), an appreciable rate of 0.5 - 1.0 nmoles of H_2O_2 per mg protein/minute is observed. Since the test conditions employed in some of these investigations can be considered to approximate the physiological status, we may expect incomplete oxygen reduction by the respiratory chain to be a common event occurring *in vivo*. Being unaware of any role of O_2^- and H_2O_2 beneficial to mitochondria, we can view their formation as one of the inevitable risks of living with oxygen.

REFERENCES

1. Weisiger, R.A. and Fridovich, I. (1973). *J. Biol. Chem. 248*, 4793-4796.
2. Weisiger, R.A. and Fridovich, I. (1973). *J. Biol. Chem. 248*, 3582-3592.
3. Flohé, L., Günzler, W.A. and Schock, H.H. (1973). *FEBS Letters 32*, 132-134.
4. Flohé, L. and Schlegel, W. (1971). *Hoppe-Seyler's Z. Physiol. Chem. 352*, 1401-1410.
5. Bueding, E. and Charms, B. (1952). *J. Biol. Chem. 196*, 615-627.
6. Jensen, P.K. (1966). *Biochim. Biophys. Acta 122*, 157-166.
7. Hinkle, P.C., Butow, R.A., Racker, E. and Chance, B. (1967). *J. Biol. Chem. 242*, 5169-5173.
8. Chance, B. and Oshino, N. (1971). *Biochem. J. 122*, 225-233.
9. Loschen, G., Flohé, L. and Chance, B. (1971). *FEBS Letters 18*, 261-264.
10. Loschen, G., Azzi, A. and Flohé, L. (1974). In "Glutathione" (L. Flohé, H.Ch. Benöhr, H. Sies, H.D. Waller and A. Wendel, eds.) pp 228-236, Georg Thieme Publishers, Stuttgart.
11. Loschen, G., Azzi, A. and Flohé, L. (1973). *Hoppe-Seyler's Z. Physiol. Chem. 354*, 791-794.
12. Boveris, A., Oshino, N. and Chance, B. (1972). *Biochem. J. 128*, 617-630.
13. Boveris, A. and Chance, B. (1973). *Biochem. J. 134*, 707-716.
14. Loschen, G., Azzi, A., Richter, Ch. and Flohé, L. (1974). *FEBS Letters 42*, 68-72.
15. Rotilio, G., Bray, R.C. and Fielden, E.M. (1972). *Biochim. Biophys. Acta 268*, 605-609.
16. Misra, H.P. and Fridovich, I. (1972). *J. Biol. Chem. 247*, 3170-3175.
17. Hodgson, E.K. and Fridovich, I. (1973). *Photochem. Photobiol. 18*, 451-455.
18. Loschen, G. (1975). Ph.D. Thesis, University of Tübingen, W. Germany.
19. Hodgson, E.K. and Fridovich, I. (1975). *Biochemistry 14*, 5294-5298.
20. Hodgson, E.K. and Fridovich, I. (1975). *Biochemistry 14*, 5299-5303.
21. Loschen, G., Azzi, A. and Flohé, L. (1973). *FEBS Letters 33*, 84-88.
22. Wikström, M.K.F. (1973). *Biochim. Biophys. Acta 301*, 155-193.
23. Chance, B., Wilson, D.F., Dutton, P.L. and Erecinska, M. (1970). *Proc. Nat. Acad. Sci. 66*, 1175-1182.
24. Chance, B. and Williams, G.R. (1956). In "Advances of Enzymology" Vol.17, (F.F. Nord, ed.) p 65 ff, Interscience Publishers Inc., New York.

25. Losche, G., Azzi, A. and Flohé, L. (1974). In "Alcohol and Aldehyde Metabolizing Systems" (R.G. Thurman, T. Yonetani, J.R. Williamson and B. Chance, eds.) pp 215-219, Academic Press, New York and London.
26. Zimmerman, R., Flohé, L., Weser, U. and Hartmann, H.J. (1973). *FEBS Letters 29,* 117-120.
27. Flohé, L. and Zimmermann, R. (1974). In "Glutathione" (L. Flohé, H.Ch. Benöhr, H. Sies, H.D. Waller and A. Wendel, eds.) pp 245-260, Georg Thieme Publishers, Stuttgart.
28. Lavelle, F., Michelson, A.M. and Dimitrijevic, L. (1973). *Biochim. Biophys. Acta 55,* 350-357.
29. Fridovich, I. (1975). In "Annual Reports in Medical Chemistry" (R.V. Heinzelman, ed.) Vol.10, pp 257-264, Academic Press, New York and London.
30. Flohé, L., Günzler, W.A. and Ladenstein, R. (1976). In "Glutathione Peroxidase: Metabolism and Function" (I.M. Arias and W.B. Jakoby, eds.) pp 115-138, Raven Press, New York.
31. Patterson, C.O.P. and Myers, J. (1973). *Plant Physiol. 51,* 104-109.
32. Telfer, A., Cammack, R. and Evans, M.C.W. (1970). *FEBS Letters, 10,* 21-24.

SUPEROXIDE AND HYDROXYLATION REACTIONS

B. HALLIWELL

Department of Biochemistry
King's College
Strand, London WC2R 2LS, U.K.

INTRODUCTION

Why is O_2^- dangerous? O_2^- is a good nucleophile - indeed, it has been used as a nucleophile in several chemical laboratories (San Filippo *et al.*, 1975; Corey *et al.*, 1975) - and should therefore react readily with electrophilic sites on biological molecules. For example, it attacks the carbonyl group of 2-ketoglutaric acid, giving succinate as the eventual product (San Filippo *et al.*, 1976). Also, since E_o for the system O_2/O_2^- is -0.33V (Muir-Wood, 1974; Ilan *et al.*, 1976), O_2^- can reduce molecules such as cytochrome *c*. E_o' for the system O_2^-/H_2O_2 is +0.94V and so O_2^- can oxidise molecules such as ascorbic acid (Nishikimi, 1975a). Ascorbic acid is widely distributed in plants and animals, and the products of its oxidation seem to be highly toxic, especially if traces of metal ions are present (Orr, 1967a,b; Wong *et al.*, 1974; Karnovsky, 1975). O_2^- slowly oxidises NADH and NADPH, but its reaction with NADH is greatly accelerated by lactate dehydrogenase (Bielski and Chan, 1973).

Nevertheless, O_2^- does not seem to be especially reactive: it does not, for example, attack alkanes or alkenes.

Dismutation of O_2^-, either spontaneously or by the action of superoxide dismutase, gives H_2O_2. H_2O_2 is a rather unreactive molecule, which is probably always present at low concentrations in aerobic cells (Halliwell, 1974). However, in the presence of certain transition metal ions, H_2O_2 breaks down to give highly reactive radical species. A mixture of Fe^{2+} and H_2O_2 ("Fenton's reagent") is an extremely reactive system, attacking both aromatic and aliphatic compounds (Walling, 1975a).

Haber and Weiss (1934) proposed that O_2^- reacts with H_2O_2 (eqn. 1):

$$H_2O_2 + O_2^- \longrightarrow .OH + OH^- + O_2 \qquad (1)$$

As Fee has stated, there is no *direct* evidence that the Haber-Weiss reaction occurs in aqueous systems at physiological pH. Nevertheless, there is considerable circumstancial evidence for the production of .OH from H_2O_2 and

O_2^- in that many reactions inhibited by catalase, superoxide dismutase and scavengers of the hydroxyl radical have been reported (Fridovich, 1974, 1975; Hodgson and Fridovich, 1976a,b; Halliwell and Ahluwalia, 1976; also see Bors *et al.*, 1976). The hydroxyl radical is extremely reactive and may be one of the major species responsible for the properties of Fenton's reagent (Walling, 1975a).

The non-enzymic dismutation of O_2^- has been reported to generate O_2 in the singlet state (Khan, 1970). Singlet O_2 is reactive enough to attack molecules such as alkenes (e.g. the polyunsaturated fatty acids found in membrane lipids) and aromatic compounds. However, singlet O_2 generation from O_2^- has not been confirmed by other workers (Nilsson and Kearns, 1974; Bors *et al.*, 1974; King *et al.*, 1975). To complicate the matter further, O_2^- itself may quench singlet O_2 by electron transfer, an observation which might account for the different results mentioned above, (Guiraud and Foote, 1976). Also, Kellogg and Fridovich (1975) have proposed that the O_2 generated in the Haber-Weiss reaction is in the singlet state. Clearly, further studies of this latter reaction are required.

APPROACHES TO THE STUDY OF THE ROLE OF O_2^- IN HYDROXYLATION REACTIONS

Essentially two types of approach have been used. First, chemical model systems have been studied to investigate means by which hydroxylation might occur. Criteria which such systems have to meet in order to be considered as suitable models for enzyme-catalysed reactions have been discussed in detail by Ullrich *et al.* (1972) and Hamilton (1974).

One of the most important of these criteria is that the non-enzymic system should show the "NIH shift" when hydroxylating aromatic molecules. It has been found for many enzyme-catalysed hydroxylations that a hydrogen atom at the position on the aromatic ring which eventually gains the hydroxyl group becomes shifted to an adjacent carbon atom, as shown in eqn. 2 (Daly *et al.*, 1972).

$$C_6H_5\text{-}{}^*H \xrightarrow[O_2]{\text{enzyme}} C_6H_4(OH)\text{-}{}^*H \quad (2)$$

Secondly, direct studies on enzymic hydroxylating systems may be carried out. Both enzymic and non-enzymic systems might be dependent on species such as O_2^-, H_2O_2, $.OH$ or singlet O_2 for hydroxylation. How well can we detect the involvement of such species?

H_2O_2 is broken down by catalase, so inhibition observed on addition of catalase to a system is frequently taken as evidence that H_2O_2 is required. Catalase should first, of course, be checked for possible contamination by superoxide dismutase (Halliwell, 1973). Some systems might also produce intermediates which react with proteins, e.g. addition of catalase to autoxidising ascorbic acid decreases the rate of autoxidation, but this cannot be attributed solely to removal of H_2O_2, as there is a reaction between the protein and some species produced during autoxidation (Orr, 1967a,b). Autoxidation of ascorbate is inhibited by superoxide dismutase also, but dismutase apoenzyme or bovine serum albumin are equally effective (Halliwell and Foyer, 1976). *Controls*, using inactive enzymes and bovine serum albumin, should therefore always be carried out. A similar caution has been made by Amano *et al.* (1975).

An inhibition of a reaction on addition of superoxide dismutase is usually taken to indicate the involvement of O_2^-, provided that the above controls give negative results. The assumption is that this enzyme *specifically* catalyses breakdown of O_2^-. This has been challenged by Finazzi-Agro *et al.* (1972) and Paschen and Weser (1973), who report that superoxide dismutase also quenches singlet O_2. However, the quenching of singlet O_2 seems to be a non-specific effect of the protein rather than an enzymic reaction: the metal-free apoenzyme is equally effective (Matheson *et al.*, 1975). This again emphasises the need for controls before drawing conclusions from inhibitions by superoxide dismutase or catalase.

Most studies have used the copper-zinc superoxide dismutase from erythrocytes as a probe for the involvement of O_2^-. However, copper-zinc enzymes are inactivated by the H_2O_2 usually present in such systems, especially at alkaline pH values (t1/2 for inactivation of bovine erythrocuprein [3.8 x 10^{-7} M] by 0.13 mM H_2O_2 is 3 min at pH 10.8, but 27 min at pH 9.6) (Bray *et al.*, 1974; Hodgson and Fridovich, 1975a). Further, the enzyme is itself capable of bringing about certain peroxidations such as those of linoleic acid, diphenylisobenzofuran and bilirubin, reagents often used as "specific" scavengers of singlet O_2 (Hodgson and Fridovich, 1975b). In view of these results, it may be seen that an observation of a short-lived inhibition on addition of copper-zinc superoxide dismutase to a reaction mixture might merely mean that the enzyme has become inactivated. Such results should always be checked using a superoxide dismutase which is not inhibited by H_2O_2, such as the manganoenzyme from *Plectonema boryanum* (Asada *et al.*, 1975) or *Rhodopseudomonas spheroides* (Lumsden, 1975).

EVIDENCE THAT O_2^- IS INVOLVED IN HYDROXYLATION REACTIONS

In Non-Enzymic Systems

Addition of O_2^-, dissolved in dimethylsulphoxide, to pure aqueous solutions of aromatic compounds failed to induce hydroxylation (Fee and Hildenbrand, 1974). However, when O_2^- was generated continuously in aqueous solution by use of a mixture of NADH and phenazine methosulphate, hydroxylation of added aromatic compounds was observed (Ravindranath *et al.*, 1974; Ashok Kumar *et al.*, 1975). Hydroxylation is inhibited by superoxide dismutase, but not by the heat-denatured enzyme nor by catalase, and so formation of .OH by the Haber-Weiss reaction does not appear to be involved. The pH optimum for hydroxylation is 4.5 - 5.0, which suggests that $HO_2^{\cdot}$ rather than O_2^- may be required, as the pK_a for the reaction:

$$O_2^- + H^+ \longrightarrow HO_2^{\cdot}$$

is 4.88 (Behar *et al.*, 1970). Traces of metal ions stimulate the hydroxylation, but it is not clear if they are absolutely necessary for it to proceed. It would be interesting to know if hydroxylation by the NADH/PMS system shows the NIH shift.

Hydroxylation of aromatic compounds *in vitro* can also be achieved by a mixture of $FeSO_4$, EDTA and ascorbic acid (Udenfriend *et al.*, 1954). However, hydroxylation by this system is not inhibited by superoxide dismutase (Halliwell and Ahluwalia, 1976) and it does not usually show the NIH shift (Hamilton, 1974).

Enzyme-Catalysed Reactions

Enzyme-catalysed hydroxylation reactions derive the oxygen attached to the substrate from molecular O_2. In many cases, only 1 atom from the O_2 molecule enters the substrate and the other appears in H_2O. Enzymes catalysing this type of reaction may be classified as "mono-oxygenases". Many mono-oxygenases require an external reductant (e.g. NAD(P)H, ascorbate, tetrahydropteridines) for hydroxylation to proceed, and the general reaction catalysed may be represented as in eqn. 3, where SH is the substrate and RH_2 the reductant:

$$SH + O_2 + RH_2 \longrightarrow SOH + H_2O + R \tag{3}$$

Some hydroxylases, however, are "dioxygenases", inserting both atoms of oxygen from O_2 into the substrates: typical are the lysyl and prolyl hydroxylases, which are discussed later.

Flavoproteins. Several hydroxylases are flavoprotein mono-oxygenases, including p-hydroxybenzoate hydroxylase (eqn. 4), orcinol hydroxylase (eqn. 5) and salicylate hydroxylase (eqn. 6).

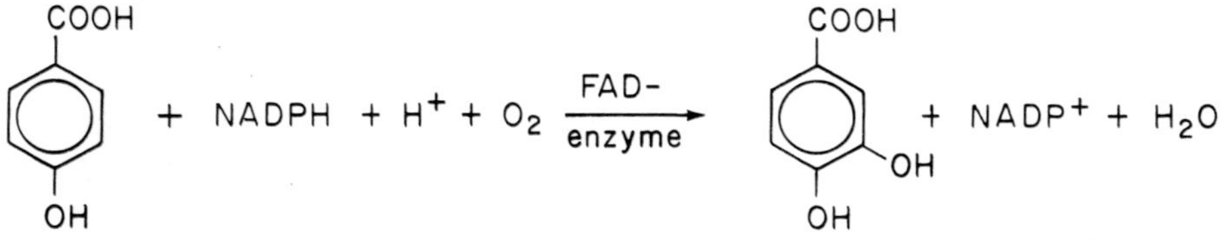

$$\text{p-hydroxybenzoate} + NADPH + H^+ + O_2 \xrightarrow[\text{enzyme}]{\text{FAD-}} \text{protocatechuic acid} + NADP^+ + H_2O \tag{4}$$

protocatechuic acid

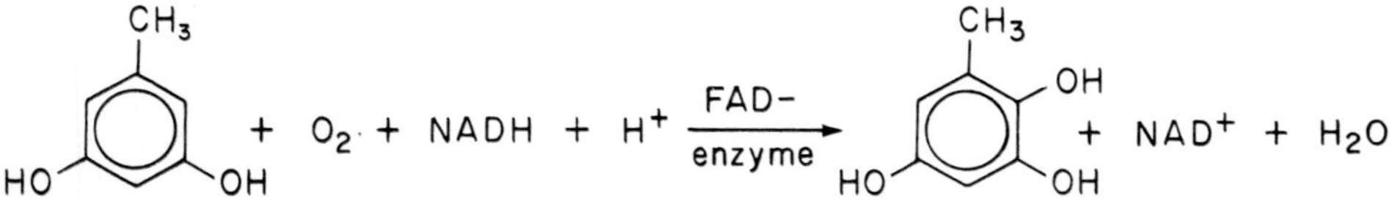

$$\text{orcinol} + O_2 + NADH + H^+ \xrightarrow[\text{enzyme}]{\text{FAD-}} \text{2,3,5-trihydroxytoluene} + NAD^+ + H_2O \tag{5}$$

2,3,5-trihydroxytoluene

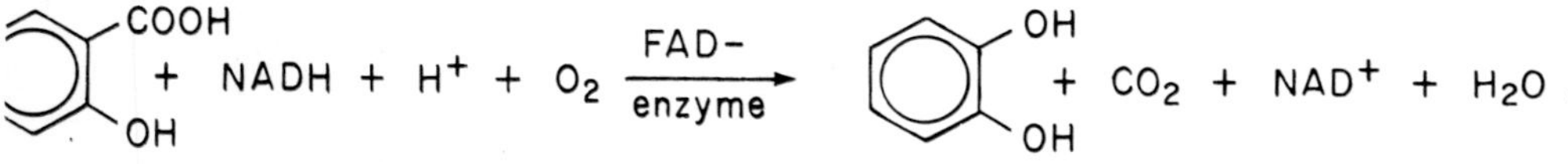

$$\text{salicylate} + NADH + H^+ + O_2 \xrightarrow[\text{enzyme}]{\text{FAD-}} \text{catechol} + CO_2 + NAD^+ + H_2O \tag{6}$$

catechol

The reduced flavoproteins interact with O_2 and substrate to give a ternary complex, which breaks down to yield oxidised enzyme, H_2O and hydroxylated product. The flavin prosthetic group is then reduced again by interaction with NAD(P)H (see Gunsalus *et al.* (1975) for a detailed discussion of the order of substrate addition to different flavoprotein hydroxylases).

Since reduced flavins in aqueous solution convert O_2 to O_2^- (Massey *et al.*, 1969; Ballou *et al.*, 1969), it may be that O_2^-, generated at the active site, is involved in hydroxylation. Strickland and Massey (1972) reported that O_2^- generated by a flavin in a simple model system converts p-hydroxybenzoate to polyhydroxybenzoates and other products. However, Massey *et al.* (1969) did not detect production of O_2^- by the reduced forms of several flavoprotein hydroxylases, using cytochrome *c* as a probe. This suggests that O_2^- or $HO_2^{\cdot}$, if involved, is not released into free solution, where it could reduce cytochrome *c*, but remains bound to the active site, perhaps as a flavin-oxygen adduct (e.g. a flavin peroxide).

Prema Kumar *et al.* (1972) reported that hydroxylation catalysed by the fungal enzyme m-hydroxybenzoate-4-hydroxylase (eqn. 7)

$$\text{m-hydroxybenzoic acid (COOH, OH)} + NADPH + H^+ + O_2 \xrightarrow[\text{enzyme}]{\text{FAD-}} \text{protocatechuic acid (COOH, OH, OH)} + NADP^+ + H_2O \quad (7)$$

protocatechuic acid

is abolished by superoxide dismutase but not by heat-denatured dismutase. This observation strongly suggests that O_2^- is somehow involved, but also means that it is accessible to the superoxide dismutase, i.e. released free into the solution. Further studies on this enzyme are clearly required.

Tetrahydropteridine-Dependent Hydroxylases. Some monoxygenases use a reduced pteridine as reductant, well known examples being phenylalanine hydroxylase from liver and tyrosine hydroxylase from adrenal medulla. These enzymes also require Fe^{2+}. The overall reaction may be written as follows, where R-H is the substrate:

$$\text{R-H} + \text{tetrahydropteridine} + O_2 \xrightarrow{Fe^{2+}} \text{R-OH} + \text{dihydropteridine} + H_2O \quad (8)$$

Tetrahydropteridine-requiring hydroxylases have many characteristics, suggesting that their mechanism is similar to that of flavoprotein hydroxylases (see Hamilton, 1974). Indeed, autoxidation of tetrahydropteridines generates O_2^- (Kaufman and Fisher, 1974; Nishikimi, 1975b; Heikkila and Cohen, 1975). Although superoxide dismutase does not inhibit hydroxylation by phenylalanine or tyrosine hydroxylases (Kaufman and Fisher, 1974; Petrack and Chertock, 1974), this might merely mean that O_2^-, generated at the active site, is inaccessible to the dismutase and might perhaps exist as a pteridine-O_2^- adduct or even as $Fe^{3+}\text{-}O_2^- \rightarrow Fe^{2+}\text{-}O_2$. Catechols such as Tiron are very powerful inhibitors of pteridine-dependent hydroxylases, and this might be consistent with an ability to penetrate to the active site and react with bound O_2^- (Miller, 1970): indeed, Tiron has been used as a probe for the involvement of O_2^- in reactions not inhibited by superoxide dismutase (e.g. Miller and Macdowall, 1975). Of course, it must not be assumed that O_2^- is the only oxidant which can act upon catechols.

Incubation of phenylalanine in air with 6,7-dimethyltetrahydropterin causes formation of some tyrosine (Woolf *et al.*, 1971). Also, reduced pteridines can replace ascorbic acid in the Udenfriend system (Bobst and Viscontini, 1966), although, as we have seen, this system is not a good model for enzymic hydroxylations.

Peroxidase. It has been known for some time that a mixture of peroxidase and dihydroxyfumaric acid catalyses the hydroxylation of aromatic compounds (e.g. Buhler and Mason, 1961). This hydroxylation does not show the NIH shift

(Daly and Jerina, 1970). Work in my laboratory has demonstrated that hydroxylation by this system is completely inhibited by superoxide dismutase and by Cu^{2+}, Mn^{2+} or Fe^{2+}-EDTA at concentrations which scavenge O_2^- (Halliwell, 1975a; Halliwell and Ahluwalia, 1976). Superoxide dismutase or Cu^{2+} only partially inhibit dihydroxyfumarate oxidation in these reaction mixtures, and Mn^{2+} increases it, so inhibition of hydroxylation by these reagents cannot be attributed to an inhibition of dihydroxyfumarate oxidation.

Solutions of dihydroxyfumarate slowly oxidise and always contain traces of H_2O_2. Apparently, the peroxidase uses this H_2O_2 to catalyse further oxidation (Yamazaki and Piette, 1963) and one of the products so obtained reduces O_2 to O_2^-, which may have an autocatalytic effect on the breakdown of dihydroxyfumarate and can also produce more H_2O_2 by non-enzymic dismutation. Indeed, Yamazaki and Yamazaki (1973) observed that oxidation of dihydroxyfumarate by peroxidase is transiently inhibited by superoxide dismutase (a copper-zinc enzyme was used - see the cautionary note above) and Nilsson *et al.* (1969) provided electron spin resonance evidence for O_2^- production in the peroxidase-dihydroxyfumarate system at alkaline pH values. Further, it is known that peroxidase, in the presence of dihydroxyfumarate, becomes largely converted into compound III ("oxyperoxidase"), which appears to be an enzyme-O_2^- adduct ($Fe^{3+}—O_2^- \longleftrightarrow Fe^{2+}—O_2$).

Oxyperoxidase can be formed by a direct reaction of O_2^- with the enzyme (Sawada and Yamazaki, 1973) and can break down again to give free O_2^- (Rotilio *et al.*, 1975). Breakdown of oxyperoxidase is not increased by superoxide dismutase but is greatly enhanced by addition of Cu^{2+} (Yokota and Yamazaki, 1965), which may be able to penetrate to the active site and react with the bound O_2^-.

The complete inhibition by superoxide dismutase, but not by the heat-denatured enzyme, shows that O_2^- is essential for hydroxylation by the peroxidase-dihydroxyfumarate system. This reaction is also partially inhibited by scavengers of .OH, such as Tris, formate, ethanol and mannitol, at concentrations having little effect on dihydroxyfumarate oxidation (Halliwell and Ahluwalia, 1976). Hence, a Haber-Weiss reaction may be involved, to produce .OH as the true hydroxylating species. The lack of the NIH shift in hydroxylations catalysed by this system perhaps suggests that .OH is unlikely to be an intermediate in other enzyme-catalysed hydroxylations.

A mixture of $FeSO_4$ and dihydroxyfumarate also catalyses hydroxylation of aromatic compounds (Goscin and Fridovich, 1972) and superoxide dismutase partially inhibits this reaction.

Cytochrome P_{450}-Dependent Hydroxylating Systems. The haemoproteins cytochrome P_{450}, together with various electron-donating systems, are involved in a large number of hydroxylations, including 11β hydroxylation of steroids in adrenal mitochondria, hydroxylation of "foreign compounds" by microsomes and hydroxylations in some micro-organisms, especially studied being the camphor methylene hydroxylase system from *Pseudomonas putida*.

The liver microsomal enzyme system, which catalyses hydroxylation of alkanes, fatty acids and many drugs, has been resolved into three components: a cytochrome P_{450}, an NADPH-dependent cytochrome P_{450} reductase (which also reduces cytochrome *c*) and lecithin. These purified components may be reconstituted to give an active hydroxylating system. Whereas hydroxylation by isolated, intact microsomes is not usually significantly inhibited by superoxide dismutase, hydroxylation by the reconstituted system is (Strobel and

Coon, 1971): it is also strongly inhibited by Tiron. Further, NADPH and the reductase can be replaced, although at diminished efficiency, by a O_2^--generating system, and it is known that the NADPH-cytochrome *c* reductase from rat liver can generate O_2^- (Pedersen and Aust, 1972). These results imply that O_2^-, bound to cytochrome P_{450}, might be the true hydroxylating species, and Fig.1 shows a scheme of the type proposed by many workers to account for these results (Symposium, 1972; Colloquium, 1976).

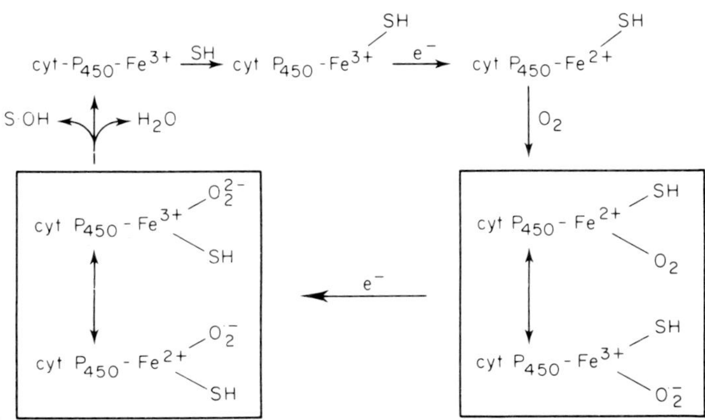

Fig.1. *A scheme accounting for hydroxylation by cytochrome P_{450}. The reductants and reductase enzymes supplying electrons (e^-) are not shown. SH is the substrate for hydroxylation and S.OH the product. An O_2^--generating system presumably acts to convert cyt P_{450} - Fe^{2+} - SH directly into cyt P_{450} - Fe^{2+} - SH - O_2^-, but it could also be argued that it merely acts as a reductant ($O_2^- \longrightarrow O_2 + e^-$) - see Strobel and Coon (1971).*

In the camphor methylene hydroxylase system from *Pseudomonas putida*, electrons are supplied to a cytochrome P_{450} (P_{450} cam) by the iron-sulphur protein putidaredoxin, which is kept reduced by an NADH-linked flavoprotein dehydrogenase. Sliger *et al.* (1974) showed that breakdown of the P_{450} cam-O_2-substrate ternary complex produces O_2^-, in full agreement with the scheme in Fig.1.

In adrenal mitochondria, cytochrome P_{450} is reduced by electrons from adrenodoxin (an iron-sulphur protein): a flavoprotein dehydrogenase which catalyses reduction of adrenodoxin by NADPH is also present. The complex of adrenodoxin and dehydrogenase has been shown to produce O_2^- when supplied with NADPH and O_2 (Chu and Kimura, 1973). This observation, like that of O_2^- generation by the liver NADPH-cytochrome *c* reductase, merely means that the reduced forms of the respective enzymes *can* interact with O_2 to yield O_2^-, and not that they actually do so *in vivo*: the electrons may well be passed on directly to the cytochrome P_{450}, which then themselves complex with O_2 (see Fig.1). The somewhat "looser" conditions of the reconstituted microsomal systems used by Strobel and Coon (1971), as compared with intact microsomes, might have made direct electron transfer more difficult and allowed O_2^- to become an intermediate. Nevertheless, it cannot be ruled out that, in some cases, the reducing system might donate O_2^- directly to the cytochrome P_{450} *in vivo*.

It is also interesting to note that cytochrome P_{450} can react with $NaIO_4$, some peroxides or $NaClO_2$ to give intermediates active in hydroxylation (Hrycay *et al.*, 1975; Estabrook *et al.*, 1976).

Richter *et al.* (1976) have observed an inhibition of some demethylation reactions catalysed by the liver microsomal cytochrome P_{450} system upon addition of a copper-tyrosine complex, which is known to catalyse the breakdown of O_2^- (Joester *et al.*, 1972). They proposed that this complex was able to penetrate to the (hydrophobic) active site of the cytochrome P_{450} and break down bound O_2^-. Clearly, these simple complexes, and even free Cu^{2+} itself might be of great use in probing the mechanisms of hydroxylation reactions. It must be borne in mind, however, that Cu^{2+} can inhibit many enzymes by simple reaction with -SH groups.

Rubredoxin-Dependent Systems. The organism *Pseudomonas oleovorans* contains an "ω-hydroxylating system" which catalyses the terminal hydroxylation of alkanes and fatty acids: this system consists of an iron-sulphur protein (rubredoxin), an NADH-linked flavoprotein, rubredoxin reductase, and an ω-hydroxylase. Although this system seems to resemble some of the cytochrome P_{450}-dependent hydroxylating systems discussed above, it has been found that the ω-hydroxylase contains only non-haem iron (Ruettinger *et al.*, 1974). Nevertheless, a mixture of NADH, rubredoxin and the reductase generates O_2^-, which appears to be able to interact with the ω-hydroxylase (May *et al.*, 1973). These preliminary observations are clearly in agreement with a hydroxylation mechanism involving enzyme-bound O_2^-, perhaps similar to that in Fig.1.

2-Ketoglutarate-Dependent Hydroxylases. Several hydroxylases which require 2-ketoglutarate for activity are known. These enzymes are dioxygenases, incorporating both atoms of oxygen from O_2 into the substrates, and the reaction they catalyse may be represented as in eqn. 9, where S. is the substrate:

$$\text{2-ketoglutarate} + S + O_2 \longrightarrow \text{succinate} + CO_2 + S.OH \qquad (9)$$

These enzymes usually require Fe^{2+} and a reductant, such as ascorbate, for activity. (Further details of the properties of these enzymes may be found in a recent review by Abbot and Udenfriend (1974).)

Perhaps the best known members of this group are the proline and lysine hydroxylases, which are responsible for the hydroxylation of protocollagen to give collagen. Hydroxylation by prolyl hydroxylase is not inhibited by superoxide dismutase (Cardinale and Udenfriend, 1974), but this does not rule out the involvement of bound O_2^- in hydroxylation. It is interesting to note that O_2^- converts 2-ketoglutarate into succinate *in vitro* (San Filippo *et al.*, 1976): persuccinic acid is a possible intermediate in this reaction. Hamilton (1974) has proposed that the initial stage of hydroxylation catalysed by these enzymes is the attack of an iron-O_2 complex on the carbonyl group of 2-ketoglutarate to give persuccinate, which then hydroxylates the substrate. Some evidence is available to support this assertion (Hamilton, 1974). Obviously, a complex such as Fe^{2+}-O_2 would have considerable resonance contribution from Fe^{3+}-O_2^-, and so the reaction might well be regarded as nucleophilic attack by a form of superoxide. The reductant may function to regenerate Fe^{2+} from Fe^{3+}.

Copper-Containing Hydroxylases. A. Dopamine-β-Hydroxylase. Dopamine-β-hydroxylase is a mono-oxygenase which converts dopamine to noradrenaline (eqn. 10). It contains copper and also requires ascorbate for optimal activity. Liu *et al.* (1974) reported that hydroxylation catalysed by this enzyme is decreased by addition of large amounts of superoxide dismutase to the reaction mixture, but no control with heated enzyme was carried out. In view of the interactions of ascorbic acid with superoxide dismutase (Halliwell and Foyer, 1976), further studies on this enzyme need to be performed. Inhibition of the hydroxylation by superoxide dismutase has not been confirmed by other workers.

$$\text{HO-C}_6\text{H}_3\text{(OH)-CH}_2\cdot\text{CH}_2\cdot\overset{+}{\text{N}}\text{H}_3 + \text{ascorbate} + O_2 \longrightarrow \text{HO-C}_6\text{H}_3\text{(OH)-CHOH}\cdot\text{CH}_2\cdot\overset{+}{\text{N}}\text{H}_3 + \text{dehydroascorbate} + H_2O \qquad (10)$$

B. Phenolases. Phenolases are enzymes which catalyze two reactions: hydroxylation of a phenol (eqn. 11) and oxidation of o-dihydric phenols to quinones (eqn. 12).

$$\text{R-C}_6\text{H}_4\text{-OH} + O_2 + \text{red}\cdot H_2 \longrightarrow \text{R-C}_6\text{H}_3\text{(OH)}_2 + H_2O + \text{red}_{ox} \qquad (11)$$

$$\text{R-C}_6\text{H}_3\text{(OH)}_2 + \tfrac{1}{2}O_2 \longrightarrow \text{R-C}_6\text{H}_3\text{O}_2 \text{ (o-quinone)} + H_2O \qquad (12)$$

The reductant (red. H_2) used by these enzymes for hydroxylation (eqn. 11) seems to be a diphenol, which becomes converted to a quinone: hence the continuance of hydroxylation may be dependent upon the availability of diphenol. Hydroxylation of a monophenol, in the absence of added diphenol, proceeds only after a lag period in which the diphenol required arises by some mechanism as yet unclear (see Vanneste and Zuberbuhler, 1974). The lag period is shortened by adding reductants such as NAD(P)H, ascorbate or tetrahydropteridines, which appear to act by reducing quinones back to diphenols rather than by interacting directly with the enzyme.

Phenolases contain copper, but its valency state during catalysis is not yet clear. A scheme originally proposed in 1955 by Mason (eqns. 13 - 15) still seems capable of accounting for most of the properties of phenolases: the form presented below assumes an active site containing 2 atoms of Cu:

$$E(Cu^{2+})_2 \longrightarrow + \text{ o-diphenol} \quad E(Cu^{+})_2 + \text{o-quinone} + 2H^{+} \qquad (13)$$

"resting enzyme"

$$E(Cu^{+})_2 + O_2 \rightleftarrows E(Cu^{+})_2O_2 \qquad (14)$$

$$E(Cu^{+})_2O_2 + \text{monophenol} \longrightarrow E(Cu^{2+})_2 + \text{o-diphenol} + 2H_2O \qquad (15)$$

Jolley *et al.* (1974) showed that mushroom phenolase ("tyrosinase") reacts with H_2O_2 to give a complex, called "oxytyrosinase". The spectrum of this complex is abolished when H_2O_2 is removed by adding catalase, and it also disappears under anaerobic conditions, but appears again on admitting O_2, suggesting that an equilibrium exists, i.e.

$$H_2O_2 + E \rightleftharpoons (E\text{-}H_2O_2 \overset{?}{\rightleftharpoons} E\text{-}O_2) \rightleftharpoons E + O_2 \qquad (16)$$

Addition of mono- or diphenol substrates causes a breakdown of oxytyrosinase, suggesting that it may be a catalytic intermediate. However, since catalase does *not* inhibit hydroxylation by tyrosinase, the significance of the H_2O_2-induced complex is not very clear. Similar oxygenated complexes have been observed when phenolases from spinach-beet (Vaughan and McIntyre, 1975) and *N. crassa* (Gutteridge and Robb, 1975) are treated with H_2O_2.

An intermediate hydroxylating species such as $Cu^+\text{-}O_2$ (eqn. 14) might have considerable resonance contribution from $Cu^{2+}\text{-}O_2^-$. We therefore reasoned that it might be possible to detect involvement of O_2^- in the hydroxylation catalysed by phenolases. Preliminary studies showed that superoxide dismutase does not inhibit hydroxylation by either mushroom or spinach-beet enzymes (e.g. see Table I), but this could merely be due to its inability to remove O_2^- bound to the active site of the enzymes.

TABLE I

Effect of Superoxide Dismutase on Hydroxylation of 4-Hydroxycinnamic Acid (P-Coumaric Acid) by Mushroom Phenolase

Reductant added	μmol caffeic acid formed/30 min	
	no SOD added	SOD present
None	0.05	0.03
Ascorbate	0.18	0.20
NADH	0.17	0.21
Dihydroxyfumarate	0.15	0.22

The reaction mixtures contained, in a total volume of 1 ml, p-coumaric acid (2.5 μmol), reductant (10 μmol), KH_2PO_4 (40 μmol) and sufficient KOH to adjust the pH to 7.5. 10 units (McCord and Fridovich, 1969) of bovine superoxide dismutase were also added where indicated. Reaction was started by adding tyrosinase purified from *Agaricus sp.* (Jolley *et al.*, 1974) and reaction mixtures incubated at 30°C for 0.5 h. 10 μl conc. HCl was then added and the caffeic acid (3,4-dihydroxycinnamic acid) produced extracted into ether and assayed as described by Halliwell and Ahluwalia (1976). Similar results were obtained using phenolase purified from spinach-beet leaves, except that $(NH_4)_2SO_4$ (500 μmol) was also included in the reaction mixture (Vaughan and Butt, 1969). It may be seen that SOD had no inhibitory effect on hydroxylation. The slight stimulation observed in the presence of reductants may be due to a decrease in the rate of their autoxidation.

If bound O_2^- is an intermediate in hydroxylation, one might then expect to promote this reaction by exposing the enzyme to a monophenol in the presence of a O_2^--generating system. Illuminated chloroplasts catalyse hydroxylation of 4-hydroxycinnamic acid, apparently by the action of a phenolase. This hydroxylation is completely inhibited by superoxide dismutase (Halliwell, 1975b). Since chloroplasts generate O_2^- on illumination (Asada *et al.*, 1974), it seemed likely that O_2^- was interacting with the phenolase to promote hydroxylation.

Recent experiments in my laboratory have begun to investigate this further, using phenolases purified from a variety of sources. Table II shows a typical result obtained when mushroom phenolase in the presence of 4-hydroxycinnamic acid is exposed to an illuminated mixture of FMN and EDTA, which is known to generate O_2^-.

TABLE II

Effect of Superoxide Dismutase and a O_2^--Generating System on Hydroxylation by Mushroom Phenolase

	μmol caffeic acid formed/h	μmol p-coumaric acid used up/h
Complete	0.18	0.52
Omit FMN	0.10	0.42
Omit tyrosinase	0	0
Dark*	0.11	0.42
Complete, + 10 units superoxide dismutase⁺	0.01	0.19
Omit FMN, + 10 units SOD⁺	0.09	-
Complete, + 10 units superoxide dismutase°	0.01	0.19
Omit FMN, + 10 units SOD°	0.09	-
Complete, + 10 units heat-denatured dismutase⁺	0.20	-
Complete, + 100 μg bovine serum albumin	0.20	-
Complete, + 10^3 units catalase	0.19	-

*Tube wrapped in aluminium foil to exclude light
⁺Purified bovine erythrocuprein: units as defined by McCord and Fridovich (1969).
°Purified mangano-dismutase from *B. stearothermophilus*.

The reaction mixtures contained, in a total volume of 1.00 ml, p-coumaric acid (2.5 μmol) FMN (0.05 μmol), EDTA (0.1 μmol), KH_2PO_4 (40 μmol) and sufficient KOH to adjust the pH to 7.5. They were incubated with tyrosinase at 30°C for 1 h while illuminated by a 150 w Osram filta-lite at a distance of 22 cm. After 1 h, 10 μl conc. HCl was added and the products extracted into ether. Caffeic acid was assayed as described by Halliwell and Ahluwalia (1976), p-coumaric acid as described by Vaughan and Butt (1969).

Note that, although caffeic acid formation in the absence of FMN is almost unaffected by superoxide dismutase, this enzyme almost completely inhibits its formation when FMN is present. The reason for this is unknown.

Where indicated, SOD was heated at 100°C for 30 min and cooled before use.

Accumulation of caffeic acid (3,4-dihydroxycinnamic acid) is increased, and this is prevented by superoxide dismutases, but not by catalase, denatured dismutase, nor by bovine serum albumin. Oxidation of added caffeic acid by tyrosinase was found to be unaffected by the O_2^--generating system or by superoxide dismutase, so the increased formation of caffeic acid from p-coumarate cannot be due to an inhibition of the catechol oxidase activity of the enzyme (eqn. 12). Hence the O_2^--generating system appears to promote hydroxylation directly and, indeed, an increased disappearance of 4-hydroxy-cinnamic acid can be detected in its presence. This is *consistent* with bound O_2^- being an intermediate in hydroxylation. However, it is also possible to argue that O_2^-, which is known to reduce Cu^{2+} to Cu^+ (Rabani *et al.*, 1973) is merely acting to replace o-diphenol in reducing the enzyme's copper (eqn. 13) and allowing it to complex with O_2.

CONCLUSIONS

Perhaps the simplest way in which an enzyme might be envisaged as using O_2^- to bring about hydroxylation would be that O_2^- is formed by the enzyme in opposition to the appropriate part of the substrate molecule and then brings about hydroxylation. Evidence consistent with this has been obtained for many hydroxylases. However, it seems likely that, in most cases, the O_2^- does not exist "as such", but as a complex exhibiting a resonance structure, to which O_2^- contributes (e.g. for cytochrome P_{450}, phenolase and prolyl hydroxylase, and possibly for flavoprotein and pteridine-dependent hydroxy-lases). Direct involvement of *free* O_2^- in an enzyme-catalysed hydroxylation remains to be rigorously demonstrated.

REFERENCES

1. Abbott, M.T. and Udenfriend, S. (1974). In "Molecular Mechanisms of O_2 Activation" (Hayaishi, O., ed.), Academic Press, New York.
2. Amano, D., Kagosaki, Y., Usui, T., Yamamoto, S. and Hayaishi, O. (1975). *Biochem. Biophys. Res. Commun. 66*, 272-279.
3. Asada, K., Kiso, K. and Yoshikawa, K. (1974). *J. Biol. Chem. 249*, 2175-2181.
4. Asada, K., Yoshikawa, K., Takahashi, M., Maeda, Y. and Enmanji, K. (1974). *J. Biol. Chem. 250*, 2801-2807.
5. Ashok Kumar, A., Rao, B.S.S.R., Vaidyanathan, C.S. and Appaji Rao, N. (1975). *Ind. J. Biochem. Biophys. 12*, 163-167.
6. Ballou, D., Palmer, G. and Massey, V. (1969). *Biochem. Biophys. Res. Commun. 36*, 898-904.
7. Behar, D., Czapski, G., Rabani, J., Dorfman, L.M. and Schwarz, H.A. (1970). *J. Phys. Chem. 74*, 3209-3213.
8. Bielski, B.H.J. and Chan, P.C. (1973). *Arch. Biochem. Biophys. 159*, 873-879.
9. Bobst, A. and Viscontini, M. (1966). *Helv. Chim. Acta 49*, 884-888.
10. Bors, W., Saran, M., Lengfelder, E., Spottl, R. and Michel, C. (1974). *Curr. Top. Radiat. Res. Q. 9*, 247-309.
11. Bors, W., Lengfelder, E., Saran, M., Fuchs, C. and Michel, C. (1976). *Biochem. Biophys. Res. Commun. 70*, 81-87.
12. Bray, R.C., Cockle, S.A., Fielden, E.M., Roberts, P.B., Rotilio, G. and Calabrese, L. (1974). *Biochem. J. 139*, 43-48.
13. Buhler, D.R. and Mason, H.S. (1961). *Arch. Biochem. Biophys. 92*, 424-437.

14. Cardinale, G.J. and Udenfriend, S. (1974). *Adv. Enzymol. 41,* 245-300.
15. Chu, J.W. and Kimura, T. (1973). *J. Biol. Chem. 248,* 5183-5187.
16. Colloquium on Cytochrome P_{450} (1976). *Biochem. Soc. Trans. 3,* 803-835.
17. Corey, E.J., Nicolaou, K.C., Shibasaki, M., Machida, Y. and Shiner, C.S. (1975). *Tetrahedron Lett. 37,* 3183-3186.
18. Daly, J.W. and Jerina, D.M. (1970). *Biochim. Biophys. Acta 208,* 340-342.
19. Daly, J.W., Jerina, D.M. and Witkop, P. (1972). *Experientia 28,* 1129-1149.
20. Estabrook, R.W., Werningloer, J., Hrycay, E.G., O'Brien, P.J., Rahimtula, A.D. and Peterson, J.A. (1976). *Biochem. Soc. Trans. 3,* 811-813.
21. Fee, J.A. and Hildenbrand, P.G. (1974). *FEBS Lett. 39,* 79-82.
22. Finazzi-Agro, A., Giovagnoli, C., De Sole, P., Calabrese, L., Rotilio, G. and Mondovi, B. (1972). *FEBS Lett. 21,* 183-185.
23. Fridovich, I. (1974). *Adv. Enzymol. 41,* 35-97.
24. Fridovich, I. (1975). *Ann. Rev. Biochem. 44,* 147-159.
25. Goscin, S.A. and Fridovich, I. (1972). *Arch. Biochem. Biophys. 153,* 778-783.
26. Guiraud, H.J. and Foote, C.S. (1976). *J. Amer. Chem. Soc. 98,* 1984-1986.
27. Gunsalus, J.C., Pederson, T.C. and Sligar, S.G. (1975). *Ann. Rev. Biochem. 44,* 377-407.
28. Gutteridge, S. and Robb, D. (1975). *Eur. J. Biochem. 54,* 107-116.
29. Haber, F. and Weiss, J. (1934). *Proc. Roy. Soc. London Ser.A 147,* 332-351.
30. Halliwell, B. (1973). *Biochem. J. 135,* 379-381.
31. Halliwell, B. (1974). *New Phytol. 73,* 1075-1086.
32. Halliwell, B. (1975a). *FEBS Lett. 56,* 34-38.
33. Halliwell, B. (1975b). *Eur. J. Biochem. 55,* 355-360.
34. Halliwell, B. and Ahluwalia, S. (1976). *Biochem. J. 153,* 513-518.
35. Halliwell, B. and Foyer, C. (1976). *Biochem. J. 155,* 697-700.
36. Hamilton, G.A. (1974). In "Molecular Mechanisms of O_2 Activation" (Hayaishi, O., ed.), Academic Press, New York.
37. Heikkila, R.E. and Cohen, G. (1975). *Experientia 31,* 169.
38. Hodgson, E.K. and Fridovich, I. (1975a). *Biochemistry 14,* 5294-5299.
39. Hodgson, E.K. and Fridovich, I. (1975b). *Biochemistry 14,* 5299-5303.
40. Hodgson, E.K. and Fridovich, I. (1976a). *Biochim. Biophys. Acta 430,* 182-188.
41. Hodgson, E.K. and Fridovich, I. (1976b). *Arch. Biochem. Biophys. 172,* 202-205.
42. Hrycay, E.G., Gustafsson, J.A., Ingelman-Sundberg, M. and Ernster, L. (1975). *Biochem. Biophys. Res. Commun. 66,* 209-216.
43. Ilan, Y.A., Czapski, G. and Meisel, D. (1976). *Biochim. Biophys. Acta 430,* 209-224.
44. Joester, K.E., Jung, G., Weber, U. and Weser, U. (1972). *FEBS Lett. 25,* 25-28.
45. Jolley, R.L., Evans, L.H., Makino, N. and Mason, H.S. (1974). *J. Biol. Chem. 249,* 335-345.
46. Karnovsky, M.L. (1975). In "The Phagocytic Cell in Host Resistance" (Bellanti, J.A. and Dayton, R.H., eds.), Raven Press, New York.
47. Kaufman, S. and Fisher, D.B. (1974). In "Molecular Mechanisms of O_2 Activation" (Hayaishi, O., ed.), Academic Press, New York.
48. Kellog, E.W. and Fridovich, I. (1975). *J. Biol. Chem. 250,* 8812-8817.
49. Khan, A.U. (1970). *Science 168,* 476-477.

50. King, M.M., Lai, E.K. and McCay, P.B. (1975). *J. Biol. Chem. 250,* 6496-6502.
51. Liu, T.Z., Shen, J.T. and Ganong, W.F. (1974). *Proc. Soc. Exp. Biol. Med. 146,* 37-40.
52. Lumsden, J. (1975). D. Phil. Thesis, University of London, England.
53. Mason, H.S. (1955). *Adv. Enzymol. 16,* 105-184.
54. Massey. V., Strickland, S., Mayhew, S.G., Howell, L.G., Engel, P.C., Matthews, R.G., Schuman, M. and Sullivan, P.A. (1969). *Biochem. Biophys. Res. Commun. 36,* 891-897.
55. Matheson, I.B.C., Etheridge, R.D., Kratowich, N.R. and Lee, J. (1975). *Photochem. Photobiol. 21,* 165-171.
56. May, S.W., Abbott, B.J. and Felix, A. (1973). *Biochem. Biophys. Res. Commun. 54,* 1540-1545.
57. Miller, R.W. (1970). *Can. J. Biochem. 48,* 935-939.
58. Miller, R.W. and Macdowall, F.D.H. (1975). *Biochim. Biophys. Acta 387,* 176-187.
59. Muir-Wood, P. (1974). *FEBS Lett. 44,* 22-24.
60. Nilsson, R., Pick, F.M. and Bray, R.C. (1969). *Biochim. Biophys. Acta 192,* 145-148.
61. Nilsson, R. and Kearns, D.R. (1974). *J. Phys. Chem. 78,* 1681-1683.
62. Nishikimi, M. (1975a). *Biochem. Biophys. Res. Commun. 63,* 463-468.
63. Nishikimi, M. (1975b). *Arch. Biochem. Biophys. 166,* 273-279.
64. Orr, C.M.W. (1967a). *Biochemistry 6,* 2995-3000.
65. Orr, C.M.W. (1967b). *Biochemistry 6,* 3000-3006.
66. Paschen, W. and Weser, U. (1973). *Biochim. Biophys. Acta 327,* 217-222.
67. Pederson, T.C. and Aust, S.D. (1972). *Biochem. Biophys. Res. Commun. 48,* 789-795.
68. Petrack, B. and Chertock, H.R. (1974). *Fed. Proc. 33,* 535 Abs.
69. Prema Kumar, B., Ravindranath, S.D., Vaidyanathan, C.S. and Appaji Rao, N. (1972). *Biochem. Biophys. Res. Commun. 49,* 1422-1426.
70. Ravindranath, S.D., Ashok Kumar, A., Prema Kumar, R., Vaidyanathan, C.S. and Appaji Rao, N. (1974). *Arch. Biochem. Biophys. 165,* 478-484.
71. Richter, C., Azzi, A. and Wendel, A. (1976). *FEBS Lett. 64,* 332-337.
72. Rotilio, G., Falcioni, G., Fioretti, E. and Brunori, M. (1975). *Biochem. J. 145,* 405-407.
73. Ruettinger, R.T., Olson, S.T., Boyer, R.F. and Coon, M.J. (1974). *Biochem. Biophys. Res. Commun. 57,* 1011-1017.
74. San Filippo, J., Chern, C.I. and Valentine, J.S. (1975). *J. Org. Chem. 40,* 1678-1680.
75. San Filippo, J., Chern, C.I. and Valentine, J.S. (1976). *J. Org. Chem. 41,* 1077-1078.
76. Sawada, Y. and Yamazaki, I. (1973). *Biochim. Biophys. Acta 327,* 257-265.
77. Sligar, S.G., Lipscomb, J.D., Debrunner, P.G. and Gunsalus, I.C. (1974). *Biochem. Biophys. Res. Commun. 61,* 290-296.
78. Strickland, S. and Massey, V. (1971). In "Second International Symposium on Oxidases and Related Oxidation Reduction Systems" (King, T.E., Mason, H.S. and Morrison, M. eds.), Academic Press, New York.
79. Strobel, H.W. and Coon, M.J. (1971). *J. Biol. Chem. 246,* 7826-7829.
80. Symposium on Biological Hydroxylation Mechanisms (1972). Boyd, G.S. and Smellie, R.M.S. (eds.), Academic Press, London and New York.
81. Udenfriend, S., Clark, C.T., Axelrod, J. and Brodie, B.B. (1954). *J. Biol. Chem. 208,* 731-739.

82. Ullrich, V., Ruf, H.H. and Mimoun, H. (1972). In "Symposium on Biological Hydroxylation Mechanisms" (Boyd, G.S. and Smellie, R.M.S., eds.), Academic Press, New York.
83. Vannester, W.H. and Zuberbuhler, A. (1974). In "Molecular Mechanisms of O_2 Activation" (Hayaishi, O., ed.), Academic Press, New York.
84. Vaughan, P.F.T. and Butt, V.S. (1969). *Biochem. J. 113,* 100-115.
85. Vaughan, P.F.T. and McIntyre, R. (1975). *Biochem. J. 151,* 759-762.
86. Walling, C. (1975). *Acc. Chem. Res. 8,* 125-131.
87. Walling, C. and Johnson, R.A. (1975). *J. Amer. Chem. Soc. 97,* 363-367.
88. Wong, K., Morgan, A.R. and Paranchyck (1975). *Can. J. Biochem. 52,* 950-958.
89. Woolf, L.I., Jakubovic, A. and Chan-Henry, E. (1971). *Biochem. J. 125,* 569-574.
90. Yamazaki, I. and Piette, L.H. (1963). *Biochim. Biophys. Acta 77,* 47-64.
91. Yamazaki, H. and Yamazaki, I. (1973). *Arch. Biochem. Biophys. 154,* 147-159.
92. Yokota, K. and Yamazaki, I. (1965). *Biochem. Biophys. Res. Commun. 18,* 48-53.

IN VIVO SCAVENGING OF SUPEROXIDE RADICALS BY CATECHOLAMINES

G. COHEN and R.E. HEIKKILA

Department of Neurology
Mount Sinai School of Medicine
New York 10029, U.S.A.

INTRODUCTION

One of the first assay methods described for superoxide dismutase (SOD) was based upon the ability of epinephrine, a catecholamine, to scavenge superoxide radicals ($O_2^{\bar{\cdot}}$) (McCord and Fridovich, 1969). In this reaction, epinephrine was transformed to a colored product, adrenochrome (Fig.1). SOD intercepted the $O_2^{\bar{\cdot}}$ (generated by xanthine oxidase) and suppressed development of the red color.

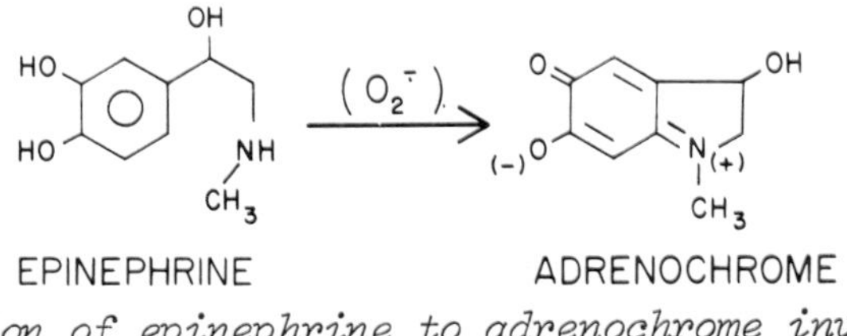

EPINEPHRINE $\xrightarrow{(O_2^{\bar{\cdot}})}$ ADRENOCHROME

Fig.1. *The conversion of epinephrine to adrenochrome involves a complex series of reactions. Epinephrine can be oxidized to a semiquinone by superoxide radicals. Transfer of a second electron (oxidation of the semiquinone) produces adrenaline quinone, which undergoes an intramolecular cyclization to form a substituted dihydroindole. This product is oxidized by oxygen or, perhaps, by superoxide to yield a dihydroindolequinone which tautomerizes to the zwitterion (adrenochrome) shown in the figure.*

Earlier, the co-oxidation of epinephrine during the xanthine-xanthine oxidase reaction had been noted (Valerino and McCormack, 1969). It was subsequently established (McCord and Fridovich, 1969) that the xanthine oxidase reaction generated $O_2^{\bar{\cdot}}$ as an intermediate and that the formation of adrenochrome was the result of scavenging of $O_2^{\bar{\cdot}}$ by epinephrine. Later, an assay system based on the autoxidation of epinephrine was described (Misra and Fridovich, 1972). In this method, the spontaneous reaction between epinephrine and molecular oxygen was catalyzed by base (pH 10.2) and probably by trace metals (that is, the rate was sensitive to the metal-chelating agent, EDTA). Superoxide radicals were generated during the reaction between epinephrine and molecular oxygen. It was proposed that $O_2^{\cdot -}$ catalyzed the overall rate by acting as a propagating species in a chain reaction mechanism. SOD suppressed the catalytic phase and, thereby, suppressed the rate of color development. These studies, in addition to providing a simple analytic method for measuring SOD, established epinephrine as a scavenger of superoxide radicals.

This current report describes the capacity of another catecholamine, norepinephrine (NE), to act as a scavenger of $O_2^{\cdot -}$ within terminals of the sympathetic nervous system. Additionally, evidence will be presented for the generation of hydroxyl radicals (·OH) mediated by $O_2^{\cdot -}$ within sympathetic nerves. The generation of these radicals is associated with a toxic phenomenon, namely the destruction of sympathetic nerve terminals. Scavenging of superoxide by NE, as well as scavenging of ·OH by certain exogenously administered agents (to be described), is associated with protection of sympathetic nerve terminals.

THE NEUROTOXIC ACTIONS OF 6-HYDROXYDOPAMINE (6-OHDA) AND 6-AMINODOPAMINE (6-ADA)

6-OHDA (see Fig.2) has gained widespread prominence in neurological and behavioral research (see review by Kostrzewa and Jacobowitz, 1974). Injected intravenously or intraperitoneally into experimental animals, 6-OHDA selectively destroys the nerve terminals of the peripheral sympathetic system. Injected directly into the brain, 6-OHDA produces relatively selective lesions of NE-containing and dopamine-containing nerve terminals. In higher dose, the cell bodies of catecholamine-neurons are also destroyed. 6-ADA (Fig.2) appears to share many of these actions with 6-OHDA (Heikkila *et al.*, 1973; Jonsson and Sachs, 1973a).

An illustration of the damage produced by 6-OHDA is shown in Fig.3. The noradrenergic nerve plexus in a normal mouse iris is shown in Fig.3A. Twenty-four hours after the injection of 6-OHDA, the nerve plexus has been completely destroyed (Fig.3B). Intravenous or intracerebral injections of 6-OHDA provide powerful research tools for removing the sympathetic nervous system or central catecholamine nerve tracts, respectively, from experimental animals. The molecular mechanisms of action of 6-OHDA and 6-ADA have been the focus of research and debate (Jonsson, Malmfors and Sachs, 1975).

The reason for the specificity of action of 6-OHDA and 6-ADA lies in their structural similarity to the catecholamine neurotransmitters. The axonal membranes of catecholamine neurons contain energy-dependent mechanisms that "recognize" NE and dopamine and transport them into the terminals against a concentration gradient. The transport system also operates with certain other phenolic phenylethyl~~amines~~, ~~including 6-OHDA and 6-ADA~~. As a result, 6-OHDA and 6-ADA are selectively accumulated by catecholamine neurons. Drugs such as cocaine or desmethylimipramine, which inhibit the catecholamine transport systems, prevent the uptake of 6-OHDA and 6-ADA and thereby prevent neuronal damage. The intracellular concentrations of 6-OHDA required to initiate damage have been estimated by Sachs and Jonsson (1975): the toxic levels appear to be in the range of 25 mM.

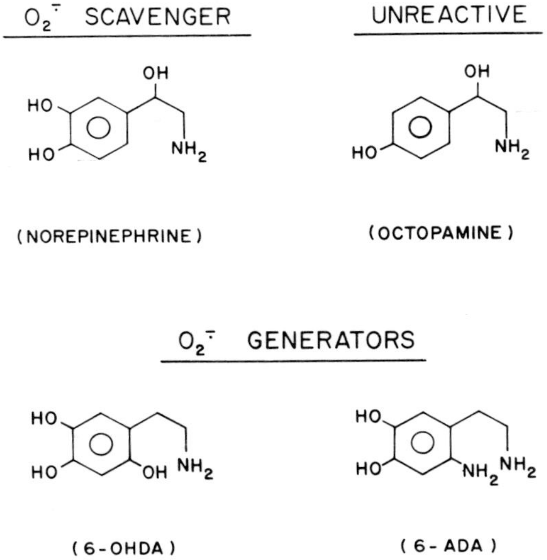

$O_2^{\bar{\cdot}}$ SCAVENGER

UNREACTIVE

(NOREPINEPHRINE)

(OCTOPAMINE)

$O_2^{\bar{\cdot}}$ GENERATORS

(6-OHDA)

(6-ADA)

Fig.2. *The catecholamine, norepinephrine (NE), is the principal neurotransmitter of the peripheral sympathetic nervous system. NE is also the transmitter in certain neuronal tracts of the brain; additionally, NE coexists with epinephrine in the adrenal medulla. Another catecholamine, dopamine (3,4-dihydroxyphenylethylamine), is a neurotransmitter in other neuronal tracts of the brain. Octopamine is a phenolic amine which, like NE, can be taken up, stored and released by catecholamine nerve terminals but, unlike NE, cannot react with superoxide radicals. 6-Hydroxydopamine (2,4,5-trihydroxyphenylethylamine) and 6-aminodopamine (2-amino-4,5-dihydroxyphenylethylamine) are neurotoxins that can destroy catecholamine-containing nerve tracts in the brain as well as nerve terminals in the peripheral sympathetic nervous system.*

GENERATION OF SUPEROXIDE RADICALS BY 6-HYDROXYDOPAMINE AND 6-AMINODOPAMINE

SOD has been used to detect the generation of $O_2^{\bar{\cdot}}$ during the spontaneous autoxidation of 6-OHDA (Heikkila and Cohen, 1973). The reaction between 6-OHDA and molecular oxygen at physiologic pH (pH 7.4) is remarkably rapid. The quinoidal products of the reaction are intensely colored and provide a convenient spectrophotometric tool to follow the reaction. Fig.4 shows that the overall rate is sensitive to the presence of SOD. This observation forms the basis of a spectrophotometric assay system for SOD (Heikkila and Cabbat, 1976; see also article by Heikkila and Cohen, this volume); the 6-OHDA assay system is similar to the epinephrine-adrenochrome system but differs from the latter in that measurements are carried out in a neutral pH range (pH 6.0 - 7.4). Oxygen uptake (formation of H_2O_2) during the autoxidation of 6-OHDA is similarly slowed by the presence of SOD (Heikkila and Cohen, 1973). The inhibitory actions of SOD on the rates of formation of quinones and H_2O_2 indicate that $O_2^{\bar{\cdot}}$ exerts a catalytic role in the overall reaction scheme.

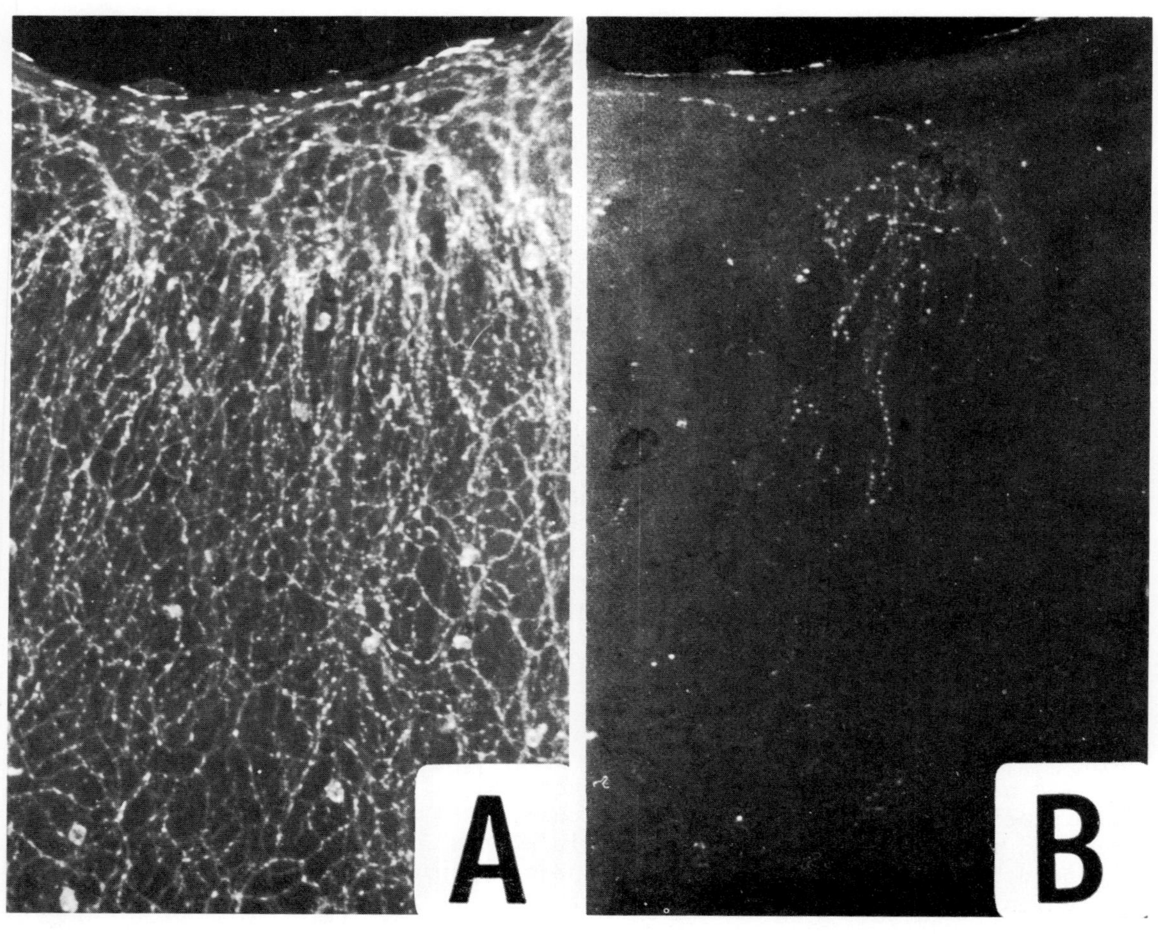

Fig.3.

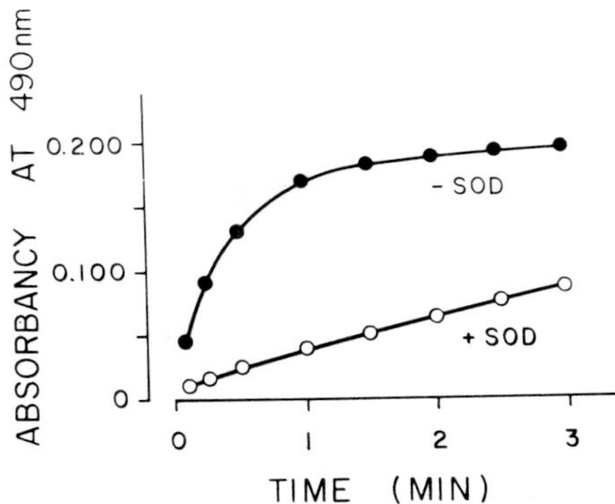

Fig.4. *The spontaneous autoxidation of 6-OHDA (0.1 mM) in Krebs Ringer phosphate buffer at pH 7.4 results in the rapid accumulation of quinoidal products which absorb light maximally at 490 nm. The reaction rate is slowed by the addition of SOD (1.0 μg/ml).*

The mechanism for catalysis may be similar to that proposed by Misra and Fridovich (1972) for the base-catalyzed oxidation of epinephrine. A suggested reaction sequence is shown below:

$$6\text{-OHDA} + O_2 \longrightarrow SQ\cdot + O_2^{\overline{\cdot}} \quad \text{(relatively slow)} \qquad (1)$$

$$6\text{-OHDA} + H^+ + O_2^{\overline{\cdot}} \longrightarrow SQ\cdot + HO_2^- \quad \text{(fast)} \qquad (2)$$

$$SQ\cdot + O_2 \longrightarrow Q + O_2^{\overline{\cdot}} \quad \text{(fast)} \qquad (3)$$

where SQ· is the semiquinone of 6-OHDA, Q is the quinone and HO_2^- is the ionized form of H_2O_2. In this sequence, the reaction of 6-OHDA with $O_2^{\overline{\cdot}}$ (reaction 2) is presumed to be much faster than the corresponding reaction with molecular oxygen (reaction 1). Therefore, after $O_2^{\overline{\cdot}}$ has been initially generated, reaction 2 bypasses the slow step. Reactions 2 and 3 represent a chain mechanism in which $O_2^{\overline{\cdot}}$ is consumed and regenerated, while 6-OHDA is oxidized to its quinone and $O_2^{\overline{\cdot}}$ is reduced to H_2O_2. SOD (which catalyzes the reaction: $O_2^{\overline{\cdot}} + O_2^{\overline{\cdot}} + H^+ \longrightarrow O_2 + HO_2^-$) interrupts the chain and slows the overall rate by transforming the mechanism to the sequence shown in reactions 1 and 3.

Supportive evidence for this proposed reaction mechanism has emerged from experiments in which ascorbic acid was added to the system (Heikkila and Cohen, 1975). Ascorbate bleaches (reduces) the quinone of 6-OHDA and simultaneously stimulates oxygen consumption (Table I).

Fig.3 (opposite). *The sympathetic nerve plexus in a normal mouse iris (Fig. 3A) arises from cell bodies located in the superior cervical ganglion. The bright dots (varicosities) are nerve terminals; they are sites of large numbers of NE-storage vesicles. This nerve plexus was visualized with the Falck-Hillarp method (reviewed by Jonsson, 1971) which transforms NE into a highly fluorescent derivative by means of a condensation reaction with formaldehyde. This photograph was taken in a fluorescence microscope. The iris was exposed to ultra-violet light; the NE-fluorophore emitted a yellow-green light which visualized the nerve plexus. Twenty-four hours after the injection of 6-OHDA·HBr (20 mg/kg, i.v.) the nerve plexus has been completely destroyed (Fig. 3B).*

TABLE I

Oxygen Consumption During the Spontaneous Autoxidation of 6-Hydroxydopamine at pH 7.4: Effect of Ascorbic Acid (10^{-2} M) and Superoxide Dismutase (100 μg/ml)

	μMoles O_2 per Minute Without SOD	μMoles O_2 per Minute With SOD	Inhibition by SOD (%)
6-OHDA Alone	28 ± 6	< 2	95
6-OHDA + Ascorbate	67 ± 7	44 ± 3	35

A phosphate buffer (0.05 M) was employed. The initial rate of reaction was measured at 15 seconds. The minimal detectable rate was 2 nmoles per min. Data from Heikkila and Cohen (1975). The consumption of oxygen by ascorbate alone was insignificant.

Under these circumstances, SOD exerts a much weaker inhibition of the overall rate of oxygen consumption. The reaction between 6-OHDA quinone and ascorbate would be expected to generate SQ· as follows:

$$Q + AH_2 \longrightarrow SQ\cdot + AH\cdot \quad (4)$$

where AH_2 is ascorbate and AH· is the monodehydroascorbate radical. Formation of SQ· via reaction 4 would facilitate oxygen consumption (reaction 3), bypassing the slower reaction 1 and making the overall rate of oxygen consumption less dependent on reaction 2 (less dependent upon $O_2^{\overline{\cdot}}$).

The autoxidation reaction of 6-ADA exhibits different properties. Again a remarkably rapid reaction with molecular oxygen at physiological pH has been noted (Cohen and Heikkila, 1974). However, the rate of reaction is not sensitive to SOD. Nonetheless, the $O_2^{\overline{\cdot}}$ formed as a reaction intermediate has been detected by its ability to reduce ferricytochrome *c* (that is, SOD inhibited the 6-ADA-mediated reduction of ferricytochrome *c*) (Cohen and Heikkila, 1974). It is likely that the relative rates for reactions 1 and 2 are different for 6-ADA and 6-OHDA. The $O_2^{\overline{\cdot}}$-mediated catalysis of the autoxidation rate for 6-OHDA and the lack of catalysis for corresponding autoxidation of 6-ADA presents an important distinction which will be utilized to evaluate the role of intraneuronal NE as a superoxide radical scavenger.

REACTION OF NOREPINEPHRINE WITH SUPEROXIDE RADICALS *IN VITRO*

The autoxidation reaction of 6-OHDA can serve as a convenient tool to study scavenging of $O_2^{\overline{\cdot}}$ by NE. NE, added to the reaction mixture *in vitro*, suppresses both the rate of oxygen consumption and the rate of formation of quinoidal products (Fig.5). Thus, NE appears to act in much the same manner as SOD (see Fig.4). These observations are not surprising in light of the known capacity of epinephrine and other catechol compounds to scavenge $O_2^{\overline{\cdot}}$ (Rapp, Adams and Miller, 1973).

In similarly designed experiments with 6-ADA, it was shown that NE did not slow the rate of production of quinoidal products or H_2O_2 (Sachs *et al.*, 1975). This latter observation is in keeping with the previously mentioned failure of SOD to slow the autoxidation of 6-ADA.

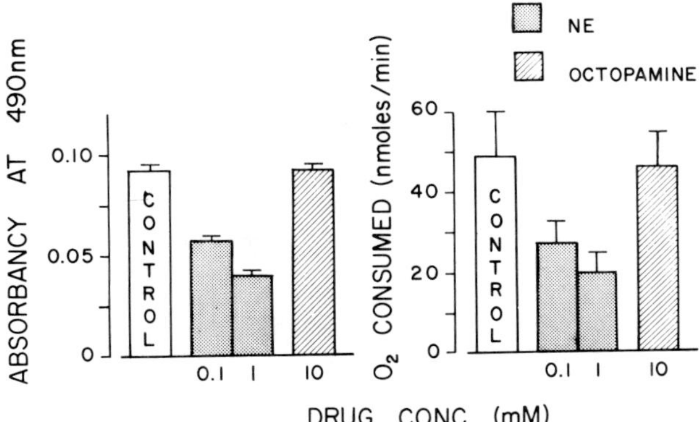

Fig.5. *NE at 0.1 mM or 1.0 mM suppresses the formation of quinoidal products and the consumption of oxygen by 0.1 mM 6-OHDA in a Krebs Ringer phosphate buffer at pH 7.4 and 37°C (data from Sachs* et al., *1975). The monophenolic amine, octopamine (10 mM), has no effect. The absorbancy change was measured at 15 sec. Oxygen consumption was measured with a Clark electrode; oxygen consumption can be equated with the formation of H_2O_2. In similarly designed experiments with 6-ADA (0.1 mM), neither NE (1 mM) nor octopamine (10 mM) slowed the course of the autoxidation reaction (data not shown).*

Dopamine acts similarly to NE; that is, dopamine suppresses the autoxidation rate of 6-OHDA, but not 6-ADA. On the other hand, the monophenolic amine, octopamine (Fig.2), does not suppress the autoxidation of either 6-OHDA (Fig.5) or 6-ADA; similarly, tyramine (p-hydroxyphenylethylamine) does not suppress the autoxidation of either 6-OHDA or 6-ADA (Sachs *et al.*, 1975). These latter results indicate that monophenolic amines are not effective scavengers of superoxide radicals.

EVIDENCE FOR THE REACTION OF NOREPINEPHRINE WITH SUPEROXIDE RADICALS *IN VIVO*

Experiments with 6-Hydroxydopamine

It has been generally assumed that the autoxidation of 6-OHDA is a critical event for the destruction of catecholamine nerve terminals (Kostrzewa and Jacobowitz, 1974; Sachs and Jonsson, 1975). The reactive superoxide radicals, generated as intermediates, could play an indirect role by catalyzing the formation of reaction products (Fig.5) or, alternatively, $O_2^{\overline{\cdot}}$ could exert a direct toxic action (e.g. see Fridovich, 1972). Therefore, NE should be capable of protecting neurons. The concentration of NE in nerve terminals of the rat has been estimated to be in the range of 6 mM (Dahlstrom, Haggendal and Hokfelt, 1966). Reference to Fig.5 shows that concentrations in this range can scavenge $O_2^{\overline{\cdot}}$ and slow the overall autoxidation rate of 6-OHDA. Additionally, Jonsson and Sachs (1973b) have shown that peripheral sympathetic nerves become sensitized to the toxic action of 6-OHDA after depletion of endogenous NE. An example of this phenomenon is shown in Fig.6.

The following experiments were designed to test the ability of NE to scavenge $O_2^{\overline{\cdot}}$ *in vivo* (Sachs *et al.*, 1975). We used both normal rats and NE-depleted (sensitized rats). One hour prior to the injection of 6-OHDA, 50 ng of NE in 5 μl isotonic saline was injected into the fluid of the anterior chamber of the right eye of each rat. The left eye received 5 μl of saline vehicle and served as a control. NE injected in this manner can enter the nerve terminals of the iris via the catecholamine transport system of the axonal membrane. Thus, an experimental model is achieved in which each rat serves as its own control. Previously sensitized rats possess one iris with an NE-depleted nerve plexus and one iris with NE levels restored; in normal rats, one iris is normal and one iris contains more than the normal complement of NE. The question to be answered: does the sensitivity of the iris to injected 6-OHDA depend upon the neuronal stores of NE?

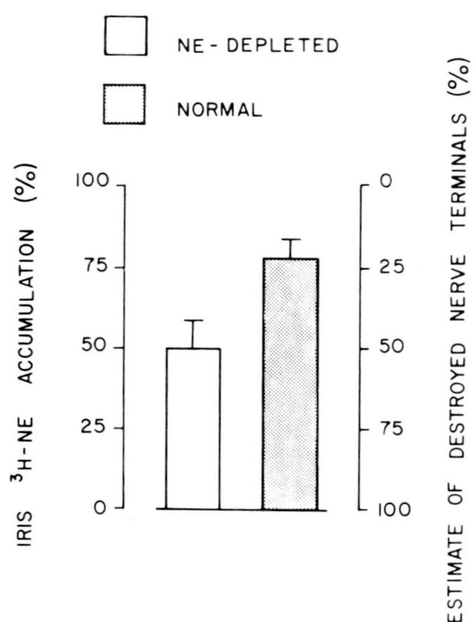

Fig.6. *Normal rats and rats depleted of endogenous NE received intravenous injections of 6-OHDA (5 mg/kg). Depletion of NE was achieved by pretreating the animals 16 hours earlier with a tyrosine hydroxylase inhibitor (alpha-methyl-p-tyrosine methyl ester). The animals with NE-depleted nerve terminals were more sensitive to the destructive action of 6-OHDA. The measurement of ^{3}H-NE accumulation by the iris serves to estimate the relative numbers of nerve terminals compared to rats not injected with 6-OHDA (from Sachs* et al.*, 1975).*

The results of these experiments are presented in Fig.7. An increased level of neuronal NE exerted a strong protective action on the nerve plexus. This was true both in rats previously depleted of NE and in control rats. These results are consistent with scavenging of $O_2^{\cdot-}$ by NE within intact neurons[1].

A source of concern in the interpretation of these experiments was that the 6-OHDA may have competed with injected NE for entry into the nerve terminals. Decreased damage by 6-OHDA might be explained by decreased entry of 6-OHDA into the nerve plexus. However, control experiments eliminated this possibility (Sachs *et al.*, 1975): NE or saline was injected into the anterior chambers of the right and left eyes as usual. Then ^{3}H-6-OHDA was administered intravenously and the accumulation of ^{3}H by each iris was measured. Results of these studies showed that the accumulation of 6-OHDA by the nerve plexus[2] was unchanged by intraocular injection of NE.

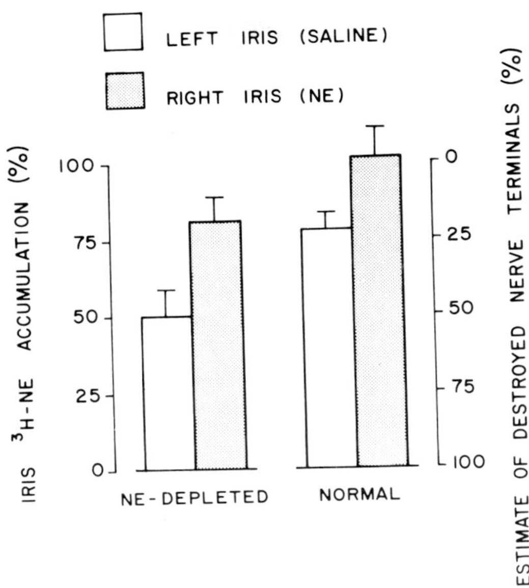

Fig.7. *Normal rats and rats depleted of endogenous NE received intraocular injections of NE (50 ng in saline; right eye) or saline alone (left eye) one hour before intravenous injections of 6-OHDA (5 mg/kg). Prior injection of NE protected the irides in both normal and sensitized (NE-depleted) rats. The measurement of ^{3}H-NE accumulation serves to estimate the relative numbers of nerve terminals compared to rats not injected with 6-OHDA (from Sachs* et al., *1975).*

In other control experiments, octopamine (Fig.2) was used in place of NE. Octopamine, like NE, enters nerve terminals via the catecholamine transport system, but as mentioned previously, it does not react with superoxide radicals. If protection by injected NE were the result of an action other than scavenging of $O_2^{\overline{\cdot}}$, it would be expected that a "false neurotransmitter" such as octopamine could be substituted for NE. But octopamine failed to alter the neurotoxic potential of 6-OHDA (Sachs *et al.*, 1975).

Experiments with 6-Aminodopamine

The neurotoxin 6-ADA provides a valuable contrast to 6-OHDA (see Table II). Unlike 6-OHDA, the autoxidation of 6-ADA is not catalyzed by $O_2^{\overline{\cdot}}$ nor suppressed by NE. Therefore, the autoxidation rate of 6-ADA *in vivo* should not be sensitive to intraneuronal NE levels. In experiments with rats, similar in design to those described for 6-OHDA, the intraocular injection of NE had no effect on the neurotoxic properties of 6-ADA; that is, the same level of neuronal destruction was observed when nerve terminals were normal, or depleted of NE, or filled with NE by intraocular injection (Sachs *et al.*, 1975). The contrasting effects of NE on the autoxidation rates for 6-OHDA and 6-ADA, and the correlation of these effects with similar contrasting actions of NE on neurotoxic potential *in vivo* (see Table II), support the view that NE can serve as a potent scavenger of $O_2^{\overline{\cdot}}$ within intact neurons.

TABLE II

A Comparison of 6-OHDA and 6-ADA:
Autoxidation Rates *in Vitro* and Neurodegenerative Action *in Vivo*

	6-OHDA	6-ADA
Autoxidation Catalyzed by Superoxide Radicals	Yes	No
Autoxidation Slowed by NE	Yes	No
Neurotoxicity Diminished by NE	Yes	No
Neurotoxicity Enhanced by NE Depletion	Yes	No
Autoxidation Slowed by Octopamine	No	No
Neurotoxicity Diminished by Octopamine	No	No

The data with 6-ADA would appear to indicate that $O_2^{\overline{\cdot}}$ does not play a direct toxic role in the destruction of adrenergic neurons, but rather that $O_2^{\overline{\cdot}}$ regulates the formation of toxic byproducts during the autoxidation of 6-OHDA. However, it is difficult to estimate the effects of NE on levels of $O_2^{\overline{\cdot}}$ within neurons exposed to 6-OHDA versus 6-ADA. Because $O_2^{\overline{\cdot}}$ catalyzes its own rate of production from 6-OHDA (but not from 6-ADA), $O_2^{\overline{\cdot}}$ levels may be differentially altered by NE in neurons exposed to either 6-OHDA or 6-ADA. Therefore, the role of $O_2^{\overline{\cdot}}$ in 6-OHDA toxicity cannot be fully evaluated from the current data.

GENERATION OF HYDROXYL RADICALS BY 6-HYDROXYDOPAMINE AND 6-AMINODOPAMINE

Beauchamp and Fridovich (1970) described a system in which the highly reactive hydroxyl radical (·OH) was detected by means of its reaction with methional to form ethylene; the ethylene was measured by gas chromatography. These investigators noted a strongly oxidizing species that was generated during the xanthine-xanthine oxidase reaction. This agent oxidized ferrocytochrome *c* and also reacted with methional to form ethylene. Scavengers of ·OH such as ethanol and benzoate blocked ethylene formation and prevented oxidation of ferrocytochrome *c*. So did catalase and SOD. These and other observations led Beauchamp and Fridovich to suggest that the oxidizing species was ·OH, formed via the reaction proposed by Haber and Weiss (1934):

$$H_2O_2 + O_2^{\overline{\cdot}} \longrightarrow \cdot OH + OH^- + O_2 \qquad (5)$$

More recently, Kellogg and Fridovich (1975) provided evidence that singlet oxygen, rather than oxygen in the ground state, was a product of this reaction.

In our studies (Cohen and Heikkila, 1974), the autoxidation reaction of 6-OHDA or 6-ADA resulted in the generation of ethylene from methional. Ethylene production was inhibited by ethanol or benzoate. It was also blocked by catalase and partially inhibited by SOD[3]. Some studies were conducted in an acetate buffer at pH 6.4 containing EDTA: these conditions prevented the formation of ethylene from H_2O_2 alone, a reaction which may have been catalyzed by trace metals:

$$H_2O_2 + Fe^{2+} \longrightarrow \cdot OH + OH^- + Fe^{3+} \qquad (6)$$

Results of these experiments indicated that formation of ·OH was probably via reaction 5. Because hydroxyl radicals are considered to be toxic to cells (e.g. Myers, 1973; Pryor, 1973), ·OH should be considered as a contributory damaging species produced during the autoxidation of 6-OHDA and 6-ADA.

HYDROXYL RADICAL SCAVENGERS AS PROTECTIVE AGENTS AGAINST 6-OHDA and 6-ADA *IN VIVO*

Thiourea is a powerful scavenger of ·OH *in vitro* (Dorfman and Adams, 1973). We tested a substituted thiourea, namely 1-phenyl-3-(2-thiazolyl)-2-thiourea (PTTU), as an ·OH scavenger *in vitro* and as a protective agent *in vivo* for sympathetic nerves against 6-OHDA and 6-ADA.

In studies *in vitro*, PTTU appeared to scavenge ·OH as effectively as did thiourea (Table III). Thiourea and PTTU were closely comparable in their relative abilities to block ethylene production; they were each better than ethanol or benzoate in this regard. Since ·OH is probably produced by reaction 5, it was theoretically possible that PTTU had suppressed ethylene by reacting either with $O_2^{\cdot-}$ or H_2O_2. However, control experiments showed that PTTU did not react with either of these substances (Cohen *et al.*, 1976). Therefore, a direct reaction of PTTU with ·OH was supported.

TABLE III

Hydroxyl Radical Scavenging by Various Agents [a]

Drug Additions (mM)		Ethylene Evolved (% of Control ± SEM)
Ethanol	0.5	90.1 ± 2.3
Benzoate	0.5	84.1 ± 3.7
Thiourea	0.5	66.8 ± 1.8
PTTU	0.5	57.9 ± 3.0
Ethanol	5.0	52.2 ± 1.5
Thiourea	5.0	15.5 ± 0.5
Controls		
Complete System (no drug addition)		100 ± 5.8[b]
No 6-ADA		4.4 ± 0.2
SOD, 50 μg/ml		51.3 ± 2.5
Catalase, 33 μg/ml		3.9 ± 0.1

a. The system consisted of 0.2 M phosphate buffer containing 6-ADA (0.1 mM), H_2O_2 (1 mM) and EDTA (0.1 mM) at pH 8.4. 2-Keto-4-thiomethylbutyric acid (1 mM) was used in place of methional for the ethylene-forming reaction with hydroxyl radicals. The sensitivity of the system to both SOD and catalase implies that the source of ·OH was probably the Haber-Weiss reaction. Data from Cohen *et al.* (1976).

b. Control generation of ethylene was 1.30 ± 0.08 nmoles (N = 86) in 6 minutes.

PTTU is known to penetrate into NE-containing neurons *in vivo* (Johnson, Boukma and Kim, 1970). When tested *in vivo*, PTTU (200 mg/kg, i.p.) blocked the neurodegenerative actions of both 6-OHDA (Cohen *et al.*, 1975) and 6-ADA (Cohen *et al.*, 1976). Additionally, methimazole (another thiourea analogue) was protective, and ethanol and n-butanol each exerted a lesser protective action (Fig.8). All of these agents are ·OH scavengers, while none blocks the catecholamine transport system of sympathetic nerves (Cohen *et al.*, 1976). These results are consistent with a protective mechanism based on scavenging of hydroxyl radicals.

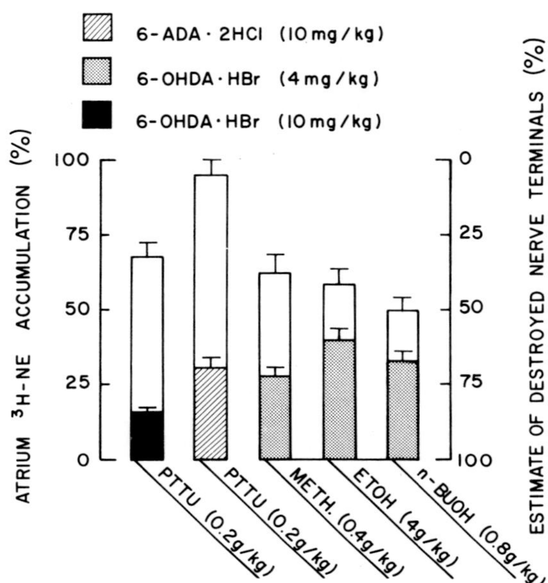

Fig.8. *Protection by PTTU, methimazole, ethanol and n-butanol against nerve terminal destruction induced by 6-OHDA or 6-ADA. The accumulation of* ^{3}*H-NE by isolated mouse atria* in vitro *was used to estimate the degree of damage to adrenergic nerve terminals in the atrium. Data from Cohen* et al. *(1976). The shaded bars show data with 6-OHDA or 6-ADA alone; the clear bars show data from corresponding groups of mice in each experiment treated with protective agent one hour before 6-OHDA or 6-ADA.*

EXPERIMENTS WITH ALLOXAN AND DIALURIC ACID

Alloxan and dialuric acid are, respectively, the oxidized (quinoidal) and reduced (polyhydric) forms of a pair of agents that induce a diabetic state in experimental animals (Bruckmann and Wertheimer, 1945). This diabetes is the result of selective destruction of the insulin-producing beta cells of the pancreas.

In vitro, dialuric acid reacts spontaneously with oxygen to form alloxan and H_2O_2; during this reaction, $O_2^{\cdot-}$ and ·OH are generated (Cohen and Heikkila, 1974). Alloxan, can be recycled to dialuric acid by ascorbate and perhaps by other tissue reducing agents (Heikkila and Cohen, 1975; Heikkila, Barden and Cohen, 1974). Hence, the molecular mechanism underlying destruction of the beta cells of the pancreas by alloxan may be similar, in many respects, to that for destruction of sympathetic nerve terminals of 6-OHDA and 6-ADA.

In vivo experiments with mice showed that an intoxicating dose of ethanol

(4 g/kg; approximately 0.09 M if uniformly distributed) prevents the diabetogenic action of alloxan (Heikkila, Barden and Cohen, 1974). Similarly, other ·OH scavengers, such as thiourea, methanol, n-propanol and n-butanol, prevent the development of a diabetic state (Heikkila *et al.*, 1976). The greater efficacy of n-propanol and n-butanol compared to ethanol or methanol (Heikkila *et al.*, 1976) may be the result of faster rate constants of the longer-chain alcohols for reaction with hydroxyl radicals (Dorfman and Adams, 1973). These and other observations support the point of view that ·OH, generated in the pancreas from dialuric acid (perhaps via reaction 5), plays a prominent role in the cytotoxic process. This conclusion is further supported by the observation that PTTU also proved to be a very effective protective agent against alloxan (Cohen *et al.*, 1975).

CONCLUSIONS

A. Norepinephrine (the natural transmitter substance of the peripheral sympathetic nervous system) is present at a sufficiently high intracellular concentration to contribute to the scavenging of superoxide radicals within sympathetic nerves.

B. Superoxide radicals catalyze the autoxidation of 6-OHDA (a neurotoxin). The neurodegenerative action of 6-OHDA *in vivo* on sympathetic nerves can be modulated towards either increased or decreased damage, respectively, by lowering or raising the neuronal levels of norepinephrine.

C. Although 6-ADA (another neurotoxin) also generates $O_2^{\cdot-}$, its rate of autoxidation *in vitro* is not catalyzed by $O_2^{\cdot-}$, and its neurodegenerative action on peripheral sympathetic nerves is unaltered by neuronal levels of norepinephrine.

D. Both 6-OHDA and 6-ADA generate hydroxyl radicals (·OH), ostensibly via the Haber-Weiss reaction between $O_2^{\cdot-}$ and H_2O_2. Scavengers of ·OH, such as 1-phenyl-(2-thiazolyl)-2-thiourea (PTTU), methimazole, ethanol, and n-butanol, can prevent, either wholly or in part, the neurodegenerative actions of 6-OHDA or 6-ADA.

E. Alloxan (a toxin to the insulin-producing cells of the pancreas), upon reduction by ascorbate to dialuric acid, generates $O_2^{\cdot-}$, ·OH and H_2O_2 during a spontaneous reaction with molecular oxygen. Scavengers of ·OH, such as PTTU, ethanol or n-butanol prevent the development of a diabetic state.

F. The generation of $O_2^{\cdot-}$ and ·OH (as well as H_2O_2) in cells appears to be intimately involved in the destruction of sympathetic nerves by 6-OHDA or 6-ADA, and in the destruction of the beta cells of the pancreas by alloxan.

FOOTNOTES

[1]SOD levels within sympathetic nerves have not been measured. However, it seems likely that SOD is present. First, SOD levels have been measured in the adrenal medulla (Peeters-Joris *et al.*, 1975) which arises from the same embryological cell type (sympathoblasts) as the nerves of the sympathetic nervous system. Secondly, SOD levels are rather uniformly distributed in brain areas (Heikkila and Cohen, unpublished observations).

[2]The dose of ^3H-6-OHDA (5 mg/kg) was identical to that used in neurotoxicity studies. Desmethylimipramine, a drug that blocks the axonal membrane pump for catecholamines, was used to distinguish ^3H within nerve terminals from ^3H present in non-neuronal areas.

[3]SOD produced a biphasic effect on the production of ethylene from 6-OHDA: Ethylene production was inhibited during the early, vigorous phase of the reaction, but, after 40 min, some augmentation of ethylene production was observed. At this late stage, the autoxidation reaction of 6-OHDA had been completed and ethylene production possibly reflected reactions of indoles and other products (Blank *et al.*, 1976) with the accumulated H_2O_2 and with oxygen.

REFERENCES

1. Beauchamp, C. and Fridovich, I. (1970). *J. Biol. Chem. 245*, 4641-4646.
2. Blank, C.L., McCreery, R.L., Wightman, R.M., Chey, W. and Adams, R.N. (1976). *J. Med. Chem. 19*, 178-180.
3. Bruckman, G. and Wertheimer, E. (1945). *Nature (London) 155*, 267-268.
4. Cohen, G. and Heikkila, R.E. (1974). *J. Biol. Chem. 249*, 2447-2452.
5. Cohen, G., Allis, B., Winston, B., Mytilineou, C. and Heikkila, R. (1975). *Eur. J. Pharmacol. 33*, 217-221.
6. Cohen, G., Heikkila, R.E., Allis, B., Cabbat, F., Dembiec, D., MacNamee, D., Mytilineou, C. and Winston, B. (1976). *J. Pharmacol. Exp. Therap.*, in press.
7. Dahlstrom, A., Haggendal, J. and Hokfelt, T. (1966). *Acta Physiol. Scand. 67*, 289-294.
8. Dorfman, L.M. and Adams, G.E. (1973). *NSRDS-National Bureau of Standards 46*, Washington, D.C.
9. Fridovich, I. (1972). *Accts. Chem. Res. 5*, 321-326.
10. Haber, F. and Weiss, J. (1934). *Proc. Roy. Soc. A147*, 332-351.
11. Heikkila, R.E. and Cabbat, F. (1976). *Anal. Biochem.*, in press.
12. Heikkila, R.E. and Cohen, G. (1973). *Science (Wash.) 181*, 456-457.
13. Heikkila, R.E. and Cohen, G. (1975). *Ann. N.Y. Acad. Sci. 258*, 221-230.
14. Heikkila, R.E., Mytilineou, C., Cote, L.J. and Cohen, G. (1973). *J. Neurochem. 21*, 111-116.
15. Heikkila, R.E., Barden, H. and Cohen, G. (1974). *J. Pharmacol. Exp. Therap. 190*, 501-506.
16. Heikkila, R.E., Winston, B., Cohen, G. and Barden, H. (1976). *Biochem. Pharmacol. 25*, 1085-1092.
17. Johnson, G.A., Boukma, S.J. and Kim, E.G. (1970). *J. Pharmacol. Exp. Therap. 171*, 80-87.
18. Jonsson, G. (1971). *Progr. in Histochem. Cytochem. 2*, 299-334.
19. Jonsson, G. and Sachs, C. (1973a). *J. Neurochem. 21*, 117-124.
20. Jonsson, G. and Sachs, C. (1973b). *Res. Commun. Chem. Pathol. Pharmacol. 5*, 287-296.
21. Jonsson, G., Malmfors, T. and Sachs, C. (eds.) (1975). "Chemical Tools in Catecholamine Research, Vol.I: 6-Hydroxydopamine as a Denervation Catecholamine Research", North-Holland Publ. Co., Amsterdam.
22. Kellogg, E.W. III, and Fridovich, I. (1975). *J. Biol. Chem. 25*, 8812-8817.
23. Kostrzewa, R.M. and Jacobowitz, D.M. (1974). *Pharmacol. Revs. 26*, 199-288.
24. McCord, J.M. and Fridovich, I. (1969). *J. Biol. Chem. 25*, 6049-6055.
25. Misra, H.P. and Fridovich, I. (1972). *J. Biol. Chem. 247*, 3170-3175.
26. Myers, L.S. Jr. (1973). *Fed. Proc. 32*, 1882-1894.
27. Peeters-Joris, C., Vandevoorde, A.-M. and Baudhuin, P. (1975). *Biochem. J. 150*, 31-39.

28. Pryor, W.A. (1973). *Fed. Proc. 32,* 1862-1869.
29. Rapp, U., Adams, W.C. and Miller, R.W. (1973). *Can. J. Biochem. 51,* 158-173.
30. Sachs, C. and Jonsson, G. (1975). *Biochem. Pharmacol. 24,* 1-8.
31. Sachs, C., Jonsson, G., Heikkila, R. and Cohen, G. (1975). *Acta Physiol. Scand. 93,* 345-351.
32. Valerino, D.M. and McCormack, J.J. (1969). *Fed. Proc. 28,* 545.

THE INACTIVATION OF COPPER-ZINC SUPEROXIDE DISMUTASE BY DIETHYLDITHIOCARBAMATE

R.E. HEIKKILA and G. COHEN

Department of Neurology
Mount Sinai School of Medicine
Fifth Avenue and 100th Street
New York 10029, U.S.A.

INTRODUCTION

Many distinct superoxide dismutases (SOD) including manganese containing, iron containing and copper-zinc containing have been isolated and characterized, (see other chapters in this volume). These superoxide dismutases, although differing in the metal ion which they contain, as well as several other parameters, all catalyze the same reaction, the dismutation of the superoxide radical to hydrogen peroxide and oxygen. The copper-zinc containing SOD is widely distributed in mammalian tissues and contains the copper at the active site of the enzyme in close proximity to histidine (Richardson, 1977). It was previously shown that long term administration of the copper chelating agent penicillamine caused modest decrements in SOD content of various tissues in the rat including liver, brain and erythrocytes (Albergoni *et al.*, 1975). Additionally, long-term penicillamine treatment in patients with Wilson's disease resulted in a lowering of erythrocyte SOD levels (Alexander and Benson, 1975); in this latter study, other tissues were not analyzed. Although penicillamine was active *in vivo* both in experimental animals as well as in people, it had no effect on SOD activity *in vitro* (Albergoni *et al.*, 1975). This observation may indicate that a metabolite of penicillamine is the active inhibitory species *in vivo*. In contrast to penicillamine, the copper chelating agent diethyldithiocarbamate (DDC) has been shown to cause a marked inactivation of SOD both *in vitro* and *in vivo* in mice (Heikkila *et al.*, 1976). In the present report we will describe the inhibitory effects of DDC on SOD and report on the inactivation of SOD activity *in vitro* by structural analogs of DDC.

MATERIALS AND METHODS

Superoxide dismutase activity was measured by a number of different assay procedures. Among them were assays based on the inhibitory effects of SOD on the reduction of nitroblue tetrazolium (Beauchamp and Fridovich, 1971) or cytochrome *c* (McCord and Fridovich, 1969). Additionally, assays based on the inhibitory effects of SOD on the autoxidation of pyrogallol (Marklund and Marklund, 1974), epinephrine (Misra and Fridovich, 1972) or 6-hydroxydopamine (6-OHDA, Heikkila and Cabbat, 1976) were used. The 6-OHDA assay, which was used most frequently, was done in 10 ml 0.05 M potassium phosphate buffer containing 10^{-4} M EDTA, at pH 6.5 or pH 7.4, and at 37°C. The rate of formation of quinoidal oxidation products was measured spectrophotometrically at 490 nm. Pure copper-zinc superoxide dismutase, obtained from bovine erythrocytes, was used as a standard and was purchased from Truett Laboratories (Dallas, Texas). Diethyldithiocarbamate (sodium salt) was purchased from Sigma and 4-methyl-1-homopiperazinyldithiocarboxylic acid (FLA-57) from Regis Chemicals (Morton Grove, Illinois).

IN VITRO INACTIVATION OF SOD BY DIETHYLDITHIOCARBAMATE

6-OHDA is a catecholamine analog, which when injected into experimental animals, is accumulated by catecholamine containing nerve terminals and causes their degeneration (see Kostrzewa and Jacobowitz, 1974; Jonsson *et al.*, 1975). 6-OHDA is a very unstable compound which rapidly consumes O_2 and forms quinoidal products (Saner and Thoenen, 1971) as well as several highly reactive oxygen metabolites, among them H_2O_2, O_2^- and ·OH (Cohen and Heikkila, 1974). It is generally felt that the intraneuronal formation of one or more of these reactive oxidation products is the cause of the neurotoxicity (see references in Jonsson *et al.*, 1975).

The autoxidation of 6-OHDA was strongly inhibited by superoxide dismutase (Heikkila and Cohen, 1973; see also Fig.1). This result indicated that O_2^- was generated during the autoxidation of 6-OHDA and that O_2^- promoted the overall rate of autoxidation. The incubation of 10^{-3} M diethyldithiocarbamate with the copper-zinc containing bovine erythrocyte SOD for 1.5 hours resulted in a total inactivation of the enzymatic activity; that is, SOD lost its capacity to retard the autoxidation of 6-OHDA. Shorter periods of incubation resulted in lesser degrees of inactivation of SOD, while DDC alone had no effect on the oxidation of 6-OHDA (data not shown). The *in vitro* inactivation of SOD by DDC was confirmed with several other assays, namely epinephrine-adrenochrome autoxidation, pyrogallol autoxidation, cytochrome *c* reduction and nitroblue tetrazolium reduction. Results with the nitroblue tetrazolium assay system are also shown in Fig.1. It was previously reported that DDC partially inactivated the copper-zinc spinach superoxide dismutase *in vitro* (Asada *et al.*, 1974). Additionally, Misra (1975) reported that DDC formed spectrophotometrically distinct complexes with SOD *in vitro*.

Brain and liver homogenates, like copper-zinc containing bovine erythrocyte SOD slowed the rate of autoxidation of 6-OHDA. This observation was consistent with the presence of SOD in brain and liver. Brain or liver homogenates incubated with 10^{-3} M DDC for 1.5 hours, also completely lost their capacity to slow the autoxidation of 6-OHDA. This result indicated that DDC inactivated tissue SOD *in vitro*.

EFFECTS OF DIALYSIS AND $CuSO_4$ ADDITION ON SOD INACTIVATED BY DDC

Dialysis of the DDC-inactivated SOD did not bring about a restoration of enzymatic activity (Fig.2). Additionally, passage of the inactivated SOD through a Sephadex G-25 column did not restore activity. However, the incubation of the dialyzed, DDC-inactivated SOD with 5×10^{-4} M $CuSO_4$ resulted in a total restoration of enzymatic activity (Fig.2).

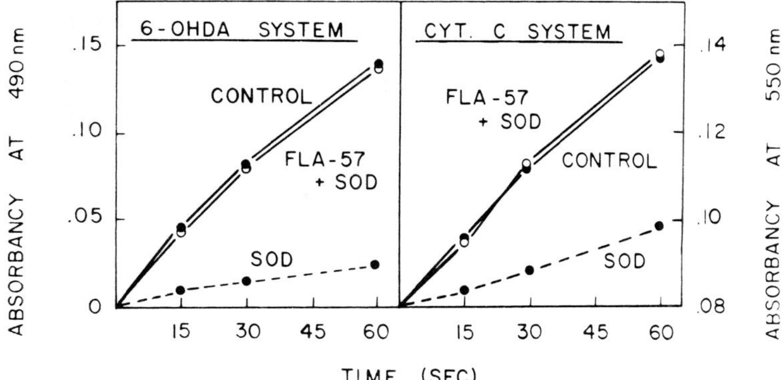

Fig.1. *The inhibition of superoxide dismutase* in vitro *by diethyldithiocarbamate. Left: 6-hydroxydopamine (6-OHDA) autoxidation system. The reaction mixture consisted of 10 ml of 0.05 M phosphate buffer at pH 6.5 containing 10^{-4} M EDTA at 37°C. SOD was present at 200 ng/ml and DDC at 10^{-3} M. Samples were incubated at 37°C for 1.5 hrs. and then the 6-OHDA was added to 2 x 10^{-4} M and the absorbancy of quinoidal products measured at 490 nm. DDC alone had no effect on the autoxidation of 6-OHDA. Right: Nitroblue tetrazolium (NBT) reduction system. SOD (1 mg/ml) and DDC (10^{-2} M) were incubated alone or together in distilled water at 37°C for 1.5 hrs. Aliquots of 10 μl were then removed and SOD assayed (Beauchamp and Fridovich, 1971) at 37°C in 10 ml of 0.05 M phosphate buffer at pH 7.8, containing 10^{-4} M EDTA. Final concentrations in the assay system were SOD, 1 μg/ml; NBT, 2 x 10^{-5} M; xanthine, 10^{-4} M; xanthine oxidase, 6.8 μg/ml. The reaction was started by the addition of the xanthine oxidase and was monitored at 560 nm.*

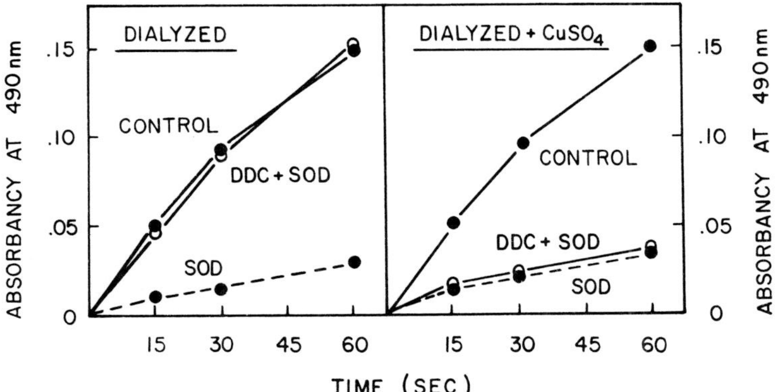

Fig.2. *Inhibition of superoxide dismutase (SOD) by diethyldithiocarbamate. Left: Lack of effect of dialysis. Right: Reversal of inhibition by $CuSO_4$. SOD (1 mg/ml) and DDC (10^{-2} M) were incubated alone and together in distilled water at 37°C for 1.5 hrs. The samples were then dialyzed against distilled water at 4°C for 16 hrs. and then against fresh distilled water for 4 hrs. Aliquots (10 μl) were removed and assayed for SOD activity with the 6-OHDA autoxidation system (left). Final SOD concentration in the assay system was 1 μg/ml. The dialyzed samples were then incubated with 5 x 10^{-4} M $CuSO_4$ for 1 hr. at 37°C and 10 μl aliquots assayed (right). The final concentration of SOD in the assay for again 1 μg/ml. The concentration of $CuSO_4$ was 5 x 10^{-7} M; this concentration of $CuSO_4$ by itself had no effect on the assay.*

The incubation of SOD at 1 mg/ml with 10^{-2} M DDC (e.g. in experiments in Fig.2) led to the formation of an intense yellow color which was not lost upon dialysis and was not extractable with organic solvents. In contrast, the similar intense yellow color of the Cu^{2+}-DDC complex was extractable into organic solvents. The yellow color of the dialyzed DDC-inactivated enzyme became extractable after incubation with $CuSO_4$. These data were consistent with the formation of a complex between DDC and the copper of SOD with a concomitant loss of enzymatic activity. The data argue against a removal of enzyme copper by DDC as the mechanism of inactivation. Most likely the addition of $CuSO_4$ broke the SOD-DDC complex and resulted in the formation of a Cu^{2+}-DDC complex (which was extractable into organic solvents) and the release of a fully active SOD no longer containing bound DDC.

IN VITRO INACTIVATION OF SOD BY STRUCTURAL ANALOGS OF DDC

FLA-57 is a structural analog of DDC (Fig.3). Like DDC, FLA-57 is a potent copper chelator and a commonly used inhibitor of the enzyme dopamine beta hydroxylase (Florvall and Corrodi, 1970). Incubation of SOD with FLA-57 resulted in a loss of superoxide dismutase activity as demonstrated by a loss in ability to slow down the autoxidation of 6-OHDA (Fig.4). The inactivation of SOD by FLA-57 was confirmed with other assays, including cytochrome *c* reduction (Fig.4).

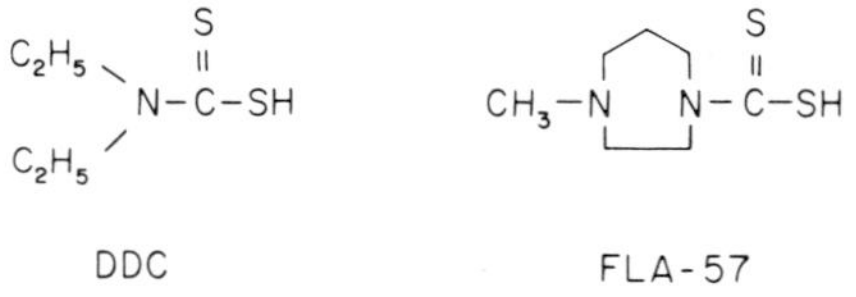

$(C_2H_5)_2N-C(=S)-SH$ — DDC

$CH_3-N\langle\ \rangle N-C(=S)-SH$ — FLA-57

Fig.3. *The structures of DDC and FLA-57.*

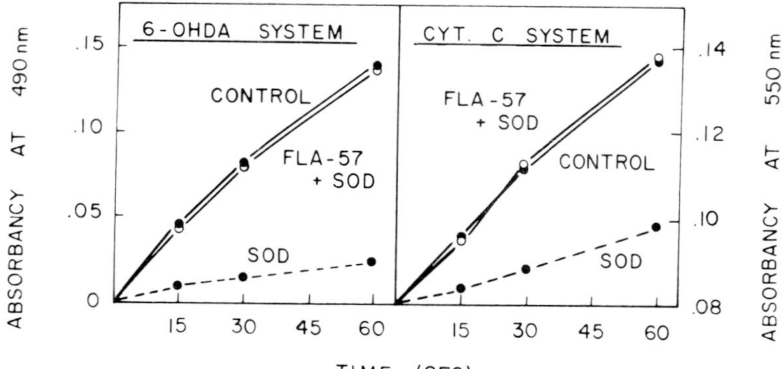

Fig.4. *The inhibition of superoxide dismutase by FLA-57. Superoxide dismutase (1 mg/ml) and FLA-57 (10^{-2} M) were incubated alone or together at 37°C in distilled water for 2 hours and then 10 μl aliquots were removed and assayed for SOD activity. Left: 6-OHDA assay system, as described in Fig.1. Final concentrations of SOD and FLA-57 in the assay were 1 μg/ml and 10^{-5} M respectively. FLA-57 alone had no effect on the autoxidation of 6-OHDA. Right: Cytochrome* c *assay system. The assay was run as described by McCord and Fridovich (1969) at 25°C in 10 ml of 0.05 M phosphate buffer at pH 7.8, containing 10^{-4} M EDTA. Final concentrations in the assay system were SOD, 1 μg/ml; cytochrome* c, *10^{-5} M; xanthine, 5 x 10^{-5} M and xanthine oxidase, 6.8 μg/ml. The reaction was started by the addition of xanthine oxidase and the absorbancy increase measured at 550 nm.*

The inactivation *in vitro* of SOD by FLA-57 appeared similar to that by DDC. The FLA-57 inactivated enzyme, like the DDC-inactivated enzyme was yellow in color and showed no return of activity nor loss of color upon dialysis. Moreover, the enzymatic activity was completely restored by incubation with $CuSO_4$. These results appeared to indicate that the -NCSSH grouping (see Fig.3) was the important structural parameter for the inactivation of SOD. And in fact we have observed that three other compounds containing this grouping, among them dimethyldithiocarbamate, were potent SOD inhibitors *in vitro*.

LACK OF EFFECT OF OTHER COPPER-CHELATING AGENTS ON SUPEROXIDE DISMUTASE ACTIVITY *IN VITRO*

It was previously reported (Asada *et al.*, 1974) that several metal chelating agents in addition to DDC, namely cyanide, axide and o-phenanthroline, inhibited spinach leaf superoxide dismutase *in vitro*. In the same study, several copper chelating agents including neocuproin, cuprizone, and αα-dipyridyl had no inhibitory effect on SOD. In the present study, we tested several copper chelating agents as inhibitors of SOD. All of the compounds in Table I, some of them chelators of cupric copper, others chelators of cuprous copper, showed no inhibitory effect on superoxide dismutase activity *in vitro*. What is there about these potent copper chelators that makes them ineffective as inhibitors of SOD? The available evidence suggests that they fail to chelate the protein-bound copper in SOD. This is indicated by the fact that several of the compounds yield intensely colored complexes with inorganic copper but under similar conditions fail to give any color with superoxide dismutase. This is in marked contrast to results obtained with DDC and FLA-57, which give an intense yellow color, both with $CuSO_4$ as well as with superoxide dismutase.

TABLE I

Compounds tested and found to be totally lacking as *in vitro* inhibitors of bovine erythrocyte superoxide dismutase. SOD (0.5 mg/ml, alone or together with the compounds, 10^{-2} M) was incubated for 1.5 hrs. at 37°C in distilled water or dimethylformamide (those compounds not soluble in water). Aliquots (10 μl) were then removed and assayed for SOD activity in the 6-OHDA system at pH 6.5. It should be emphasized that SOD incubated in either water or dimethylformamide has the same effect in the assay system; that is, incubation with dimethylformamide alone caused no loss of SOD activity. Additionally, DDC (10^{-2} M) totally inactivated SOD when the incubation was carried out in dimethylformamide.

Inactive Compounds
Methimazole
Penicillamine
Propylthiouracil
Phenylthiazolylthiourea
Pyridyldiphenyltriazine
Cuprizone
Bathocuproine
Diethyldithiophosphate

EFFECTS OF *IN VIVO* DDC ADMINISTRATION TO MICE

After it was found that DDC completely inactivated SOD *in vitro*, it was a logical extension to attempt to inhibit SOD activity *in vivo*. In this regard it was fortunate that DDC could be given to experimental animals in large doses (Deitrich and Erwin, 1971). DDC (1.5 g/kg) caused large decreases in SOD activity of mouse brain, liver and blood at 3 hours (Table II).

TABLE II

The superoxide dismutase content of various tissues of mice treated with diethyldithiocarbamate. Tissues were removed 3 hours after the intraperitoneal injection of DDC (1.5 g/kg) or the saline vehicle and assayed for their SOD content with the 6-OHDA assay (n = 8 to 12 per group). Brain and liver were assayed at pH 6.5 and blood at 7.4. Since serum contains very little SOD, the erythrocyte SOD can be calculated by dividing by the hematocrit, which in the mouse is approximately 45%. Data are calculated from a standard curve run with bovine erythrocyte SOD.

Tissue	SOD Content (μg/g ± S.D.) Control	DDC Treated	% Loss
Brain	100 ± 6	52 ± 10	48
Liver	697 ± 80	200 ± 32	71
Blood	88 ± 12	12 ± 12	86

This dose of DDC, if uniformly distributed throughout the entire animal, represents an average concentration of 9×10^{-3} M. In contrast to the large losses in SOD activity, liver and blood catalase measured at 3 hours after 1.5 g/kg DDC were 85 ± 5% (mean ± SEM, $p < 0.025$, n = 20) and 97 ± 7% (n = 11, $p > 0.3$) of their respective control values. Thus the large losses in SOD activity brought about by DDC *in vivo* cannot be the result of non-specific tissue damage. DDC at 0.5 and 1.0 g/kg also caused large decrements in liver SOD activity (Table III). These *in vivo* inhibitory effects of DDC on SOD were confirmed with other assays for SOD besides the 6-OHDA assay.

TABLE III

The effect of various doses of DDC on the SOD content of mouse liver. Mice were injected intraperitoneally with DDC and livers removed 3 hours later and assayed for their SOD content with the 6-OHDA assay at pH 6.5 (n = 8).

DDC Dose (g/kg)	Liver Superoxide Dismutase (μg/g ± S.D.)	% Loss
0	632 ± 48	-
0.5	368 ± 80	42
1.0	200 ± 48	68
1.5	168 ± 32	74

The only compound previously shown to affect SOD activity in experimental animals *in vivo* is penicillamine (Albergoni *et al.*, 1975), a compound interestingly with no action on SOD *in vitro* (Albergoni *et al.*, 1975; see also Table I). Penicillamine administration twice daily for 20 days at 150 mg/kg furthermore caused only relatively modest losses in SOD activity compared to the large losses caused by a single injection of DDC. A single injection of a large dose of DDC represents a simple method for lowering tissue SOD levels; this experimental approach may help investigators to evaluate the biological significance of SOD in various tissues *in vivo*.

CONCLUSIONS

A. Diethyldithiocarbamate (DDC) is a potent *in vitro* inhibitor of copper-zinc containing superoxide dismutase.
B. The inactivation of superoxide dismutase by DDC is not reversed by dialysis but is reversed by incubation with copper sulfate.
C. 4-methyl-1-homopiperazinyldithiocarboxylic acid (FLA-57), a structural analog of DDC, similarly inactivates superoxide dismutase *in vitro*. This inactivation is also not reversed by dialysis, but is reversed by copper sulfate.
D. DDC is a potent inhibitor of tissue superoxide dismutase in mice *in vivo*. DDC (1.5 g/kg) caused large losses in SOD activity in the brain, liver and whole blood.
E. The use of DDC may help to elucidate the biological significance of SOD.

REFERENCES

1. Albergoni, V., Cassini, A., Favero, A. and Rocco, G.P. (1975). *Biochem. Pharmacol. 24*, 1131-1133.
2. Alexander, N.M. and Benson, G.D. (1975). *Life Sciences 16*, 1025-1032.
3. Asada, K., Takahashi, M. and Nagate, M. (1974). *Agr. Biol. Chem. 38*, 471-473.
4. Beauchamp, C. and Fridovich, I. (1971). *Analyt. Biochem. 44*, 276-287.
5. Cohen, G. and Heikkila, R.E. (1974). *J. Biol. Chem. 249*, 2447-2452.
6. Deitrich, R.A. and Erwin, V.G. (1971). *Mol. Pharmacol. 7*, 301-307.
7. Florvall, L. and Carrodi, H. (1970). *Acta Pharm. Suecica. 7*, 7-22.
8. Fridovich, I. (1972). *Accts. Chem. Res. 5*, 321-326.
9. Heikkila, R.E. and Cabbat, F. (1976). *Analyt. Biochem.*, in press.
10. Heikkila, R.E. and Cohen, G. (1973). *Science, 181*, 456-457.
11. Heikkila, R.E., Cabbat, F. and Cohen, G. (1976). *J. Biol. Chem. 251*, 2182-2185.
12. Jonsson, G., Malmfors, T. and Sachs, C. (Eds.) (1975). "Chemical Tools in Catecholamine Research, Vol.I: 6-Hydroxydopamine as a Denervation Tool in Catecholamine Research", North-Holland Publ. Co., Amsterdam.
13. Kostrzewa, R.M. and Jacobowitz, D.M. (1974). *Pharmacol. Revs. 26*, 199-288.
14. Marklund, S. and Marklund, G. (1974). *Eur. J. Biochem. 47*, 469-474.
15. McCord, J.M. and Fridovich, I. (1969). *J. Biol. Chem. 244*, 6049-6055.
16. Misra, H.P. (1975). *Fed. Proc. Abstr. 34*, 624.
17. Misra, H.P. and Fridovich, I. (1972). *J. Biol. Chem. 247*, 3170-3175.
18. Richardson, D.C. (1977). This volume.
19. Saner, A. and Thoenen, G. (1971). *Molec. Pharmacol. 7*, 147-154.

THE ACTION OF $Cu(TYROSINE)_2$ AS A SUPEROXIDE DISMUTASE MODEL ON HEPATIC MICROSOMAL DEMETHYLATION

C. RICHTER and A. AZZI

C.N.R. Unit for the Study of Physiology of Mitochondria
and Institute of General Pathology
University of Padova, Italy

U. WESER and A. WENDEL

Physiologisch-Chemisches Institut
der Universität Tübingen
D-74 Tübingen, Hoppe-Seylerstrasse 1, West Germany

In the presence of molecular oxygen and suitable reducing systems hydroxylation reactions of many hepatic microsomal substrates are known to be cytochrome p-450 dependent (Orrenius and Ernster, 1974). In the absence of both oxygen and a reducing system $NaIO_4$, $NaClO_2$ or organic hydroperoxides are capable of catalysing hydroxylation reactions *in vitro* (Hrycay *et al.*, 1976). Attempts were made to elucidate the nature of the involved oxygen species or possible intermediates. At present several species are discussed including superoxide anions (Strobel and Coon, 1971; Coon *et al.*, 1972), oxenoid compounds (Hamilton, 1964; 1969; 1974; Ullrich and Staudinger, 1966; Sih, 1969) and peroxide complexes bound to the differently oxidized iron states of cytochrome p-450 (Staudt *et al.*, 1974; Rahimtula, 1974; Hrycay and O'Brien, 1971).

Due to the supporting action of a superoxide generating system in microsomal hydroxylation reactions in the absence of cytochrome p-450 reductase it was argued that superoxide may participate as an intermediate (Strobel and Coon, 1971; Coon *et al.*, 1972). The direct $O_2^{\cdot-}$ mediated reduction of cytochrome p-450 rather than a distinct intermediate in the hydroxylation process was not fully excluded. The fact that superoxide dismutase was unable to inhibit the hydroxylation reaction successfully could also be ascribed to steric inhibition at the active site of the hydroxylating enzyme.

The catalysis of superoxide dismutation by both low molecular weight copper chelates and enzyme bound copper (Brigelius *et al.*, 1974; 1975) prompted us to use these small copper complexes as monitors of superoxide mediated microsomal hydroxylation reactions. Provided there is accumulation of superoxide ions near the active site of the membrane bound hydroxylation system then only small hydrophobic compounds may successfully reach this active site. For this purpose $Cu(tyr)_2$ was employed to examine the microsomal hydroxylations, both for the NADPH and for the organic hydroperoxide dependent hydroxylation. The involvement of superoxide for cytochrome p-450 reduction is supported by the observation that $Cu(tyr)_2$ is a potent inhibitor of hydroxylation reactions in the presence of cumene hydroperoxide.

REACTIVITY OF $Cu(tyr)_2$ ON O-DEMETHYLATION OF p-NITROANISOL AND N-DEMETHYLATION OF AMINOPYRINE

Liver microsomes were prepared from phenobarbital pretreated rats (Remmer *et al.*, 1967). The initial rate of p-nitrophenol formation from p-nitroanisol catalysed by microsomes was monitored in a dual wavelength spectrophotometer at 417 minus 480 nm (Netter and Seidel, 1964). The NADPH regenerating system isocitrate isocitrate-dehydrogenase-NADP served as a source for reducing equivalents. The rate was sufficient to reduce 60 nmoles NADP per min per ml. Aminopyrine demethylation was assayed after 10 min incubation by estimation of formaldehyde (Nash, 1953). It can be seen that $Cu(tyr)_2$ inhibits at fairly low concentrations both p-nitrophenol and aminopyrine demethylation. The inhibitory effect was more pronounced at low than at high substrate concentrations (Table I).

TABLE I

$Cu(tyr)_2$ Induced Inhibition of Microsomal Demethylation

	Substrate conc. (mM)	$Cu(tyr)_2$ (μM)	Inhibition (%)
a) Demethylation of p-nitroanisol	3	10	39
		20	57
		30	73
		40	76
		50	80.5
	0.034	10	42
		20	75
		30	93
b) Demethylation of aminopyrine	10	50	90

Incubation conditions: a) 150 mM KCl, 50 mM Tris-HCl pH 7.75, 10 mM $MgCl_2$, 7 mM isocitrate, 10 mM nicotinamide, 1 mM NADPH, 60 μM units isocitrate dehydrogenase, and 3 mg microsomal protein in a total of 2 ml. b) 150 mM KCl, 50 mM Tris-HCl pH 7.75, 10 mM $MgCl_2$, 10 mM nicotinamide, 1 mM semicarbazide, 1 mM NADPH, 1 mM glucose-6-phosphate, 2.6 U glucose-6-phosphate dehydrogenase, and 15 mg microsomal protein in a total of 6 ml. 0% inhibition corresponds to 1.46, 0.79 and 8 nmoles product formed per min and mg of protein for 3 mM p-nitroanisol, 34 μM p-nitroanisol, and 10 mM aminopyrine, respectively.

Double reciprocal plotting of the rate of p-nitrophenol formation versus the amount of added p-nitroanisol reveals that the presence of $Cu(tyr)_2$ affected V_{max} but not K_m of the system (Fig.1). A non-competitive type of

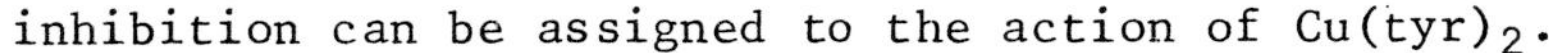
inhibition can be assigned to the action of $Cu(tyr)_2$.

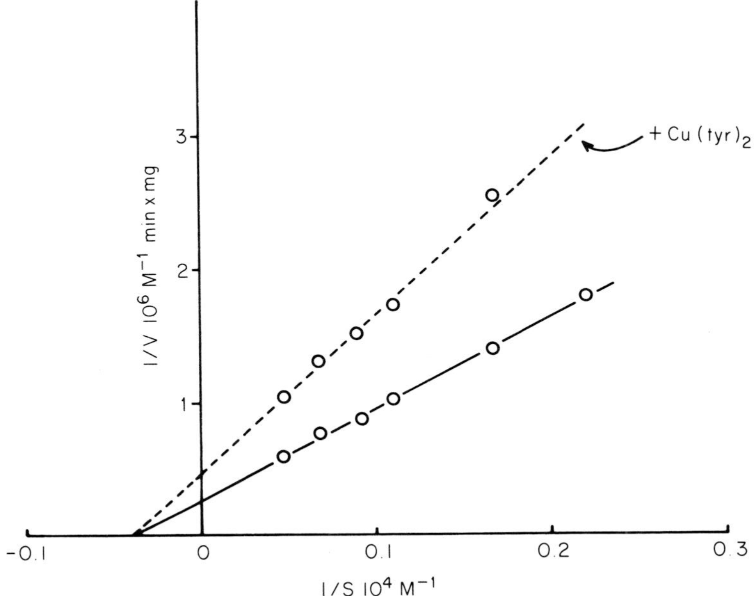

Fig.1. *Double-reciprocal plot of the rate of p-nitrophenol formation as a function of p-nitroanisol concentration in the presence and absence of* $Cu(tyr)_2$. *The incubation mixture was the same as described in Table Ia. The concentration of* $Cu(tyr)_2$ *was 25 μM.* Σ_{mM} *= 13.2 at 417 - 480 nm was used for p-nitrophenol formation.*

REDOX STATE OF MICROSOMAL PIGMENTS IN THE PRESENCE OF $Cu(tyr)_2$

The addition of an appropriate substrate to microsomes stimulates the oxidation of microsomal pigments and of the required NADPH (Estabrook, 1972), reducing equivalents being necessary for substrate hydroxylation. The steady state reduction of microsomal pigments following the initial addition of NADPH and in the presence of $Cu(tyr)_2$ is shown in Fig.2A. The action of Cu $(tyr)_2$ proved rather similar but more distinct than that of the substrate aminopyrine. The steady state of reduction triggered by the addition of NADPH is diminished followed by an accelerated rate of reoxidation. It was attractive to conclude that $Cu(tyr)_2$ stimulates the withdrawal of electrons from the microsomal electron transfer chain. It was of interest to examine the action of $Cu(tyr)_2$ under aerobic conditions. Anaerobiosis was maintained by adding a small aliquot of actively respiring mitochondrial fragments and 5 mM succinate (Beyer, 1967). In this case no Cu(II)-dependent difference was seen in the reduction of the microsomal pigments (Fig.2B). Thus molecular oxygen is probably required for the Cu(II) mediated oxidation of microsomal pigments, and $Cu(tyr)_2$ *per se* is unable to act as an electron acceptor under the above conditions.

Assuming that the microsomal pigments (measured at 557 minus 540 nm) are predominantly due to cytochrome b_5, and knowing the concentration of this cytochrome (measured separately according to Omura and Sato, 1964), the steady state reduction and the half time of its reoxidation after NADPH exhaustion, the "turnover number" can be calculated (Chance, 1943; Modirzadeh and Kamin, 1965). The oxygen dependent numerical value was 0.045 which rose to 0.077 in the presence of $Cu(tyr)_2$. No turnover at all was observed in the absence of oxygen.

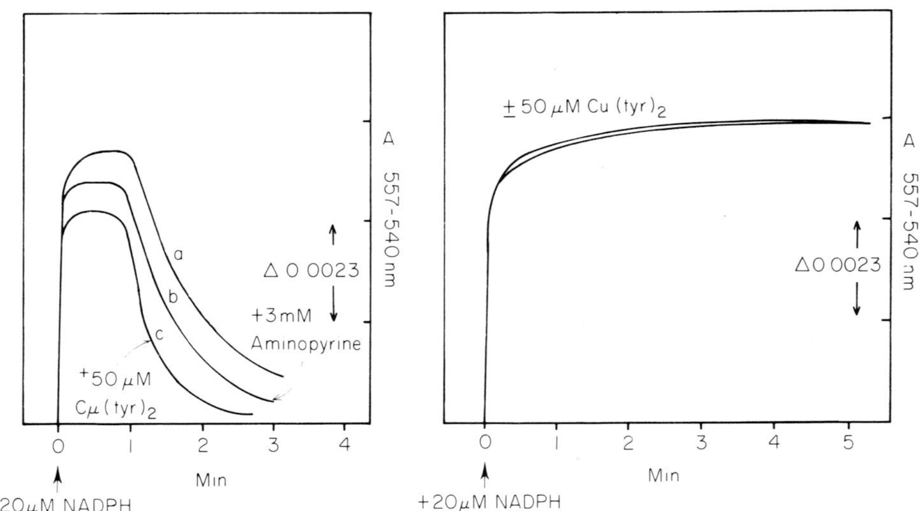

Fig.2. *Steady state kinetics of microsomal pigments. Microsomes from phenobarbital induced rats were diluted to 1 mg x ml^{-1} into buffer containing 150 mM KCl, 50 mM Tris-HCl pH 7.8, 10 mM $MgCl_2$, and 10 mM nicotinamide. A. Extent of microsomal pigment reduction in air-saturated buffer upon addition of 20 μM NADPH in a total of 2 ml (Nash, 1953): curve* a, *substrate omitted; curve* b, *3 mM aminopyrine added; curve* c, *in the presence of 50 μM $Cu(tyr)_2$. B. Extent of microsomal pigment reduction in anaerobic buffer upon addition of 20 μM NADPH in the presence and absence of $Cu(tyr)_2$ in a total of 3.5 ml. Anaerobiosis was obtained by the presence of 0.9 mg x ml^{-1} of mitochondrial particles and 3 mM succinate. Control experiments showed that the addition of NADPH to mitochondrial particles in the absence of microsomes did not result in spectral changes under these conditions.*

A series of experiments was devised and performed to support the initial observation that the action of $Cu(tyr)_2$ is near the oxygen site in the microsomal electron transport chain. The oscilloscopic traces of the kinetics of microsomal pigment reduction using the stopped flow apparatus are depicted in Fig.3. The addition of NADPH induced a fast reduction of microsomal pigments which proved copper independent. A pseudo first-order rate constant was calculated to be 331 s^{-1} in the absence and 301 s^{-1} in the presence of 50 μM $Cu(tyr)_2$. NADPH-cytochrome *c* reductase was also not inhibited by $Cu(tyr)_2$ (150 nmoles x min^{-1} x mg^{-1}, control, and 140 nmoles min^{-1} x mg^{-1} in presence of 50 μM $Cu(tyr)_2$.) Furthermore, it was ascertained that the NADP-isocitrate isocitrate-dehydrogenase regenerating system was not affected by $Cu(tyr)_2$. It was unlikely that $Cu(tyr)_2$ may have acted as a substrate for the same binding sites as indicated by: *1* the non competitive type of inhibition (see Fig.1), *2* no significant spectrum of a cytochrome p-450 substrate complex was seen after the addition of 50 μM $Cu(tyr)_2$, *3* no spectral changes of the aminopyrine-cytochrome-p-450 complex in the presence of $Cu(tyr)_2$ occur.

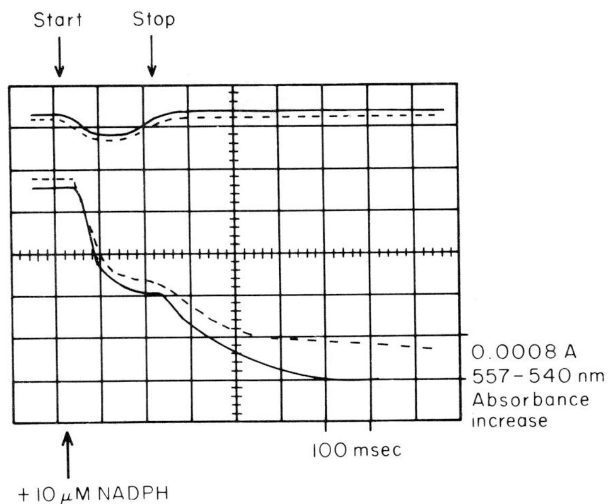

Fig.3. *Kinetics of reduction of microsomal pigments in the presence and absence of $Cu(tyr)_2$. The reaction was measured in a stopped-flow apparatus as described by Chance* et al. *(1967) using a wavelength pair of 557 and 540 nm. Microsomal protein was diluted to a final concentration of 2 mg x ml^{-1} into a buffer containing 150 mM KCl, 50 mM Tris-HCl pH 7.75, 10 mM $MgCl_2$, 10 mM nicotinamide, and placed into the large syringe of the apparatus. The small syringe contained NADPH to give a concentration of 10 μM after mixing. The first order rate constant was determined according to Chance* et al. *(1967). A. Control experiment (——). B. In the presence of 50 μM $Cu(tyr)_2$ (---).*

$Cu(tyr)_2$ DEPENDENT RATE OF OXYGEN UPTAKE BY RAT LIVER MICROSOMES

Provided that $Cu(tyr)_2$ is acting as a superoxide dismutase on microsomal demethylations a decreased rate of oxygen uptake should be observed, i.e. during hydroxylation for every two electrons delivered to cytochrome-p-450 one water molecule and one molecule of hydroxylated substrate is formed with the concomitant uptake of one molecule of oxygen (Orrenius and Ernster, 1974). Added $Cu(tyr)_2$ should completely inhibit substrate hydroxylation attributable to superoxide dismutation into H_2O_2 and O_2, the H_2O_2 being further decomposed into 1/2 O_2 by traces of microsomal catalase. In this case the ratio between the oxygen uptake in the presence and absence of $Cu(tyr)_2$ should be 0.5. Indeed the average ratio of the inhibited and uninhibited values at the aminopyrine concentrations used was 0.54 ± 0.029 (standard error) which agrees with the predicted mode of action of the $Cu(tyr)_2$.

REACTION OF DIFFERENT SUPEROXIDE DISMUTASE ACTIVE COMPOUNDS ON MICROSOMAL O-DEMETHYLATION OF p-NITROANISOL AND CYTOCHROME p-450 TRANSFORMATION

An alternative explanation of the inhibition of demethylase activity by $Cu(tyr)_2$ could be given by a transformation reaction of cytochrome p-450 into cytochrome p-420, the latter being inactive in hydroxylation. A number of superoxide dismutase active compounds were examined for their ability to transform cytochrome p-450. At the same time the inhibitory effect on p-nitroanisol demethylation was examined. A detectable transformation reaction by $CuSO_4$ (Brigelius *et al.*, 1974; 1975) was seen (9.5%) while demethylation was 11% (Table II). This is strongly contrasted when $Cu(tyr)_2$ was used: 3.4% transformation and 100% inhibition of demethylase activity. Tyrosine alone and 2Cu,2Zn-superoxide dismutase (erythrocuprein) was inactive in either assay. Fe(II)EDTA (Halliwell, 1975) caused 9.5% transformation and a somewhat increased demethylase activity (21%). The rather hydrophilic $Cu(lys)_2$ (220 μM) complex inhibited p-nitrophenol formation by 42% and caused 4.1% cytochrome transformation.

From the above data we can conclude rather high specificity of the $Cu(tyr)_2$ complex in the inhibitory action on microsomal demethylation. The enzyme 2Cu2Zn-superoxide dismutase (erythrocuprein) was unable to show any inhibition, and the other hydrophilic low molecular weight transition metal chelates were only slightly active. Furthermore, it appears that the site of action $Cu(tyr)_2$ is in the region of cytochrome b_5 and/or cytochrome p-450, affecting the oxidation but not the reduction kinetics of microsomal pigments.

ACTIVITY OF $Cu(tyr)_2$ ON CUMENE HYDROPEROXIDE SUPPORTED MICROSOMAL DEMETHYLATION AND SUPEROXIDE FORMATION

Cumene hydroperoxide supported both microsomal demethylation and hydroxylation reactions in the absence of NADPH by acting directly at the cytochrome p-450 level (Hrycay *et al.*, 1976). The initial rate and extent of p-nitrophenol formation were inhibited by $Cu(tyr)_2$. This inhibition was not attributed to any measurable decomposition of cumene hydroperoxide produced by the copper complexes, as monitored spectrophotometrically at 250 nm (Table III). Furthermore, it was shown that addition of cumene hydroperoxide to rat liver microsomes facilitates the formation of $O_2^{\cdot-}$ as detected by production of nitroform from tetranitromethane (Czapski and Bielski, 1963; Hodgson and Fridovich, 1973). The generation of $O_2^{\cdot-}$ is inhibited in the presence of Cu $(tyr)_2$. The action of the copper complex, however, was not detectable at high tetranitromethane concentrations (100 μM), but such a lack of inhibition is consistent with more efficient trapping of $O_2^{\cdot-}$ by tetranitromethane. Since the generation of $O_2^{\cdot-}$ in NADPH enriched microsomes was previously shown to occur at the reductase level, their involvement in the hydroxylation mechanism could not be evaluated. In the present experiments only cytochrome p-450 appears to be active in the reaction supported by cumene hydroperoxide (Hrycay *et al.*, 1976). $O_2^{\cdot-}$ is probably generated at this level. The microsomal mediated formation of $O_2^{\cdot-}$ in the presence of cumene hydroperoxide appears to be related to hydroxylation in the same way as the substrate p-nitroanisol inhibits their formation (Table III).

It was of interest to perform a rapid flow analysis of the velocity of cumene hydroperoxide-induced formation of a 440 nm species, most likely a ferryl-oxygen intermediate (Hrycay *et al.*, 1976). The formation rate of the 440 nm species was not affected by 50 μM $Cu(tyr)_2$ while the extent was diminished by 50%. The observed 440 nm species, as previously indicated appears to be an important intermediate in the sequence of hydroxylation reactions. From the decreased extent of the 440 nm species it was suggested that this reduction may be a consequence of the diminution of a precursor bound to $O_2^{\cdot-}$. A biphasic formation of the 440 nm species was proposed with a rate constant of $k = 6.9\ s^{-1}$ for the fast, and $k = 0.19\ s^{-1}$ for the slow phase. Neither the fast nor the slow constant were affected by the presence of 50 μM $Cu(tyr)_2$ (Fig.4). Similarly, a biphasic reduction of cytochrome p-450 in the presence of NADPH was observed (Gigon *et al.*, 1969; Matsubara *et*

TABLE II

The Action of Various Compounds with Superoxide Dismutase Activity on Microsomal O-Demethylation and on Transformation of Cytochrome p-450 to Cytochrome p-420

Compound	nmoles p-nitrophenol x min^{-1} x mg^{-1}	Inhibition (%)	Cytochrome p-450-CO (nmoles x mg^{-1})	Transformation into cytochrome p-420 (%)
No inhibitor	0.62	-	1.79	-
50 μM $CuSO_4$	0.55	11.3	1.62	9.5
2.5 μM erythrocuprein	0.64	0.0	1.79	0.0
50 μM $FeSO_4$ + 100 M EDTA	0.49	21.0	1.62	9.5
55 μM $Cu(lys)_2$	0.64	0.0	n.d.	n.d.
115 μM $Cu(lys)_2$	0.42	32.2	n.d.	n.d.
220 μM $Cu(lys)_2$	0.36	42.0	1.70	4.1
50 μM $Cu(tyr)_2$	0.00	100.0	1.73	3.4
Tyrosine (saturated)	0.62	0.0	1.79	0.0

The formation of p-nitrophenol from p-nitroanisol (34 μM) was determined in the presence of the indicated compounds as described in the legend of Table Ia. For the determination of the transformation of cytochrome p-450 to cytochrome p-420, 8 mg of microsomal proteins were diluted into 6 ml of buffer containing 150 mM KCl, 50 mM Tris-KCl pH 7.8, and 10 mM $MgCl_2$, and reduced by the addition of a few grains of dithionite. The suspension was divided equally into cuvettes and the sample cuvette was gassed for 1 min with CO. The content of cytochrome p-450 was calculated using a Σ_{mM} of 91 at 450 - 490 nm (Omura and Sato, 1964).

al., 1976). In the presence of cumene hydroperoxide, extrapolation of the slow phase to zero using semi-logarithmic plotting showed the same amount of slowly reacting component (54%) of the total compared with that of Gigon *et al.* (1969), suggesting that the observed intermediate is formed with participation by cytochrome p-450.

TABLE III

Action of $Cu(tyr)_2$ on Cumenehydroperoxide-Supported Microsomal Demethylation and $O_2^{\bar{\cdot}}$ Formation

	Control	50 μM $Cu(tyr)_2$ added	Inhibition (%)
a) Demethylation of p-nitroanisol Initial rate (nmoles x min^{-1} x mg^{-1})	3.4	1.5	56
extent (nmoles)	1.7	0.4	76
b) Formation of $O_2^{\bar{\cdot}}$ (nmoles nitroform formed min^{-1} x mg^{-1})	4.7	2.4	51
c) Formation of $O_2^{\bar{\cdot}}$ (nmoles nitroform formed min^{-1} x mg^{-1}) in the presence of 3 mM p-nitroanisol	0.0	0.0	-

Assay conditions in a) were as described in Table Ia) except that NADPH and the reducing system were omitted, and that the reaction was instead started by the addition of 55 μM cumenehydroperoxide. In b) and c) the formation of nitroform from tetranitromethane was measured as described by Czapski and Bielski (1963) and Omura and Sato (1964). The concentration of tetranitromethane was 10 μM. The reaction was started by the addition of 70 μM cumene hydroperoxide.

It was demonstrated that $Cu(tyr)_2$ is a very convenient model of superoxide dismutase and quite useful in the elucidation of the mode of action of microsomal demethylation. It was proposed that $O_2^{\bar{\cdot}}$ should be the activated oxygen species involved in microsomal hydroxylation reactions. No decision was possible favouring or rejecting this assumption. The data by Strobel and Coon (1971), Coon *et al.* (1972) obtained with a reconstituted hydroxylation system in which the superoxide generating component could substitute for NADPH were discussed with caution. The inhibitory action of superoxide dismutase on such a system could also be attributed to a role of superoxide radicals in cytochrome p-450 reduction and not their involvement in the hydroxylation process.

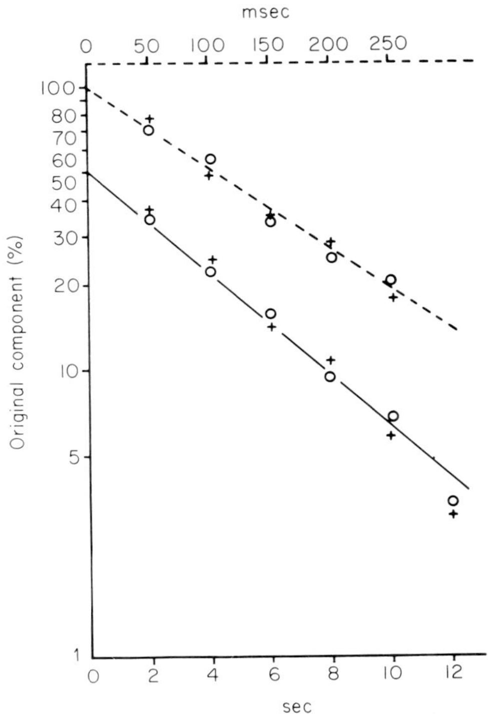

Fig.4. *Semilogarithmic plot of the rate of formation of the "activated oxygen" intermediate of cytochrome p-450 induced by cumene hydroperoxide. The reaction was measured in a stopped-flow apparatus as described in Fig.3 except that the small syringe contained cumene hydroperoxide to give a concentration of 330 μM after mixing. (+) in the presence, (o) in the absence of 50 μM Cu(tyr)$_2$. A. Slow phase. B. Fast phase.*

One attractive explanation for the inability of superoxide dismutase to react in intact microsomes during hydroxylation was that $O_2^{\overline{\cdot}}$ produced at the level of cytochrome p-450 was not accessible to the enzyme. A number of small molecular weight copper complexes have proven to catalyse the spontaneous superoxide dismutation quite efficiently. They were used in this study as monitors to examine the possibility that $O_2^{\overline{\cdot}}$ is actually formed as an intermediate in microsomal hydroxylations. Cu(tyr)$_2$ was the only complex among those tested which at low concentrations was a potent inhibitor of microsomal demethylations.

The observed inhibition proved to be non-competitive with regard to the substrate used, and could not be ascribed to any other effect on the reactions leading to the reduction of cytochrome p-450. NADPH-cytochrome *c* reductase activity as well as the rate of cytochrome b_5 reduction by NADPH were not significantly diminished, nor did the complex act as a "bad" substrate or alter the molecular characteristics of cytochrome p-450 in its transformation to cytochrome p-420. The Cu(tyr)$_2$ acted rather as an electron sink, stimulating the oxidation of both NADPH and microsomal pigments. This phenomenon was only seen in the presence of molecular oxygen, thus excluding the possibility that the sink was due to Cu(II) redox reactions. Provided this reaction also occurred under anaerobic conditions a more oxidized state of microsomal pigments should have been observed with a simultaneous increased rate of NADPH oxidation. The necessity of molecular oxygen in the NADPH-supported reaction for the activity of $Cu(tyr)_2$ as well as the known literature data is in accord with the proposal shown in Fig.5.

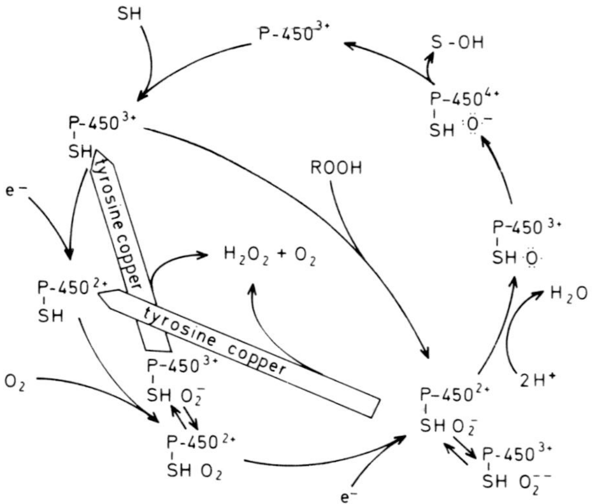

Fig.5. *Proposed mode of action of $Cu(tyr)_2$ during reductase- or cumene-hydroperoxide-mediated microsomal demethylation.*

In this scheme the inhibitory action of the complex is explained via the catalysis of superoxide dismutation into H_2O_2 and oxygen, completely blocking substrate hydroxylation. The hemoprotein-superoxide complexes are formed either *via* the reductase-mediated sequential addition of two electrons (Estabrook *et al.*, 1972), or *via* cumene hydroperoxide.

On addition of $Cu(tyr)_2$ an increased rate of hydrogen peroxide generation is to be expected (Fig.5). The quantitation of this phenomenon is rather complex because hydrogen peroxide is also produced at sites in the microsomal membrane other than cytochrome p-450. Further complications are expected due to contaminating quantities of catalase and/or glutathione peroxidase.

From the presented data the conclusion can be drawn that O_2^- is generated by cytochrome p-450 in the presence of hydroperoxide, and that these radicals play an active role in substrate hydroxylation. From the stopped-flow experiments it was deduced that cumene hydroperoxide-supported hydroxylation is $Cu(tyr)_2$ dependent, and the extent but not the rate of generation of the so called "oxygenated intermediate" is also diminished when $Cu(tyr)_2$ is present. We propose a modification of the hydroxylation reaction (Hrycay *et al.*, 1976) catalysed by cytochrome p-450 in the presence of cumene hydroperoxide. The data suggest that superoxide bound to hemoprotein is either a precursor of the observed intermediate or the intermediate itself, and that this species can be considered to be directly involved in substrate hydroxylation.

Additional experimental support is derived from the result that hemoglobin can hydroxylate aniline in a way similar to the action of a terminal oxidase enzyme. Mieyal *et al.* (1976) showed that this reaction can be inhibited by catalase and to some extent by superoxide dismutase. It was proposed that in this reaction a hemoglobin-bound form of hydrogen peroxide (aniline-hemoglobin^{3+}-OOH^-) is an active participant, a species which appears formally identical with the SH-P-450^{3+}-O_2^{2-} species depicted in Fig.5, which is thought to be in resonance with SH-P-450^{2+}-O_2^-.

REFERENCES

1. Beyer, R.E. (1967). *Methods Enzymol. 10,* 186-194.
2. Brigelius, R., Spöttl, R., Bors, W., Lengfelder, E., Saran, M. and Weser, U. (1974). *FEBS Lett. 47,* 72-75.
3. Brigelius, R., Hartmann, H.J., Bors, W., Saran, M. Lengfelder, E. and Weser, U. (1975). *Z. Physiol Chem. 356,* 739-746.
4. Chance, B. (1943). *J. Biol. Chem. 151,* 553-577.
5. Chance, B., DeVault, D., Legallais, V., Mela, L. and Yonetani, T. (1967). In "Nobel Symposium 5" (Claesson, S. ed.), Interscience Publ., New York.
6. Coon, M.J., Strobel, H.W., Autor, A.P., Heidema, J. and Duppel, W. (1972). In "Biological Hydroxylation Mechanisms"(Boyd, G.S. and Smellie, R.M.S., eds.) pp. 45-54, Academic Press, New York.
7. Czapski, G.H. and Bielski, M.J. (1963). *J. Phys. Chem. 67,* 2180-2187.
8. Estabrook, R.W., Baron, J., Franklin, M, Mason, I., Waterman, M. and Peterson, J. (1972). In "Miami Winter Symposium" (Schultz, J. and Cameron, B.F., eds.) *4,* pp 197-230, Academic Press, New York.
9. Gigon, P.L., Gram, T.E. and Gilette, J.R. (1969). *Mol. Pharmacol. 5,* 109-122.
10. Halliwell, B. (1975). *FEBS Lett. 56,* 34-38.
11. Hamilton, G.A. (1964). *J. Amer. Chem. Soc. 86,* 3391-3392.
12. Hamilton, G.A. (1969). *Adv. Enzymol. 32,* 55-96.
13. Hamilton, G.A. (1974). In "Molecular Mechanisms of Oxygen Activation" (Hayaishi, O., ed.) pp 405-451, Academic Press, New York.
14. Hodgson, E.K. and Fridovich, I. (1973). *Biochem. Biophys. Res. Comm. 54,* 270-274.
15. Hrycay, E.G. and O'Brien, P.J. (1971). *Arch. Biochem. Biophys. 147,* 14-27.
16. Hrycay, E.G. and O'Brien, P.J. (1972). *Arch. Biochem. Biophys. 153,* 480-494.
17. Hrycay, E.G., Gustafsson, J.A., Ingelmann-Sunderberg, M. and Ernster, L. (1976). *Eur. J. Biochem. 61,* 43-52.
18. Matsubara, T., Baron, J., Peterson, L.L. and Peterson, J.A. (1976). *Arch. Biochem. Biophys. 172,* 463-469.
19. Mieyal, J.J., Ackerman, R.S., Blume, J.L. and Freeman, L.S. (1976). *J. Biol. Chem. 251,* 3436-3441.
20. Modirzadeh, J. and Kamin, H. (1965). *Biochim. Biophys. Acta 99,* 205-226.
21. Nash, T. (1953). *Biochem. J. 55,* 416-421.
22. Netter, K.J. and Seidel, G. (1964). *J. Pharmacol. Exp. Ther. 146,* 61-65.
23. Omura, T. and Sato, R. (1964). *J. Biol. Chem. 239,* 2370-2378.
24. Orrenius, S. and Ernster, L. (1974). In "Molecular Mechanisms of Oxygen Activation" (Hayaishi, O., ed.) pp 215-244, Academic Press, New York.
25. Rahimtula, A.D., O'Brien, P.J., Hrycay, E.G., Peterson, J.A. and Estabrook, R.W. (1974). *Biochem. Biophys. Res. Comm. 60,* 695-702.
26. Remmer, H., Greim, H., Shenkman, J.B. and Estabrook, R.W. (1967). *Methods Enzymol. 10,* 703-708.
27. Sih, G.J. (1969). *Science (Wash. D.C.) 163,* 1297-1300.
28. Staudt, H., Lichtenberger, F. and Ullrich, V. (1974). *Eur. J. Biochem. 46,* 99-106.
29. Strobel, H.W. and Coon, M.J. (1971). *J. Biol. Chem. 246,* 7826-7829.
30. Ullrich, V. and Staudinger, H.J. (1966). In "Biological and Chemical Aspects of Oxygeneases" (Bloch, K. and Hayaishi, O., eds.) pp 235-249, Maruzen, Tokyo.

2 Cu, 2 Zn-SUPEROXIDE DISMUTASE IN COPPER DEPLETED RATS

W. BOHNENKAMP and U. WESER

Physiologisch-Chemisches Institut
der Universität Tübingen, D-7400 Tübingen
Hoppe-Seyler-Strasse 1

With the development of a method for measuring the 2 Cu, 2 Zn-superoxide dismutase activity conveniently in limited amounts of crude biological material, this enzyme can be measured over long periods under changing conditions. It was found that competing superoxide dismutase activities derived from other components, including Fe- and/or Mn-enzymes as well as low molecular weight chelates (Joester *et al.* 1972; Brigelius *et al.* 1975), can be successfully excluded. This is easily done by gel electrophoresis followed by activity staining (Beauchamp and Fridovich, 1971; Bohnenkamp and Weser, 1975). Measurements with radioactive copper showed that the 2 Cu, 2 Zn-superoxide dismutase obtains its copper from ceruloplasmin (ferroxidase) (Marceau and Aspin, 1973; Hsieh and Frieden, 1975). Thus, it is not surprising that those copper proteins are related to each other. In this contribution, we summarize the observed activities of 2 Cu, 2 Zn-superoxide dismutase, ferroxidase, catalase and glutathione peroxidase in blood and brain of copper depleted rats.

Ferroxidase was determined by measuring the formation of Fe^{3+}-transferrin as described by Johnson *et al.*, 1967. Copper depletion was performed by feeding young rats (35 - 40 g body weight) a diet containing less than 0.2 mg Cu/kg dry weight. This is about 1% of the copper in a normal diet. For drinking water deionized and quartz distilled water was provided. Two control groups of animals were used, one receiving the regular diet and tap water, and the other group maintained on a low copper diet and quartz distilled drinking water containing 35 mg Cu $(Leu)_2/1$ (Kirchgessner *et al.*, 1967). No significant differences could be measured between these control groups.

Blood samples (about 100 - 200 μl/animal) were collected from the tail vein and were rapidly mixed with EDTA or, in the case of ferroxidase determination, with heparin.

MEASUREMENTS IN BLOOD

During copper depletion, the copper concentration in the plasma decreased faster than the copper content of the blood cells (Figs.1 and 2). Compared to the controls, the copper content was diminished by about 75%. This data is in contrast with that published by Dreosti and Quicke (1968), who found a constant copper concentration in the erythrocytes which was independent of the copper status of their rats. This probably is due to a lower copper intake in the present studies. The absolute copper content of the controls was 0.97 ± 0.05 μg Cu/ml of packed blood cells and 0.36 ± 0.03 μg Cu/ml plasma.

Only insignificant changes in blood cells were detected between normal and copper depleted rats. At the beginning of copper depletion the number of lymphocytes increased from 75% to 85% of the total white blood cell count. However three weeks later the lymphocyte count returned to normal. Monocytes and rhabdoid nuclear leucocytes were unaffected by copper depletion. The polymorphonuclear leucocytes showed more segments in their nuclei during copper deficiency than under normal nutritional conditions. Staining and counting was according to Hallmann (1960).

Figs.1 and 2 show that the copper content of plasma and ferroxidase activity decrease more rapidly during copper depletion than the total copper content or the 2 Cu, 2 Zn-superoxide dismutase activity in erythrocytes.

Figs.1 and 2 also show that the ferroxidase activity correlates more closely with the copper content of the plasma than does 2 Cu, 2 Zn-superoxide dismutase activity with the copper content of the blood cells. Thus the decrease in plasma copper (down 78%) and ferroxidase activity (down 75%) are comparable, while the decrease in blood cell copper (down 72%) and 2 Cu, 2 Zn-superoxide dismutase activity (down 56%) are quite different. This is in accord with the fact that 95% of the plasma copper is ferroxidase-bound but only 60% of the blood cell copper is represented by the 2 Cu, 2 Zn-superoxide dismutase.

It appears that in copper deficiency, copper in 2 Cu, 2 Zn-superoxide dismutase is less readily removed than all other erythrocyte copper and/or blood cell copper is used to maintain the activity of this enzyme. This follows from the observed difference between the decrease in erythrocyte copper and the decrease in 2 Cu, 2 Zn-superoxide dismutase activity and the assumption that the normal copper content in the blood cells is not in excess of that needed to satisfy bodily requirements (Evans, 1973).

2 Cu, 2 Zn-SUPEROXIDE DISMUTASE IN BRAIN

Porter and Folch (1957) and Porter and Ainsworth (1959) have described the copper containing "cerebrocuprein" and Carrico and Deutsch (1969) have demonstrated the identity with erythrocuprein. We have measured the 2 Cu, 2 Zn-superoxide dismutase activity in brain during copper depletion and have found that all conclusions from the data in blood can also be applied to the conditions in brain (Fig.3).

The differences between the relative decrease in brain tissue copper and the 2 Cu, 2 Zn-superoxide dismutase activity is even more pronounced. On the 45th day of depletion the copper content of the tissue decreased to 35% of the control value, while the 2 Cu, 2 Zn-superoxide dismutase activity decreased only to 62% of the control value.

As the 2 Cu, 2 Zn-superoxide dismutase activity was determined in a blood cell lysate and in an 85 000 g-supernatant of the brain homogenate it could be argued that the differences between the findings in blood and brain are

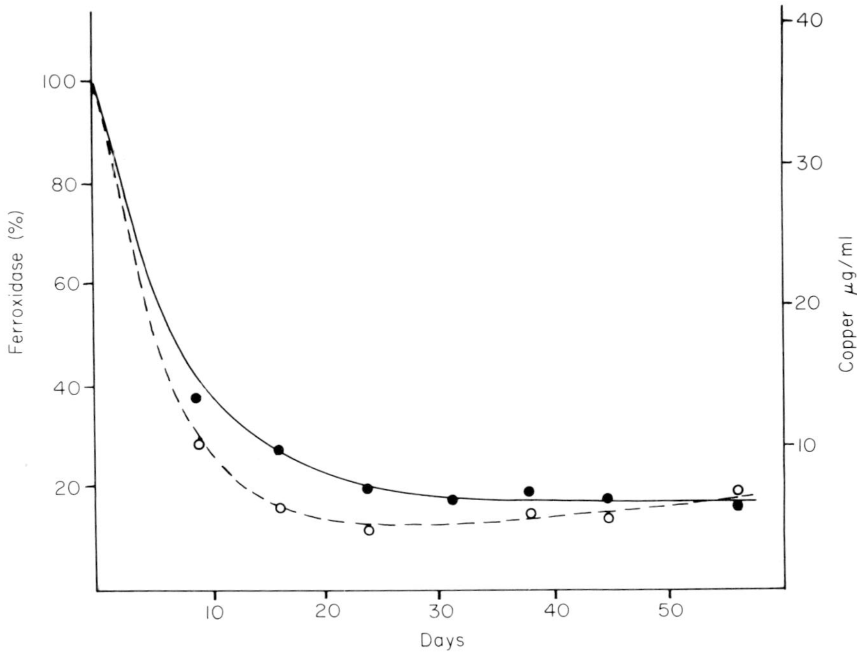

Fig.1. (o) *copper content during copper depletion of rats as μg/ml plasma. Measurements were performed in a Varian AA 6 DB equipped with the mini Massman-cuvette CRA 63.* (o) *ferroxidase activity of plasma from rats as percent of the control.*

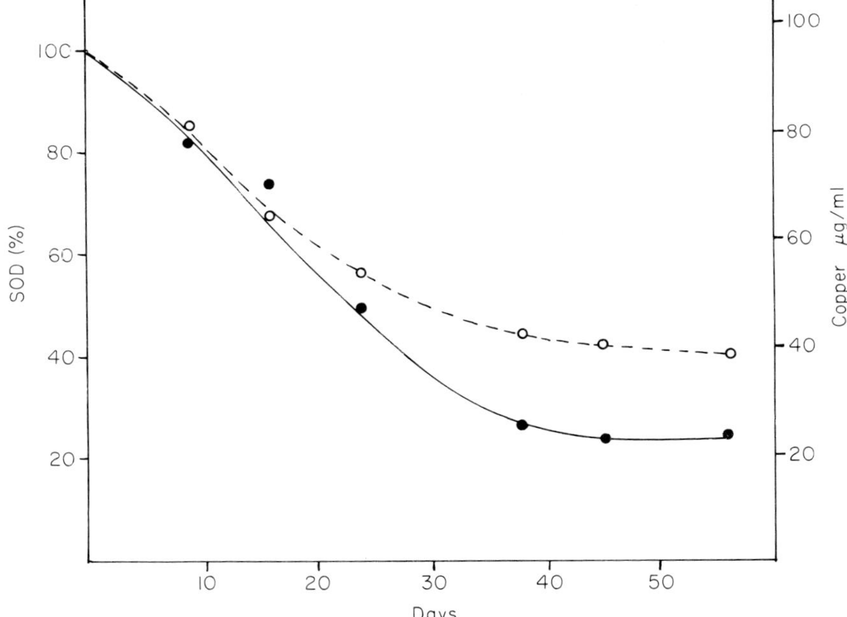

Fig.2. (o) *copper content in copper deficient rats as μg/ml of packed blood cells. Measurements were performed as in Fig.1.* (o) *2 Cu, 2 Zn-superoxide dismutase activity in blood cells of copper depleted rats as percent of the control.*

attributable to different absorption effects in the course of preparation. However, if there are very different preparation effects on the 2 Cu, 2 Zn-superoxide dismutase in brain, the myelin-rich glia cell and myelin-poor neuron should show such effects. No more than 9.5% overall difference could be detected between the 2 Cu, 2 Zn-superoxide dismutase activity/mg protein derived from neuronal or glial cells. This is in good accord with the activities measured in telencephalon and cerebellum (Bohnenkamp and Weser, 1975).

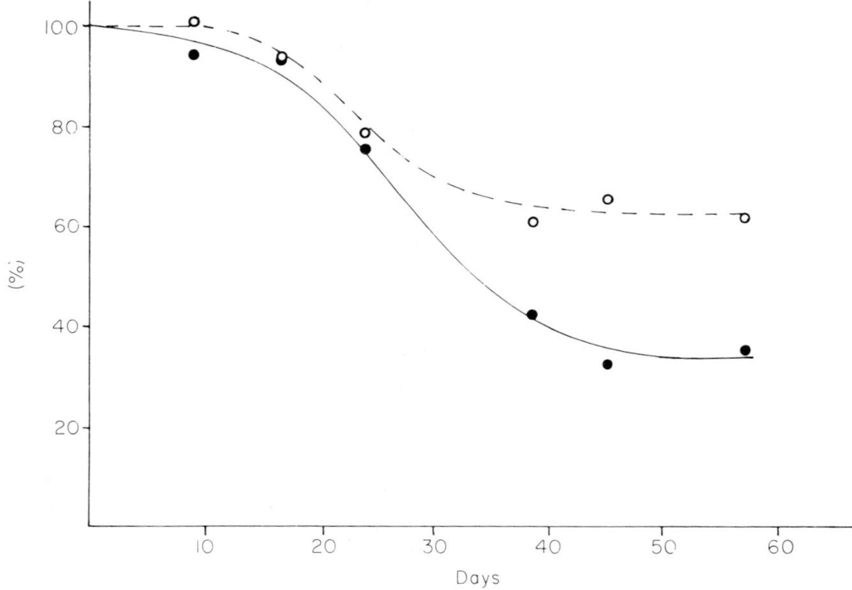

Fig.3. (●) *copper content and* (o) *2 Cu, 2 Zn-superoxide dismutase activity in the 85 000 g-supernatant of rat brain homogenates during copper depletion as percent of the control.*

In the course of this experiment rat brain was dissected by passage through a series of polyamide sieves with mesh sizes from 900 - 35 µm and centrifuged in a sucrose gradient (Rose, 1969). The washed gradient sections were tested for purity by microscopic examination and by the carbonicanhydrase activity which is found only in glial cells (Giacobini, 1962). The almost identical 2 Cu, 2 Zn-superoxide dismutase activities indicate once more the broad distribution of this enzyme.

The intracellular distribution of the 2 Cu, 2 Zn-superoxide dismutase activity in forebrain and cerebellum rarely changes during copper depletion. Table I shows the method of preparing cell fractions and the measured 2 Cu, 2 Zn-superoxide dismutase activities of the final preparation from normal fed rats. The numerical values refer to the activity displayed by 10^{-12} mol bovine erythrocuprein. During copper depletion the 2 Cu, 2 Zn-superoxide dismutase activity decreases in the supernatant to 34 ± 7 and in the mitochondria to 47 ± 8 per mg protein.

Thus the distribution of 2 Cu, 2 Zn-superoxide dismutase activity between 85 000 g-supernatant and mitochondria (Clark and Nicklas, 1970) in brain is always 1:1.3 using the preparation method described above, even though total mitochondrial 2 Cu, 2 Zn-superoxide dismutase activity corresponds well with the degree of copper depletion.

TABLE I

Course of preparation for brain cell nuclei, mitochondria and 85 000 g-supernatant. For these three fractions the 2 Cu, 2 Zn-superoxide dismutase activity is indicated per mg total protein in numbers which refer to the superoxide dismutase activity of 10^{-12} mol bovine erythrocuprein.

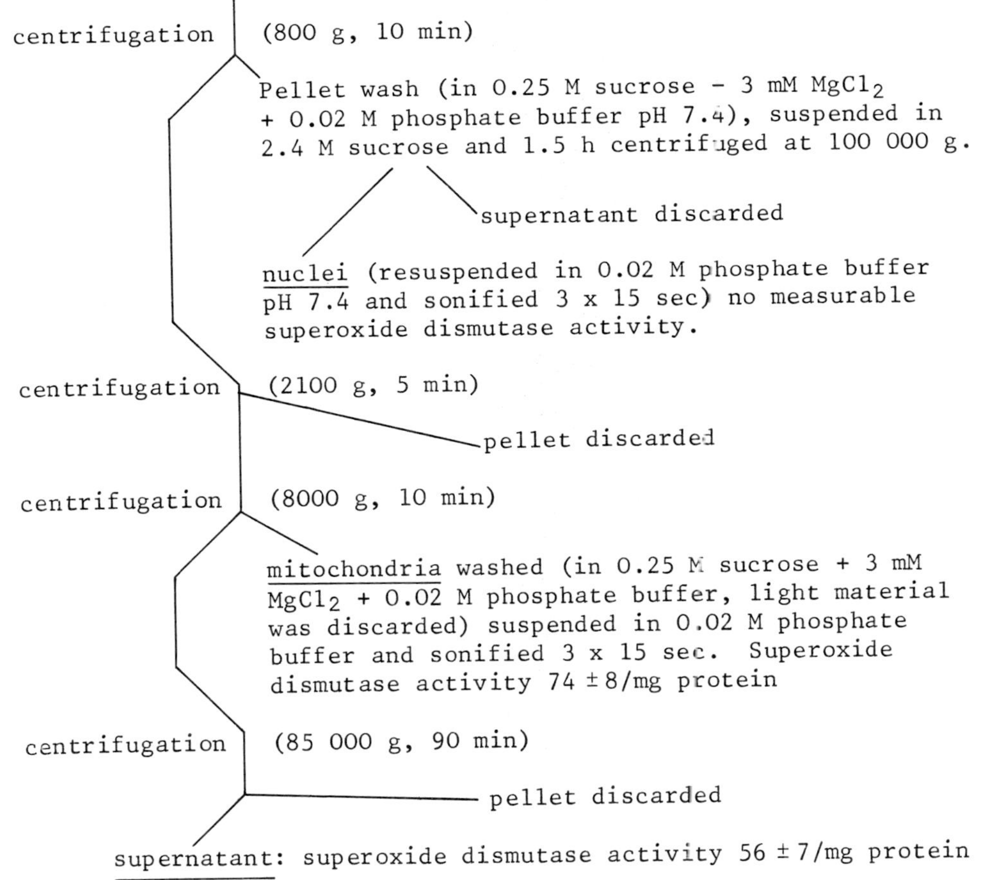

CATALASE AND GLUTATHIONE PEROXIDASE

Provided the 2 Cu, 2 Zn-superoxide dismutase *in vivo* reacts in the following way (McCord and Fridovich, 1969)

$$2H^+ + 2\ O_2^- \xrightarrow{\text{SOD}} O_2 + H_2O_2$$

it would be necessary for the organism to scavenge the produced peroxide. This is preferably done by catalase and/or glutathione peroxidase.

In the case of catalase the enzyme activity shows rather complex changes during copper depletion (Fig.4).

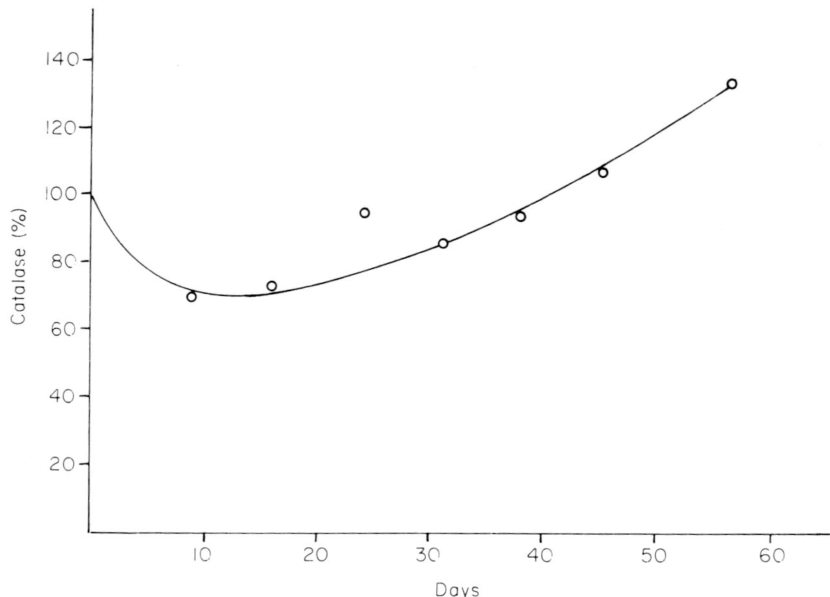

Fig.4. *Catalase activity in blood cells of copper deficient rats as percent of the control.*

Compared to the control the lowest catalase activity was reached within two weeks then the activity proceeded to rise. The control value was exceeded after the 6th week and by the 8th week a stiumlation of 30% above the control value was observed. This is in contrast to the prediction that catalase, a heme enzyme, should decrease during copper deficiency as the ferroxidase does not provide enough Fe^{3+} for the heme synthesis. It is intriguing to note that after the initial drop in catalase activity during copper depletion a rise in this activity is seen at a stage where the remaining 2 Cu, 2 Zn-superoxide dismutase activity is only 70% of the control and the ferroxidase activity displays only 20% of the control.

In contrast to the catalase activity, glutathione peroxidase activity does not show such an effect. This is of special interest, since in brain, the most oxygen-dependent tissue, no catalase activity can be detected (Sorgato *et al.*, 1974). The deleterious action of uncontrolled oxidation reactions (Gilbert, 1972; Weser *et al.*, 1975) might occur much faster in this tissue.

The glutathione peroxidase activity remains unaffected in blood as well as in brain regardless of the copper concentration present in the organism. This is in contrast to Marchena *et al.* (1974) who did not find any glutathione peroxidase in brain.

EXCESSIVE COPPER AND COPPER REPLETION

Giving a surplus of copper to normal fed rats over a period of 4 weeks by 7 intraperitoneal injections of 0.13 mg $CuSO_4$ in saline solution per kg body weight it is not possible to raise the enzyme activities markedly over the control values, even though the copper content of plasma is increased by 62% and in blood cells by 26% (Table II).

TABLE II

Copper concentrations and enzyme activities of rats with excess copper and during the course of repletion after copper deficiency, in percent of control

Treatment	Copper		Enzyme activities		
			Ferroxidase	2 Cu, 2 Zn-superoxide dismutase	
	Plasma	Packed blood cells	Plasma	Blood	Brain
7 Injections (0.31 mg $CuSO_4$/kg)	162	126	108	112	104
Copper repletion					
1.5 Days	96	81	77	93	98
12 Days	108	97	100	96	99

When copper depleted rats (43rd day of depletion) received normal dietary copper concentrations in the drinking water for repletion, the enzyme activities of ferroxidase and 2 Cu, 2 Zn-superoxide dismutase in blood and brain were found to rise in 36 hours to about 90% of the control values and to reach normal values in a period of 12 days (Table II).

Catalase activities show no significant rise after 36 hours of copper repletion but a 66% stimulation after 12 days.

REFERENCES

1. Beauchamp, C. and Fridovich, I. (1971). *Anal. Biochem. 44*, 267-287.
2. Bohnenkamp, W. and Weser, U. (1975). *Hoppe-Seylers Z. Physiol. Chem. 356*, 747-748.
3. Brigelius, R., Hartmann, H.J., Bors, W., Lengfelder, E., Saran, M. and Weser, U. (1975). *Hoppe-Seylers Z. Physiol. Chem. 356*, 739-745.
4. Carrico, R.J. and Deutsch, H.F. (1969). *J. Biol. Chem. 244*, 6087-6093.
5. Clark, J.B. and Nicklas, W.J. (1970). *J. Biol. Chem. 245*, 4724-4731.
6. Dreosti, I.E. and Quicke, G.V. (1968). *Br. J. Nutr. 22*, 1-7.
7. Evans, G.W. (1973). *Physiol. Rev. 53*, 535-570.
8. Giacobini, E. (1962). *J. Neurochem. 9*, 169-177.
9. Gilbert, D.L. (1972). *Anesthesiology 37*, 100-121.
10. Günzler, W.A., Kremers, H. and Flohé, L. (1974). *Z. Klin. Chem. Klin. Biochem. 12*, 444-448.
11. Hallman, L. (1960). *Klinische Chemie und Mikroskopie*, pp. 396-404 Thieme Verlag, Stuttgart.
12. Hsieh, S.H. and Frieden, E. (1975). *Biochem. Biophys. Res. Commun. 67*, 1326-1331.
13. Joester, K.E., Jung, G., Weber, U. and Weser, U. (1972). *FEBS Lett. 25*, 25-28.
14. Johnson, D.A., Osaki, S. and Frieden, E. (1967). *Clin. Chem. 13*, 142-150.
15. Kirchgessner, M., Weser, U. and Müller, H.L. (1967). *Z. Tierphysiologie, Tierernährung und Futtermittelkunde 22*, 76-81.
16. McCord, J.M. and Fridovich, I. (1969). *J. Biol. Chem. 244*, 6049-6055.
17. Marceau, N. and Aspin, N. (1973). *Biochim. Biophys. Acta 328*, 351-358.
18. Marchena, O. de, Guarnieri, M. and McKhann, G. (1974). *J. Neurochem. 22*, 773-776.
19. Porter, H. and Ainsworth, S. (1959). *J. Neurochem. 5*, 91-102.
20. Porter, H. and Folch, J. (1957). *J. Neurochem. 1*, 260-267.
21. Rose, S.P.R. (1969). In "Handbook of Neurochemistry"(Lajtha, A. Ed.) Vol. 2, pp. 183-193, Plenum Press, New York.
22. Sorgato, M.C., Sartorelli, L., Loschen, G. and Azzi, A. (1974). *FEBS Lett. 45*, 92-95.
23. Weser, U., Paschen, W. and Younes, M. (1975). *Biochem. Biophys. Res. Commun. 66*, 769-777.

SUPEROXIDE ANION AS AN INTERMEDIATE OR A SUBSTRATE FOR CERTAIN OXYGENASES

F. HIRATA and O. HAYAISHI

Department of Medical Chemistry
Kyoto University Faculty of Medicine
Kyoto 606, Japan

INTRODUCTION

Superoxide anion (O_2^-) is known to be produced by a number of biologically significant oxidations and is probably generated in all oxygen metabolizing cells (Fridovich, 1975). Superoxide dismutase has been used by numerous investigators to demonstrate the involvement of O_2^- in these processes. If O_2^- is an intermediate, superoxide dismutase will inhibit the reaction by intercepting O_2^-. For example, superoxide dismutase has been used to investigate the involvement of O_2^- in a number of oxygenase-catalyzed reactions. Thus, cytochrome P-450 (Strobel and Coon, 1971), anthranilate hydroxylase (Kumar *et al.*, 1972), metahydroxybenzoate-4-hydroxylase (Kumar *et al.*, 1972), cysteamine oxygenase (Federici *et al.*, 1973), ethionamine sulfoxidase (Prema and Gopinathan, 1974), dopamine β-hydroxylase (Liu *et al.*, 1974), *p*-coumaric acid hydroxylase (Halliwell, 1975) and 2-nitropropane oxygenase (Kido *et al.*, 1975) have all been reported to be inhibited by dismutase.

In this communication, we wish to present evidence indicating the participation of O_2^- in the indoleamine 2,3-dioxygenase catalyzed reaction and to provide some insight into the interaction of the enzyme with O_2^-.

GENERAL PROPERTIES OF INDOLEAMINE 2,3-DIOXYGENASE

Distribution and Purification

Indoleamine, 2,3-dioxygenase is a heme containing dioxygenase that catalyzes the oxygenative ring cleavage of substituted and unsubstituted indoleamines such as tryptophan, 5-hydroxytryptophan, tryptamine and serotonin (Hirata and Hayaishi, 1972). When the tissue distribution of apparent enzyme activity was determined with the high speed supernatants from various organs of rabbit, the enzyme activity was ubiquitously distributed. Much higher activities are associated with the ileum, lung and colon (Hayaishi *et al.*, 1975). Low but significant activities are also detected in the brain, kidney and heart.

Starting from crude extracts of rabbit intestine, the best source of enzyme, the enzyme has been purified about 200-fold with an overall yield of about 15% (Hayaishi *et al.*, 1975). The most highly purified enzyme preparation is essentially homogeneous when examined by acrylamide gel electrophoresis and ultracentrifugation. The enzyme has one molecule of protoheme IX as a prosthetic group per protomer (mol. wt. = 58,000) and does not contain significant amounts of copper.

Substrate Specificity and Assay Method

A highly purified preparation of intestinal indoleamine 2,3-dioxygenase can degrade both the D- and L-isomers of tryptophan to form the corresponding isomers of formylkynurenine (Higuchi and Hayaishi, 1967; Yamamoto and Hayaishi, 1967). As shown in Table I, it also catalyzes the oxygenative ring cleavage of D- and L-5-hydroxytryptophan, tryptamine, serotonin, and melatonin (Hirata *et al.*, 1972; 1974). Skatol, indole, indoleacetic acid and 5-hydroxyindoleacetic acid do not serve as substrates. It is evident from these studies that a substrate molecule must contain an indole nucleus and a substituted or unsubstituted amino group before it can serve as a substrate. Hence, the enzyme has been termed "indoleamine 2,3-dioxygenase".

TABLE I

Substrate Specificity

Substrate	V_{max} (munits/mg protein)	K_m (M)
D-Tryptophan	140	300
L-Tryptophan	140	20
D-5-Hydroxytryptophan	9	100
L-5-Hydroxytryptophan	11	6
Tryptamine	5	250
Serotonin	3	150
Melatonin	0.1	40

Three methods have been employed for the enzyme assay. First, any of the previously mentioned indoleamine substrates can be incubated with enzyme. When the indole ring is cleaved, the absorption maximum shifts about 40 nm to a longer wavelength. When formamidase is included in the reaction mixture and the formyl group is hydrolyzed, the absorption maximum shifts by another 40 nm. These spectral properties have been utilized for spectrophotometric determination of these products (Hirata and Hayaishi, 1972). Second, radioactive substrates can be incubated with the enzyme. The products may be isolated by thin layer or paper chromatography and the radioactivity of the product determined. The third and most convenient method involves the use of ring-2-carbon labeled substrate as originally described by Peterkofsky (1968) for a tryptophan 2,3-dioxygenase assay. In the presence of excess formamidase, the radioactive carbon is eventually released as formate, which is then easily isolated and quantitated.

PARTICIPATION OF O_2^- IN THE CATALYTIC PROCESS OF INDOLEAMINE 2,3-DIOXYGENASE

Requirement of Ascorbic Acid and Methylene Blue as Cofactors

The purified enzyme requires both methylene blue and ascorbic acid for maximum activity. Fig.1 illustrates the dependence of enzyme activity on methylene blue in the presence of an excess amount of ascorbic acid and on ascorbic acid in the presence of an excess amount of methylene blue. The concentrations required for half maximal activity are 2 μM and 0.2 mM, respectively. Methylene blue could not be replaced by any natural cofactors, metals, or other compounds, except for toluidine blue and some related dyes. On the other hand, ascorbic acid could be replaced by some enzyme systems known to generate O_2^- (Table II). With the standard assay mixture containing substrate, enzyme, buffer and methylene blue, the reaction can proceed in the presence of either ascorbic acid or xanthine oxidase or glutathione reductase with their respective substrates but not with glucose oxidase or amino acid oxidase. Ascorbate, the xanthine oxidase system and the glutathione reductase system are all known to generate O_2^- as well as H_2O_2. However, the glucose oxidase and amino acid oxidase systems have been shown by Massey *et al.* (1969) to generate H_2O_2 but not O_2^-. These results provide the first clue that O_2^- and not H_2O_2 is involved in this reaction.

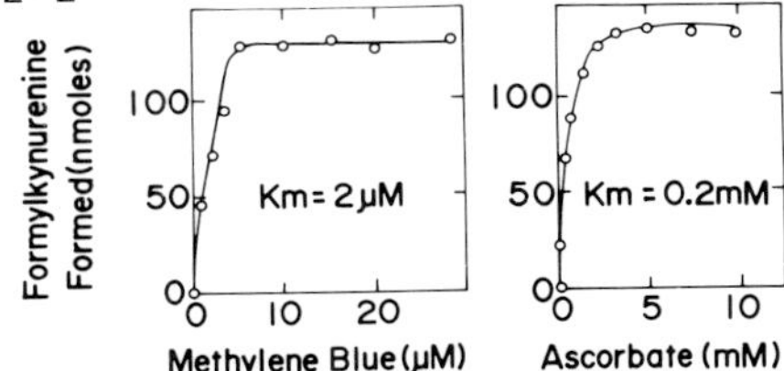

Fig.1. *Dependency of indoleamine 2,3-dioxygenase on methylene blue and ascorbic acid.*

TABLE II

Replacement of Ascorbic Acid by Superoxide and Hydrogen Peroxide Generating Systems

Addition	Product Formation (nmoles)
None	0
Ascorbate	28.0
Xanthine oxidase system	30.0
Glutathione reductase system	25.4
H_2O_2	0
Glucose oxidase system	0
L-Amino acid oxidase system	0

Inhibition by Superoxide Dismutase

In order to intercept O_2^-, superoxide dismutase was added to the reaction mixture and its effect on indoleamine 2,3-dioxygenase was examined. As shown in Fig.2, indoleamine 2,3-dioxygenase was instantaneously inhibited by a highly purified preparation of superoxide dismutase. The degree of inhibition was essentially identical whether it was added during the steady state of the reaction or prior to the start of the reaction. This inhibition was a function of dismutase concentration and of the pH of the medium. At pH 7.5 and 10.0, a half maximal inhibition was obtained with dismutase concentrations of 1.0 and 0.5 μM, respectively. The pH profile of the inhibition was similar to that of the inhibition by dismutase of aerobic cytochrome *c* reduction as previously reported by Klug *et al.* (1972), with the maximum inhibition being around pH 10.

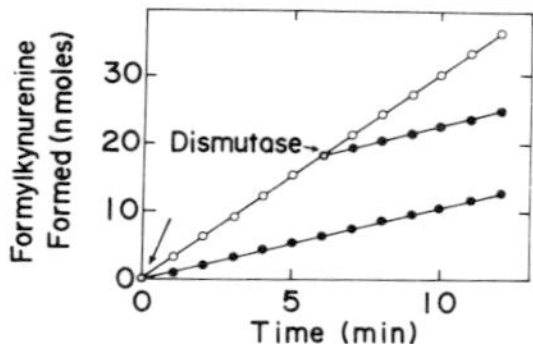

Fig.2. *Inhibition of indoleamine 2,3-dioxygenase by superoxide dismutase from bovine erythrocytes.*

In order to ascertain whether the observed inhibition was specifically due to dismutase, various preparations of bovine erythrocyte dismutase and dismutases from different origins were examined (Fig.3). Under conditions in which the native enzyme caused 50% inhibition, boiled enzyme, apodismutase, and copper ion, which is a component of erythrocyte dismutase, failed to inhibit the reaction, while the reconstituted dismutase was inhibitory to the same degree as the native dismutase. The inhibition observed could be completely abolished by inhibitors of the dismutase, that is, chelating agents for copper such as diethyldithiocarbamate, xanthogenate and diphenylthiocarbazone. Since these inhibitors neither activated nor inhibited the dioxygenase under these conditions, the inhibitory action of dismutase appeared to be due to its catalytic activity. Furthermore, dismutase preparations from bovine erythrocytes, green peas, spinach leaves and *E. coli* were all inhibitory to the same extent regardless of their origins or the nature of the prosthetic groups (Fig.3). These results, taken together, strongly indicate that the inhibition of the dioxygenase by dismutase was due to its specific catalytic decomposition of O_2^- rather than to some non-specific protein-protein interaction or to any toxic effect of copper in the dismutase preparations.

		μM	% Inhibition (0, 25, 50, 75)
I	Dismutase	2	
	Boiled Enzyme	5	
	Apoenzyme	5	
	Cu^{++}	5	
	Reconstituted	2	
		units	
II	Erythrocyte	200	
	Green Pea	200	
	Spinach	220	
	E.coli	170	

Fig.3. *Inhibition of indoleamine 2,3-dioxygenase by various preparations of superoxide dismutase.*

Tiron, another scavenger for O_2^-, also inhibited the dioxygenase at any stage of the reaction. The concentration and pH dependencies of the inhibition by Tiron were very similar to those observed for superoxide dismutase inhibition. On the other hand, compounds which are reported to scavenge singlet oxygen, such as guanosine (Clagett and Galen, 1971), β-carotene (Foote and Denny, 1968) and 1,4-diazobicyclo-(2,2,2)-octane (Ouannes and Wilson, 1968) hardly inhibited the dioxygenase reaction, even though the concentrations used were at least ten times higher than those previously reported (Fig.4). These results indicated that O_2^- itself participates in the reaction of indoleamine 2,3-dioxygenase rather than single oxygen, which may be generated secondarily from O_2^- (Kahn, 1970).

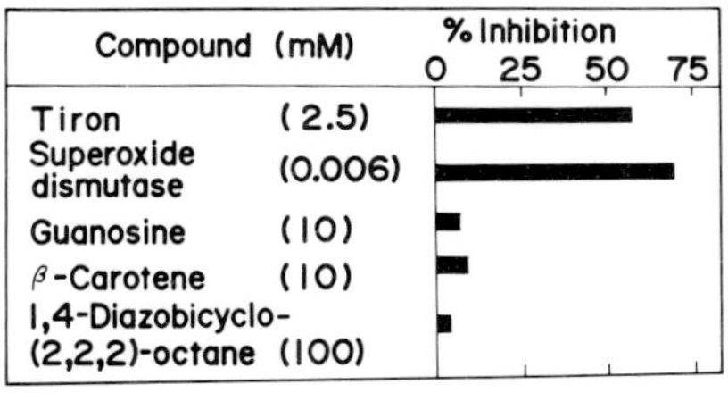

Compound	(mM)	% Inhibition (0 25 50 75)
Tiron	(2.5)	
Superoxide dismutase	(0.006)	
Guanosine	(10)	
β-Carotene	(10)	
1,4-Diazobicyclo-(2,2,2)-octane	(100)	

Fig.4. *Effect of scavengers for O_2^- and for singlet oxygen on the indoleamine 2,3-dioxygenase reaction.*

Factors Involved in Inhibition of Dioxygenase by Dismutase

The degree of inhibition of dioxygenase by dismutase was independent of the concentration of the substrate, D-tryptophan, but was inversely dependent on the concentrations of cofactors, ascorbic acid and methylene blue, indicating that these compounds might be involved in the generation of O_2^-. In fact, it has been reported that O_2^- is produced by the autoxidation of reduced methylene blue (McCord and Fridovich, 1970). Under steady-state conditions of the reaction, approximately one-third of the methylene blue was found to be reduced by ascorbic acid. When the concentration of O_2^- in the reaction mixture was varied as described later, the inhibition of the dioxygenase reaction by a constant amount of dismutase decreased as the concentration of O_2^- increased, indicating that O_2^- is a common substrate for both indoleamine 2,3-dioxygenase and dismutase. In accord with this interpretation, the degree of inhibition was inversely related to the amount of dioxygenase present in the reaction mixture. Furthermore, when the amount of the dioxygenase was increased, the rate of the reaction became constant, indicating that the generation of O_2^- was a rate-limiting step under these conditions.

It was previously reported by Brady *et al.* (1971) and independently by us (Hirata and Hayaishi, 1971) that O_2^- can reductively activate tryptophan 2,3-dioxygenase during the initial stage of the reaction. Superoxide dismutase failed to inhibit tryptophan 2,3-dioxygenase from rat liver or pseudomonad. However, with these enzymes, a prolonged lag period was observed if superoxide dismutase were added prior to the start of the reaction. In order to determine whether O_2^- is required only in the initial stage of the indoleamine 2,3-dioxygenase reaction or is required to maintain the steady state, experiments were carried out to inhibit the xanthine oxidase system by allopurinol, a specific inhibitor for xanthine oxidase, during the course of the reaction (Fig.5). Ascorbic acid was eliminated from the reaction mixture and the time course of product formation was determined spectrophotometrically. There was no measurable activity until xanthine oxidase and its substrate was added. When allopurinol was added to this system, the reaction was instantaneously and completely inhibited. However, the addition of ascorbic acid restored the dioxygenase reaction, indicating that the inhibition was not due to the direct effect of allopurinol on the dioxygenase. These results led us to the tentative conclusion that O_2^- must be generated during the reaction and participates intimately in the catalytic process as an intermediate or substrate rather than in reductive activation, as in the case of pseudomonad and hepatic tryptophan 2,3-dioxygenase.

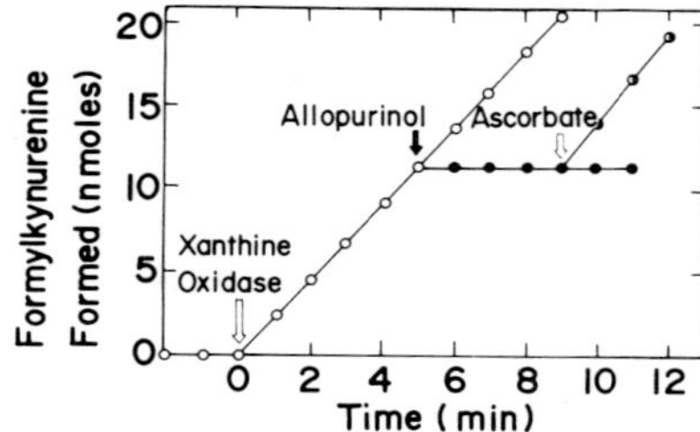

Fig.5. *Effect of allopurinol, a specific inhibitor for xanthine oxidase, on the indoleamine 2,3-dioxygenase reaction in the presence of the xanthine oxidase system.*

Evidence for the Participation of Superoxide Anion

In order to confirm the interpretation that O_2^-, generated either chemically by autoxidation of ascorbic acid or enzymatically by the xanthine oxidase system, is utilized by indoleamine 2,3-dioxygenase, O_2^- was directly introduced into the reaction mixture in place of ascorbic acid. Tetrabutylammonium superoxide can be produced by electrolytic reduction of molecular oxygen (McCord and Fridovich, 1969), or commercial potassium superoxide can be dissolved in dimethylsulfoxide. When O_2^-, thus prepared, was added to the reaction mixture containing a large amount of enzyme, a burst of activity was observed which lasted only for several seconds presumably because of the extreme instability of O_2^- in water (Fig.6). Product formation was not observed in the absence of dioxygenase or in the presence of a boiled enzyme preparation, indicating that this was an enzymic process. When twice as much O_2^- was added, almost twice as much product was produced. This process could be repeated a number of times. The stoichiometry between O_2^- added and the product formed is shown in Fig.6 (right). With relatively small amounts of O_2^-, the amount of formylkynurenine formed corresponded approximately to about 70% of the O_2^- added. However, when the amount of O_2^- was increased, the yield of product as a percent of O_2^- added decreased correspondingly, possibly because a larger fraction of O_2^-, which was not bound to the dioxygenase, decomposed rapidly and was not available to the enzyme. It is noteworthy that the amount of the product formed depended exclusively on the amount of the O_2^- added and not on the presence or absence of oxygen. These results strongly suggest that the enzyme preferentially utilizes O_2^- rather than molecular oxygen as an oxidizing agent.

Affinity of Indoleamine 2,3-Dioxygenase For Superoxide Anion

In order to establish further the participation of superoxide anion in the catalytic process and to clarify its role in this reaction, a new assay procedure was developed for the dioxygenase (Fig.7). The new assay is based on the continuous infusion of O_2^- into a reaction mixture containing DL-[ring-2-^{14}C]tryptophan as substrate, formate as a carrier, catalase to decompose

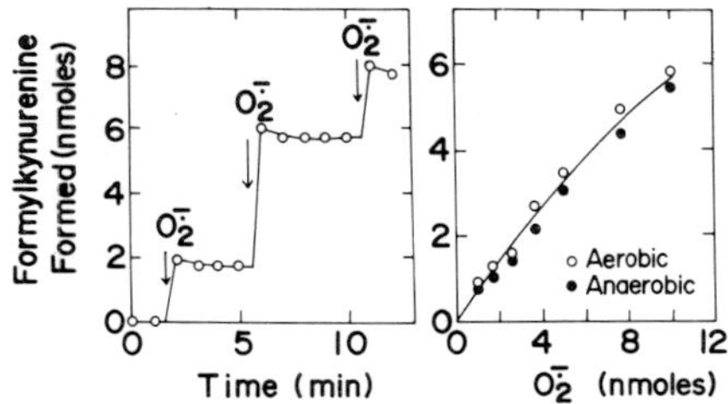

Fig.6. *Left: Effect of chemically prepared superoxide anion on the reaction of indoleamine 2,3-dioxygenase. Right: Stoichiometry between O_2^- added and product formed.*

H_2O_2 formed by spontaneous decomposition, formamidase to complete the hydrolysis of the formyl group, and indoleamine 2,3-dioxygenase in a total volume of 0.2 ml. Potassium superoxide dissolved in dimethylsulfoxide is infused at a constant rate into the reaction mixture which is vigorously mixed with a vortex mixer or a magnetic stirrer. The reaction is terminated with 5% trichloroacetic acid and the acidified reaction mixture is passed through a short column of Dowex-50. ^{14}C-labelled formate was eluted with water and was counted by a scintillation counter.

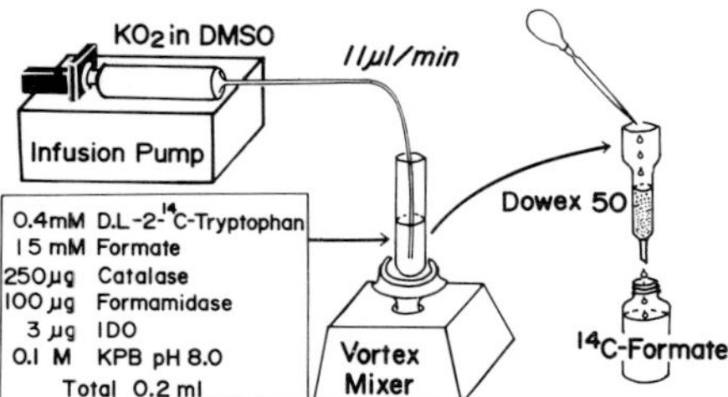

Fig.7. *A new assay procedure for indoleamine 2,3-dioxygenase.*

The time course of the reaction is illustrated in Fig.8. When methylene blue was not included in the reaction mixture, the rate of the reaction rapidly declined after about 1 min. However, in the presence of 25 µM methylene blue, the reaction proceeded linearly for at least 6 min. When 25 µM methylene blue was added 2 min after the initiation of the reaction, the reaction accelerated and became essentially identical to that observed when 25 µM methylene blue was present from the outset of the reaction. These results suggest that the initial velocity was not significantly affected by the presence of methylene blue, but that methylene blue was essential for maintaining the steady state of the reaction even when O_2^- was continuously supplied to the incubation mixture. The role of methylene blue in this reaction and search for a natural cofactor which can replace methylene blue *in vivo* are now under investigation in our laboratory.

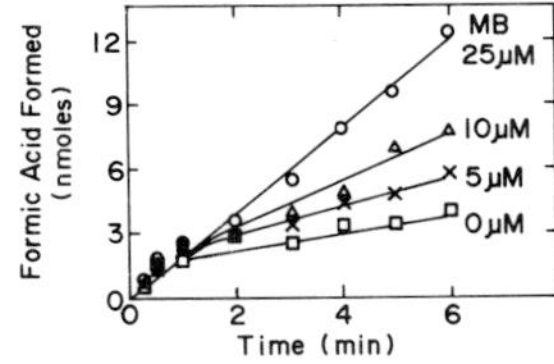

Fig.8. *Time course of the reaction accompanying direct infusion of O_2^-.*

Under the newly developed assay conditions, the maximal rate of the reaction, V_{max}, was almost identical with that observed under the previous assay conditions in the presence of ascorbate and methylene blue. Furthermore, K_m values for tryptophan, were also identical under both assay conditions. On the other hand, the pH optimum of the initial velocities lies in the neighbourhood of 8.5 with the new assay procedure, whereas the optimum pH was found to be around 7.0 with the conventional assay procedure in the presence of methylene blue and ascorbic acid. The reason for this discrepancy is not clearly understood, but it may be attributable to the pH dependency of the rate of generation and the stability of O_2^-.

The steady-state concentration of O_2^- cannot be measured without special equipment such as an electron-spin-resonance spectroscope and mass spectrometer. Furthermore, these physical methods of measuring O_2^-, although direct and unequivocal, lack sensitivity at the low concentrations of O_2^- being studied. Using the new assay procedures, the steady-state concentration of O_2^- is calculable, because a known amount of O_2^- is constantly infused and the decay constant of O_2^- has been previously reported (Behar *et al.*, 1970). When the amount of O_2^- was varied, the concentration of O_2^- for half maximal activity was estimated to be approximately 1.3 μM (Fig.9). This value was much lower than that reported for superoxide dismutase (355 μM) by Rigo *et al.* (1975).

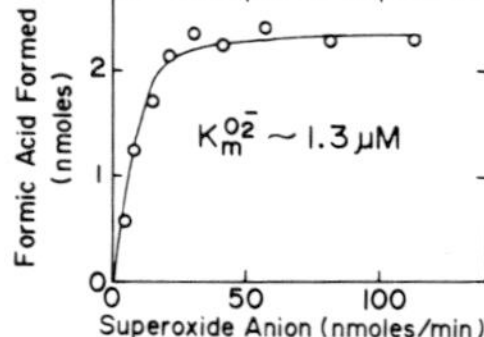

Fig.9. *K_m of indoleamine 2,3-dioxygenase for O_2^-.*

INTERACTION OF O_2^- WITH INDOLEAMINE 2,3-DIOXYGENASE

Formation of Oxygenated Indoleamine 2,3-dioxygenase

The native indoleamine 2,3-dioxygenase has absorption maxima at 406, 499 and 630 nm, typical of a high spin ferric hemoprotein (Fig.10). It could be reduced to a ferrous form enzymatically by methemoglobin reductase and an NADH generating system or chemically, by sodium borohydride or sodium dithionite. However, the ferrous form is unstable and undergoes autoxidation, regenerating the ferric form. When potassium superoxide dissolved in dimethylsulfoxide was infused into a buffered solution containing the native ferric enzyme, a new spectral species with absorption maxima at 418, 542 and 576 nm appeared. The observed spectrum was distinct from those of the ferric or ferrous form of enzyme. Essentially identical results were obtained when O_2^- was generated in the reaction mixture by the xanthine oxidase system. The formation of this new spectral species was completely abolished in the presence of superoxide dismutase, indicating that this process involves the direct binding of O_2^- to the ferric form of enzyme. The positions of the α-, β- and Soret bands of this new spectral species were similar to those of the oxygenated form of known heme proteins, such as oxyhemoglobin (Sidwell *et al.*, 1938) oxymyoglobin (Yamazaki *et al.*, 1964) and compound III of peroxidase (Yokota and Yamazaki, 1965). Hence, it may be presumed that this new spectral species represents the so-called oxygenated form of indoleamine 2,3-dioxygenase, namely, the binary complex of oxygen-enzyme. This new spectral species was also observed during the steady state of the reaction in the presence of D- or L-tryptophan, and resembled the spectrum of the enzyme-substrate-oxygen complex, an obligatory intermediate which appears during the course of the reaction catalyzed by tryptophan 2,3-dioxygenase (Ishimura *et al.*, 1970).

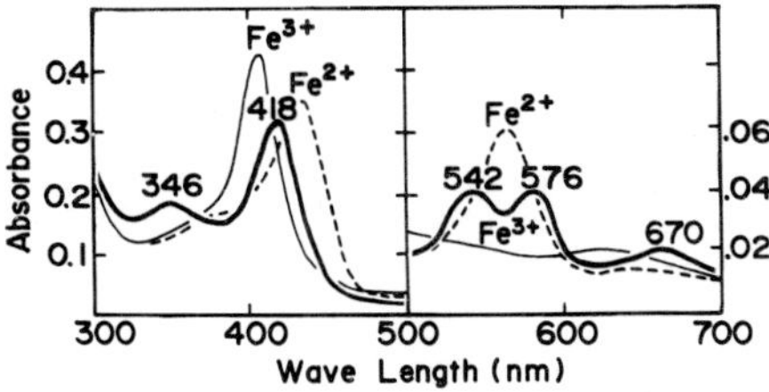

Fig.10. *Absorbance spectra of ferrous, ferric and oxygenated forms of indoleamine 2,3-dioxygenase.*

Properties of Oxygenated Indoleamine 2,3-dioxygenase

Since the initial observation of the oxygenated intermediate of tryptophan 2,3-dioxygenase, it has been pointed out that the valence state of the heme iron in such complexes is in the state of $Fe^{2+}O_2 \rightleftharpoons Fe^{3+}O_2^-$ (Hayaishi, 1964). Misra and Fridovich (1972) have reported that O_2^- is generated during the autoxidation of oxymyoglobin and oxyhemoglobin, by direct dissociation of $Fe^{3+}O_2^-$, whereas it is well known that molecular oxygen is normally released when the partial pressure of oxygen is lowered. Therefore, in order to investigate the valence state of the heme iron of the oxygenated indoleamine 2,3-dioxygenase, the effects of potassium cyanide, a ligand for ferric heme, and carbon monoxide, a ligand for ferrous heme, were examined. Although 10 mM potassium cyanide did not react with the ferrous enzyme, the addition of 1 mM potassium cyanide resulted in immediate disappearance of the oxygenated form. Furthermore, when carbon monoxide was bubbled vigorously through a solution of the oxygenated form obtained by the infusion of O_2^-, a spectrum resembling that of the ferrous enzyme-carbon monoxide complex appeared. These results indicate that the bound oxygen molecule can be replaced by potassium cyanide and carbon monoxide. In fact, the oxygenated form of indoleamine 2,3-dioxygenase underwent spontaneous decomposition to produce the ferric form, whereas the ferrous form was obtained by reduction of oxygen pressure. When oxygen was introduced to the ferrous indoleamine 2,3-dioxygenase, the new spectral species was again observed and was not distinguishable from that obtained by the reaction of the ferric enzyme with O_2^-. These results suggest that the new spectral species observed is the $Fe^{3+}O_2^-$ complex which, in turn, is in equilibrium with the $Fe^{2+}O_2$ complex.

When D- or L-tryptophan was added to the incubation mixture containing the oxygenated form of the enzyme, the ferric form of enzyme and formylkynurenine were immediately released, indicating that the new spectral species is catalytically active. The specificity of the oxygenated form for indoleamine derivatives were next examined. Metabolizable substrate analogues such as tryptophan, 5-hydroxytryptophan, tryptamine and serotonin were cleaved, whereas indoleacetic acid and skatol, non-metabolizable compounds, were inactive. The amount of the product produced from these substrates was proportional to the amount of the oxygenated enzyme employed, but the stoichiometry was not unity. Investigations on the oxygenated form of indoleamine 2,3-dioxygenase are now in progress. However, the above results appear consistent with, but do not necessarily prove, the sequence of reactions shown in Fig.11. The native ferric form of the enzyme reacts first with O_2^- to form the oxygenated form. Alternatively such an oxygen adduct may be formed by the addition of oxygen to the ferrous form of enzyme. This oxygenated form of enzyme then reacts with the substrate to form the product. The overall stoichiometry indicates that one electron should be released, if the enzyme utilizes O_2^- as an oxidizing agent rather than molecular oxygen. It is most likely that the electron is transfered to methylene blue, because methylene blue accelerates the autoxidation of the oxygenated form of enzyme as well as the ferrous form to the ferric form. The ferrous form in the terminal stage of the reaction could be demonstrated in the presence of carbon monoxide.

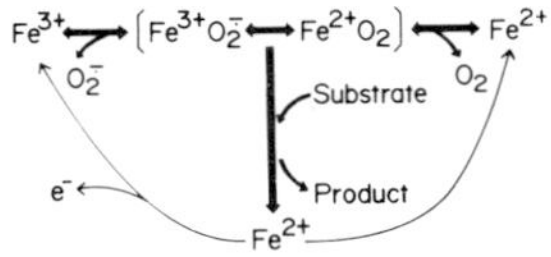

Fig.11. *Hypothetical reaction sequence of indoleamine 2,3-dioxygenase.*

SUMMARY AND PERSPECTIVES

Oxygen is one of the most abundant elements and is essential for practically all forms of life. The principle process in biological oxidation proceeds exclusively through the transfer of hydrogen atoms of electrons through various carriers ultimately to oxygen. The complete reduction of oxygen requires four electrons and the intermediates that would be produced if this reduction were to proceed by univalent steps would be superoxide anion (O_2^-), hydrogen peroxide, and hydroxyl radical ($OH^\cdot$). Among these, the most stable and the only one that accumulates appreciably in neutral aqueous media is hydrogen peroxide. Hydrogen peroxide is primarily degraded by catalase or is utilized for the oxidation of organic compounds by peroxidases.

Recently, it has been shown that O_2^- is a common intermediate in the biological reduction of oxygen by oxidases and dehydrogenases, and that superoxide dismutase is the primary enzyme for catalyzing the disproportionation of this species. Analogous to the utilization of hydrogen peroxide by peroxidases, there might be enzymes that utilize O_2^- for oxidizing organic compounds. As described in this communication, indoleamine 2,3-dioxygenase might be such a novel enzyme. Furthermore, in the case of oxygenases such as cytochrome P.450 (Strobel and Coon, 1971), dopamine β-hydroxylase (Liu *et al.*, 1974) and 2-nitropropane oxygenase (Kido *et al.*, 1976), it has been reported that superoxide dismutase inhibits the reactions catalyzed by these enzymes and that O_2^- generating systems involving xanthine oxidase or phenazine methosulfate replaced their cofactors without any loss of enzyme activity. Chemical model systems of oxygenases have been extensively studied because they afford a less complicated approach to the study of enzyme mechanisms. Among these studies has been the use of bis(salicylidene)ethylenediaminocobalt (II) in place of tryptophan 2,3-dioxygenase for catalyzing oxygenation reactions. This metal complex can catalyze the oxygenation of 3-substituted indoles to produce the corresponding *o*-formylaminoacetophenone (Nishinaga, 1975). The oxygen molecule incorporated in this complex has been found by X-ray crystallography and by electron spin resonance spectrophotometry to be mostly (80 to 90%) in the O_2^- state (Nishinaga, 1974). Furthermore, the preparation of the complex of iron porphyrin with O_2^- also supported the formulation of the presumably analogous complexes of heme proteins as superoxide-ferric iron derivatives (Allen *et al.*, 1974). Such speculation is also supported by Collman *et al.* (1974). The migration of an electron among metal, substrate, and oxygen might activate oxygen and substrate, although the detailed chemistry and the exact nature of the activated oxygen complex remains uncertain. If such is the case, other oxygenases may be found to utilize O_2^- as an intermediate or a substrate.

ACKNOWLEDGEMENT

This manuscript was prepared during the tenure of O.H. as a Fogarty International Scholar at the National Institutes of Health, Bethesda, Maryland, U.S.A.

REFERENCES

1. Allen, H., Hill, O., Turner, D.R. and Pellizer, G. (1974). *Biochem. Biophys. Res. Commun. 56*, 739-744.
2. Behar, D., Czapski, G., Rabani, J., Dorfman, L.M. and Schwartz, H.A. (1970). *J. Phys. Chem. 74*, 3209-3213.
3. Brady, F.O., Forman, H.J. and Feigelson, P. (1971). *J. Biol. Chem. 246*, 7119-7124.
4. Clagett, D.C. and Galen, T.J. (1971). *Arch. Biochem. Biophys. 146*, 196-201.
5. Collman, J.P., Gagne, R.R., Reed, C.A., Robinson, W.T. and Rodley, G.A. (1974). *Proc. Nat. Acad. Sci. U.S.A. 71*, 1326-1329.
6. Fridovich, I. (1975). *Ann. Rev. Biochem. 44*, 147-149.
7. Federici, G., Fiori, A., Costa, M., Barboni, E. and Dupré, S. (1973). Abstract of Ninth International Congress of Biochemistry, Stockholm, p. 331.
8. Foote, C.S. and Denny, R.W. (1968). *J. Amer. Chem. Soc. 90*, 6233-6235.
9. Halliwell, B. (1975). *Eur. J. Biochem. 55*, 355-360.
10. Hayaishi, O. (1964). Proceedings of the Plenary Sessions Sixth International Congress of Biochemistry, New York, p. 31.
11. Hayaishi, O., Hirata, F., Fujiwara, M., Ohnishi, T. and Nukiwa, T. (1975). Proceedings of FEBS Meeting, p. 131-144.
12. Higuchi, K. and Hayaishi, O. (1967). *Arch. Biochem. Biophys. 120*, 397-403.
13. Hirata, F. and Hayaishi, O. (1971). *J. Biol. Chem. 246*, 7825-7826.
14. Hirata, F. and Hayaishi, O. (1972). *Biochem. Biophys. Res. Commun. 47*, 1112-1119.
15. Hirata, F., Hayaishi, O., Tokuyama, T. and Senoh, S. (1974). *J. Biol. Chem. 249*, 1311-1313.
16. Ishimura, Y., Nozaki, M., Hayaishi, O., Nakamura, T., Tamura, M. and Yamazaki, I. (1970). *J. Biol. Chem. 245*, 3593-3602.
17. Khan, A.U. (1970). *Science 168*, 476-477.
18. Kido, T., Soda, K., Suzuki, T. and Asada, K. (1976). *J. Biol. Chem.*, in press.
19. Klug, D., Rabani, J. and Fridovich, I. (1972). *J. Biol. Chem. 247*, 4839-4842.
20. Kumar, R.P., Ravindranath, S.D., Vaidyanathan, C.S. and Rao, N.A. (1972). *Biochem. Biophys. Res. Commun. 49*, 1422-1426.
21. Liu, T.Z., Shen, J.-T. and Ganong, W.F. (1974). *Proc. Exp. Biol. Med. 146*, 37-40.

22. Massey, V., Strickland, S., Mayhew, S.G., Howell, L.G., Engel, P.C., Mathews, R.G., Schuman, M. and Sullivan, P.A. (1969). *Biochem. Biophys. Res. Commun. 36*, 891-897.
23. McCord, J.M. and Fridovich, I. (1969). *J. Biol. Chem. 244*, 6049-6055.
24. McCord, J.M. and Fridovich, I. (1970). *J. Biol. Chem. 245*, 1374-1377.
25. Misra, H.P. and Fridovich, I. (1972). *J. Biol. Chem. 247*, 6960-6962.
26. Nishinaga, A. (1974). *Kagakuzohkan 61*, 179-191.
27. Nishinaga, A. (1975). *Chem. Letters*, 273-276.
28. Ouanness, C. and Wilson, T. (1968). *J. Amer. Chem. Soc. 90*, 6527-6528.
29. Peterkofsky, B. (1968). *Arch. Biochem. Biophys. 128*, 637-645.
30. Prema, K. and Gopinathan, K.P. (1974). *Biochem. J. 143*, 613-624.
31. Rigo, A., Viglino, P. and Rotilio, G. (1975). *Biochem. Biophys. Res. Commun. 63*, 1013-1018.
32. Sidwell, A.E. Jr., Munch, R.H., Barron, E.S.G. and Hogness, T.R. (1938). *J. Biol. Chem. 123*, 335-343.
33. Strobel, H.W. and Coon, M.J. (1971). *J. Biol. Chem. 246*, 7826-7829.
34. Yamamoto, S. and Hayaishi, O. (1967). *J. Biol. Chem. 242*, 5260-5266.
35. Yamazaki, I., Yokota, K. and Shikawa, K. (1964). *J. Biol. Chem. 239*, 4151-4156.
36. Yokota, K. and Yamazaki, I. (1965). *Biochem. Biophys. Res. Commun. 18*, 48-53.

LACK OF INHIBITION OF DOPAMINE-β-HYDROXYLASE BY SUPEROXIDE DISMUTASE

E.J. DILIBERTO, Jr. and S. KAUFMAN

Laboratory of Neurochemistry, NIMH
Bethesda, Maryland 20014, U.S.A.

Dopamine-β-hydroxylase (3,4-hydroxyphenylethylamine, ascorbate: oxygen oxidoreductase (β-hydroxylating); EC 1.14.17.1), a mixed-function oxidase, catalyzes the ascorbate-dependent conversion of dopamine to norepinephrine (Levin *et al.*, 1960). Ascorbate functions in this system by reducing enzyme-bound Cu^{2+} to Cu^{1+}. Even in the absence of ascorbate, this reduced form of the enzyme reacts with substrate and molecular oxygen to form the hydroxylated product and to regenerate the oxidized form of the enzyme (Friedman and Kaufman, 1965).

Recently, it was suggested that superoxide anion, generated from ascorbate, was responsible for reduction of the enzyme; evidence for this proposal was provided in part by the observation that partially purified dopamine-β-hydroxylase is inhibited by superoxide dismutase (Liu *et al.*, 1974).

In numerous experiments in our laboratory we have been unable to confirm the report that superoxide dismutase inhibits dopamine-β-hydroxylase. Typical results are shown in Table I, where it can be seen that purified superoxide dismutase in amounts as high as 200 μg/ml do not inhibit the hydroxylation of tyramine. These results are to be contrasted with those of Liu *et al.* who reported 80% inhibition of hydroxylase by 200 μg/ml of superoxide dismutase. In separate experiments in which dopamine-β-hydroxylase activity was measured by a spectrophotometric assay (Levin and Kaufman, 1961), we were also unable to show inhibition of hydroxylase activity by dismutase in concentrations as high as 1 mg/ml.

Our results, therefore, provide no support for the idea that superoxide anion is an intermediate in the reduction of dopamine-β-hydroxylase by ascorbate. It should be noted that our earlier observation that ascorbate can reduce the hydroxylase to an active form *in the absence of oxygen* (Friedman and Kaufman, 1965) is also inconsistent with the proposal of Liu *et al.* that superoxide anion is an intermediate in the reduction of the hydroxylase by ascorbate.

TABLE I

Effects of Superoxide Dismutase on Dopamine-β-Hydroxylase Activity

Superoxide Dismutase	Octopamine Formed μmol/min/mg protein
None	5.51
50 μg/ml	5.30
100 μg/ml	5.84
200 μg/ml	5.40

Dopamine-β-hydroxylase activity was assayed by the procedure of Friedman and Kaufman (1966). The reaction mixture contained (in micromoles): fumarate, 5; sodium phosphate, pH 6.0, 12.5; ascorbate, 2.5; [*side chain*-2-^{14}C]tyramine (0.14 μCi/μmol), 1.25; catalase, 15 μg; and purified dopamine-β-hydroxylase, 1 μg. The final volume was 0.25 ml. The activity of the superoxide dismutase preparation (Miles Lab., bovine erythrocytes) was tested and found to have a specific activity comparable to that of purified dismutase (Misra, H.P. and Fridovich, I. (1972)).

REFERENCES

1. Friedman, S. and Kaufman, S. (1965). *j. Biol. Chem. 240*, 4763.
2. Levin, E.Y., Levenberg, B. and Kaufman, S. (1960). *J. Biol. Chem. 235*, 2080.
3. Levin, E.Y. and Kaufman, S. (1961). *J. Biol. Chem. 236*, 2043.
4. Liu, T.Z., Shen, J.-T. and Ganong, W.F. (1974). *Proc. Soc. Exp. Bio. Med. 146*, 37.
5. Misra, H.P. and Fridovich, I. (1972). *J. Biol. Chem. 247*, 3170.

ENZYME-CATALYZED CHAIN OXIDATION OF NICOTINAMIDE ADENINE DINUCLEOTIDE BY SUPEROXIDE RADICALS

B.H.J. BIELSKI

Department of Chemistry
Brookhaven National Laboratory
Upton, New York 11973, U.S.A.

P.C. CHAN

Department of Biochemistry
State University of New York Downstate Medical Center
Brooklyn, New York 11203, U.S.A.

INTRODUCTION

Superoxide free radicals are involved in many biological reactions some of which are deleterious to the living cell. At present there is a controversy regarding the reactivity of O_2^-. Extensive research in recent years on the mechanisms of involvement of O_2^- in various biological systems has led to the belief that other more "active oxygen species" derived from the superoxide radical may be the immediate causes of the observed effects. The suggested species include the hydroxyl radical, singlet molecular oxygen, metal-oxygen complexes and peroxides.

This presentation is intended to show that under suitable conditions the superoxide radical itself can be the active oxidizing agent. For example in the presence of a biological catalyst such as lactate dehydrogenase, the rate of reaction between the superoxide radical and NADH is increased by four orders of magnitude. Since this sytem is free of metal ions to form complexes with the radical (or to convert it into some other reactive species) the enhancement of this reactivity must be due to some form of activation of the NADH molecule. Experimental evidence suggests that in the presence of the enzyme this reaction is facilitated by the resonance stabilization of the transient intermediate, enzyme-NAD· radical. As the nucleotide radical reduces molecular oxygen to form another superoxide radical at a diffusion controlled rate, a chain reaction is thus initiated.

METHODS OF SUPEROXIDE RADICAL GENERATION

In these studies superoxide free radicals were generated either by the xanthine oxidase system or by high energy ionizing radiation. Both of these methods have inherent advantages and limitations.

Xanthine Oxidase Method

It has been demonstrated that O_2^- radicals and H_2O_2 are produced during the xanthine oxidase catalyzed oxidation of xanthine to uric acid in the presence of molecular oxygen (1,2). The mechanism proposed by Fridovich (1) assumes that after the oxidation of xanthine, the reduced form of the enzyme (EH_2) is reoxidized via two alternative pathways. The univalent electron transfer pathway yields superoxide radicals:

$$EH_2 + 2O_2 \longrightarrow E + 2O_2^- + 2H^+ \quad (1)$$

The other pathway of reoxidation of EH_2, apparently leads directly to hydrogen peroxide formation by a two-electron transfer to molecular oxygen:

$$EH_2 + O_2 \longrightarrow E + H_2O_2 \quad (2)$$

The percentage of EH_2 oxidation by Reaction 1 is dependent on the concentrations of molecular oxygen and xanthine as well as on the pH of the system, and it is determined by the quantitative reduction of ferricytochrome *c* by O_2^- (1).

This method is convenient for generation of O_2^- at a very slow rate for optimal chain length. It is, however, not suitable for studies on mechanisms and kinetics.

Ionizing Radiation Method

Superoxide radicals can be generated by high energy ionizing radiation either under steady state conditions in a ^{60}Co gamma-ray source, under flow-radiolysis conditions where O_2^- is formed in isolation and transported to a mixing chamber to encounter the scavenger, or by pulse radiolysis where it is formed on a microsecond scale in a single burst of energy deposition. While each of these radiation techniques has its limitations, combined they constitute a powerful research tool with a wide range of experimental possibilities. With the exception of the flow-radiolysis technique, these methods require the addition of chemicals to protect enzymes and other biological compounds from reacting directly with the primary radicals generated by the radiation.

We have found that sodium formate which is necessary for converting all primary radicals to superoxide radicals, is also an effective protector for the system. Moreover, formate does not affect the catalytic activity of the enzyme on NADH oxidation by pyruvate (3).

Based upon results from a number of studies applicable to both steady state and pulse radiolysis, the overall mechanism of superoxide radical formation (which refers to O_2^- and HO_2 in equilibrium) in oxygenated formate solutions can be described by the following reactions:

$$H_2O \rightsquigarrow OH\ (2.74,\ e^-_{aq}\ (2.76),\ H\ (0.55),$$

$$H_3O^+\ (2.76),\ H_2\ (0.55),\ H_2O_2\ (0.72) \quad (I)$$

$$e^-_{aq} + O_2 \longrightarrow O_2^- \quad (3)$$

$$H + O_2 \longrightarrow HO_2 \quad (4)$$

$$HO_2 \rightleftharpoons H^+ + O_2^- \quad (5,-5)$$

$$OH + HCOO^- \longrightarrow \cdot COO^- + H_2O \quad (6)$$

$$\cdot COO^- + O_2 \longrightarrow O_2^- + CO_2 \quad (7)$$

The numerical value in parenthesis in Equation (I) represents the G value, the number of molecules formed or transformed per 100 eV of energy dissipated in the system. From the established $G(O_2^-) = G(OH + e_{aq}^- + H) = 6.05$ in oxygenated formate solutions and appropriate ferrous dosimeter calibrations, absolute superoxide radical concentrations can be determined to within a few percent. The quantitative generation of superoxide radicals per unit dose input is pH independent in the range from 2 - 11.

In the absence of appropriate scavengers, the superoxide and perhydroxyl radical disproportionate by second order kinetics to molecular oxygen and hydrogen peroxide:

$$HO_2 + O_2^- + H_2O \longrightarrow O_2 + H_2O_2 + OH^- \quad (8)$$

Rigorous treatments of the kinetics of this system as a function of pH can be found in References 4 - 6.

KINETICS AND THE CHAIN MECHANISM

The kinetic parameters of the chain oxidation of enzyme bound NADH initiated by superoxide radicals (O_2^- and HO_2) and propagated by molecular oxygen was studied as a function of pH by pulse radiolysis (7,8). The reaction steps of the chain mechanism were confirmed by studies with the other methods: flow radiolysis and steady-state generation of the radicals by either γ-ray radiation (3) or xanthine oxidase (9). Since lactate dehydrogenase (LDH) appeared to be the most efficient catalyst for the chain oxidation, it was used preferentially in all mechanistic studies.

Taking into account the existing equilibrium between superoxide and perhydroxyl radicals (Reactions, 5,-5; pK 4.8 (6)) and the equilibria between free and enzyme bound NADH/NAD$^+$ (Reactions 9,-9; 13,-13), the following chain reaction mechanism was proposed

$$\text{Enzyme} + \text{NADH} \rightleftharpoons \text{Enzyme-NADH} \quad (9,-9)$$

$$\text{Enzyme-NADH} + O_2^- \longrightarrow \text{Enzyme-NAD}\cdot + HO_2^- \quad (10)$$

$$\text{Enzyme-NADH} + HO_2 \longrightarrow \text{Enzyme-NAD}\cdot + H_2O_2 \quad (11)$$

$$\text{Enzyme-NAD}\cdot + O_2 \longrightarrow \text{Enzyme-NAD}^+ + O_2^- \quad (12)$$

$$\text{Enzyme-NAD}^+ \rightleftharpoons \text{Enzyme} + \text{NAD}^+ \quad (13,-13)$$

$$O_2^- + X \longrightarrow \text{Product(s)} \quad (14)$$

The chain oxidation is initiated in Reactions 10 and 11, and then propagated by Reaction 12. Reaction 14 is the chain termination step in which X represents some unidentified component(s).

Experiments for studying the rate of Reaction 10 were carried out above neutral pH, where the contribution of Reaction 11 is negligible. Experimentally, the rates were determined spectrophotometrically by observing NADH concentration changes which were initiated by a given quantity of superoxide radicals formed by an electron pulse. The rate of decrease of NADH is an indirect measurement of the rate of decrease of superoxide radicals, which can be described by Equation (II):

$$-\frac{d[O_2^-]}{dt} = k_{10}[\text{E-NADH}][O_2^-] + k_{14}[X][O_2^-] - k_{12}[\text{E-NAD}\cdot][O_2] \quad (II)$$

Since Reaction 12 is diffusion controlled (7,10,11), that is $k_{12} >> k_{10}$, the rate of disappearance of O_2^- in Reaction 10 is equal to its rate of regeneration by Reaction 12 and hence Equation (II) is reduced to Equation (III):

$$-\frac{d[O_2^-]}{dt} = k_{14}[X][O_2^-] \qquad \text{(III)}$$

The rate of disappearance of NADH by Reaction 10 is given by:

$$-\frac{d[E\text{-}NADH]}{dt} = k_{10}[E\text{-}NADH][O_2^-] \qquad \text{(IV)}$$

Substitution of Equation (III) into Equation (IV) and integration between the time limits t = 0 (for all practical purposes taken at the end of the electron pulse) and t = ∞ (end of experimental run) yields

$$k_{10} = \left[\frac{k_{14}[X]\Delta[NADH]}{[E\text{-}NADH][O_2^-]_o}\right] \qquad \text{(V)}$$

where $[O_2^-]_o$ is the quantity of superoxide radicals generated by the electron pulse, k_{14} X is the experimentally observed pseudo first order rate constant for NADH disappearance and $\Delta[NADH]/[O_2^-]_o = \nu$ is the chain length, i.e. the number of NADH molecules oxidized per superoxide radical generated initially in the given run.

The second order rate constant $k_{10} = (1.0 \pm 0.2) \times 10^5\ M^{-1}\ s^{-1}$ for O_2^- oxidation of lactate dehydrogenase-bound NADH was calculated from experimental data by Equation (V). This rate constant is four orders of magnitude higher than the value reported by Land and Swallow (10) for the oxidation of free NADH by O_2^-, $k_{15} << 27\ M^{-1}\ s^{-1}$:

$$O_2^- + NADH + H^+ \longrightarrow NAD\cdot + H_2O_2 \qquad (15)$$

For the pH range where Reaction 11 contributes to the overall disappearance of NADH, it can be shown that by taking the ($HO_2 \rightleftharpoons O_2^- + H^+$) equilibrium into consideration the rate follows Equation (VI):

$$k_{cal} = \frac{1}{\nu}\ \frac{k_{11} + k_{10}(K_{HO_2}/[H^+])}{1 + K_{HO_2}/[H^+]} \qquad \text{(VI)}$$

Corresponding rate measurements over the pH range between 4.5 and 9.0, yielded an estimated rate constant for HO_2 oxidation of LDH-NADH ($k_{11} \sim 2.0 \times 10^6\ M^{-1}\ s^{-1}$, Fig.1, ref. 8).

The profile of the chain length *vs* pH shows a maximum at pH 7.2 (7). Besides the characteristic of the enzyme, the other factors which may also affect the chain length as a function of pH are the decay of superoxide radicals and the binding of NADH and NAD^+ to lactate dehydrogenase.

Initiation Reaction

Independent of the source of origin (xanthine oxidase system or high energy ionizing radiation), superoxide radicals initiate the chain reaction as described by Reactions 10 and 11. The following observations are in support of the postulation that the superoxide radical is the initiator of the lactate dehydrogenase catalyzed chain oxidation of NADH:

a) Addition of superoxide dismutase effectively inhibited the chain reaction (3,9).

b) Addition of known amounts of ferricytochrome *c*, a good superoxide scavenger, resulted in a quantitative lowering of the chain length.

c) Ascorbate is another scavenger of superoxide radicals (12). Addition of small concentrations of ascorbate lowered the rate of NADH oxidation in a competitive fashion. When all ascorbate was consumed, the chain reaction accelerated to that of control runs in the absence of ascorbate (3). High ascorbate concentrations inhibited altogether the chain reaction.

d) Any significant role of hydrogen peroxide, a by-product of both the radiation and the xanthine oxidase system, was ruled out. In control experiments, in which either catalase or hydrogen peroxide was added, only a negligible change in NADH oxidation was observed (9).

Propagation Reaction

Earlier reports (10,11) have established that the NAD· radical reacts with molecular oxygen at a diffusion-controlled rate, ($k_{16} = 1.9 \times 10^9\ M^{-1}\ s^{-1}$):

$$NAD\cdot + O_2 \longrightarrow NAD^+ + O_2^- \quad . \qquad (16)$$

In our pulse radiolysis studies of the LDH-NADH system, the LDH-NAD· was not detectable in the presence of molecular oxygen, which suggests that Reaction 12 proceeds at a very rapid rate similar to that of Reaction 16. Hence in an air saturated solution (250 μM O_2) it may be assumed that the superoxide radical is regenerated from molecular oxygen in Reaction 12 at a rate equal to its consumption in Reactions 10 and 11.

Experiments carried out in the absence of molecular oxygen showed that lactate dehydrogenase stabilizes the NAD· radical. The stabilized radical complex LDH-NAD·, which could be observed for several minutes at 405 nm, yielded the enzymatically inactive dimer $(NAD)_2$ as a final product. This suggests a slow dissociation step in Reaction 17 followed by the relatively rapid dimerization in Reaction 18, ($k_{18} = 5.6 \times 10^7\ M^{-1}\ s^{-1}$, (14)):

$$LDH\text{-}NAD\cdot \longrightarrow LDH + NAD\cdot \qquad (17)$$

$$NAD\cdot + NAD\cdot \longrightarrow (NAD)_2 \qquad (18)$$

In view of the slow dissociation of LDH-NAD· one may postulate that during the chain oxidation process in an aerobic system, molecular oxygen must diffuse to the active site of lactate dehydrogenase to react with the bound nucleotide radical, Reaction 12. This is supported by the fact that the final product NAD^+ was tested to be enzymatically active (3).

Termination Reaction

The termination step, Reaction 14, is still not well understood, since X may be any unknown functional group(s) on the enzyme capable of interacting with the superoxide radical. As presented in the kinetic analysis, it is Reaction 14 which is being monitored indirectly when one observes the oxidation of NADH in the chain reaction. This is the only process which diminishes the superoxide radical concentration in this system above neutral pH where the spontaneous decay of O_2^- is relatively slow. At lower pH, however, dismutation of superoxide radicals as shown in Reaction 8 will play an increasingly important role in the termination process.

An alternative termination with radical-radical interaction in Reaction 19

$$\text{LDH-NAD}\cdot + O_2^- + 2H^+ \longrightarrow \text{LDH-NAD}^+ + H_2O_2 \qquad (19)$$

can be ruled out because of the magnitude of the rate constant for Reaction 12 and the high ratio of $[O_2]/[O_2^-]$ in these studies.

Enzyme Catalysis

Although nicotinamide adenine dinucleotide is a co-factor in a large number of biological oxidation-reduction reactions, the mechanism of its involvement on a molecular level is still not thoroughly understood. At present it is generally assumed that when the co-factor and a specific substrate are bound at the active site of a dehydrogenase, a hydride transfer takes place during the reaction process.

In non-enzymatic systems recent studies have revealed that the co-factor is also capable of undergoing one-electron transfer reactions with the formation

$$\text{NADH} \underset{+H}{\overset{-H}{\rightleftharpoons}} \text{NAD}\cdot \underset{+e}{\overset{-e}{\rightleftharpoons}} \text{NAD}^+ \qquad (20)$$

(NADH) (NAD·) (NAD^+)

The physico-chemical properties of this free radical (NAD·) which have been studied mainly by pulse radiolysis, indicate that its formation from the corresponding parent compounds in Reaction 20 requires relatively strong oxidizing agents such as OH (13) and Br_2^- (10) or strong reducing agents such as e^-_{aq} and CO_2^- (14).

The results in this study indicate that a two-step mechanism with an obligatory nucleotide radical intermediate can also take place at the active site of an enzyme like lactate dehydrogenase. Moreover these findings demonstrate that when NADH is bound to lactate dehydrogenase, accompanied probably by conformational changes, it is made much more reactive toward the relatively weak oxidizing radicals O_2^- (Reaction 10) and HO_2 (Reaction 11). A structural change in the co-factor upon binding to lactate dehydrogenase was first deduced from spectral shifts by Chance and Nielands (15), and later confirmed by X-ray crystallographic studies (16).

The experimental observations in support of the postulation that the free radical chain oxidation of NADH (Reactions 9 - 13) are taking place at the active site of lactate dehydrogenase are:

a) NADPH cannot substitute for NADH in the chain oxidation, suggesting that a certain structural specificity of the nucleotide is required.

b) Addition of oxamate, which is a well known competitive inhibitor for pyruvate reduction by NADH and which has been shown to be bound at the substrate site of lactate dehydrogenase in X-ray crystallography studies (17), inhibitis also the superoxide-induced chain oxidation of NADH (9).

c) According to the proposed chain mechanism (Reactions 9 - 14) the observed rates of the reaction are proportional to the total number of active sites. When O_2^- was generated at a constant rate in the presence of varying amounts of LDH and NADH, the chain length increased with increasing ratio of [NADH]/[LDH] and remained unchanged after all the active sites on the enzyme were saturated by NADH (Fig.2 in Ref. 3).

With an excess of NADH the concentration of LDH-NADH can be kept constant throughout the run. This is possible because of the favorable association rate constant for LDH-NADH, $k_{22} = 1 \times 10^8\ M^{-1}\ s^{-1}$ (18):

$$\text{LDH} + \text{NADH} \longrightarrow \text{LDH-NADH} \tag{21}$$

and the dissociation rate constant for LDH-NAD^+, $k_{23} = 3 \times 10^2\ s^{-1}$ (19):

$$\text{LDH-NAD}^+ \longrightarrow \text{LDH} + \text{NAD}^+ \tag{22}$$

These favorable rate constants facilitate a rapid turnover in NADH oxidation and consequently also influence the chain length. Under favorable conditions a chain length of 70 has been observed (20). Studies on other enzymes (glyceraldehyde-3-phosphate dehydrogenase and glyoxylate reductase) with less favorable conditions for the turnover yielded much shorter chain lengths, and still others (malate dehydrogenase and isocitrate dehydrogenase) showed no detectable catalytic activity for NADH or NADPH oxidation by superoxide radicals at all (20).

One unique characteristic of a dehydrogenase-catalyzed oxidation of NADH is the stereospecificity of the hydrogen atoms on C-4 of the pyridine ring, in contrast to the unbound NADH in which the two hydrogen atoms are not distinguishable. This difference can serve to ascertain whether a reaction actually takes place at the active site of an enzyme. In the oxidation of LDH-NADH by O_2^- the hydrogen atom from the nucleotide is transferred to the final product H_2O_2 from which it exchanges freely with water. Consequently a stereospecificity in the reaction process cannot be examined. Therefore, we have selected a suitable radical of an organic acid to interact with the bound NADH to test further the postulation that a free radical reaction can be catalyzed at the active site of a dehydrogenase (21).

The malate radical ($^-$OOC-ĊH-CH(OH)-COO$^-$) was prepared by interacting OH with furmarate. The oxidation of NADH by malate radicals to form NAD^+ and malate showed a stoichiometry of one to two, indicating a two-step reaction with a nucleotide radical intermediate. Tritium-labeled NADH was prepared enzymatically so that all the labels were placed on the A side of the pyridine ring. When the (A-T)NADH was bound to lactate dehydrogenase and then oxidized by malate radicals, all the tritium was transferred to the product malate. Experiments with flow radiolysis indicate that in the first oxidation step a hydrogen atom was transferred from the bound nucleotide (LDH-NADH) to a malate radical, generating a nucleotide radical intermediate (LDH-NAD·), which subsequently transferred an electron to a second malate radical to form LDH-NAD^+ (21). These findings provide further support that lactate dehydrogenase is capable of catalyzing oxidation of NADH at the active site by free radical reactions similar to those in the chain oxidation of LDH-NADH by O_2^- (Reactions 9 - 14).

SUMMARY

Superoxide radicals react very slowly with free NADH. When the nucleotide is bound to lactate dehydrogenase the reaction rate is increased by four orders of magnitude. Moreover the enzyme bound nucleotide radical intermediate reacts rapidly with molecular oxygen to produce NAD^+ and another O_2^-, thereby propagating a chain reaction. Under favorable conditions a chain length of 70 has been observed, i.e. 70 molecules of NADH are oxidized to enzymatically active NAD^+ for every O_2^- radical generated by either xanthine oxidase or high energy ionizing radiation. The chain oxidation of NADH is inhibitied by superoxide dismutase, sodium oxamate (competitive inhibitor for lactate dehydrogenase) and ascorbate (a scavenger for O_2^-). NADPH cannot substitute for NADH. As a further support for the postulation that a free radical reaction can be catalyzed at the active site of lactate dehydrogenase, a malate radical is substituted for O_2^- as the hydrogen atom acceptor. The results show that the transfer of a hydrogen atom from LDH-NADH to a free radical is stereospecific. Some other dehydrogenases tested have similar catalytic activity for the chain oxidation of NADH by O_2^- but with much shorter chain lengths, while still others are not capable of catalyzing the oxidation process at all. This study shows that the superoxide radical itself, without being converted to some other species, can be a reactive oxidizing agent in a biological system when the other reactant is suitably activated.

ACKNOWLEDGEMENT

This study was supported in part, by Grant No. PCM 76-00505 from the National Science Foundation, and it is carried out partly under the auspices of the Energy Research and Development Administration.

REFERENCES

1. Fridovich, I. (1970). *J. Biol. Chem. 245*, 4053-4057.
2. Nilsson, R., Pick, F.M. and Bray, R.C. (1969). *Biochim. Biophys. Acta 192*, 145-148.
3. Bielski, B.H.J. and Chan, P.C. (1973). *Arch. Biochem. Biophys. 159*, 873-879.
4. Czapski, G. and Bielski, B.H.J. (1963). *J. Phys. Chem. 67*, 2180-2184.
5. Behar, D., Czapski, G., Rabani, J., Dorfman, L.M. and Schwarz, H.A. (1970). *J. Phys. Chem. 74*, 3209-3213.
6. Rabani, J. and Nielsen, S.O. (1969). *J. Phys. Chem. 73*, 3736-3744.
7. Bielski, B.H.J. and Chan, P.C. (1975). *J. Biol. Chem. 250*, 318-321.
8. Bielski, B.H.J. and Chan, P.C. (1976). *J. Biol. Chem. 251*, 3841-3844.
9. Chan, P.C. and Bielski, B.H.J. (1974). *J. Biol. Chem. 249*, 1317-1319.
10. Land, E.J. and Swallow, A.J. (1971). *Biochim. Biophys. Acta 234*, 34-42.
11. Willson, R.L. (1970). *Chem. Commun.* 1005.
12. Nishikimi, M. (1975). *Biochem. Biophys. Res. Comm. 63*, 463-468.
13. Schellenberg, K.A. and Hellerman, L. (1958). *J. Biol. Chem. 231*, 547-556.
14. Land, E.J. and Swallow, A.J. (1968). *Biochim. Biophys. Acta 162*, 327-337.
15. Chance, B. and Neilands, J.B. (1952). *J. Biol. Chem. 199*, 383-387.
16. Chandrasekhar, K., McPherson, A. Jr., Adams, M.J. and Rossmann, M.G. (1973). *J. Mol. Biol. 76*, 503-518.
17. Rossman, M.G., Adams, M.J., Buehner, M., Ford, G.C., Hackert, M.L., Lentz, P.J. Jr., McPherson, A. Jr., Schevitz, R.W. and Smiley, I.E. (1971). Cold Spring Harbor Symp. Quant. Biol. *36*, 179-191.
18. Heck, H.D'A., McMurray, C.H. and Gutfreund, H. (1968). *Biochem. J. 108*, 793-796.
19. Borgmann, U., Moon, T.W. and Laidler, K.J. (1974). *Biochemistry 13*, 5152-5158.
20. Chan, P.C. and Bielski, B.H.J., Manuscript in preparation.
21. Chan, P.C. and Bielski, B.H.J. (1975). *J. Biol. Chem. 250*, 7266-7271.

SUPEROXIDE AND PHOTOSYNTHETIC REDUCTION OF OXYGEN

J.F. ALLEN

Botany School
University of Oxford
South Parks Road
Oxford OX1 3RA, U.K.

I. INTRODUCTION

In plant-type photosynthesis sunlight is able to bring about the transfer of reducing equivalents from water to an oxidant, with a consequent liberation of molecular oxygen. This process involves formation of NADPH (by reduction of the oxidant NADP) and of ATP (by phosphorylation of ADP). NADPH and ATP are respectively oxidised and hydrolysed in the subsequent pathway of CO_2-assimilation, a process which itself has no direct requirement for light.

Oxygen evolution is by no means a necessary condition for the conversion of radiant energy into chemical potential energy. In the most abundant form of photosynthesis in the biosphere, water is the electron donor and so oxygen is produced. The alternatives represented by anaerobic, bacterial photosynthesis and by *Halobacterium* photophosphorylation are interesting biochemically and as evolutionary relics, but they today make a negligible contribution to global primary production.

While evolution of oxygen is the prominent feature of at least that form of photosynthesis found in plants, light-dependent reduction of oxygen may also occur in certain circumstances. In the absence of other electron acceptors, isolated chloroplasts may actually consume oxygen in the light. Photosynthetic oxygen uptake by chloroplasts is termed the "Mehler reaction", and hydrogen peroxide is its first stable product.

The relatively recent identification of the superoxide anion as the initial product of photosynthetic oxygen reduction may be a valuable one for certain areas of photosynthesis research, because chloroplast oxygen uptake is widely used as a measure of photosynthetic electron transport *in vitro*. The theme of the present chapter is that several discrete categories of Mehler reaction can now be described, each in terms of the way in which superoxide functions as an intermediate.

II. REDUCTION AND CONSUMPTION OF OXYGEN BY ISOLATED CHLOROPLASTS

A. *The Mehler Reaction: Oxygen as a Hill Oxidant*

In the "Hill reaction" (Hill, 1939) oxygen is produced by illuminated chloroplasts when electrons are transferred from water to an electron acceptor which consequently may be termed a "Hill oxidant" (X in equation 1)

$$H_2O \longrightarrow 2H + \tfrac{1}{2}O_2$$

$$2H + X \longrightarrow XH_2$$

$$\text{Overall: } H_2O + X \longrightarrow \tfrac{1}{2}O_2 + XH_2 \qquad (1)$$

Artificial Hill oxidants include ferric oxalate, potassium ferricyanide and indophenol dyes. The natural electron acceptor, NADP, is reduced by chloroplasts only in the presence of the soluble electron carrier ferredoxin. NADP/NADPH has a low redox potential (E'_o = -320 mV) but NADPH is not autoxidisable; when NADP is reduced energy is stored in a relatively stable form. This energy is consumed in the conversion of diphosphoglycerate to triose phosphate in the "dark" pathway of CO_2-fixation.

If NADP is absent from the chloroplast preparation, electrons may be transferred to oxygen instead. Mehler (1951a) identified hydrogen peroxide as a product of this reaction by using catalase in its peroxidative capacity in order to show that ethanol could be converted by illuminated chloroplasts to acetaldehyde. In this situation oxygen is consumed (equation 2), and net oxygen consumption has the same stoichiometric relationship to electron transport as oxygen evolution has when some other electron acceptor is used.

$$H_2O \longrightarrow 2H + \tfrac{1}{2}O_2$$

$$O_2 + 2H \longrightarrow H_2O_2$$

$$\text{Overall: } H_2O + \tfrac{1}{2}O_2 \longrightarrow H_2O_2 \qquad (2)$$

That oxygen is simultaneously consumed and evolved in this reaction was confirmed, in experiments using ^{18}O-oxygen as a tracer, by Mehler and Brown (1952) and by Brown and Good (1955). Dismutation of the hydrogen peroxide by catalase results in zero net oxygen exchange. In measurements of chloroplast oxygen uptake it is now customary to add sodium azide as an inhibitor of endogenous catalase.

Good and Hill (1955) showed that the rate of oxygen uptake in the Mehler reaction could be increased by addition of various autoxidisable electron acceptors such as flavin derivatives, dyes (Janus green and litmus) and the bipyridyl compounds methyl viologen and benzyl viologen. In the presence of saturating concentrations of methyl viologen, rates of oxygen uptake of several hundreds of micromoles per milligram of chlorophyll per hour can be observed; these represent rates of electron transport as high or higher than those which must occur *in vivo*. ATP synthesis is, of course, coupled to photosynthetic electron transport, and "pseudocyclic" is the term used to describe photophosphorylation that is driven by non-cyclic electron transport with oxygen as the terminal electron acceptor (Arnon *et al.*, 1961; Arnon *et al.*, 1967).

Good and Hill (1955) also confirmed earlier observations that chloroplast oxygen uptake could be stimulated by manganese ions (Mn^{2+}) (Gerretsen, 1950) and by ascorbate (Mehler, 1951b). Net oxidation of such reagents was observed, while this was not the case for the flavins, quinones or viologens which merely serve catalytically as electron carriers between the chloroplast and oxygen. Mehler (1951b) nevertheless ascribed to manganese ions a catalytic role similar to that of quinones, while to account for ascorbate's effect he suggested a reduction by ascorbate of the more oxidised product of the photolysis of water (HO·). The terminal steps of oxygen evolution would then be replaced by oxidation of ascorbate and net oxygen uptake would thereby be enhanced. The same proposition is contained within the more recent statement by Bohme and Trebst (1969) and by Ben-Hayyim and Avron (1970a), that ascorbate can by itself competitively replace water as an electron donor to photosystem II.

Non-cyclic electron transport in plant photosynthesis involves two light reactions which operate in series (Hill and Bendall, 1960). Fig.1 depicts an outline of the "Z-scheme" for photosynthetic electron transport, and includes some of the oxygen-reducing and other reactions which are discussed in this chapter. The Z-scheme is produced by plotting the sequence of electron transfer reactions on a scale of the standard redox potential of its components.

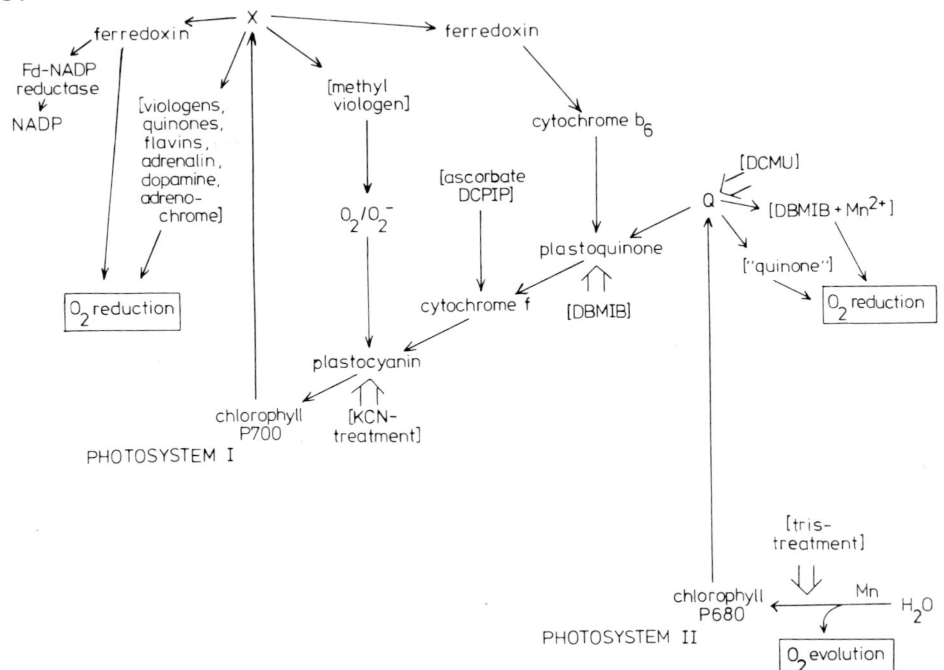

Fig.1. *Photosynthetic electron transport and associated reactions that involve oxygen. Artificial treatments or additions are in brackets.*

If the oxygen-evolving mechanism is inactivated by some treatment such as incubation of the chloroplasts with tris buffer, then ascorbate will indeed serve as an alternative electron donor (Yamashita and Butler, 1968), though without such treatment it now seems that ascorbate will not by itself compete with water. Like ascorbate, Mn^{2+} has been thought to be able to displace water as a donor to photosystem II (Ben-Hayyim and Avron, 1970b; Walker *et al.*, 1970).

With the replacement of manometric methods of measuring oxygen exchange by polarographic ones, oxygen uptake has become widely used in studies of photosynthetic electron transport. If electron transport in photosystem II is inhibited (e.g. by Tris-treatment or by addition of DCMU) then oxygen evolution can no longer be observed, though electron transport which is associated with photosystem I may still easily be measured as oxygen uptake, provided electrons are supplied by a reductant such as ascorbate together with catalytic amounts of a dye such as DCPIP (Vernon and Zaugg, 1960). Despite the obvious utility of the Mehler reaction and its modifications, care must be taken in interpretation of the oxygen electrode traces obtained. Some experimental conditions (e.g. catalase activity) which must be taken into account have been described by Whitehouse *et al.* (1971), and, specifically for the photosystem I reaction, by Allen and Hall (1974). The fact that ferredoxin is an effective mediator of the Mehler reaction (Telfer *et al.*, 1970) raises the possibility that what is apparently a straightforward Hill reaction with NADP as electron acceptor may actually involve oxygen uptake as well (Allen, 1975b).

B. *Chloroplast Oxygen Uptake With Methyl Viologen as the Mediator*

1. *Stimulation by Ascorbate and Manganese.* If methyl viologen (MV) is used as the carrier of electrons between the chloroplast and oxygen, a special case of the Mehler reaction described by equation 2 can be written as follows:

$$H_2O + 2MV^{2+} \longrightarrow 2MV^{+} + \tfrac{1}{2}O_2 + 2H^{+} \qquad (3)$$

$$2MV^{+} + O_2 + 2H^{+} \longrightarrow 2MV^{2+} + H_2O_2 \qquad (4)$$

$$\text{Overall: } H_2O + \tfrac{1}{2}O_2 \longrightarrow H_2O_2 \qquad (2)$$

Both light reactions (see Fig.1) are involved in the electron transfer represented by equation 3.

If ascorbate (AH_2) were to replace water as the electron donor (equation 5), the stoichiometry of the overall reaction would be altered from that described by equation 2 to that described by equation 6.

$$AH_2 + 2MV^{2+} \longrightarrow 2MV^{+}\ A + 2H^{+} \qquad (5)$$

$$2MV^{+} + O_2 + 2H^{+} \longrightarrow 2MV^{2+} + H_2O_2 \qquad (4)$$

$$\text{Overall: } AH_2 + O_2 \longrightarrow A + H_2O_2 \qquad (6)$$

For any given rate of electron transport, ascorbate would increase the rate of oxygen uptake by no more than a factor of two. Bohme and Trebst (1969) used anthraquinone as the mediator, and found that the rate of oxygen uptake was more than doubled on addition of ascorbate. To account for this observation they suggested that ascorbate, in acting as an electron donor to photosystem II, by-passed a phosphorylation site which was situated between water and photosystem II, and which was also limiting to the overall rate of electron transport. The fact that the rate of photophosphorylation was apparently unaffected by ascorbate is consistent with this explanation provided that it is assumed that a second phosphorylation site exists and is not by-passed in the same way. DCMU-sensitivity and susceptibility to a "red-drop" effect (Ben-Hayyim and Avron, 1970a) made it necessary to assume that the ascorbate-stimulated rate of oxygen uptake was driven by electron transport involving both photosystems. The same mechanism was invoked for similar effects of manganese ions (Mn^{2+}) on chloroplast oxygen uptake mediated by diquat (Ben-Hayyim and Avron, 1970b). One difficulty with this explanation was that NADP reduction was unaffected by ascorbate, while ascorbate oxidation was much slower with NADP as the electron acceptor than it was with oxygen as the electron acceptor (Bohme and Trebst, 1969). A second problem was that ascorbate was able to increase the rate of methyl viologen-mediated oxygen uptake by more than a factor of two even in uncoupled chloroplasts, where, by definition, electron transport cannot be limited in rate by phosphorylation reactions (Allen and Hall, 1973).

Though NADP reduction itself was unaffected by ascorbate, net oxygen evolution with NADP (with added ferredoxin) as the electron acceptor was, under certain conditions, inhibited by ascorbate (Ben-Hayyim and Avron, 1970a). This effect was interpreted as additional evidence for ascorbate's ability to suppress oxidation of water and itself become the electron donor to photosystem II.

2. *The Effect of Superoxide Dismutase.* An alternative solution to the problem of ascorbate's effect on chloroplast oxygen uptake was proposed by Epel and Neumann (1972). Following Elstner *et al.* (1970), Epel and Neumann suggested that superoxide was the initial product of autoxidation of methyl viologen, and that the reduction of oxygen to hydrogen peroxide (equation 4) could be regarded as an overall reaction which involved both univalent reduction of oxygen (equation 7) and dismutation of superoxide (equation 8). The methyl viologen-mediated Mehler reaction may then be written:

$$H_2O + 2MV^{2+} \longrightarrow 2MV^+ + \tfrac{1}{2}O_2 + 2H^+ \qquad (3)$$

$$2MV^+ + 2O_2 \longrightarrow 2MV^{2+} + 2O_2^- \qquad (7)$$

$$2O_2^- + 2H^+ \longrightarrow O_2 + H_2O_2 \qquad (8)$$

$$\text{Overall: } H_2O + \tfrac{1}{2}O_2 \qquad H_2O_2 \qquad (2)$$

Ascorbate's effect in stimulating oxygen uptake may then be attributed to a replacement of the dismutation step (equation 8) with oxidation of ascorbate by superoxide (equation 9). The overall reaction then becomes that shown in equation 10, with three times as much oxygen being consumed as is the case in the absence of ascorbate (equation 2).

$$H_2O + 2MV^{2+} \longrightarrow 2MV^+ + \tfrac{1}{2}O_2 + 2H^+ \qquad (3)$$

$$2MV^+ + 2O_2 \longrightarrow 2MV^{2+} + 2O_2^- \qquad (7)$$

$$AH_2 + 2O_2^- + 2H^+ \longrightarrow 2A + 2H_2O_2 \qquad (9)$$

$$\text{Overall: } H_2O + AH_2 + 1\tfrac{1}{2}O_2 \longrightarrow A + 2H_2O_2 \qquad (10)$$

Accordingly the effect of ascorbate is merely a consequence of its reaction with superoxide, and the electron transport reactions which lead to reduction of methyl viologen (equation 3) are quite unaffected. Thus the scheme is consistent with the evidence for the involvement of both photosystems in ascorbate-stimulated oxygen uptake, and with the unchanged rate of phosphorylation.

A three-fold increase in methyl viologen-mediated oxygen uptake in broken, washed chloroplasts is accompanied by an equivalent effect of ascorbate in which the ADP/O ratio in the same system is decreased (Allen and Hall, 1973).

Epel and Neumann (1972) also pointed out that their explanation of ascorbate's effect on oxygen uptake could accommodate an explanation of inhibition by ascorbate of net oxygen evolution with NADP as the electron acceptor. If a ferredoxin-mediated, oxygen-consuming component of this reaction were susceptible to stimulation by ascorbate, then ascorbate's reaction with superoxide would cause an inhibition of net oxygen evolution without affecting electron transport to NADP.

The significant advantage that Epel and Neumann's (1972) hypothesis (that ascorbate stimulates oxygen uptake because of its reaction with superoxide) has over its alternative (that ascorbate replaces water as the electron donor to photosystem II) is that the former hypothesis is the one more vulnerable to experimental test. A disproof of Epel and Neumann's hypothesis (though not of its alternative) would be the persistence of ascorbate-stimulation of oxygen uptake in a situation where the ascorbate-superoxide reaction (equation 9) could not occur. In fact the addition of superoxide dismutase, which accelerates the dismutation reaction (equation 8), not only prevents ascorbate from stimulating chloroplast oxygen uptake; it also reverses the effect of ascorbate which has previously been added (Epel and Neumann, 1973; Allen and Hall, 1973). Superoxide dismutase is known to be antagonistic to the effect of ascorbate in all the cases that were quoted previously as evidence for ascorbate's suppression of oxygen evolution associated with Photosystem II. Ascorbate's stimulatory effect on oxygen uptake (Epel and Neumann, 1973), its inhibitory effect on the ADP/O ratio (Allen and Hall, 1973), and its inhibitory effect on NADP-dependent net oxygen evolution (Epel and Neumann, 1973; Allen, 1975c) are all reversed in a cyanide-sensitive fashion by addition of superoxide dismutase to the appropriate reaction.

Epel and Neumann (1973) used a chloroplast-free model system in which methyl viologen was reduced enzymatically by ferredoxin-NADP reductase in the presence of NADPH. By adding limiting amounts of NADPH they were able to show directly that the stoichiometry of oxygen uptake was changed by ascorbate in the way predicted by their scheme; from $NADPH/O_2 = 1$ (without ascorbate) to $NADPH/O_2 = \frac{1}{2}$ (with ascorbate). Epel and Neumann (1973) also demonstrated that Mn^{2+} and dithiothreitol will each substitute for ascorbate in causing a superoxide dismutase-sensitive enhancement of oxygen uptake in the chloroplast-free system.

Dithiothreitol was subsequently shown by Marchant (1974) to accelerate oxygen uptake in chloroplasts, albeit at roughly fifty-fold higher concentrations than ascorbate. Cysteine, another reductant for superoxide, is similar to dithiothreitol in this respect (Allen, 1975c). Marchant (1974) also obtained results relating to ascorbate's effect on the ADP/O ratio of oxygen-consuming chloroplasts. Marchant's data gave some statistically rigorous support to the supposition that ascorbate, dithiothreitol (at least at lower concentrations) and superoxide dismutase have no effect on either electron transport or photophosphorylation *per se*, and that their effects on the ADP/O ratio are merely consequences of their effects on the stoichiometry of oxygen uptake in the Mehler reaction.

Another observation by Allen and Hall (1973), Epel and Neumann (1973) and Marchant (1974) was that in order to achieve a given degree of inhibition of ascorbate's effect on oxygen uptake, a greater amount of superoxide dismutase had to be added to broken, washed chloroplasts than had to be added to intact chloroplasts. This clearly represents evidence for chloroplast-associated superoxide dismutase activity. From the magnitude of this effect Allen and Hall (1973) estimated that such an activity in spinach chloroplasts would correspond to about five hundred superoxide dismutase units per milligram of chlorophyll.

Any reagent that will reduce superoxide to peroxide should be a suitable substitute for ascorbate in such experiments. Thus Mn^{2+} and glutathione also produce enhanced chloroplast oxygen uptake which can then be inhibited by superoxide dismutase. With cysteine or glutathione the concentration-dependence of the effect resembles that with dithiothreitol. In broken chloroplasts, which present fewer problems of permeability, the principal factors involved should be the rate constant of the reaction of the reductant with superoxide and any residual or membrane-associated superoxide dismutase activity. Spinach superoxide dismutase, which can be present as a contaminant of ferredoxin-NADP reductase preparations, may also have to be taken into account in the chloroplast-free model system (Allen, 1975c).

The identity of the autoxidisable electron acceptor is to a certain extent immaterial, provided that it mediates only univalent oxygen reduction as outlined above. A list of such mediators can now include the bipyridyl derivatives methyl viologen (dimethyl), benzyl viologen (dibenzyl) and diquat (ethylene); FMN and adrenochrome (Allen, 1975c); and anthraquinone and phenazine methosulphate (Epel and Neumann, 1973). Autoxidisable compounds which seem to mediate a different type of Mehler reaction include adrenalin (section IIC of the present chapter) and ferredoxin (seection IID).

3. *Superoxide, Ascorbate and Stoichiometries of Oxygen Uptake.* It is clear that in situations where oxygen exchange is to be taken as an absolute measure of photosynthetic electron transport, the number of oxygen molecules evolved or consumed per electron pair transferred must be known. In normal Hill (equation 1) or Mehler (equation 2) reactions $O_2/2e^- = \frac{1}{2}$, while for the methyl viologen-mediated Mehler reaction in the presence of ascorbate (according to equation 10) $O_2/2e^- = 1\frac{1}{2}$.

Electron transport in photosystem I can also be measured as methyl viologen-mediated oxygen uptake, provided that a suitable electron donor such as DCPIP is present. The site at which electrons are donated lies between the two photosystems, as shown in Fig.1. In practice DCPIP is used in catalytic amounts, being reduced by a compound which consequently provides the pool from which electrons are drawn by photosystem I; this compound is invariably ascorbate. The relationship of oxygen uptake to electron transport in such a system has widely been assumed to differ from that of the complete non-cyclic chain only in that oxygen is not evolved where ascorbate instead of water is the ultimate source of electrons (Izawa, 1968; Strotmann and von Gosseln, 1972; Ort and Izawa, 1973; Gould and Izawa, 1973a). The overall reaction would, in this view, be that of equation 6, the value $O_2/2e^-$ = 1 would therefore hold, and in measuring the stoichiometry of photophosphorylation the relationship $P/2e^- = P/2O$ would be expected to apply.

If, however, the generation by methyl viologen of superoxide is recognised (equation 7), the overall reaction becomes that of equation 11 (below), as a consequence of the oxidation of ascorbate by superoxide (equation 9).

$$AH_2 + 2MV^{2+} \longrightarrow 2MV^+ + A + 2H^+ \qquad (5)$$

$$2MV^+ + 2O_2 \longrightarrow 2MV^{2+} + 2O_2^- \qquad (7)$$

$$AH_2 + 2O_2^- + 2H^+ \longrightarrow A + 2H_2O_2 \quad (9)$$

$$\text{Overall: } 2AH_2 + 2O_2 \longrightarrow 2A + 2H_2O_2 \quad (11)$$

An important implication is that the ratio $O_2/2e^-$ for the photosystem I reaction (equation 11) has a value of two, and so the adoption of equation 6 and its $O_2/2e^-$ value of one leads to a corresponding overestimate of the rate of electron transport in this sytem, and hence to an underestimate of the stoichiometry of photosystem I phosphorylation. Equation 11 implies the relationship $P/2e^- = P/4O$ rather than $P/2e^- = P/2O$.

Support for this interpretation is the observation that superoxide dismutase is inhibitory to oxygen uptake by photosystem I (Epel and Neumann, 1973; Allen and Hall, 1974; Ort and Izawa, 1974). Allen and Hall found that with saturating concentrations of superoxide dismutase the rate of oxygen uptake was half that of the control, while Ort and Izawa obtained a significantly smaller percentage inhibition. Addition of catalase also halves the rate of oxygen uptake, both in the presence and in the absence of superoxide dismutase (Allen and Hall, 1974). Oxygen uptake in the presence of both superoxide dismutase and catalase therefore proceeds at only a quarter of the rate that is observed with neither enzyme present.

Fig.2 depicts a more general scheme for oxygen uptake in the Mehler reaction, and the accompanying Table I shows the stoichiometry of oxygen uptake that is associated with each set of conditions. Five possible values for the $O_2/2e^-$ ratio are given. For measurable O_2 uptake that results from electron transport through both photosystems the $O_2/2e^-$ ratio has a value of ½, or, in the presence of ascorbate but of neither superoxide dismutase nor catalase, of 1½. For oxygen uptake driven only by photosystem I, the $O_2/2e^-$ ratio has three possible values; 1 with either superoxide dismutase or catalase, 2 with neither enzyme, and ½ with both. Less than saturating concentrations of ascorbate, superoxide dismutase or catalase (or of the enzymes' inhibitors) will, of course, produce $O_2/2e^-$ ratios which are intermediate between any two of the five values in Fig.2.

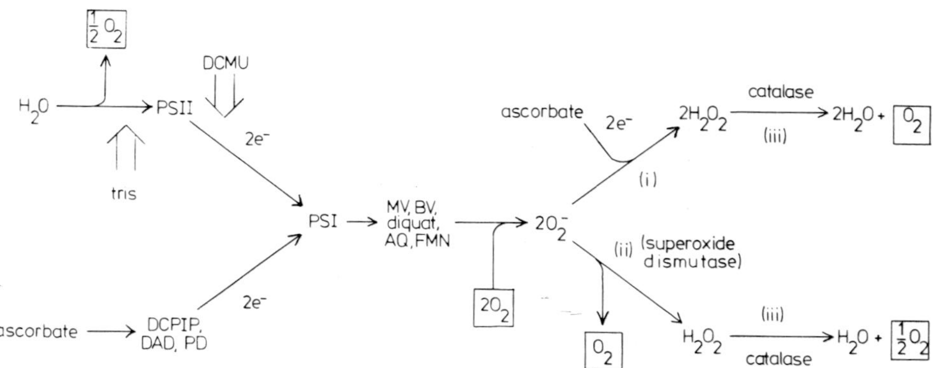

Fig.2. (with an accompanying Table). *Oxygen-evolving and oxygen-consuming reactions that determine the relationship of chloroplast oxygen uptake to photosynthetic electron transport: effects of ascorbate, superoxide dismutase, and catalase.*

TABLE I

	Enzyme Additions	Reactions Involved	$O_2/2e^-$	$P/2e^-$ Equals
Water as electron donor; PSI and PSII				
without ascorbate	None	(ii)	½	*P/O*
	SOD	(ii)	½	*P/O*
	catalase	(ii) + (iii)	0	-
	SOD + catalase	(ii) + (iii)	0	-
with ascorbate	None	(i)	1½	*P/3O*
	SOD	(ii)	½	*P/O*
	catalase	(i) + (iii)	½	*P/O*
	SOD + catalase	(ii) + (iii)	0	-
Ascorbate as electron donor; PSI only				
	None	(i)	2	*P/4O*
	SOD	(ii)	1	*P/2O*
	catalase	(i) + (iii)	1	*P/2O*
	SOD + catalase	(ii) + (iii)	½	*P/O*

It follows from this interpretation that the effects of ascorbate, superoxide dismutase, and catalase are effects only on the fate of the oxygen which is reduced initially to superoxide. Electron transport, when measured as NADP reduction, is indeed unaffected by these treatments (Allen and Hall, 1974).

4. *Further Evidence for Superoxide as an Intermediate.* Elstner *et al.* (1970) were the first to suggest that the superoxide radical is the initial product of photosynthetic oxygen reduction and that photosynthetic ascorbate oxidation can therefore result from ascorbate's reaction with superoxide. Elstner and Kramer (1973) agreed with Epel and Neumann's (1972) explanation of ascorbate's effect on the chloroplast oxygen uptake resulting from autoxidation of low-potential electron acceptors such as methyl viologen or anthraquinone. At the same time they proposed a different scheme to account for ascorbate-stimulated phosphorylation *per se* in the absence of artificial cofactors.

Evidence that superoxide is the initial product of autoxidation of diquat has been provided by Stancliffe and Pirie (1971). In a study of the reaction of methyl viologen (paraquat) with oxygen, Farrington *et al.* (1973)

identified superoxide as the initial product, and suggested that diffusion of superoxide from its site of formation in the chloroplast could result in peroxidation of membrane lipids in the chloroplast envelope and tonoplast; these are sites which are particularly vulnerable to the herbicidal action of paraquat.

By following the EPR signal of the tiron semiquinone Greenstock and Miller (1975) have shown that tiron can be used as a specific probe for superoxide in chloroplast reactions. With this technique, Miller and Macdowall (1975) showed that chloroplast superoxide production was considerably enhanced by addition of methyl viologen. Both ascorbate and adrenalin competitively obscured formation of the tiron semiquinone. A similar effect of superoxide dismutase was observed only when photosystem I sub-chloroplast particles were pre-incubated with the enzyme. This led Miller and Macdowall (1975) to conclude that tiron gains access to the site of superoxide production in undamaged thylakoids whereas superoxide dismutase does not. Evidence for enhancement by methyl viologen of chloroplast superoxide production has also been obtained in an EPR spin-trapping study by Harbour and Bolton (1975).

Asada and Kiso have identified production of superoxide in chloroplasts by a number of more commonly encountered techniques. Initiation of sulphite oxidation that was sensitive to both DCMU and superoxide dismutase was found to occur in chloroplasts even without the addition of autoxidisable cofactors (Asada and Kiso, 1973a); similar results were also obtained for oxidation of adrenalin (Asada and Kiso, 1973b). Asada *et al.* (1974) demonstrated a superoxide dismutase-sensitive reduction of cytochrome *c* by chloroplasts. In the presence of cytochrome *c*, hydrogen peroxide production was absent, indicating that chloroplast oxygen reduction is entirely univalent under these conditions. Though superoxide production was apparently stimulated by methyl viologen, benzyl viologen, diquat and triquat, a significant "endogenous" level of both adrenalin oxidation and cytochrome *c* reduction led Asada *et al.* (1974) to conclude that a membrane-bound chloroplast component, possibly the primary electron acceptor of photosystem I, is able to bring about univalent reduction of oxygen.

C. *Chloroplast Oxygen Uptake Mediated by Adrenalin or Dopamine and by Adrenochrome*

1. *Inhibition by Superoxide Dismutase and by Ascorbate: Two-Step Oxygen Reduction.* In the methyl viologen-mediated Mehler reaction superoxide dismutase has no effect on oxygen uptake. Elstner and Heupel (1974a) described a photosynthetic chloroplast oxygen uptake which was sensitive to inhibition by superoxide dismutase. They were able to demonstrate this effect both on the ferredoxin-mediated reaction and by using dopamine as the autoxidisable cofactor. That increasing the rate of superoxide dismutase (by adding superoxide dismutase) inhibits the overall reaction suggests that superoxide normally participates in some reaction other than dismutation in oxygen uptake mediated by ferredoxin or by dopamine. Elstner and Heupel (1974a) proposed that superoxide is reduced to peroxide by one (reduced) state of the autoxidising cofactor, while a second (intermediate) state is the one responsible for generation of superoxide. A third (oxidised) state then accepts from the photosynthetic chain two electrons per molecule, hence completing the cycle. For the reactions mediated by dopamine or adrenalin, this scheme is endorsed by the present author, and evidence for it will be reviewed in this section. The reaction in the presence of ferredoxin is thought by

Elstner and Heupel (1974a and 1974b; Elstner *et al.*, 1975) to be mediated by a putative chloroplast component with properties similar to dopamine's, while in the present chapter the ferredoxin-mediated reaction will be assumed to proceed by a different mechanism, and hence to merit a section (IID) of its own.

Among the compounds that were tested by Elstner and Heupel (1974a) as cofactors for a Mehler reaction, adrenalin and dopamine gave the highest rates of oxygen uptake. Either cofactor, at catalytic concentrations, gives a light-dependent, DCMU-sensitive and catalase-sensitive chloroplast oxygen uptake, which is directly proportional to chloroplast concentration at saturating light intensity (Allen, 1975c). To this extent each of these compounds resembles any other mediator of the Mehler reaction. They nevertheless have a number of peculiar properties, one of which is the aforementioned sensitivity to inhibition by superoxide dismutase.

Using the terminology of Misra and Fridovich (1972), a two step cycle (equations 12 and 13) for the chain-oxidation of adrenalin can be used to explain the sensitivity to superoxide dismutase of adrenalin-mediated chloroplast oxygen uptake. An initial oxidation of adrenalin (RH_3^-) by superoxide (equation 12) is followed by univalent reduction of oxygen by RH_3 (equation 13). Similar steps are likely to be involved in the oxidation of dopamine (Heikkila and Cohen, 1973). RH_2 may then function as a Hill oxidant (equation 14) to regenerate adrenalin, and the overall reaction (equation 2) would be the Mehler reaction as it has previously been defined.

$$O_2^- + RH_3^- + 2H^+ \longrightarrow RH_3 + H_2O_2 \qquad (12)$$

$$RH_3 + O_2 \longrightarrow RH_2 + O_2^- + H^+ \qquad (13)$$

$$H_2O + RH_2 \longrightarrow RH_3 + \tfrac{1}{2}O_2 + H^+ \qquad (14)$$

$$\text{Overall: } H_2O + \tfrac{1}{2}O_2 \longrightarrow H_2O_2 \qquad (2)$$

The observed inhibition by superoxide dismutase of overall oxygen uptake can be understood to be a result merely of competitive replacement of the reaction of equation 12 by the reaction of superoxide dismutation. This would prevent regeneration of the electron acceptor (RH_2), and both oxygen uptake and electron transport would cease.

Initiation of the cycle would require production of superoxide in order for the first reaction (equation 12) to take place; this could be achieved by a slow univalent reduction of oxygen by endogenous chloroplast components such as the membrane-bound iron-sulphur proteins of photosystem I (Asada *et al.*, 1974). The time taken for accumulation of the intermediates RH_3 and RH_2 would represent the lag-phase which Elstner and Heupel (1974a) reported.

Both the overall rate of oxygen uptake and the degree of sensitivity to superoxide dismutase are also dependent on pH. The rate of oxygen uptake has a maximum value in the region pH 8.0 to pH 8.5, while sensitivity to inhibition by superoxide dismutase is most marked at pH 7.5, and decreases with increasing pH (Allen, 1975c). At higher pH the oxygen uptake, though perhaps initiated by formation of superoxide in the chloroplast, becomes largely a consequence of the chain-oxidation of adrenalin to adrenochrome, a process which has been shown to have a pH-dependence consistent with this interpretation (Misra and Fridovich, 1972). In this view the cycle would break down at higher pH because of replacement or regeneration of adrenalin (equation 14) by the following reactions:

$$RH_2 + O_2^- + H^+ \longrightarrow RH + H_2O_2 \qquad (15)$$

$$RH + O_2 \longrightarrow R + O_2^- + H^+ \qquad (16)$$

Both dopamine-mediated (Elstner and Heupel, 1974a) and adrenalin-mediated (Allen, 1975c) Mehler reactions may be either stimulated by ascorbate or inhibited by it. Inhibition occurs if ascorbate is added to the chloroplasts before the mediator, and is likely to be a result of the ascorbate-superoxide reaction (equation 9) preventing initial oxidation of adrenalin (equation 12) and, as a consequence, accumulation of intermediates of the cycle. Like superoxide dismutase, ascorbate would then competitively inhibit both oxygen uptake and electron transport. Stimulation occurs if ascorbate is added after the mediator and during the linear phase of the sigmoidal oxygen uptake trace. Here saturating concentrations of RH_3 and RH_2 would already have accumulated, and disappearance of superoxide caused by its reaction with ascorbate would not, at first, interrupt the cycle. It could, however, cause a transient increase in the rate of oxygen uptake by preventing oxygen evolution from a residual spontaneous dismutation of superoxide. In any event the stimulation is transient, and the rate subsequently declines to less than that observed before addition of ascorbate.

2. *Adrenochrome-Mediated Single-Step Oxygen Reduction.* Of the five redox states of adrenalin that are thought to be involved in the chain-reaction of adrenalin oxidation (Misra and Fridovich, 1972), only three (RH_3^-, RH_3 and RH_2) are included in the foregoing explanation of adrenalin's mediation of a Mehler reaction. In support of this explanation, Allen (1975c) showed that the end-product of adrenalin oxidation, adrenochrome (R), mediates chloroplast oxygen uptake but does so by the single-step mechanism described in section IIB.

Adrenochrome at catalytic concentrations produces a light-dependent chloroplast oxygen uptake which is completely inhibited by DCMU and by catalase. The oxygen uptake appears promptly on addition of the mediator, and any lag-phase is shorter than the response-time (a few seconds) of the oxygen electrode. Oxygen uptake is unaffected by addition of superoxide dismutase, but acceleration of oxygen uptake resulting from addition of ascorbate is prevented. The reaction is clearly analogous to the methyl viologen-mediated one and is therefore likely to involve one-electron reduction of adrenochrome (R) by a component of the chloroplast, and one-electron reduction of oxygen by RH. The superoxide produced will either dismute or be reduced by ascorbate.

D. *Ferredoxin-Mediated Chloroplast Oxygen Uptake*

1. *Superoxide Production on Autoxidation of Ferredoxin.* In an investigation of the reaction of reduced iron-sulphur proteins with oxygen, Orme-Johnson and Beinert (1969) found that autoxidation of ferredoxin from *Clostridium pasteurianum* led to the production of an EPR-detectable concentration of superoxide, while autoxidation of spinach ferredoxin did not. Misra and Fridovich (1971) reported that both clostridial and spinach ferredoxins, when reduced enzymatically with NADPH and spinach ferredoxin-NADP reductase, bring about an aerobic oxidation of adrenalin which is sensitive to inhibition by superoxide dismutase. Nakamura and Kimura (1972) reported a similar result

for spinach ferredoxin. Allen (1975c) produced confirmatory evidence (oxidation of adrenalin and reduction of nitroblue tetrazolium) that superoxide can be produced by autoxidation of spinach ferredoxin. In all experiments involving enzymatic reduction of ferredoxin, ferredoxin-NADP reductase by itself seems either not to reduce oxygen or to do so very slowly.

Leaving aside for the moment the experiment of Orme-Johnson and Beinert (1969), there seems to exist good evidence that ferredoxins reduce oxygen univalently. Misra and Fridovich (1971) found, however, that adrenochrome production was relatively slower than NADPH oxidation, and concluded that a variable proportion of total oxygen reduction by both bacterial and plant-type ferredoxins proceeded by a two-electron transfer. For spinach ferredoxin they showed that the "percent univalent flux" increases with oxygen concentration, and was greater at pH 7.8 than at pH 6.8. Asada *et al.* (1974) found that while methyl viologen stimulated production of superoxide by illuminated chloroplasts, spinach ferredoxin did not. Since ferredoxin is known to stimulate chloroplast oxygen uptake, it might be concluded that a ferredoxin-mediated Mehler reaction, unlike the methyl viologen-mediated one, proceeds by some mechanism which involves at least some divalent reduction of oxygen.

2. *Inhibition by Superoxide Dismutase: Two-Step Oxygen Reduction by Ferredoxin.* Although hydrogen peroxide is its product, and it has the usual overall stoichiometry

$$H_2O + \tfrac{1}{2}O_2 \longrightarrow H_2O_2 \qquad (2)$$

(Telfer *et al.*, 1970), the ferredoxin-mediated Mehler reaction is indeed unusual in a number of respects. The rate of oxygen uptake is not, for example, appreciably increased by addition of ascorbate (Allen, 1975a and 1975c). It may, in fact, be inhibited by it (Elstner and Heupel, 1974a and 1974b).

A more significant observation is that chloroplast oxygen uptake in the presence of spinach ferredoxin is inhibited by superoxide dismutase on its own (Elstner and Heupel, 1974a; Allen, 1975a). Such an inhibition, observed with either adrenalin or dopamine as the mediator, results from a necessary involvement of superoxide in catalysis of oxygen reduction by three valence states of the mediator. If ferredoxin itself is the component which reduces oxygen in the ferredoxin-mediated chloroplast oxygen uptake, then the adrenalin-type mechanism cannot apply, since, under physiological conditions plant-type ferredoxins have only two valence states (Tagawa and Arnon, 1968). How, then, can superoxide dismutase cause an inhibition? Elstner and Heupel's (1974a and 1974b) solution to this problem is the proposal that ferredoxin itself does not reduce oxygen directly, but instead transfers electrons to an "oxygen reducing factor" which confers sensitivity to ascorbate and to superoxide dismutase on the overall reaction. The oxygen reducing factor is then assumed to have three valence states, and to participate in an oxygen-reducing cycle which, like that involving adrenalin or dopamine, has superoxide as an essential intermediate.

Allen (1975a) has proposed an alternative explanation to account for inhibition by superoxide dismutase of the ferredoxin-mediated Mehler reaction. This explanation rests on the proposal that ferredoxin itself reduces oxygen in the Mehler reaction, and that it does so by a "two-step" mechanism in which reduced ferredoxin reduces both oxygen to superoxide (equation 17) and superoxide to peroxide (equation 18). Re-reduction of ferredoxin by non-cyclic photosynthetic electron transport (equation 19) gives the familiar overall reaction (equation 2) which is the one required by the catalase-sensitivity of net oxygen uptake by chloroplasts in the presence of ferredoxin.

$$Fd_{red} + O_2 \longrightarrow Fd_{ox} + O_2^- \quad (17)$$

$$Fd_{red} + O_2^- + 2H^+ \longrightarrow Fe_{ox} + H_2O_2 \quad (18)$$

$$H_2O + 2Fd_{ox} \longrightarrow 2Fd_{red} + \tfrac{1}{2}O_2 + 2H^+ \quad (19)$$

$$\text{Overall: } H_2O + \tfrac{1}{2}O_2 \longrightarrow H_2O_2 \quad (2)$$

Addition of superoxide dismutase would then eliminate one route of ferredoxin oxidation, by replacing the reaction of equation 18 with superoxide dismutation (equation 8).

$$2Fd_{red} + 2O_2 \longrightarrow 2Fd_{ox} + 2O_2^- \quad (17)$$

$$2O_2^- + 2H^+ \longrightarrow H_2O_2 + O_2 \quad (8)$$

$$H_2O + 2Fd_{ox} \longrightarrow 2Fd_{red} + \tfrac{1}{2}O_2 + 2H^+ \quad (19)$$

$$\text{Overall: } H_2O + \tfrac{1}{2}O_2 \longrightarrow H_2O_2 \quad (2)$$

If this were to occur when ferredoxin concentration limits the rate of the overall reaction, then the effect of superoxide dismutase would be to inhibit the rate of regeneration of oxidised ferredoxin, and hence competitively to inhibit both oxygen uptake and photosynthetic electron transport.

The introduction of superoxide dismutase would not prevent regeneration of oxidised ferredoxin by the first step (equation 17), and so addition of superoxide dismutase would neither totally inhibit autoxidation of ferredoxin (as it would with adrenalin) nor change the stoichiometry of oxygen uptake to electron transport (as it would with methyl viologen in the presence of ascorbate). The overall stoichiometry (equation 2) would remain unchanged, while the overall process would be subject to a partial feedback inhibition caused by a decrease in the steady-state concentration of oxidised ferredoxin.

The hypothesis requires reduced ferredoxin to have a greater affinity for superoxide than it has for oxygen. The relative rates of the reactions described by equations 17 and 18 will, of course, depend on oxygen concentration. At lower oxygen concentrations superoxide as an intermediate would have a lower steady-state concentration than it would at higher oxygen concentrations. This could help to explain the observations of Misra and Fridovich (1971) that univalent oxygen reduction, as measured by adrenalin oxidation, constitutes a greater proportion of total oxygen reduction at higher oxygen concentrations. An ability of the putative reaction of ferredoxin with superoxide (equation 18) to compete with a superoxide-detecting reaction such as adrenalin oxidation would give the appearance of direct reduction of oxygen to hydrogen peroxide. This may well account for the "divalent flux" of Misra and Fridovich, and for ferredoxin's apparent failure to enhance superoxide production by illuminated chloroplasts (Asada *et al.*, 1974). A comparatively low steady-state concentration of superoxide could also explain the absence of EPR-detectable superoxide production by plant ferredoxin in the experiment of Orme-Johnson and Beinert (1969). Mehler reactions mediated by any of a variety of both plant and bacterial ferredoxins show sensitivity to inhibition by superoxide dismutase (Allen, 1975c), though at equal concentrations (10 μM) the bacterial proteins seemed to mediate significantly (approximately two-fold) faster reactions than did the plant-types. Despite a common mechanism of oxygen reduction, relative affinities for oxygen may in part explain the generation of higher concentrations of superoxide by clostridial ferredoxin than by spinach ferredoxin (Orme-Johnson and Beinert, 1969; Misra and Fridovich, 1971).

The two-step hypothesis makes no specific prediction about the effect of ascorbate on chloroplast oxygen uptake, since replacement of the ferredoxin-superoxide reaction (equation 18) by the ascorbate-superoxide reaction (equation 9), should result in inhibition of electron transport itself as well as an increase in the stoichiometry of oxygen uptake to electron transport.

3. *Evidence For Two-Step Oxygen Reduction*. Allen (1975a) obtained a double reciprocal plot of the rate of chloroplast oxygen uptake against ferredoxin concentration, in which the effect of superoxide dismutase appeared to increase the ferredoxin concentration required for half saturation from 30 μM to 50 μM, while V_{max} was unchanged at about 70 μmoles (O_2 (mg chl)$^{-1}$h^{-1}. Superoxide dismutase may therefore compete with ferredoxin for an intermediate (superoxide) of the overall reaction. A similar plot of the rate of ascorbate-stimulated oxygen uptake against methyl viologen concentration shows superoxide dismutase to be a non-competitive inhibitor. The "K_m" was unchanged at 3 μM methyl viologen, and the rate of oxygen uptake at infinite methyl viologen concentration would be decreased by superoxide dismutase from 670 to 280 μmoles (mg chl)$^{-1}$h^{-1}.

In the chloroplast-free model system, in which ferredoxin is reduced by NADPH via ferredoxin-NADP reductase, there is a one-to-one molar ratio between oxygen consumption and NADP oxidation, regardless of the presence of superoxide dismutase or ascorbate, while the rate of both NADPH oxidation and oxygen uptake is slower in the presence of superoxide dismutase than in its absence (Allen, 1975c). Elstner and Heupel (1974b) found no effect of superoxide dismutase in this system. They concluded that a chloroplast membrane-bound factor is required for such inhibition to occur. Spinach superoxide dismutase present as a contaminant of ferredoxin-NADP reductase (Shin, 1971) will also produce this result (Allen, 1975c). In any event the pattern of inhibition of overall reaction rate but not of oxygen uptake relative to NADPH oxidation contrasts markedly with that obtained in a similar experiment performed with methyl viologen. Here NADPH oxidation is unaffected by ascorbate and by superoxide dismutase. In the presence of ascorbate however, superoxide dismutase inhibits both the rate of oxygen uptake and the number of oxygen molecules reduced per NADPH molecule oxidised (Epel and Neumann, 1973; Allen, 1975c).

On addition of ferredoxin to chloroplasts or to the chloroplast-free model system, oxygen uptake appears promptly (Allen, 1975c), suggesting that an adrenalin-like mediator is not involved.

The inhibition by superoxide dismutase of ferredoxin-mediated oxygen uptake also occurs in chloroplasts depleted of ferredoxin-NADP reductase (Allen, 1975c), indicating that this enzyme plays no part in the reaction.

Another line of evidence against the assumption that a factor other than ferredoxin plays some part in superoxide dismutase's inhibition of a ferredoxin-mediated Mehler reaction is that the autoxidation of chemically-reduced

Spirulina ferredoxin was shown to be inhibited by superoxide dismutase (Allen, 1975a). When superoxide dismutase was present from the start, the rate of ferredoxin autoxidation was stimulated by addition of cyanide.

E. *Alternative Schemes*

Elstner *et al.* (1970) suggested that the superoxide radical could be considered an active intermediate in ascorbate oxidation by chloroplasts. They found that photosynthetic oxidation of hydroxylamine, ascorbate or glycollate required the presence of an autoxidisable acceptor (anthraquinone or triquat), and occurred only in the presence of cyanide. The requirement for cyanide can now be viewed as the need to inhibit a chloroplast-associated superoxide dismutase (Elstner *et al.*, 1975), though in the earlier paper (Elstner *et al.*, 1970) a "cyanoperoxidase" was invoked as a catalyst of these reactions.

Elstner and Kramer (1973) found no effects of ascorbate or superoxide dismutase on ATP synthesis by chloroplasts in the presence of a low potential, autoxidisable electron acceptor. Their conclusions about the mechanism of ascorbate-stimulation of oxygen uptake in such circumstances agree with those of Epel and Neumann (1973); oxidation of ascorbate was found to be sensitive to inhibition by superoxide dismutase. Elstner and Kramer (1973) also found that low "endogenous" rates of phosphorylation (i.e. those occurring in the absence of added electron acceptors or cofactors) were, however, stimulated by addition of ascorbate, and that the stimulation was sensitive to inhibition by superoxide dismutase. They observed no net oxidation of ascorbate accompanying the stimulated rate of phosphorylation, and concluded that a catalytic cycle which involved three valence states of ascorbate was responsible for the stimulation. According to this scheme, ascorbate is oxidised by superoxide to monodehydroascorbate, which in turn becomes oxidized by donating electrons to the photosynthetic chain. Dehydroascorbate is then assumed to accept electrons from photosystem I, thus completing the cycle. Apart from the *ad hoc* character of the assumptions which the scheme involves, its chief weakness as a candidate for a physiological process is that ferredoxin is likely to be a more effective electron acceptor than dehydroascorbate.

If the "ascorbate-cycle" of Elstner and Kramer (1973) were to function *in vivo*, the necessary concentration of ascorbate within the chloroplast would be expected to inhibit electron flow to oxygen via "oxygen reducing factor" (ORF), another process held to occur *in vivo* by Elstner and Heupel (1974a, 1974b). Here "observed rates of oxygen reduction by chloroplast lamellar systems in the presence of ferredoxin are not due to an autoxidation of reduced ferredoxin but to the catalysis by a membrane-bound factor (ORF_{bound})" (Elstner and Heupel, 1974b). Stated in this way it is contradictory to the proposal that ferredoxin itself is the autoxidising component of the ferredoxin-mediated Mehler reaction (Telfer *et al.*, 1970; Allen, 1975a, and section IID of the present chapter). ORF is assumed by its proponents to have three valence states which participate in a catalytic cycle of the type outlined by them for dopamine (Elstner and Heupel, 1974a), and discussed in section IIC of this chapter. ORF is thereby thought to confer sensitivity on ferredoxin-mediated oxygen uptake to inhibition by superoxide dismutase and ascorbate. The two-step mechanism of ferredoxin autoxidation already discussed (Allen, 1975a) explains these observations and is both more economical and vulnerable to experimental test.

The ORF which Elstner and Heupel (1974b) obtained in the supernatant from a heat-treated sugar beet or spinach chloroplast preparation is clearly active in mediating superoxide dismutase-sensitive chloroplast oxygen uptake and glyoxylate decarboxylation. The small volumes in which the chloroplasts were washed prior to heat-treatment may not entirely exclude the possibility of cytoplasmic contamination. Heat-activation of the glyoxylate-decarboxylating properties of the factor is another possibility. In other respects the factor resembles the incompletely characterized "cytochrome reducing substance" of Fujita and Myers (1971).

A low-molecular weight factor which catalyses photosynthetic production of both hydrogen peroxide and superoxide has been isolated from *Euglena gracilis* by Elstner and Heupel (1976). In spinach chloroplasts this compound causes an oxygen uptake which is stimulated by addition of ferredoxin. In a similar system with *Euglena* chloroplasts or detergent-treated spinach chloroplasts, ferredoxin has an inhibitory effect on oxygen uptake (Elstner *et al.*, 1976). A role *in vivo* for Elstner and Heupel's catalyst of univalent oxygen reduction has yet to be described.

F. *Oxygen Uptake and Photosystem II*

Addition of DBMIB, an inhibitor at the plastoquinone site (Bohme *et al.*, 1971), to chloroplasts makes possible a separation of two sites of electron transport and coupled phosphorylation (see Fig.1); one site is associated with each of the two photosystems (see Trebst, 1974). Photophosphorylation normally occurs only when plastoquinone is able to function as the hydrogen carrier of a proton-translocating loop, though in the presence of DBMIB an artificial, hydrophobic hydrogen carrier (such as PMS or DCPIP) may replace phastoquinone and so restore phosphorylation.

Gould and Izawa (1973b) reported that in the absence of ferricyanide DBMIB may also mediate a Mehler reaction. The chloroplast oxygen uptake resulted in formation of hydrogen peroxide, as indicated by release of oxygen on subsequent addition of catalase. Insensitivity of the oxygen uptake to treatment of the chloroplasts with cyanide or polylysine (both of which inhibit electron flow through plastocyanin) led Gould and Izawa (1973b) to conclude that photosystem I was not involved. The stoichiometry of photosystem II phosphorylation which they obtained with the DBMIB-mediated Mehler reaction was equivalent to a $P/2e^-$ ratio of about 0.35.

The role of superoxide in photosystem II oxygen uptake has not yet been established, but a significant stimulation of the oxygen uptake by Mn^{2+} has been reported (Miles, 1976). However, superoxide dismutase inhibits only the initial phase of a biphasic oxygen consumption which follows addition of $MnCl_2$ (Allen, unpublished results). It is possible that Mn^{2+}/Mn^{3+} catalyses an overal dismutation of superoxide (Lumsden and Hall, 1975), though slow post-illumination evolution of oxygen (which is prevented by prior addition of catalase) suggests strongly that net oxidation of Mn^{2+} has occurred during oxygen uptake by photosystem II. This oxygen evolution probably results from oxidation of hydrogen peroxide by Mn^{3+}; a similar dark evolution of oxygen was reported by Walker *et al.* (1970) to be associated with the now more clearly understood effect of manganese on FMN-mediated chloroplast oxygen uptake.

Trebst *et al.* (1976) have recently introduced a cofactor of DBMIB-insensitive and KCN-insensitive chloroplast oxygen uptake; 2,3-dimethyl-5,6-methylenedioxy-p-benzoquinone. This compound ("quinone" in Fig.1) functions as a Mehler reaction cofactor both in the absence of DBMIB, and in the presence of DBMIB as a concentration (1 μM) which is inhibitory to conventional non-cyclic electron flow but which is also too small for DBMIB itself to function as the cofactor.

In all examples of chloroplast oxygen uptake which have been assigned to photosystem II alone, the precise mechanism of oxygen reduction remains to be established. Caution must therefore be exercised in interpreting such oxygen uptake as an absolute measure of electron transport in photosystem II.

III. CONCLUSION

There is as yet no compelling evidence that reduction of oxygen plays an essential role in photosynthesis. The Mehler reaction which occurs *in vitro* in broken, washed chloroplasts in the presence of an autoxidisable electron acceptor is perhaps more fully understood as a result of identification of superoxide as an intermediate, though any genuinely biological advance represented by the work discussed here must rest on whatever help it gives to an understanding of photosynthesis *in vivo*. The next step in this direction is likely to take the form of a demonstration that hydrogen peroxide or superoxide either does or does not participate in the "complete" photosynthetic process as it occurs in an intact, CO_2-fixing chloroplast. Evidence for hydrogen peroxide production during CO_2-fixation has already been provided for chloroplasts by Egneus *et al.* (1975) and Kaiser (1976) and, for algal suspensions, by Patterson and Myers (1973) and Radmer and Kok (1976). Enhancement by superoxide dismutase of CO_2-fixation by a chloroplast preparation has been reported (Ziegler and Libera, 1975), but the authors' interpretation of the effect is that broken chloroplasts in the preparation produce superoxide which, if not dismuted catalytically, impairs in some way the CO_2-fixing activity of the unbroken chloroplasts remaining.

The envelope of an intact chloroplast represents simultaneously a necessary condition for complete chloroplast photosynthesis and a barrier to the relatively crude experimental manipulation that is carried out on broken chloroplasts and on chloroplast particles. To some extent this problem is resolved by the reconstituted chloroplast system of Walker and Lilley (1974), in which soluble components of the chloroplast stroma are added back to the chloroplast lamellae. Generation of hydrogen peroxide or superoxide in the reconstituted system should be relatively easy to identify.

Even if photosynthetic reduction of oxygen were to occur only in unusual physiological circumstances, such as during the onset of poised cyclic phosphorylation or during pseudo-cyclic phosphorylation when demands for ATP are exceptionally high, a purely univalent reduction of oxygen would certainly be detrimental to chloroplast function. Thus reduced ferredoxin's relatively high affinity for oxygen and the mechanism of its reaction with oxygen are both quite likely to be of some adaptive significance.

ACKNOWLEDGEMENTS

I am indebted to Prof. D.O. Hall for enabling and encouraging me to work on chloroplast oxygen reduction, and to Dr. J. Lumsden and others who, at King's College London, helped to introduce me to this field. I am most grateful to Prof. F.R. Whatley FRS and Dr. P. John for reading the manuscript meticulously and for drawing my attention to a number of errors and ambiguities which otherwise would have found their way into the completed chapter. Financial support for my research comes from the U.K. Science Research Council.

REFERENCES

1. Allen, J.F. (1975a). *Biochem. Biophys. Res. Commun. 66*, 36-43.
2. Allen, J.F. (1975b). *Nature 256*, 599-600.
3. Allen, J.F. (1975c). Ph.D. Thesis, University of London, King's College.
4. Allen, J.F. and Hall, D.O. (1973). *Biochem. Biophys. Res. Commun. 52*, 856-862.
5. Allen, J.F. and Hall, D.O. (1974). *Biochem. Biophys. Res. Commun. 58*, 579-585.
6. Arnon, D.I., Losada, M., Whatley, F.R., Tsujimoto, H.Y., Hall, D.O. and Horton, A.A. (1961). *Proc. Nat. Acad. Sci. (USA) 47*, 1314-1334.
7. Arnon, D.I., Tsujimoto, H.Y. and McSwain, B.D. (1967). *Nature 214*, 562-566.
8. Asada, K. and Kiso, K. (1973a). *Eur. J. Biochem. 33*, 253-257.
9. Asada, K. and Kiso, K. (1973b). *Agr. Biol. Chem. 37*, 453-454.
10. Asada, K., Kiso, K. and Yoshikawa, K. (1974). *J. Biol. Chem. 249*, 2175-2181.
11. Ben-Hayyim, G. and Avron, M. (1970a). *Eur. J. Biochem. 15*, 155-160.
12. Ben-Hayyim, G. and Avron, M. (1970b). *Biochim. Biophys. Acta 205*, 86-94.
13. Bohme, H. and Trebst, A. (1969). *Biochim. Biophys. Acta 180*, 137-148.
14. Bohme, H., Reimer, S. and Trebst, A. (1971). *Z. Naturforsch. 266*, 341-352.
15. Brown, A.H. and Good, N. (1955). *Arch. Biochem. Biophys. 57*, 340-354.
16. Egneus, H., Heber, U., Matthieson, U. and Kirk, N. (1975). *Biochim. Biophys. Acta 408*, 252-268.
17. Elstner, E.F. and Heupel, A. (1974a). *Z. Naturforsch 29c*, 559-563.
18. Elstner, E.F. and Heupel, A. (1974b). *Z. Naturforsch 29c*, 564-571.
19. Elstner, E.F. and Heupel, A. (1976). *Arch. Biochem. Biophys. 173*, 614-622.
20. Elstner, E.F. and Kramer, R. (1973). *Biochim. Biophys. Acta 314*, 340-353.
21. Elstner, E.F., Heupel, A. and Vaklinova, S. (1970). *Z. Pflanzenphysiol. 62*, 184-200.
22. Elstner, E.F., Stoffer, C. and Heupel, A. (1975). *Z. Naturforsch 30*, 53-56.
23. Elstner, E.F., Wildner, G.F. and Heupel, A. (1976). *Arch. Biochem. Biophys. 173*, 623-630.
24. Epel, B.L. and Neumann, J. (1972). 6th International Congress of Photobiology, Germany. Abstract No. 237.
25. Epel, B.L. and Neumann, J. (1973). *Biochim. Biophys. Acta 325*, 520-529.
26. Farrington, J.A., Ebert, M., Land, E.J. and Fletcher, K. (1973). *Biochim. Biophys. Acta 314*, 372-381.
27. Fujita, Y. and Myers, J. (1971). In "Methods in Enzymology" (San Pietro, A., ed.) 23A, 613-618, Academic Press.
28. Gerretsen, F.C. (1950). *Plant Soil 2*, 323-343.
29. Good, N. and Hill, R. (1955). *Arch. Biochem. Biophys. 57*, 355-366.
30. Gould, J.M. and Izawa, S. (1973a). *Biochim. Biophys. Acta 314*, 211-223.
31. Gould, J.M. and Izawa, S. (1973b). *Eur. J. Biochem. 37*, 185-192.
32. Greenstock, C.L. and Miller, R.W. (1975). *Biochim. Biophys. Acta 396*, 11-16.
33. Harbour, J.R. and Bolton, J.R. (1975). *Biochem. Biophys. Res. Commun. 64*, 803-807.
34. Heillila, R.E. and Cohen, G. (1973). *Science 181*, 456-457.
35. Hill, R. and Bendall, F. (1960). *Nature 186*, 136-137.

36. Izawa, S. (1968). In "Comparative Biochemistry and Biophysics of Photosynthesis" (Shibata *et al.* eds.) p. 140, Tokyo.
37. Kaiser, W. (1976). *Biochim. Biophys. Acta 440,* 476-482.
38. Lumsden, J. and Hall, D.O. (1975). *Biochem. Biophys. Res. Commun. 64,* 595-602.
39. Marchant, R.H. (1971). *Biochem. Soc. Trans. 2,* 532-534.
40. Mehler, A.H. (1951a). *Arch. Biochem. Biophys. 33,* 65-77.
41. Mehler, A.H. (1951b). *Arch. Biochem. Biophys. 33,* 339-351.
42. Mehler, A.H. and Brown, A.H. (1952). *Arch. Biochem. Biophys. 38,* 365-370.
43. Miles, C.D. (1976). *FEBS Letts. 61,* 251-254.
44. Miller, R.W. and Macdowall, F.D.H. (1975). *Biochim. Biophys. Acta 387,* 176-187.
45. Misra, H.P. and Fridovich, I. (1971). *J. Biol. Chem. 246,* 6886-6890.
46. Misra, H.P. and Fridovich, I. (1972). *J. Biol. Chem. 247,* 3170-3175.
47. Nakamura, S. and Kimura, T. (1972). *J. Biol. Chem. 247,* 6462-6468.
48. Orme-Johnson, W.H. and Beinert, H. (1969). *Biochem. Biophys. Res. Commun. 36,* 905-911.
49. Ort, D.R. and Izawa, S. (1973). *Plant Physiol. 52,* 595-600.
50. Patterson, C.O.P. and Myers, J. (1973). *Plant Physiol. 51,* 104-109.
51. Radmer, R.J. and Kok, B. (1976). *Plant Physiol. 58,* 336-340.
52. Shin, M. (1971). In "Methods in Enzymology" (San Pietro, A., ed.) 23A, pp 440-447, Academic Press.
53. Stancliffe, T.C. and Pirie, A. (1971). *FEBS Letts. 17,* 297-299.
54. Strotmann, H. and von Gosseln, C. (1972). *Z. Naturforsch. 27,* 445-455.
55. Tagawa, K. and Arnon, D.I. (1968). *Biochim. Biophys. Acta 153,* 602-613.
56. Telfer, A., Cammack, R. and Evans, M.C.W. (1970). *FEBS Letts. 10,* 21-24.
57. Trebst, A. (1974). *Ann. Rev. Plant Physiol. 25,* 423-458.
58. Trebst, A., Reimer, S. and Dallacker, F. (1976). *Plant Sci. Letts. 6,* 21-24.
59. Vernon, L.P. and Zaugg, W.S. (1960). *J. Biol. Chem. 235,* 2728-2733.
60. Walker, D.A. and Willey, R.McC. (1974). *Plant Physiol. 54,* 950-952.
61. Walker, D.A., Ludwig, L.J. and Whitehouse, D.G. (1970). *FEBS Letts. 6,* 281-284.
62. Whitehouse, D.G., Ludwig, L.J. and Walker, D.A. (1971). *J. Exp. Bot. 22,* 772-791.
63. Yamashita, T. and Butler, W.L. (1968). *Plant Physiol. 43,* 1978-1986.
64. Ziegler, I. and Libera, W. (1975). *Z. Naturforsch. 30c,* 634-637.

ABBREVIATIONS

AQ : anthraquinone
BV : benzyl viologen
DAD : diaminodurene
DBMIB : 2,5-dibromo-3-methyl-6-isopropyl-p-benzoquinone
DCMU : 3-(3,4-dichlorophenyl)-1,1-dimethylurea
DCPIP : 2,6-dichlorophenolindophenol
FMN : flavin mononucleotide
MV : methyl viologen
PD : p-phenylenediamine
PMS : phenazine methosulphate
PS : photosystem

SUPEROXIDE DISMUTASE IN PHOTOSYNTHETIC ORGANISMS

J. LUMSDEN, L. HENRY and D.O. HALL

University of London
King's College
68 Half Moon Lane
London SE24 9JF, U.K.

THE TYPES OF PHOTOSYNTHETIC ORGANISMS

An evolutionary hierarchy of photosynthetic organisms can be constructed primarily on the basis of biochemical composition and of the structural and functional aspects of the photosynthetic process - e.g. the comparative study of ferredoxin amino acid sequences by Hall *et al.* (1975). On these bases, the most primitive photosynthetic organisms are the Chlorobacteriaceae or green bacteria (e.g. *Chlorobium*) and the Thiorhodaceae or purple, sulphur bacteria (e.g. *Chromatium*). These two groups of organisms are strictly anaerobic and obligately photosynthetic, utilising sulphide, thiosulphate or hydrogen as electron donors.

More versatile are the Athiorhodaceae or purple, non-sulphur bacteria (e.g. *Rhodopseudomonas*). These are reduced organic compounds as electron sources but are not obligately photosynthetic - the presence of oxygen suppresses the formation of photosynthetic pigments and these organisms have an aerobic respiratory metabolism in the dark.

The Cyanophyceae, or blue-green algae, embody a major evolutionary advance - the capability for utilising water as a source of electrons with the concomitant release of molecular oxygen - and they are the only prokaryotes with oxygenic photosynthesis. Although some blue-greens can grow chemoheterotrophically, others can only maintain a *status quo* in the dark and there is some evidence that the Krebs cycle may be incomplete (Uzzell and Spolsky, 1974; Stanier, 1975).

Currently, majority opinion favours the hypothesis that the next major advance in the evolution of photosynthetic organisms was the development of an endosymbiotic relationship between a blue-green algal-like cell and a large prokaryote, with a capacity for phagocytosis, for which no close analogue is extant - the "protoeukaryote" (Margulis, 1970). As Stanier *et al.* (1971) point out, the further evolution of the eukaryotic (nucleated) algae has a "two dimensional" aspect in that varying degrees of multicellular organisation occur within a given basic cellular architecture of which there are a number of types. While the progressive nature of the developments within a given cell type is apparent, the interrelationships of the different types are comparatively obscure.

The Rhodophyceae or red algae (e.g. *Porphyridium*) is the only group with chloroplasts which show a close affinity with present-day blue-green algae (Stanier, 1974) - both contain phycocyanin and phycoerythrin as auxiliary pigments which moreover exhibit considerable amino acid sequence homology (Glazer *et al.*, 1976).

By contrast, the chloroplasts of the green algae - such as the Chlorophyceae (e.g. *Scenedesmus, Codium*) and the Charophyceae (e.g. *Chara, Spirogyra*) - have the same pigment systems as the higher plants (i.e. chlorophylls a and b), which are generally regarded as deriving from a green algal ancestor (but see Stewart and Mattox, 1975 for a critical review). To account for the diversity of algal chloroplasts within the framework of the endosymbiotic hypothesis, there must either have been considerable evolution of the established organelle (Lee, 1972) or one must assume a polyphyletic origin for chloroplasts (Raven, 1970), postulating rare or extinct prokaryotic organisms with the requisite pigment contents - such as may indeed have been discovered recently by Lewin and Withers (1975).

It is to be hoped that a study of superoxide dismutases from representative photosynthetic organisms will shed some fresh light on one or more of these problems.

HIGHER PLANTS

A striking feature of the superoxide dismutases so far purified from higher plants is their close similarity to the mammalian cupreins. This has been shown in some detail for the enzymes from green peas (Sawada *et al.*, 1972) and spinach (Asada *et al.*, 1973). The latter enzyme was found to be localised primarily within the chloroplast; it has been crystallised and preliminary studies have been reported (Morita and Asada, 1974).

Using the technique of polyacrylamide gel electrophoresis in conjunction with a specific stain for superoxide dismutase activity, Beauchamp and Fridovich (1973) studied the occurrence of the enzyme in wheat. Wheat germ extracts produced three bands of activity, two of which were cyanide-sensitive (these were purified and shown to be cupreins) and one, cyanide-insensitive. This latter isozyme was presumed to be of mitochondrial origin, by analogy with the mammalian situation and because it was absent from the grain heads of the wheat plant, which have few of these organelles. The wheat heads contained both of the cupreins but only one was found in extracts of stalks and leaves.

One of the purified cupro-zinc isozymes exhibited the unusual property of remaining active in the presence of 1% sodium dodecyl sulphate; gel electrophoresis showed that this activity was attributable to the persistent dimeric form of the enzyme and subsequent conversion to the monomer resulted in the loss of enzymatic activity. Evidence suggestive of an intrachain disulphide bridge was presented (cf. erythrocuprein) and this was thought to be an important contributory factor towards the unusual stability of the enzyme.

Three distinct isozymes of superoxide dismutase have also been found in roots of Jerusalem artichoke (*Helianthus tuberosus*) - unpublished results of Henry *et al*. The cyanide-insensitive enzyme is indeed localised within the mitochondrial matrix while one of the cyanide sensitive isozymes appears to be confined to the mitochondrial intermembrane space and the other is cytoplasmic (Fig.1).

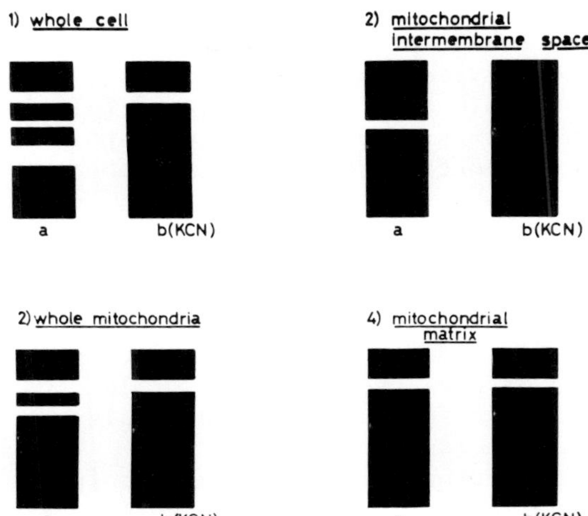

Fig.1. *Gel electrophoresis of superoxide dismutase isozymes from roots of Jerusalem artichoke (Henry* et al., *unpublished). Gel 1: total cell extract; gel 2: whole mitochondrial extract; gel 3: mitochondrial intermembrane space; gel 4: mitochondrial matrix. Gels labelled (b) were stained in the presence of 1 mM cyanide.*

SUPEROXIDE DISMUTASE IN CHLOROPLASTS

Copper-Zinc SOD

The presence of SOD in chloroplasts isolated from spinach was first reported by Asada *et al.* (1973). They noted that spinach leaves contained a single copper-zinc SOD of which at least 40% was localised in the chloroplast; of this, over half could be removed by osmotic shock but the rest remained membrane-bound, being released only on extraction of the lamellae with acetone or butanol. Schmid (1975) also reported that treatment with 1 mM EDTA caused either removal or inactivation of the lamellar SOD activity. Using carefully prepared "intact" chloroplasts, Lumsden and Hall (1974) measured the specific activity of SOD in the chloroplast stroma (the soluble material) at 6 units/mg protein, ranking among the higher values recorded for a variety of aerobic bacteria (1.4 → 7.0 units/mg) by McCord *et al.* (1971) and implying an important role for SOD within the chloroplast. The suggestion that the stromal enzyme was a different isozyme from that in the cytoplasm has not been confirmed by other groups.

Elstner and Heupel (1975) fractionated chloroplast membranes with the detergent digitonin. This solubilised only 10% of the membrane-bound activity and most of the remainder was shown to be associated with photosystem I-derived particles, probably reflecting the principal site of superoxide production in the chloroplast. The close membrane association of much of the SOD activity may be of great importance: Miller and MacDowall (1975) reported that inhibition of the reaction between superoxide and tiron by exogenous SOD could be achieved only if the (digitonin) chloroplast fragments used were incubated with an excess of the enzyme for one hour prior to illumination. This suggests that the site of superoxide production within the particles was not readily accessible either to the added macromolecule or to the endogenous, membrane-bound pool. Similarly, Takahama and Nishimura

(1976) found that exogenous SOD was only partially effective in suppressing the peroxidation of membrane lipids in illuminated chloroplasts while a low molecular weight singlet oxygen scavenger was almost completely effective in this respect. It is probable that such a role is fulfilled *in vivo* by lipid-soluble compounds such as β-carotene and α-tocopherol, the latter of which has been shown to be an important constituent of photosystem I (Baszynski, 1974).

It is necessary to ask whether superoxide production by photosystem I is likely to occur *in vivo*. Allen (1975) studied the isolated chloroplast system using the physiological electron acceptor of photosystem I (ferredoxin) and concluded that, although ferredoxin caused the univalent reduction of oxygen, the superoxide so produced remained bound, being released as H_2O_2 only after the transfer of a second electron; Miller and MacDowall (1975) have also presented evidence for such a bound complex of superoxide with an endogenous chloroplast reductant.

The precise role of chloroplastic SOD thus remains to be clarified and there is another combination of circumstances which deserves to be investigated: one product of superoxide dismutation is hydrogen peroxide, chloroplasts apparently lack catalase activity (Gregory, 1968) and hydrogen peroxide causes the rapid inactivation of cupreins (Bray *et al.*, 1974; Hodgson and Fridovich, 1975), including that from spinach (Asada *et al.*, 1975). We must presume that other pathways exist for the disposal of hydrogen peroxide within the chloroplast.

Chloroplast Manganese

"It is intriguing to consider ... that manganese, which is essential for photosynthetic oxygen evolution, may be part of a chloroplast superoxide dismutase" (Fridovich, 1974).

Manganese is the only lamellar redox component which has definitely been associated with the oxygen-evolving site of photosystem II. Cheniae and Martin (1970) demonstrated two distinct pools of manganese, the larger of which is directly involved in the water-splitting reaction. Unfortunately, it is very labile and easily lost from the thylakoid membrane; fractionation of chloroplast lamellae by detergent treatment generally causes the loss of all water-splitting ability and to date no "partial reactions", which might be ascribed to this manganese complex, have been observed. Takahama *et al.* (1974) found that, in chloroplasts where the electron flow from water was blocked, H_2O_2 could act as an electron donor to photosystem II. However, added SOD had no effect and there is no evidence for the generation of free superoxide during this reaction.

Aqueous Mn^{2+} can react with superoxide anion (Epel and Neumann, 1973) and it exhibits a superoxide dismutase-like effect in the common assay systems for that enzyme (Fong *et al.*, 1973; Albergoni and Cassini, 1974; Lumsden and Hall, 1975a). It therefore seemed promising to attempt to use this readily accessible property as a probe of the state of chloroplast-bound manganese and to investigate whether treatments of chloroplasts which disrupt the water-splitting system would reveal a superoxide dismutase capability of bound manganese in excess of that expected for an equivalent quantity of free Mn^{2+}. Lumsden and Hall (1974, 1975a) prepared subchloroplast particles by treatment of chloroplast lamellae with the detergent Triton X-100 followed by aqueous phase partition and discontinuous sucrose gradient. One class of particles was obtained which had a consistently higher SOD-like effect than the others: the inhibitory effect of these particles on the SOD assay (photochemical assay of Beauchamp and Fridovich, 1971) had two components. About 60% of the activity was destroyed by heating at 85°C and this fraction was also inhibited by micromolar concentrations of hydrogen peroxide in the presence of 1 mM cyanide to inhibit peroxidase activity (Fig.2). Washing the particles with 1 M Tris or with 2 mM EDTA produced complete inactivation of the activity thus implicating bound metal ions and in fact, the SOD-like effect could be restored by Cu^{2+} or Mn^{2+}. However, only the latter was peroxide-inhibitable and the "reconstituted" particles were in either case no longer heat sensitive.

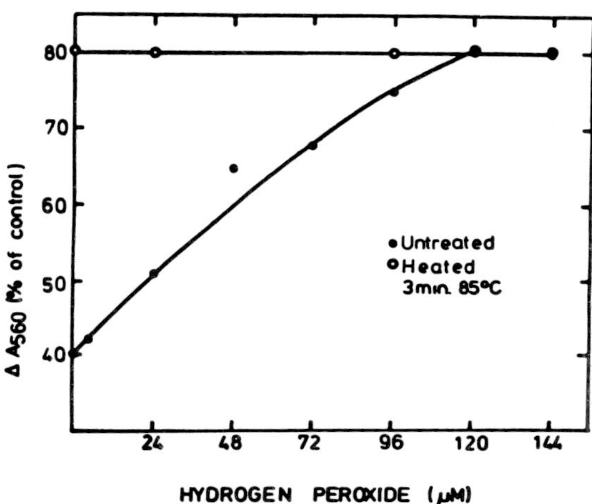

Fig.2. *Effect of hydrogen peroxide on the superoxide dismutase activity of subchloroplast particles (see Lumsden and Hall, 1975a).*

The inhibition by peroxide might be explained in terms of three different redox states for the bound manganese, with the Mn(IV)/Mn(III) couple catalysing the removal of superoxide more effectively than the Mn(III)/Mn(II) couple (Fig.3) and, in this connection, it is worth recalling the apparent involvement of higher valence states of manganese in the water-splitting process (Radmer and Cheniae, 1971).

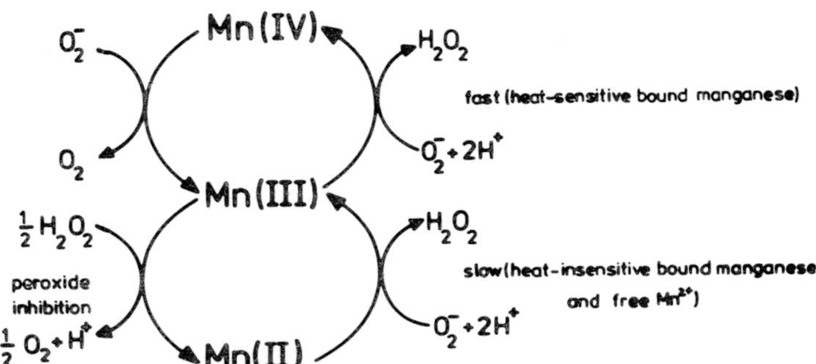

Fig.3. *Proposed mechanism of action of free and particle-bound manganese (see Lumsden and Hall, 1975a).*

It remains unclear whether the bound manganese of the original particles is in a "native" binding site or whether it has been displaced by the detergent and readsorbed at another site on the membrane. So far, it has been assumed that the particle-bound manganese must necessarily have been derived from the water-splitting complex of photosystem II. The absence of cyanide-insensitive SOD activity in intact lamellae need not mean that the activity of the particles is induced by the detergent treatment. 90% of lamellar manganese is located on the inner surface of the thylakoid membranes (Blankenship and Sauer, 1974) and treatment with Triton would remove this permeability barrier between the manganese site and the components of the SOD assay. In a careful study, Blankenship *et al.* (1975) found that Tris-washed chloroplasts could be reactivated so as to regain completely the original rate of oxygen evolution although they had lost 35% of the original manganese content. They suggested that part or all of this fraction might be accounted for by a lamellar superoxide dismutase, a point which might be resolved by the preparation of particles from Tris-washed chloroplasts and Tris-washed, reactivated chloroplasts. Obviously, much work remains to be done in this important area.

EUKARYOTIC ALGAE

No superoxide dismutase has yet been purified from a eukaryotic alga although the enzyme has been studied in crude extracts from a number of these organisms by gel electrophoresis. Lumsden and Hall (1975b) reported the occurrence of cyanide-insensitive superoxide dismutase in the red alga *Porphyridium cruentum* (one band of activity) and in the green algae *Scenedesmus obliquus* (2 major and 2 - 3 minor bands) and *Codium fragile* (2 bands); one of the *Codium* enzymes was shown to occur in the chloroplast stroma and the other in the cytoplasm.

Further studies of this kind (L. Henry, unpublished results) have confirmed the exclusive occurrence of the cyanide-insensitive enzyme in other members of the Chlorophyceae (*Ulva*; *Cladophora*) but *Spirogyra* was found to contain only a single cyanide-sensitive dismutase and extracts of *Chara* and *Nitella* exhibited both kinds of activity - one cyanide-sensitive and two cyanide-insensitive bands (Fig.4). The last two named organisms are members of the Charophyceae and the significance of this distinction will be discussed in a subsequent section.

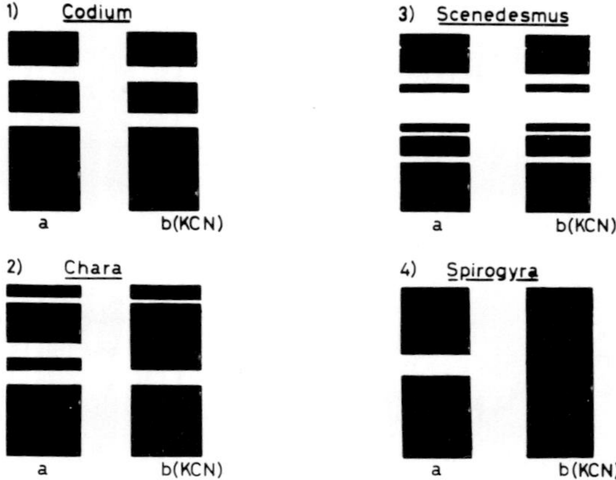

Fig.4. *Gel electrophoresis of superoxide dismutase from eukaryotic algae (Henry and Hall, unpublished). Gel 1:* Codium; *gel 2:* Chara; *gel 3:* Scenedesmus; *gel 4:* Spirogyra. *Gels were stained in the absence (a) or presence (b) of 1 mM cyanide.*

BLUE-GREEN ALGAE

It is probable that atmospheric oxygen was initially derived from oxygenic photosynthesis by blue-green algal-like cells and some of the oxygen so produced would inescapably have become reduced by electron transport components of low redox potential within these cells. Patterson and Myers (1973) have recorded *in vivo* production of hydrogen peroxide by the blue-green alga *Anacystis nidulans* especially under conditions of impaired CO_2 fixation which would result in a "build-up" of reduced ferredoxin. Subsequently, Abeliovich *et al.* (1974) demonstrated that the superoxide dismutase activity in crude extracts of the same organism was substantially increased under aerobic conditions of cell growth.

Thus, the superoxide dismutases of this class of organisms is of particular interest and, to date, the enzyme has been purified from two species - *Plectonema boryanum* (Asada *et al.*, 1975; Misra and Keele, 1975) and *Spirulina platensis* (Lumsden and Hall, 1975b; Lumsden *et al.*, 1976). All three groups isolated a dismutase of the iron-containing type although Asada *et al.* also reported the occurrence of a minor manganodismutase in *Plectonema* without presenting any detailed characterisation of this enzyme.

The physico-chemical properties of the *Spirulina* ferridismutase are representative - it has a molecular weight of 37 - 38,000, is composed of two identical subunits associated non-covalently and has an iron content of 1 g atom per mole. This unusual stoicheiometry may result from the specific conditions of cell growth or enzyme isolation employed - it seems probable that both subunits have metal-binding sites as the *Plectonema* enzyme isolated by Asada's group contains 2 Fe per mole. Concentrated solutions of the *Spirulina* enzymes are yellow in colour and exhibit a broad absorption band between about 320 and 550 nm, the only feature being a shoulder at 350 nm of molar extinction coefficient 2180 $M^{-1}cm^{-1}$; the circular dichroism spectrum has one negative band centred at 395 nm.

The ESR spectrum of the *Spirulina* dismutase is shown in Fig.5: the three resonances around g = 4.3 can be ascribed to transitions in the middle Kramers doublet (g_z = 4.88, g_x = 3.95, g_y = 3.60) but the assignment of the low field resonances are less obvious (Lumsden *et al.*, 1976). From the g-values of the middle Kramers doublet transition, an estimate of the "rhombicity" of the metal ion binding site can be obtained by the method of Wickman *et al.* (1965) in terms of the ratio E/D (E = 0 for axial symmetry and E/D = 0.33 for the completely rhombic case). The value of E/D is calculated to be 0.22, the same as for the *Plectonema* enzyme, while the *E. coli* enzyme, by comparison, appears to be more nearly rhombic (Yost and Fridovich, 1973).

The blue-green algal dismutases, like the cupreins, are inactivated by hydrogen peroxide; the second order rate constant for inactivation is 0.88 $M^{-1}s^{-1}$ for the *Spirulina* enzyme; they are not sensitive to cyanide, but Asada *et al.* (1975) found azide to be an effective inhibitor of the *Plectonema* enzyme.

An apparently similar type of superoxide dismutase has been detected in crude extracts of the related blue-green algae *Nostoc muscorum*, *Anabaena flos-aquae* and *Anabaena cylindrica* (Lumsden and Hall, 1975b). In the case of the last organism, it has proved possible to obtain homogeneous populations of the two cell types, namely the vegetative cells and the heterocysts (Tel-Or and Stewart, 1976). The heterocysts have an essentially anaerobic metabolism - they lack the oxygen-evolving photosystem II and are the repository for the important enzyme nitrogenase, which is rapidly inactivated by oxygen.

Preliminary studies (on material kindly supplied by Dr. E. Tel-Or) have demonstrated the presence of superoxide dismutase activity in both cell types (L. Henry, unpublished observations), a result which is perhaps not so surprising in view of the similar findings for anaerobic bacteria described in the following section.

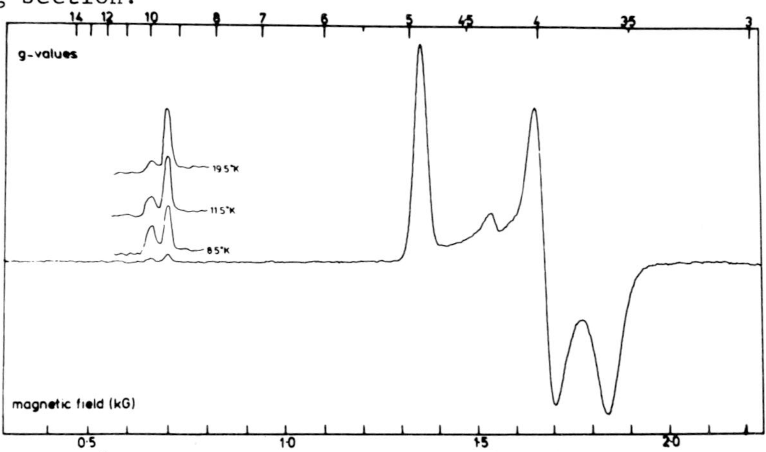

Fig.5. *Electron spin resonance spectrum of* Spirulina platensis *iron-containing superoxide dismutase at 8.5°K (from Lumsden* et al.*, 1976).*

PHOTOSYNTHETIC BACTERIA

Obligate Anaerobes

The presence of superoxide dismutase in a range of obligately anaerobic bacteria was first reported by Hewitt and Morris (1975), contrary to the findings of an earlier study by McCord *et al.* (1971). Among the organisms tested were the green bacteria *Chlorobium thiosulphatophilum* and the purple, sulphur bacterium *Chromatium* sp., crude extracts of which had 30% and 1 - 2% respectively of the specific activity of aerobic *Escherichia coli* extract. We have confirmed the occurrence of superoxide dismutase in the two photosynthetic bacteria and have demonstrated that the enzymes have virtually identical isoelectric points (Lumsden and Hall, 1975b).

Hewitt and Morris (1975) pointed out that oxygen has a bacteriostatic rather than a bactericidal effect on these organisms and it is further suggested that, although the possession of superoxide dismutase activity may serve to protect the genetic material from oxidative damage, thus allowing a resumption of growth when the oxygen is removed, it cannot prevent the destruction of certain important enzymes, such as the complex iron-sulphur proteins. Indeed, the aerobic inactivation of *Chromatium* nitrogenase was not significantly retarded in the presence of an excess of superoxide dismutase (S.L. Albrecht and J. Lumsden, unpublished observations). An alternative possibility is that molecular oxygen might "drain-off" too much reducing power from the photosynthetic electron transport chain, so preventing the generation of reduced pyridine nucleotide. Boucher and Gingras (1975) have reported that isolated reaction centres of *Rhodospirillum rubrum* can generate superoxide and this probably applies generally to photosynthetic bacteria.

Facultative Aerobes

These studies have been extended to the purple, non-sulphur bacterium

Rhodopseudomonas spheroides which has the ability to grow either anaerobically (photosynthetically) or as an aerobic heterotroph. In both cases, the same major isozyme of superoxide dismutase is present but with distinct patterns of minor isozymes (Lumsden *et al.*, 1976) (Fig.6). It is not clear whether this reflects the different requirements of the two conditions of growth or whether it is simply a nutritional phenomenon: the similar, non-photosynthetic organism *Pseudomonas ovalis* gives a different pattern of isozymes depending on whether the growth medium is supplemented with iron or manganese (Yamakura, 1976).

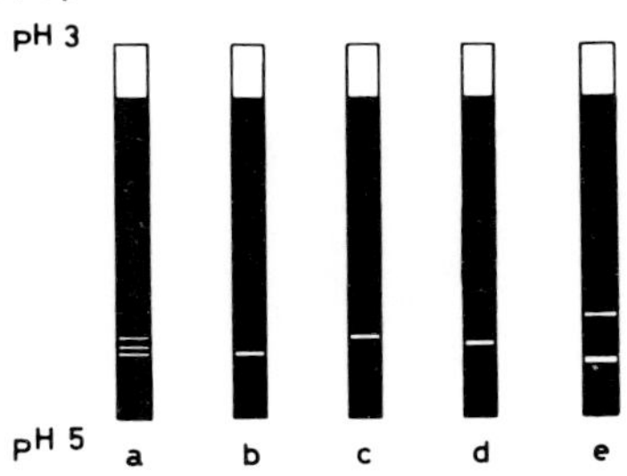

Gel electrofocusing of Rps. spheroides isozymes

a. crude extract (aerobic)
b. SOD I
c. SOD II
d. SOD III
e. crude extract (anaerobic)

Fig.6. *Gel isoelectric focusing of superoxide dismutase isozymes from* Rhodopseudomonas spheroides. *Gel (a): crude extract of aerobically-grown cells; gel (b): major isozyme; gels (c) and (d): minor isozymes; gel (e): crude extract of anaerobically-grown cells (from Lumsden* et al.*, 1976).*

The principal *Rps. spheroides* dismutase is a manganoenzyme with properties generally similar to those of the *E. coli* enzyme (Keele *et al.*, 1970). It is ESR silent but has characteristic visible absorbance (Fig.7) and circular dichroism spectra (Fig.8).

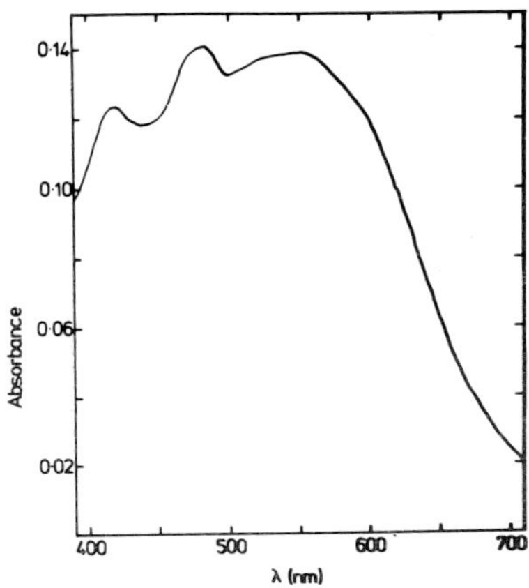

Fig.7. *Visible spectrum of* Rhodopseudomonas spheroides *manganese-containing superoxide dismutase recorded at 77°K (from Lumsden* et al.*, 1976).*

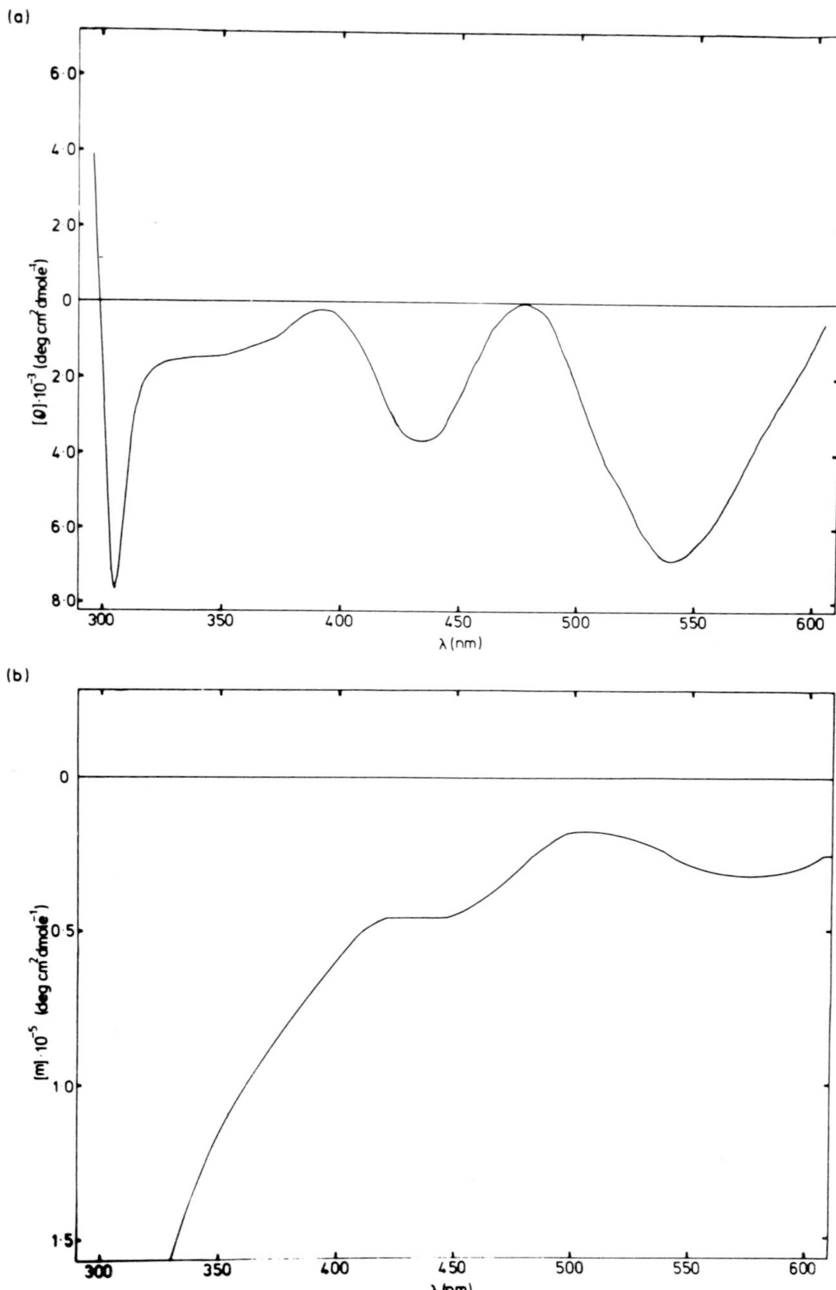

Fig.8. *Circular dichroism and optical rotatory dispersion of* Rps. spheroides *superoxide dismutase (Lumsden, 1975).*

Unlike the *E. coli* enzyme (Keele *et al.*, 1975), but in common with the *Plectonema* manganodismutase (Asada *et al.*, 1975), it is resistant to inactivation or inhibition by hydrogen peroxide, a property which probably derives from the structure of the metal-binding site.

EVOLUTIONARY IMPLICATIONS

Some ways in which superoxide dismutase can be used as a basis for speculations on the origin and evolution of photosynthetic organisms have been set down in detail elsewhere (Lumsden, 1975; Lumsden and Hall, 1975b) and here we will simply record them briefly, expanding only on points where recent work has shed new light.

Anaerobic Bacteria

To explain the presence of superoxide dismutase in anaerobic bacteria we must postulate either (i) the independent evolution of SOD in these organisms or (ii) that SOD antedates oxygenic photosynthesis by blue-green algae; perhaps evolving in response to the ultraviolet photolysis of water to oxygen (Berkner and Marshall, 1965) or (iii) the transfer of genetic information for SOD from an aerobe to an anaerobe or (iv) that all primitive anaerobes which could not adapt to an aerobic mode of existence were killed and present-day anaerobes represent secondary adaptations of once-aerobic organisms. We must await the publication of amino acid sequence data in order to narrow the field of possibilities.

Possibility (iii) is a disturbing one to anyone interested in tracing prokaryotic phylogeny from protein sequences and one which should have a restraining effect on the drawing of too detailed conclusions from such data (see comments in Dickerson and Timkovich, 1975).

Aerobic Bacteria

Aerobic prokaryotes divide into three categories with respect to photosynthesis, i.e. the purple, non-sulphur bacteria, the blue-green algae and the non-photosynthetic, oxygen-respiring bacteria; one school of thought has it that the last (and largest) group originally evolved from photosynthetic ancestors. Uzzell and Spolsky (1974), on the principle of "minimising the amount of evolution inferred to have taken place", favour blue-green algal-like organisms for this role while Dickerson *et al.* (1976) leave the question open and suggest that some aerobic respirers evolved from purple, non-sulphur bacteria (cf. *Paracoccus* and *Rhodopseudomonas*) while others derive from blue-green algal origins (cf. *Saprospira* and *Spirulina*). Whether the purple, non-sulphur bacteria themselves evolved from blue-green algae or whether they were the result of the independent acquisition of respiration by more primitive photosynthetic bacteria is another point of controversy.

Following the purification of superoxide dismutase from two representative photosynthetic prokaryotes (Lumsden *et al.*, 1976), N-terminal amino acid sequence studies have been carried out on these enzymes (J.I. Harris *et al.*, unpublished results) and the findings of this study are presented in detail elsewhere in this volume (Steinman and Harris), showing a comparison between the SODs of photosynthetic and non-photosynthetic organisms.

Eukaryotic Algae

Osterberg (1974) has pointed out that, under the general reducing conditions of the primitive atmosphere, the sea-water concentrations of iron and manganese would have been high while copper would have been trapped in insoluble cuprous sulphides and would thus have been unavailable for incorporation into metalloenzymes until the net build-up of atmospheric oxygen occurred. This is frequently assumed to have followed closely upon the advent of the primitive blue-green algae but Hall (1971) considered that these organisms were too restricted in their possible environments to have achieved a biomass sufficient to cause this net increase which would thus require the evolution of the more versatile eukaryotic algae.

In this way, Lumsden and Hall (1975b) attempted to account for the apparent absence of copper-containing superoxide dismutase in green algae but a difficulty of this hypothesis was how to account for the appearance of cupreins in higher plants. As mentioned earlier, the algae examined in this survey were members of the Chlorophyceae; however a recent extension to the Charophyceae (L. Henry, unpublished results) has demonstrated the occurrence of cyanide-sensitive superoxide dismutase in a number of these organisms. This result is in agreement with previous biochemical and cytological studies which suggested that the Charophyceae were on the evolutionary line leading to the higher plants while the Chlorophyceae were more remote from this line (see review by Stewart and Mattox, 1975).

Thus, the Charophyceae, as well as possessing cyanide-sensitive superoxide dismutase, possess these other properties which are also characteristic of higher plants:

(a) glycollate oxidase as the key enzyme of glycollate metabolism, compared with glycollate dehydrogenase in the Chlorophyceae (Frederick *et al.*, 1973);
(b) a phragmoplast type of spindle and microtubular arrangement during cytokinesis and
(c) motile cells in which laterally-attached flagellae are associated with a single broad band of microtubules (Pickett-Heaps and Marchant, 1972). Again, by comparison, the Chlorophyceae have the phycoplast system of microtubules and motile cells have anterior flagellae with four cruciate microtubular roots.

CONCLUSIONS AND FUTURE PROSPECTS

The study of the superoxide dismutases from photosynthetic organisms has only just commenced but it already promises to yield much useful information on many aspects of their evolution. As the possession of this enzyme appears to be indispensible to all living things which come into contact with oxygen, it should be able to afford us insights into such ancient processes as the building-up of atmospheric oxygen, the development of photosystem II, the relationships between photosynthetic and non-photosynthetic prokaryotes, the endosymbiotic origin of organelles and the evolution of algae into higher plants. A number of hypotheses have been put forward on the basis of work with crude or partially purified enzymes and, as more of these become available in a pure form, metal analyses and sequence determinations should put these to a rigorous test.

Finally, the precise roles of superoxide dismutase within the chloroplast, both at the reducing side of photosystem I and (if any) in the inner workings of the oxygen evolving complex, are challenging questions which would seem to demand great biochemical skill in their solution.

REFERENCES

1. Abeliovich, A., Kellenberg, D. and Shilo, M. (1974). *Photochem. Photobiol. 19*, 378-382.
2. Albergoni, V. and Cassini, A. (1974). *Comp. Biochem. Physiol. 47*B, 767-777.

3. Allen, J.F. (1975). *Biochem. Biophys. Res. Commun. 66*, 36-43.
4. Asada, K., Urano, M. and Takahashi, M. (1973). *Eur. J. Biochem. 36*, 257-266.
5. Asada, K., Yoshikawa, K., Takahashi, M.-A., Maeda, Y. and Enmanji, K. (1975). *J. Biol. Chem. 250*, 2801-2807.
6. Baszynski, T. (1974). *Biochim. Biophys. Acta 347*, 31-35.
7. Beauchamp, C.O. and Fridovich, I. (1971). *Anal. Biochem. 44*, 276-287.
8. Beauchamp, C.O. and Fridovich, I. (1973). *Biochim. Biophys. Acta 317*, 50-64.
9. Berkner, L.V. and Marshall, L.C. (1965). *J. Atmos. Sci. 22*, 225-261.
10. Blankenship, R.E. and Sauer, K. (1974). *Biochim. Biophys. Acta 357*, 252-266.
11. Blankenship, R.E., Babcock, G.T. and Sauer, K. (1975). *Biochim. Biophys. Acta 387*, 165-175.
12. Boucher, F. and Gingras, G. (1975). *Biochem. Biophys. Res. Commun. 67*, 421-426.
13. Bray, R.C., Cockle, S.A., Fielden, E.M. Roberts, P.B., Rotilio, G. and Calacrese, L. (1974). *Biochem. J. 139*, 43-48.
14. Cheniae, G.M. and Martin, I.F. (1970). *Biochim. Biophys. Acta 197*, 219-239.
15. Dickerson, R.E. and Timkovich, R. (1975). In "The Enzymes" 3rd edn. (Boyer, P.D. ed.), Vol.11, pp 397-547, Academic Press, London and New York.
16. Dickerson, R.E., Timkovich, R. and Almassy, R.J. (1976). *J. Mol. Biol. 100*, 473-491.
17. Elstner, E.F. and Heupel, A. (1975). *Planta (Berl.) 123*, 145-154.
18. Epel, B.L. and Neumann, J. (1973). *Biochim. Biophys. Acta 325*, 520-529.
19. Fong, K.-L., McCay, P.B., Poyer, J.L., Keele, B.B. and Misra, H.P. (1973). *J. Biol. Chem. 248*, 7792-7797.
20. Frederick, S.E., Gruber, P.J. and Tolbert, N.E. (1972). *Plant Physiol. 52*, 318-323.
21. Fridovich, I. (1974). *Advan. Enzymol. 41*, 35-97.
22. Glazer, A.N., Apell, G.S., Hixson, C.S., Bryant, D.A., Rimon, S. and Brown, D.M. (1976). *Proc. Natl. Acad. Sci. U.S. 73*, 428-431.
23. Gregory, R.P.F. (1968). *Biochim. Biophys. Acta 159*, 429-439.
24. Hall, D.O., Rao, K.K. and Cammack, R. (1975). *Sci. Prog. Oxf. 62*, 285-317.
25. Hall, J.B. (1971). *J. Theor. Biol. 30*, 429-454.
26. Hewitt, J. and Morris, J.G. (1975). *FEBS Letts. 50*, 315-318.
27. Hodgson, E. and Fridovich, I. (1975). *Biochemistry 14*, 5294-5299.
28. Keele, B.B., McCord, J.M. and Fridovich, I. (1970). *J. Biol. Chem. 245*, 6176-6181.
29. Keele, B.B., Giovagnoli, C. and Rotilio, G. (1975). *Physiol. Chem. and Phys. 7*, 1-6.
30. Lee, R.E. (1972). *Nature 237*, 44-46.
31. Lewin, R.A. and Withers, N.W. (1975). *Nature 256*, 735-737.
32. Lumsden, J. (1975). Ph.D. Thesis, University of London.
33. Lumsden, J. and Hall, D.O. (1974). *Biochem. Biophys. Res. Commun. 58*, 35-41.
34. Lumsden, J. and Hall, D.O. (1975a). *Biochem. Biophys. Res. Commun. 64*, 595-602.
35. Lumsden, J. and Hall, D.O. (1975b). *Nature 257*, 670-672.

36. Lumsden, J., Cammack, R. and Hall, D.O. (1976). *Biochim. Biophys. Acta*, in press.
37. Margulis, L. (1970). "Origin of Eukaryotic Cells", Yale University Press.
38. McCord, J.M., Keele, B.B. and Fridovich, I. (1971). *Proc. Natl. Acad. Sci. U.S.A. 68*, 1024-1027.
39. Miller, R.W. and MacDowall, F.D.H. (1975). *Biochim. Biophys. Acta 387*, 176-187.
40. Misra, H.P. and Keele, B.B. (1975). *Biochim. Biophys. Acta 379*, 418-425.
41. Morita, Y. and Asada, K. (1974). *J. Mol. Biol. 86*, 685-686.
42. Osterberg, R. (1974). *Nature 249*, 382-383.
43. Patterson, C.O.P. and Myers, J. (1973). *Plant Physiol. 51*, 104-109.
44. Pickett-Heaps, J.D. and Marchant, H.J. (1972). *Cytobios 6*, 255-264.
45. Radmer, R. and Cheniae, G.M. (1971). *Biochim. Biophys. Acta 253*, 182-186.
46. Raven, P.H. (1970). *Science 169*, 641-646.
47. Sawada, Y., Ohyama, T. and Yamazaki, I. (1972). *Biochim. Biophys. Acta 268*, 305-312.
48. Scmid, R. (1975). *FEBS Letts. 60*, 98-102.
49. Stanier, R.Y. (1975). *Biochem. Soc. Trans. 3*, 352-357.
50. Stanier, R.Y., Doudoroff, M. and Adelberg, E.A. (1971). "General Microbiology" 3rd edn., MacMillan, London.
51. Stewart, K.D. and Mattox, K.R. (1975). *Bot. Rev. 41*, 104-135.
52. Takahama, U. and Nishimura, M. (1976). *Plant Cell Physiol. 17*, 111-118.
53. Takahama, U., Inoue, H. and Nishimura, M. (1974). *Plant Cell Physiol. 15*, 971-978.
54. Tel-Or, E. and Stewart, W.D.P. (1976). *Biochim. Biophys. Acta 423*, 189-195.
55. Uzzell, T. and Spolsky, C. (1974). *Amer. Sci. 62*, 334-343.
56. Wickman, H.H., Klein, M.P. and Shirley, D.A. (1965). *J. Chem. Phys. 42*, 2113-2117.
57. Yamakura, F. (1976). *Biochim. Biophys. Acta 422*, 280-294.
58. Yost, F.J. and Fridovich, I. (1973). *J. Biol. Chem. 248*, 4905-4908.

ROLE OF OXYGEN IN COELENTERATE BIOLUMINESCENCE: EVIDENCE FOR ENZYME-SUBSTRATE, OXYGEN-CONTAINING INTERMEDIATES

M.J. CORMIER, W.W.WARD and H. CHARBONNEAU

Bioluminescence Laboratory
Department of Biochemistry
University of Georgia
Athens, GA 30602, U.S.A.

Similarities in the chemical mechanisms leading to light emission among the two major classes of bioluminescent coelenterates, the Anthozoans and Hydrozoans, have been documented in a series of recent papers and reviews (Cormier *et al.*, 1975; Shimomura and Johnson, 1975a; Hori *et al.*, 1975; Ward and Cormier, 1975; Cormier *et al.*, 1974; Hori *et al.*, 1973; Cormier *et al.*, 1973). The substrates and products in these systems are essentially identical and the mechanisms leading to excited state product formation are apparently the same as well. The most interesting differences are at the enzymological level. Thus, both the Anthozoans and Hydrozoans trigger their bioluminescence by Ca^{2+} but only the Hydrozoans possess the ability to form unusually stable luciferin-protein-oxygen complexes which are referred to as photoproteins. In both cases the reactions catalyzed are of the oxygenase type and considerable information on the role of oxygen in these reactions may be obtained by a careful study of these systems. The role of oxygen in coelenterate bioluminescence will be the major focus of this chapter and the principal findings are briefly outlined below.

CHEMISTRY AND ENZYMOLOGY OF *IN VITRO* ANTHOZOAN BIOLUMINESCENCE

Bioluminescence in the Anthozoans proceeds via a classical luciferin-luciferase reaction. Thus from the Anthozoans, such as *Renilla reniformis*, one may isolate a luciferase which catalyzes the bioluminescent oxidation of a substrate luciferin in the presence of dissolved oxygen (Hori *et al.*, 1973; Mathews *et al.*, 1976).

Renilla luciferase has recently been purified to homogeneity and some of its characteristics determined (Matthews *et al.*, 1976). The enzyme, which is active as a single polypeptide chain monomer of 35,000 daltons, is a glycoprotein containing 3 free SH groups but no disulfide linkages. This enzyme catalyzes the reaction illustrated in Fig.1 with a turnover number of 111 μmoles luciferin min^{-1} $\mu mole^{-1}$ enzyme.

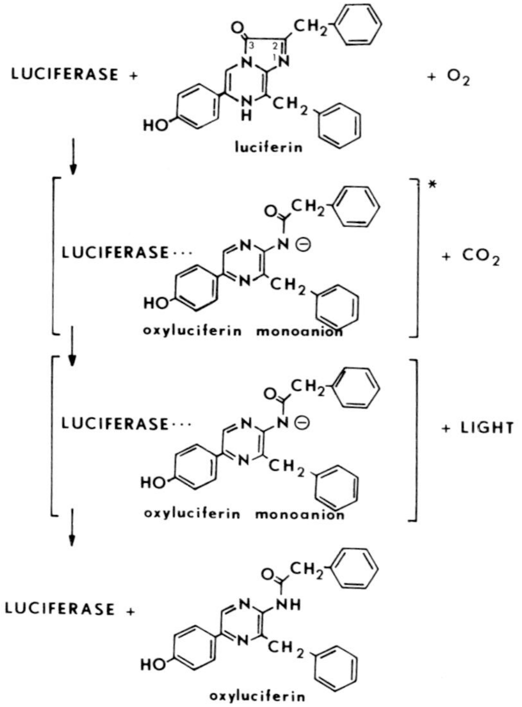

Fig.1. *Chemical path to Anthozoan* in vitro *bioluminescence. The luciferase catalyzed reaction illustrated here was elucidated for the Anthozoan,* Renilla reniformis *(Hori* et al.*, 1973).*

Initiation of the reaction presumably occurs via the formation of a luciferin carbanion which attacks molecular oxygen to yield a linear hydroperoxide intermediate as follows:

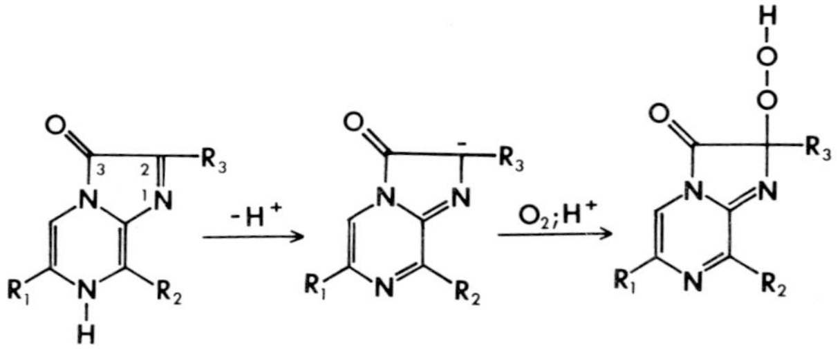

As may be deduced from Fig.1 and the above scheme, molecular oxygen is apparently incorporated into one of the products, oxyluciferin, while a second product, CO_2 is derived from carbon 3 of luciferin. Thus *Renilla* luciferase can be classified as an oxygenase. The initial product of the reaction is an electronic excited state of a luciferase-oxyluciferin monoanion complex which relaxes to the ground state with the production of blue light (λ_{max} = 490 nm). The product excited state yield during the reaction approaches 100% (Ward and Cormier, 1976).

Another protein, luciferin binding protein, has recently been discovered which is important in the nerve-linked control of the bioluminescence flash in *Renilla* and other Anthozoans (Anderson *et al.*, 1974). This protein (mol. wt. = 24,000) has also been isolated to homogeneity and many of its properties determined (Charbonneau and Cormier, unpublished data). When isolated in the absence of Ca^{2+} this protein exhibits a deep yellow color due to its content of non-covalently bound luciferin. The luciferin binding protein also contains a Ca^{2+} binding site and the binding of Ca^{2+} induces a conformational change in the protein which allows it to discharge reversibly its bound luciferin (Anderson *et al.*, 1975).

When the luciferin binding protein is mixed in solution with luciferase and oxygen, in the absence of Ca^{2+}, no light is produced. However, upon the addition of Ca^{2+}, a brilliant flash of light is observed. These relationships are illustrated below where BP and LF refer to binding protein and luciferin respectively.

$$\text{BP-LH} + Ca^{2+} \rightleftarrows \text{Ca-BP-LH}$$

$$\text{Ca-BP-LH} + \text{Luciferase (E)} \rightarrow \text{Ca-BP} + \text{E-LH}$$

$$\text{E-LH} + O_2 \rightarrow \text{E-oxyluciferin*} + CO_2$$

$$\text{E-oxyluciferin*} \rightarrow \text{E-oxyluciferin} + \text{Light}$$

Thus, in the Anthozoans, the Ca^{2+}-linked control of bioluminescence and the oxidation of luciferin to produce light require the participation of two distinct proteins. As outlined below, both of these functions are achieved by a single protein in the Hydrozoans.

CHEMISTRY AND ENZYMOLOGY OF *IN VITRO* HYDROZOAN AND CTENOPHORE BIOLUMINESCENCE

Chemistry of Light Emission in Photoproteins

In 1962 Shimomura *et al.* reported the isolation of a protein from the luminous jellyfish *Aequorea* which produces blue light in the presence of Ca^{2+}. They termed this protein a photoprotein and referred to it specifically as aequorin. Photoproteins have since been isolated from other Hydrozoans and Ctenophores. For example, from the ctenophores *Mnemiopsis* and *Beroë*, Ca^{2+}-triggered photoproteins have been isolated which are referred to as mnemiopsin and berovin in keeping with the original nomenclature (Ward and Seliger, 1974a, 1974b).

One of the most interesting observations on the mechanism of photoprotein luminescence is the fact that dissolved oxygen is not required for the Ca^{2+}-triggered luminescence reaction (Shimomura *et al.*, 1962). This has been confirmed with other photoproteins as well and for some time the aequorin bioluminescence reaction was viewed simply as:

$$\text{Photoprotein} + Ca^{2+} \rightarrow \text{Ca-protein} + \text{Products} + \text{Light}$$

Thus the aequorin luminescence reaction appeared to be distinctly different from other bioluminescent systems that were being investigated, such as *Renilla*. However, when sufficient data became available on both the *Renilla* and *Aequorea* systems it became apparent that the chemistry of their luminescence reactions might be similar or identical (Cormier *et al.*, 1973; Hori *et al.*, 1973; Cormier *et al.*, 1974; Shimomura and Johnson, 1973). Recent evidence suggests that this is correct. This is based, in part, on the quantitative extraction of luciferin, indistinguishable from *Renilla* luciferin, from each of the three photoproteins aequorin, mnemiopsin and berovin (Ward and Cormier, 1975; Hori *et al.*, 1975). In addition, the products of the aequorin luminescence reaction are CO_2 and oxyluciferin as illustrated in Fig.1 except that *Aequorea* oxyluciferin is reported to contain a p-hydroxybenzyl group at carbon 2 (Shimomura and Johnson, 1973; Shimomura *et al.*, 1974). Thus, the photoprotein luminescent reaction can be viewed as:

$$\text{Protein-Luciferin} + Ca^{2+} \rightarrow \text{Ca-Protein-Oxyluciferin}^* + CO_2$$

$$\text{Ca-Protein-Oxyluciferin}^* \rightarrow \text{Ca-Protein-Oxyluciferin} + \text{Light (469 nm)}$$

Once Ca^{2+} is bound to a photoprotein the reaction would be expected to proceed in a manner similar to that shown in Fig.1 especially since the structures of the starting substrate (luciferin) and products (oxyluciferin and CO_2) are essentially identical. Thus, as previously suggested (Anderson *et al.*, 1974), photoproteins play a dual role. In the absence of Ca^{2+} they function as a luciferin binding protein whereas, in the presence of Ca^{2+}, they function as a *Renilla*-like luciferase. However, during single turnover conditions, photoproteins accomplish this latter function in the absence of dissolved oxygen indicating that they contain some form of bound oxygen. Evidence for this is given below.

Evidence That Photoproteins are Stable Enzyme-Substrate, Oxygen-containing Intermediates.

We recently provided evidence that photoproteins are protein-luciferin complexes which also contain an oxygenated species (different from O_2) that is not bound to luciferin, but rather to some site on the protein (Hori *et al.*, 1975; Ward and Cormier, 1975). It was shown that *Renilla* luciferin can exist in several tautomeric forms, one with an absorption maximum at 435 nm (I) and another at 454 nm (IV).

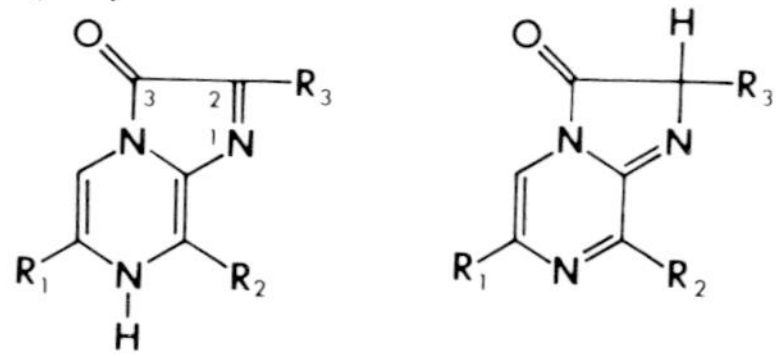

I (λ max = 435 nm) IV (λ max = 454 nm)

In water and methanol I predominates, while in aprotic solvents IV predominates. Presence of a peroxide group at position 2 of luciferin would fix the electronic structure into a tautomeric form resembling IV, the 454 nm species which corresponds to the absorption maximum of aequorin. Therefore, if photoproteins were viewed as a protein-luciferin hydroperoxide complex, the visible absorption band for all photoproteins would be centered near 454 nm. However, mnemiopsin absorbs maximally at 435 nm and in this respect resembles tautomer I. Since tautomer I cannot accommodate a peroxide function at position 2, the mnemiopsin chromophore is viewed as an exception to the luciferin-hydroperoxide model. Furthermore, the *Renilla*-like luciferins, which can be extracted from photoproteins, react with *Renilla* luciferase to produce blue light (490 nm) but *only* in the presence of dissolved oxygen. This fact, and the stability of the extracted luciferins, clearly indicates that they do not exist as linear hydroperoxides in photoproteins.

Based on the evidence cited above we have proposed a new model for photoprotein luminescence. This model, illustrated in Fig.2, involves a free *Renilla* luciferin noncovalently bound to the protein. Luciferin assumes configurations I in mnemiopsin and IV in aequorin, indicating that these proteins confer different environments on the chromophore. This is supported by the fact that aequorin and mnemiopsin emit different colors of light from the same excited state oxyluciferin species (see Fig.2). To account for the lack of a dissolved oxygen requirement for photoprotein luminescence, a site on the protein (not luciferin) is presumed to contain an oxygenated species, possibly a hydroperoxide group. Calcium ion binding would then induce a conformational change in the protein resulting in transfer of the oxygenated species to luciferin to form the key luciferin hydroperoxide intermediate III. In mnemiopsin, this conformational change must also induce a change in luciferin tautomers (from I to IV) prior to the formation of III.

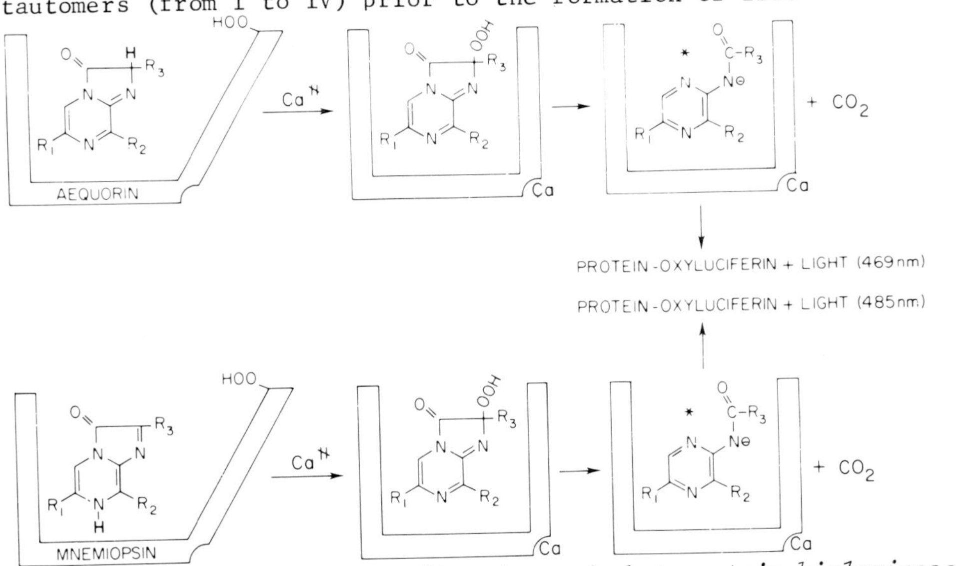

Fig.2. *Proposed scheme for the Ca^{2+}-triggered photoprotein bioluminescence of aequorin and mnemiopsin.*

According to the model illustrated in Fig.2 photoproteins are oxygenases which are unique since, in the absence of Ca^{2+}, they exist as oxygen-containing, enzyme-substrate intermediates that are remarkably stable. The fact that oxygen is incorporated into photoproteins during their formation was demonstrated recently (Shimomura and Johnson, 1975). Aequorin luminescence was first initiated with Ca^{2+} and the apoaequorin was separated from Ca^{2+} and products by gel filtration. The photoprotein could then be 90% regenerated by incubating a mixture of apoaequorin, luciferin, dissolved oxygen and 2-mercaptoethanol (2-ME) for 3 hours. Thus:

$$\text{Apoaequorin} + O_2 + \text{Luciferin} \xrightarrow{\text{2-ME}} \text{Aequorin}$$

The luciferin used in the above regeneration experiment contained a p-hydroxybenzyl group at carbon 2 but we can also regenerate aequorin from apoaequorin using the fully active luciferin analogue shown in Fig.1.

If apoaequorin is incubated as outlined above with O_2, luciferin and 2-ME, but in the *presence* of Ca^{2+}, a long lasting luminescence is observed that is characterized by turnover of aequorin (Shimomura and Johnson, 1975). Thus under these conditions apoaequorin is functioning as a *Renilla*-like luciferase as originally suggested by Anderson *et al.* (1974).

Taken together the data outlined above lends strong support for the photoprotein model illustrated in Fig.2.

The photoprotein model proposed by Shimomura *et al.* (1974) is considerably more complex and we suggest that it is unnecessary. They have proposed the existance of a second chromophore in aequorin, different from luciferin, which they refer to as yellow compound. The existence of yellow compound is based primarily on the fact that pretreatment of aequorin with $NaHSO_3$ allows the extraction of a chromophore which absorbs near 440 nm when dissolved in ethyl ether. They further suggest that this chromophore is responsible for the 454 nm absorption of aequorin and that it functions as an internal oxidation-reduction reagent during the incorporation of oxygen into photoproteins.

Based on our findings outlined above it is clear that the 454 nm absorption of aequorin, as well as the 435 nm absorption of mnemiopsin, can be explained by the solvent-dependent tautomeric equilibrium that occurs with luciferin involving configurations I and IV. We suggest that yellow compound may be either a degradation product of luciferin, or a bisulfite addition product of either luciferin or one of its products.

Therefore an alternative proposal is required in order to obtain a satisfactory mechanism for the incorporation of oxygen into photoprotein during the regeneration experiments, and at least two such proposals can be made. One of these represents an analogy to the mechanism proposed for superoxide dismutase (Fridovich, 1975) in which a metal in the protein is alternately oxidized and reduced by sequential interaction with two molecules of superoxide anion. For example, if photoproteins contained a metal such as copper, iron or manganese it would be possible to view the regeneration as shown in scheme B, Fig.3. In this scheme, O_2 would bind to the metal and temporarily extract an electron forming a superoxide anion adduct. Then during the luminescent reaction (scheme A, Fig.3) O_2 would be transferred to a luciferin carbanion with reduction of the metal to its original state. Superoxide anion has recently been shown to be the substrate for indoleamine 2,3-dioxygenase by Hirata and Hayaishi (1975) lending precedence for involvement of superoxide anion as an intermediate in this type of oxygenase reaction. Supporting evidence for this proposal derives from the fact that reagents which would be expected to react with superoxide anion, such as tetrazolium blue and p-benzoquinone, at final concentrations of 10^{-3} M, will completely inactivate aequorin and mnemipsin if preincubated with the photoprotein prior to Ca^{2+} addition. The incubation time required varies from minutes to hours and depends on the photoprotein. Superoxide dismutase and cytochrome *c* have no effect but there is likely to be an accessibility problem in these cases. *Renilla* luciferase is also inhibited by scavengers of superoxide anion. In this case, immediate inhibitions of 50 - 80% are observed with ferricytochrome *c* (10^{-5} M), p-benzoquinone (10^{-4} M) and tetrazolium blue (10^{-4} M) while superoxide dismutase had no effect. These reagents do not react with free luciferin under the conditions used. The inhibition of both *Renilla* luciferase and photoprotein activities by scavengers of superoxide anion might be expected since these proteins all catalyze essentially the same reactions. In any event, these experiments suggest that a careful examination of metal content in both *Renilla* luciferase and the photoproteins should be made.

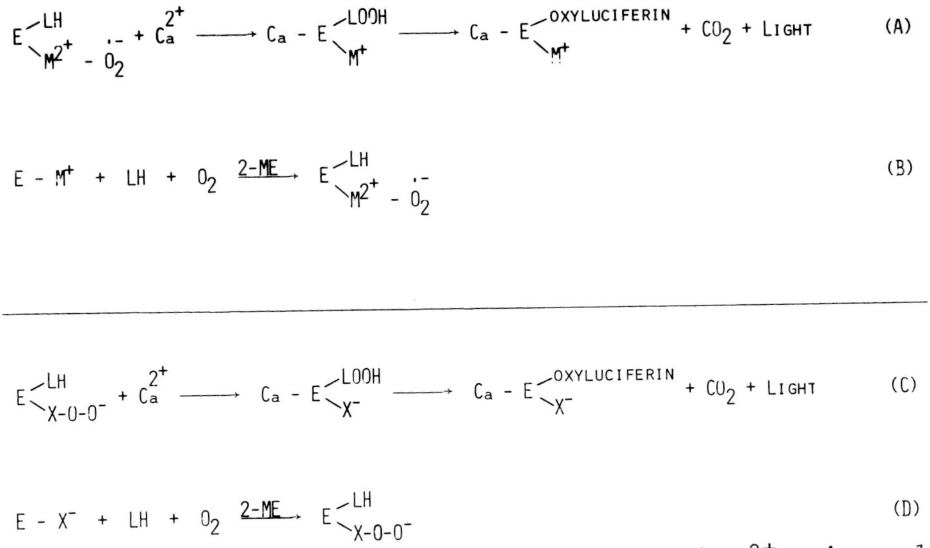

$$E{<}^{LH}_{M^{2+}-\dot{O}_2^-} + Ca^{2+} \longrightarrow Ca - E{<}^{LOOH}_{M^+} \longrightarrow Ca - E{<}^{OXYLUCIFERIN}_{M^+} + CO_2 + LIGHT \quad (A)$$

$$E - M^+ + LH + O_2 \xrightarrow{2\text{-}ME} E{<}^{LH}_{M^{2+}-\dot{O}_2^-} \quad (B)$$

$$E{<}^{LH}_{X\text{-}O\text{-}O^-} + Ca^{2+} \longrightarrow Ca - E{<}^{LOOH}_{X^-} \longrightarrow Ca - E{<}^{OXYLUCIFERIN}_{X^-} + CO_2 + LIGHT \quad (C)$$

$$E - X^- + LH + O_2 \xrightarrow{2\text{-}ME} E{<}^{LH}_{X\text{-}O\text{-}O^-} \quad (D)$$

Fig.3. *Hypothetical schemes for the regeneration, and Ca^{2+}-triggered luminescence, of photoproteins. Abbreviations are: LH, luciferin; $E\text{-}M^+$, apophotoprotein containing a metal M; $E\text{-}X^-$, apophotoprotein containing a stable carbanion X^-; 2-ME, 2-mercaptoethanol; LOOH, luciferin hydroperoxide.*

A second alternative mechanism does not assume the presence of a metal in photoproteins but rather the existence of a stabilized anion derived perhaps from an imidazoyl or sulfhydryl moiety in the protein. Thus during photoprotein regeneration (scheme D, Fig. 3) O_2 would simply add to this stabilized anion, designated by X^-, forming a hydroperoxide. During the Ca^{2+}-triggered luminescence, shown in scheme C, Fig.3, O_2 would be transferred from one anion (X^-) to another (luciferin carbanion) to form the protein bound luciferin hydroperoxide intermediate III.

To be sure the schemes in Fig.3 are hypothetical but they provide viable alternatives to the one proposed by Shimomura and Johnson (1975) and may also form the basis for new experimental approaches.

ACKNOWLEDGEMENTS

This work was supported in part by grants from the National Science Foundation (BMS 74-06914) and the Energy Research and Development Administration (AT 38-1-635). W.W. Ward is an NIH Postdoctoral Fellow (1 F 32-EY-05104-01). We thank Mr. R. McCann for technical assistance.

REFERENCES

1. Anderson, J.M., Charbonneau, H. and Cormier, M.J. (1974). *Biochemistry 13*, 1195.
2. Anderson, J.M., Charbonneau, H. and Cormier, M.J. (1975). *Fed. Proc. 34*, 1610.
3. Anderson, J.M. and Cormier, M.J. (1973). *J. Biol. Chem. 248*, 2937.

4. Cormier, M.J., Hori, K. and Anderson, J.M. (1974). *Rev. Bioenerg. 346,* 137.
5. Cormier, M.J., Hori, K., Karkhanis, Y.D., Anderson, J.M., Wampler, J.E., Morin, J.G. and Hastings, J.W. (1973). *J. Cell. Physiol. 81,* 291.
6. Cormier, M.J., Lee, J. and Wampler, J.E. (1975). *Ann. Rev. Biochem. 44,* 255.
7. Fridovich, I. (1975). *Ann. Rev. Biochem. 44,* 147.
8. Hirata, F. and Hayaishi, O. (1975). *J. Biol. Chem. 250,* 5960.
9. Hori, K., Anderson, J.M., Ward, W.W. and Cormier, M.J. (1975). *Biochemistry 14,* 2371.
10. Hori, K., Wampler, J.E., Matthews, J.C. and Cormier, M.J. (1973). *Biochemistry 12,* 4463.
11. Matthews, J.C., Hori, K. and Cormier, M.J. (1976). *Biochemistry,* in press.
12. Shimomura, O. and Johnson, F.H. (1972). *Biochemistry 11,* 1602.
13. Shimomura, O. and Johnson, F.H. (1973). *Tetrahedron Lett. 31,* 2963.
14. Shimomura, O. and Johnson, F.H. (1975a). *Proc. Nat. Acad. Sci., U.S.A. 72,* 1546.
15. Shimomura, O. and Johnson, F.H. (1975b). *Nature 256,* 236.
16. Shimomura, O., Johnson, F.H. and Morise, H. (1974). *Biochemistry 13,* 3278.
17. Shimomura, O., Johnson, F.H. and Saiga, Y. (1962). *J. Cell. Comp. Physiol. 59,* 223.
18. Wampler, J.E., Hori, K., Lee, J. and Cormier, M.J. (1971). *Biochemistry 10,* 2903.
19. Ward, W.W. and Cormier, M.J. (1975). *Proc. Nat. Acad. Sci., U.S.A. 72,* 2530.
20. Ward, W.W. and Cormier, M.J. (1976). *J. Phys. Chem.,* in press.
21. Ward, W.W. and Seliger, H.H. (1974a). *Biochemistry 13,* 1491.
22. Ward, W.W. and Seliger, H.H. (1974b). *Biochemistry 13,* 1500.

SOD GENES IN HUMANS : CHROMOSOME LOCALIZATION AND ELECTROPHORETIC VARIANTS

PIERRE-MARIE SINET

Hôpital Necker, Service de Biochimie Génetique
149 rue de Sèvres, 75015 Paris, France

INDOPHENOL OXIDASE AND TETRAZOLIUM OXIDASE: FIRST TERMS FOR SUPEROXIDE DISMUTASES

The reduction of tetrazolium salts in the presence of phenazine methosulfate (PMS) is currently used for the staining of certain enzymatic activities after starch gel electrophoresis. These enzymes are either dehydrogenases (glucose-6-phosphate dehydrogenase, 6 phosphogluconate dehydrogenase) or they can be coupled to a dehydrogenase (hexokinase, phosphoglucomutase, adenylate kinase) for the reduction of tetrazolium in the presence of PMS.

After electrophoresis of tissue extracts from human or other organisms, when the staining of one of these enzymes is carried out, in addition to the blue bands of the enzyme under investigation, there appear achromatic areas (the background of the gel is light blue because the reaction is not carried out in total darkness). For a long time, these achromatic bands were ignored.

In 1967, Brewer who had observed the presence in thousands of human hemolysates of one achromatic area of remarkable constancy, found in the red cells of one individual three achromatic bands. The observation of this electrophoretic variant which he found also in the propositus family, allowed him to suggest a protein support to these achromatic bands, the gene of which is autosomal.

Furthermore, he showed that: - in tissues other than the red cells, there are two achromatic areas: the area A (IPOA) identical to that of the red cell band and another, the area B (IPOB) with slower electrophoretic velocity, both determined by separate genetic loci.

- these achromatic bands appear on the gel after addition of PMS and tetrazolium only (and no coenzymes or substrates of the enzymes studied) followed by light exposure.

These proteins, catalysing the "Nadi reaction" (Yonetani, 1963) - a test for idophenol oxidase - were consequently called *indophenol oxidase* (IPO), and also *tetrazolium oxidase* (TO) because they prevent the reduction of tetrazolium to formazan.

Thus study of some properties of IPOs, such as tissue distribution, genic support and genetic polymorphism in human and other species (Baur and Show, 1969) began before knowledge of their enzymatic function.

After the discovery of superoxide dismutase (SOD) (McCord and Fridovich, 1969), Beauchamp and Fridovich (1971) described a specific staining procedure for localizing SOD activity on polyacrylamide gels: SOD inhibits the nitro blue tetrazolium reduction (NBT) by O^-_2 generated by reoxidation of photochemically reduced flavins. Because of the analogy of this method with the staining of IPO by Brewer, these authors suggested that SOD and IPO may be identical. This hypothesis was confirmed by Lippitt and Fridovich (1973). Table I summarizes the properties of the two SODs found in human tissues (Brewer, 1967; Beckman *et al.*, 1973a; Salin and McCord, 1974; McCord *et al.*, 1977).

TABLE I

Summary of Properties of Human SODs

	Metal	No. of subunits	Inhibition by CN^-	Tissue Distribution	Intracellular localization
SOD A	Cu-Zn	2	+	all tissues	essentially cytoplasmic
SOD B	Mn	4	-	all tissues except red cells	mitochondrial and cytoplasmic

CHROMOSOME ASSIGNMENT OF GENES FOR SOD A AND SOD B BY CELL HYBRIDIZATION

The use of cell hybridization techniques (Braski *et al.*, 1960; Migeon and Miller, 1968) has already allowed the chromosome assignment of about 100 genes in humans. After the fusion stimulated by inactivated viruses of somatic cells (fibroblasts or leucocytes) of two different species (human x mouse or human x hamster) and culture in a selective medium of the hybrid cells, one can isolate a number of hybrid clones each identifiable by its human chromosome content. The hybrid cells preferentially lose the human chromosomes, which allows correlation of the presence or absence in different clones of a particular human gene product with the presence or absence of the chromosome which carries this gene. This methodology is possible only if the gene products of human and other species are easily distinguishable. This is the case for SOD A and SOD B after starch gel electrophoresis of human x mouse hybrid cell extracts (Fig.1).

In this way, the human SOD genes were assigned as follows:

- SOD A to chromosome 21 (Tan *et al.*, 1973)
- SOD B to chromosome 6 (Creagan *et al.*, 1973) this being one of the first assignments to a particular human chromosome of a mitochondrial enzyme.

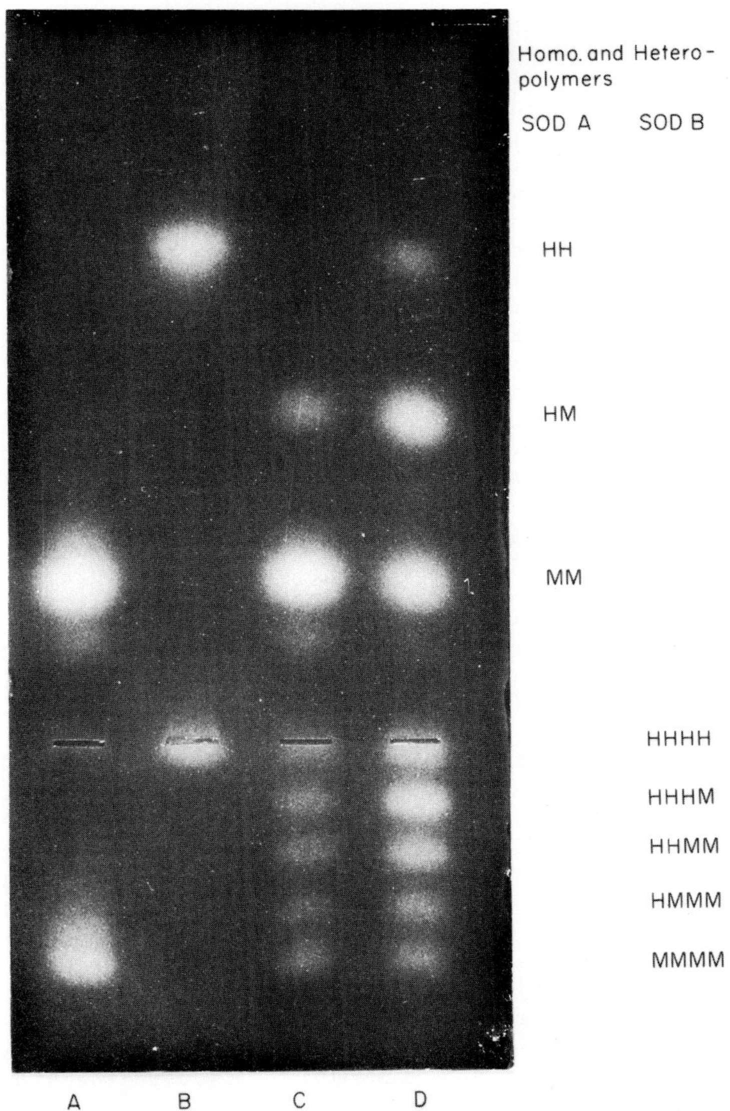

Fig.1. *Starch gel electrophoresis patterns from fibroblast extracts stained for SOD. A: mouse, B: human, C: human x mouse hybrids. (Rebourcet, R., Nguyen, V.C., Weil, D. and Frezal, J.*

SOD A: GENE-DOSAGE EFFECT IN TRISOMY 21 (DOWN SYNDROME)
AND ACCURATE GENE LOCALIZATION ON CHROMOSOME 21.

Before the development of techniques of cell hybridization, the attempts to assign genes to none sexual human chromosomes (autosomes) were limited to the survey of rare individuals with chromosomal abnormality (de Grouchy, 1974). For example, the observation of a quantitative change of a particular enzyme in patients bearing a deletion or a trisomy allowed the supposition that the gene coding for this enzyme was located on the deleted or trisomic chromosome. Although demonstrated in some organisms (bacteria, plants, insects), the correlation between gene dose and gene product in mammalians had little experimental support. In XX and XO mice, Epstein (1969) had demonstrated a direct proportionality between the glucose-6-phosphate dehydrogenase activity and the number of X chromosomes in oocytes in which X inactivation does not occur. In humans, the heterozygotes of certain recessive metabolic diseases have a 50% decrease in the activity of the enzyme which is completely deficient in homozygotes.

Due to gene assignments by cell hybridization, the examples of gene-dosage effects are now greatly increasing, and new techniques of chromosome banding (Caspersson *et al.*, 1970; Dutrillaux and Lejeune, 1971) allow more accurate gene localization on chromosomes. Thus Mayeda *et al.* (1973) observed a decrease of LDH-B (E.C. 1.1.1.27) level in red cells of a patient with a deletion in the short arm of chromosome 12 (12 p); similarly, Rethoré *et al.* (1975) observed an increase of LDH-B in a patient showing a trisomy 12 p. Analysis of cell hybrids has indeed established the assignment of the LDH-B gene to chromosome 12. Another interesting example concerns cell antiviral response to human interferon. The gene for antiviral protein (AVP) induced by interferon has been assigned to chromosome 21 (Tan *et al.*, 1973). In fibroblasts, from monosomic 21, normal and trisomic 21 subjects, it was shown that the sensitivity of antiviral response to human interferon increases with the number of chromosomes 21 (Tan *et al.*, 1974; Tan, 1975).

Because the SOD A gene has been assigned to chromosome 21, it was interesting to detect in trisomy 21 (Down Syndrome) a similar gene-dosage effect. An increase of about 50% in the erythrocyte SOD A activity of these patients has been observed (Sinet *et al.*, 1974; Sichitiu *et al.*, 1974). The ratio of activities $\frac{\text{SOD Tri.21}}{\text{SOD controls}}$, does not differ from 1.5, value of the gene ratio between trisomic 21 and normal subjects (Crosti *et al.*, 1976; Michelson *et al.*, 1977). This increase in activity reflects a higher level of SOD protein (Frantz *et al.*, 1975). In these patients the hematological values and maturity of red cells are apparently normal (Sparkes and Baughan, 1969; Layser and Epstein, 1972). A similar increase of SOD A activity is also observed in blood platelets (Sinet *et al.*, 1975) in which, however, a decrease of SOD B activity is noted. To explain the 50% excess of SOD A in trisomy 21, the most probable hypothesis is that of a gene-dosage effect.

In order to obtain a more accurate localization of the SOD A gene on chromosome 21, we have measured this enzyme in red cells from several cases of partial monosomy and trisomy 21 resulting from malsegregation of a parental translocation (Sinet *et al.*, 1976) (Fig.2). The relationship between chromosome constitution and enzymatic activity indicates a highly probable localization of SOD A gene in sub-band 21 q 22.1, the length of which is one tenth to one fifth of the whole chromosome 21.

Furthermore, the phenotypic analysis of these cases indicates that the trisomy of this single sub-band is responsible for the majority of clinical features of trisomy for the whole chromosome. Nevertheless, it is not possible to conclude at present that excess of SOD A plays a role in the pathogenesis of this disease. There is an apparent discrepancy between the protective role of SOD against O^-_2 and some clinical features of trisomy 21 such as frequent occurrence of leukemia, premature senility (Burger and Vogel, 1973) comparable to Alzheimer's Disease (Schochet *et al.*, 1973), and the short lifetime of cultured fibroblasts (Schneider and Epstein, 1973; Segal and McCoy, 1974). Increased knowledge of the modalities and levels of production of different radical species in the cell (O^-_2 may be a scavenger of other more reactive radicals) will undoubtedly aid the resolution of these problems.

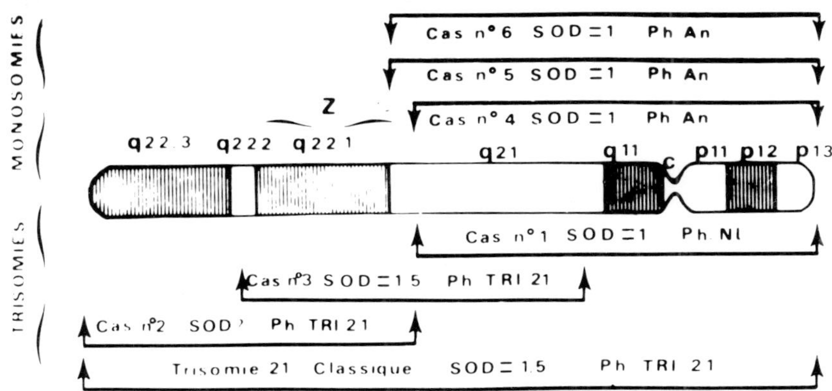

Fig.2. *Chromosome 21 (human) with schematic representation of results of the analysis from cases of monosomies and trisomies. The arrows limit chromosome segments present once in partial monosomies, and in triplicate in partial trisomies. Z, probable localization of the SOD A gene. Ph.N1: normal phenotype; pH.An.: abnormal phenotype; pH Tri 21: trisomy 21 phenotype. SOD = 1 in controls, SOD = 1.5 in complete trisomy 21.*

ELECTROPHORETIC VARIANTS OF SOD A

For numerous enzymes, rare alleles give rise to electrophoretic variants (Harris *et al.*, 1974). The observation of such an SOD A variant led Brewer (1967) to study IPO. Other investigators (Ritter and Wendt, 1971; Welch and Mears, 1972; Kirjarinta *et al.*, 1973; Beckman, 1973b), from routine examination of different populations, have also described an electrophoretic variant of SOD A. The electrophoretic properties of these variants being apparently comparable, Beckman and Parkinen (1973c) proposed the existence of two SOD A genes: SOD^1 and SOD^2 with associated phenotypes: SOD 1, SOD 2-1, SOD 2. In starch gel electrophoresis, at pH 8.6, the phenotype SOD 2 shows a slower anodic mobility than does SOD 1. The phenotype SOD 2-1 shows, in addition to SOD 1 and SOD 2 bands, a hybrid enzyme with intermediate mobility (Fig.3).

The rare SOD^2 variant allele has been found with a relatively high frequency (1 to 3%) only in northern Sweden, in northern Finland, in the Orkney Islands and in Newfoundland (Beckman, 1973b; Beckman and Parkinen, 1973c; Beckman *et al.*, 1973d; Welch and Mears, 1972; Carter *et al.*, 1976). A hypothesis of successive population migrations from a single Scandinavian source could explain these observations. The strict identity of the SOD A variants observed in these different populations is presently under investigation.

Finally, the fact that frequencies of SOD 1, SOD 2-1, and SOD 2 in areas of relatively high frequency of the SOD^2 gene follow the Hardy Weiberg law, and the clinical analysis of families in which this allele segregates (Beckman *et al.*, 1975) indicate that the SOD^2 gene (or genes) is not the cause of particular diseases or selective disadvantage for SOD 2-1 heterozygotes and SOD 2 homozygotes.

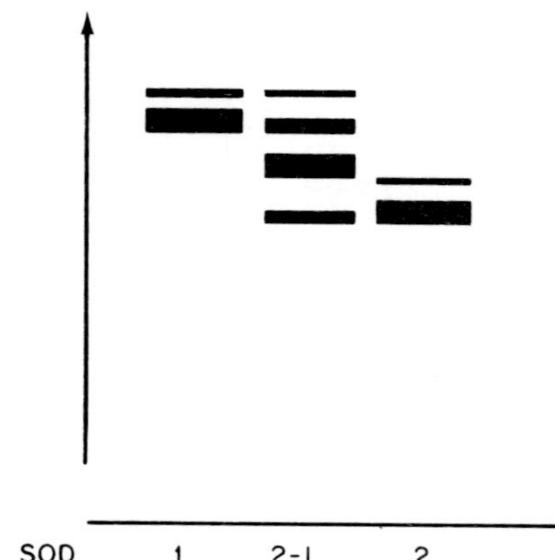

Fig.3. *Schematic figure showing the three phenotypes SOD 1, 2-1 and 2 from red cell extracts. (From Beckman and Parkinen, 1973c).*

REFERENCES

1. Baur, E.W. and Schorr, R.T. (1969). *Science 166,* 1524-1525.
2. Beauchamp, C. and Fridovich, I. (1971). *Anal. Biochem. 44,* 276-287.
3. Beckman, G., Lundgren, E. and Tärnvik, A. (1973a). *Hum. Hered. 23,* 338-345.
4. Beckman, G. (1973b). *Hereditas 73,* 305-309.
5. Beckman, G. and Parkinen, A. (1973c). *Hum. Hered. 23,* 346-351.
6. Beckman, G., Beckman, L. and Nilsson, L.O. (1973d). *Hereditas 73,* 305-310.
7. Beckman, G., Beckman, L. and Nilsson, L.O. (1975). *Hereditas 79,* 43-46.
8. Braski, G., Sorieul, S. and Cornefert, F. (1960). *C.R. Acad. Sci. (Paris) 251,* 1825.
9. Brewer, G.J. (1967). *Amer. J. Hum. Genet. 19,* 674-680.
10. Burger, P.C. and Vogel, F.S. (1973). *Amer. J. Pathol. 73,* 457-468.
11. Carter, N.D. Auton, J.A., Welch, S.G. Marshall, W.H. and Fraser, G.R. (1976). *Hum. Hered. 26,* 4-7.
12. Caspersson, T., Zech, L. and Johansson, C. (1970). *Exp. Cell Res. 62,* 490-492.
13. Creagan, R., Tischfield, J., Ricciuti, F. and Ruddle, F.H. (1973). *Humangenetik 20,* 203-209.
14. Crosit, N., Serra, A., Rigo, A. and Viglino, P. (1976). *Hum. Genet. 31,* 197-202.
15. Dutrillaux, B. and Lejeune, J. (1971). *C.R. Acad. Sci. (Paris) 272,* 2638-2640.
16. Epstein, C.J. (1969). *Science 163,* 1078-1079.
17. Frants, R.R., Eriksson, A.W., Jongbloet, P.H. and Hamers, A.J. (1975). *Lancet 2,* 42-43.
18. Grouchy, J. de (1974). *Ann. Biol. Clin. 32,* 169-178.
19. Harris, H., Hopkinson, D.A. and Robson, E.B. (1974). *Am. Hum. Genet. 37,* 237-253.
20. Kirjarinta, M., Fellman, J., Gustafson, C., Keisala, E. and Eriksson, A.W. (1969). *Scand. J. Clin. Lab. Invest. 22, suppl. 108,* 46.
21. Layser, R.B. and Epstein, C.J. (1972). *Amer. J. Hum. Genet. 24,* 533-543.

22. Lippitt, B. and Fridovich, I. (1973). *Arch. Biochem. Biophys. 159,* 738-741.
23. Mayeda, K., Weiss, L., Lindahl, R. and Dully, M. (1974). *Amer. J. Hum. Genet. 26,* 59-64.
24. McCord, J.M. and Fridovich, I. (1969). *J. Biol. Chem. 244,* 6049-6055.
25. MCord, J.M., Boyle, J., Day, E., Rizzdo, L. and Salin, M., this volume.
26. Michelson, A.M., Puget, K., Durosay, A. and Bonneau, J.C. (1977), this volume.
27. Migeon, B.R. and Miller, C. (1968). *Science 162,* 1005-1007.
28. Réthoré, M.O., Kaplan, J.C., Junien, C., Cruveiller, S., Lafourcade, J. and Lejeune, J. (1975). *Ann. Genet. 18,* 81.
29. Ritter, H. and Wendt, G.G. (1971). *Humangenetik 14,* 72.
30. Salin, M.L. and McCord, J.M. (1974). *J. Clin. Invest. 54,* 1005-1009.
31. Schneider, E.L. and Epstein, C.J. (1972). *Proc. Soc. Exp. Biol. Med. 141,* 1092-1094.
32. Schochet, S.S. Jr., Lambert, P.W. and McCormick, W.F. (1973). *Acta Neuropath. (Berl.) 23,* 342-346.
33. Segal, D.J. and McCoy, E.E. (1974). *J. Cell. Physiol. 83,* 85-90.
34. Sichitiu, S., Sinet, P.M., Lejeune, J. and Frézal, J. (1974). *Humangenetik 23,* 65-72.
35. Sinet, P.M., Allard, D., Lejeune, J. and Jérome, H. (1974). *C.R. Acad. Sci. (Paris) 278,* 3267-3270.
36. Sinet, P.M. Lavelle, F., Michelson, A.M. and Jérome, H. (1975). *Biochem. Biophys. Res. Commun. 67,* 904-909.
37. Sinet, P.M. Couturier, J., Dutrillaux, B., Poissonnier, M., Raoul, O., Réthoré, M.O., Allard, D., Lejeune, J. and Jérome, H. (1976). *Exp. Cell Res. 97,* 47-55.
38. Sparkes, R.S. and Baughan, M.A. (1969). *Amer. J. Hum. Genet. 21,* 430-439.
39. Tan, Y.H., Tischfield, J. and Ruddle, F.H. (1973). *J. Exp. Med. 137,* 317-330.
40. Tan, Y.H., Schneider, E.L., Tischfield, J., Epstein, C.J. and Ruddle, F.M. (1974). *Science 186,* 61-63.
41. Tan, Y.H. (1975). *Nature 253,* 280-282.
42. Welch, S.G. and Mears, G.W. (1972). *Hum. Hered. 22,* 38-41.
43. Yonetani, T. (1963). In "The Enzymes", 2nd Ed. (P.D. Boyer, H. Lardy and K. Myrbäch, Eds.) Chapter 2, pp. 41-70. Academic Press, New York and London.

CLINICAL ASPECTS OF THE DOSAGE OF ERYTHROCUPREIN

A.M. MICHELSON, K. PUGET and P. DUROSAY

Institut de Biologie Physico-Chimique
Service de Biochimie-Physique
13, rue P. et M. Curie
75005 Paris, France

J.C. BONNEAU

Centre Régional de Transfusion Sanguine
et de Génétique Humaine
B.P. 5 - 609, Chemin de la Bretèque - Bois-Guillaume
France

The chemical and biochemical properties of the superoxide radical anion which is produced by all aerobic cells as a normal metabolic intermediate of molecular oxygen during oxidative processes suggest that control of this highly reactive species is a primary necessity. This control is exercised by a class of metallo-proteins, the superoxide dismutases. It is clear that if too little SOD were present in the cell, the steady state level of $O_2^{\overline{\cdot}}$ would rise above normal and in the human this could give rise to increased oxidation of various neuromediators for example, leading to mental aberrations. In addition, damage to membranes both of cells and sub-cellular structures such as mitochondria, particularly in the liver, could cause various hepatic problems.

Conversely, if too much SOD is present in the cell, this would inhibit certain enzymatic oxidations in which $O_2^{\overline{\cdot}}$ is a necessary intermediate, perhaps blocking the normal biosynthesis of neuromediators and thus causing mental debility and other aberrations. Further, the substrate for indolamine dioxygenase appears to be $O_2^{\overline{\cdot}}$ (Hirata and Hayaishi, 1975) and one of the roles of this enzyme is to degrade dimethyltryptamine (a hallucinogen normally present as a metabolic intermediate in the brain) as well as serotonin.

Apart from direct oxidative or reductive degradation of biological molecules by $O_2^{\overline{\cdot}}$, this radical can directly or indirectly (via formation of other free radicals) induce mutations. Indeed, the low level of mutation in humans (40 per day, corresponding to 10^6 nucleotide base changes in 70 years, that is 0.34 mm of nucleic acid modified in 10^{10} kilometers of DNA) may well be due to the protective action of SOD. In addition, we have shown (Cooperman *et al.*, 1977) that an $O_2^{\overline{\cdot}}$ producing system is extremely efficient in inducing covalent cross-linking of the factor of protein initiation IF 3 with the ribosome. Lack of reversibility thus induced affects protein synthesis in general and leads to diminished cell viability (Michelson and Buckingham, 1974). Hence $O_2^{\overline{\cdot}}$ is toxic for the cell at many different levels of cell components and structures - small molecules, macromolecules, the genetic and translational apparatus, and membranes in general. Levels of SOD must therefore be very strictly controlled in order to avoid hepatic and neurological lesions due to excessive quantities of $O_2^{\cdot-}$, without seriously inhibiting essential processes involving this radical as an intermediate.

We have therefore begun a survey of levels of SOD in normal humans and in a wide variety of subjects suffering from various pathological syndromes. In principle, a biochemical explanation for certain types of degenerative processes in the human exists, but has yet to be associated with a specific illness or illnesses. Since red cell erythrocuprein is extremely important in view of the production of $O_2^{\cdot-}$ by autoxidation of hemoglobin (Misra and Fridovich, 1972) in the erythrocyte, we have first considered erythrocyte SOD levels, though in certain cases the study has been extended to the mitochondrial SOD in other cells such as platelets. In addition, it is necessary to consider other enzymes which are included in the group of defense mechanisms against uncontrolled oxidation, such as catalase and glutathione peroxidase. In the erythrocyte, both catalase and erythrocuprein are cytoplasmic (no SOD activity is found in erythrocyte ghosts) whereas glutathione peroxidase is generally considered a membrane bound enzyme.

We have previously presented preliminary results on the level and distribution of erythrocyte SOD in normal humans (Lavelle *et al.*, 1974). These showed a Gaussian distribution of erythrocuprein levels over a very narrow range of values. Subsequently, confirmation of these results was reported by Winterbourn *et al.* (1975), who also indicated that in Wilson's disease (abnormal serum copper) a normal erythrocuprein value was observed in the one case examined. Levels of SOD and catalase in various human tissues (determined immunochemically) obtained by autopsy of three subjects have been described by Hartz *et al.* (1973) but extremely important variations among the three cases were observed, perhaps due to degradative processes subsequent to death.

METHODS

Hemoglobin was estimated as cyanomethemoglobin by addition of 20 μl of lysed erythrocytes to 5 ml of Drabkin's solution. The optical density at 540 nm was measured and compared with a standard curve, using Drabkin solution as blank (Drabkin and Austin, 1935).

Superoxide Dismutase (Erythrocuprein) Estimation

The blood sample (2 ml) was centrifuged for 10 min at 3000 g and the serum removed by suction. Cold water (1 ml) was added to the packed blood cells (volume, 1 ml) to effect lysis. After at least 2 hours lysis at 4°C with gentle stirring, 0.1 ml was removed for estimation of catalase activity and hemoglobin. To the remaining 1.9 ml was added 0.8 ml of a mixture of chloroform and ethanol (3 : 5 v/v) at 0°C to precipitate the hemoglobin, followed by 0.3 ml H_2O, with agitation. The mixture was then centrifuged for 10 min at 3000 g and the clear pale yellow supernatant containing the erythrocuprein was decanted and used directly for estimation of dismutase activity.

This was measured by inhibition of the reduction of nitroblue tetrazolium by $O_2^{\cdot-}$ produced via photoreduction of riboflavin (Beauchamp and Fridovich, 1971).

The incubation mixture contained 2 x 10^{-6} M riboflavin, 10^{-2} M methionine,

2×10^{-5} M KCN and 1.67×10^{-4} M nitroblue tetrazolium in 5×10^{-2} M phosphate buffer, pH 7.8. To this solution (2 ml) was added 10 μl or 5 μl of the extract of erythrocuprein described above. The solution was then irradiated at 365 nm for 2.0 min with a Mineralight B 100 A lamp at about 20 cm distance, the distance being adjusted to give an increase in optical density at 560 nm (after 2 min irradiation) of 0.200 to 0.220 in a corresponding blank in the absence of superoxide dismutase. Measurements are linear up to 50% inhibition.

Taking into consideration the various dilutions and a standard curve established with pure human erythrocuprein the quantity of enzyme as μg of superoxide dismutase per ml of blood was obtained by multiplication of the percent inhibition by 1.1935 for an aliquot of 10 μl extract or by 2.387 for an aliquot of 5 μl.

It may be noted that photoreduction of riboflavin is not necessary to produce $O_2^{\cdot-}$ since NBT is itself photosensitive at 365 nm with production of superoxide radicals. When the riboflavin was omitted from the incubation mixture, and the intensity slightly increased (distance of lamp 0.8 times that used in presence of flavin) to give the same control increase in optical density at 560 nm after 2 min irradiation in absence of SOD, identical results for the dosage of erythrocuprein were obtained (Table I).

TABLE I

Estimation of Erythrocuprein

References	SOD μg/ml With Rb	SOD μg/ml Without Rb	% Difference
149512	69.2	68.2	- 1.4
149514	60.0	59.1	- 1.4
149516	68.1	69.8	+ 2.6
149517	72.7	70.5	- 3.0
MIC	115.3	112.6	- 2.3

In the case of platelets the SOD activities were measured by a more sensitive method involving inhibition of the chemiluminescent reaction produced in the system O_2/hypoxanthine/xanthine oxidase/luminol (Puget and Michelson, 1974). Activity was measured in absence and in presence of 2 mM cyanide (to inhibit the Cu-SOD). About 0.2 ng of erythrocuprein can be readily estimated by this technique.

Estimation of Catalase Activity

An aliquot of the lysate of red cells described above was diluted 100 fold in water and 10 μl of this solution was added to 5 ml of 2.1×10^{-5} M H_2O_2 in 10^{-2} M K_2 HPO_4 pH 7.8. The solution was incubated at 25°C for 20 min and then residual H_2O_2 (and hence catalase activity) was determined by measurement of light emitted by oxidation of luminol catalysed by horse radish peroxidase, using the following technique.

The cuvette contained 10 μg of horse radish peroxidase (Boehringer) in 3 ml of 10^{-4} M EDTA, 10^{-4} M luminol in 0.1 M phosphate pH 7.8. Reaction was initiated by injection of 1 ml of the above solution of H_2O_2 incubated 20 min with the erythrocyte extract and values of Imax and total light (L) emitted (integration with time) measured with the apparatus previously described (Henry *et al.*, 1970). Calibration curves (Fig.1) were established by using 1 to 15 ng of crystalline beef liver catalase (Boehringer, 50 000 units/mg) per ml of 2.1×10^{-5} M H_2O_2.

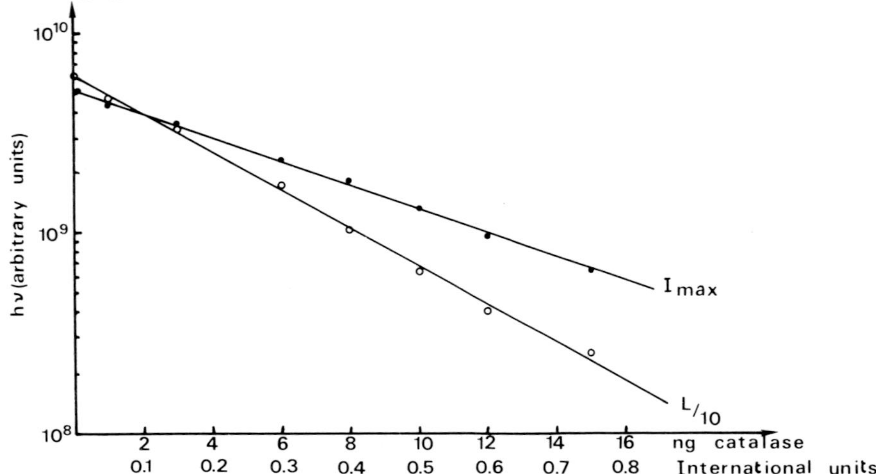

Fig.1. *Calibration curves (Imax and total light, L) for the estimation of catalase activity.*

Imax and L are expressed in arbitrary, but constant units, and catalase as μg per ml of blood is given by the expressions 50 000 [(1.6532 - log. I) x 0.0167 + 0.001] for Imax and 50 000 [(1.6902 - log. L/10) x 0.0105 + 0.001] for total light emitted. The average of the two estimates (which should not differ by more than 5%) was used and calculations were effected with a suitable programme for the Hewlett Packard 65. The results are correct within a range of 1 to 15 ng of catalase/ml, that is, to femtomole quantities of catalase.

Preliminary experiments indicated that hemoglobin in amounts equivalent to the blood sample used (0.02 μl) had no effect on the determinations. No variation was observed after storage of the erythrocyte lysate for 4 days at 4°C. The results are expressed in terms of μg amounts of equivalent crystalline beef liver catalase rather than human. In principle an expression of the type

$$K = \frac{1}{t} \log_{10} \frac{A}{A - x}$$

where A represents initial concentration of H_2O_2 and A - x the residual concentration, should be used. However, in the region of catalase concentration measured, the relationship between residual H_2O_2 (expressed in logarithmic terms) and activity is sufficiently linear to be used directly, since log A - x is a linear function of log $\frac{1}{A - x}$. It is to be noted that the initial concentration of H_2O_2 is sufficiently low that catalase is not destroyed, as occurs at higher concentrations of H_2O_2. The results are generally expressed in μgs of erythrocuprein or catalase per ml of blood and per gm of hemoglobin. Other expressions such as enzyme per erythrocyte are less satisfactory due to variations of hemoglobin per cell. Since the major source of $O_2^{\bar{\cdot}}$ in the erythrocyte is probably autoxidation of hemoglobin we prefer the above system, despite the possibility of variations in cellular hemoglobin. This problem has been discussed by Winterbourn *et al.* (1975) and by Beutler (1971) and there is general agreement that expression of the results per gm of hemoglobin is not only easier, but most accurate.

RESULTS AND DISCUSSION

Normal Subjects

Levels of SOD and catalase were determined in blood samples from 200 white French rural subjects. The results are shown in Fig.2. It can be seen that a very sharp distribution of SOD/ml blood (average 58.1 μg) or SOD/g Hb (average 461.4 μg) is present, as reported earlier (Lavelle *et al.*, 1974), with a standard deviation of less than one tenth the average value. The value of 461.4 μg SOD/g Hb is close to that estimated by Winterbourn *et al.* (1975) of 500 μg, but lower than that reported (700 μg) by Stansell and Deutsch (1965). The average value for catalase was 4.043 mg/g Hb with a standard deviation of about 16% of this average. One case of partial acatalasia was found (198 μg/ml of blood, 1.554 mg/g Hb), not shown in Fig.2c and d.

Superoxide dismutase is calculated in terms of μg of a preparation of human erythrocuprein. An earlier preparation, though pure with respect to protein, was of low specific activity. This led to an earlier published figure of one molecule of erythrocuprein per 600 molecules of hemoglobin (Lavelle *et al.*, 1974). The present work indicates that the true value is one molecule of erythrocuprein per 1127 molecules of hemoglobin. The results of catalase determinations of some 200 normal subjects indicate that in the erythrocytes there is on average one molecule of catalase per 959 molecules of hemoglobin. Given the relationship that an average erythrocyte contains about 290 x 10^6 molecules of hemoglobin it may thus be calculated that about 257 000 molecules of erythrocuprein and 302 000 molecules of catalase are present per normal erythrocyte.

As shown in Table II no significant variations with respect to sex or age[1] were detected for levels of SOD/g Hb within the population studied, though it may be noted that the average of seven subjects older than 60 years was 420 μg/g Hb compared with 461 μg for the total population (average Hb/100 ml of blood was 12.76 gms compared with a global average of 12.67 gms).

Since hemoglobin itself is perhaps the principle source of $O_2^{\bar{\cdot}}$, correlation of SOD levels with the amount of hemoglobin per 100 ml of blood was attempted. The results are shown in Table III. It can be seen that there is a tendency for the catalase/g Hb to increase with decrease in hemoglobin and a similar relative increase in SOD/g Hb with decrease in hemoglobin is even more marked (431.2 μg SOD/g Hb at 14.37 g Hb/100 ml and 506.6 μg SOD/g Hb at 10.83 g Hb/100 ml).

Hemodialysis

Since hemodialysis reduces the hemoglobin level it was of interest to study such cases. Blood samples from 24 subjects were examined; the results are shown in Fig.3. The distribution of SOD levels in the population remains remarkably sharp but there is a significant increase to 540.3 μg/g Hb coincident with the low hemoglobin values (mean 8.01 g Hb/100 ml, compared with the figure for the normal population $\bar{x}$ = 12.67, Sx = 1.31, $S\bar{x}$ = 0.09) whereas catalase/g Hb is reduced (Table III).

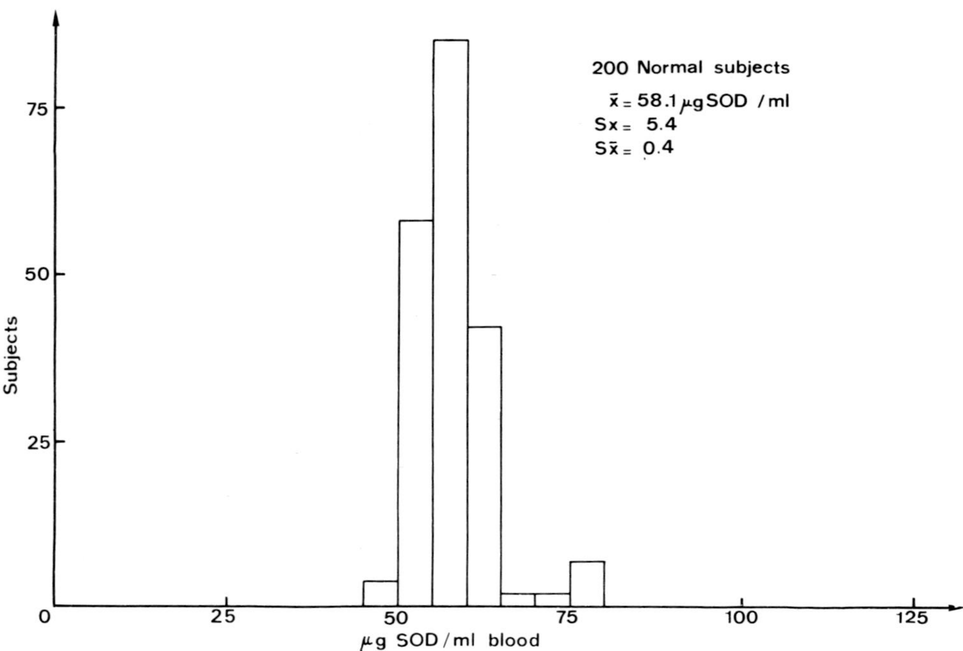

Fig.2a.

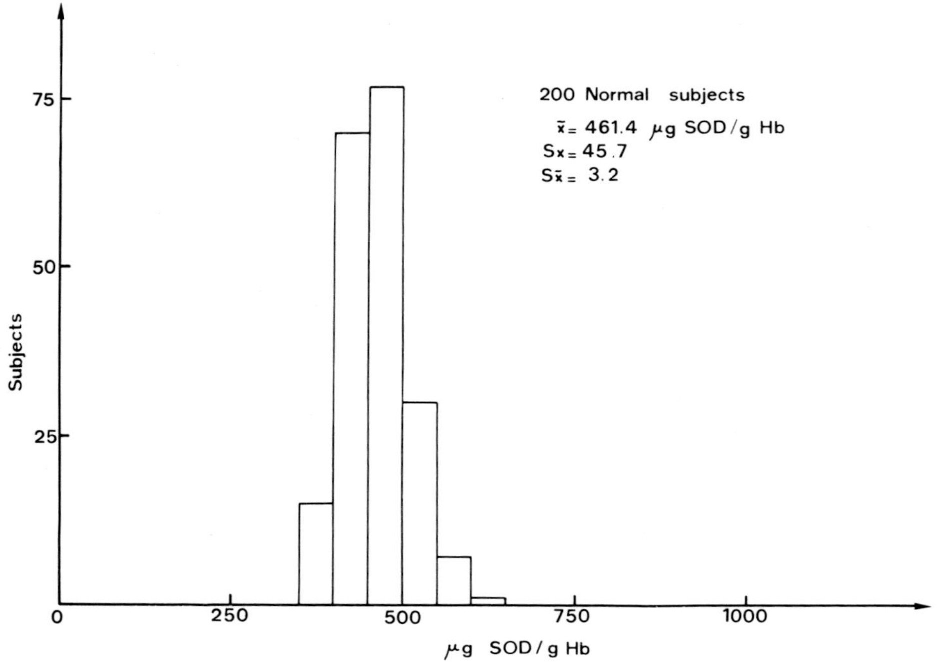

Fig.2b.

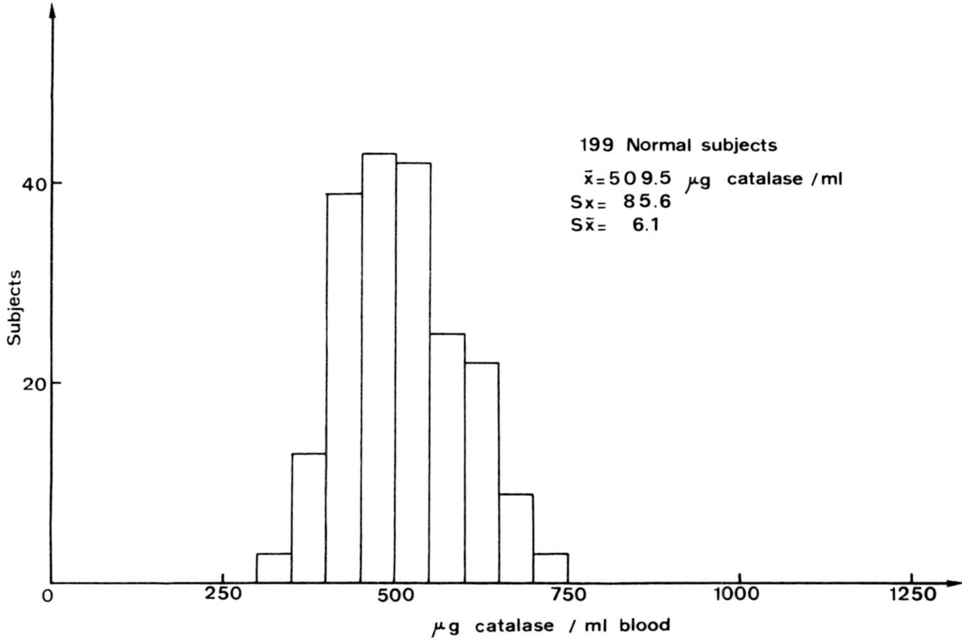

Fig.2c.

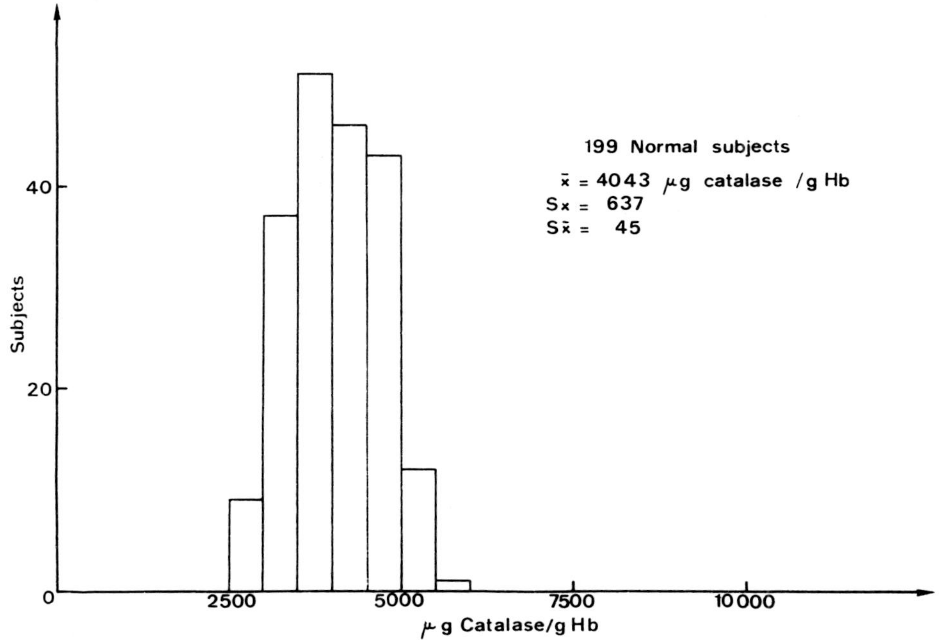

Fig.2d.

Fig.2 - *Distribution of values of erythrocyte SOD a) per ml of blood, b) per g of hemoglobin; and of catalase, c) per ml of blood, and d) per g of hemoglobin in a normal French rural population (200 subjects).*

24 Hemodialysis cases
x̄ = 43.2 μg SOD/ml
Sx = 5.5
Sx̄ = 5.1
Subjects
10
5
0
25
50
75
100
125
μg SOD / ml blood

Fig.3a.

24 Hemodialysis cases
x̄ = 540.3 μg SOD / g Hb
Sx = 37.5
Sx̄ = 7.7
Subjects
15
10
5
0
250
500
750
1000
μg SOD/ gHb

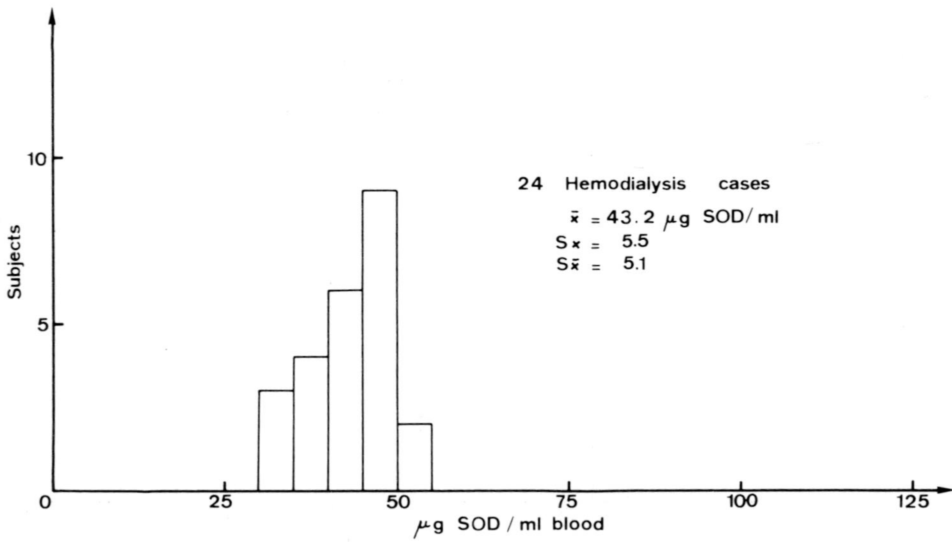

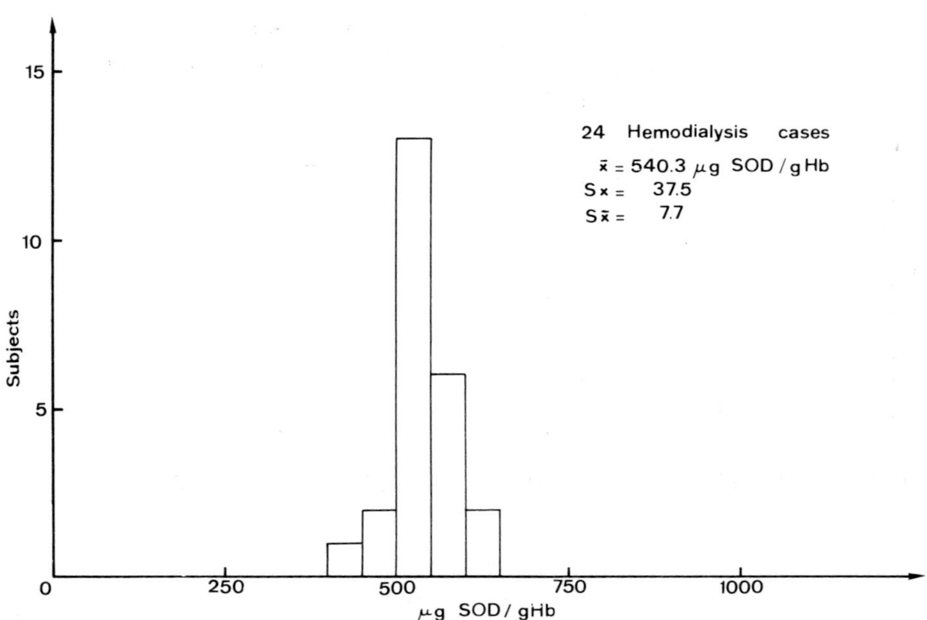

Fig.3b.

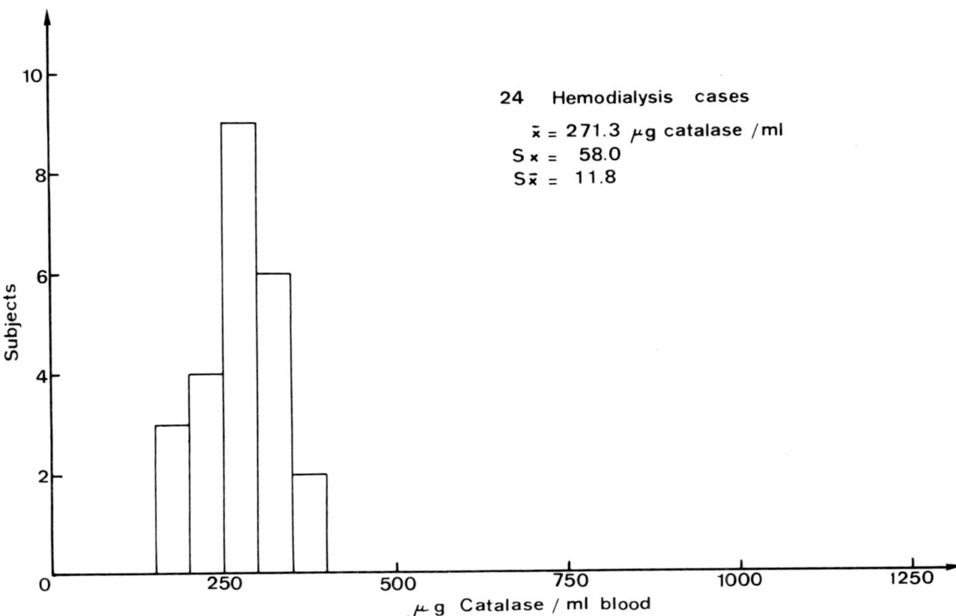

Fig.3c.

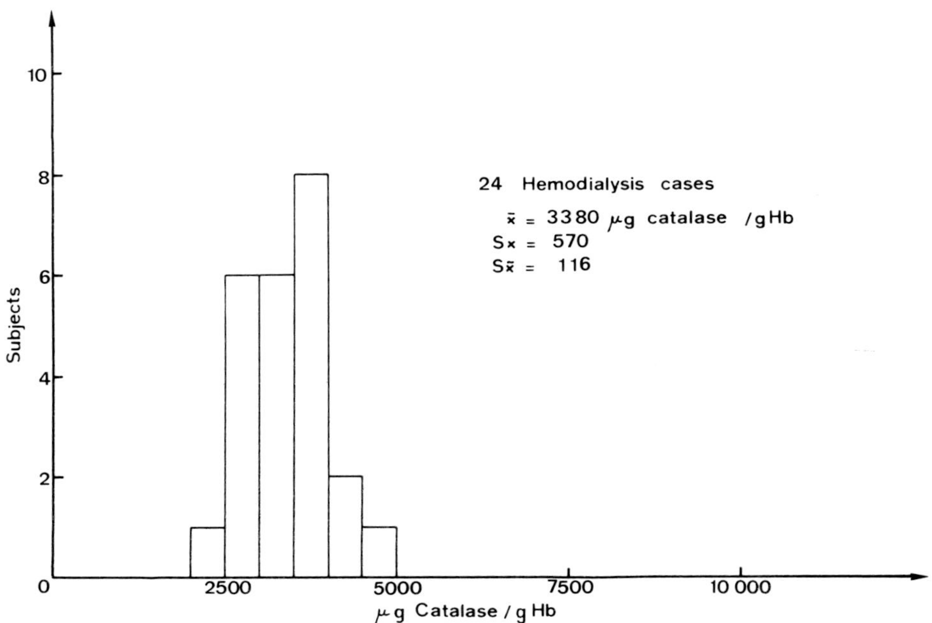

Fig.3d.

Fig.3 - *As for figure 2 for 24 hemodialysis cases.*

TABLE II

Average values μg SOD/g Hb				
	$\bar{x}$	Sx	$S\bar{x}$	No. of Subjects
Males	459.1	50.3	4.7	114
Females	467.4	39.6	4.3	85
Age group (years)				
10 - 20	466.0	-	-	1
20 - 29	467.8	43.3	5.3	68
30 - 39	463.3	51.6	7.0	54
40 - 49	467.0	50.2	8.25	37
50 - 59	454.6	36.3	6.4	32
60 - 69	419.6	26.7	10.1	7
Total population	461.4	45.8	3.2	200

TABLE III

Influence of Hemoglobin on Erythrocyte SOD and Catalase Levels

	Average Hb per 100 ml		μg SOD/g Hb		μg Catalase/g Hb		No. of
	$\bar{x}$	Sx	$\bar{x}$	Sx	$\bar{x}$	Sx	Subjects
Hb	12.67	1.3	461.4	45.8	4033	638	200
≥ 13.7	14.37	0.9	431.2	44.6	3860	654	46
≥ 12.7	13.64	0.9	442.9	41.9	3920	649	103
≤ 12.6	11.64	0.8	481.0	41.6	4172	603	97
≤ 11.6	10.83	0.6	506.6	45.6	4254	572	39

Standard error of the mean lies between 3.2 and 7.3 μg for the SOD levels and 45 - 97 μg for the catalase

Hemodialysis patients	8.01	1.1	540.3	38.3	3380	582	24

Hemochromatosis

Five patients were examined. Both erythrocuprein and catalase were completely normal (Table IV).

Cancer Patients

Estimation of erythrocyte erythrocuprein in the blood of 22 cancer patients before surgical treatment showed a completely normal pattern of enzyme levels.

TABLE IV

	g Hb/100 ml		μg SOD/ml		μg SOD/g Hb		μg Catalase/ml		μg Catalase/g Hb		No. of Subjects
	$\bar{x}$	Sx	$\bar{x}$	Sx	$\bar{x}$	Sx	$\bar{x}$	Sx	$\bar{x}$	Sx	
Hemochromatosis	11.72	0.9	54.7	1.8	468.0	28.4	452.8	65.2	3856	396	5
Splenomegaly myeloid	11.09	3.06	51.1	5.4	480.0	91.7	502.0	135.3	4535	226	4
Alcoholic cirrhosis	12.60	1.41	55.1	5.3	439	29.3	588.9	83.1	4702	660	8
Mothers	11.12	1.18	55.8	5.2	504.2	44.1	389.4	56.8	3614	472	6 SOD }
Children	11.95	1.74	59.9	8.4	501.2	18.9	404.0	79.8	3480	408	5 catalase }
Normal population	12.67	1.31	58.1	5.4	461.4	45.8	509.5	85.8	4043	638	200

Splenomegaly Myeloid

Blood from four patients was examined. No significant differences of SOD and catalase from normal levels were observed (Table IV).

Alcoholic Cirrhosis

Eight cases were examined. As shown in Table IV the levels of SOD and catalase per gm Hb were not significantly different from those of a normal population.

Porphyria

Two cases of porphyria were available. Both presented completely normal erythrocyte SOD and catalase levels (435, 431 μg SOD and 3216, 4147 μg catalase per g Hb).

Vaquez's Disease (Polycythemia Rubra Vera)

Examination of two cases showed a large increase in SOD/ml of blood (96.4 and 94.8 μg/ml) corresponding to the large increase in number of erythrocytes. The SOD/g Hb was also increased (to 634 and 585 μg/g Hb) though with a normal population, increase in hemoglobin per ml of blood (15.2 and 16.2 g/100 ml in these two cases) is accompanied by a decrease in SOD/g Hb. It thus appears that whereas the quantity of SOD per erythrocyte is normal, relative to hemoglobin there is an increase. Catalase levels were essentially normal (4347 and 4581 μg/g Hb and 661, 742 μg/ml corresponding to the increased number of erythrocytes).

Trisomy 18

A single case of trisomy 18 was examined. Despite the fact that the gene for erythrocuprein is on chromosome 21 (and for the mitochondrial Mn-SOD on chromosome 6) and hence a normal erythrocyte SOD level would be expected, the value was excessively high when expressed per g Hb (691 μg SOD) though due to the low hemoglobin (7.1 g/100 ml) in this case, the SOD/ml of blood (49.1 μg) was quite normal. Catalase (323 μg/ml; 4546 μg/g Hb) fell within normal limits.

Mothers and Children

Blood samples were taken from six mothers and from the six new born children (umbilical cord), all cases being completely normal. No major differences between mother and child in SOD or catalase levels were observed and indeed the figures were in each case remarkably similar. Mean values for the mothers and for the children are shown in Table IV. It has been reported (Autor and Roberts, 1974) that fetal lung tissues (and lung tissue from infants succumbing to hyaline membrane disease) contain a lower level of SOD activity than normal human adult lung tissue. Erythrocyte SOD levels in normal neonatal infants were slightly higher than those of premature infants (Autor and Roberts, 1974).

Trisomy 21

Since the gene for erythrocuprein is located on chromosome 21 in humans it was of considerable interest to study trisomy 21 mongoloids, to see whether a direct gene-dosage effect exists (though considerable caution must be exercised in interpretation of any results due to the control and regulation of what is in fact a very complex system involved in oxygen metabolism). Forty two subjects were available but one showed completely normal erythrocyte SOD and catalase levels (No. 148536; 12.5 g Hb/100 ml; 56.2 μg SOD/ml; 449 μg SOD/g Hb; 504 μg catalase/ml and 4028 μg catalase/g Hb) and was not taken into consideration since this case is possibly a partial trisomy 21 (the caryotype has not as yet been determined), in which the third chromosome lacks that part containing the gene for Cu-SOD. Ages ranged from 6 - 48 years. Hemoglobin levels were normal ($\bar{x}$, 12.55 g Hb/100 ml; Sx 1.13; S$\bar{x}$ 0.18).

The results are presented in Fig.4. It can be seen that erythrocyte catalase levels both with respect to mean values and range and distribution of levels in the population are completely normal (in confirmation of Pantekalis *et al.*, 1970) and indeed Figs.4c and 4d can be essentially superimposed on Figs.2c and 2d, thus providing an internal control. (Ratio of mean values of catalase in trisomy 21 to normal = 0.95). However, the erythrocuprein levels are completely displaced (Figs.4a and 4b) to higher values, with ratios of the mean values $\frac{\text{Trisomy 21}}{\text{Normal}}$ = 1.47 both for SOD/ml and SOD/g Hb, in accord with a gene dosage relationship. In view of other cases of high SOD levels (reported in this chapter) this result cannot be taken as proof of such a relationship. However, at least some of the pathological effects may be due to high levels of SOD since in cases of partial trisomy 21 in which the segment of chromosome containing the gene for erythrocuprein is absent, many of the morphological and mental anomalies are considerably reduced (Sinet, private communication). It would be of interest to attempt to correlate severity of the symptoms with the level of SOD in each case (since this ranges from the extreme upper limit of that of a normal population to a value of 1.8 to 1.9 times the normal mean) but for various technical reasons this has not been possible. As with a normal population, no sex-linked character in trisomy 21 cases was observed (for 24 female subjects $\bar{x}$ = 691.2 μg SOD/g Hb, standard deviation = 64.3 and for 15 male subjects $\bar{x}$ = 653.0 μg SOD/g Hb, Sx = 57.0.

During the course of this work similar results (based on fewer cases) have been reported by Sinet *et al.* (1974), by Rotilio (private communication), and by Frants *et al.* (1975). It may be noted that hemoglobin values, erythrocyte numbers, red cell volumes and maturity of erythrocytes in Trisomy 21 cases are essentially normal (Sparkes and Baughan, 1969; Layzer and Epstein, 1972).

These results have been further extended to estimation of superoxide dismutase activities (cytoplasmic and mitochondrial) in platelets from nine normal subjects and eleven trisomic 21 patients of similar age (Sinet *et al.*, 1975a). Activity stained gel electrophoresis of extracts of human platelets show identical profiles from normal or trisomic 21 subjects. Two distinct bands are present and have Rf values corresponding to those of copper-zinc and manganese SOD from human leukocytes (Salin and McCord, 1974). In addition, we found that all the SOD activity is fully retained after the organic solvent treatment, both enzymes being resistant to chloroform-ethanol. Human manganese SOD thus appears to be different from other mitochondrial Mn-SODs, such as the chicken liver enzyme, which are definitely destroyed by such treatment (Weisiger and Fridovich, 1973). (See also Autor and Roberts, 1974, for a similar observation with human lung tissue). We have also observed this resistance to both cyanide and to organic solvent treatment[2] with SOD isolated from the protozoa, *Tetrahymena pyriformis* (Lavelle and Michelson, unpublished observations).

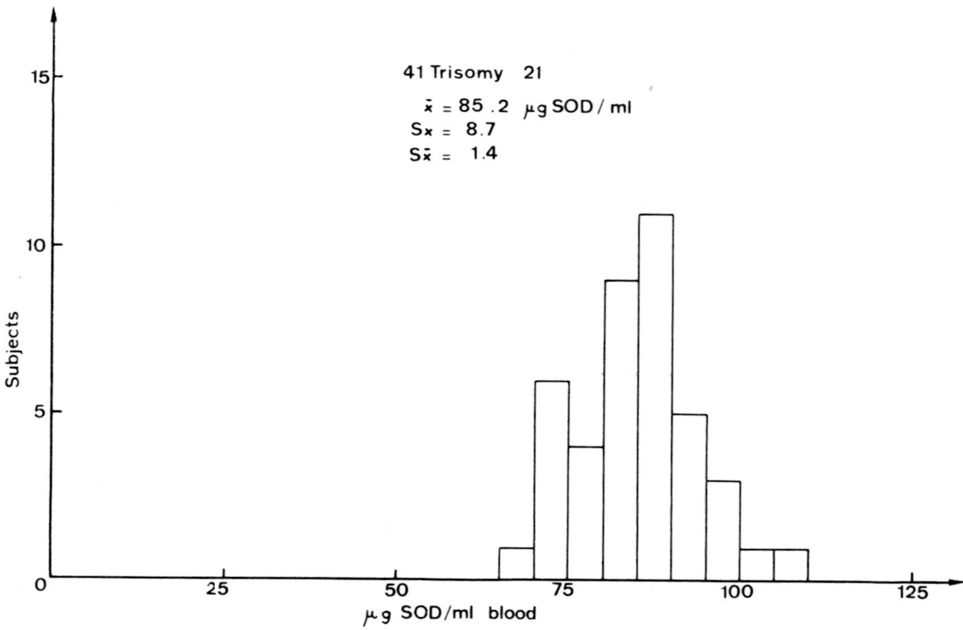

Fig.4a.

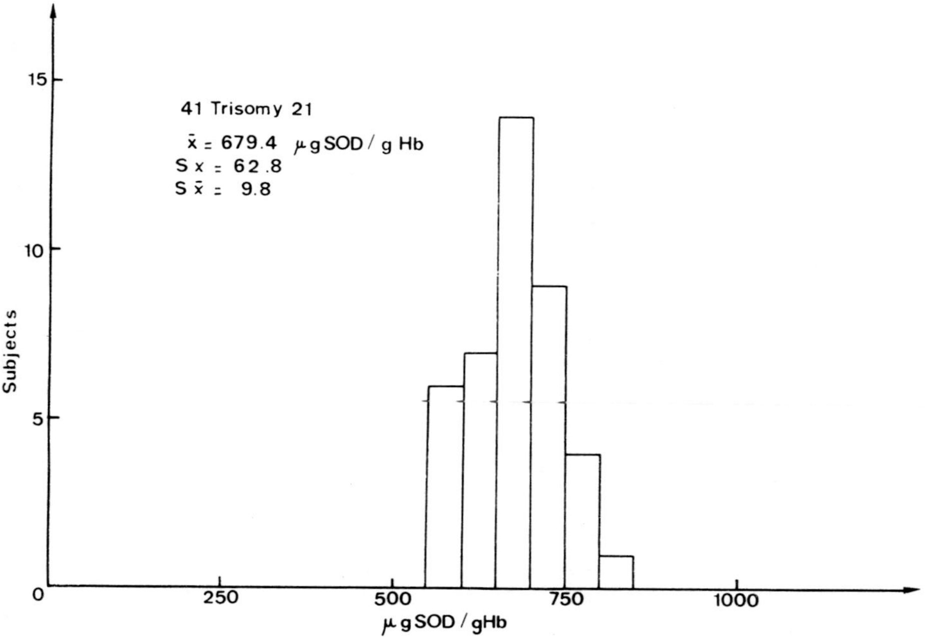

Fig.4b.

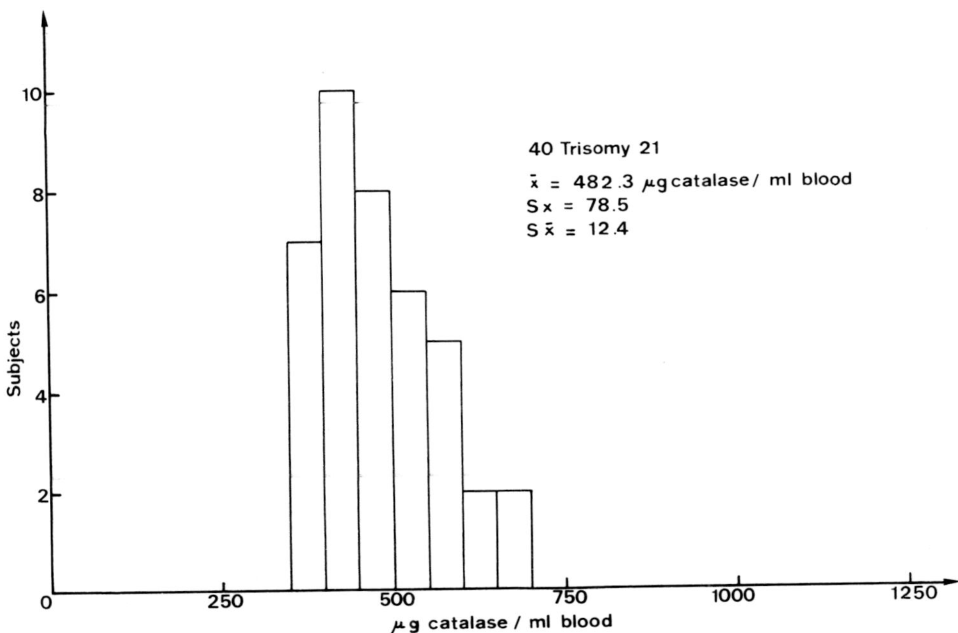

Fig.4c.

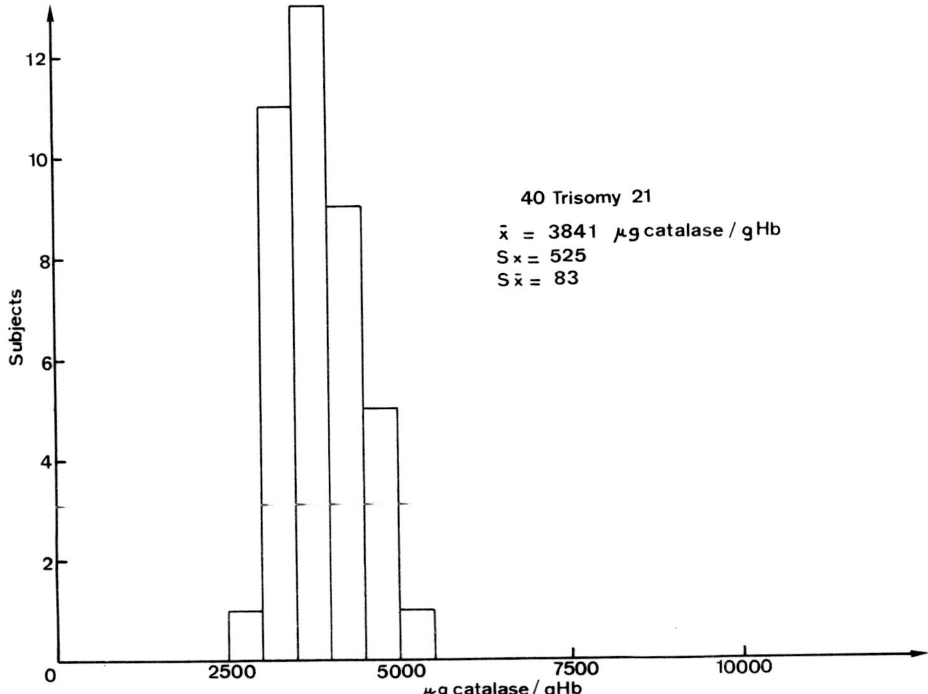

Fig.4d.

Fig.4 - *As for figure 2 for 41 cases of trisomy 21.*

The results of estimations of SOD activity in platelets from controls and from trisomy 21 patients are shown in Table V. The cytoplasmic SOD activities are significantly increased in platelets of trisomic 21 patients. This finding might be related to modifications of platelet properties in trisomy 21 such as their density (Kamoun and Jerôme, 1975) and perhaps their life span. Thus, in leucocytes which have a modified life span in trisomy 21, enzymes such as alkaline phosphatase and even glucose-6-phosphate dehydrogenase which is sited on the X-chromosome, are increased. However, the levels of these enzymes are unaltered in the platelets (Hsia *et al.*, 1971). On the other hand, the ratio of the average values of trisomic 21 SOD (Cu)/ control SOD (Cu) is 1.56, which is comparable with the ratio observed in erythrocytes, cells in which no difference in maturity characteristics have been discerned between normal subjects and trisomy 21 patients (Layser and Epstein, 1972; Sparkes and Baughan, 1969). Thus, in platelets as in the erythrocytes, the increase of the SOD (Cu) activity in trisomy 21 is probably due to a direct gene-dosage effect.

TABLE V

SOD Activities in Platelets From Normal Controls and Trisomic 21 Patients

	SOD activities in units/mg protein (mean ± standard deviation)		P	Ratio: $\frac{\text{Trisomy 21}}{\text{Controls}}$
	Controls n = 9	Trisomic 21 Patients n = 11		
Total SOD	46.73 ± 11.5	67.10 ± 13.52	<0.01	1.44
SOD (Mn) (mitochondrial)	6.44 ± 1.65	4.31 ± 1.70	<0.02	0.67
SOD (Cu) (cytoplasmic) = Total SOD minus SOD (Mn)	40.30 ± 10.90	62.79 ± 12.93	<0.001	1.56

The mitochondrial (manganese) SOD activity is significantly decreased in platelets from trisomic 21 subjects. The analysis of this finding is made difficult by the lack of reports relating to the properties of mitochondria in trisomy 21. Only the platelet mono-amine oxidase has been studied (Benson and Southgate, 1971; Lott *et al.*, 1972; Paasonen *et al.*, 1964) but the results are conflicting, and depend on the type of substrates used for assays. Yet these data, which may appear surprising because the SOD (Mn) gene is located on chromosome 6, raise the hypothesis of a possible regulation of this enzyme by the intra-cytoplasmic levels of SOD (Cu) and/or superoxide anions. Moreover, an increase in the superoxide radical concentration in mitochondria could well occur as a consequence of the reduced SOD (Mn) level and this might be the origin of damage in the mitochondria and consequently in cell function in trisomy 21. As part of a general programme on the pathological aspects of oxygen metabolism, we have also studied erythrocyte glutathione peroxidase activities in trisomy 21 cases (Sinet *et al.*, 1975b). In the red blood cell, H_2O_2 produced biochemically or otherwise, can be eliminated by the action of various enzymes such as catalase and glutathione peroxidase (GPX). It is generally considered (Cohen and Hochstein, 1963) that in erythrocytes, at low concentrations of H_2O_2 the second enzyme catalyses reduction of H_2O_2 preferentially by the reaction

$$2\ GSH + H_2O_2 \xrightarrow[GPX]{} GSSG + 2\ H_2O$$

Catalase activity in erythrocytes from cases of trisomy 21 is completely normal and therefore levels of glutathione peroxidase in such patients were examined.

Twelve trisomy 21 subjects free of congentical cardiopathy were compared with eighteen normal individuals of similar age, all older than three years. The difference between average levels of glutathione peroxidase in the two populations is highly significant (Table VI) and the ratio of the averages for trisomy 21 compared with normal subjects is 1.55, closely similar to that observed in the case of determinations of erythrocuprein. In contrast with the clear cut difference in distribution for erythrocuprein, the levels of glutathione peroxidase acitivity for trisomy 21 cases and normal subjects partially overlap.

Two hypotheses may be presented at the moment to explain the concurrent increase of activity to the same extent for glutathione peroxidase and erythrocuprein. The first implies that the gene for expression of glutathione peroxidase is located on chromosome 21, as is that for erythrocuprein, while the second invokes participation on intracellular $O_2^{\bar{\cdot}}$ or H_2O_2, or of superoxide dismutase, in the regulation of erythrocyte glutathione peroxidase activity. With respect to the effect of superoxide dismutase, this enzyme can in fact increase intracellular levels of H_2O_2 by catalysed dismutation of $O_2^{\bar{\cdot}}$ as opposed, not to spontaneous dismutation (which gives the same quantity of H_2O_2), but to elimination of $O_2^{\bar{\cdot}}$ by oxidative processes or by diffusion (given the relatively long life time of $O_2^{\bar{\cdot}}$ and the kinetics of dismutation by superoxide dismutase).

TABLE VI

Comparison of Glutathione Peroxidase Activity in Erythrocytes From Normal Subjects and Trisomy 21 Patients

	Number of Subjects	µM NADPH/min/g Hb Average	Standard Deviation	Standard Error of the Mean	P
Normal	18	5.73	1.54	0.36	<0.001
Trisomy 21	21	8.89	1.91	0.55	

Ratio T 21/Normal = 1.55

(The variation of glutathione peroxidase levels in various cases of hepatic illnesses generally leading to severe hemolysis is currently under investigation in collaboration with Dr. Najman, Hôpital St. Antoine, Paris.

No significant relationship[3] between very low levels of erythrocyte glutathione peroxidase - less than one third normal - and values for SOD or catalase is observed.)

With respect to trisomy 21 cases a final point may be mentioned. A preliminary report (Corberand *et al.*, 1974) indicates that spontaneous production of $O_2^{\overline{\cdot}}$ (as measured by reduction of nitroblue tetrazolium) by resting trisomy 21 phagocytes is significantly higher than in normal subjects (ratio 1.46) but that during active phagocytosis this difference disappears (24 children, 3 - 19 years old). Hence the possibility that increased sensitivity of trisomy 21 infants to infectious diseases is due to lack of $O_2^{\overline{\cdot}}$ during phagocytosis is eliminated. There is an increased reduction of NBT by neutrophils of new born infants (Humbert *et al.*, 1970).

Psychiatric Cases

Since in cases of trisomy 21, high SOD levels are associated with mental debility, it was of interest to examine various mental disorders. Again, as in all the previous cases, subjects were rural whites (Normandy), and all suffered severe mental illness. Estimations of erythrocyte SOD and catalase of 30 patients are presented in Fig.5. A significant increase in SOD was noted (27% increase of the mean value compared with the mean of a normal population) together with a smaller increase (18%) in catalase levels.

The highest value of SOD (714 µg/g Hb) was found in a case of paranoid psychosis, which also revealed a very high catalase level (5872 µg/g Hb). Schizophrenics (6 cases) showed levels of 504 - 682 µg SOD/g Hb while various other psychoses (manic depressive, chronic hallucinations, Korsakoff syndrome, paranoic delirium, mental debility, etc. ...) gave values in the range 541 - 672 µg SOD/g Hb and 4158 - 5799 µg catalase/g Hb. The age range was 23 - 84 years. No sex linked differences were detected and hemoglobin levels were normal ($\bar{x}$ 12.54 g Hb/100 ml; Sx 0.85; S$\bar{x}$ 0.15).

It thus appears that various mental diseases are related to increased SOD levels, suggesting a possible novel therapeutic approach using inhibitors of SOD. However, this survey requires considerable amplification and a thorough clinical study of SOD levels related to specific forms of mental illness.

Age and Senility

Estimation of erythrocyte SOD and catalase in 23 old people (70 - 99 years) in excellent, robust health gave results indistinguishable from those of the normal rural population examined as control as shown in Fig.6. Hemoglobin levels were normal ($\bar{x}$ 11.81 g Hb/100 ml; Sx 1.57; S$\bar{x}$ 0.33) though slightly low in certain cases.

A somewhat different aspect was seen when 34 old people (54 - 94 years, 7 males and 34 females) in very bad health, requiring clinical care and generally in pathological condition, were examined. The hemoglobin levels were normal (mean value 12.08 g Hb/100 ml lysate; Sx 1.45; S$\bar{x}$ 0.25) as were catalase values (Fig.7) but the distribution of erythrocuprein levels was much more dispersed. Although mean values were normal, standard deviations were more than 2.5 times larger than those of the normal rural Normandy population. Indeed there are indications of a bimodal distribution of low values of SOD and high values about the normal mean. The various pathological disorders included arterial hypertension, coronary insufficiency, myocardial necrosis, arteriosclerosis, arthrosis, respiratory insufficiency, osteoporosis, senile dementia, coxarthrosis, etc. At the highest levels, coronary insufficiency showed values of 615 and 695 μg SOD/g Hb, while in a case of hyperlipidic arterial hypertension both SOD and catalase were elevated (702 μg SOD and 5400 μg catalase/g Hb). Two cases of hemiplegia showed very low SOD levels - 232 and 330 μg/g Hb - whereas catalase was normal. The first subject, a 75 year old female, also suffering from hyperlipidemia, is the only example so far found with a value of SOD as low as 50% of the normal mean (461.4 μg/g Hb) which is also reflected in the value of SOD/ml of blood (25.5 μg compared with 58.1 μg/ml).

It would thus appear that high values or particularly, very low values of SOD are often associated with senility symptoms, some of which may well result (in the case of low SOD levels) either directly or indirectly from a diminished protection against aging processes caused by $O_2^{\cdot-}$ or other free radicals derived via superoxide. Highest values of SOD/g Hb were found in the age group 80 - 91 (20 subjects) with a mean of 504.1 μg and lowest values, mean 382.7 μg SOD/g Hb, in the age group 71 - 80 (9 subjects) with rather low figures also found in the group 54 - 70 years. This suggests that longevity is not associated with low SOD levels, but that aging processes occur more rapidly under these conditions. At the molecular level this could be due to modification of repair enzymes, protein-protein cross-linking and covalent linkage of lipids or nucleic acids to proteins caused by an excess of $O_2^{\cdot-}$, as well as direct modification of DNA.[4]

Ethnic Differences

We were able to obtain blood samples from seven Saharan touaregs. No significant differences in catalase or SOD levels were observed if the samples are compared with a normal French rural population with the same hemoglobin level, as shown in Table VII, though it must be noted that the samples were not in good condition.

A completely different result was obtained with 40 francophone African negroes resident in Paris. As shown in Fig.8 both SOD and catalase are extremely high, particularly when compared with a control white rural population with about the same average hemoglobin per ml of blood, since hemoglobin values were generally high (Table VII). Thus erythrocyte SOD in these negroes is about 55% higher than that of the control rural white population, whereas the catalase is some 33% higher, indicating a very strong ethnic difference.

However, these results must be accepted with caution since these negro subjects were resident for various times in a large city (Paris) and cannot be compared directly with a French rural population, and in addition the African ethnic origin was extremely varied. It would be of considerable interest to pursue this work with a specific ethnic group in an African country. As shown in the next section, residence in a large city may have an effect on the levels of erythrocuprein.

Urban and Rural

With the exception of the touaregs and francophone Africans, all comparisons have hitherto been effected among a rural population. Somewhat to our surprise, estimations of erythrocyte SOD in members of this department showed a curious distribution of very high levels of activity. Since one explanation could be the difference between a city population compared with rural, a

30 Psychiatric patients
x̄ = 73.7 μg SOD/ml
Sx = 10.1
Sx̄ = 1.9
Subjects
μg SOD/ml blood
0
25
50
75
100
125
2
4
6
8

Fig.5a.

30 Psychiatric patients
x̄ = 587.0 μg SOD/gHb
Sx = 66.4
Sx̄ = 12.1
Subjects
μg SOD/gHb
0
250
500
750
1000
2
4
6
8

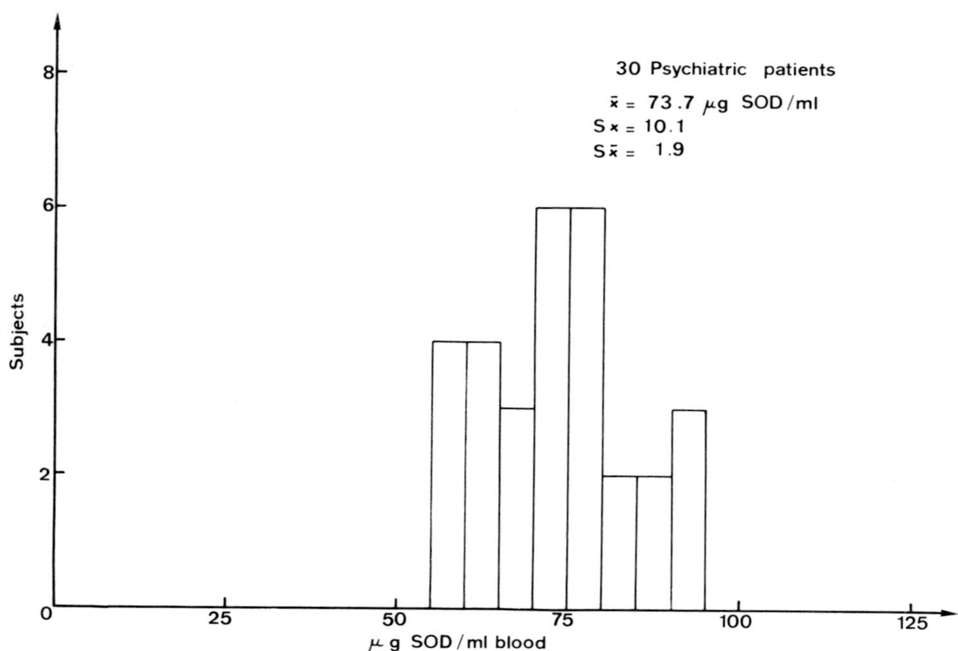

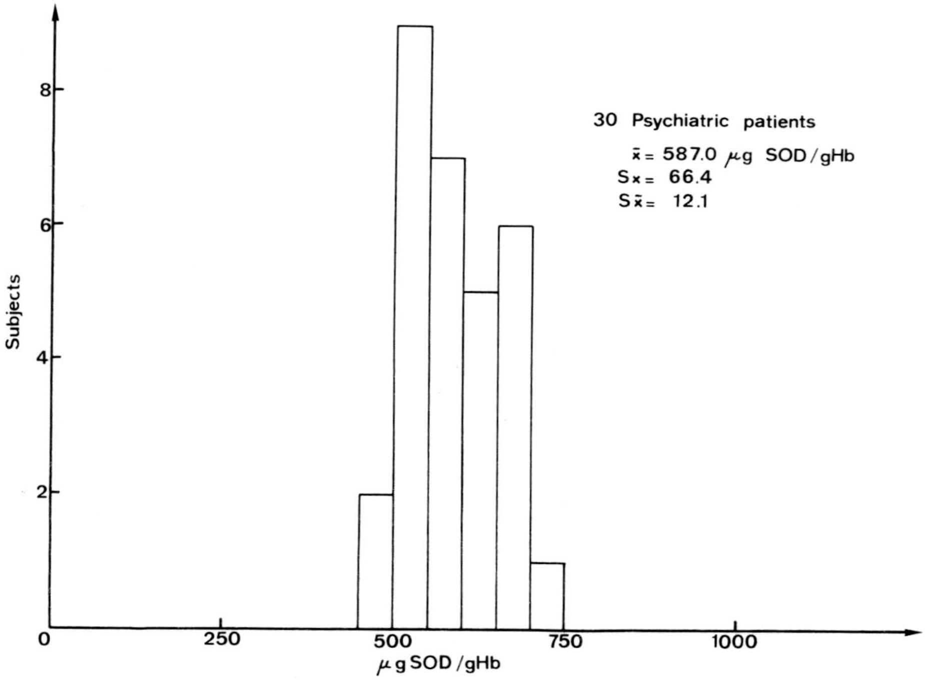

Fig.5b.

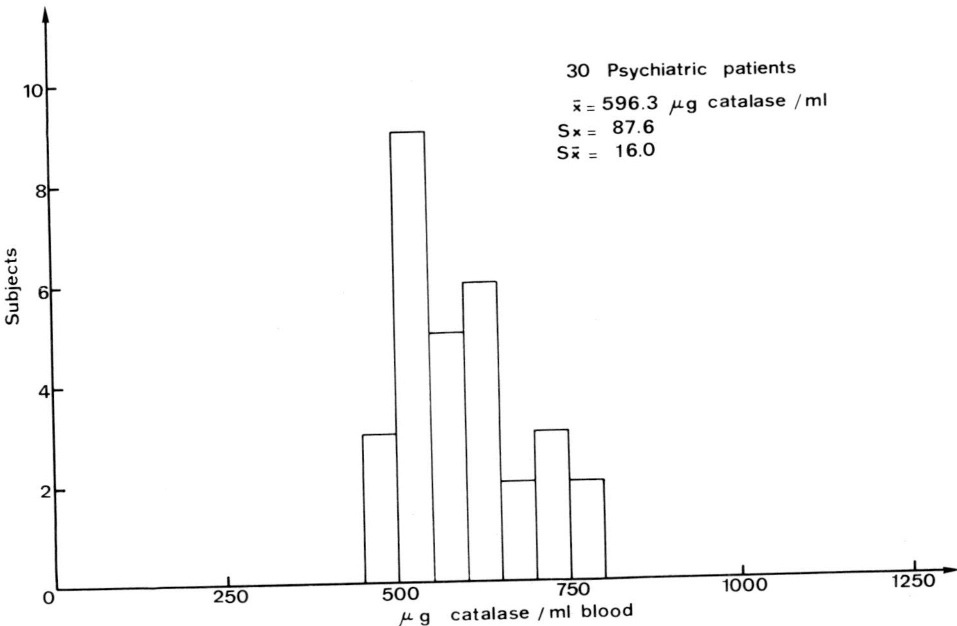

Fig.5c.

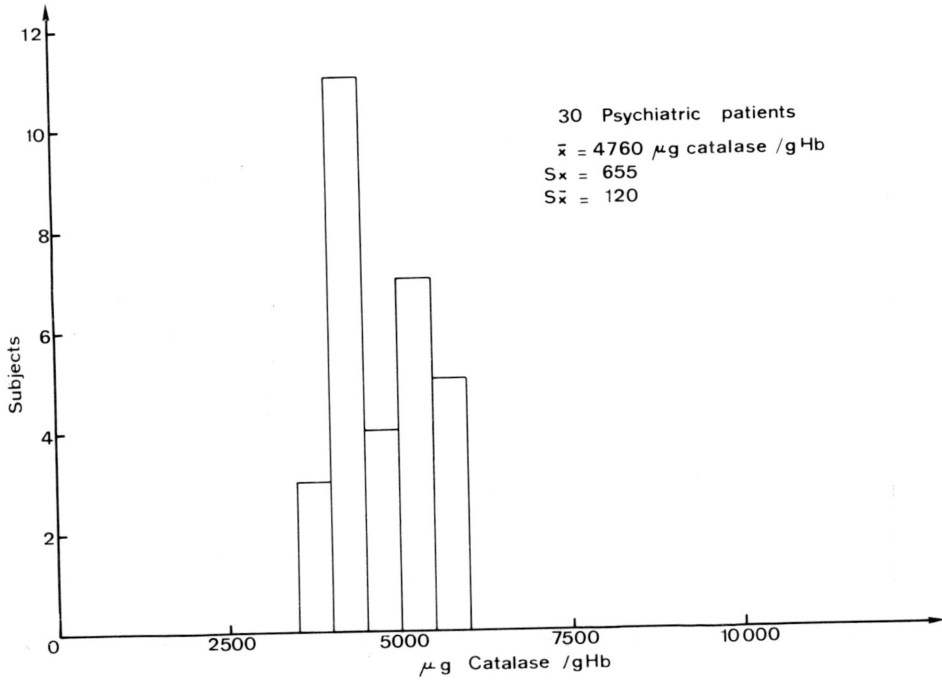

Fig.5d.

Fig.5 - *As for figure 2 for 30 psychiatric patients.*

15
10
5
0
Subjects
25
50
75
100
125
μg SOD/ml blood
23 subjects > 70 years
in excellent health
x̄ = 54.2 μgSOD/ml
Sx = 4.5
Sx̄ = 0.9

Fig.6a.

10
5
0
Subjects
250
500
750
1000
μgSOD / gHb
23 subjects > 70 years
in excellent health
x̄ =463.3 μgSOD / g Hb
Sx = 44.3
Sx̄ = 9.2

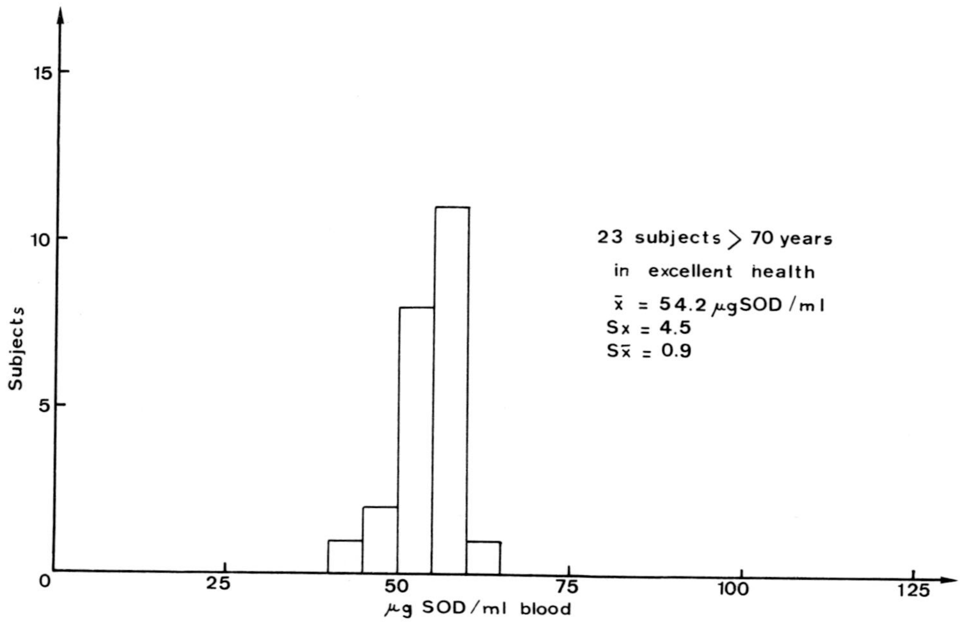

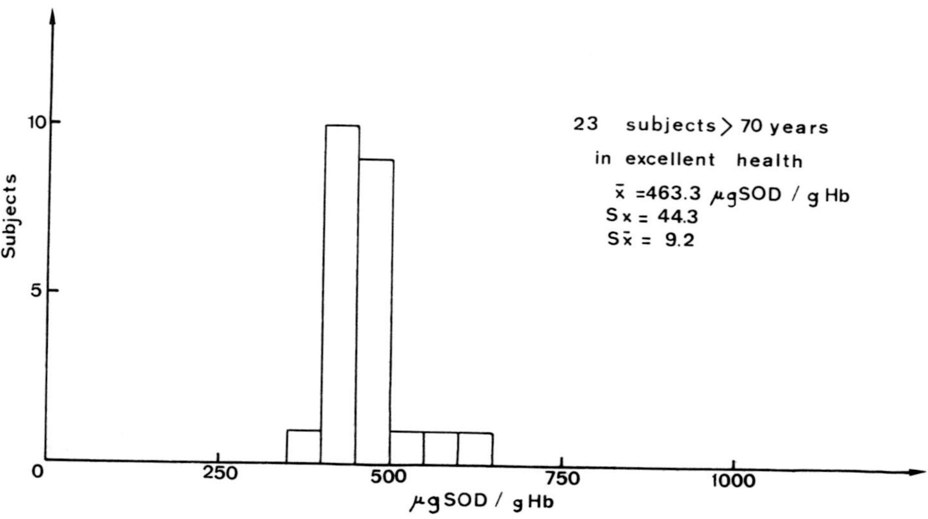

Fig.6b.

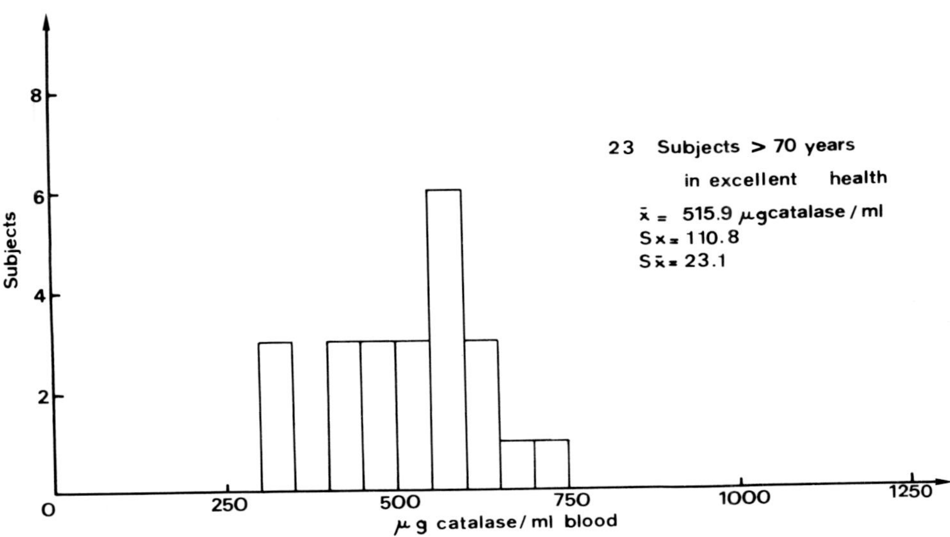

Fig.6c.

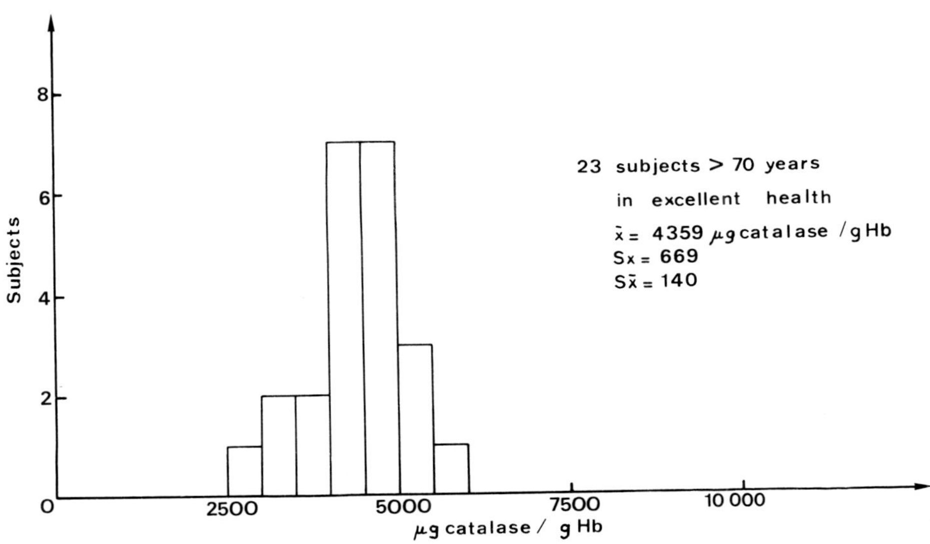

Fig.6d.

Fig.6 - *As for figure 2 for 23 old people in excellent health.*

34 Subjects > 70 years
in very poor health
x̄ = 55.6 μg SOD/ml
Sx = 15.1
Sx̄ = 2.6
Subjects
15
10
5
0
25
50
75
100
125
μg SOD/ml blood

Fig.7a.

34 Subjects > 70 years
in very poor health
x̄ = 460.9 μg SOD/gHb
Sx = 113.4
Sx̄ = 19.4
Subjects
15
10
5
0
250
500
750
1000
μg SOD/gHb

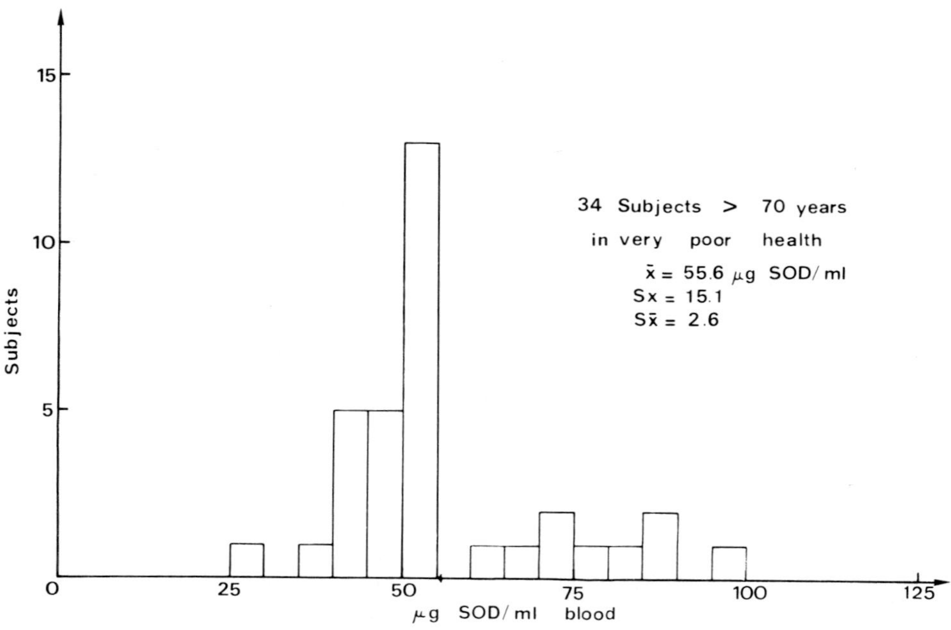

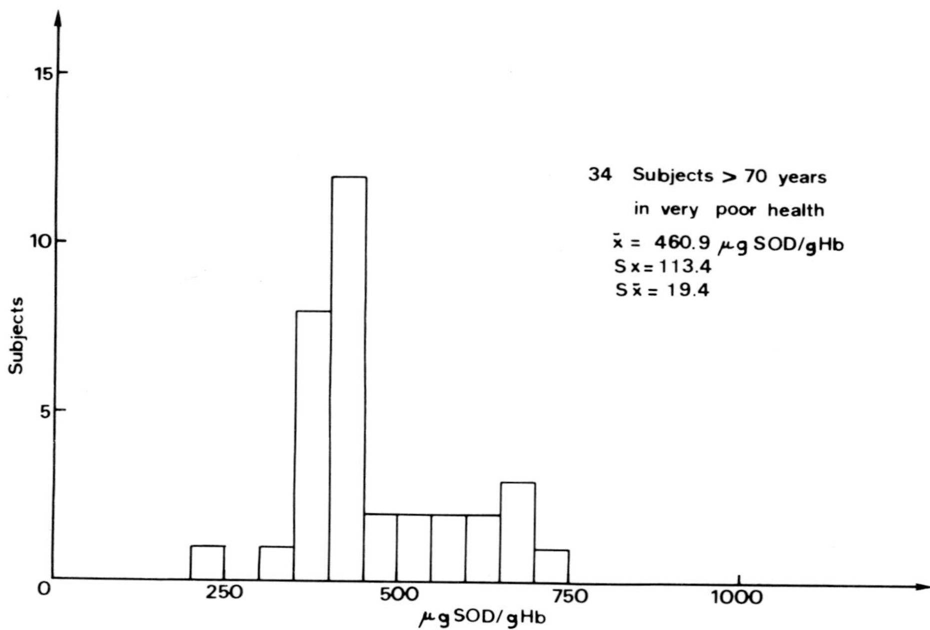

Fig.7b.

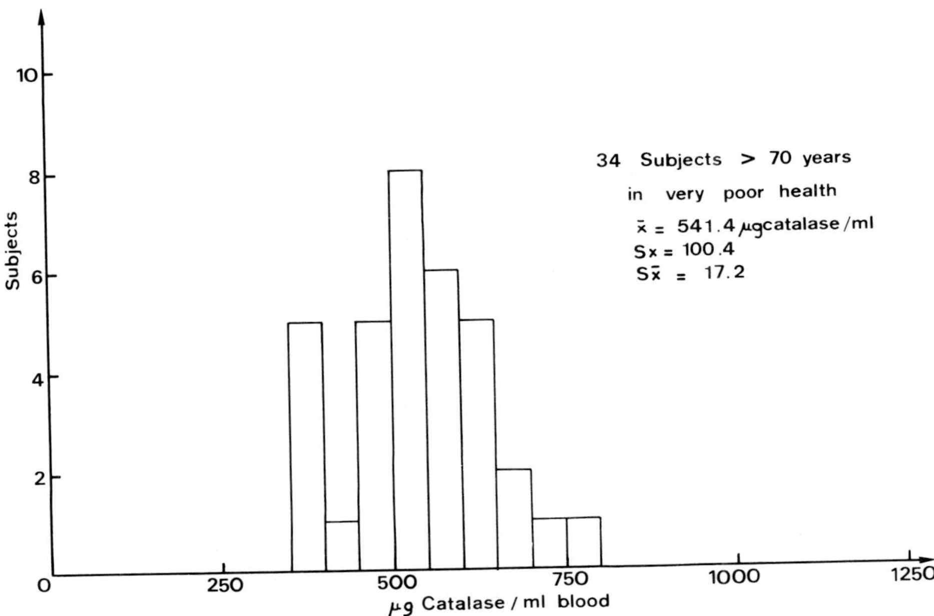

Fig.7c.

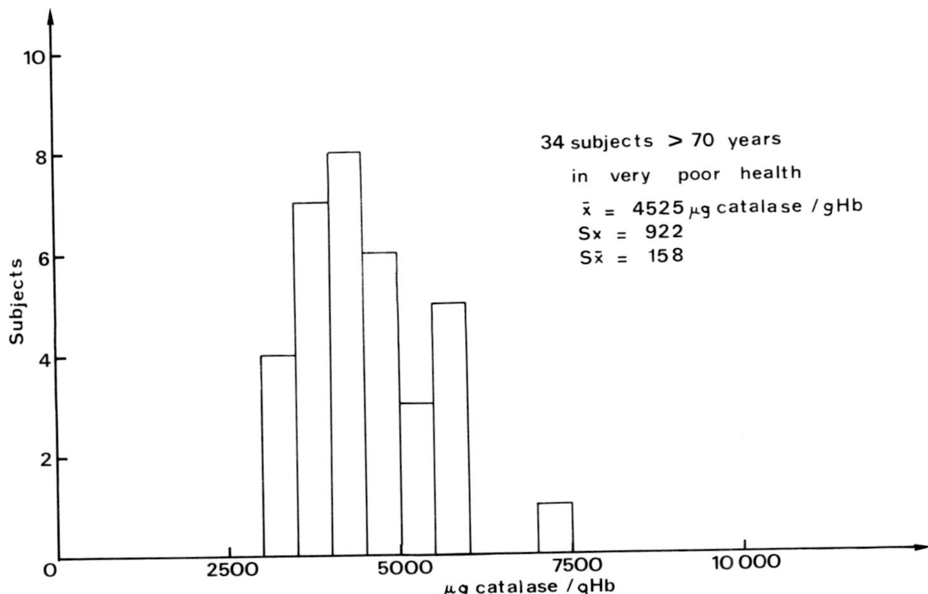

Fig.7d.

Fig.7 - *As for figure 2 for 34 old people in very poor health.*

40 Negroids
x̄ = 95.5 μg SOD /ml
Sx = 13.6
Sx̄ = 2.2
Subjects
0
5
10
25
50
75
100
125
μg SOD /ml blood

Fig.8a.

40 Negroids
x̄ = 672.5 μg SOD / g Hb
Sx = 80.9
Sx̄ = 12.8
Subjects
0
5
10
15
250
500
750
1000
μg SOD / g Hb

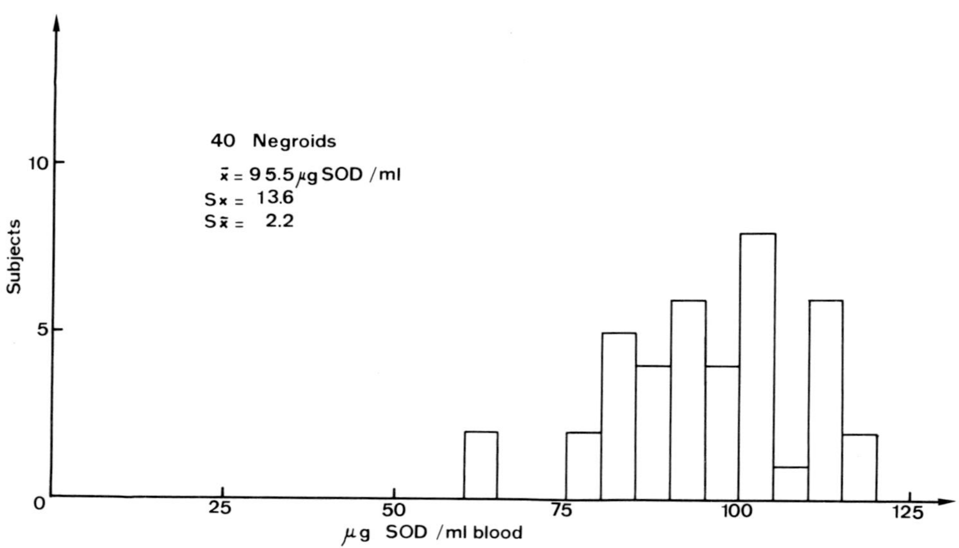

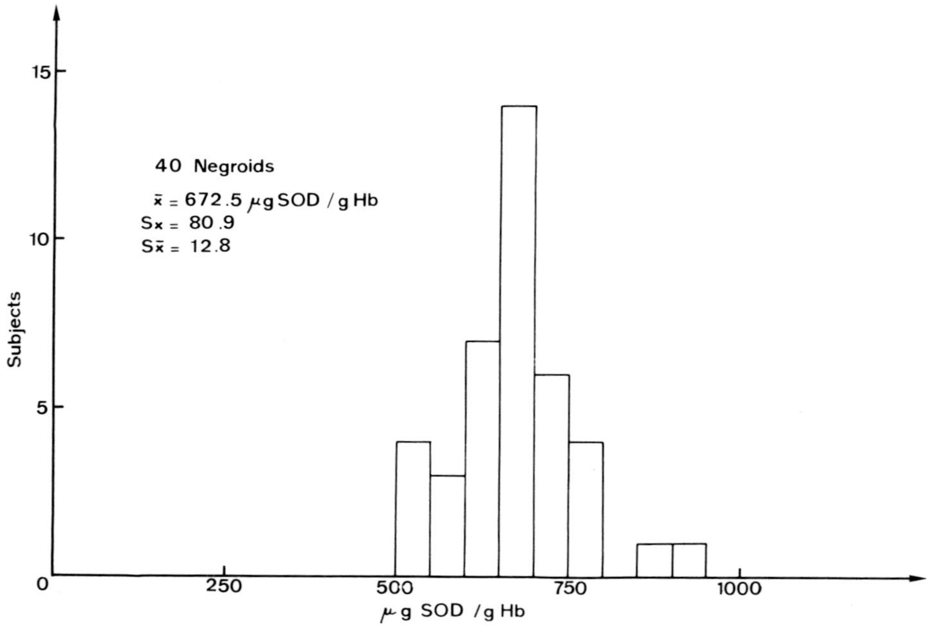

Fig.8b.

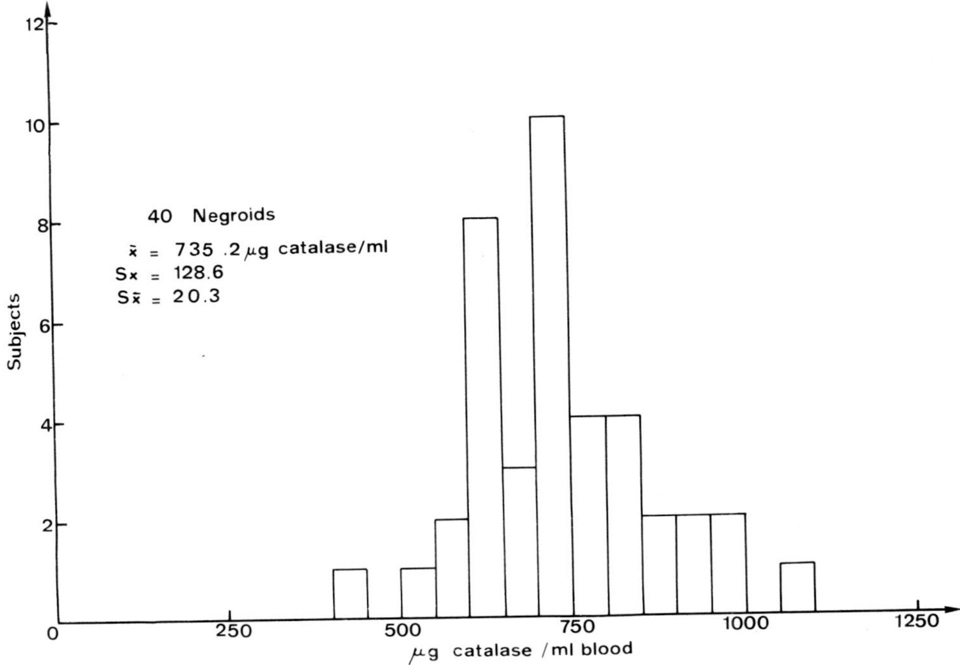

Fig.8c.

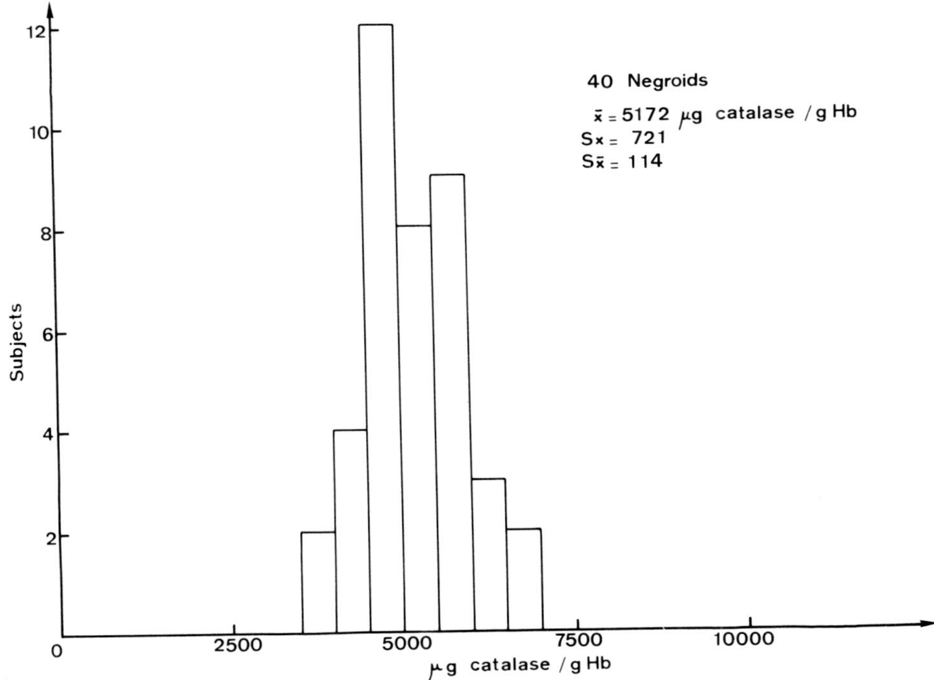

Fig.8d.

Fig.8 - *As for figure 2 for 40 negroes.*

TABLE VII

		Normal French Rural Subjects (39) Low Hemoglobin	Touaregs (7 subjects)	Normal French Rural Subjects (46) High Hemoglobin	African Negroes (40) Resident in Paris	White Parisiens (20)
Hemoglobin	$\bar{x}$	10.83	10.46	14.37	14.26	15.57
	Sx	0.58	1.86	0.88	1.85	1.40
	S$\bar{x}$	0.09	0.70	0.13	0.29	0.31
μg SOD/ml	$\bar{x}$	54.9	52.3	62.0	95.5	92.3
	Sx	-	6.6	-	13.8	9.8
	S$\bar{x}$	-	2.5	-	2.2	2.2
μg SOD/g Hb	$\bar{x}$	506.6	506.7	431.2	672.5	593.6
	Sx	45.6	67.8	44.6	81.9	43.1
	S$\bar{x}$	7.3	25.6	6.6	13.0	9.6
μg catalase/ml	$\bar{x}$	460.7	379.6	554.7	735.2	760.0 (11 subjects)
	Sx	-	70.3	-	130.2	169.9
	S$\bar{x}$	-	26.6	-	20.6	51.2
μg catalase/g Hb	$\bar{x}$	4254	3645	3860	5172	4726
	Sx	572	388	654	730	902
	S$\bar{x}$	92	147	97	115	272

series of 20 white (normal) Parisien subjects was examined. The results are shown in Table VII. This group showed much higher hemoglobin levels than the rural population (mean 15.57 g/100 ml compared with 12.67 g/100 ml), and in comparison with white rurals with highest hemoglobin levels erythrocuprein was significantly increased by 38% (per g Hb) whereas catalase gave a mean value of 22% (per g Hb) greater than that of the rural population.

These results suggest profound differences in oxygen metabolism between rural and city populations and need to be further developed. We have followed one subject over three years and there is a notable constancy of erythrocyte SOD (limits 614 - 632 μg/g Hb with a variation in hemoglobin from 16.4 to 18.3 g/100 ml). No difference was seen between a fasting state and after a copious meal.

Hepatic Diseases

Preliminary results suggest a bimodal distribution of levels of SOD in a series of patients suffering from various hepatic and hemolytic diseases with a bias towards low values (Puget, Michelson and Najman, 1976).

Conclusions

The above survey of SOD levels, although limited with respect to numbers of subjects examined, and hence necessarily to be regarded with a certain caution, nevertheless indicates that relatively high erythrocyte SOD can be associated with various pathological states. Low levels of SOD appear to be even more harmful and up till now excessively low values have been very rarely encountered (two cases of old people in very bad health). It may well be that levels of less than 50% of the normal mean for SOD are more or less lethal due to the increased toxicity of uncontrolled $O_2^{\overline{\cdot}}$. In normal humans, the extremely strict control of SOD with a very narrow distribution of values in a population, indicates that the regulation of synthesis of this enzyme in mammalian cells is worth study.

It may well be that SOD plays an important role in the inhibition of certain aging processes and somatic mutations; we have begun work on possible effects of SOD on the longevity of eucaryotic cells in culture.

The clinical value of knowledge of the various enzymes connected with protection against activated forms of oxygen, particularly by SOD, remains to be verified by much larger trials. It is clear that if the results presented in this chapter are confirmed then a new specific kind of treatment can at least be envisaged (e.g. use of SOD inhibitors with Mongoloid or psychiatric cases). Certain experimental approaches are alas technically impossible. Thus a plausible explanation for cataract formation is cross-linking of proteins of the crystallin induced by $O_2^{\overline{\cdot}}$ (cf Cooperman and Michelson, 1977). We have already shown that cataract lens proteins contain SOD concentrated in particular in the exterior zone (Puget and Michelson, 1976). It is possible that diminished protection against $O_2^{\overline{\cdot}}$ by a low level of SOD ultimately results in cataract formation (which could well be inhibited by external application of the enzyme). For ethical and technical reasons comparison between normal and cataract human eyes is not feasible and we have only been able to show small quantities of SOD in the latter, perhaps considerably lower than the normal.

Finally, the possible application of $O_2^{\overline{\cdot}}$ and SOD with respect to cancer cannot be neglected. Application of an $O_2^{\overline{\cdot}}$ producing system causes considerable protection against tumour formation after injection of new born mice with murine sarcoma virus (Michelson and Cherman, 1976) while injection of SOD into transplanted melanoma tumours in hamsters causes an increase in survival time (Puget, Michelson and Lacour, 1976) though in this case it is not clear that the SOD is acting enzymatically. Thus the bivalent possibilities of too much or too little biologically available $O_2^{\cdot-}$ (or too little or too much SOD) again become apparent. It is clear that we are at the very beginning of the possible medical aspects of $O_2^{\cdot-}$ and of superoxide dismutases and very much work yet remains to be developed both in fundamental exploration and in pragmatic practical applications.

FOOTNOTES

[1]We have recently shown (F. Lavelle, unpublished results) that erythrocuprein content of young and old erythrocytes from a given individual is constant; no variation was observed within the life span of the mature red cell.

[2]Cyanide and chloroform/ethanol insensitivity is also shown by the three distinct SODs present in paramecia, and by the three major Mn-SOD activities (pIs, 7.30, 7.55, 7.75) present in human platelets (K. Puget, P. Durosay and A.M. Michelson, unpublished results). A fourth band, pI 7.80, can also be detected. While certain of these SODs, which together make up about 20% of the total SOD activity, are certainly mitochondrial, others may well be cytoplasmic in addition to the 80% Cu-SOD activity. With respect to resistance to chloroform/ethanol, it may be noted that all the above mentioned Mn-SODs have a molecular weight of 80,000 to 90,000 and thus differ from the two subunit, one Mn, *E. coli* SOD (M. wt. 40,000). The multiplicity of bands of activity on electrofocalisation of human platelet Mn-SODs, may be real, or an artifact due to spontaneous degradation of the original *in vivo* SOD. We have noted a change in activity of this fraction (Mn-SOD) of the total cell SOD (in contrast with the Cu-SOD) with time of storage under physiological conditions (except temperature, $4^{\circ}C$). Activity increases for several days, then decreases, indicating conformational or structural changes. A similar formation of multiple protein bands, both active and inactive has been noted on storage of a bacteriocuprein pI, 8.15, separated from the main component (source *P. leiognathi*). This instability *in vitro* (Michelson *et al.*, 1977) may reflect an *in vivo* situation and could have considerable consequences for the metabolic turnover of superoxide dismutase and the control and regulation of the cellular levels of the differently located SODs. The *in vivo* rate of turnover of SOD has not yet been investigated.

[3]It has nevertheless been noted that in all cases which show marked improvement, the glutathione peroxidase level increases and SOD decreases. This suggests that normally an inverse relationship exists between SOD and the peroxidase scavenger of H_2O_2 (in contrast with the parallel increase found in trisomy 21). Similarly, in human diabetics, we have found that low glutathione peroxidase is associated with high SOD with respect to erythrocytes.

[4]For the effects of high SOD levels, other processes must be invoked. It may be noted that in cases of trisomy 21 (50% increase in SOD) there is frequent occurrence of premature senility (Burger and Vogel, 1973) as well as leukemia, and reduced life times of cultured fibroblasts (Segal and McCoy,

1974; Schneider and Epstein, 1972). We have already mentioned the possibility of inhibition of certain enzymic oxidations by excessive quantities of SOD, which may also interfere at some step in the respiratory chain. An alternate type of explanation could implicate the following considerations.

The role of $O_2^{\cdot-}$ as a radical trap for $HO^{\cdot}$ (a much more reactive and dangerous activated form of oxygen) may be important via the reaction $O_2^{\cdot-} + HO^{\cdot} \longrightarrow HO^- + O_2$. Excessive levels of SOD could thus increase the concentration of $HO^{\cdot}$ in the cell, and indeed in several systems studied, for example lysis of erythrocytes, a slight acceleration rather than inhibition is observed when increased SOD is present (Michelson and Durosay, 1977; Bartosz *et al.*, 1977). Oxidation of lysosomal membrane lipids by the xanthine oxidase system leading to lysis is stimulated by SOD rather than inhibited (Fong *et al.*, 1973). It may also be noted that increase in amount of intravenously injected SOD nullifies the radioprotective effect seen at low concentrations (Petkau *et al.*, 1975). While the Haber-Weiss reaction, $O_2^{\cdot-} + H_2O_2 \longrightarrow O_2 + HO^{\cdot} + HO^-$, is frequently postulated as the source of biological $HO^{\cdot}$ radicals, other mechanisms such as the reaction of Fe^{2+} (or oxyhemoglobin, reduced cytochrome *c* and other heme proteins) or flavin radicals with H_2O_2 cannot be excluded, i.e.

$$Fe^{2+} + H_2O_2 \longrightarrow HO^{\cdot} + HO^- + Fe^{3+}$$

$$FMNH^{\cdot} + H_2O_2 \longrightarrow HO^{\cdot} + H_2O + FMN$$

Such reactions can of course be initiated by $O_2^{\cdot-}$, either through dismutation to H_2O_2 or by creation of both H_2O_2 and the intermediate free radical:

$$H^+ + O_2^{\cdot-} + RH_2 \longrightarrow RH^{\cdot} + H_2O_2$$

$$RH^{\cdot} + H_2O_2 \longrightarrow R + H_2O + HO^{\cdot}$$

$$H^+ + O_2^{\cdot-} + RH_2 \longrightarrow R + H_2O + HO^{\cdot}$$

Another source of $HO^{\cdot}$ could well be the *reduction* of ferri-proteins (e.g. oxidised cytochrome *c*) or even of the various SODs by H_2O_2.

In certain cases, only catalase will inhibit, whereas in others both SOD and catalase will be effective, as in the postulated Haber-Weiss reaction. The existence of dual inhibition by both enzymes is thus not absolute proof of the Haber-Weiss reaction, but of a rather more generalised concept of the production of $HO^{\cdot}$ radicals. As mentioned above, excessive SOD could in certain circumstances increase the amount of $HO^{\cdot}$ available. The same is true of catalase (which also sometimes stimulates instead of inhibiting) by interference with another mechanism of destruction of $HO^{\cdot}$ radicals, in which less reactive $O_2^{\cdot-}$ is produced:

$$H_2O_2 + HO^{\cdot} \longrightarrow H_2O + H^+ + O_2^{\cdot-} \quad k = 4.5 \times 10^7\ M^{-1}sec^{-1}.$$

It is clear that all assumptions with respect to the direct effects of different levels of such enzymes in biological systems are necessarily oversimplified at present.

ACKNOWLEDGEMENTS

We thank Dr. Najman and Prof. Dreyfus (Paris) for blood samples, and Prof. C. Ropartz, Director of the Centre Regional de Transfusion Sanguine et de Génétique Humaine, Bois-Guillaume, France, for his interest and encouragement, as well as the vital role he played in the development of this project.

REFERENCES

1. Autor, A.P. and Roberts, R.J. (1974). *Federation Proceedings 33,* 1505.
2. Beauchamp, C. and Fridovich, I. (1971). *Anal. Biochem. 44,* 276-287.
3. Benson, P.F. and Southgate, J. (1971). *Amer. J. Hum. Genet. 23,* 211-214.
4. Beutler, E. (1971). "Red Cell Metabolism: A Manual of Biochemical Methods", Grune and Stratton, New York, p 30.
5. Bartosz, G., Leyko, W., Kedziora, J. and Jeske, J. (1977). *Int. J. Radiat. Biol. 31,* 197-200.
6. Burger, P.C. and Vogel, F.S. (1973). *Amer. J. Pathol. 73,* 457-468.
7. Cohen, G. and Hochstein, P. (1963). *Biochemistry 2,* 1420-1428.
8. Cooperman, B.S., Dondon, J., Finelli, J., Grunberg-Managa, M. and Michelson, A.M. (1977). *FEBS Letters 76,* 59-63.
9. Corberand, J., de Larrard, B., Pris, J. and Colombies, P. (1974). *Nouvelle Revue Française d'Hématologie 14,* 298-301.
10. Drabkin, D.L. and Austin, J.H. (1935). *J. Biol. Chem. 112,* 51-57.
11. Fong, K.L., McCoy, P.F., Foyer, J.L., Keele, B.B. and Misra, H. (1973). *J. Biol. Chem. 248,* 7792-7797.
12. Frants, R.R., Eriksson, A.W., Jongbloet, P.H. and Hamers, A.J. (1975). *Lancet ii,* 42-43.
13. Hartz, J.W., Funakoshi, S. and Deutsch, H.F. (1973). *Clinica Chimica Acta 46,* 125-132.
14. Henry, J.P., Isambert, M.F. and Michelson, A.M. (1970). *Biochim. Biophys. Acta 205,* 437-450.
15. Hirata, F. and Hayaishi, O. (1975). *J. Biol. Chem. 250,* 5960-5966.
16. Hsia, D.Y.Y., Justice, P., Smith, G.F. and Dowben, R.M. (1971). *Amer. J. Dis. Child 121,* 153-161.
17. Humbert, J.R., Kurtz, M.L. and Hathaway, W.E. (1970). *Pediatrics 45,* 125-128.
18. Kamoun, P. and Jérôme, H. (1975). In "Serotonin in Mental Disorders" (D.I. Boullin, ed.), John Wiley, London (in press).
19. Lavelle, F., Puget, K. and Michelson, A.M. (1974). *Compte Rendu Acad. Sci.* (Paris) *278,* 2695-2698.
20. Layser, R.B. and Epstein, C.J. (1972). *Amer. J. Hum. Genet. 24,* 533-543.
21. Lott, I.T., Chase, T.N. and Murphy, D.L. (1972). *Pediat. Res. 6,* 730-735.
22. Michelson, A.M. and Buckingham, M.E. (1974). *Biochem. Biophys. Research Commns. 58,* 1079-1086.
23. Michelson, A.M. and Cherman, S. (1976). In preparation.
24. Michelson, A.M. and Durosay, P. (1977). *Photochem. Photobiol. 25,* 55-63.
25. Michelson, A.M., Puget, K., Durosay, P. and Golse, B. (1977). In preparation.
26. Misra, H.P. and Fridovich, I. (1972). *J. Biol. Chem. 247,* 6960-6962.
27. Paasonen, M.K., Solatuntari, E. and Kivalo, E. (1964). *Psycho-Pharmacologia 6,* 120-124.
28. Pantekalis, S.M., Karalis, A.G., Alexion, D., Vardas, E. and Valaes, T. (1970). *Amer. J. Hum. Genet. 22,* 184-193.
29. Petkau, A., Kelly, K., Chelack, W.S., Pleskach, S.D., Barefoot, C. and Meeker, B.E. (1975). *Biochem. Biophys. Research Commns. 67,* 1167-1174.
30. Puget, K. and Michelson, A.M. (1974). *Biochimie 56,* 1255-1267.
31. Puget, K. and Michelson, A.M. (1976). Unpublished results.
32. Puget, K., Michelson, A.M. and Lacour, F. (1976). Unpublished results.
33. Puget, K., Michelson, A.M. and Najman, A. (1976). Unpublished results.

34. Salin, M.L. and McCord, J.M. (1974). *J. Clin. Invest. 54,* 1005-1009.
35. Schneider, E.L. and Epstein, C.J. (1972). *Proc. Soc. Exp. Biol. Med. 141,* 1092-1094.
36. Segal, D.I. and McCoy, E.E. (1974). *J. Cell Physiol. 83,* 85-90.
37. Sinet, P.M., Allard, D., Lejeune, J. and Jérôme, H. (1974). *Compte Rendu Acad. Sci.* (Paris) *278,* 3267-3270.
38. Sinet, P.M., Lavelle F., Michelson, A.M. and Jérôme, H. (1975a). *Biochem. Biophys. Research Communs. 67,* 904-909.
39. Sinet, P.M., Michelson, A.M., Bazin, A., Lejeune, J. and Jérôme, H. (1975b). *Biochem. Biophys. Research Communs. 67,* 910-915.
40. Sparkes, R.S. and Bauthan, M.A. (1969). *Amer. J. Hum. Genet. 21,* 430-439.
41. Stansell, M.J. and Deutsch, H.F. (1965). *J. Biol. Chem. 240,* 4299-4305.
42. Weisiger, R.A. and Fridovich, I. (1973). *J. Biol. Chem. 248,* 4793-4796.
43. Winterbourn, C.C., Hawkins, R.E., Brian, M. and Carrell, R.W. (1975). *J. Lab. Clin. Med.* 337-341.

EFFECTS OF PLANT AND ANIMAL TISSUE LESIONS ON SUPEROXIDE DISMUTASE ACTIVITIES

B. MATKOVICS

Biological Isotop Laboratory
"A.J." University
H-6701 Szeged, Hungary

INTRODUCTION

All aerobic organisms, plants, animals and even microorganisms, possess an enzymatic defense system which is able to protect them from radicals formed in the course of the metabolism. Radicals produced during intermediate metabolism could give rise to side-reactions, and these might result in products different from those of the main trends of the metabolism.

The past few years has seen a resurgence in publications dealing with radicals formed from molecular oxygen and with the role of these in aerobic metabolism (Fridovich, 1975; Matkovics, 1976a,b). There have been a significant number of papers connected with the superoxide anion (Fridovich, 1975).

In the present paper we deal with the quantitative levels of plant and animal SOD under normal conditions and then report our results on the quantitative variations in SOD with changes in age. Finally, we shall discuss the quantitative variations in SOD after induction of some plant lesions, e.g. a virus infection, or some metabolic disease.

MATERIALS AND METHODS

The material can be classified into two main groups: (i) plant material, (ii) animal and human material.

(i) Bean plants were grown in water culture in a light-thermostat in our laboratory (Simon *et al.*, 1974). The seeds and vegetables were bought in spring.

Tomato plants of various ages, either normal or infected with strain TMV O, were obtained from the Vegetable-Production Research Institute in Kecskemét, Hungary; the plants were cultivated there under controlled conditions, and the virus infection was effected by J. Farkas (to whom we are indebted for his exact and conscientious work and cooperation) using the conditions described by Mészöly and Vidéki (1963).

(ii) Human material: normal and pathological placentas were obtained from

A. Jakoboviks of the Department of Obstetrics and Gynaecology, University Medical School, Szeged, Hungary. If the tissues were not processed immediately, they were stored at -20°C until use. Our examinations to date show that the tissue SOD activity does not essentially alter under these conditions.

Human diabetic blood samples were provided by clinics specializing in diabetes.

In general, Wistar, mottled and Gödöllo (Hungary) white rats about 2 months old were used in experiments on streptozotocin-induced diabetes. After starvation for 12 hr, the rats were injected twice with 50 mg streptozotocin/kg, and kept for about 2 months on a normal diet (rat pellets LATI Nr II. Gödöllo, Hungary). Diabetes was regarded as definite if the rats excreted glucose in the urine. In addition, the diabetic rats were checked on several occasions by blood sugar determinations; a blood sugar level higher than 150 mg% in the fasting state was considered a manifestation of diabetes, and only the results of enzymatic measurements on such rats were included in the diabetic group.

SOD determinations were also carried out on washed red blood cells from human diabetics in whom blood sugar was controlled with oral antidiabetics. The erythrocytes were separated from the plasma by centrifugation. They were washed three times with cold physiological saline solution. The erythrocytes obtained after the final centrifugation were hemolyzed with an equal volume of distilled water, and enzymatic activity measurements were made on aliquots of the hemolyzates.

Of the methods available for the quantitative determination of SOD, we selected the SOD-dependent inhibition of the spontaneous autocatalytic oxidation of L-adrenaline at pH 10.2, first described by Misra and Fridovich (1972) with a graphic method for determination of the activities (Matkovics *et al.*, 1976). This technique can sometimes give rise to erroneous results when used with crude extracts, particularly in presence of ascorbate.

L-adrenaline was obtained from Boehringer (Mannheim, FRG), and glutathione peroxidase from Calbiochem (Basel, Switzerland). All other compounds and buffer materials used were commercial products of Reanal (Budapest, Hungary), of analytical purity.

General Preparation of Materials

(a) Tomato plants were planted out on the 12th day, and were subjected to TMV infection on the 14th day. Plants were examined for enzymatic activities on the 15th, 30th and 50th days after virus infection.

The plants were washed free of earth and dried between filter papers and 1 g wet weight taken from the roots, the stems and the leaves. Each part was ground with quartz, then washed into a centrifuge tube with 5 ml 0.05 M phosphate buffer (pH 7.8), followed by centrifugation; the supernatant was used for measurement of SOD. In addition to the SOD determinations, simultaneous measurements were made of the dry weight, the amounts of protein and ascorbic acid and the peroxidase and catalase activities.

The air-dried plant seeds were also ground with quartz sand, and here too the supernatant obtained by centrifugation was used for enzymatic activity measurements (Do Quy Hai *et al.*, 1975).

Growth of the bean plant, Phaseolus vulgaris, was similar. Enzymes were again measured in the supernatant from the centrifugation of a buffered extract of the plant part after grinding with quartz sand (Simon *et al.*, 1974)

(b) Animal tissues were generally pretreated in a Potter glass homogenizer or ground with quartz sand. Here too the ratio of wet tissue to phosphate buffer was 1:5. Besides the earlier-mentioned enzymes (SOD, peroxidase and catalase), quantitative determinations were often made of transaminases (GOT and GPT), tissue free glucose and glycogen, as well as cholesterol and protein in the animal tissue homogenizates.

The protein was always determined quantitatively with the photometric variant of the Folin method described by Lowry *et al.* (1951). Bovine plasma albumin was employed to prepare the standard curve.

Quantitative measurement of glutathione peroxidase was performed by the method of Paglia and Valentine (1967).

The rats, which were otherwise maintained on a normal diet, received 0.5% aqueous H_2O_2 ad libitum rather than water. This treatment was begun at the age of 2 months; after a further 2 months had passed, the SOD activities of the larger organs were measured. Initially, the H_2O_2-drinking rats lagged behind water-drinking rats in their development, but later they caught up with the controls. The most striking difference in comparison with the controls when the animals were killed was the thinness of the viscera and large vessels.

The error in the quantitative determination of SOD was ± 5-10%. The Figures always show the means of a large number of measurements (in general of more than ten).

In the calculation of the data, the PL/1 programme was used on an R 10 (Robotron, Magdeburg, GDR) computer.

Spectrophotometric measurements were made on Spektromom 201 or 360 photometers (MOM, Budapest, Hungary).

THE ROLE OF SOD IN PLANTS

This question is rather general for, as will be seen later, SOD can be found in dormant plant seeds, in the fruit, and in every part of the plant in the most varied stages of development.

The review of Halliwell (1974) deals with the role of SOD in plant and animal tissues. He states that SOD protects the plant parts against excess of O_2^- anion radical and other radicals formed from molecular oxygen. In his view, however, important roles are similarly played by catalase and glutathione peroxidase in the enzymatic protection of plants. We shall provide data complementing this conception, and analyze the importance of SOD in overcoming TMV infection in plants.

SOD Values of Plant Seeds and Fruits: Early Development of the Bean Plants and Ageing of the Leaves

(a) In 1974 (Do Quy Hay *et al.* 1970) we described an enzymatic examination of the peroxide metabolism in monocotyledonous and dicotyledonous plant seeds (the enzymes in question both here and subsequently are SOD, peroxidase and catalase). It can be observed from Figs.1 and 2 that SOD occurs in fewer monocotyledonous than dicotyledonous seeds and that its concentration per g air-dried seed is lower in the former than in the latter group. The number of varieties of seeds examined were almost the same in the two groups but SOD activities less than 10 U/g air-dried seed are not shown in the Figures. Our seed SOD activity measurements extended to all the important cultivated plants. It appears that of the dicotyledonous seeds only soya possesses an

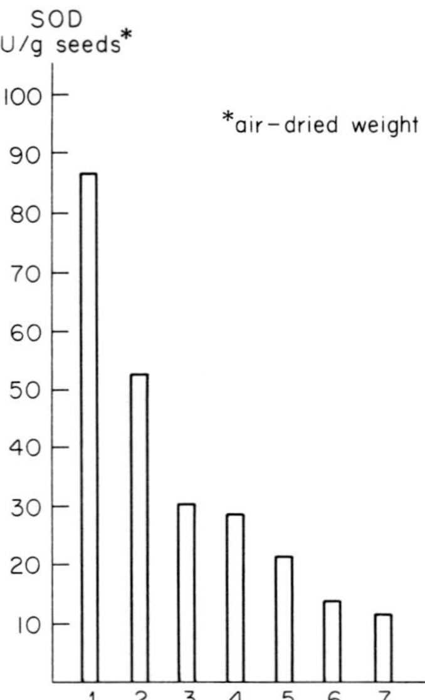

Fig.1. *SOD values of monocotyledonous seeds, calculated in U/g air-dried weight. 1. Panicum miliaceum L. 2. Triticum aestivum. 3. Triticale. 4. Zea mays L. 5. Hordeum vulgare L. 6. Festuca rubra. 7. Sorgum vulgare.*

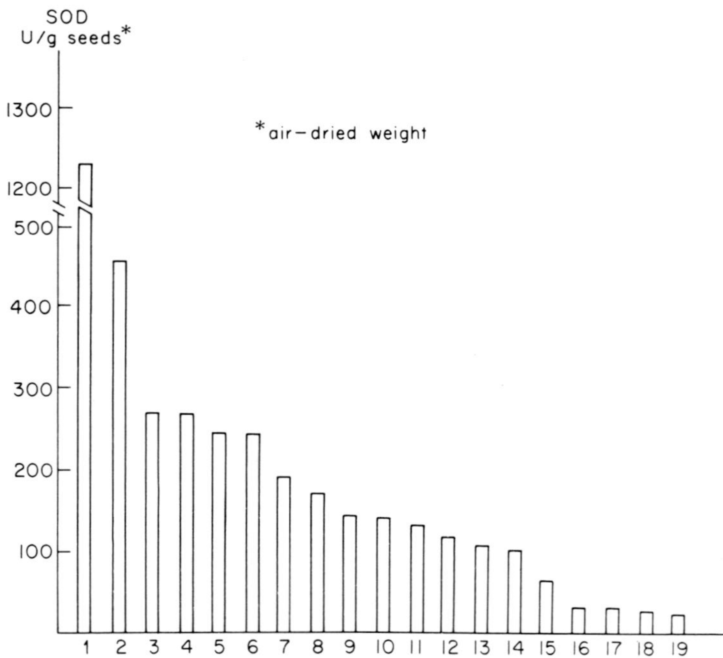

Fig.2. *SOD values of dicotyledonous seeds, calculated in U/g air-dried weight. 1. Glycine soya. 2. Papaver somniferum. 3. Pisum sativum ssp. hortense. 4. Brassica oleracea. 5. Raphanus sativus L. 6. Citrullus lanatus thumb. 7. Phaseolus coccineus. 8. Hatthiola bicornis. 9. Brassica napus L. 10. Lycopersicon esculentum. 11. Phaseolus vulgaris L. 12. Cucumis sativus. 13. Linum usitatissimum. 14. Cucurbita pepo L. 15. Hibiscus cannabinus L. 16. Ricinus communis L. 17. Cannabis sativa. 18. Petroselium hortense. 19. Medicago sativa L.*

outstanding SOD activity, the values for the other seeds being only 1/6 - 1/3 of that for soya in general.

(b) Bean was selected as a dicotyledonous plant of moderate SOD activity. When this begins to germinate it is observed that by the age of 2 days after being planted out the SOD activity of the small plantlets developing in the wet chamber in a light-thermostat is nearly three times that of the seed; by the 8th day it is 500 - 600 U/g wet tissue, this value persisting subsequently. (Fig.3).

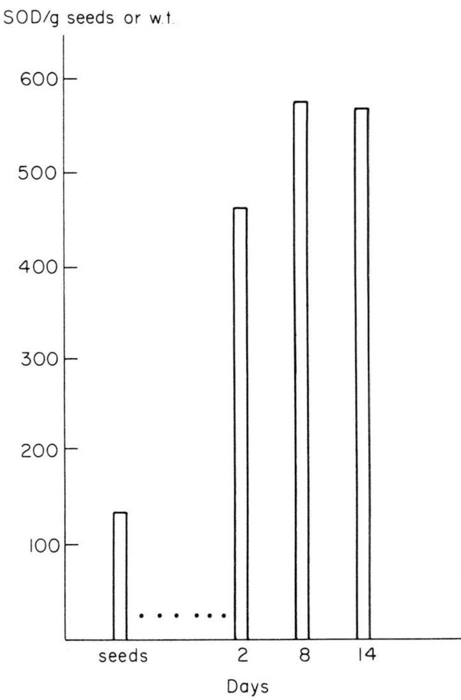

Fig.3. *Total SOD activity values of bean and bean seedlings. Column 1: air-dried white bean. Columns 2 - 4: seedlings on the 2nd, 8th and 14th days after planting-out.*

If the root, stem and leaf SOD activities of the bean seedling are examined separately, however, the picture presented in Fig.4 is obtained. From the 8th day on the SOD predominates in the leaf.

The above results and conclusions are not always so clear-cut, and depend to a large extent on the species and the conditions under which the plants are kept.

The role of SOD in photosynthesis has not been satisfactorily clarified. In our view, however, SOD is of great importance in the photosynthetic processes of both normal and diseased plants.

(c) Fig.5 presents the SOD activities of some common vegetables, among them paprika.

(d) Fig.6 shows the changes with age in the SOD content of paprika leaves. The SOD activity is strongly decreased in the leaves of paprika at the age of ca. 2 months. The activity in the fruit, however, is the highest as regards the total plant (compare Fig.5).

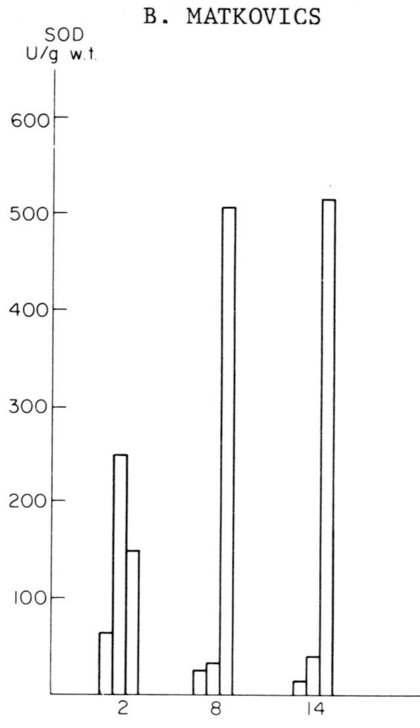

Fig.4. *SOD activity values of bean seedlings with ages as given in Fig.3, in the sequence root, stem and leaf.*

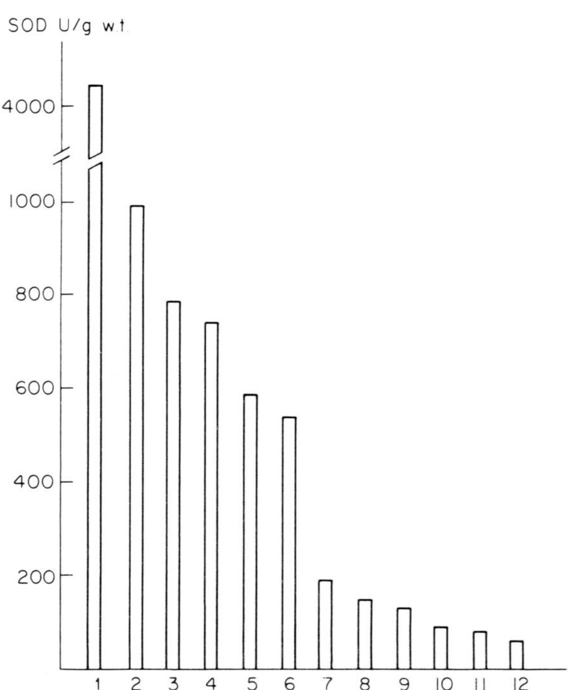

Fig.5. *SOD activities of various plant fruits, calculated per g wet tissue. 1. Cecei paprika. 2. Radish root. 3. Red cabbage. 4. Tomato paprika. 5. White cabbage. 6. Pea. 7. Spinach. 8. Garlic. 9. Lemon peel. 10. Onion. 11. Orange peel. 12. Potata tuber.*

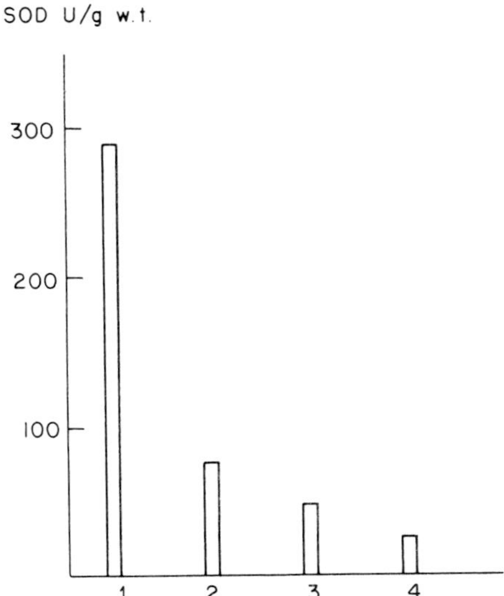

Fig.6. *SOD values of paprika leaf as a function of age. 1. 2-leaf stage. 2. 30-day leaf. 3. 50-day leaf. 4. 80-day leaf.*

It appears from the above that most SOD is contained in those parts of the plant with highest metabolic activity.

Changes of SOD in Normal and TMV Infected Tomato Plant Varieties

Fig.7 compares the SOD values in various parts of the 44-day tomato plant, 30 days after infection with TMV. In the tomato varieties, SOD could be detected only in the 30-day plants.

The high SOD content of the stem is obvious at first glance. This is related to the high dry matter content of the stem. It is also striking that the SOD values are lower in the infected than in the non-infected plant parts. This picture exhibits a slight deviation only in the case of the sensitive plant Mm. The SOD values of the resistant strain change most significantly with infection.

For a clearer picture of the effect of TMV infection on the SOD, our results were first calculated in terms of SOD/mg protein dry weight. It should be mentioned that the protein values of all the TMV-infected plant parts were lower than those of the non-infected plant parts.

The resulting data are shown in Fig.8. It is again apparent that the SOD values of the stem are high. At the time of examination the SOD values were lower in every case at the site of infection in the leaves. The same holds for the SOD activities of the roots. It seems as if the SOD is stored in the stem until the plant overcomes the TMV infection. (With the exception of the Mm-Tml and resistant Mm-Tm2^2 strains, the SOD values were higher in the stem of the virus-infected plants than in the normal plants; cf. Fig.8).

Fig.9 compares the total SOD values of normal and infected tomato varieties. This is supporting evidence that the SOD values per mg protein dry weight do not differ substantially from one another. With the exception of two sensitive plants (columns 3 and 4), the SOD content of the virus-infected plant is less than that of the normal plant. The most significant difference can be observed in the tolerant and resistant strains (columns 1 and 6).

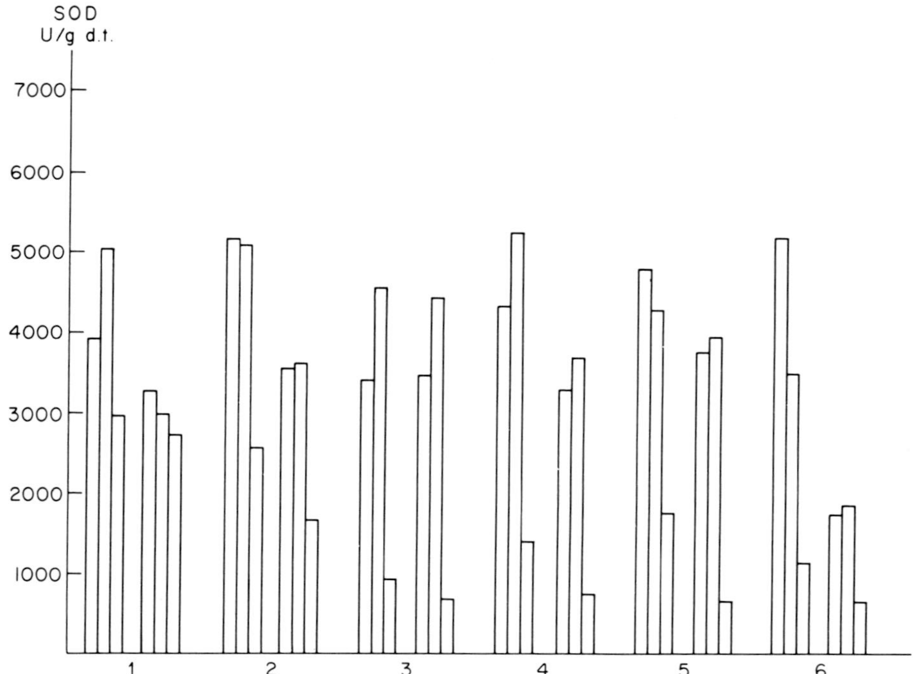

Fig.7. *Comparison of normal and TMV-infected forms of various Lycopersicon esculentum varieties at the age of 44-days. (Column sequence: root, stem, leaf. Comparative values given in SOD U/g dry tissue weight.) 1. Money-maker (Mm)-Tml normal and infected. 2. Mm-Tml/+ normal and infected. 3. Mm normal and infected. 4. K 363 normal and infected. 5. S.A.D. normal and infected. 6. Mm-Tm2² normal and infected.*

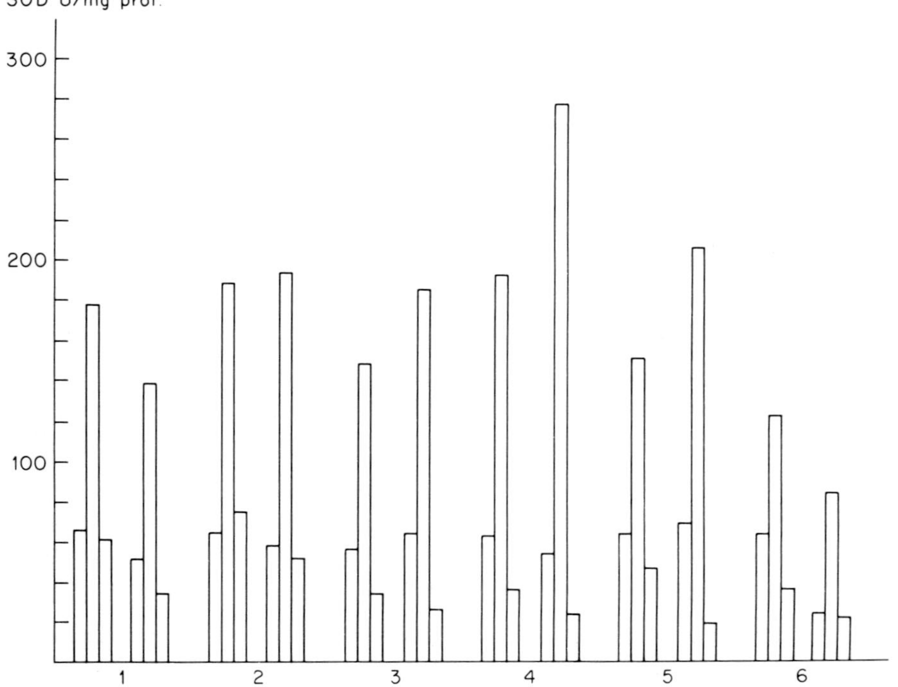

Fig.8. *The same 30-day varieties as in Fig.7, but here the SOD activities are calculated in U/mg protein dry weight.*

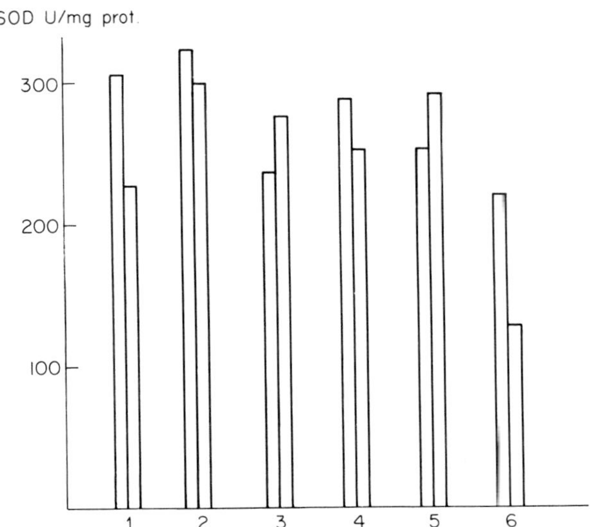

Fig.9. *Total SOD values of normal and infected individuals of the 30-day tomato plants referred to U/mg protein dry weight.*

SOD cannot be used as an indicator enzyme for the detection of TMV infection, for it is late in appearing in normal and infected outdoor tomato plant parts. The SOD values of infected plant parts exhibit significant differences, but normal and infected plants cannot be differentiated from one another solely on the basis of the differences in the SOD values. There is only one period in the plant development when SOD can be detected and can yield well-measurable quantitative values. SOD disappears from the 50-day plant parts, irrespective of whether the infection is withstood or not.

The appearance of SOD in the plant parts and its disappearance at the end of the development require further, more detailed investigations, extending to more sampling days, in both normal and TMV-infected plants; in our view this will permit a more exact understanding of the role of SOD in the metabolism of normal and TMV-infected plants. We are carrying out such examinations at the moment.

CHANGES OF SOD IN ANIMAL TISSUES

In this section we shall deal with the quantitative changes of SOD in rat tissues from several aspects, and also include some human SOD measurements.

(a) Fig.10 illustrates the SOD values of normal and pathological human placentas averaged according to age. The pathological placentas originated from spontaneous abortions or from the interruption of pregnancies in the event of foetal death, abortion, Rh incompatibility, mola hydatidosa, etc. It can be observed clearly that the SOD values of the pathological placentas are about half those of the normal placentas. The placentas which we term normal originated from normal births or from the interruption of normal pregnancies by request. Quantitative determination of SOD in human placentas provides supporting evidence retrospectively for the existence of disease since a placental SOD value lower than normal is always the consequence of some pathological process. (Kovács *et al.*, 1975).

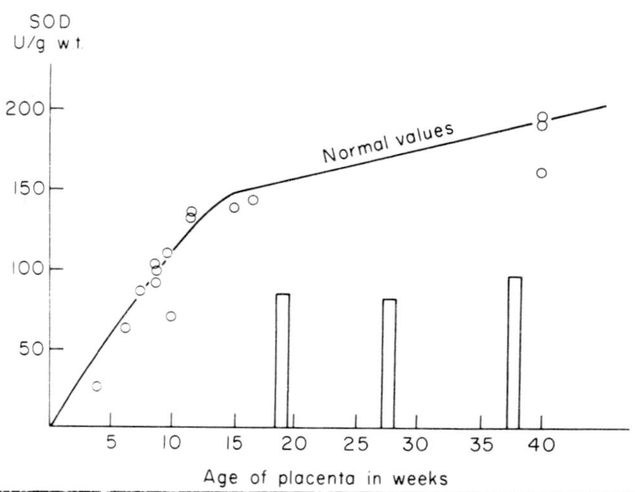

Fig.10. *SOD values of normal and pathological placentas. The variation of the normal SOD values is shown by the curve, while the SOD averages for the various pathological placentas are given in column graphs.*

(b) Fig.11 shows the SOD values of the following organ homogenates in developed (4-month) normal rats: liver, kidney, testis, spleen, brain, lung, pancreas, myocardium, skeletal muscle and RBC hemolyzate (Matkovics *et al.*, 1976). It can be seen that the SOD activity of the liver is outstandingly high, while the SOD values of the kidney and erythrocyte hemolyzate also exceed 1000 U/g wet tissue. If similar comparative studies are carried out on other vertebrates, it can be stated that the SOD proportions are roughly the same in the various organ homogenizates as in the rat, i.e. the SOD value is highest in the liver, followed by the kidney, etc. (Matkovics *et al.*, 1976; Pham Van Hien *et al.*, 1974; Hartz *et al.*, 1973).

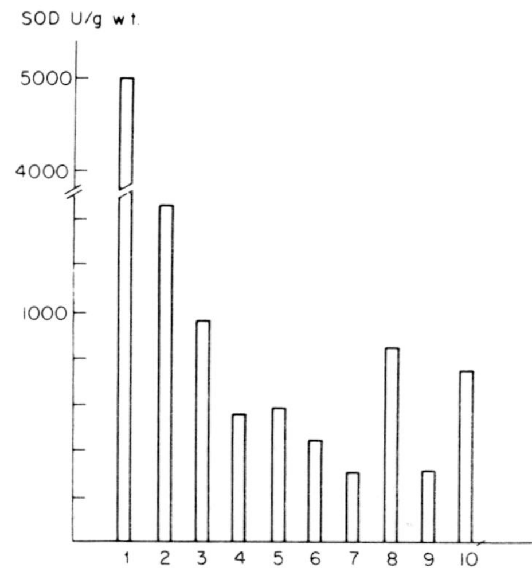

Fig.11. *SOD values of the main tissue and organ homogenizates of 4-month-old rats. (Scatter of measurements: ± 5 – 10%.) 1. Liver. 2. Kidney. 3. Testis. 4. Spleen. 5. Brain. 6. Lung. 7. Pancreas. 8. Heart. 9. Muscle. 10. Haemolyzate.*

(c) Fig.12 presents the age-dependent variations of the SOD in homogenates of the six largest organs. These organs can be easily prepared in appropriate amounts from newborn rats. Thus, comparisons were made of the SOD contents of the liver, kidney, brain, lung, myocardium and erythrocyte hemolyzate. The SOD value for the erythrocyte hemolyzate of 8-day-old rats could not be evaluated because it is very low (7 - 10 U/g w.t.). In Fig.12 the low nature of the first and sixth columns is striking. This means that the SOD contents of the organs increase progressively up to the second month, i.e. until the sexual maturity of the rats, and subsequently gradually decrease.

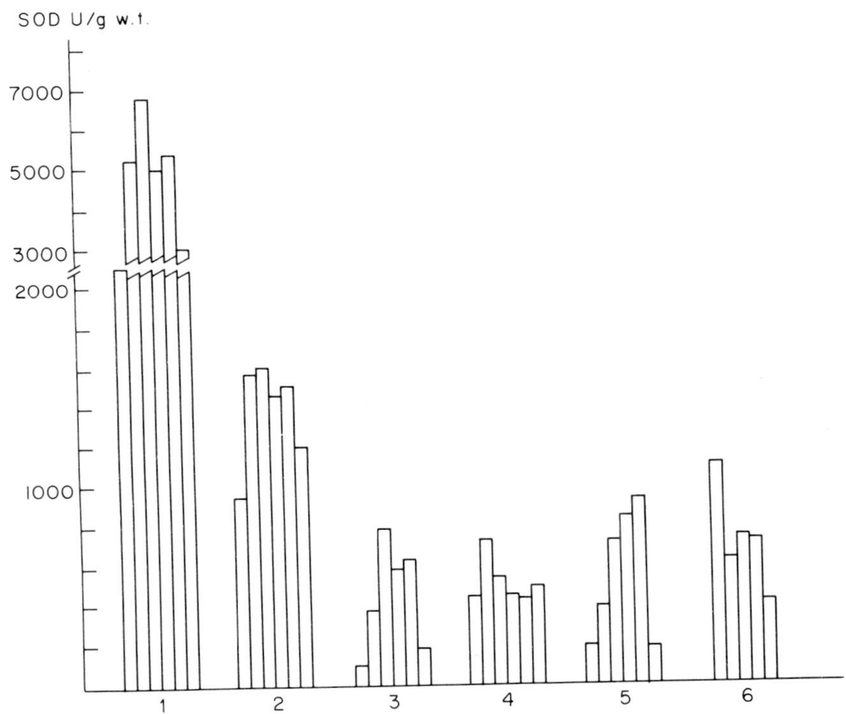

Fig.12. *SOD values of organ homogenizates from rats of various ages, in U/g wet tissue. Columns 1-6: 8-day, 1-month, 2-month, 4-month, 6-month and 14-month rats. 1. Liver. 2. Kidney. 3. Brain. 4. Lung. 5. Heart. 6. RBC haemolyzate (here the 8-day value is very low, from 7 - 10 U/g wet tissue).*

For easier comparison, the SOD values of the organ homogenates from rats of various ages are also tabulated numerically in Table I.

It can be clearly seen that the variation in SOD follows the development, insofar as an increase in the values can be observed during the development, with a subsequent stagnation and finally a regression. The independence of the organs has not yet developed in practice in 8-day suckling rats; this is well reflected in the low SOD values. The SOD decrease in older rats can be attributed in our view to secondary, radical-caused damage of the SOD molecule, which results in a decrease in activity and hence radical-caused damage of the tissues is enhanced with advance in age. It is obvious that the superoxide radical is not the only radical damaging the enzyme structure; the hydroxyl radical, $OH^{\cdot}$, formed in spontaneous reactions of H_2O_2 and of H_2O_2 and O_2^- (Haber-Weiss reaction), must also be included.

TABLE I

SOD Values of Organ Homogenates from Rats of Various Ages, in the Sequence of Figure 12, in U/g w.t.

Tissue	8 days	1 month	2 month	4 month	6 month	14 month
Liver	2100	5200	6800	5000	5350	3000
Kidney	950	1585	1600	1450	1505	1200
Whole Brain	100	395	795	590	635	180
Lung	450	720	545	450	430	490
Heart	200	395	720	850	940	190
Erythrocytes	-	1100	635	750	720	420

(d) There is a striking SOD decrease in rats made diabetic with streptozotocin, if these rats are maintained on the same diet as the others after the development of diabetes. This change, which occurs for all organs, is illustrated in Fig.13. The decrease is sometimes so marked that the SOD values measured in the diabetic rat organ homogenates are as little as 1/3 or less of the normal values.

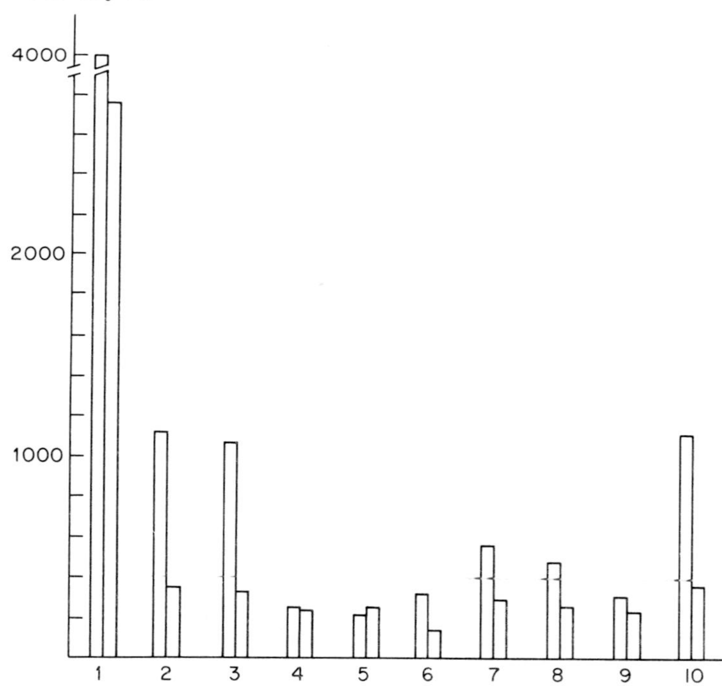

Fig.13. *Comparison of SOD values of 10 main organ homogenates from rats of the same age, in normal cases and ca. 2 months after streptozotocin treatment, when the high blood sugar level had become constant. 1. Liver. 2. Kidney. 3. Testis. 4. Brain. 5. Lung. 6. Pancreas. 7. Spleen. 8. Heart. 9. Muscle. 10. RBC haemolyzate. (In every case the first column gives the normal values, and the second those for streptozotocin-induced diabetes, in SOD U/g wet tissue.)*

Our investigations into the factor causing the SOD decrease were continued by providing sexually mature rats not with drinking water, but with 0.5% H_2O_2 ad libitum for about 2 months. The controls received water. Both groups were fed with a healthy rat diet. The animals were then killed, and the SOD values of the largest organs and the erythrocyte hemolyzate were examined. Our results are presented in Fig.14. The Figure clearly reveals that H_2O_2 is not an SOD inhibitor, but an important SOD inducer, in that various increases in the SOD values could be observed in all the organs, the value being nearly doubled in the liver. As with the SOD decreases in diabetes, these increases were significant, in most cases being well in excess of the standard deviations of the measurements.

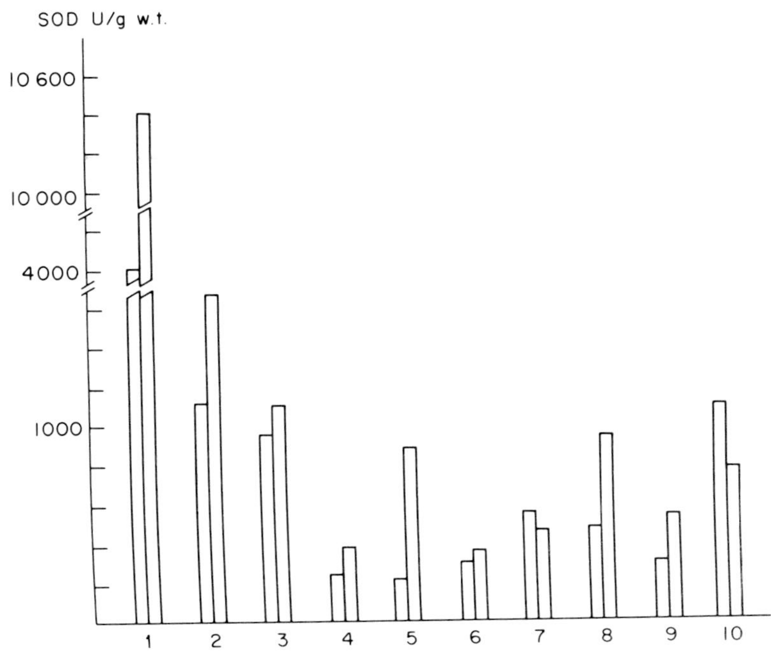

Fig.14. *Effect of drinking 0.5% H_2O_2 for ca. 2 months on the SOD activities of the main organs. 1. Liver. 2. Kidney. 3. Testis. 4. Brain. 5. Lung. 6. Panceas. 7. Spleen. 8. Heart. 9. Skeletal muscle. 10. RBC haemolyzate. (The first column gives the normal, second the treated SOD values.)*

It may be noted that replacement of water by dilute H_2O_2 significantly increased not only the SOD values in the tissue homogenates of the large organs, but also the catalase values.

(e) Let us now return to human examinations, as the aim of all the animal experiments described was to provide experimental support for our observations in our earlier SOD measurements with normal and diabetic human hemolyzates in a large number of cases. The fasting blood sugar values of the diabetics examined lay in the range 110-160 mg%. The patients regulated their blood sugar values with oral antidiabetics, and blood samples were taken at their half-yearly control check-ups. The SOD values of distilled water hemolyzates of erythrocytes washed with physiological saline are listed in Table II, together with comparative values obtained in an analogous manner from normal individuals.

TABLE II

SOD and GP of normal and treated diabetic erythrocyte hemolyzates, and the mean errors of the measurements. For SOD, the activities of the supernatants of the chloroform:ethanol-treated hemolyzates are also given.

	Enzyme Sources	$\bar{X} \pm S$
SOD:	Normal hemolyzate	1169 ± 273
	Diabetic hemolyzate	41 ± 57
	$CHCl_3$: EtOH extracts of normal hemolyzate	1173 ± 98
	$CHCl_3$: EtOH extracts of diabetic hemolyzate	223 ± 160
GP:	Normal hemolyzate	6.43 ± 0.65
	Diabetic hemolyzate	16.45 ± 0.58

There is a significant difference between the SOD values of the erythrocyte hemolyzates from normal and diabetic individuals. This difference, although smaller, can still be observed in the supernatants after chloroform: ethanol treatment of the hemolyzates.

Table II also shows gluthathione peroxidase (GP) values for erythrocyte hemolyzates of normal individuals and treated diabetic cases. Here too a significant difference is found between the normal and diabetic enzymatic activity values, those for the erythrocyte hemolyzates of diabetics being about 3 times higher than for normal individuals.

In the present stage of our work it can be said that the amount of SOD is appreciably decreased in organ and tissue homogenates of diabetic rats (see Fig.13). In our view, based on the animal experiments, the decrease is not caused by an increase in the blood and tissue H_2O_2 concentrations, but is the consequence of other changes, as shown by the tissue SOD increase observed when the animals drank H_2O_2 solution. We assume that the H_2O_2 concentration increase in the organs and tissues reverses to some extent the SOD-catalyzed reaction and an increase in the amount of O_2^- induces synthesis of SOD.

In experimentally-induced diabetes, therefore, the tissue SOD decrease occurs for other reasons. It may be mentioned that as assumed by Halliwell (1974), an important role is played by GP in human diabetes, partly in tissue H_2O_2 decomposition, and partly in the oxidation of reduced glutathione; according to Eggleston and Krebs (1974), the change in the concentration of reduced glutathione regulates tissue oxidation directly (pentose phosphate cycle).

We hope to continue studies on the changes in diabetic oxidoreductases with particular reference to SOD and hope to elucidate the role of these factors in diabetic metabolism changes.

The following workers took part in the measurement and evaluation of the data, and in the active solution of the problems: Dr. Katalin Kovács, Pham Van Hien, R. Novák, Hoang Duc Hanh and L. Szabó. The streptozotocin was kindly provided by Dr. I Zador (Ayerst Lab., New York). Special thanks are due to Mrs. Sz. I. Varga for her untiring efforts in helping to prepare this manuscript.

REFERENCES

1. Do Quy Hay, Kovács, K., Matkovics, I. and Matkovics, B. (1975). *Biochem. Physiol. Pflanzen (BPP), 167,* 357-359.
2. Eggleston, L.V. and Krebs, H.A. (1974). *Biochem. J. 138,* 425-435.
3. Flohé, L. (1971). *Klin. Wschr. 49,* 669-683.
4. Fridovich, I. (1975). In "Free Radicals in Biology" (W.A. Pryor, ed.) Vol. I, Academic Press, New York and London.
5. Halliwell, B. (1974). *New Phytol. 73,* 1075-1085.
6. Hartz, J.W., Funakoshi, S. and Deutsch, H.F. (1973). *Clin. Chim. Acta 46,* 125-132.
7. Keele, B.B. Jr., McCord, J.M. and Fridovich, I. (1971). *J. Biol. Chem. 246,* 2875-2880.
8. Kovács, K., Szónyi, D.G. and Matkovics, B. (1975). *Acta Biol. Szeged 21,* 99-11.
9. Lowry, O.H., Rosebrough, H.J., Lewis Farr, A. and Randall, R.I. (1951). *J. Biol. Chem. 193,* 265-275.
10. Matkovics, B., Fodor, I. and Kovács, K. (1975). *Enzyme 19,* 285-293.
11. Matkovics, B. and Varga, Sz.I. (1976a). In "A Biológia Aktuális Problémái" (Gy. Csaba, ed.), in press, Vol.7 (in Hungarian), Medicina Könyvkiadó V. Budapest, Hungary.
12. Matkovics, B. (1976b). *Acta Univ. Lodziensis, Folia Biochimica et Biophysica, Sera III, Nr.1,* in press.
13. Matkovics, B., Novák, R., Hoang Duc Hanh, Szabó, L., Varga, Sz.I. and Zalesna, G. (1976). *Comp. Biochem. Physiol. Ser. 56B,* in press.
14. Matkovics, B., Novák, R., Hoang Duc Hanh, Szabó, L. and Varga, Sz.I. (1976). *Comp. Biochem. Physiol., Ser. B.,* under publication.
15. Misra, H.P. and Fridovich, I. (1972). *J. Biol. Chem. 247,* 3170-3174.
16. Mészöly, Gy. and Vidéki, L. (1963). *Duna-Tiszaközi Mezogazdasági Kisérleti Intézet Évkönyve,* Yearbook of the Duna Tiszakozi Agricultural Experimental Institute, pp. 17-37 (in Hungarian).
17. Paglia, D.E. and Valentine, W.N. (1967). *J. Lab. Clin. Med. 70,* 158-169.
18. Pham Van Hien, Kovács, K. and Matkovics, B. (1974). *Enzyme 18,* 341-347.
19. Simon, L.M. Fátrai, Zs., Jónás, D.E. and Matkovics, B. (1974). *Biochem. Physiol. Pflanzen (BPP) 166,* 387-392.

ORGOTEIN, THE DRUG VERSION OF BOVINE Cu-Zn SUPEROXIDE DISMUTASE:

I. A SUMMARY ACCOUNT OF SAFETY AND PHARMACOLOGY IN LABORATORY ANIMALS

W. HUBER and M.G.P. SAIFER

Diagnostic Data Inc.
518 Logue Avenue, Mountain View
California 94043, U.S.A.

HISTORY

The beginning of our work some 12 years ago was motivated by the recognition that successful treatment of chronic inflammatory disease remained one of the great unsolved problems in medical therapy. Then, as now, all effective anti-inflammatories had side effects that become more serious the longer the treatment. Treatment of chronic disease, to remain useful, has to be both efficacious and safe for however long therapy is necessary. Thus, a molecule of mammalian rather than synthetic origin appeared to us an attractive solution if it were safe in target species and could be produced in sufficient amounts and economically.

Of the three major groups: proteins, lipids and carbohydrates, the first seemed the most attractive in view of a considerable body of literature data indicating their pharmacodynamic efficacy, even though their safety was questioned, at least for the enzymes that had been used, i.e. proteases, amylases and lipases (Innerfield, 1960; Abderhalden, 1961).

While many of the early enzyme preparations were shown to be immunogenic in man, we hoped that an anti-inflammatory protein could be found with minimal immunogenicity and also be toxicologically safe.

A review of the literature indicated that the effects described when different proteins were used as drugs were mostly anti-inflammatory. This literature has been reviewed recently by Wolf and Ransberger (1972). We were impressed that these effects were germane to a great diversity of protein preparations from numerous tissues, (liver, testes, placenta, kidney, etc.) of various mammals. This led us to the concept that the effective molecule could well be the same protein present as the active component in many such preparations. The isolation of a minor component from a complex mixture appeared in the early 1960s to have come within reach, in view of developments in gel filtration, ion exchange chromatography, gel electrophoresis, and other preparative methods for protein isolation.

During 1965 we isolated a protein from a bovine liver extract which proved identical with a protein isolated from a pharmaceutical preparation of bovine testicular hyaluronidase (Wydase, Wyeth Laboratories, Inc.). The latter, when given in high doses, had been found to possess anti-inflammatory activity in urologic disorders (Schulte, 1964). Using its characteristic three-band pattern in thin-gel agarose electrophoresis for identification, the protein was further purified and found to contain both copper and zinc, in addition to a rather unusual amino acid composition. An identity with erythro-, hepato- and cerebro-cuprein was first suspected but later discarded, as none of the cupreins at that time had been described to contain zinc and because of significant differences from reported amino acid compositions. Our purified protein was shown to possess anti-inflammatory activity in several models of induced inflammation in laboratory animals, and preliminary toxicity and sensitization studies showed promising safety.

Thus, in early 1966 we had a promising substance (Huber *et al.*, 1968). How we determined that we had an effective and safe drug which could meet exacting pharmaceutical standards of identity, purity, and potency is described below.

We first saw the characterization of superoxide dismutase (SOD) by McCord and Fridovich (1969) in early 1970. Comparison of their analtyical data with ours indicated that orgotein and superoxide dismutase were probably identical, a fact which was established beyond doubt by our assaying orgotein, which we had obtained from ten species of mammals, birds and fish, in their cytochrome *c* reduction assay and confirmed by an exchange of samples. This also established the identity of orgotein congeners with the various cupreins, the first member having been isolated by T. Mann and D. Keilin (1939) and described by them and later investigators during 30 years as copper-containing proteins of unknown function.

THE PRODUCT

As drug, orgotein[†] is supplied in single-dose vials of sterile, non-pyrogenic, lyophilized powder, stabilized with 2 mg of sucrose per milligram of protein. Product quality is controlled by the procedures listed in Table I which insure that at least 80% of orgotein protein is Cu-Zn SOD.

In this dosage form, the product is highly soluble in aqueous solvents (to over 100 mg/ml) and is remarkably heat resistant. The shelf life at 40°C is considerably in excess of five years (Fig.1).

SAFETY

Extensive safety evaluations have been carried out with orgotein. These include acute, subacute and chronic toxicity studies, with assessment of hematology, biochemistry and histopathology, as well as reproduction and teratology studies in a variety of animal species by several routes of administration (Carson *et al.*, 1973). Single doses and a variety of repeated dose schedules up to 56 weeks have been used to determine whether any toxic

[†]Orgotein is the non-proprietary name assigned by the United States Adopted Names Council. Under the trade name Palosein it is available for use in horses and dogs from Diagnostic Data, Inc., Mountain View, California 94043. For human use, it is prepared by the same company under the trade name Ontosein and is presently available in the U.S.A. only for investigational use. (Huber *et al.*, 1971 - 1974).

effects would be elicited (Table II). Reproduction studies included observation of sexual maturation, mating success, gestation, organogenesis, parturition and lactation. No toxic effects have been seen in any of these studies.

TABLE I

Orgotein Quality Control

Procedure	Specification of Orgotein for Administration to Humans
Atomic Absorption Analysis for Cu and Zn	1.9 ± 0.4 gram atoms per mole
Thin Film Agarose Gel Electrophoresis	>80% of protein in position of major SOD charge isomers
Specific Absorbance	A_{280}/mg = 0.23 ± 0.05 at pH 8.5
Fluorometric Measurement of Albumin	<0.5% of protein, using crystalline bovine serum albumin as standard
Superoxide Dismutase Activity according to McCord and Fridovich (1969)	3,100 ± 500 U/mg protein at pH 7.8
Pyrogenicity per USP XIX	Non-pyrogenic at 2.5 mg/kg in rabbits

Orgotein has been safely administered in man and animals by a large variety of routes, i.e. intravenous, intra-arterial, intraperitoneal, intramuscular, subcutaneous, intra-articular, subconjunctival, intraocular, intrathecal, intramural, intrapulmonary and topical.

Aspects of sensitization have been explored in guinea pigs using intradermal and intraperitoneal sensitization routes with intradermal and intravenous challenge. Challenge doses given to 54 horses and 40 dogs after termination of one or more courses of orgotein therapy showed no local or systemic allergic reactions, immediate or delayed.

The weak immunogenicity of orgotein may result from some general principles, such as low content of aromatic amino acids, excellent solubility, compact size, and favorable clearance kinetics. In addition, the degree to which the structure of the Cu-Zn SOD proteins has been conserved during evolution is impressive. Difference index calculations (Metzger *et al.*, 1968) for orgotein from 11 animal species (cow, sheep, rabbit, pig, guinea pig, man, dog, chicken, horse, mouse and rat) place it among the more highly conserved proteins studied. Beyond amino acid composition, aspects of secondary and tertiary structure conservation are also remarkable. Immunologic cross reactivity could be demonstrated for all mammalian species tested (Table III) and molecular hybridization occurs for all species pairs tested, including mammal-fish pairs (Fig.2) and even fish-fungi pairs (Tegelström, 1975). While immunologic cross reaction measures surface alterations, molecular hybridization measures compatibility of the intersubunit interfaces. The natural function of this enzyme must be critcal, indeed, for selection pressure to have maintained such evolutionary stability (Steinman and Hill, 1973; Tegelström, 1975).

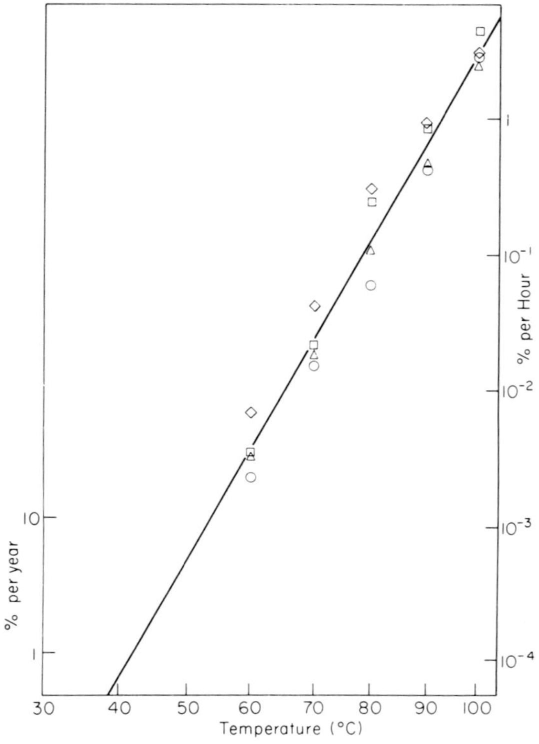

Fig.1. *Temperature dependence of rate of disappearance of orgotein band protein. Vials from four commercial lots of Palosein brand of veterinary orgotein designated □ ; ○ ; △ ; and ◇ ; were heated at the temperature indicated for various times. Their contents were then dissolved and electrophoresed on pH 8.4 tris-glycine buffered agorose gels. After staining the protein with 0.2% amido black in 5% acetic acid, the gel strips were scanned at 590 nm and integrated. The rate of disappearance of the three principal charge isomers (bands 1, 2 and 3) as a percentage of total protein was determined during the initial 20% change for each lot at each temperature and is here plotted as log-rate $^1/T$. The fraction of protein in bands 1, 2 and 3 decreases more rapidly during heating than does SOD activity, solubility or anti-inflammatory potency. It is thus the most sensitive of the stability-indicating parameters used for quality control.*

TABLE II

Orgotein Toxicity Studies

Study	Species	No. of Animals Treated with Orgotein	Test Dose (mg/kg) Route	Dosage Schedule
Single dose (acute)	Mouse	10	100 i.v.	Once
		10	60 i.p.	Once
		10	4,000 s.c.	Once
	Rat	5	60 i.p.	Once
	Monkey	2	60 i.m.	Once
		7	10 i.v.	Once
	Rabbit	10	10 i.v.	Once
Ten dose Series (subacute)	Guinea Pig	20	0.4 i.m.	2 x/week for 5 weeks
	Monkey	4	0.4 i.m.	2 x/week for 5 weeks
Chronic	Monkey	10	0.2 i.m.	5 x/week for first 5 weeks; 3 x/week for last 47 weeks
		10	2.0 i.m.	2 x/week for first 5 weeks; 3 x/week for last 47 weeks
	Rat	56	0.2 i.m.	5 x/week for 56 weeks
		56	2.0 i.m.	2 x/week for 56 weeks
Reproduction and Teratology	Rat	30 males	0.2 and 2.0 i.m.	Daily from 4 weeks age to 13 weeks (mated at 11 weeks)
		60 females	0.2 and 2.0 i.m.	Daily from 2 weeks prior to mating to parturition[a]
		66 females	0.2 and 2.0 i.m.	Day 6 to 16 of gestation
		66 females	0.2 and 2.0 i.m.	Day 15 gestation to day 21 post-partum
	Rabbit	40 females	0.2 and 2.0 i.m.	Day 6 to 18 of gestation

[a]One-half by natural delivery, one-half by cesarean section at day 13 of gestation.

TABLE III

Immunologic Cross-Reaction of Some Orgotein Congeners

SOD Source	Anti-Bovine SOD Serum	Anti-Rat SOD Serum	Anti-Human SOD Serum
Bovine	100%	24%	3%
Rat	10%	100%	1%
Man	10%	24%	100%
Horse	2%	32%	6%
Sheep	ND	ND	14%
Dog	2%	ND	ND
Guinea Pig	2%	ND	ND
Mouse	2%	ND	ND
Chicken	<1%	ND	ND

Sera were prepared in New Zealand White rabbits. Cross-reaction was measured by titration of diluted serum with SOD to equivalence as measured on zymograms. Free SOD in antigen excess was well separated from soluble complexes.

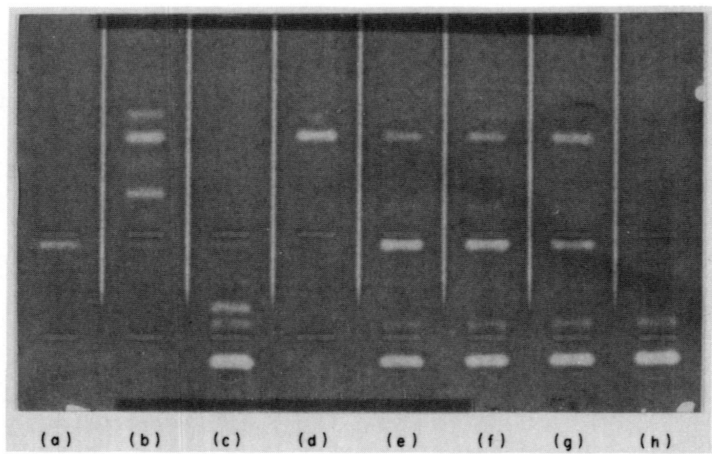

Fig.2. *Molecular hybridization of orgotein congeners. Orgotein preparations or mixtures at 5 mg/ml of each were incubated at 50°C in 1 mM tris-HCl (pH 8). Samples were diluted 1:200 after 150 min unless otherwise indicated and subjected to agarose gel electrophoresis for 20 min at 250 V, using pH 8.4 tris-glycine buffer. The wells across the center of the figure were used for sample application. Zymograms were developed with NBT-riboflavin stain. During incubation, the quantity of native SOD decreased while the amount of hybrid activity increased. The orgotein preparations used had been isolated from the erythrocytes of guinea pig and human and from the pancreas of the swordfish. Swordfish orgotein (a); swordfish-human mixture (b); swordfish-guinea pig mixture (c); human orgotein (d); human-guinea pig mixture (e); human-guinea pig mixture after 60 min (f); human-guinea pig mixture after 15 min (g); guinea pig orgotein (h).*

A critical test of immunologic "tolerance" to orgotein was carried out in mice. Adult animals were subjected to heminephrectomy and then given a series of 12 injections (0.04 or 2.0 mg/kg in 6 weeks; 3/wk, 2/wk and 1/wk for 2 weeks each). Twenty-four hours after the last injection the remaining kidney was removed and examined for immunopathology by light and fluorescence microscopy. There was no increase in IgG or C'3 deposited in the basement membranes, nor was any histopathology observed using the methods of Oldstone (1973).

PHARMACOLOGY

Using standard methods of evaluation and a wide range of doses, the effects of orgotein on different organ systems has been explored as part of an extensive preclinical evaluation program.

General

Central and Autonomic Nervous System. In all of the toxicity studies in mice, rats, guinea pigs and monkeys using doses from 0.2 to 100 mg/kg, there have been no effects on behaviour which would suggest any effect of orgotein on the central or autonomic nervous system. Orgotein up to 10 mg/kg intravenously in mice did not effect either the performance of the mice on a rotating rod nor the sleeping time under the influence of a barbiturate (Thompson, 1974).

Cardiovascular and Respiratory Systems. Orgotein has been given intravenously in doses up to 10 mg/kg to dogs, cats, squirrel monkeys and stump-tailed Macaque monkeys. The compound was found to be essentially free of effects in these species on all the parameters measured, including blood pressure, heart rate, ECG, respiration rate, tidal volume, and the response to noradrenaline, isoprenaline, acetylcholine, vagal stimulation, and occlusion of the carotid artery.

Gastrointestinal System. In all our acute and chronic toxicity studies there has been no evidence of any adverse effect. Gross and microscopic histology of the gastrointestinal tract was unaffected. Food intake and growth were unaffected and there was no overt evidence of an effect on gastrointestinal motility.

Immune System. The remarkable immunological tolerance to orgotein seen in cross-species experiments led to investigations of its possible immunosuppressive effects. These included differentiation of anti-body producing cells *in vitro*, rejection of godfish scale homografts, lymphochoriomeningitis pathology, and the graft-vs-host reaction. No suppression by orgotein was observed in any of these systems at doses ranging up to 50 mg/kg. The effect of treatment with orgotein at doses of 10 mg/kg given intraperitoneally on the development of Leishmaniasis lesions in infected guinea pigs also has been tested. Orgotein was given three days prior to and for two weeks after infection. The development of the delayed hypersensitivity reaction in the ears of the infected animals was monitored by ear thickness measurement. Continuous orgotein treatment had no effect on the onset or on the severity of symptoms of Leishmaniasis in guinea pigs (Thompson, 1974). This reaction is reported to be a manifestation of cell-mediated delayed hypersensitivity and these results are consistent with the lack of effect of orgotein on the delayed hypersensitivity reaction of guinea pigs to tuberculin (purified protein derivative) which has been observed in our laboratory.

Defense Against Infection. The effect of orgotein on acute bacterial infection in mice has been investigated. Orgotein was given to mice at 15 mg/kg subcutaneously six hours before and immediately prior to infection with *Diplococcus pneumoniae* Type 1 organisms at different dilutions. Orgotein had no effect on the survival time of the infected animals. In addition, orgotein was found to have no deleterious effect on the capacity of vaccinated mice to withstand infection with *D. pneumoniae* (Thompson, 1974).

Drug Interaction. In a typical screen, orgotein in doses up to 10 mg/ml was found not to interfere with the antibacterial efficacy of antibiotics and sulfa drugs.

In man and animal, orgotein was found not to interfere with the effect of steroidal and nonsteroidal anti-inflammatories. In fact, an additive effect is observed when orgotein is given together with suboptimal doses of these drugs. In no case did orgotein administration increase the side effects of other drugs.

Metabolism

Studies with orgotein to elucidate bioavailability, pharmacokinetics, tissue distribution and biotransformation in animals have been carried out with mice, rats, hamsters, guinea pigs, and dogs. Since safety considerations do not limit the administered doses, high doses and many routes of administration have been included (i.v., s.c., i.m., i.p., subconjunctival, intratracheal and intracisternal). Our studies to identify the fate of injected orgotein have been performed using its superoxide dismutase activity, as well as with radioisotopes of technetium (^{99m}Tc) and iodine (^{125}I) as tracers.

Each tracer has advantages for specific purposes. The six hour $t_{\frac{1}{2}}$ of ^{99m}Tc allows one to work in comparative safety with the millicurie quantities of labeled orgotein required for producing pinhole camera images of rapidly changing distributions (Fig.3). Superoxide dismutase assays make it possible to follow the course of absorption from injection sites, the kinetics of development of serum concentrations, and the distribution into tissues and excretions of the enzymatically active drug. ^{125}I-Iodo-orgotein is especially well suited for longer-term kinetic studies of distribution of the protein after the injected enzyme activity becomes undetectable. The results obtained with these techniques complement each other to produce a pattern of orgotein metabolism from the first minutes after injection until 10 to 14 days later when an iodine-labeled dose of orgotein has been essentially cleared from the body (Table IV).

Intravenously injected Tc-orgotein begins to accumulate in the kidneys of rats within two minutes after administration and by 15 minutes the label has been extensively transferred from the blood-pool to the kidneys (Fig.3a-d). In dogs, the process is somewhat slower, but the same pattern results. The larger kidney of the dog allows one to see that the label is shifting into the cortex of the kidneys as it is cleared from the circulation (Fig.3f). In neither rats nor dogs is there extensive excretion of the kidney-accumulated label into the urine during the first hours following injection of low doses of orgotein. Control experiments have indicated that the pattern of kidney

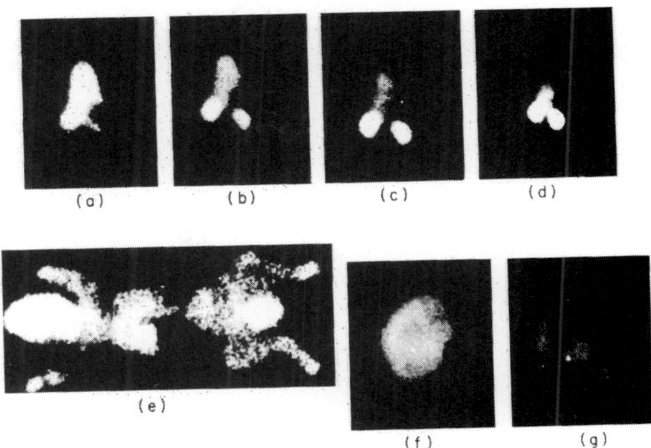

Fig.3. *Pinhole camera scintigrams of the distribution of ^{99m}Tc-labeled orgotein after injection. Orgotein was labeled with ^{99m}Tc in an acid solution of pertechnetate using Sn(II) as reducing agent (Lin, 1975) and injected without further purification. A rat was injected intravenously with 1.2 mg orgotein containing 2 mCi of ^{99m}Tc and imaged after 2 minutes (a); same rat after 5 minutes (b); after 10 min (c); 14 min (d). A rat was injected subcutaneously at the left side of the neck with a similar dose of labeled protein and was kept unanesthetized. After 3 hours it was sacrificed, the kidneys were removed, and the animal was skinned. The sensitivity of the scintillation camera was increased in order to image the distribution of ^{99m}Tc in the skinned carcass (e). A dog was injected intravenously with 5 mCi of Tc-orgotein and close-up scintigrams of one kidney were made after less than 1 hr (f); and after 2 hrs (g).*

accumulation observed for orgotein is not that of free technetium. Furthermore, SOD assays by a zymogram technique employing thin-film agarose gels in Tris-glycine buffer at pH 8.5 and a NBT-Riboflavin staining technique based on the method of Beauchamp and Fridovich (1971) show that the kidneys do contain about 75% of an i.v. dose of orgotein 35 to 45 minutes after injection in rats (Fig.4) and that bovine SOD is removed from the serum with a half-time of about six minutes (Fig.5). Dissection of canine kidneys two hours after i.v. injection revealed a selective distribution of bovine SOD in the canine kidney resulting in a 100:1 ratio of enzyme activity between the cortex and medulla, also confirming the Tc-pattern. That these clearance kinetics are not unique to the bovine molecule was established by injecting a mixture of 1 mg each of bovine and rat orgotein i.v. into a rat and following the simultaneous clearance of the two SODs by the zymogram technique. Bovine and rat SOD were cleared from rat plasma at indistinguishable rates until over 95% of each had been removed from the circulation.

When orgotein is administered by parenteral routes of administration other than i.v., bovine SOD appears in the serum in concentrations which increase during the first one to two hours and then gradually decline (Fig.5). The integral of serum concentration over time for a given route of administration, divided by that integral for the intravenous route, can be taken as a measure of the bioavailability of a drug administered by that route. Measured as SOD, the plasma bioavailability of orgotein has been found to be about 60% subcutaneously, 55% intracisternally, 50% intraperitoneally, 40% intratracheally, and 30% intramuscularly. These results have been obtained

TABLE IV

Excretion of Radioiodine
After Subcutaneous Injection of ^{125}I-Orgotein[a] in Dogs

Days Since Injection	Percentage of Radioactivity Recovered: Dog A Urine	Dog A Feces	Dog B Urine	Dog B Feces
1	26	0.4	39	0.6
2	28	0.2	34	1.1
3	24	0.1	6	0.1
4	12	0.1	8	0.2
5	4	0.01	3	0.2
6	1	0.05	2	0.03
7	0.7	0.02	0.6	0.03
8	0.3	0.00	0.4	0.00
9	0.3	0.00	0.3	0.00
10	0.2	0.00	0.1	0.00
11	0.1	0.00	0.1	0.00
12	0.1	0.00	0.1	0.00
TOTAL	96.7	0.9	93.6	2.3
Residual in Body[b]	2.5		4.8	

[a]Each dog received 2.4 mg of orgotein labeled with ^{125}I by New England Nuclear Corp. at 1.2 mCi/mg using Chloromine-T as oxidant.

[b]The thyroids accounted for 88 and 95% of the ^{125}I remaining in dogs A and B at sacrifice (2.2 and 4.7% of the recovered ^{125}I).

with rats except for the intratracheal route for which hamsters were employed for technical reasons. In the intracisternal series, it was possible to measure the rate of clearance of bovine SOD from the cerebrospinal fluid. Clearance was pseudo-first order with a half-time of about 75 minutes. Conversely, i.v. administered orgotein did not appear in the cerebrospinal fluid.

Since the bioavailabilities of orgotein administered parenterally by routes other than i.v. are all less than 100%, we have looked at the sites of injection for residual bovine SOD activity. Essentially complete clearance from peritoneum, CSF, synovial fluid, lungs and s.c. or i.m. injection sites does occur. In order to rule out the possibility that enzymatically active orgotein accumulates in some unanticipated tissue, whole-body recovery experiments have been performed. The incomplete calcification of bones in suckling mice (up to three days after birth) permits their complete homogenization for extraction of SOD. In these neonates, the total body content of soluble bovine SOD decreased with a half-time of three to four hours while their kidney contents rose during the first four hours after s.c. injection (Fig.6). The maximum recovery from kidneys accounted for 25 to 35% in various individuals, and urinary excretion accounted for another few percent. Thus, incomplete bioavailability of orgotein's SOD must result from either inactivation or insolubilization of the enzyme. These possibilities can be experimentally differentiated and appropriate experiments have been initiated. Incubation of orgotein at 37° with blood, kidney homogenates or in intact kidneys does not result in inactivation comparable to that seen *in vivo*. The natural mechanism of SOD inactivation is certainly worthy of further attention.

From the point of tissue injection, SOD spreads slowly by diffusion and more rapidly by the movement of interstitial fluid. Its activity can be traced by sectioning around an injection site and extracting each section for SOD assay. The gradual increase in the area infiltrated is accompanied by a progressive loss of total SOD recoverable from the area of the injection site (Table V). After injection under the conjunctiva of guinea pig eyes, bovine SOD was found to penetrate over the course of several hours into the inner structures of the eye, including the uvea, lens, retina, vitreous humor, and optic nerve (Table VI).

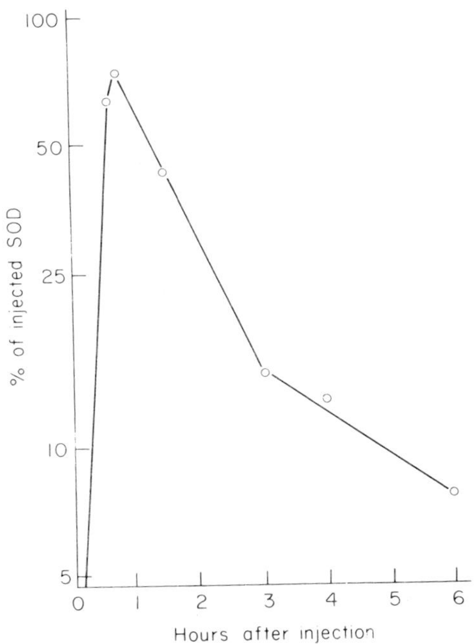

Fig.4. *Orgotein content of rat kidneys after intravenous injection. The kinetics of accumulation and elimination of bovine SOD in the kidneys of rats after intravenous injection of 1 mg of orgotein were followed for 6 hours. Pairs of animals were sacrificed at the indicated times and kidney homogenates were prepared and diluted serially for the quantitation of bovine SOD with NBT-riboflavin stained zymograms. Only a few per cent of the injected dose appeared in the urine during the first hour. No bovine SOD was recovered from urine more than 2 hours after injection.*

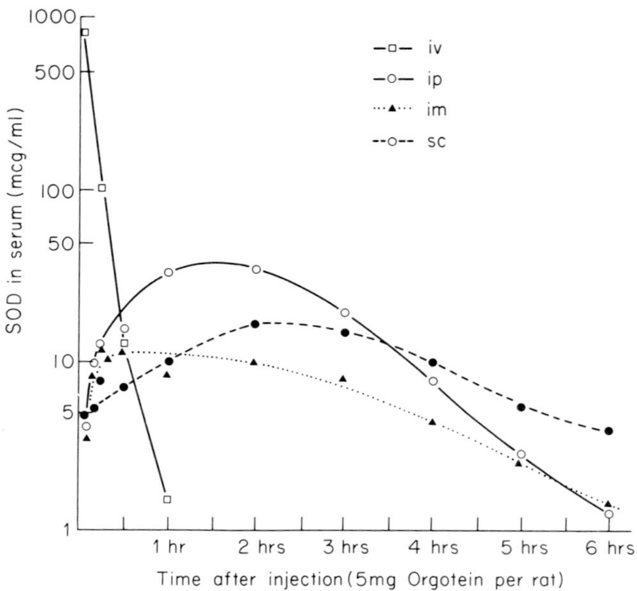

Fig.5. *Superoxide dismutase levels in serum after injection of orgotein by four routes. Groups of 5 to 8 rats were injected with 5 mg of orgotein by the indicated routes. Blood samples were taken from the tail at the intervals shown. SOD was measured in serum by the zymogram method after dilution to levels suitable for quantitation.* □ *i.v.; 0 i.p.;* ● *s.c.;* ▲ *i.m. Results are presented in micrograms of SOD per ml of serum.*

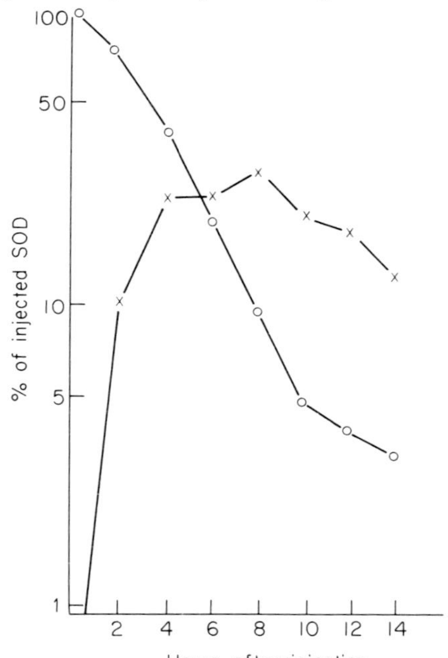

Fig.6. *Distribution kinetics of orgotein in suckling mice. The accumulation of orgotein in the tissues of 2 to 3 gm mice after subcutaneous injection of 100 mcg of orgotein in 0.01 ml saline was followed by zymograms. The fraction of the SOD injected which was recovered in homogenates of the kidneys and the rest of the body at various times is shown. Kidney recovery, X; Body recovery,*

TABLE V

Injection Site Distribution of Orgotein

Route of Injection	Time Since Injection (hrs)	Distance from Injection Site for 50%[a] of Residual SOD (cm)	Total % of SOD Recovered
i.m.	0.5	0.7	66
guinea pig	1.0	1.0	45
(thigh)	2.0	1.2	36
s.c.	0	1.0	100
rat	0.5	1.0	96
(dorsal skin)	1.0	1.2	80
	2.0	1.0	76
	4.0	1.0	63
	6.0	2.5	32
	8.0	2.0	17

The recovery of SOD from tissue sections near the site of orgotein injection was measured in zymograms. Total recovery from whole thigh muscle (i.m.) or 3 cm diameter dorsal skin area (s.c.) is expressed relative to the injected dose.
[a]80% for s.c. distribution study

TABLE VI

Bovine Superoxide Dismutase Recovery (mcg)[a] From Eye Tissues After Subconjunctival Injection of 1 mg Orgotein[b] in Guinea Pigs

Tissue[c]	Hours Since Injection		
	1	4	6.5
Conjunctiva	150	25	20
Sclera	20	10	10
Cornea	25	3	2
Uvea	10	5	5
Lens	<<0.2	0.2	0.4
Retina and Vitreous	7	1.0	0.2
Optic Nerve	12	0.8	1
Blotting Paper	20	15	12
Total in both eyes	244	60	50

[a]Measured on NBT-riboflavin stained zymograms which separated bovine and guinea pig SODs completely.
[b]Administered to each eye in 0.1 ml saline.
[c]Dissected tissues from both eyes were combined for homogenization.

Despite the primitive state of knowledge of the biotransformation of orgotein, we expect that the ultimate fate of the protein is the same as for other proteins, i.e. breakdown to amino acids and resynthesis, metabolism, or excretion. The events occurring between its injection and its elimination somehow result in profound influences upon the course of certain disease processes. Our metabolism work to date has demonstrated tissue distributions and bioavailabilities compatable with its use as a drug, using a variety of parenteral routes of administration. The remarkable biological stability of orgotein is thus another property making it ideally suited for drug use.

Pharmacodynamics

The outstanding safety of orgotein, at least in part, is due to the fact that it remains extracellularly after administration and thus does not interfere with intracellular functions. It appears to be a pure anti-inflammatory without the anti-pyretic, analgesic or immunosuppressive actions of other drugs used in inflammatory diseases.

The initial evidence for the anti-inflammatory efficacy of orgotein came from models of induced inflammation in laboratory animals. None of these models is in any way specific for any drug. Their common denominator is an inflammatory reaction which is to be inhibited or suppressed by the test drug. The models listed in Table VII, among others, have been used by us to explore the efficacy of orgotein. In most of the induced inflammations mentioned in Table VII, parameters used for assessment of efficacy included reduction of weight and volume of inflamed tissue.

TABLE VII

Experience With Orgotein in Models of Inflammation

Model	Species	Total No. of Animals Used	Effective Dose Range (mg/kg)	Reference to Methods
Carrageenan-Induced	Rats	380	≥1.0	Winter *et al.*
Foot Edema	Mice	900	≥1.0	(1962)
Carrageenan-Induced	Rats	325	≥0.5	Benitz and
Abscess	Mice	3400	≥0.5	Hall (1963)
Carrageenan-Induced Pleurisy	Rats	270	≥1.0	Vinegar *et al.* (1973)
Reversed Arthus	Guindea Pigs	3000	≥0.05	Ungar *et al.*
Reaction	Rats	200	≥0.5	(1959)
$ZnCl_2$ Necrosis	Dogs	40	≥0.5	Elgars (1975)
Hypodermin Reaction	Horses	28	≥0.01	Cushing *et al.* (1973)
Adjuvant-Induced	Rats	200	≥3.0	Beck and
Foot Edema	Mice	50	≥2.0	Whitehouse (1974)
Autoimmune Glomerulo-Nephritis	Mice $(NZB \times W)F_1$	24	2.0	Oldstone and Dixon (1969)

Induced pleurisy in rats provides a more sophisticated model of inflammation, as both exudate volume and cellular infiltration can be quantitated simultaneously. When assessed in blind experiments at 24 hours after carrageenan injection into the pleural cavity, orgotein produces statistically significant decreases in these parameters. In the dose ranges studied, inhibition of leukocyte infiltration parallels that of exudate volume. The reduction in leukocyte infiltration due to orgotein is apparently due to the reduction of polymorphonuclear neutrophil leukocytes (PMNs) in the exudate (Table VIII). This shows that orgotein can influence the PMN response always seen as an early signal of an inflammatory event.

TABLE VIII

Range of Effects of Orgotein in the Carrageenan-Induced Pleurisy Model in Rats

Treatment	No.	Effusion Volume (ml)	Total Leukocytes in Pleural Cavity ($X10^{-6}$)	% PMN
Normal Rats	4	<0.1	5 ± 2	1 ± 1
Carrageenan Rats				
Saline Control	34	0.9 - 1.4	127 - 173	68 - 82
Orgotein treated				
2 - 3 mg/kg	28	0.65 - 0.75[a]	101 - 118	60 - 72[a]
8 mg/kg	12	0.42 - 0.76[a]	92 - 109[a]	57 - 66[b]
27 - 32 mg/kg	28	0.36 - 0.56[b]	74 - 100[b]	50 - 66[b]

The procedure used was based on Vinegar *et al.* (1973). The ranges shown are for the averages obtained with groups of 6 or 8 rats per experiment. Carrageenan rats received 0.15 ml of 1% carrageenan in saline into the pleural cavity 24 hours prior to sacrifice. 3 ml of 0.1% Na_2 EDTA in saline was used to harvest cells from the thoracic cavity after 24 hours. Orgotein treated rats were dosed subcutaneously with 0.5 ml orgotein in saline at 1 hour before and 3 hours after the injection of carrageenan. The tabulated dose is the total dose administered.

[a]Difference from saline control significant ($p < 0.05$)
[b]Difference from saline control significant ($p < 0.005$)

Our longest experience with orgotein has been in the reversed Arthus reaction, which we have developed as a quantitative 3-point parallel line bioassay modeled after the guidelines of the United States Pharmacopeia (1975). It yields quantitative comparisons with a reference standard. The results are programmed for analysis of variance, testing the difference in slopes, and the significance of difference in position of the parallel lines. From the common slope and the difference between the midpoints of the lines, the activity of the unknown as a percentage of the reference standard is calculated. The effect of orgotein in this assay also expresses itself histopathologically, as the saline controls and prednisolone recipients differ from the orgotein-treated animals by the presence of severe vascular damage, microthrombosis, PMN infiltration, nuclear debris, and hemorrhages. This difference is particularly distinct at 90 to 120 minutes after induction of inflammation.

In models of induced acute inflammation, administration of orgotein either prior to the induction event or during the early phases of development of the response produces measurable inhibition of the inflammation. In most of our recent experiments, the dose has been divided equally between a prophylactic administration at one to three hours prior to induction and a "therapeutic" administration at one to three hours after induction. The interval between the first dosing and induction of inflammation can be increased to at least six hours without markedly decreasing the effectiveness of the drug. When the predosing interval is increased to 24 or more hours a time-dependent diminution of potency is observed.

Orgotein efficacy as an inhibitor of induced inflammation can be reliably demonstrated with doses above 1.0 mg/kg; a dose level which is below the effective dose range for most non-steroidal anti-inflammatories in the same models (Ungar *et al.*, 1959). The response to orgotein in models of induced inflammation is often characterized by a dose-response curve with a zero-effect intercept in the range of 0.1 to 1 mg/kg (Fig.7). Similar effects of SOD on carrageenan-induced edema have been reported by Oyanagui (1976). Ceiling effects have not been observed even at doses of several hundred mg/kg in some models. The safety of orgotein allows such high-dose experiments to be performed without the complicating features of toxicity regularly seen with other anti-inflammatories.

Models of chronic inflammation have been regularly produced by utilizing immune-related events. Adjuvant-induced polyarthritis in the rat is considered a model for rheumatoid arthritis. The arthritic response follows injections of 1% *Mycobacterium tuberculosis* suspended in perhydrosqualene (Beck and Whitehouse, 1974) into the foot-pad of Sprague-Dawley rats. Subcutaneous administration of orgotein in a blind experiment, for a total of 10 treatments starting on the day of adjuvant injection, reduced the increase in foot size by about 27% (injected foot), and about 30% (contralateral foot) on day 8 and 35% in both feet on day 14. A similar decrease was seen in the arthritic score for both the injected and the contralateral feet, but the reductions were statistically significant only for the injected foot. The lack of significance in the contralateral foot was mostly due to large variations in the control group.

Another immune-related model used was the (NZB x W)F_1 mouse which develops autoimmune disorders involving antibodies to nuclear antigens (ANA, ADNA) and red blood cells (ARBC) which lead to immune complex glomerulonephritis and hemolytic anemia, respectively. This model has been considered as comparable to autoimmune diseases in the human, especially lupus erythematosis. Orgotein administration (2 mg/kg) every other day, starting at $1\frac{1}{2}$ months of age, caused pronounced reductions of antibody titers, deposits of IgG in the glomeruli and plaque-forming cells in the spleen when evaluated by the procedures of Oldstone and Dixon (1969), (Table IX).

Whether the effects of orgotein on these immune events are related to the same action mechanism(s) operating in non-immune related models of induced inflammation is open to question. A selective immunosuppression is shown by experiments in which DNA complexed with methylated BSA was used to immunize (Balb/c x C57Bl/6)F_1 mice. When orgotein was administered to such mice according to the same schedule as used for the NZB studies, the immune

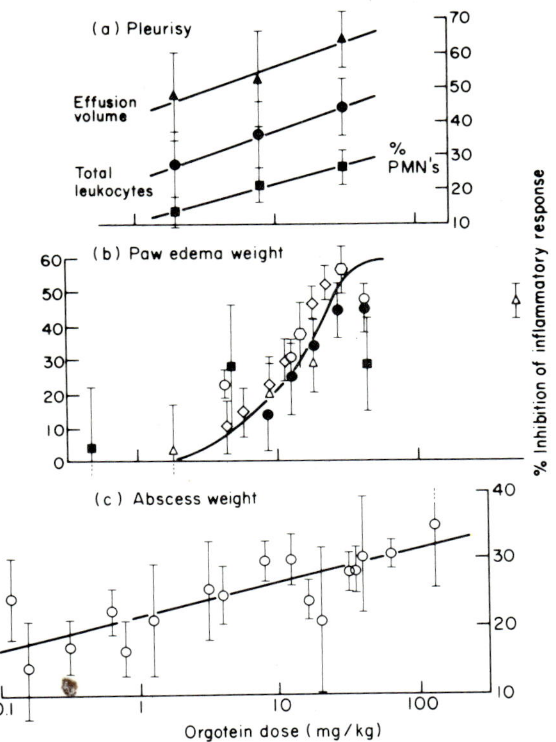

Fig.7. *Effects of orgotein in three models of induced inflammation. Inflammatory responses to carrageenan injection were measured 24 hours after induction in control animals (dosed with saline instead of orgotein). The inhibitory effect of orgotein is expressed for each response as % inhibition ± S.E. vs. the log of the orgotein dose. Dosing was done with blind-labeled samples of Palosein brand of orgotein or saline, usually at 1 hr before and 3 hrs after induction with carrageenan. The dose plotted is the total dose administered.*

A. Carrageenan-induced pleurisy in rats. Method as in Table VIII.

B. Carrageenan-induced paw edema in mice. In analogy with Winter et al. *(1962), paw edema was induced in 20 - 23 gm mice by injection of 0.03 ml of 1% carrageenan in saline into the right hind paw. Orgotein or saline was administered s.c. in .1 ml at times between -4 and +3 hours to 8 to 10 Swiss-Webster mice per dose. Edema was measured by subtracting the weights of the amputated uninjected left feet from the carrageenan injected feet. The different symbols represent results obtained on different days.*

C. Carrageenan-induced subcutaneous abscess in mice. Abscesses were induced by s.c. injection of 0.2 ml of 1% or 1½% carrageenan in saline in analogy with the procedure of Benitz and Hall (1963). Groups of 5 - 10 Swiss-Webster mice were dosed s.c. with orgotein or saline in divided doses. The mice were sacrificed and abscesses were carefully dissected and weighed after 24 hours.

response to DNA was significantly damped. The combined data suggests that in autoimmune disorders orgotein treatment may influence events leading to the formation of antibody producing cells (Table X).

TABLE IX

Orgotein Treatment of NZB x W Mice

	Treatment 6 - 7 Months Control	6 - 7 Months Orgotein	8 - 9 Months Control	8 - 9 Months Orgotein
Incidence of ANA	96%	100%	100%	100%
ANA Titre, Arithmetric	8.1	2.6	ND	ND
Plaque Forming Cells to DNA (per 10^6 spleen cells)				
Direct	55	6	99	10.3
Indirect	210	31.3	483	61
IgG Eluted from Glomeruli (mcg)	55	9.1	275	9.9
Mortality	0	0	72%	40%

The development of immune complex disease was followed in groups of control and orgotein treated F_1 mice. The orgotein group received 2 mg/kg 3 times each week.

TABLE X

Orgotein Effect on DNA Immunization

Experiment	Treatment	No. Mice	Direct PFC[a] per 10^6 Cells	p<
1	Orgotein	14	3.6	0.05
	Saline	14	7.9	
2	Orgotein	14	5.4	0.25
	Saline	14	9.3	
3	Orgotein	6	1.2	0.01
	Saline	6	5.1	

DNA-Methylated BSA complex was injected into (Balb/c x C57B1/6)F_1 mice once. Orgotein at 2 mg/kg was injected 3 x/day until sacrifice at day 4. The procedures were those of Lampert and Oldstone (1973)

[a]Plaque Forming Cells to DNA were detected in a modified Jerne assay.

At present, we can only speculate on the mechanisms of action responsible for the anti-inflammatory effects of orgotein in animal models. The available evidence comes from *in vitro* experiments, as the models of induced inflammation do not yet permit valid assessment because of their complexity. As documented extensively in this volume, extracellular SOD protects PMNs and macrophages against lysis induced by phagocytosis. At this time, the most attractive working hypothesis to us for the mechanism by which orgotein produces its anti-inflammatory effects is stabilization of membranes of cells or cell organelles involved in inflammatory events. Should such events occur *in vivo*, extracellular orgotein could reduce the spillage of lysosomal and other cellular inflammants and thereby interrupt a cycle maintaining inflammation.

We have found that orgotein can stabilize isolated lysosomes against autolysis during incubation at 37°C in buffered, isotonic sucrose. This effect, while requiring high concentrations of orgotein, is not a nonspecific protein effect, since BSA is inactive in the system. These observations are supported by findings that orgotein at 0.01 mg/ml (3×10^{-7} M) significantly inhibits naphthylamidase release from human skin lysosomes upon exposure to 10^{-4} M histamine (Bitensky, 1975). This model has been proposed as representative of lysosome labilization in human epidermis. Other non-steroidal anti-inflammatories require significantly higher doses to produce the same effect (Chayen *et al.*, 1972).

One is hard-pressed to explain orgotein efficacy demonstrated in the NZB and Balb/c x C57/B16 mouse models on the basis of the *in vitro* models of O_2^- related events. This applies also to the fact that demonstrable enzyme activity of exogenous Cu-Zn SOD in serum and in tissue extracts decays *in vivo* at rates measured in hours while its clinical effects last for days, weeks, or even months (Menander-Huber and Huber, 1976). Thus, to date we have demonstrated that orgotein is effective and has a high margin of safety. It is ideally suited for use as a drug. The proof of its action mechanism(s) remains a challenging problem for the future.

REFERENCES

1. Abderhalden, R. (1961). "Clinical Enzymology" (P. Oesper, trans.) Vol.2, pp 225-325, Van Norstrand, Princeton and Toronto.
2. Beauchamp, C. and Fridovich, I. (1971). *Anal. Biochem. 44*, 276-287.
3. Beck, F.W.J. and Whitehouse, M.W. (1974). *Proc. Soc. Exp. Biol. Med. 146*, 665-669.
4. Benitz, K.F. and Hall, L.M. (1963). *Arch. Int. Pharmacodyn. Thera. 144*, 185-195.
5. Bitensky, L. (1975). Personal communication.
6. Carson, S., Vogin, E.E., Huber, W. and Schulte, T.L. (1973). *Tox. Appl. Pharm. 26*, 184-202.
7. Chayen, J., Bitensky, L. and Ubhi, G.S. (1972). *Beitr. Path. 147*, 6-20.
8. Cushing, L.S., Decker, W.E., Santos, F.K., Schulte, T.L. and Huber, W. (1973). *Modern Vet. Pract. 54*, 17-20.
9. Elars Bioresearch, Inc. (1975). Unpublished data.
10. Huber, W. (1971a and b). U.S. Patent Nos. 3,579,495 and 3,624,251.
11. Huber, W. (1972a and b). U.S. Patent Nos. 3,637,640 and 3,687,927.
12. Huber, W. (1973). U.S. Patent No. 3,781,414.
13. Huber, W., Chow, S.H. and Saifer, M.G. (1973). U.S. Patent No. 3,763,136.

14. Huber, W., Chow, S.H. and Saifer, M.G. (1974a and b). U.S. Patent Nos. 3,806,411 and 3,813,289.
15. Huber, W., Huebner, A. and Saifer, M.G. (1973). U.S. Patent No. 3,763,137.
16. Huber, W. and Schulte, T.L. (1972). U.S. Patent No. 3,637,641.
17. Huber, W. and Schulte, T.L. (1973a,b and c). U.S. Patent Nos. 3,758,682, 3,773,928 and 3,773,929.
18. Huber, W. and Schulte, T.L. (1974). U.S. Patent No. 3,832,338.
19. Huber, W., Schulte, T.L., Carson, S., Goldhamer, R.E. and Vogin, E.E. (1968). *Tox. Appl. Pharmacol. 17,* 308-309.
20. Innerfield, J. (1960). "Enzymes in Clinical Medicine", McGraw Hill Book Co., New York.
21. Lampert, P.W. and Oldstone, M.B.A. (1973). *Science 180,* 408-410.
22. Lin, M.S. (1974). In "Radiopharmaceuticals" (G. Subramanian *et al.* eds.) pp 36-48, Society of Nuclear Medicine, New York.
23. McCord, J.M. and Fridovich, I. (1969). *J. Biol. Chem. 244,* 6049-6055.
24. Mann, T. and Keilin, D. (1939). *Proc. Roy. Soc. Biol. Sci. 126,* 303-315.
25. Menander-Huber, K.B. and Huber, W. (1976). In "Superoxide and Superoxide Dismutases" (A.M. Michelson, J.M. McCord and I. Fridovich, eds.) in press, Academic Press, London.
26. Metzger, H., Shapiro, M., Mosimann, J. and Vinton, J. (1968). *Nature 219,* 1166-1168.
27. Oldstone, M.B.A. (1973). Personal communication.
28. Oldstone, M.B.A. and Dixon, F.J. (1969). *J. Exp. Med. 129,* 483-488.
29. Oyanagui, Y. (1976). *Biochem. Pharm. 25,* 1465-1472.
30. Schulte, T.L. (1964). Personal communication.
31. Steinman, H.M. and Hill, R.L. (1973). *Proc. Nat. Acad. Sci. U.S.A. 70,* 3725-3729.
32. Tegelström, H. (1975). *Hereditas 81,* 185-198.
33. Thompson, S. (1974). Unpublished observations.
34. Ungar, G., Kobrin, S. and Sezesby, B.R. (1959). *Arch. Int. Pharmacodyn. 123,* 71-76.
35. United States Pharmacopeia XIX (1975). "Design and Analysis of Biological Assays", pp 601-611, United States Pharmacopeial Convention, Inc., Rockville, MD 20852.
36. Vinegar, R., Traux, J.F. and Selph, J.L. (1973). *Proc. Soc. Exp. Biol. Med. 143,* 711-714.
37. Winter, C.A., Risley, E.A. and Nuss, G.W. (1962). *Proc. Soc. Exp. Biol. Med. 111,* 544-547.
38. Wolf, M. and Ransberger, K. (1972). "Enzyme Therapy", Vantage Press, New York.

ORGOTEIN, THE DRUG VERSION OF BOVINE Cu-Zn SUPEROXIDE DISMUTASE II. A SUMMARY ACCOUNT OF CLINICAL TRIALS IN MAN AND ANIMALS

K.B. MENANDER-HUBER and W. HUBER

Diagnostic Data Inc.
58, Logue Avenue, Mountain View
California 94043, U.S.A.

INTRODUCTION

The formation of the superoxide radical (O_2^-) as an important step in physiological and pathological reactions has been described in several chapters of this book (Babior; Johnston and Lehmeyer; Salin and McCord; Michelson). When the O_2^- formation occurs in the cytoplasm and the mitochondria, endogenous intracellular superoxide dismutase (SOD) acts as a scavenger. If, however, O_2^- radicals are produced outside the cell, the body does not seem to have any defense mechanism against the damage these radicals and their secondary products can cause.

As exogenous orgotein does not penetrate the cell membrane (Huber and Saifer, 1972; Petkau *et al.*, 1976) it should be able to prevent pathology related to extracellular O_2^- formation by virtue of its SOD activity. The intimate relationship between phagocytosing, i.e. O_2^- producing, PMNs and macrophages and inflammation certainly is a clue why orgotein can control acute and chronic inflammatory conditions. But as the inflammatory cascade consists of many events (Berry, 1976), inhibition of O_2^- radicals and their derivatives need not be the only action mechanism of orgotein in pathology.

The following account gives the status of our clinical experience in man and animal to date. Our work includes double-blind trials in conditions such as rheumatoid arthritis and osteoarthritis commonly used to test the clinical efficacy and safety of anti-inflammatory drugs. In addition orgotein efficacy has been explored in diseases such as radiation induced inflammation.

CLINICAL STUDIES IN MAN

Inflammations of the Musculoskeletal System

Rheumatoid arthritis is an autoimmune condition considered as a prime model for an inflammatory disease. The criteria for the diagnosis and for active disease have been standardized (Am. Rheumatism Assoc. (ARA), 1973; U.S. Food and Drug Administration (FDA), 1976). The 1976 FDA guidelines state, "the major emphasis in testing anti-inflammatory drugs has been put on their efficacy in rheumatoid arthritis as it is a good example of a musculo-skeletal inflammation that tends to be non-self-limiting, less placebo responsive and less variable from day to day than are most disease processes that may be considered. It is also a prevalent condition and under study by many physicians and clinics and historically anti-inflammatory drugs have been shown to have substantial effects in rheumatoid arthritis".

In our work, orgotein efficacy in rheumatoid arthritis has been explored in trials where it has been compared with placebo and also with other anti-rheumatic drugs, i.e. penicillamine and gold. In our U.S. double-blind placebo-controlled studies all patients had definite rheumatoid arthritis (ARA 1973, FDA 1976), and though stabilized on a constant regimen of aspirin or aspirin plus corticosteroids all still had active disease (Table I).

TABLE I

Criteria for Active Rheumatoid Arthritis
(Three criteria have to be present)

Parameter	Required Activity
No. of Painful or Tender Joints	> 6
No. of Swollen Joints	> 3
Duration of Morning Stiffness (Minutes)	> 45
Westergren Sedimentation Rate (mm/hr.)	> 28

Such types of patients can be considered as "non-responders" to their basic drug regimen and therefore represent a particularly challenging sample for the evaluation of a new drug. They also represent the group of patients in which orgotein probably would be used first after registration. The details for our double-blind placebo-controlled studies are given in Table II.

This will give us about 96 (four trials) patients in the corticosteroid plus aspirin group and about 72 (three trials) patients in the aspirin group by fall of 1976. During May, 1976, we were able to conduct a preliminary statistical analysis of the results obtained from patients who had completed the study (49 in each group). The results are presented in Table III and demonstrate that in the total available sample orgotein was markedly more effective than placebo. Most importantly the efficacy of orgotein over placebo was in the same direction, i.e. improvement in *all* the parameters tested.

As demonstrated (Huber and Saifer, 1976) orgotein is not an analgesic in standard animal models. The clinical experience seems to confirm this as the onset of orgotein effects, at least with the dose regimen used, appears to take several weeks in some of the patients. However, once established the orgotein effect seems to last at least one month beyond termination of therapy as shown by the observations at week 16, four weeks after the end of orgotein dosage.

The remarkable safety of orgotein demonstrated in animals (Carson *et al.*, 1973) was confirmed in these sensitive patients. Only one patient dropped

TABLE II

Protocol Parameters of Double-Blind Placebo-Controlled Study in Rheumatoid Arthritis

Parameter	
No. of Trials	7
No. of Patients/Trial	24
Orgotein or Placebo Dosage	8 mg/inj. x 4/week for 4 weeks, 8 mg/inj. x 3/week for next 8 weeks
Concomitant Medication (Maintenance Therapy)	Corticosteroid + ASA, or ASA (To be Kept Constant)
Therapy Excluded Before & During Trial (Time w/o Therapy Before Entering)	Non-steroidal Anti-inflammatories (2 weeks), Antimalarials (6 mos.) Immunosuppressive (6 mos.), Gold Except as Maintenance Therapy (6 mos.), Corticosteroids Except as Maintenance Therapy (2 mos.).
Efficacy Parameters Evaluated	Morning Stiffness, Grip Strength, Walking Time, Pain, Tenderness, Swelling, Use of Analgesics, Patient's Evaluation of Status, Doctor's Evaluation of Status.
Safety Parameters Evaluated	CBC, SMA-12, Urinalysis

TABLE III

Results of Double-Blind Placebo-Controlled Studies in Rheumatoid Arthritis

Evaluated Parameter	Mean Change in Patients (49) On Corticosteroids + Aspirin				Mean Change in Patients (49) On Aspirin			
	at 12 weeks		at 16 weeks		at 12 weeks		at 16 weeks	
	Orgotein	Placebo	Orgotein	Placebo	Orgotein	Placebo	Orgotein	Placebo
Morning Stiffness (%)	-51***	+37	-32	+71	-41	- 2	-37*	- 7
Grip Strength (%)	+13.2**	- 3.4	+17.8*	- 0.3	+12.6*	+ 3.0	+14.8*	+ 2.4
No. of Analgesics per 14 days	+ 1.6*	+12.0	+ 0.6**	+13.4	- 5.8*	+ 5.1	- 2.3**	+14.2
Pain (Units)	-16.1*	+ 3.1	-15.2*	+ 1.0	-14.0	-12.2	-14.8	-10.3
Tenderness (Units)	-12.6*	+ 2.5	-16.8	- 6.8	-15.0	- 7.4	-13.8*	- 2.0
Active Joints (No.)	- 4.0*	+ 2.4	- 5.2*	+ 1.7	- 4.9	- 2.6	- 4.7	- 3.0

* $p < .05$ (Student's t-test)

** $p < .01$ (Student's t-test)

*** $p < .001$ (Student's t-test)

out of the study because of side-effects attributable to orgotein, i.e. injection site reaction. Interestingly, in the orgotein groups the gastro-intestinal side-effects, including ulceration, were less than in the placebo groups even though both groups received comparable dose regimens of corticosteroid and/or aspirin.

To provide safety data upon long term administration of orgotein, all patients concluding the double-blind trials are given the opportunity to enter a maintenance program extending to at least one year. The orgotein regimen is kept constant or increased during this period while the steroid and/or aspirin dosage is being reduced. Results to date appear to confirm prior observations that orgotein is safe and effective as a steroid and aspirin sparing drug.

Evaluation of orgotein efficacy against reference drugs, i.e. gold and penicillamine, has been carried out in Europe. Entry into the trial required diagnosis of definite or classical rheumatoid arthritis (Ropes *et al.*, 1959). The patients were allowed to continue treatment with the same dosage of non-steroidal anti-inflammatories and/or low dose corticosteroids that they were taking on admission. The duration of the trials was 16 weeks. The dose regimens are listed in Table IV. Parameters evaluated included severity of pain, duration of morning stiffness, grip strength, joint circumference and articular index. Other details of the trial and a summary of the results are given in Table IV.

These trials indicate that orgotein in rheumatoid arthritis is comparable in effectiveness to gold and in some parameters to penicillamine, both of which are considered as anti-rheumatic drugs. This occurred even though orgotein dosage was decreased while penicillamine dosage was increased and gold dosage remained constant. Gold and penicillamine both have serious side-effects (Girdwood, 1974; Golding, 1973) which tend to limit treatment periods severely for many patients.

Degenerative Joint Disease (Osteoarthroses, Osteoarthritis)

Osteoarthritis represents a disease with localized inflammatory pathology. The disease is progressive and periods of spontaneous remission are rare (FDA, 1976). For this reason it can be used to explore the efficacy and safety of locally administered anti-inflammatory drugs.

An open drug study (Lund-Olesen and Menander, 1974) had indicated that orgotein can be used safely and effectively when given intra-articularly into knees and hips in single or multiple doses. A double blind controlled trial designed for intra-group comparison of orgotein and placebo has been completed (Lund-Olesen *et al.*, 1976). The details for the trial are summarized in Table V. The design was set up to provide proof of both efficacy and long term safety.

The results (Table VI) show that orgotein injected intra-articularly is a safe and effective durg. Intra-articular injections into the knee joint cause a pronounced improvement in parameters such as pain, function, use of aids, use of analgesics and doctor's assessment. The changes are statistically in favor of orgotein.

During the course of this trial it was established that orgotein could be injected repeatedly intra-articularly with safety, provided the endotoxin content is kept at levels below 2 nanogram/mg protein. These are endotoxin levels which are lower than those specified in the prevailing USP test for pyrogenicity (U.S. Pharmacopeia XIX, 1975).

TABLE IV

Controlled Trials in Rheumatoid Arthritis Comparing Orgotein Against a Reference Drug

Country	No. of Patients	Reference Drug	Orgotein**	Efficacy	Safety
England	9 + 9	Penicillamine*	12 mg/inj.	Same in Morning Stiffness Swelling Grip Strength	Orgotein Much Better
France	11 + 10 + 10	Gold (100 mg/week)	4 or 12 mg/inj.	Same	Orgotein Much Better
Belgium	18 + 16	Gold (75 mg/week)	8 mg/inj.	Same	Orgotein Much Better

* 250 mg/day for 2 weeks, 500 mg/day for next 2 weeks, 750 mg/day for next 2 weeks, 1,000 mg/day for next 2 weeks, 1,250 mg/day for next 2 weeks, 2,500 mg/day for the next 16 weeks.

** 3 inj/week for 4 weeks, 2 inj/week for next 12 weeks, 2 inj/week for next 10 weeks.

TABLE V

Protocol Parameters of Double-Blind Placebo-Controlled Trial in Osteoarthritis

	Protocol Parameters
Patient Selection	X-ray Evidence of Osteoarthritis of Knee Joints Combined with Active Inflammation
No. of Patients	44
Dosage of Experimental Drugs (Orgotein, Placebo)	2 mg Injected into Each Knee Joint Every 2 weeks for 6 months
Parameters Followed:	
Pain	Day Pain, Night Pain, Pain Walking, Use of Analgesics
Function	Longest Distance Walked, Limp, Use of Aids, Stair Climbing
Overall Evaluation	Patient's Evaluation, Doctor's Evaluation
Safety	Patients Asked About Possible Side-effects

TABLE VI

Results of Double-Blind Controlled Trial in Osteoarthritis

Evaluated Parameter	Mean Change	
	Orgotein	Placebo
Change in Pain (%)	-80.0**	-31.3
Change in Use of Analgesics (%)	-60.7*	-30.0
Change in Impairment of Function (%)	-42.6*	-15.9
Doctor's Evaluation (Units)[a]	+120.0**	+57

a 0 = No change; 100 = Improved; 200 = Much Improved.

* p <.05 (Student's t-test)

** p <.01 (Student's t-test)

Inflammatory Conditions as a Sequel to Radiation Therapy

Acute Inflammation. Acute inflammation is an unavoidable sequelae of radiation therapy, and is probably the composite of several events.

Cell destruction, be it by radiation or other means, results in the release of substances which attract phagocytes including PMN's and macrophages, which when activated exhibit a respiratory burst which initially appears to be largely converted into O_2^- by generating systems located on the plasma surface of mammalian cells (Weening *et al.*, 1975; Petkau *et al.*, 1976).

Extracellular O_2^- could no doubt damage the activated phagocytes as well as adjacent cells and connective tissue. This promotes on-site lysis and release of cellular inflammants causing additional breakdown, influx of more phagocytosing cells and with it an ever widening spread of inflammation. During radiation therapy this sequence of events is further stimulated by the daily radiation session, thus maintaining cellular breakdown. Based on considerations that extracellular orgotein prevents or modifies such events without interfering with the intracellular efficacy of tumor cell destruction by direct or indirect radiation events, a double-blind placebo-controlled trial was undertaken (Edsmyr, 1976).

Radiation therapy of pelvic tumor involves the use of large doses (5,000 - 8,000 rad) of high voltage radiation, administered daily in increments of about 200 rad. Such dose regimens cause undesirable acute side effects on the bowel, the bladder and the urethra. That these are inflammatory in character has long been recognized (Moss and Ackermann, 1965). The patients were injected with placebo or orgotein directly after each radiation session throughout the entire course of the radiation therapy. The details of the trial are presented in Table VII.

TABLE VII

Protocol Parameters of Double-Blind Placebo-Controlled Study on the Orgotein Efficacy in Ameliorating Side-Effects Due to Radiation Therapy of Bladder Tumors

	Protocol Parameters
Patient Selection	Patients with Bladder Tumors T2 - T4
No. of Patients	38
Orgotein or Placebo Dosage	4 mg Subcutaneously after Completion of Daily Radiation Therapy
Radiation Dosage	6,400 or 8,400 rad.
Parameters Evaluated	Maximal Voiding Volume, Voiding Frequency, Pain, Proctitis, Medication Used

The results showed that orgotein can be used safely and effectively to ameliorate and/or prevent the side-effects of high dose radiation as evaluated by signs and symptoms caused by the inflammation of the bladder and the bowel (Table VIII). Long term radiation side-effects also were ameliorated.

The good results observed in this double-blind trial indicate that with orgotein a therapeutic rather than symptomatic treatment approach of radiation side effects is possible. Other double-blind trials using the same protocol are in progress at different centers in Europe and the U.S. including also patients who receive radiation therapy for carcinomas of the prostate and the cervix of the uterus.

Chronic Inflammation. Chronic inflammation of the bladder as a consequence of radiation therapy for carcinoma of the cervix is an incapacitating disease in which a wide spectrum of drugs and surgical procedures have been used over the years, generally with but temporary success (Frick and Hittmair, 1967).

TABLE VIII

Patient Distribution and Results of Double-Blind Trial on the Efficacy of Orgotein in Ameliorating Side-Effects Due to Radiation Therapy of Bladder Tumors

	Level of Statistical Significance
Maximal Voiding Volume	$p < .05$[a]
Interval Between Voidings During Day	$p < .05$[a]
Severity of Signs and Symptoms from Bladder	$p < .05$[a]
Percent Visits With Diarrhea	$p < .025$[a]
Percent of "Diarrhea Visits" Requiring Medication	$p < .001$[a]
Dose of Antidiarrheal Medication	$p < .0025$[b]

[a] Chi-square test

[b] Student's t-test

An open drug study on 22 patients (Marberger *et al.*, 1975) treated with orgotein, injected intramurally into the bladder wall by needle catheter, showed that orgotein was effective and safe. Effective mean-dose regimen was found to be 30.4 (10 - 90) mg total and 11.3 (7 - 30) mg per injection.

The clinical response to orgotein was first seen in symptoms such as pain, spasms, dysuria, incontinence, and hematuria. The cystoscopic findings indicated an anti-inflammatory effect, followed by disapperance of edema, softening of the bladder wall and increase in maximal voiding volume. The mean duration of clinical benefit experienced by the patients after a prior orgotein injection was 22 (3.5 - 55.4) months.

Exploratory Studies

Open drug trials to explore further the clinical effect and safety of orgotein have been and are being carried out in parallel with double-blind studies (Table IX).

These studies demonstrate that orgotein can be used with complete safety when administered subconjunctively, intramurally, intrathecally and topically. The extent of orgotein efficacy in these conditions will have to be substantiated by double-blind, controlled trials. This is particularly important for multiple sclerosis since this disease is known for its fluctuations in severity and has also been described to respond well to placebo (McAlpine, 1972).

CLINCIAL STUDIES IN ANIMALS

The efficacy and safety of orgotein has been explored in musculoskeletal diseases in animals particularly horse and dog. The scope of these investigations is presented in Table X.

For systematic administration and depending upon severity, doeses of 5 to 10 mg are given daily till improvement is noted. Thereafter injections every

TABLE IX

Miscellaneous Drug Studies With Open Label Orgotein in Man

Disease	Route of Administration	No. of Patients
Chronic Post-operative Cystitis[1]	Intramural	25
Peyronie's Disease[1]	Intraplaque	30
Herpes Simplex[2]	Topical	10
Multiple Sclerosis[3]	Intrathecal & Subcutaneous	35
Uveitis, Choroiditis[4]	Subconjunctival	12

[1]Marberger *et al.* (1974

[2]Sellman and Menander-Huber (1975)

[3]Lund-Olesen and Menander-Huber (1976)

[4]Sellman and Menander-Huber (1975)

other day can be continued to maximal response. A regimen of no more than ten injections has been found effective in most cases.

Many of the disorders (Table X) are localized. Therefore orgotein administration into and around the affected sites has been explored and found to be especially effective as well as safe. Doses of 5 mg or less were administered intra-articularly, or by instillation into and around the affected area once weekly for three to five weeks.

In the horse the results show that orgotein has a potent enhancing effect on the resolution of certain injuries commonly affecting this species. These are principally synovitis, periostitis, and arthritis with the most marked responses being observed in the inflammatory phase. Orgotein compared very favourably with existing treatments and while its activity may often be slower in onset it is considered to be superior at the end of the course of treatment in most cases. The relapse rate following treatment was observed to be low and in addition the recovery time was significantly reduced over that expected (Linton, 1976; Faull *et al.*, 1976).

The results (Table X) of a multi-center double-blind study in dogs (Diagnostic Data, Inc., 1974) involving 154 animals clearly demonstrated both the safety of orgotein and its superior efficacy relative to placebo ($p < 0.001$). One of the significant findings in the study was the lasting effect of orgotein. Evaluation at eight weeks after end of therapy demonstrated that there was no tendency towards relapse in any of the animals that had been treated with orgotein.

In horses and dogs orgotein has been proven extremely safe even with multiple repeat injections, both systemic and local, irrespective of the interval between treatments. No systemic side-effects were seen with injection into muscle, traumatized tissue, around bone or into joints. Swelling and tenderness were seen occasionally on local injection but this passed off usually in a matter of days (Linton, 1976; Faull *et al.*, 1976). This lack of side-effects in combination with the fact that orgotein does not inhibit

TABLE X

Clinical Studies with Orgotein in Animals

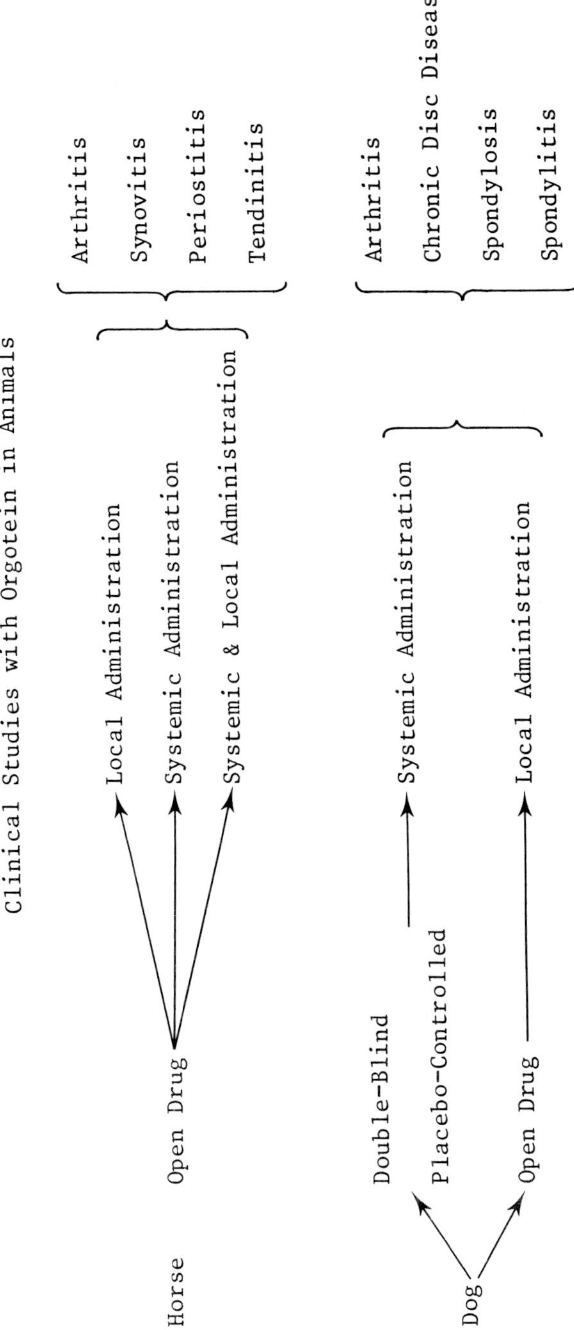

wound healing should make its substitution for the anti-inflammatory steroids desirable.

CONCLUSION

In conclusion, orgotein appears to be a unique anti-inflammatory drug by virtue of a combination of the following characteristics:

1. It is an effective anti-inflammatory.
2. Its effects are long lasting.
3. It can be used in combination with other anti-inflammatory drugs, such as aspirin and corticosteroids and provides additional improvement.
4. Treatment can be adjusted to the localization of the pathology, i.e. systemic for systemic disease and local for local disease.
5. It is unique in its lack of tissue-toxicity and adverse reactions.
6. Its immunogenic potential is of a very low order.
7. In combination with other anti-inflammatories it does not increase the incidence and/or severity of their side-effects.

These make orgotein a drug of promising potential, as inflammation is an important part of many more diseases than have been explored by us so far.

ACKNOWLEDGEMENTS

W. Huber would like to acknowledge the good fortune which brought him into contact with men like M.G. Smith and T.L. Schulte, M.D. who have accompanied him through many tortuous research trials with steadfast support and enthusiasm. Special thanks are due to the small group of farsighted investors, among them W.T. Brady, T.J. Davis, Jr., P.W. Hewitt, J. Martin, Jr., T.I. Mosley, A.M. Poniatoff, A. Rock and Brooks Walker, Sr., who provided the initial capital, as well as seasoned advice, to get the orgotein project started and have been in our corner ever since.

W. Huber, K.B. Menander-Huber and M.G.P. Saifer want to acknowledge the help of their research associates, especially P.H.C. Dang, Ph.D. and L.D. Williams, Ph.D., whose enthusiasm and competence contributed materially to our knowledge and understanding over the years.

We thank M.B.A. Oldstone, M.D. of Scripps Clinic and Research Foundation for performing the immunopathology studies with NZB x W and Balb/c w C57Bl mice which are described here and B. Gerstl, M.D. of Stanford University for expert histopathology consultations and stimulating counsel.

K.B. Menander-Huber would like to thank our medical and veterinary investigators, especially K. Lund-Olesen, M.D. and F. Edsmyr, M.D., Ph.D., for their extraordinary and valuable cooperation and help.

REFERENCES

1. American Rheumatism Assoc. (1973). "Primer on the Rheumatic Diseases" pp 137-138. The Arthritis Foundation, New York, N.Y. 10036.
2. Babior, B.M. (1977). In "Superoxide and Superoxide Dismutases" (A.M. Michelson, J.M. McCord and I. Fridovich, eds.), in press, Academic Press, London.
3. Berry, C.L. (1976). In "The Cell in Medical Science" (F. Beck and J.B. Lloyd, eds.), Vol.IV, pp 291-319, Academic Press, New York and London.
4. Carson, S., Vogin, E.E., Huber, W. and Schulte, T.L. (1973). *Tox. Appl. Pharm. 26*, 184-202.
5. Diagnostic Data, Inc. (1974). "Supplemental New Animal Drug Application,

45-863V, for use of Orgotein in Ankylosing Spondylitis, Spondylosis and Disc Disease of Dogs". Approved, Van Houweling, C.D. (1976). Fed. Reigster *41*, 151, p 32583.

6. Edsmyr, F., Huber, W. and Menander, K.B. (1976). *Curr. Therap. Res. 19*, 198-211.
7. Faull, G.L., Baker, de B.B., Walt, V.D. and Hofmeyer, C.F.B. (1976). *J. So. African Vet. Assoc. 47*, 39-40.
8. Frick, J. and Hittmair (1967). *Urol. Int. 22*, 429-439. See also for earlier references.
9. Girdwood, R.H. (1974). *Brit. Med. J. 1*, 501-518.
10. Golding, D.N. (1973). *Brit. J. Hosp. Med. June*, 805-813.
11. Huber, W. and Saifer, M.G.P. (1972). Unpublished observations.
12. Huber, W. and Saifer, M.G.P. (1976). In "Superoxide and Superoxide Dismutases" (A.M. Michelson, J.M. McCord and I. Fridovich, eds.) in press, Academic Press, London.
13. Johnston, R.B. and Lehmeyer, J.E. (1977). In "Superoxide and Superoxide Dismutases" (A.M. Michelson, J.M. McCord and I. Fridovich, eds.) in press, Academic Press, London.
14. Linton, J.A.M. (1976). *Irish Vet. J. 30*, 53-56.
15. Lund-Olesen, K. and Menander, K.B. (1974). *Curr. Therap. Res. 16*, 706-717.
16. Lund-Olesen, K., Menander-Huber, K.B. and Huber, W. (1976). Unpublished Observations.
17. Marberger, H., Huber, W., Bartsch, G., Schulte, T. and Swobada, P. (1974). *Int. Urol. and Nephrol. 6*, 61-74.
18. Marberger, H., Bartsch, G., Huber, W., Menander, K.B. and Schulte, T.L. (1975). *Curr. Therap. Res. 18*, 466-475.
19. McAlpine, D., Lumsden, C.E. and Acheson, E.D. (1972). "Multiple Sclerosis, a Reappraisal", pp 127-244, The Williams and Wilkins Co., Baltimore.
20. Michelson, A.M. (1977). In "Superoxide and Superoxide Dismutases" (A.M. Michelson, J.M. McCord and I. Fridovich, eds.), in press, Academic Press, London.
21. Moss, W.T. and Ackermann, L.V. (1965). "Therapeutic Radiology", Mosby, St. Louis.
22. Petkau, A., Kelly, K., Chelack, W.S. and Barefoot, C. (1976). *Biochem. Biophys. Res. Commun. 70*(2), 452-458.
23. Roper, M.W., Bennett, G.A. and Cobb, S. (1959). *Arth. Rheum. 2*, 16-20.
24. Salin, M.L. and McCord, J.M. (1977). In "Superoxide and Superoxide Dismutases" (A.M. Michelson, J.M. Mcord and I. Fridovich, eds.), in press, Academic Press, London.
25. Sellman, A. and Menander-Huber, K.B. (1975). Unpublished observations.
26. U.S. Food and Drug Administration (1976). "Guidelines for the Clinical Evaluation of Nonsteroidal Anti-Inflammatory Drugs", FDA Public Records and Documentation Center, Rockville, MD 20852.
27. U.S. Pharmacopeia XIX (1975). "Pyrogen Test" p 613, U.S. Pharmacopeial Conv. Inc., Rockville, MD 20852.
28. Weening, R.S., Wever, R. and Roos, D. (1975). *J. Lab. Clin. Med. 85*, 245-252.

EPILOGUE AND PROSPECTS

I. FRIDOVICH, J.M. McCORD and A.M. MICHELSON

During the past decade there has been an exponential increase in our knowledge of the biological production and scavenging of the superoxide radical. Thus, the first symposium devoted to superoxide and superoxide dismutases bears eloquent witness to the present vigorous state of this young field of investigation. We can briefly enumerate much of what has been learned. It is now apparent that O_2^- is produced by numerous, biochemically-relevant oxidations. The oxidations of hydroquinones (1), leucoflavins (1,2), catecholamines (3,4), ferrohemoproteins (5-7) and iron-sulfur proteins (8-13) have all been shown to generate O_2^-. There are several oxidative enzymes which carry out the univalent reduction of oxygen (14-19). Indeed, such subcellular organelles as mitochondria (20-24) and chloroplasts (25-33) have been shown to produce O_2^- and one specialized type of cell, the phagocytic neutrophil, seems to use O_2^- as part of its antimicrobial armamentarium (38-41).

The biological production of O_2^- is thus a reality and it has called forth the evolution of enzymes devoted to the catalytic scavenging of this potentially deleterious free radical. These enzymes are called superoxide dismutases, although the proper nomenclature would be superoxide: superoxide oxidoreductases. They are unique in that their substrate is a reactive and unstable free radical; they are ubiquitous among respiring organisms and are even found, although sporadically, among organisms whose usual life style is anaerobic (42). Superoxide dismutases have been isolated from bacteria, protozoa, algae, fungi, higher plants, insects, fishes, birds and mammals, and their physical, chemical and catalytic properties have been explored (43-45). Three distinct types of superoxide dismutases have been found. One type contains both copper and zinc, while the second type contains iron and the third type manganese at the active site. The copper and zinc enzyme from bovine erythrocytes has been studied by X-ray crystallography and a structure at 3.0 Å resolution has been attained (46). The other types of superoxide dismutases are currently being analyzed by crystallographic methods in several laboratories. We may confidently expect that detailed structural information will soon be available for at least one specific example of each of the three major classes of superoxide dismutases.

The all important question of the biological function of the superoxide dismutases has begun to be explored. There has been controversy, but some things are clear. These enzymes do scavenge O_2^- with nearly perfect catalytic efficiency, operating at, or close to, the diffusion limit (47,48). No other catalytic activity has yet been associated with these enzymes, save for a low level peroxidase activity exhibited by the copper-zinc superoxide dismutases, but not by the manganese or iron enzymes (49). We must therefore assume for the superoxide dismutases, as we do for all enzymes, that the known catalytic activity is the biologically-significant catalytic activity. Superoxide dismutases have been found in all respiring organisms thus far surveyed and they have been shown to be absent from many sensitive obligate anaerobes (50). Several functional anaerobes have been found to contain low levels of these enzymes (42). This distribution itself suggests that these enzymes serve as a defense against oxygen toxicity.

Support for this proposal comes from several quarters. Thus the level of superoxide dismutase in several microorganisms and in rat lung has been shown to be increased by exposure to elevated concentrations of oxygen and the increase in superoxide dismutase has been found to correlate with an enhanced resistance towards oxygen toxicity (51-55). A mutant of *Escherichia coli* has been obtained with a temperature sensitive defect in superoxide dismutase (56). This mutant exhibited a parallel temperature-sensitive intolerance for oxygen. Human neutrophils secrete O_2^- in the microbicidal phase of their life cycle and in so doing destroy themselves as they kill the ingested bacteria. Superoxide dismutase, added to the medium, protects the neutrophils against this suicide (57) and bacteria which have an elevated level of the periplasmic superoxide dismutase are more resistant to this phagocytic kill (58). In blue-green algae, conditions of growth such as low CO_2 and high light intensity, which lead to a decrease in superoxide dismutase, render the cells prone to photo-oxidative death (59-60). The conclusion that O_2^- is cytotoxic and that superoxide dismutases serve to minimize this cytotoxicity, seems inescapable.

Superoxide radical has long been known to radiation chemists as one of the products of the radiolysis of oxygenated water. The realization that O_2^- is a species with biological relevance and the availability of such efficient and specific scavengers as the superoxide dismutases, has sparked a new interest in the chemistry of superoxide. Superoxide is being explored as a new reagent for organic chemical transformations and the spontaneous dismutation of O_2^- is being re-explored in ultrapure systems and in the presence of chelating agents to eliminate the catalytic action of trace metal contaminants. All of this work is creating a foundation of known superoxide chemistry from which it will be easier to approach the question of the biological actions of O_2^-. The production of O_2^- during the radiolysis of water provides a rational basis for the observed radiation protection by superoxide dismutases (61-67) and the anti-inflammatory action of these enzymes (68-70) can be understood in terms of the secretion of O_2^- by activated neutrophils. The mechanism of activation of the membrane bound, NADPH dependent, "superoxide synthetase", is unknown as yet but will surely be uncovered in the near future, and perhaps will prove of importance in cases of chronic granulomatous disease.

When a light is lit in a formerly dark space we are able to see things of which we were formerly unaware. At the same time we gain an appreciation of the great extent of the remaining darkness and we wish for an ever brighter light, to extend the area of the known and to decrease the hidden and feared unknown. We have been discussing some of those new things which we have recently been privileged to see clearly. What of those shapes which are still in the darkness? What are the questions and the possibilities we may hope to perceive in the near future?

We know many reactions which generate O_2^- but we do not know, in any specific organism, what are the quantitatively significant sources of O_2^-. We do not know how changes in pO_2 or in physiological or nutritional status change the rate of production of O_2^- within cells. We do not know the components of the cell which are most susceptible to attack by O_2^- or the chronic level of this attack suffered even in the presence of superoxide dismutases, catalases and peroxidases. Superoxide and hydrogen peroxide have been seen to conspire in the generation of OH· and of singlet oxygen. We know nothing of the extent of this process inside cells. Do organisms with strong protection against O_2^- or other oxygen species have a slow evolution rate, whereas the inverse is characteristic of fast evolving species? Has O_2^- in fact played a role in evolution? Is it possible that a low level of damage is always being inflicted upon respiring cells, in spite of the enzymes that scavenge O_2^- and H_2O_2 and is it possible that this chronic oxygen toxicity is one of the causes of senescence? Will we find means of specifically manipulating the activity of superoxide dismutase in micro-organisms and of thus modifying their resistance towards oxygen toxicity and towards phagocytic kill? Will superoxide dismutases or functional mimics thereof become important in the management of inflammation or be employed in selectively modifying the resistance of normal tissue during X-ray treatment of tumours? Affirmative answers to these questions would have very useful consequences. Clearly, as the light grows brighter the landscape coming into view seems ever more promising. In these days when society demands a direct social significance for the efforts of fundamental research, will superoxide dismutases be useful in prolonging the shelf life of oxidizable products, will they replace the abusive use of chemical antioxidants in food stuffs? Will SOD prevent the deplorable change in texture of frozen lobster, by inhibiting cross-linking reactions? Apart from a possible use as protection against high energy radiation will SOD, or functional mimics, diminish the inflammatory effects of excessive exposure to sunlight?

While a certain amount of satisfaction results from the application of pre-existing technology to new items of interest such as superoxide and superoxide dismutases, and while a wealth of solid information results from this approach, many of us are most intrigued by those questions which, at this stage, can barely be perceived, can barely be phrased in scientific terms. What are the precise roles of superoxide and superoxide dismutases in the properly functioning human organism? While evidence now exists which strongly supports our original hypothesis that the radical is a cytotoxic species, and is therefore something to be eliminated, is that the whole story? Does the radical play more subtle roles as well - roles of a beneficial nature? The bactericidal action of the phagocytic leukocyte appears to rely, in part, on superoxide production in the phagosome, but a large fraction of the superoxide produced by the leukocyte is released into the surrounding extracellular fluid where it appears to serve merely as a chemical insult to surrounding healthy cells. Why is there no superoxide dismutase in extracellular fluids? Is this an oversight on the part of Mother Nature, or is she smarter than we currently realize? The latter seems more likely. It now appears that there are pathological circumstances in man under which the artificial introduction of superoxide dismutase into extracellular fluids produces a beneficial effect. It is obviously crucial, however, to understand fully what metabolic balances we may, at the same time, be tilting in an unfavorable direction.

Certain contradictions exist as will be seen by an assiduous reader of this book. But conflicting view points are useful in that sterile dogma is avoided, apart from the essential role of engendering new experimental approaches. It is clear that sooner or later specialization will limit comprehension. This is the first book to cover O_2^- and superoxide dismutases from very wide-ranging approaches. It is also perhaps the last; future meetings and reports will no doubt necessarily be concentrated on certain aspects of the subject, rather than involving a global view, the price to be paid by key developments which touch an ever-widening circle of interests. Despite the impressive advances herein recorded, we are truly at the beginning of new approaches in fundamental, medical and applied research.

REFERENCES

1. Misra, H.P. and Fridovich, I. (1972). *J. Biol. Chem. 247*, 188.
2. Ballou, D., Palmer, G. and Massey, V. (1969). *Biochem. Biophys. Res. Commun. 36*, 898-904.
3. Misra, H.P. and Fridovich, I. (1972). *J. Biol. Chem. 247*, 3170-3175.
4. Cohen, G. and Heikkila, R. (1974). *J. Biol. Chem. 249*, 2447-2452.
5. Misra, H.P. and Fridovich, I. (1972). *J. Biol. Chem. 247*, 6960-6962.
6. Brunori, M., Falcioni, G., Fioretti, E., Giardina, B. and Rotilio, G. (1975). *Eur. J. Biochem. 53*, 99-104.
7. Winterbourn, C.C., McGrath, B.M. and Carrell, R.W. (1976). *Biochem. J. 155*, 493-502.
8. Nilsson, R., Pick, F.M. and Bray, R.C. (1969). *Biochim. Biophys. Acta 192*, 145.
9. Orme-Johnson, W.H. and Beinert, H. (1969). *Biochem. Biophys. Res. Commun. 36*, 905-911.
10. Nakamura, S. (1970). *Biochem. Biophys. Res. Commun. 41*, 177-183.
11. Misra, H.P. and Fridovich, I. (1971). *J. Biol. Chem. 246*, 6886-6890.
12. Nakamura, S. and Kimura, T. (1972). *J. Biol. Chem. 247*, 6462-6468.
13. Allen, J.F. (1975). *Biochem. Biophys. Res. Commun. 66*, 36-43.
14. McCord, J.M. and Fridovich, I. (1968). *J. Biol. Chem. 243*, 5753-5760.
15. McCord, J.M. and Fridovich, I. (1969). *J. Biol. Chem. 244*, 6049-6055.
16. Nakamura, S. and Yamazaki, I. (1969). *Biochim. Biophys. Acta 189*, 29.
17. Knowles, P.F., Gibson, J.F., Pick, F.M. and Bray, R.C. (1969). *Biochem. J. 111*, 53.
18. Fridovich, I. (1970). *J. Biol. Chem. 245*, 4053-4057.
19. Massey, V., Strickland, S., Mayhew, S.G., Howell, L.G. Engel, P.C., Mathews, R.G., Schuman, M. and Sullivan, P.A. (1969). *Biochem. Biophys. Res. Commun. 36*, 891-897.
20. Loschen, G., Azzi, A., Richter, C. and Flohé, L. (1974). *FEBS Lett. 42*, 68-72.
21. Forman, H.J. and Kennedy, J.A. (1974). *Biochem. Biophys. Res. Commun. 60*, 1044-1050.

22. Boveris, A. and Cadenas, E. (1975). *FEBS Lett.* *54*, 311.
23. Azzi, A., Montecucco, C. and Richter, C. (1975). *Biochem. Biophys. Res. Commun.* *65*, 597-603.
24. Boveris, A., Cadenas, E. and Stoppard, A.O.M. (1976). *Biochem. J.* *156*, 435-444.
25. Asada, K. and Kiso, K. (1973). *Eur. J. Biochem.* *33*, 253-257.
26. Allen, J.F. and Hall, D.O. (1973). *Biochem. Biophys. Res. Commun.* *52*, 856-862.
27. Epel, B.L. and Neumann, J. (1973). *Biochim. Biophys. Acta* *325*, 520-529.
28. Elstner, E.F. and Konz, J.R. (1974). *FEBS Lett.* *45*, 18-21.
29. Halliwell, B. (1975). *Eur. J. Biochem.* *55*, 355-360.
30. Miller, R.W. and Macdowell, F.D.H. (1975). *Biochim. Biophys. Acta* *387*, 176-187.
31. Harbour, J.R. and Bolton, J.R. (1975). *Biochem. Biophys. Res. Commun.* *64*, 803-807.
32. Elstner, E.F., Wildner, G.F. and Heupel, A. (1976). *Arch. Biochem. Biophys.* *173*, 623-630.
33. Kono, Y., Takahashi, M. and Asada, K. (1976). *Arch. Biochem. Biophys.* *174*, 454-462.
34. Babior, B.M. Kipnes, R.S. and Curnette, J.T. (1973). *J. Clin. Invest.* *52*, 741-744.
35. Salin, M.L. and McCord, J.M. (1974). *J. Clin. Invest.* *54*, 1005-1009.
36. Curnutte, J.T. and Babior, B.M. (1974). *J. Clin. Invest.* *53*, 1662-1672.
37. Drath, D.B. and Karnovsky, M.L. (1975). *J. Exp. Med.* *141*, 257-262.
38. Weening, R.S., Wever, R. and Roos, D. (1975). *J. Lab. Clin. Med.* *85*, 245-252.
39. Nakagawa, A. and Minakami, S. (1975). *Biochem. Biophys. Res. Commun.* *64*, 760-767.
40. Johnston, R.B., Keele, B.B., Misra, H.P., Lehmeyer, J.E., Webb, L.S., Baehner, R.L. and Rajagopalan, K.V. (1975). *J. Clin. Invest.* *55*, 1357-1372.
41. Baehner, R.L., Murrman, S.K., Davis, J. and Johnston, R.B. Jr. (1975). *J. Clin. Invest.* *56*, 571-576.
42. Hewitt, J. and Morris, J.G. (1975). *FEBS Lett.* *50*, 315-318.
43. Fridovich, I. (1972). *Accts. Chem. Res.* *5*, 321-326.
44. Fridovich, I. (1974). *Advan. Enzymol.* *41*, 35-97.
45. Fridovich, I. (1975). *Ann. Rev. Biochem.* *44*, 147-149.
46. Richardson, J.S., Thomas, K.A., Rubin, B.H. and Richardson, D.C. (1975). *Proc. Nat. Acad. Sci., U.S.A.* *72*, 1349-1353.
47. Rotilio, G., Bray, R.C. and Fielden, E.M. (1972). *Biochim. Biophys. Acta* *268*, 605-609.
48. Klug, D., Rabani, J. and Fridovich, I. (1972). *J. Biol. Chem.* *247*, 4839-4842.
49. Hodgson, E.K. and Fridovich, I. (1975). *Biochemistry* *14*, 5299-5303.
50. McCord, J.M., Keele, B.B. Jr. and Fridovich, I. (1971). *Proc. Nat. Acad. Sci. U.S.A.* *68*, 1024-1027.
51. Gregory, E.M. and Fridovich, I. (1973). *J. Bacteriol.* *114*, 543-548.
52. Gregory, E.M. and Fridovich, I. (1973). *J. Bacteriol.* *114*, 1193-1197.
53. Gregory, E.M., Yost, F.J. and Fridovich, I. (1973). *J. Bacteriol.* *115*, 987-991.
54. Gregory, E.M. Goscin, S.A. and Fridovich, I. (1974). *J. Bacteriol.* *117*, 456-460.

55. Crapo, J.D. and Tierney, D.F. (1974). *Amer. J. Physiol. 226*, 1401-1407.
56. McCord, J.M., Beauchamp, C.O., Goscin, S., Misra, H.P. and Fridovich, I. (1971). Proc. 2nd Int. Symp. Oxidases and Related Redox Systems, Memphis, Tenn., (T.E. King, H.S. Mason and M. Morrison, eds.), Univ. Park Press, Baltimore, pp 51-76.
57. Salin, M.L. and McCord, J.M. (1975). *J. Clin. Invest. 56*, 1319-1323.
58. Yost, F.J. Jr. and Fridovich, I. (1974). *Arch. Biochem. Biophys. 161*, 395-401.
59. Abeliovich, A., Kellenberg, D. and Shilo, M. (1974). *Photochem. Photobiol. 19*, 379-382.
60. Eloff, J.N., Steinitz, Y. and Shilo, M. (1976). *Appl. Environ. Microbiol. 31*, 119-126.
61. Michelson, A.M. and Buckingham, M.E. (1974). *Biochem. Biophys. Res. Commun. 58*, 1079-1086.
62. Van Hemmen, J.J. and Meuling, W.J.A. (1975). *Biochim. Biophys. Acta 402*, 133-141.
63. Petkau, A., Chelack, W.S., Pleskach, S.D., Meeker, B.E. and Brady, C.M. (1975). *Biochem. Biophys. Res. Commun. 65*, 886-893.
64. Petkau, A., Kelly, K., Chelack, W.S., Pleskach, S.D., Barefoot, C. and Meeker, B.E. (1975). *Biochem. Biophys. Res. Commun. 67*, 1167-1174.
65. Misra, H.P. and Fridovich, I. (1976). *Arch. Biochem. Biophys.*, in press.
66. Petkau, A. and Chelack, W.S. (1976). *Biochim. Biophys. Acta 433*, 445-456.
67. Petkau, A., Chelack, W.S. and Pleskach, S.D. (1976). *Int. J. Radiat. Biol. 29*, 297-299.
68. McCord, J.M. and Salin, M.L. (1975). In "Erythrocyte Structure and Function" (G.J. Brewer, ed.) A.R. Liss, Inc., New York, pp 731-746.
69. Oyanagui, Y. (1976). *Biochem. Pharmacol. 25*, 1465-1472.
70. Oyanagui, Y. (1976). *Biochem. Pharmacol. 25*, 1473-1481.

SUBJECT INDEX

D

E